Essentials of Biology & Biotechnology

Essentials of Biology & Biotechnology

Prof. Bir Bahadur

(Editor and Author)

Former Head, Dept. of Botany &
Dean Faculty of Science,
Kakatiya University, Warangal, A.P.

Contributing Authors

Dr. G. Maruthi Ram
Associate Professor, Dept. of Zoology,
Osmania University, Hyderabad, A.P.

Prof. M. A. Singhara Charya
Head Dept. of Microbiology,
Kakatiya University, Warangal, A.P.

Dr. G. Madhusudhan Rao
Reader in Zoology, A.V. College,
Gaganmahal, Hyderabad, A.P.

Dr. M. Chandraiah
Reader in Botany, Govt. Degree
College, Hanmakonda, Warangal, A.P.

C. Sirisha
Lecturer in Biotechnology, A.V. College,
Gaganmahal, Hyderabad, A.P.

PharmaMed Press

An imprint of Pharma Book Syndicate

A unit of BSP Books Pvt. Ltd.

4-4-309/316, Giriraj Lane,
Sultan Bazar, Hyderabad - 500 095.

Published by

PharmaMed Press

An imprint of Pharma Book Syndicate

A unit of BSP Books Pvt. Ltd.

4-4-309/316, Giriraj Lane, Sultan Bazar, Hyderabad - 500 095.
Phone: 040-23445605, 23445688; Fax: 91+40-23445611
E-mail: info@pharmamedpress.com

ISBN : 978-93-52300-11-2 (HB)

Preface

Biotechnology as a subject was first introduced at IARI, New Delhi, hardly 20 years ago. Due to its overwhelming popularity and demand the subject is being now offered in most Indian universities. During the last 20 years biotechnology witnessed spectacular progress due largely to intensive interdisciplinary researches in biosciences, microbial, molecular, biochemical and somatic cell, genetics, tissue culture in combination with recombinant DNA technology, cloning & PCR technologies etc. This has given new meaning to biology education since biotechnology and bioinformatics are largely instrumental in revolutionising modern biotechnology with high level of public awareness on one hand and solving wide range of day to day problems, on the other. Biotechnology revolution has had strong impact and profound implications in all spheres of life, namely food, agriculture, industry, health, environment and diagnostics; despite negative publicity about GM food, legal, moral and social issues.

The present book is intended for use in first semester biotechnology course. It presupposes a familiarity with basic biology and of course chemistry.

Our motivation for writing this text has its origin from the publishers who considering our wide experience of teaching graduate and postgraduates thought we can write basic concepts effectively and stimulate the students as they read the various chapters topic by topic.

The book essentially covers the syllabus prescribed for B.Tech. Biotechnology of JNTU and other universities and covers basic and applied aspects of plant, animal and microbial life and the various modern technologies associated with them in simple and lucid manner. The various chapters that follow attempt to reflect basic and applied concepts with emphasis on current developments and their relevance to present day needs. Students pursuing BSc, MSc,

(Biotechnology), MSc (Botany, Zoology, & Environmental Biology) will also find the book equally useful and enjoyable. A question bank of about (470) questions both essay & short questions is appended to help the students to cross check their learning process.

We thank several colleagues for their help and encouragement. We thank Mr. Nikhil Shah and Mr. Anil Shah and M/s B.S Publications for their continued encouragement, support and cooperation throughout the writing of this book. Thanks are also due to their DTP team of Mr. Naresh, Mr. Prashanth and Ms. Kalpana for their patience, cooperation and diligent work.

In preparing this book we have referred several standard text books and reviews and borrowed suitable materials. It is almost impossible to thank them individually. A list of selected references and reviews is appended for the benefit of serious readers.

Editor

Contents

CHAPTER 2

Plant Biotechnology

CHAPTER 3

Animal Science

CHAPTER 4

Biology of Parasites & Immunology

CHAPTER 5

Animal Biotechnology

CHAPTER 6

Microbial Biotechnology

Introduction

There are several books on basic and advanced aspects of biotechnology that provide information on various aspects but a book that readily and hopefully fulfills the requirements of B.Tech Biotechnology course not only in JNTU and others studying the subject at graduate and post graduate courses are few. The present text is one such attempt. There are six well designed and fairly exhaustively written chapters covering basic and applied aspects relevant to the needs of students. Needless to mention biology deals with wide variety of organisms, microbes, viruses, phages, bacteria, prious, algae, fungi, actinomycetes, plants and animals of various taxonomic groups. A study of their classification, external and internal features to relate form and function which obviously has direct bearing and relevance on current developments in plant, animal and microbial biotechnology. These are covered in chapters 2, 4 and 6. Methodologies of genetic manipulations, gene cloning, r-DNA technology, transgenic plants, animals and microbes, their applications to meet the requirements of 21st century are dealt with in detail.

The organisation of the chapters follow a logical progression of ideas.

Chapter I Introduces Plant Life, Plant classification, Cell structure tissues, Meristems, Plant growth, Life cycles of various plant groups and lichens, Flowering plants and their economic importance is also described.

Chapter II Explores Plant Biotechnology, Tissue culture Techology, Mass propagation, Phytohormones, Genetic modification methods of plants, GM (transgenic) crops, GM food, Plantibodies and their applications and Phytoremediation. Detailed account of medicinal plants and drugs & chemicals from plants is highlighted.

Chapter III Deals with Phylogeny of Invertebrates & Vertebrates, Functional morphology, Adaptive biology, Numerical taxonomy, Ecology concepts, Crop pests, Fouling organisms, etc. and also includes useful technologies like Remote sensing, World Aquaculture and its status in India, and Environmental laws as applicable in India.

Chapter IV Deals with Biology of various types of Parasites, their life cycles Parasites of fishes; parasites of livestock and man, and their control methods. Plant Parasitic Nematodes and their control are described. Immunology, Monoclonal antibodies, and Elisa techniques are also described in detail.

Chapter V Explores Animal Biotechnology, Genetic Modification in animals. Transgenic animals their applications, animal Tissue culture, r DNA methods in Vaccine development, Gene therapy, Genetic counselling, Prenatal diagnosis, Stem Cell Biology and its applications. Drugs and Chemicals from animals etc.

Chapter VI This deals with basic and applied aspects of microbial life and wide range of applications such as Microorganisms in Waste water treatment, Bioremediations, Biosensors, Biopolysters, Biomining and Bacterial leaching, Microbial enzymes, Bioindicators, Biofuels, Biofertilizers, Biopesticides, Single Cell Proteins, Mushroom culture, etc.

The reader (both the teacher and student) will find this book with historical perspectives of several important and current topics; the concepts are well illustrated for easy understanding throughout the book. Because of space limitation, we have dealt some topics rather briefly. If need be, we shall incorporate them in future edition. It is hoped that the material presented will convey to the reader both the excitement and understanding of the concepts of modern biology and leave an indelible impression that applied biology and biotechnology are not only fascinating subjects but the hope of tomorrow.

1 Plant Science

1.0 Introduction to Biology

The study of living organisms is called Biology or Natural Science. Biology is derived from two Greek words (*Bios* = life, *logos* = discourse or science). The word Biology was first used by Lamarck (France) and Treviranus (Germany) in 19th century. Ethymologically, Biology is the study of life. But life is difficult to define. According to Hungarian Nobel Laureate Szent Gyorgyi, "*life as such does not exist, no body has ever seen. The noun life has no sense, there being no such thing*". Life, therefore, may be considered as a property or as a manifestation or as a state of organism. This leads to further inquiry; what is an organism ? and what is the property of living organism ?

Organisms can be protists (*prokaryote*) composed of single cell or multicell (*eukaryote*) i.e. differentiated into tissues and organs as in plants or animals. According to Whittakar (1978) there are five kingdoms, namely, *Monera* (Prokaryotae), *Protista*, *Fungi*, *Animalia* and *Plantae*. Recently Woese *et al* (1990) on the basis of ribosomal RNA sequences distinguished, the kingdom Procaryotae comprise of two groups; Archaebacteria and Eubacteria and kingdom Eucaryotae to comprise the higher organisms. The living organisms tradionally comprise plants and animals : so biology include the study of plants (Botany) and animals (Zoology). Aristotle (384-322 BC) is considered as the father of Biology. Theophrastus, a Greek philosopher, disciple of Aristotle is credited with founding botany: Of the 200 botanical treatises written by him; only two are known to science; originally written in Greek about 300 BC their Latin translations are available. *De Causis Plantarum* and *De Historia Plantarum* contains the concepts of morphology, classification and their natural history of plants and these remained unchallenged for many centuries. He classified plants as trees, shrubs and herbs.

1.1 Origin of Life

According to the Hindu Mythology Brahma created life. Indians no less than the Greeks have shared in the work of constructing scientific concepts and methods of physical phenomena have deeply influenced the course of natural philosophy in Asia. They had their own concept of origin of life comparable to the modern theory of biochemical origin of life. Parasara (250 - 120 B.C) in his *Vrikshayuveda* (the Science of Plant life) suggests : Water transforms into a jelly-like substances ('Kalalam') in which a nucleus ('*Pindasthanukam*') is formed later and regulated by terrestrial energies it transforms into a 'germ' (bija). The Indians also agrees with the modern concept of origin of Eukaryotes from Prokaryotes. They also knew that plants preceded animals and man on this earth. Such rudiments of evolutionary thought can be seen in the writings of ancient Indian writers like Uddalaka and Yajnavolkya. It is interesting to note that Indians more than 2000 years before Darwin had thought about problems of evolution and heredity (Chaudhury, 1932). They also held that protoplasm is a colloidal system ('*Kalalibhutam*'), formed quite possibly from inorganic matter (Hora, 1953). Bible states that God created life on 5th day. The Egyptians believed that life is produced from the mud, warmed up by sun rays. Plants came first then animals and finally man. Cosmic, biological and cultural evolutions are integral parts of the story of creation. All these beliefs have no experimental basis. Then came the theory of *Abiogenesis* or Spontaneous generation. It states that life arose from non-living matter. This was supported by Francesco Redi Lazzaro Spallazani and Louis Pasteur. Cell lineage theory conceived by Turpin (1862), enlarged by Schleiden & Schwann (1838) and extended by Naegeli (1854) and Rudolf Virchow (1858) from Germany. These gave a brake to all the views, then existing, about life and proposed that cell arise from its pre-existing cell (*Omnis cellula e cellula*). This theory was proposed only after the term protoplasm was proposed. The cells are atoms of life; unit of biological structure and function.

The following are the characteristics of the living matter :

1. Chemical composition including proteins and nucleic acids
2. Definite organisation.
3. Maintenance and growth through assimilation
4. Irritability and adaptations.
5. Reproduction and heredity

Plants, animals and human as well as prokaryotes exhibit all the above characteristics.

The living world (with the exception of viruses) is characterised by two types of cells based on internal structures they contain. The prokaryte and eukaryote contain the same basic kinds of macromolecules. The prokaryote (pro = before, karyote = nucleus) comprise the bacteria, cyanophyta and the archaebacteria. Their cells are simple, devoid of plastids, mitochondria; the cell wall is made of peptiodoglycan with a nucleid; devoid of nuclear envelope; cells divide every 20 minutes. No wonder the prokaryotic cells are the most abundant cell types on the earth. The eukaryote (Eu = true, karyote = nucleus) the nucleus is membrane bound, with a distinct cell wall, which may be cellulose molecule as in plants, animal and fungi. Surrounding the cell is a plasma membrane of phospholipid bilayer whose dual function of protection barrier between the extracellular and intercellular matrices and to regulate the flow of molecules into and out of the cell. A wide variety of cell organelles are found in the cytoplasm, mitochondria, chloroplast, etc. All the eukaryotic plant and animals cells are essentially similar in most respects yet they differ in the following characters.

Plant cell wall is cellulosic hence elastic with chloroplasts, chromoplast, dictyosomes, vacuoles (constitute 90% cell volume and performs various functions of storage and excretion), the animal cell wall is not cellulosic, devoid of chloroplast and vacoules but has Golgi bodies, centrosome with 2 centrioles and furrowing as against cell plate formation in plants during cell division. Thus, there are so many kinds of organisms but at the cellular level, the living world shows unity. This unity and diversity of life is because of (i) *unity of plan*; each cell possesses a nucleus emmbeded in protoplasms (ii) *unity of function*; each cell show the same metabolism and (iii) *unity of composition* Biochemically, the main molecules of all living organism are made of same molecules. with specific structures and specific functions. For the life to be perpetuated, these macromolecules have to work in coordination and cooperation. Thus the smallest unit of integration, cooperation and reproduction is cell. The macromolecules are nucleic acids (deoxyribonucleic acid and ribonucleic acid) and proteins. The central dogma of life is : *DNA maks DNA*, (replication) *DNA makes RNA* (transcription) and *RNA makes* (translation) *protein*. The proteins are made of 20 different amino acids. Since the nucleic acids and proteins are universally present in all the organisms, therefore are called as the fundamental constituents or essential metabolites. Nature indeed has made use of few selected and limited number of building blocks to build the immense diversity of the living systems.

In 1857 Leydig first described and defined the cell as a mass of protoplasm containing a nucleus. If a comparison of various cells is made, the microbial cell has a dark central body surrounded by dull 'cytoplasm' in contrast to the cells of multicellular organisms which are specialised : therefore differentiated into the liver cells, kidney cells, brain cells etc as in most animal or human. The differentiated cells are organised into tissues and the tissues into organs and organs collectively contribute the organism. The functioning of each cell is controlled by the organism as a whole. The cells of the multicellular organisms differ from the microbial cells, which function as organism itself, an independent system of reproduction. No wonder Leeuwenhoek (1676) after discovering the bacterial cell stated, "dear god, what marvels there are in so small a creature"!

Thus, the presence of protoplasm is the most important characteristic of living organisms. It is the physical basis of life.

1.2 Plant Kingdom

Botany (Plant Kingdom) is the study of plants. Plants were of pivotal importance not only to the early man but to the present mankind, as they depended upon them as prime source of food, fuel, fodder, shelter, clothing, tools, ornaments and also as medicines. In addition to their practical and economic value, green plants are indispensable to all life on earth, through the process of photosynthesis (plants transform solar energy into chemical energy of food) which makes all life possible on the planet earth.

The behaviour of pre-Stone Age man can be inferred by studying the botany of abriginal peoples in various parts of the world. The earliest man have developed systems of plant classification. The urge to recognise different kinds of plants and to give them names is as old as the human race. They not only collected the plants but were also responsible for their domestication, development of agriculture and earlier civilization.

The invention of the optical lens during the 16th century and development of the compound microscope by Jensen (1590) and later Anthony Van Leeuwenhoek (1650) opened an era of rich discoveries about plants. Led by Robert Hooke's (1665) microscopic observations on cells of several plant tissues in his *Micrographia*, the wide spread use of the microscope during the 18th century onwards made botany an interesting and laboratory science. Carolus Linnaeus (1753) published his *Species Plantarum* which contains careful description of 6000 plant species. This was followed by his binomial nomenclature system which established the scientific naming of plants by two words, the genus name and the species name in *Latin.* This system is universally used even today. To date many systems of classification have been proposed and some classification systems even employ the application of computers.

Around the same time, Koelreuter (1760) conducted experiments in plant hybridization and obtained hybrids. During the next century, Gregor Mendel (1865) conducted experiments of hybridization using pea plants and proposed the laws of inheritance called as *Mendelism.*

With the rebirth of scientific botany during the begining of the 19th century and especially as a result of the emphasis placed by Schleiden, Nageli, Von Mohl and others on the importance of the study of developmental biology, attention was given to the sexuality in plants and in particular flowering plants. Amici (1847) demonstrated in *Orchis* that a germinal vesicle was present in the ovule before the entry of pollen tube and the vesicle in due course developed into embryo. William Hofmeister (1862) demonstrated the formation of embryo in flowering plants followed by similar observations on embryo development by Hanstein (1870) and Treub (1879).

Indian and French botanists have excelled in both classical and experimental embryology. With the advent of techniques of tissue and organ culture there has been a movement towards experimental investigations in various aspects of botany and Indians have excelled in this field also.

Plants and the immovable, amazing and highly diversified green property of the mother earth exhibit an enormous variety of form. They range from microscopic unicellular form *Chlamydomonas* to giant trees like *Sequoia*. Plants live in variety of habitats like water, marshes, plains, hills, valley, and where not ! every where, where the sun rays could penetrate. Habit of the plants changes according to habitat. Thus there are different forms of plants which include hydrophytes, mesophytes, halophytes, xerophytes etc.

Scope of Botany

Fundamentally botany is a pure science with several branches in it but over the years the frontiers of knowledge have broken and new disciplines have emerged, thus applied botany, biotechnology and bioinformatics are the craze of this century.

The various branches of botany :

1. *Morphology* : This deals with the structure and form of plants or study of shape in biological systems.
2. *Ecology* : This deals with the mutual relationship and interaction between plants and their environment.
3. *Bryology* : The study of liverworts and mosses, the bryophytes.
4. *Pteridology* : The study of ferns and their relatives, the pteridophytes.
5. *Palynology* : This deals with study of pollen, spores and allergies etc.
6. *Phycology* : (Algology) the study of algae, both marine and fresh water.
7. *Mycology* : The study of fungi both filamentous and fleshy.
8. *Ethnobotany* : The study of plants by aboriginal people now and in the distant past.
9. *Plant Pathology* : This deals with plant diseases, their causes and control measure.
10. *Physiology* : This deals with functions of plant, plant parts and organelles.
11. *Phytogeography*: The study of distribution of ranges and different kinds plants in relation to geographical regions and causes leading to divessification of floras.
12. *Cytology* : This deals with ultrastructure and functions of various components of cell.
13. *Anatomy* : The study of various tissues and their development.
14. *Taxonomy* : The study of identification, nomenclature and classification of plants.
15. *Embryology* : The study of fertilization, gametophytes, embryo sac and embryo development.
16. *Genetics* : The study of heredity and variation.
17. *Plant breeding* : This deals with crop improvement.
18. *Evolution* : The study of complex organisms from simple organisms over a period of time.
19. *Microbiology* : The study of various microorganism like bacteria, viruses etc.
20. *Taximetrics (Numerical Taxonomy)* : The relationship between plant species or those which are determined quantitatively using computers.
21. *Cytogenetics* : The science which integrates the methods and findings of cytology and genetics.
22. *Biometry* : This deals with the application of statistics to biological problems.
23. *Agriculture* : The science and art of farming, cultivating of soils, raising crops etc.
24. *Floriculture* : The study of cultivation of flowers especially decorative plants.
25. *Horticulture* : The study of growing vegetables, fruits and flowering plants.
26. *Agrostology* : The study of grasses both fodder and lawn grasses.
27. *Forestry* : The study of systematic forest management of timber and non timber plants and their conservation.
28. *Silviculture* : The study of cultivating forest.
29. *Limnology* : The study of chemical, physical and biological properties of fresh water lakes.

30. *Biotechnology* : This deals with the application of biological systems or processes in manufacturing or servicing industries and makes use of r-DNA and related technologies.

31. *Bioinformatics* : The combination of biology, computer science and mathematics. Uses computer technology to manage the quantum of biological information or Encompasses "genomics" study of genes and their functions, super computers, vast data bases and complex software to assemble and analyse the information generated by sequencing the code of life.

1.3 Plant Classification and Biodiversity

"Without classification - only chaos", remarked Linneaus to the science of biology. Therefore all organism must be classified. The term taxonomy was proposed by de Candolle (1813) and is the oldest discipline of botany and deals with three aspects of the organisms. Linnaeus adopted "species" the unit of classification in 1758 in his *Systema Naturae* and it has been used ever since.

1. Identification,
2. Nomenclature, and
3. Classification.

Taxonomy (Greek = rendering of order) and systematics (to put together) are currently used interchangely. The modern systems of classitication is the labour of several botanists especially during the last century.

Three main types of classification system of angiosperms are known based upon the morphology of the flowers, fruits and seeds.

1. Artificial system based on 1 or 2 characters of plants
2. Natural system based on as many characters as possible to group taxa seemingly related. Bentham and Hooker's (1862-1883) in their *Genera Plantarum* recognised 202 families.
3. Phylogenetic system is based on several characters to consider the inter relationship. Engler and Prantl, Hutchinson, Takhtajan and Cronquist have proposed phylogenetic systems.

In Hutchinson's system the Angiosperms are divided into predominantly herbaceous and woody groups (the Herbaceae and Lignosae, respectively). All the "*natural*" classifications recognise two series; Dicotyledonae and Monocotyedonae. Although the names are based on a difference in the number of embryonic leaves (cotyledons) as monocotyledons and dicotyledons, the ending of the names of the taxonomic ranks are consistent and follow rules of international code of Botanical nomenclature. Thus, the classification of maize, *Zea mays* be summarised as follow.

Kingdom	**Plantae**
Division	Spermatophyta
Class	Angiospermae
Sub-Class	Monocotyledones
Order	Poales
Family	Poaceae
Genus	*Zea*
Species	*Z. mays*

It is customary to indicate the name of the author who described a genus and species. The name *Zea mays* would therefore be *Zea mays* Linnaeus or *Zea mays* L. The common name corn has no scientific value. The genus and species names together comprise the binomial system of nomenclature proposed by Swedish botanist C. Linnaeus in his "*Species Plantarum*". The origin of this binomial name are basically latin, although words from many other languages including sanskrit are used and latinised. The scientific name of a plant is always written in italics and the genus name is always capitalized as per the International Botanical Code..

A brief summary of Oswald Tippo's (1942) phylogenetic classification is given in Table 1.1. The Magnoliales is considered primitive and Asteraceae as advanced and Orchidaceae more recent.

Table 1.1

Kingdom Plantae			
Sub-kingdom		*Thallophyta*	
	Division	Cyanophyta	Division Rhodophyta
		Chlorophyta	Schizophyta
		Chrysophyta	Myxomycophyta
		Pyrrophyta	Division Eumycophyta
		Phaeophyta	
Sub-kingdom		*Embryophyta*	
	Division	Bryophyta	(3 classes -Musci, Hepaticae, Anthocerotae)
		Tracheophyta	
	Sub-division	Psilopsida	
		Lycopsida	
		Sphenopsida	
		Pteropsida	
Class :		Filicinae	
		Gymnospermae	
		Angiospermae	

The functions of taxonomy are :

1. To provide a means for the communication and retrieval of information concerning all plants, animals and other organisms.
2. To facilitate the gathering of new information by permitting the prediction of characters in unfamiliar organisms, and
3. To demonstrate at once the unity and diversity of organic life by expressing the evolutionary origins of the various taxa.

Ernst Haeckel (1894-1896) published his work "*Systematic Phylogeny; a sketch of a natural system of organism based on their descent*", in which he presented what he believed to have been a genealogic tree of living world. A somewhat modified version of the phylogeny of plant kingdom is shown in Fig. 1.1. The primitive plant groups are kept at base and the advanced ones on top.

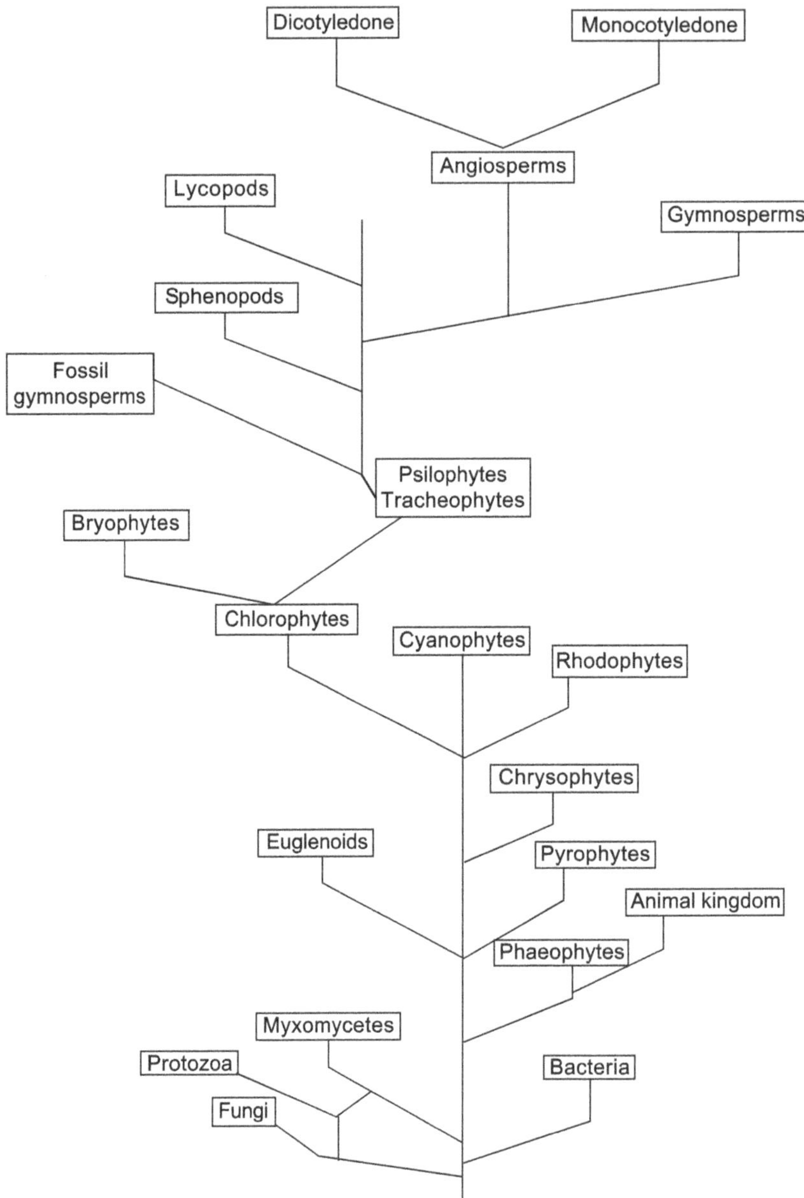

Fig. 1.1 Phylogeny of Plant Kingdom

Ancient Indians have also contributed to classification of plants based on three major considerations, namely :

1. Udhvida (botanical)
2. Virechanadi (medicinal), and
3. Annapanadi (dietic)

The systems was superficial and are similar to those of ancient Greeks and Romans, Charaka (1 century, A. D.) divided plants into the following groups and recognized two groups among *Virudhs* namely, *"Latas"* (creeper) and Gulmas (succulent herbs and shrubs).

1. *Vanaspathis* : Trees with fruits but without flowers.
2. *Vanaspathyas* : Trees bearing flowers and fruit.
3. *Aushadhis* : Herbs that wither after fructification.
4. *Virudhs* : Other herbs with spreading stems.

Prasastapada, another medical practiontioner of this period classified plants and recognised six classes, namely :

1. *Trians* : Grasses
2. *Aushadhis* : Herbs withering after fructification
3. *Latas* : Spreading and creeping herbs
4. *Avatanas* : Arboreous plants and shrubs
5. *Brikas* : Trees with flowers and fruits, and
6. *Vanaspathis* : Trees with fruits but no flowers

There is an increasing emphasis in taxonomic research of new disciplines in utilizing information other than external morphology, the structure and development of pollen grains and embryo sacs anatomy, analysis of chemical constituents, genetics, serology, etc. Studies of the numbers and morphology of chromosomes and their behaviour at meiosis have been made and have generally shown a good correlation with existing classifications. The range of basic chromosome numbers is from n = 2 (*Haplopapus gracilis*) and (*Brochycome leariloba* of Asteracae) to n = 154 *Morus nigra* of Moraceae.

The kinds of plants are inestimable. It is estimated that there are 4 lakh different kinds of plants. These are differentiated into various groups as shown in Table 1.2.

Table 1.2

Flowering plants	.....	.....	2.86.000
Gymnosperms	.....	.....	640
Ferns and allies	.....	.....	10,000
Bryophytes	.....	.....	24,000
Algae	.....	.....	9,000
Fungi	.....	.....	40-65,000
Lichens	.....	.....	1000
Blue Green Algae	.....	.....	15,000
Bacteria	.....	.....	1600-1800
Viruses	.....	.....	200

Biodiversity is variety and variability of plant and animal species on the earth. There are three broad levels of biodiversity : genetic variation within species : the variety of species within a habitat or ecosystem; and the variety of habitats on the earth. The biodiversity has two basic functions, first; it depends on the stability of the biosphere and second, it is the source of species on which mankind depends for food, fodder, fibre, shelter, medicine etc. Biodiversity is not only an important resource but also a country's strength.

Species do not occur evenly around the world and therefore there is a need to know about the distribution and in density of species in different habitats/localities. India has a very rich, variety of plants and is recognised as one of the 12 megadiversity countries (Mc Neely *et al* 1990). It is established that about 17,000 species of flowering plants belonging to 315 families are represented in India.

The green aquatic, filamentous or thalloid plants are called *algae*. This is the first distinct group of plants and hence primitive.

The non-green filamentous mycelial to fleshy mushroom are *Fungi*. It is the heterotrophic groups of plants capable of living any where, either as parasitic or saprophytic.

The thalloid group of green land plants are called *Bryophytes*. These are the first embryo bearing plants and are also called as *amphibians* of the plant kingdom. Algae, Fungi and Bryophytes are collectively called as *non-vascular cryptogams*. Bryphoytes, are highly advanced cryptogams.

The vascular plants include *pteridophytes*, *gymnosperms* and *angiosperms* and are collectively grouped under Tracheophytes because of the presence of tracheids. The Tracheophytes constitute a group of vast and richly varied assemblage of plants that dominate the earth today. They are characterised by.

1. A dominant sporophyte, and a highly reduced gametophyte (the microscopic pollen)
2. Presence of vascular elements that transport water and food
3. Development of variety of flowers and seeds

The tracheophyte has four sub-division as follow :

1. Psilopsida
2. Lycopsida
3. Sphenopsida
4. Pteropsida

The Pteropsida is sub-divided into ferns (seedless plants) and seed plants (gymnosperms and angiosperms). The ferns are the first vascular land plants and their body shows differentiation of root, stem and leaves.

The pteropsida are plants with protected seeds which produce exposed seeds not covered by fruits are the gymnosperm while the plants that bear true fruits are the angiosperms. The term angiosperm was coined in 1690 but it was Robert Brown (1827) who gave the meaning and significance to angiosperms and gymnosperms. These are sub-divided into two groups as shown below. A comparison of how they differ in several characters is shown in Table 1.3.

Model Plants System

A number of plants have been used as model system. The best example is *Arabidiopsis* because : *Arabidiopsis thaliana* a member of Brassicaeae, has a small genome with 1,20,000 kbp is extensively used because in comparison with rice 4,50,000 kbp. *Arabidopsis* can be grown easily in large numbers in plant growth chambers/greenhouse. It has a short life cycle, seed germination to production takes 6-8 weeks only. The seeds are small, 20mg each; 20,000 per plant; can be stored and large numbers of plant can be generated at will. It is easy to induce mutations by chemicals or radiations. It is easy to carry out various techniques of molecular biology. The plant genome sequencing of *Arbidopsis Genome Project* (AGP), was completed in December 2000. Other plant species for which genome sequences are being obtained include maize and rice etc.

Table 1.3

Dicotyledoneae	Monocotyledoneae
1. Two cotyledons or seed leaves in the embryo	One cotyledon in the embryo
2. Leaves dorsi ventral, principal veins of the leaves branching out from the midrib or from the base of it, forming a distinct network	Leaves isobilateral, principal veins of the leaves parallel to each other.
3. Flower parts arranged in two whorls of fours or usually fives.	Flower parts arranged in whorls of three
4. Root system characterized by a tap root, that is, a large primary root with branch roots growing from it.	Root system fibrous, that is without a principal or tap root, the roots being of about the same size.
5. Stem with vascular bundles in a cylinder (ring as seen in cross section).	Stem with the vascular bundles scattered irregularly in the pith tissues.
6. The cambium adding a new cylindrical layer ("ring" of wood each year or each growing season. *e.g.*, pea, groundnut, cotton, pulses etc.	Stem and root without a cambium, not increasing in girth by the formation of annual cylindroidal layers of wood. *e.g.*, bajra, sugar cane, palms, lillies, orchids etc.

1.4 Life History of a Typical Plant

Structure : All plants are made up of cells and tissues. The plant body in Algae, Fungi and Bryophytes is simple and is not differentiate into true root, stem and leaf therefore is called as *thallus*. The plant body in Pteridophytes, Gymnosperms and Angiosperms is differentiated into root, stem and leaf. Gymnosperms and angiosperms also show flowers, hence they are called flowering plants and kept under spermatophyta.

Nutrition : All green plants produce their own food by a process called photosynthesis. Therefore are called *Autotrophs*. Some plants like Fungi which can not prepare their food due to the absence of chlorophyll and are called *Heterotrophs*.

Metabolism : All the biochemical reactions in the plants are collectively called as metabolism. They may be constructive or destructive. The constructive process includes photosynthesis, protein synthesis etc, and are called as Anabolism and destructive process (eg. Respiration), catabolism. Complex molecules synthesised during anabolism are broken into simpler molecules in catabolism and energy is released. It is used for various body functions.

Growth : The net result of metabolic functions of the plants is manifested in growth. The growth of the plant shows several stages which are broadly divided into 1. Vegetative phase 2. Reproductive phase. A mature plant shows reproduction but not all plants reproduce sexually.

Reproduction : Life begets life. This is the unique and mysterious property of the living organisms. Every living organisms and all kinds of plants shows the capacity of reproduction. This results in continuity of species. The plants may reproduce through spores, seeds or even cuttings. Thus, there are three kinds of reproduction, namely asexual, sexual and vegetative.

Alternation of Generation

The term alternation of generations was originally coined to describe the kind of life history found in the Bryophyte and Vascular Cryptogams. (Psilopsida, Sphenopsida, Lycopsida, Pteropsida). In all these groups the sporophytic the diploid generation reproduces by means of spores. These germinate and grow into a gametophytic the haploid generation which reproduces by means of male and female reproductive organs. Fertilization of the female gamete by the male gamete produces again the diploid status of the sporophyte. Apart from certain cases where either the sporophyte or gametophyte respectively can give rise vegetatively to new comparable plants the two generations must succeed each other alternately and repeatedly (Fig. 1.2).

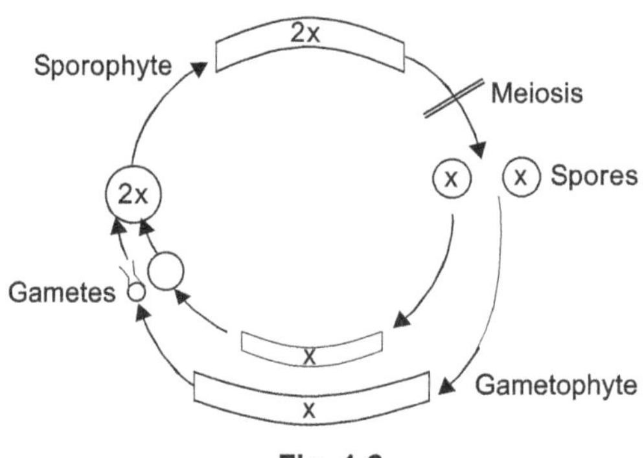

Fig. 1.2

In these plants reduction division takes place at spore formation and there is an alternation, not only of morphologically unlike generations but also of cytologically unlike generations (diploid and haploid state).

The term alternation of generations has since been applied to all members of the plant kingdom where it is relevant, and in the case of the land plants it is possible to trace a series in which at first the gametophyte (haploid) generation is the larger (or dominant) through a succession in which the gametophyte generation becomes successively more and more reduced, until in the flowering plants it is represented by only a few nuclei. In plants below the rank of Bryophyta (i.e. Algae, Fungi) alternation of generations also exists. However, the alternation of generation is not regular. It is thus evident that the phenomenon emerged before the adoption of the land habitat, but that prior to the transmigration from water to land the alternation was not inevitably regular. In the algae there is regular alternation of morphologically similar generations, as in *Dictyota, Ulva* : of morphological dissimilar generations both of which are macroscopic as in *Cutleria* or one generation microscopic and the other macroscopic as in *Laminaria*. There is thus, considerable variation in alternation of generations in red algae.

Considerable acrimony as to whether the two generations (e.g. in the Bryophyta) were completely distinct entities (antithetic theory) or whether one generation had arisen from the other and there was no such distinction (homologous theory). With present day knowledge about chromosomes and the nature of the haploid and diploid nuclei, this problem is of historical interest only.

1.5 Plant Cell - Structure and Function

Robert Hooke (1676) first observed network of cells in a thin section from the cork and this was followed by similar observations on bacteria and protozoa by Leeuwenhoek in 1683. Robert Brown (1855) proposed the term nucleus while Purkinje (1858) proposed the term protoplasm.

For convenience the plant cell is best studied under the following heads and illustrated in Fig. 1.3.

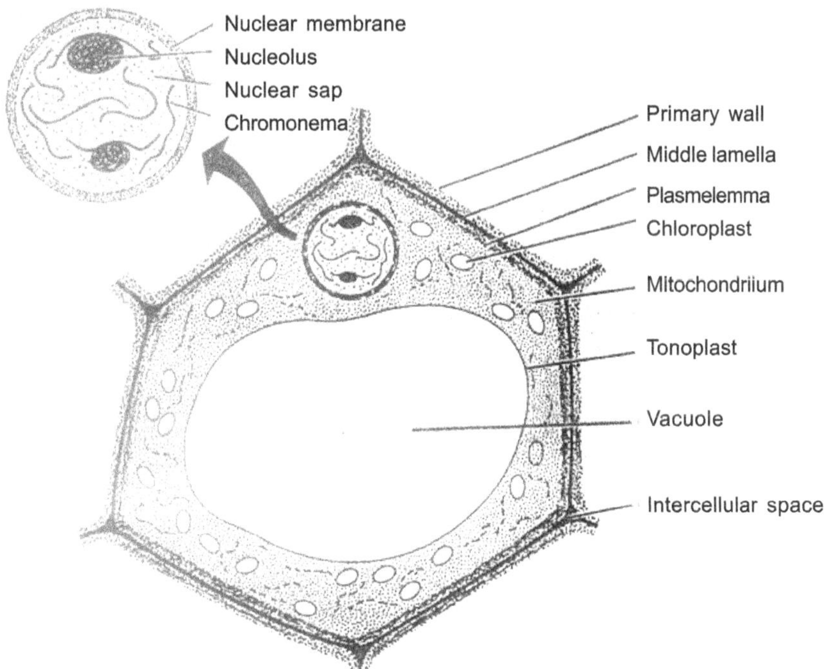

Fig. 1.3 Diagram of a generalized plant cell and enlarged view of the nucleus

1. *Protoplasmic (living) parts*
   (a) Nucleus
   (b) Nucleous
   (c) Cytoplasm
   (d) Mitochondrion
   (e) Plastids etc.

2. *Non-Protoplasmic (non-living) parts*
   (a) Vacuoles
   (b) Ergastic substances
      (i) Starch grains
      (ii) Crystals etc
   (c) Cell wall

The cytoplasm is highly organised and forms the living ground substance of the all various specialized protoplasmic bodies or organell are dispersed in it (Fig. 1.4). The cytoplasm is bounded externally by the plasma membrane or plasmalemma and internally by the tonoplast or vaculor membrane. Both the plasma membrane and the tonoplast are differentially permeable membranes and hence are of great importance in various cellular activities such as osmosis. Cytoplasm is a complex colloidal substance consisting; of water (85-900%) proteins (7-10%) carbohydrates (1-0%), lipids (1.0%) and various inorganic compounds like calcium, magnesium etc. in small amounts. Evidence from electron microscopy on cell ultrastructure provides ample proof that the cytoplasm is highly organised interwoven network of endoplasmic reticulum membrane which may be smooth at the site of lipid synthesis or rough coated with ribosomes, the site of protein synthesis the endoplasmic reticulum may be perinuclear (around the nucleus) and cortical at the cell periphery. The nucleus is large spherical bounded by a two layered nuclear membrane with distinct nuclear pores indicating continuity between the outer part of the nuclear membrane and endoplasmic reticulum within the nucleus embedded in nuclear sap (nucleoplasm) are chromonemata, which later become, chromosomes during mitotis and meiotic divisions and carries genes all along its length and biochemically made up of deoxyribonucleic acid (DNA). Nuclear chromatin of DNA threads intimately forms complex with nucleo proteins are histones low, molecular weight proteins are non histone proteins. Interphase nuclei have large spherical nucleoli that contain the machinery for the production of ribosomes and apparently involved in the synthesis of nuclear ribonucleic acid (RNA) and proteins. The main functions of a chromosome are :

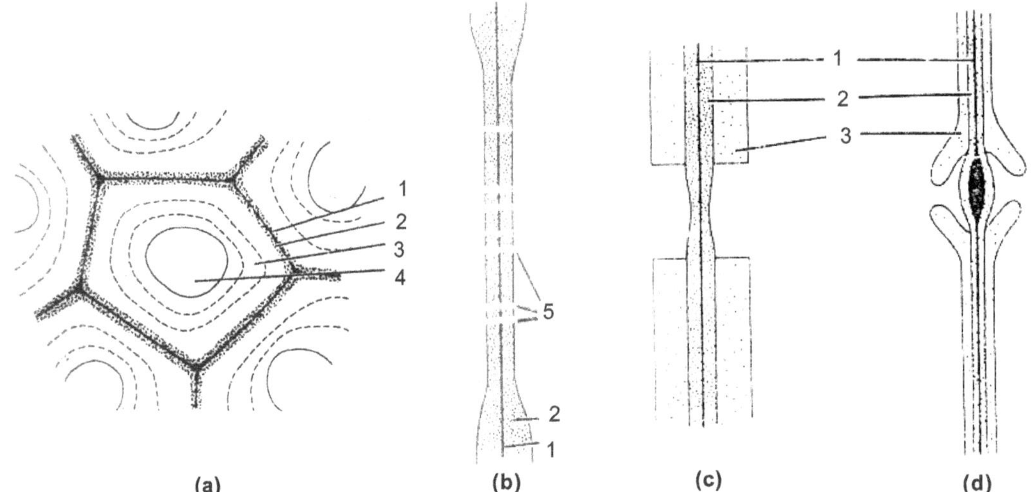

Fig. 1.4 Diagram illustrating (a) Cell wall layers, (b) A primary pit field, (c) Cross section of a simple pit, and (d) Cross section of a bordered pit

1. Middle lamella 2. Primary wall 3. Secondary wall 4. Cell lumen 5. Plasmodesmata

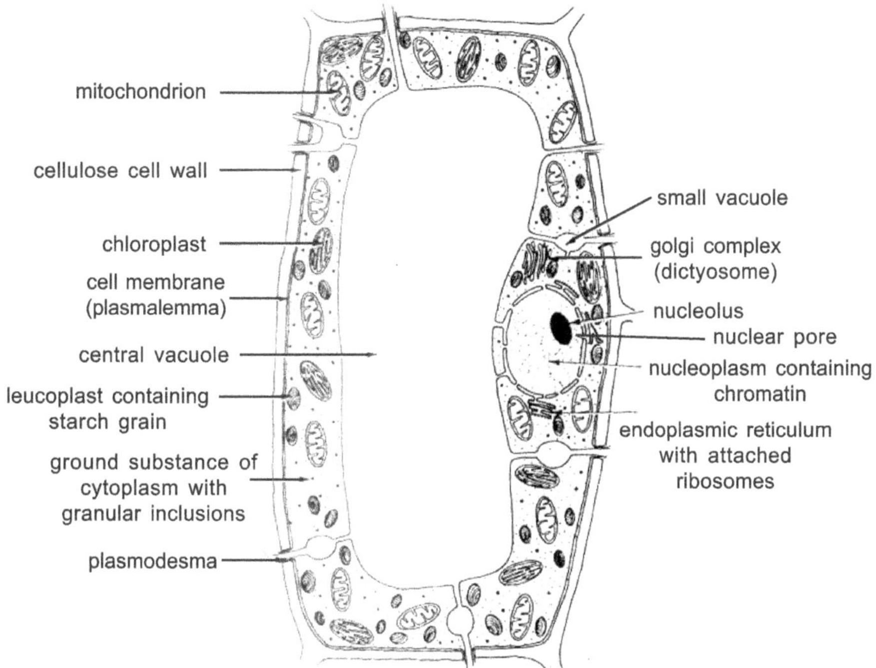

Fig. 1.5 Ultra structure of plant cell based on information obtained by light and electron microscopy

1. Transmission of genetic material from one cell to another cell and from generation to other generation
2. Control of cellular functions and development.

All living cells, both plant and animal contain small rod like or spherical double membrane structures called mitochondria referred to as the "*powerhouses*" of the cell, as they generate adenosine triphosphate (ATP) from stored food reserves carbohydrates and lipids (Fig. 1.5).

Plant cells usually contain specialised bodies called *plastids*. These can be of three main types on the basis of the colour. Green plastids are called, *chloroplasts* which contain the two green pigment chlorophyll 'A' and chlorophyll 'B'. Orange and yellow plastids are called chromoplasts and contain carotene and xanthophyll. All these pigments are necessary for photosynthesis (the process where by sunlight is converted into chemical energy), Leucoplast are colourless plastids and store food such as starch (amyloplast) and oil synthesis (elaioplasts). The Goigi complex is a membrane system similar to endoplasmic reticulum but lacks ribonucleoprotein granules and designated as "*smooth*" membrane system. Peroxisomes contain catalase that removes toxic hydrogen peroxide produced during metabolism while glyoxysomes contain lipid biosynthesis enzymes.

In Fig. 1.5 the generalised ultrastructure of cell is shown. The most important protoplasmic parts of the plant cell are vacuoles. In young cells the vacuoles are small but at maturity plant cell contains a single large vacuole which may occupy 90-95% of the total cell volume. The vacuole is seperated by a membrane the tonoplast and contains complex organic and inorganic substances. Apart from water (major component), red blue and purple anthocyanin pigments, organic acids, fats, tannins, crystals, starch, grains, amino acids, proteins and dissolved salts are the other constitutents of cell sap. This is responsible for osmotic regulation of cell.

The plant cell wall is fully permeable metabolically active wall surrounding and enclosing the protoplast as shown in Fig. 1.5. The primary wall is capable of growth in surface, varies in thickness depending upon the metabolic state and is always associated with living protoplast. Cells involved in water conduction and mechanical support have secondary walls and are often thick and consist in part of lignin or suberin. The plant cell walls are usually composed of two and rarely three distinct layer viz., 1. Middle lamella which is formed during telephase of mitosis (cell division) and act as a cementing substance between adjacent cells. 2. The primary wall, formed while the cell is still actively enlarging/elongating and 3. In some cases, the secondary wall, formed inside the primary wall at a time when the cell is nearing its maximum enlargement. The adjacent cells are interconnected by plasmodesmata pores through the wall where plasma membrane and endoplasmic reticulum connect. Everything with in the plasma membrane is termed *symplast* and everything outside it *apoplast* which is water permeable space in which hydrophilic molecules are present. The primary wall is the first finite wall layer formed by a developing cell and may be the only wall layer present in many types of all such as paenchyuma, collenchyma etc. Chemically primary wall is composed of cellulose (42%), non-cellulosic the polysaccharides (38%), hemicelluloses and pectic substances (8%) and protein (12%).

Plant cells either divide mitotically to ensure equal cell division or meiotically (reduction division) to ensure haploid condition of male/female gametes.

1.6 Biology of Algae

Algae are typically adapted for aquatic life and mostly live in water or in damp situations. They are present in all kinds of running waters, however, abundant in still waters. The aquatic algae are found either floating at the surface simply immersed in water, or attached to stones or to larger aquatic plants. The free floating algae are generally unicellular or colonial form and are called as '*plankton*'. The algae attached to the littoral zone are the '*benthos*'.

Range of plant body

The form of algae varies from a single cell to the complex giant kelps (sea weeds). Single celled *Chlorella* (5 - 8μm) are in the range of bacterial size whereas some seaweeds may attain a length of 60 m. The organisation of the thallus in algae is unicellular, colonial, filamentous (unbranched, branched, heterotrichous) membranous or foliose or tubular types. Unicellular motile (flagelletes) - *Chlamydomonas*; unicellular non-motile (coccoid habit) *Chlorella*; multicellular colonial motile - *Volvox*; multicellular aggregations - *Gleocapsa*; simple filamentous - *Ulothrix, Oedogonium*; Branched filamentous - *Cladophora*; Heterotrichous - *Coleochaete*; Siphonous - *Vaucheria, Caulerpa*; Parenchymatous or pseudoparencnymatous - *Batracospermum, Polysiphonia*. (Fig. 1.16 (a-j)).

Cell structure

Most algal cell walls are formed from cellulose sometimes impregnated with silica or calcium carbonate. Algal cells contain nuclei, mitochondria, ribosomes, golgi and chloroplasts. The internal cell structure is supported by a network of microtubules and endoplasmic reticulum. They often possess flagella that have a 9 + 2 microtubule arrangement within them, and there may be one or two per cell. The presence of eyespots near the flagella insertion point allows the cell to swim towards the light. Chloroplasts are very variable structures in the algae and can be large and single, multiple, ribbon like spiral or stellate etc.

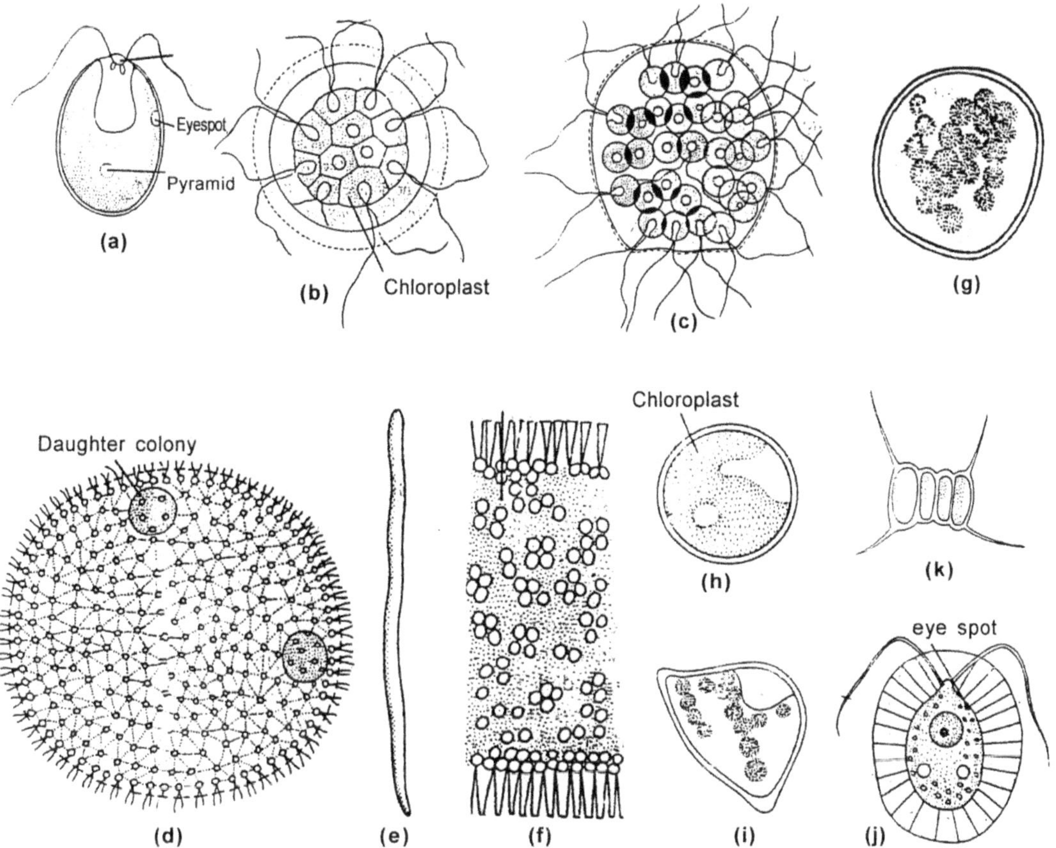

Fig. 1.6 *Range of vegetative structure*. The motile type (a)-(d) and (j); (a) motile vegetative cell of *Chlamydomonas*; (b) vegetative colony of *Pandorina* sp; (c) vegetative colony of *Eudorina unicocca*; (d) vegetative and colony of *Volvox*; (j) Motile cell of *Sphaerella*. The palmelloid type (e) and (f-e), colony of *Tetraspora cylindrica*; (f) Portion of a colony of *T. cylindrica*; the coccoid type - (g-i) and (k) (g) (h) and (i) vegetative cells of *Chlorococcum humicola*; (k) colony of *Scenedesmus* sp.

Growth

Growth in unicellular algae is synonymous with binary fission. In most unicells, haploid or diploid nuclei undergo mitosis and the cell then divides longitudinally to form two daughter cells. Coenocytic filamentous algae grow from the tip of the filament in a way very similar to that of hyphal growth.

Nutrition

Algae are auxotrophic organisms and obtain their carbon and energy requirements by the fixation of CO_2 by photosynthesis. Algae use aerobic respiration via glycolysis and citric acid cycle. Except heterocystous cyanophycean algae, other algae cannot fix nitrogen and they obtain it in a fixed inorganic or organic form. Most algae can tolerate a wide range of pH and temeperature.

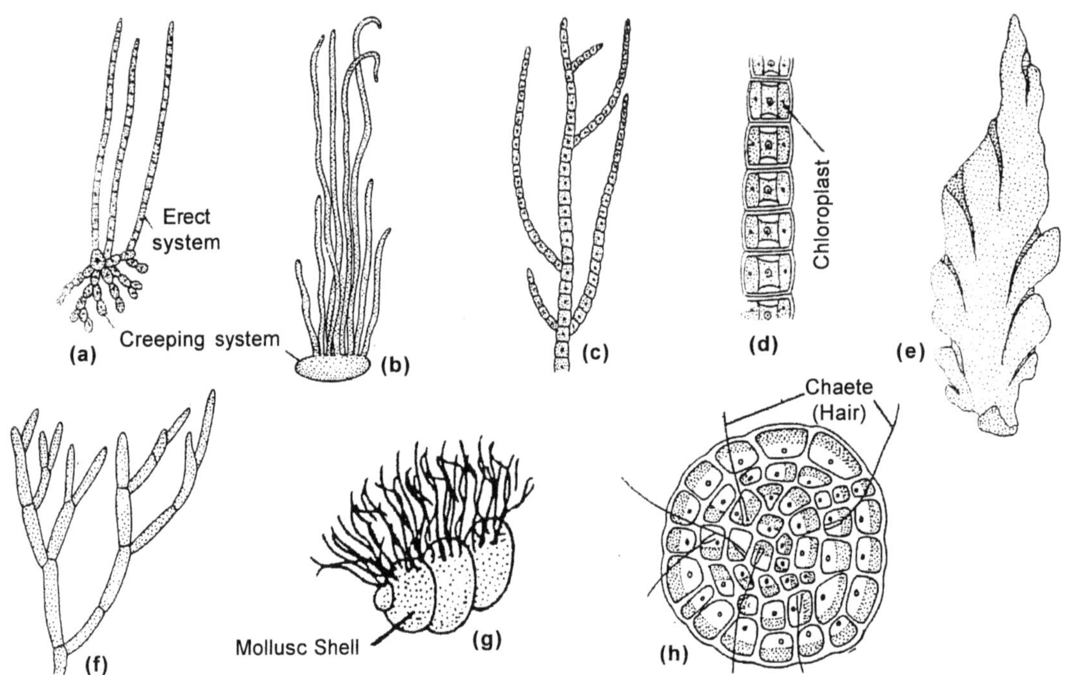

Fig. 1.7 *Range of vegetative structure.* Filamentous habit (a) heterotrichous thallus of *Stigeoclonium*; (b) portion of *Stigeoclonium*; (c) tubular thallus of *Enteromorpha*; (d) portion of the filament of *Ulothrix*; (e) foliaceous thallus of *Ulva*; (f) habit of *Cladophora* (on mullusc shell); (g) a part of *Cladophora thallus*; (h) discoid thallus of *Coleochaete*.

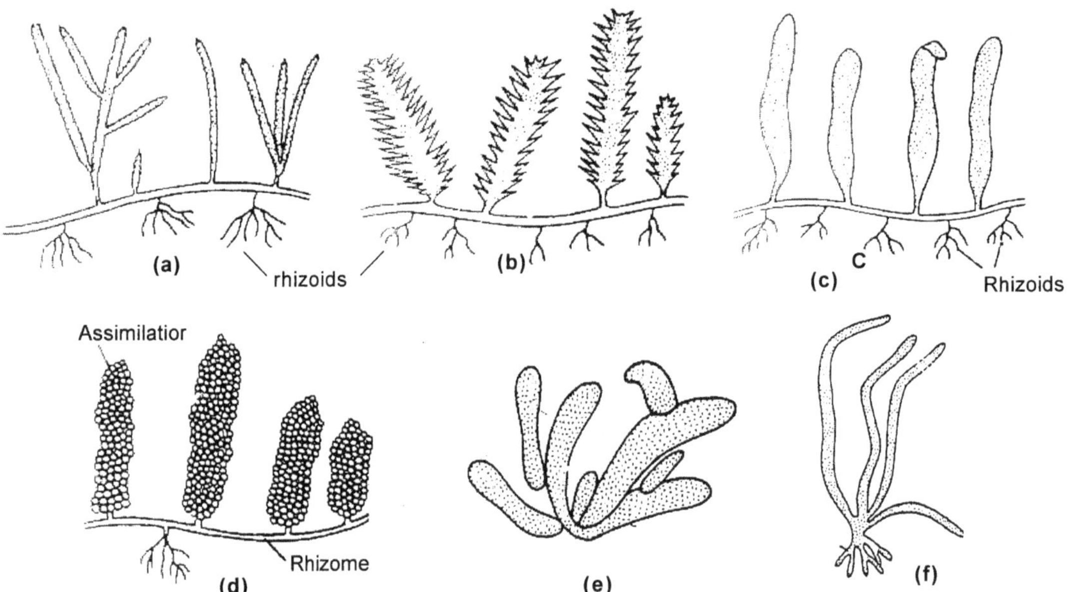

Fig. 1.8 *Range of vegetative structure.* Siphonous type. (a) thallus of *Caulerpa cupressoides*; (b) thallus of *C. crassifolia*; (c) *thallus* of *C. prolifera*; (d) thallus of *Caulerpa* sp; (e) thallus of *Valonia utricularis*; (f) thallus of *Vaucheria* sp.

Pigments

There are three kinds photoshynthetic pigments in algae : chlorophylls, carotenoids, and biloproteins (phycobilins). There are five chlorophylls : a, b, c, d and e. Chlorophyll 'a' is present in all algae. Chlorophyll 'e' is rare and has been identified in only two genera of Xanthophycophyta, *Triboneara* and *Vaucheria*. There are two kinds of carotenoids : carotenes and xanthophylls. Carotenes are linear, unsaturated hydrocarbons and xanthophylls are oxygenated derivatives of these. Biloproteins are pigment - proteins complexes and are present in two algae divisions : Rhodophycophyta and Crytophycophyta. Similarly there are two kinds of phycobilins : phycocyanin and phycoerythrin.

Motility

The motile algae have flagella occurring singly, in pairs, or in clusters at the anterior or posterior ends of the cell. There are three types of flagella : whiplash (cylinrical or smooth), tinsel (cylindrical and with hair like appendages), and ribbon or straplike. Some algae have no means of locomotion and are carried about by tides, waves and currents.

Reproduction

Reproduction in algae takes place in three ways.

1. Vegetative
2. Asexual
3. Sexual.

Vegetative

In unicellular forms the vegetative reproduction is by fission. In colonial forms it is by splitting of the mature colonies into two or more parts. In filamentous forms, parts of plant body becomes seperated and give rise to new individual by a process called *fragmentation*. The resting cells, formed to tide over a peirod, unfavourable for vegetative development is akinete formation.

Asexual

In asexual reproduction, a rejuvenation of the protoplast of algal cells is involved. This is associated with the division of protoplast and these escape from the parent cell and give rise to new plants (Fig. 1.9). The various methods are as follows :

Zoospores : Flagellated protoplasts that show distinctive features. Single (*Oedogonium*); 2 - 8 (*Ulothrix*); large numbers (*Cladophora*) compound zoospore or synzoospore (*Vaucheria*).

Aplanospores : Arrested zoospores which have skipped the motile phase. (*Ulothrix, Microspora*)

Hypnospore : Thickened spore walls of aplanospores (*Botrydium*)

Palmella stage : Under unfavourable conditions the aplanospores retained in parent cell wall and embedded in mucilaginous matrix. (*Chlamydomonas*).

Autospores : Aplanospores develop into absolute miniatures of the parent (*Scenedesmus*)

Endospores : Spores are formed within the cells (*Chamaesiphonales*)

Tetraspores : Spores formed in tetrasporangia (*Dictyota, Polysiphonia*)

Cysts : Resting stage of most flagellate forms. (*Protosiphon*).

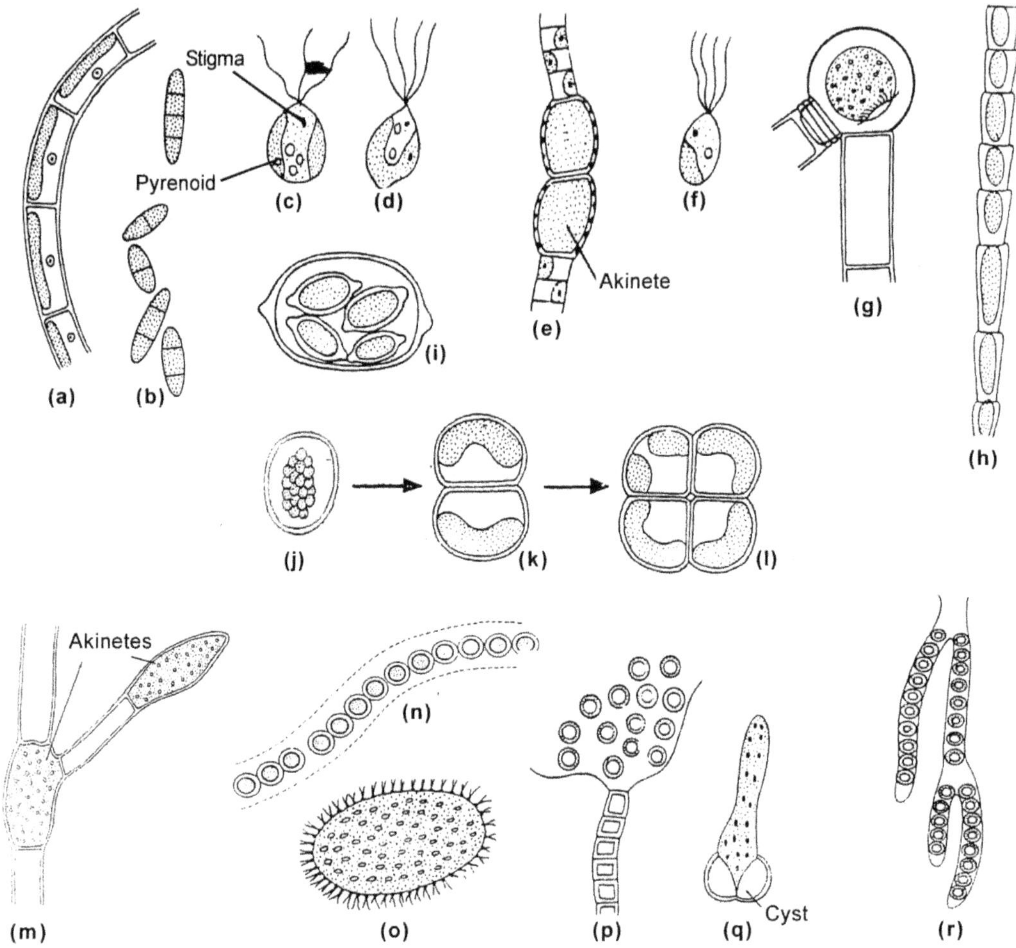

Fig. 1.9 *Reproduction Vegetative and Asexual.* (a) filament of *Hormidium*; (b) fragments of same; (c, d & f) zoospores (swarmers) of *Ulothrix*; (e) akinetes of *Ulothrix idiospora*; (g) zoospore of *Oedogonium* sp.; I. autospores of *Oocytis lacustris*; (h) akinetes of *Oedogonium*; (j, k & l). cell division in *Pleurococcus* sp.; (m) akinetes of *Pithophora*; (n) akinetes of *Ulothrix oscillarina*; (o) synzoospore of *Vaucheria*; (p) palmella stage in *Ulothrix*; (q) germination of cyst in *Vaucheria*; (r) akinetes of *Botrydium*.

Sexual

All form of sexual reproduction are found among algae. In these process there is a fusion of sex cells called *gametes*, to form a union in which blending of nuclear material occurs before new generations are formed. The union of gametes forms a zygote. (Fig. 1.10)

The gametes are identical with no visible sex determination hence, isogamous

The two gametes are unlike, differing in size (male and female) hence, heterogamous

The two sexual cells become more characteristically male and female; the large and non-motile (ovum) and small and actively motile (sperm) hence, oogamy.

Exclusive male or female thalli also exists. The two thalli may look alike but they are opposite sex types, since one produces male gametes and other ova, such plants are called unisexual (dioecious).

Plants in which gametes from the same individual can unite are said to be bisexual (monecious).

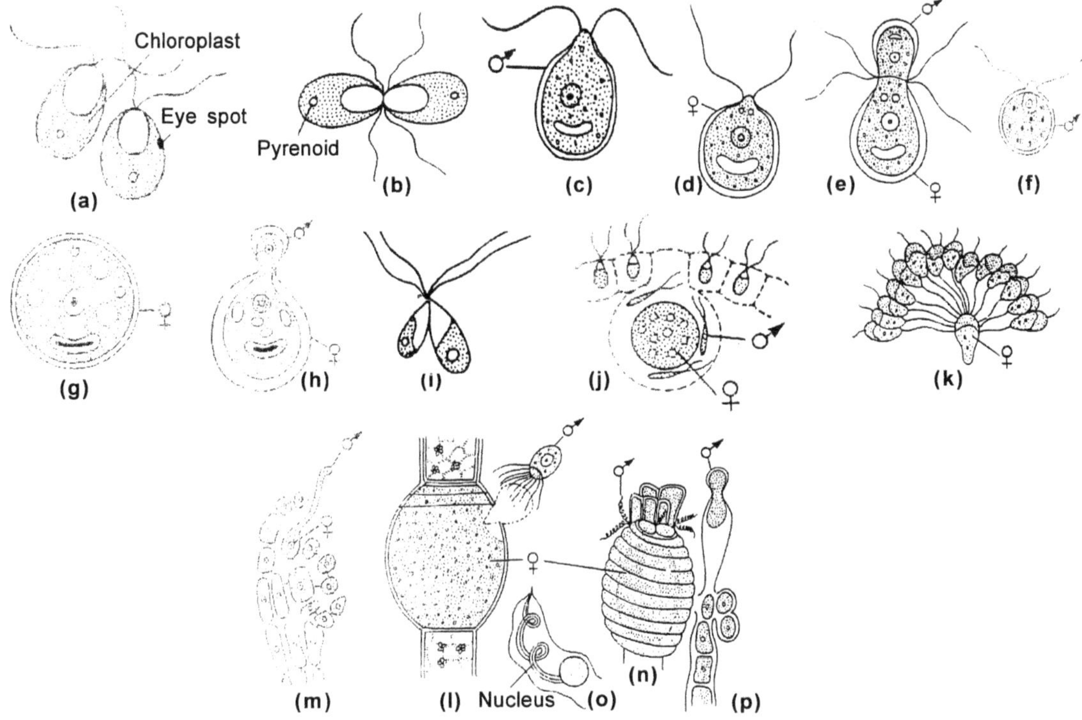

Fig. 1.10 *Reproduction Sexual.* (a) two gametes of same size (isogametes); (b) fusion of same (isogamy); (c, d) two gametes of different size (male and female); (e) fusion of same (anisogarmy); (f, g) male and female gametes; (h) fusion of same (oogamy); (i) anisogamy in *Enteromorpha intestinalis*; (j) oogamy in *Volvox*; (k) clump formation in *Ectocarpus*, large female is surrounded by male gametes; (l) oogamy in *Oedogonium*; (m) advanced oogamy in *Polysiphonia*; (n) oogamy, in *Chara*; (o) antherozoid of *Chara*; (p) oogamy in *Batrachospermum*

Life Cycles

Different types of life cycles have been recognised in the algae. (Fig. 1.11)

(a) *Haplontic :* The parent is haploid and the zygote is diploid with reduction divison occuring at the time of germination of the zygote. (eg. *Volvox, Oedogonium*)

(b) *Diplontic :* The parent is diplont and the sexual spores constitute haploid. Reduction divison at the time of gametogenesis (eg. *Diatoms*)

(c) *Diplo-haplontic :* Alternation of diploid sporotophyte with the haploid gametophyte. Reduction divison is effected at the time of formation of spores by the sporophyte (eg. *Cladophora*).

(d) *Haplo-biontic :* Two haploid generations (gametophyte and carposporophyte) alternating with a diploid one represented by the zygote. (eg. *Batracospermum*)

(e) *Diplo-biontic :* Two diploid phases (carposporophyte and tetrasporophyte) and a haploid phase (gametophyte) alternating with each other. (eg. *Polysiphonia*).

The general characteristics of major algal taxonomic groups are summarised in Table 1.4.

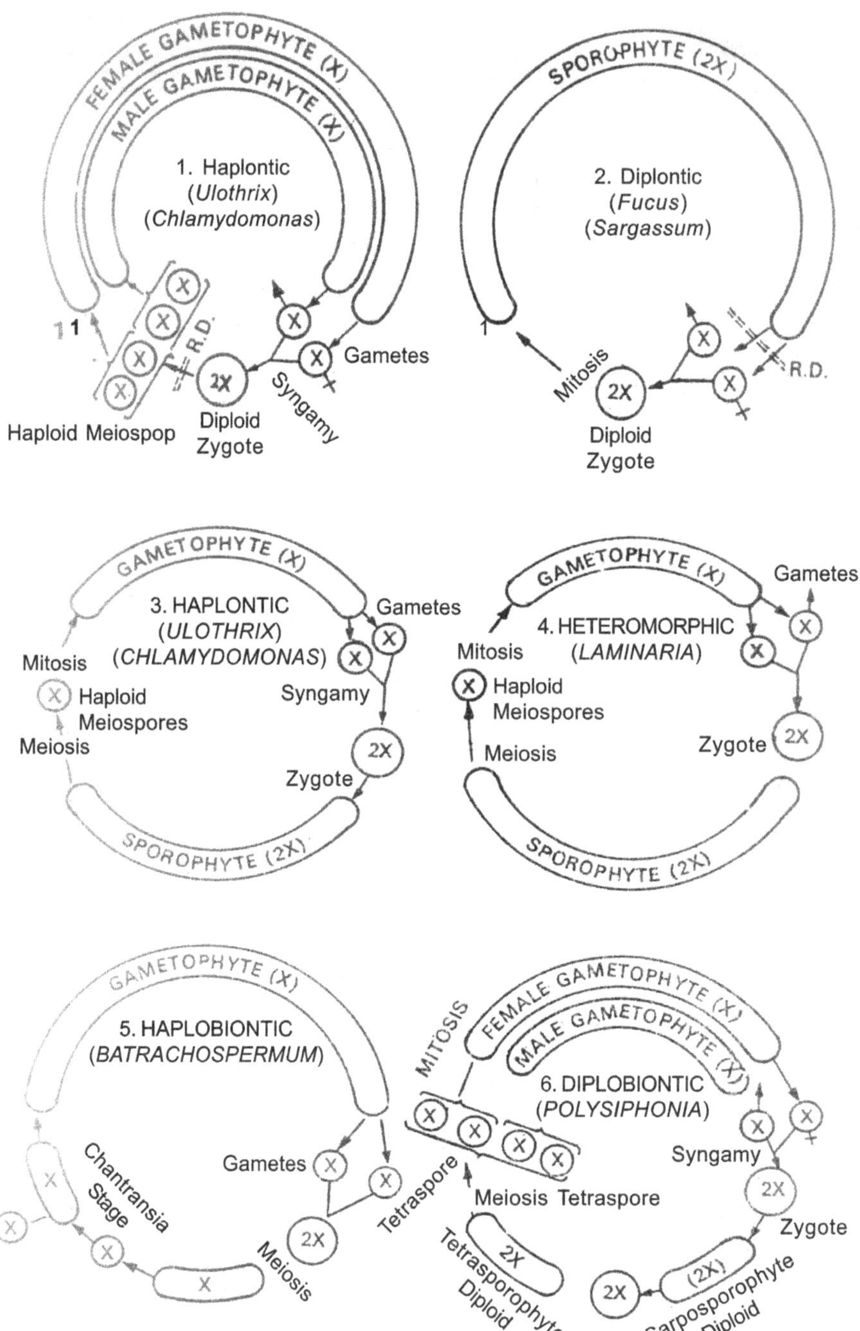

Fig. 1.11 *Life-cycle* patterns in algae. 1. haplontic type; 2. diplontic type; 3. isomorphic type; 4. heteromorphic type; 5. haplobiontic type; 6. diplobiontic type.

Table 1.4

Taxonomic group	Habitat	Pigments	Storage products	Flagella	Cell wall	Sexuality
Cyanophyta	Aquatic, terrestrial	Chlorophyll 'a' β-carotene myxoxanthin C-phycocysnin C-phycoerythrin	Cyanophycean starch	absent	a-diamino	absent pimelic acid
Chlorophyta	Aquatic terrestrial	Chlorophyll a, b, β-carotene lycopene, lutein	Starch, oils	2 or 4 anterior equal acronematic	cellulose	Isogamy to Oogamy
Charophyta	Fresh water	Chlorophyll a, b Y-carotene Lycopene, Zeaxanthin	Starch	2 subapically attached in the spermatozoid with scals on the suface	cellulose	Oogamous, male and female organs compound and complex
Bacillariophyta	Fresh water brackish material and terrestrial	Chlorophyll a, c, β-caretene ε-carotene diatoxanthin, diadinoxanthin.	Chrysolaminarin, oils.	1, pantonematic, anterior	silica, pectin.	Isogamous, anisogamous, and oogamous.
Phaeophyta	Marine	Chlorophyll a, c, β-carotene fucoxanthin, lutein, violaxanthin.	Laminarin, soluble carbohydrates, oils.	2, unequal, lateral, 1 acronematic 1 pantonematic	Cellulose, alginic acid, fucoidan.	Isogamy to oogamy
Rhodophyta	Marine (few fresh water)	Chlorophyll a, d, β-carotene lutein, Zeoxanthin, triaxnthin, R - phycoeryturin, R - phycocyasin.	Floridian starch, oils.	absent	Cellulose, polysulphate esters	Advanced Oogamous, complex, post fertilization charges.

Classification

Algae are generally classified on the basis of the following characteristics :

1. Nature and properties of pigments
2. Chemistry of reserve food products
3. Type, number and morphology of flagella
4. Morphological characteristics of cells and thalli
5. Life history, reproductive structures and methods of reproduction.

According to the scheme of classification of Fritsch (1945), the algae consists of the following eleven classes as follows : Chlorophyceae, Xanthophyceae, Chrysophyceae, Bacillariophyceae, Cryptophyceae, Dinophyceae, Chloromonadinae, Englenineae, Phaeophyceae, Rhodophyeae and Myxophyceae. Subsequently, Smith (1950) recognised eleven major groups of algae and grouped them in seven divisons as follows :

Chlorophyta, Euglenophyta, Chrysophyta, Phaeophyta, Pyrrophyta, Cyanophyta, Rhodophyta.

Klein and Cronquist (1967) reviewed the classification of algae based on chemical, structural and functional criteria and recognised six divisons and classified the blue green algae with bacteria.

Economic Importance

The algae are economically very useful and important in the agriculture fields, industry and as food for human consumption.

Agar - Agar (gellies) *Gelidium, Gracillaria, Gigartina.*
Alganic aid and Alginates - *Ascophyllum, Laminaria, Macrocystis.*
Carragaenin - *Chondrus crispus, Gigartina*
Iodine - Kelps (brown sea weeds) *Laminaria, Fucus, Eisenla, Ecklonia*
Bromine - *Rhodomela, Polysiphonia*
Glue - *Gloeopeltis furcata*
Biofertilizers - *Anabaena, Nostoc, Tolyphothrix, Aulosira.*

The following are other uses :

Algae is used as food - *Spirogyra, Oedogonium* (India) *Ulva* (Europe), *Nostoc* (Brazil) *Laminaria, Alaria* (Japan), *Durvillea, Ulva* (Chile), *Ulva lactuca* (Scotland), *Porphyra* (Korea, Japan, China) *Caulerpa* (Phyllippines), *Chlorella* (Japan, Taiwan).
Algae as fodder - *Ascophyllum, Fucus, Laminaria Rhodymenia, Sargassum*
Algae is medicine - *Laminaria, Codium* (iodine/goiter medicine)
Chlorella (chlorellin) - Antibiotic.
Cladophora, Lyngbya - antiviral and antibacterial drugs.
Digenea, Codium, Durvillea - vermifuge effect
Spirullina - antioxidant drugs and protein
Sewage treatment - *Chlorella, Scenesdesmus,*
Biological research - *Chlorella, Anacystis, Chlamydomonas, Spirullina, Synechococcus.*

Diatoms are used as oils filters, cleaning solvents, insulation of refrigerators, boiler, hollow tile bricks, sound proof rooms, tooth powders, bleaching powders and silver polishes.

Microcystis and *Aphanizomenon* produce toxic substances in to the water which are poisonous to fish, cattle etc.

1.7 Biology of Fungi

The Fungi are a group of Eukaryotic organisms devoid of chlorophyll that are of great practical and scientific interest to microbiologist. It produce green to orange color colonies on fruits, bread and cheese. Thus fungi have a diversity of morphological appearances, depending upon species. There are about 65,000 fungal species that comprise the molds, mushrooms, yeasts, rusts, smuts etc. Molds are filamentous and multicellular; yeasts are usually unicellular and eukaryotic spore bearing protists and generally reproduce both sexually and asexually.

Fungi are heterophilic organisms and require organic compounds for nutrition. When they feed on dead organic matter they are known as saprophytes.

The saprophytes decompose complex plant and animal remains breaking them by increasing its fertility, thus they can be quite beneficial to humans. They decompose, textiles, food and other materials. Saprophytic fungi are also important in industrial fermentations. eg. : brewing of beer, wine making, production of antibiotics such as penicillin etc. The making of dough and the ripening of some cheeses depend on fungal activity. As parasites, fungi cause disease in plants, humans and animals. Most of them cause plant diseases and are responsible for loss of agriculture yields.

Characteristics of Fungi

Fungi are chemoorganotrophic organisms and devoid of chlorophyll.

The fungal body (thallus) consists of single cell or filamentous structure (hence dimorphic).

Morphology

Yeast cells are larger than bacteria in size and width as well as in length. They are commonly egg shaped, elongated, have no flagella. Molds consists essentially of two components.

(a) Mycelium, and

(b) Spores

The mycelium is a complex of several filaments called "hyphae". New hyphae generally arise from spore which on germination puts out a germ tube or tubes, the germ tubes get elongated to form hyphae. The hyphae are composed of an outer tube like wall surrounding a cavity, the lumen, which is filled by protoplasm, between the protoplasm and the wall is the plasmalemma; which is a double - layered membrane. The wall matrix material in which the microfibrils are embedded consists of protein, lipids and other substances. The young hyphae may become divided into cells by cross walls which are formed by centripetal invagination. Sometimes the hyphae are separated by septa formation; but some of them are non-septate and cell structures filled by more nuclei forming coenocytic condition (Fig. 1.12 & 1.13). Mycelia can be either vegetative or reproductive but in mushrooms the compact fruiting structure is seen.

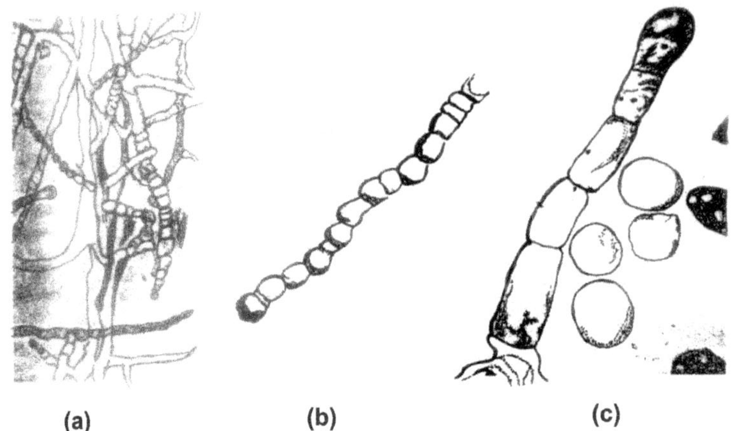

Fig. 1.12 Gross appearance of *Mucor rouxii* arthrospores. (a) Light microscopy showing extensive arthrospore formation. (b) High - magnification light microscopy showing variation in arthrospore size and shape. (c) Transmission electron micro-graph of arthrospore chain.

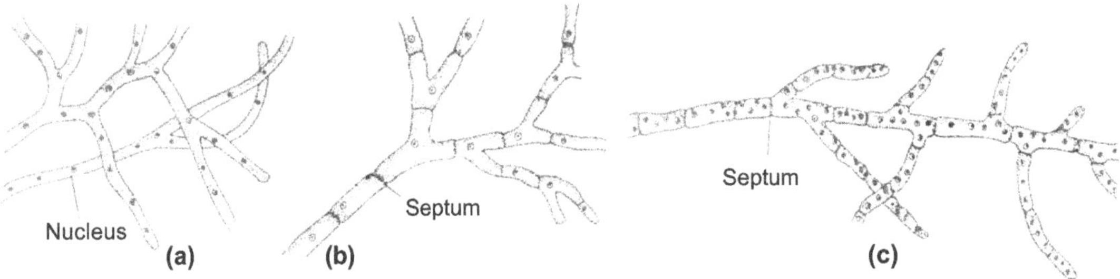

Fig. 1.13 Three types of (a) Non-septate coenocytic (b) septate uninucleate cells, (c) with multinucleate cells

The function of asexual spores is to disseminate the species, which are produced in large numbers. There are many kinds of asexual spores. (Fig. 1.14)

1. *Sporangiospores :* Single celled spores formed with in sacs (sporangium).
2. *Conidiospores or Conidia :* Small single celled conidia, called microconidia; multicelled conidia are called *macroconidia.*
3. *Oidia or orthrospores :* Single celled spores formed by disjointing of hyphal cells.
4. *Chlamydospores :* Thick walled single celled spores highly resistant to adverse condition.
5. *Blastospores :* Spores formed by budding.

Sexual Reproduction

Sexual reproduction (Fig. 1.15) is carried out by fusion of the compatible nuclei of two parent cells. The process of sexual reproduction begins with the joining of two cells and fusion of their protoplasts (plasmogamy) thus enabling the two haploid nuclei of two mating types to fuse together (karyogamy) to form a diploid nucleus (Fig. 1.16).

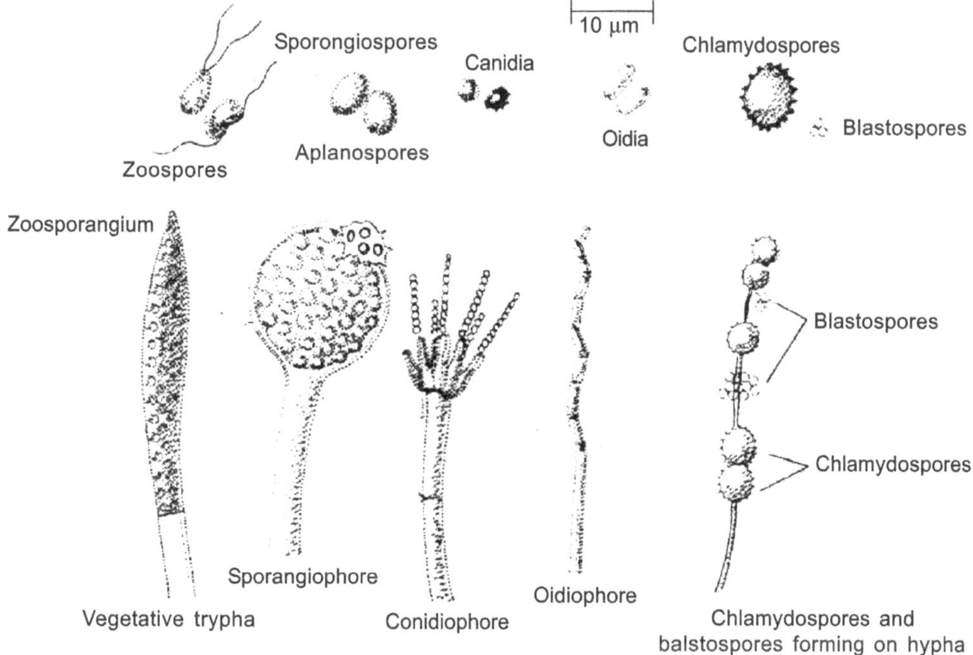

Fig. 1.14 Asexual spores in fungi

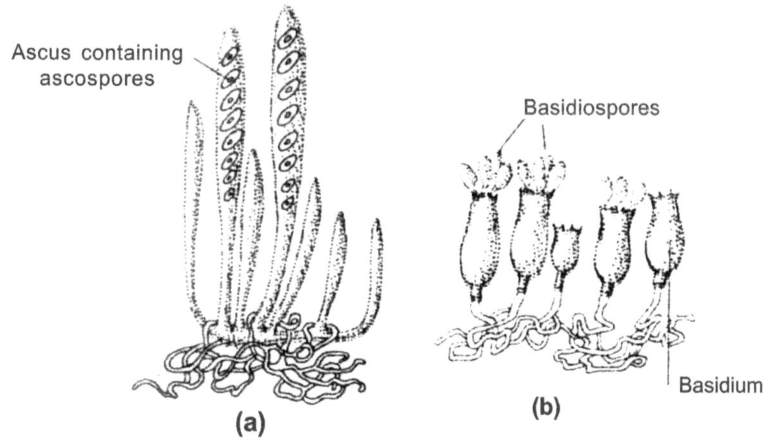

Fig. 1.15 Some sexual spore types and the structure of the corresponding reproductive mycelia in fungi. (a) Ascospores. Nuclear fusion takes place in the ascus. The diploid zygote nucleus divides by meiosis almost immediately after fusion, and produces four haploid nuclei. These haploid nuclei divide once more by mitosis, forming the eight ascospores which are typically produced in each ascus. (b) Basidiospores. Nuclear fusion and meiosis take place in the basidium. Basidiospores are then formed exogenously at the tips of the special outgrowths called sterigmata (singular, sterigma). Usually four spores are formed, one at the tip of each sterigma, but if meiosis is followed by amitotic division then eight basidial nuclei are formed (although rarely do all develop into basidiospores).

The sex organs of fungi called "*gametangia*", male gametangia is '*antheridium*' and female gametangia is oogonium. Sexual spores, which are produced by the fusion of two nuclei, occur less frequently, later and in smaller numbers than do asexual spores, they are :

1. *Ascospores :* These are single celled spores are produced in a sac called an "ascus" usually 4-8 ascospores are producecd in an ascus.
2. *Basidiospores :* These are single-celled spores, are borne of a club shaped structure called a basidium, usually 4 ascospores are produced.
3. *Zygospores :* They are large thick walled spores.
4. *Oospores :* These are formed within a special female structure, the oogonium, where fertilization of the egg takes place.

Asexual and sexual spores may be surrounded by highly organized protective structures called "fruiting bodies". Asexual fruiting bodies are called *acervulus* and *pycnidium.* The sexual fruiting bodies are called perithecium and apothecium which contains spores.

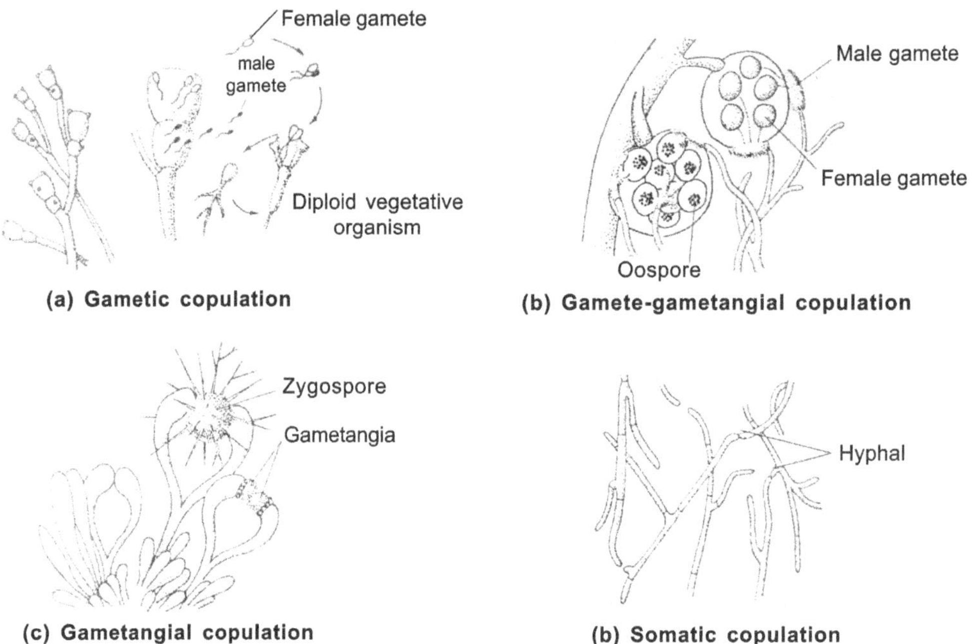

Fig. 1.16 Some sexual mechanisms in fungi moetic copulation : the fusion in pairs of sexual gametes, formed in specialized sporangia like (b) gamete-gametangial copulation : the fusion of entiated gamete-gametangial copulation: the fusion of entiated gamete of one sex with a gametangium of other sex; (c) gametangial copulation : the dired of gametangia without differentiation of gametes matic copulation : the sexual fusion of undifferential vegetative cells; (e) spermatization : spermatia with receptive hyphae of the opposite (compatible) strain.

Physiology

Fungi are able to withstand extreme environmental condition than most other microorganisms. Yeasts and molds can grow on substrate or medium containing concentrations of sugars that inhibit most bacteria. eg : jams and jellies may be spoiled by molds. Some yeasts are facultative, can grow under both aerobic and anaerobic conditions. Fungi grow over a wide range of temperature, with the optimum for saprophytic species from 22-30 °C. Pathogenic species prefer a higher temperatures optimum generally 30-37 °C. Some fungi also grow at 0 °C. Fungi are capable of using a wide variety of materials for nutrition (heterotrophic). They cannot use inorganic compounds like CO_2, but they use, ammonium salt etc.

Cultivation

Molds and yeasts can be studied by the same general cultural methods. Nearly all of them grow aerobically on the usual bacteriological culture media at temperature ranging from 20 - 30 °C. They grow slowly than bacteria on SDA (Sabourauds Dextrose Agar), and this medium is widely used to grow various pathogenic fungi.

Classification

Fungal classification is based primarily on the spore types and type of life cycles.

They are divided into perfect and imperfect fungi and they are provisionally placed in *Deuteromycetes, Ascomycetes* and *Basidiomycetes.* The classification scheme was proposed by Alexopoulus.

Some Fungi of Special Interest

The characteristics and activities of *Deutromycetes* fungi are described as under :

1. *Synchytrium :* More than 100 species of genus are parasitic on flowering plants. The most serious parasite is *S. endobioticum*. eg : black wort disease of potato. It forms warts, they are galls on which the host cells have been stimulated the divide by the fungus. The zoospores released from sporangium are capable of swimming for about 2 hrs in soil water until a new host tubes is found to infect. The zygote then encysts on the surface on the host epidermis and penetrates the host cell to infect it. The host cell reacts by undergoing "*hypertrophy*" i.e. increased cell volume.
2. *Saprolegnia :* This is the most common fungus in soil and fresh water, hence commonly called as water molds.
3. *Mucor :* This occurs abundantly on soil, fruits, vegetable and starchy foods. Some cause food spoilage; others are used for manufacture of foods. Sporangiospores may be branched, columellae are present. No stolons or rhizoids are produced.
4. *Rhizopus :* These are the common bread molds which cause much food spoilage, grow on vegetables, fruits, and other food products. Morphologically are non septate, cottony, mycelia with sporangiophores arising at the nodes where the rhizoids form.

Ascomycetes

5. *Schizosacchoromyces :* Yeasts belong to this genus. They reproduce by transverse division and also with ascospores. The best known species is *S. octosporus* which has been isolated from honey.
6. *Saccharomyces :* There are 30 species of *Saccharomyces*. The best known species is *S. cerevisiae*. This is found in nature on ripe fruits. Reproduction is by budding. The fusion normally occurs only between cells of differing types called "legitimate copulation".
7. *Neurospora :* This fungus is an Ascomycete : *N.crassa* and *N. sitophila* are common species. Some of them are responsible for food spoilage and some are used in fermentations.

Basidiomycetes

Approximately 12,000 species are known. None was implicated in human disease until recent times. *eg : Cryptococcus neoformans* now it is called *Filobasidetla neoformans* is a human pathogen, causing generalized mycotic (systemic infection) involving the blood stream as well as the lungs, and other organs.

8. *Agaricus :* Commonly known as mushroom. Pink coloration of the young gills is due to cytoplasmic pigment in the spores.
9. *Aspergillus :* This is a common fungus in nature, being found on fruits, vegetables, etc. They are economically important because they are used in industrial fermentations.
10. *Penicillium :* This is common in nature and is used in the ripening of cheese and in industrial fermentations. Fungus produces brush like conidiospores hence the name *Penicillium*.
11. *Candida : Candida albicans* is often isolated from warm blooded animals. It exists as normal flora. But sometimes this becomes pathogenic causing candidiasis called thrush.

Molds and their Association with other organisms

There are some interesting partnerships in nature involving a mold and some other organism. These associations are the partners of dependent (fungi) and independent (algae) nature of organism.

eg : Lichens, fungi and nematodes, fungi as parasites of insects, mycorrhiza.

1.8 Biology of Lichens

Lichens are the most familiar and wide spread of the stable self supporting symbioses involving an alga (phycobiont or photobiont) and fungus (mycobiont). They have a morphology, physiology and ecology which is unlike that of either the mycobiont or phycobiont component.

The lichens grown on wide ecological range soils, rocks, and tree trunks and are found in such diverse habitats as deserts, sea shores, tropical rain forests, arid deserts, mountain peaks of arctic regions, that are barred to many other life forms. Infact they occur in all continents particularly southern hemisphere. Lichen thallus structures varies from crustose forms which are tightly appressed to the substrate to leafy (foliose) forms and highly branched (fructicose) species. Reindeer moss (*Cladonia* species) a fructose lichen which grows to a height of 20-30 cm occurs over root areas of the arctic tundra and is a important for reindeer and caribou. Of the 70,000 species of fungi known approximately 25% are lichenised. Fungi that have become lichenised to diverse taxonomic groups and the algae that have been identified within them belong to 27 different genera comprising about 24,000 species. There are about 1000 lichen species.

Lichens may be divided into three distinct morphological groups based on their thallus growth habit (Fig. 1.17).

1. *Crustose type :* Crustose lichens are closely associated with stones and rocks and it is impossible to dislogic it with out breaking it. These lichens grow embedded in their substrate, the thallus is insignificant in size and without distinct lobes. The thallus lacks a lower cortex fungal layer.

   e.g., Graphic scripta, Haematoma puniceium.

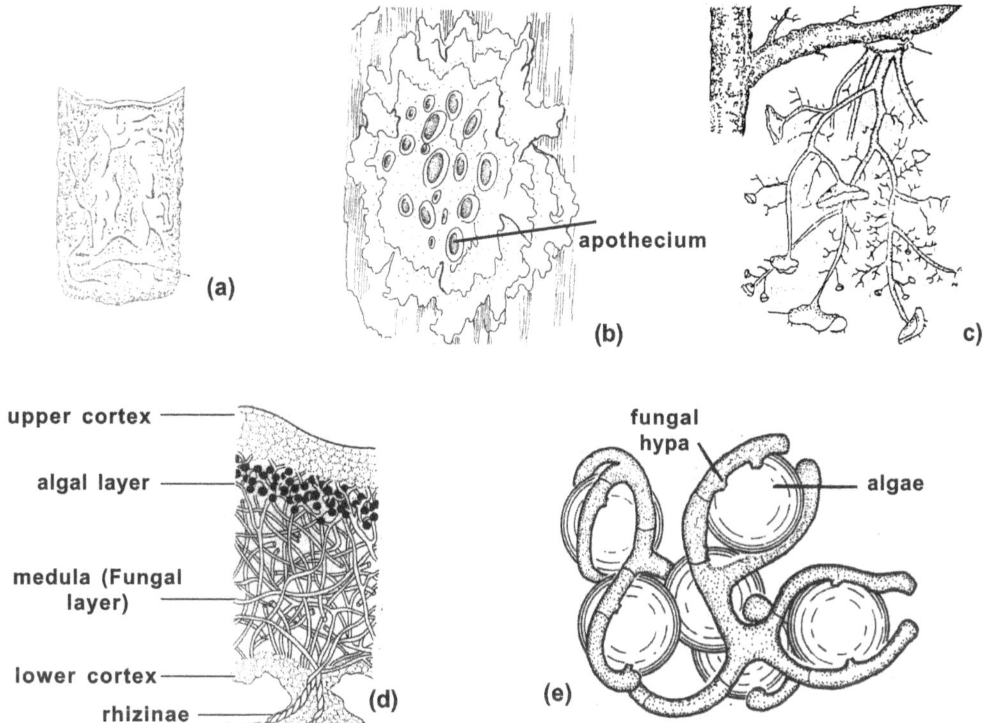

Fig. 1.17 (a-e) Lichens. Three fundamental types of lichen thalli. (a) crustose lichen (b) Foliose lichen (c) Fructicose lichen (d) Cross section of lichen thallus. Algal cells are intermingled with fungus hyphae in the upper part of the thallus (e) Fungal hyphae in close contact with the surface of the algal.

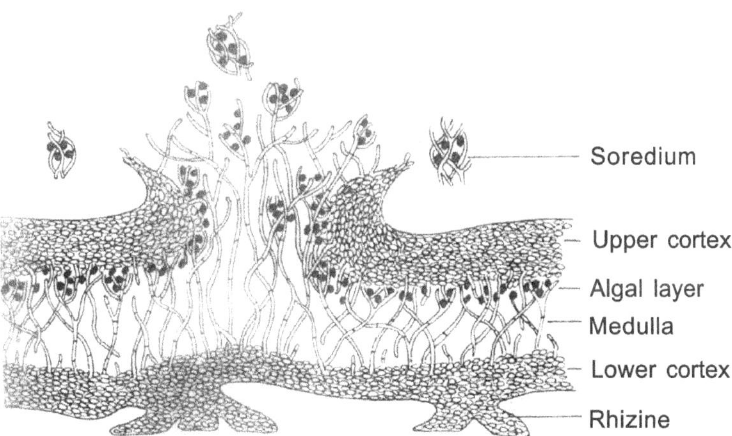

Fig. 1.17 (f) Cross-section of a typical foliose lichen showing the release of soredia. Aglal cells in the thallus are localized in the upper part of the medulla. Rhizines are bundles of fungal hyphae that anchor the thallus to the substrate and conduct water by capillary aciton.

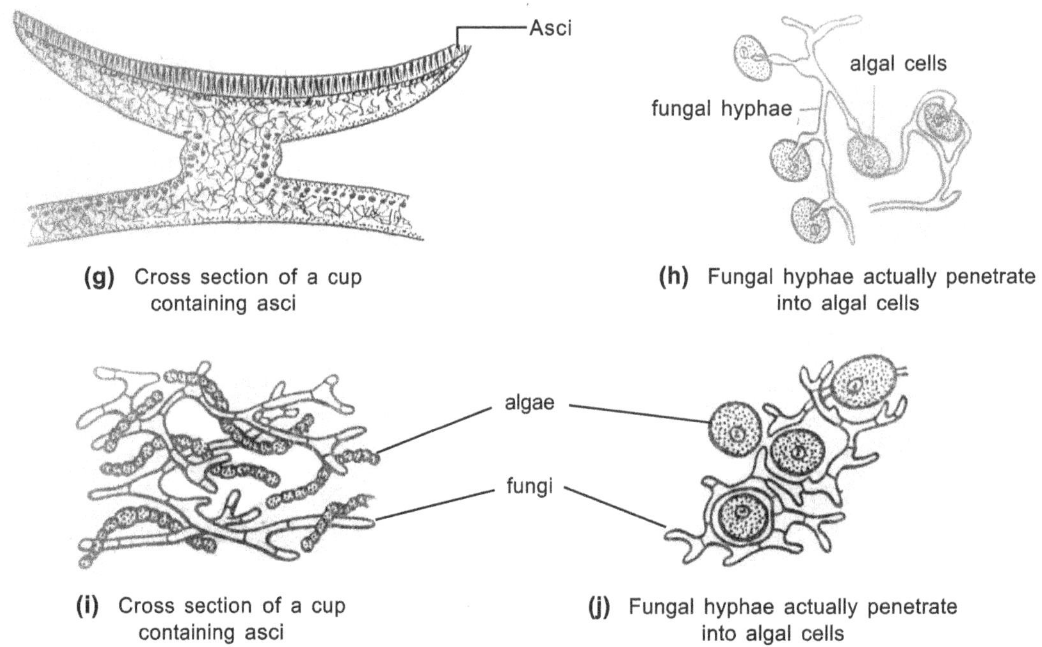

(g) Cross section of a cup containing asci

(h) Fungal hyphae actually penetrate into algal cells

(i) Cross section of a cup containing asci

(j) Fungal hyphae actually penetrate into algal cells

Fig. 1.17 (g - j) A trend in evolution from parasitism to mutalism in the lichens. In some primitive lichens the fungi actually penetrate the algal cells, as in diagram. In the more advanced species the two organism live in greater harmony for mutual benefit, as in **i** and **j** (After Odum. 1963)

2. *Foliose type :* Foliose lichens are more striking, flat, circular or lobed and grow loosely or only centrally attached to their substrates. The thallus has a lower cortex and often provided with specialized attaching filamentous structures called *rhizinae.*

   e.g., Xanthoria, Physcia, Peltigera, Parmelia, Certraria.

3. *Fructose types :* These lichens are most conspicuous, upright or pendent and attached to the substrate only at their base by a flattened disc. They may be ribbon shaped, cylindrical, ropey and shrobby in form, forming extensive and attractive growths standing out from the rocks, foliage and branches of trees.

   e.g., Cladonia, Usnea, Ramalina.

Chimeroid lichens, such as chlorophyte containing lichen *Peltigera,* have two algal components and produce cephalodia, outgrowths on the surface, which contain the cyanophytes.

Symbiotic associations of marine algae and marine fungi exist in which the fungal component grows intercellularly in the algae, but does not modify the morphology of the algae, nor lead to the productoin of chemical compounds unique to the association. These are not true lichens and have been termed as "*Mycophycobioses*". The best known of these as *Mycospherella ascophylli* which occurs in the phaeophytes *Ascophyllum nodosum* and *Pelvetia canaliculata.* It grows intercellular in the algae, without penetrating the cells and forms ascocarps and spermagonia in the host receptacles which may appear on their surfaces as black dots.

Lichens thallus is spongy and absorbs water is composed of two different organisms, a fungus and an algae. Both forming a self supporting combinations. Fungal component is mycobiont and the algal component is phycobiont. Both the phycobions and mycobiont are capable of independent growth in axentic culture and therefore no obligate relationship. Lichens are frequently cited as examples of mutulaism, but while they may occupy niches unavailable to the phycobiont in the free liching state, it is difficult to see any advantage conferred on the individual algal cells.

The fungal component (mycobiont) constitute more than 90% of the biomass of most lichens and aggregate into massive fruiting bodies. Nearly 29% of all species of fungi are found as mycobionts in lichens and most of them belong to the ascomycetes though some basidio mycetes. The algal components are chlorophytes or cyanophytes, approximately 30% genera are known of which the chlorophytes the unicellular alga, *Trebouxia* and *Pseudotrebouxia* account for over 70% of all lichen species while 10% of all lichen species contain blue-green phycobiont, the heterocyst containing filament, *Nostoc,* is the most common form.

The algae form only 3-10% by weight of the thallus body the details of thallus volume and percentage of algal wall in contact with fungal wall are shown in Table 1.5.

Table 1.5

Production of thallus volume	
Fungus	43.3% ± 0.9
Alga	6.7% ± 0.3
Extracellular matrix	34.1% ± 1.4
Air spaces	18.2% ± 1.1
Percentage of algal wall in contact with	
Funal wall	21.3% ± 2.5
Extracellular	35.1 ± 3.6

Thallus morphology is determined largely by the mycobiont although differentiation of the lichen thallus depellds on the presence of the phycobiont. Cases are known in which the same lichen fungus will develop very different morphologies if associated wilh different phycobiont. A few lichens are known to contain both green and blue green phycobionts in different parts of the same thallus. In these species the green alga is the principal phycobiont and the blue green algae is restricted to localized portions of the thallus called '*cephalodia*'.

The internal tissue of most lichens is differntiated into three or four layers, with the algae localised in a layer close to the surface of the thallus but beneath a tightly packed cortex layer as shown in Fig. 1.17 (d & e). There is always very close contact between the non-green branched fungal hyphae and the algal cells. In some cases, the fungal hyphae form a deep invagination in the algal cell wall or actually penetrate the cell wall and cause a invagination of the plasma membrane. These haustoria - like penetrations are not as common as one might imagine and are not necessary for transfer of material between symbionts.

There is one way flow of nutrient from algae to fungi and upto 90% of carbon fixed in photosynthesis move into the fungus. Carbohydrates accumulate in the primary thallus and ammonia in the cephodia. Green algae excrete polyols, erythritol, ribitol or sorbital and the cyanophytes produce glucose. The algal photosynthesis are converted to mannitol in the fungal thallus. Excretion of carbohydrates from the phycobiont cease almost immediately on their isolation into culture. The supply of carbohydrate is many times more than is required for the growlh of the fungus.

Lichens reproduction may occur by fragmentation of the vegetative thallus or by the liberation of small, wind disseminated minute spherical bodies; fungus-algae units called *soredia* which may reestablish the lichen on fresh substrate or larger corolloid outgrowths called *isidia*. The mycobiont also regularly forms typical ascomycetes fruiting bodies (*ascocarp*) which liberates ascospores. Evolution from parasitism to mutualism is clearly observed. In primitive lichen penetrate the algal cell and are thus essentially parasites of the algal. In the more advanced species two fungal mycelia or hyphae do not break into the algal cells, but the two live in close harmomy. The characteristic of the ascospores as well as fruiting structures the apothecium important in lichen identification and classification. Since these spores must germinate in the viscinity of potential phycobionts to reestablish the symbiosis; their contribution to lichen reproducrion is uncertain. (Fig. 1.17)

Sexual reproduction is morphologically similar to that found in non lichenised ascomycetes although for lichens many cytological details are lacking. The ascognia have trichogynes and there are coniodia formed in flask shaped structures the pycnia that probably function as spermatia and provides the second nucleus that takes part in the karyogomy within the ascus. The conidia do not ordinarily germinate while the ascospores germinate readily.

In lichens a trend in evolution from parasitism to mutualism is clearly seen. In primitive lichens fungi penetrate the algal cell and are thus parasites of the algae. In the more advanced species, the fungal mycelia do not break into the algal cells but the two live in close harmony (Fig. 1.17(h,j)).

The edible lichens are harvested and dried for human consumption or as fodder for cattle swine and horses. Lichens are commonly used in India and elsewhere as major components of various spices and condiments. *e.g., Cladonia rangiferina; Parmelia sp; Cerraria islandica; Lecanora esculenta.*

Lichens *e.g., Rocella montaignei* are also used in the preparation of natural dyes. Used for centuries as dye in colouring fabrics and paints. Litmus paper is used as a pH indicator in all scientific laboratories throughout the world.

Lichens are used in medicines. Few species of lichens have been used in medicines, for jaundice, fever, diarrhoea, epilepsy and various skin diseases. Several of them contain antifungal and antibacterial compounds. *e.g., Peltigera camona, Lobaria pulmonania, Everina sp. Usnea sp. Cladonia sp. Catrania islandica.*

The sweet-scented thalli of same lichens are used in the manufacturer of dhup, hawan samagris, and other perfumeries.

Lichens secrete organic acids which graudally dissolve and disintegrate the rocks over which they grow. They are the pioneers in soil formation.

About 90% of lichenised fungi associated with green algal cannot fix atmospheric nitrogen. The remaining 10% containing blue green algal genera. *Nostoc*, *Scytonema*, *Stigonema*, *Dichothrix* and *Calothrix* can fix atmospheric nitrogen. Surprisingly lichens are not especially characteristic of N_2 deficient soils/habitats, though the algal fix N_2 within the host and leak substantial quantities to it.

Lichens accumulate many distinctive organic substances (350 secondary metabolites); most of these are aromatic compounds; the most characteristics being depsides and depsidones. Other aromatic compounds are quinones pulvic acid derivative, xanthone derivatives and dibenzofuran derivatives. Various non-aromatic acids, triterpenoids and polyphenols are also found. These substances have been used in taxonomy of some lichens.

Since all lichens are very slow growing, they are the best colonisers of rock surfaces, rarely extend faster than 1-5mm per annum. They are very efficient accumulators of mineral cations, which they presumably extract from rain water and from the flow and drip down the branches of trees. Their ability to accumulate minerals makes them particularly sensitive to environmental contamination from pollution by heavy metals, flouride and for this reason they are among the most sensitive indicators of environmental pollution especially SO_2 and totally disappear in industrial areas. The '*quality*' of an environment can be best judged rather accurately from the presence of lichen growth on grave stones.

1.9 Life History of Bryophytes

The term Bryophyte was first introduced by Haeckel (1866) to plants that forms embryo from zygote by mitotic divisions. Engler (1886) suggested the term Embryophyte for the plant groups producing embryo in their life cycle which include Bryophyte, Pteridophyte, Gymnosperms and Angiosperms.

Bryophytes are the first land plants being simplest among embryophytes and highly advanced among the atracheata. They are larger than micro organisms and smaller than the higher plants. They belong to an intermediate megascopic realm. They also occupy an intermediate position with relationships on the one hand to the Chloroplyceae (green algae) and on the other hand to the higher archegoniatae; Pteridophyte. Bryophytes are represented by 960 genera and 24,000 species and comprises of liverworts, hornworts Hepaticopsida and mosses and these arose from green algae. The hornworts represents terminal point of Bryophyte evolution rather than the starting point for the higher plants.

The Bryopsida constitute small green plants called "*mosses*" that grow on walls, tree trunks or ground. There are 15000 species occurring in arctic, antarctic, temperate and tropical regions. The plant body is stem like called "caulid" and the leaflike structures the "*slyllids*". Like liverworts, mosses exhibit well defined alternations of generations between a gametophyte (n) (the protonema) which produces the adult green leafy gametophyte which on maturity produces male and female clusters. The zygote of the archegonium by the union of an egg and sperm begins the sporophyte (2n) which is embedded in the nutritive tissue of the enlarged archegonium called the calyptera. This on maturity produces a pin head like structure the capsule on a needle like seta. The sporangium contain the haploid spores which on germination produce the gametophyte.

Habitat

These are terrestrial, restricted to moist soil, shady places, damp walls, damp rocks, decaying logs in forest. Some species are purely aquatic e.g. *Riccia fluitans, Ricciocarpus natans, Riella* and *Fontinalis*. These are also called as marginal land plants. Some species like *Andraea* grows in extremely exposed conditions. Bryophytes are excellent indicators of the nature of soil, one can tell at a glance whether the substratum is acidic or basic, dry or wet or shady.

Habit

The plants are haploid, gametophytic and possess a dorsiventral thalloid body in primitive bryophytes called the liver worts. The plant body in higher bryophytes (Mosses) show erect stem, leaves and radial symmetry. These are called gametophores (*Funaria, Polytrichum* etc.). However, these structures (leaves and stem) are not homologons to those of vascular sporophytic plants. They are fixed to the soil with rhizoids, which are either unicellular or multi cellular. All bryophytes in general, show dichotomous branching. Vascular tissues are absent.

Fossil History

Fossil bryophyte spores have been reported from lower Cambrian deposits of eastern Baltic region.

Plant Size

They range from 1 mm (e.g. *Sphaerocarpus*) to 70cm (*Dawsonia*). The "*stem*" of *Polytrichum commune* from New Zealand grows to a height of 150 cms.

Nutrition

All bryophytes, with few exceptions like *Cryptothallus mirabilis* (a liver wort), are green and autotrophic. *Cryptothallus* is a saprophyte.

Reproduction

The bryophyte plants are haploid gametophytic and show,

1. Vegetative, and
2. Sexual reproduction

Vegetative Reproduction

The gametophyte undergoes vegetative reproduction and produces gametophyte. It occurs by

1. Fragmentation (eg. *Riccia, Marchantia*)
2. Gemmae (eg. *Marchantia*)
3. Adventitious branches (eg. *Targionia*)
4. Bulbils (eg. *Funaria*)
5. Tubers (eg. *Riccia, Anthoceros*)
6. Protonema (eg. *Funaria, Polytrichum*)

Sexual Reproduction

The life cycle of a bryophyte shows two distinct phases.

1. Gametophyte, and
2. Sporophyte

The gametophyte phase is long lived and therefore dominant. It shows sexual reproduction and produces the sporophyte. (Fig. 1.18).

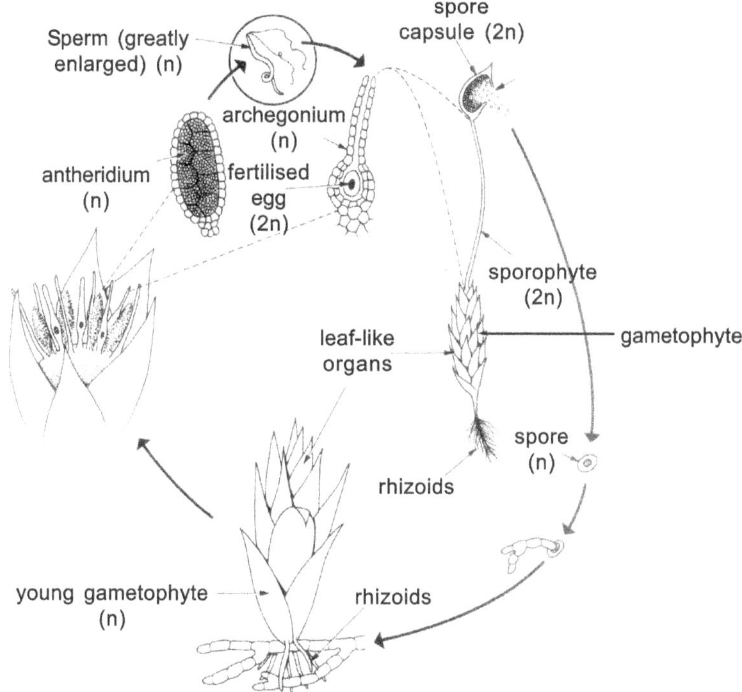

Fig. 1.18 The life cycle of a moss plant. The sporophyte is diploid and develops from the fertilised egg within the archegonium. The numberous spores in the capsule are produced by successive mitotic divisions after meiosis has occurred in the spore mother cell. The gametophyte and all its organs are haploid.

Sex Organs

Sexual reproduction is oogamous. Sex organs are multi cellular and complicate. The male sex organs is called the antheridium and the female, the archegonium. The plants may be monoecious (*Riccia, Funaria*) or dioecions (*Marchantia*).

Antheredium is club shaped and produces biflagellated (sperms) antherozoids. Archegonium is flask shaped bearing an egg cell in its venter. The antherozoids are attracted by archegonial chemicals such as sugar, salts and malic acid. Thus they show chemotactic movements. Fertilization is internal and oogamous and requires water medium for transport of the antherozoids to reach archegonia. Hence, they are called Amphibians of plant kingdom. The diploid zygote formed as a result of fertilization undergoes repeated mitotic divisions to produce embryo called sporophyte. The sporophyte never gets separated from the gametophyte.

Sporophyte

It is the diploid phase and also called *sporogonium*. This is seen endophytic (*Riccia*) or upon the gametophore. It is either parasitic on the gametophyte or semiparasitic. It is differentiated into foot seta and capsule. Its fertile part is the capsule. It consists of spore mother cells that show meiosis and produces haploid spores. It shows asexual reproduction.

The spores are unicellular, uninucleate, haploid and are the first cells of the gametophytic generation. The spores germinate and produce directly a new gametophyte or a protonema which later produces a gametophyte.

Alternation of Generations

The life cycle of a typical bryophyte shows (Fig. 1.19(a)) regular alternation of generations, Gametophytic generation alternating with the sporophytic generation. Gametophyte is independent and sporophyte being either parasitic or semi parasitic. Both differ in their morphology also. Thus Bryophytes, according to Holfmeister, posses heteromorphic alternation of generations.

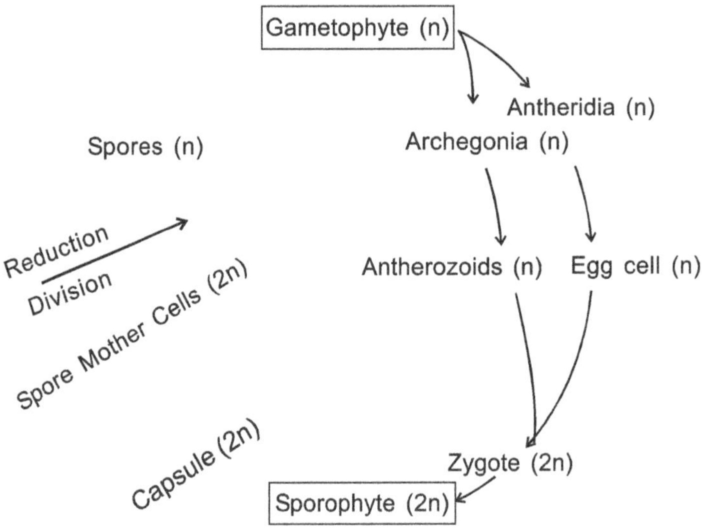

Fig. 1.19(a) Alternation of generation in a bryophyte

1.10 Life History of Pteridophytes

The term Pteridophyte has Greek origin (*Pteron* = feather, *phyta* = plants). The term was given originally to a plant group that bears feather like pinnate fronds. Tippo placed them in Tracheophyte. These are the first true land vascular plants. The plants are placed between Bryophytes and Gymnosperms.

Psilopsida (Rhyniopsida) represents the most ancient and earliest vascular plants, *Cooksonia* fossils from Europe and North America dating back to over 400 million years have been discovered from the late Silurian rocks. The sporophyte is simple, dichotomously branched, green without leaves and stomata but with rhizoids. Fossils of *Psilophyton* and *Zosterophyllum* about 395 million years old show some similarlity with the two living genera *Psilotum* and *Tmeispteris* which are also leafless dicotomously branched delicate sporophytes with rhizoids, sporangia are lateral and trilobed, the spores are homosporous.

The Lycopsida are a group of about 1000 mostly tropical small slender plants with spirally arranged microphylls. The sporphytes are prostrate and branched. Two living families Lycopodiaceae and

Selaginellaceae represent this group. The sporophytes are arranged in compact strobili; the sporophylls bear single sporangium on the abaxial surface. Spores may be similar (homosporous) as in most *Lycopodium* species, or dissimilar (heterosporous) as in some *Selaginella* species. The species germinate to form gametophytes.

The Sphenopsida (the horsetails) constitute a primitive group of vascular herbaceous plants represented by *Equisetum* the plants are less than a meter tall found in wet habitat. The sporophyte is differentiated into tall grass like hollow stems (with nodes and internodes). The branches arises in whorls, the leaves are small. The reproductive structures are terminal, compressed, strobilus, made of several sporangiophores, each sporangiophore has 5-10 sporangia, spores with 4 banded elaters that germinate to produce gametophytes. Thus, they show alternation of generations. Asexual phase is the dominant perennial sporophyte while the sexual phase is the inconspicous delicate gametophyte.

The Filicopsida is the largest group of spore bearing plants with about 15000 species of tropical to temperate wild or ornamentals ferns. The sporophyte is a well organised rhizome (under ground stem) with adventious roots, feather like leaves; the fronds which may be several meters long as in tree ferns. Fronds show circinate vernation and perform dual function of photosynthesis and reproduction. Umbrella like sporangia are formed in clusters (sori) on the lower surface and are covered with a thin tissue called the *indusium*. The spores may be more per sporangium (as in Eusporangiate ferns) or may be less in numbers (as in Leptosporangiate ferns); may be similar (homosporous) or dissimilar (heterosporous). The spores germinate to produce prothallia that bear antheridia and archegonia on the lower surface among the rhizoids. Upon fertilization, the egg of the archegonium develops into a zygote embryo and finally into mature sporophyte.

Fossil History

The ferns have a long fossil history. They flourished well during Devonian, Mississipian and Pensylvanian periods of the late Paleozoic. The late Paleozoic era can rightly be regarded as the '*Age of Pteridophytes*'.

Habit and Habitat

The sporophyte is the dominant generation and are mostly terrestial, grow in moist and shady places. The plant body shows considerable variations. Ferns like *Azolla, Salvinia, Marsilea* are aquatic, species. *Pleopeltis* is epiphytic. The plant body is divisible into root, stem and leaves. All the living Pteridophytes are herbaceous annual; leaf bearing except few woody ferns like *Angiopteris* and *Cyathea*. The leaves are megaphyllous and are called fronds which show dichotomous venation and bear sporangiam on the lower surface. Branching may be dichotomous or lateral. Growth of the plant is due to apical meristematic tissue.

Anatomy

The distinct anatomical feature of pteridophytes is the presence of vascular tissues which are well developed although it is thought to be more primitive than other members of pteropsida.

Reproduction

The plants are diploid, sporophyte and show asexual reproduction to produce gametophyte. Sporocarp (*Marsilea*) or sporophylls fronds (*Pteris, Adiantum*). They bear sporangium that contain spore mother cells which undergo reduction cell division to produce haploid spores.

The plants may be homosporous or heterosporous (*Marsilea*). In either case the spores produce gametophyte. Homosporous sporophytes produce only one kind of spores which in turn produce homothallic (Monoecious) gametophyte. Heterosporous sporophyte produces 2 kinds of spores i.e. microspores and megaspores. They form male and female gametophyte respectively.

Sexual Reproduction

The gametophytes may be independent or parasitic on the sporophyte. Pteridophytes are the only group of plants where one can find independent sporophyte and gametophyte. The gametophyte is reduced, short lived and shows distinct sexual reproduction. The sex organs are antheridium and archegonium which are embedded in the tissue of the gametophyte and are greatly reduced in size. Antheridium produces biflagellated or multi flagellated (sperms) antherozoids. Archegonium produces egg cell in the venter. Fertilisation is oogamous. The zygote thus formed undergoes numerous misotic divisions to produce sporophyte.

Alternation of Generations

The life cycle of a typical pteridophyte shows two distinct stages 1. Sporophyte 2. Gametophyte. The sporophyte is the dominant, long lived and autotrophic and undergoes asexual reproduction to produce gametophyte which is parasitic upon it. (Fig. 1.19(b)).

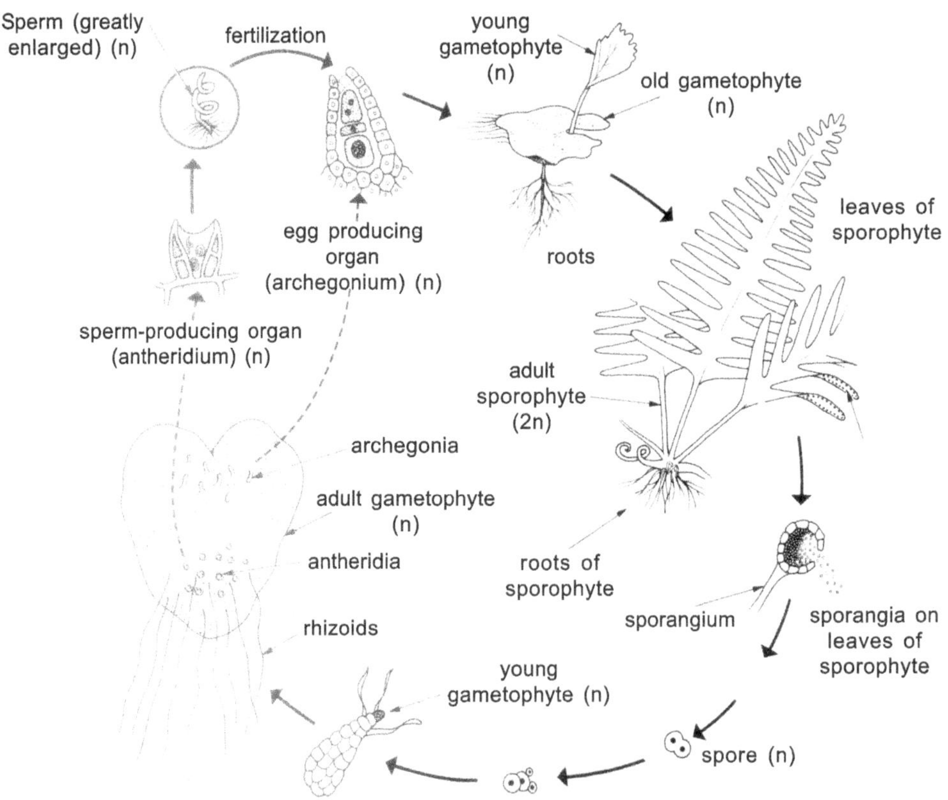

Fig. 1.19(b) The life cycle of a fern plant

The gametophyte is either autotrophic or partially parasitic on the sporophyte. It undergoes sexual reproduction to produce the sporophyte The sporophyte and gametophyte differ in both morphology and anatomy. Thus, the life cycle of a typical pteridophyte shows regular heteromorphic type of alternation of generations. (Fig. 1.19(c))

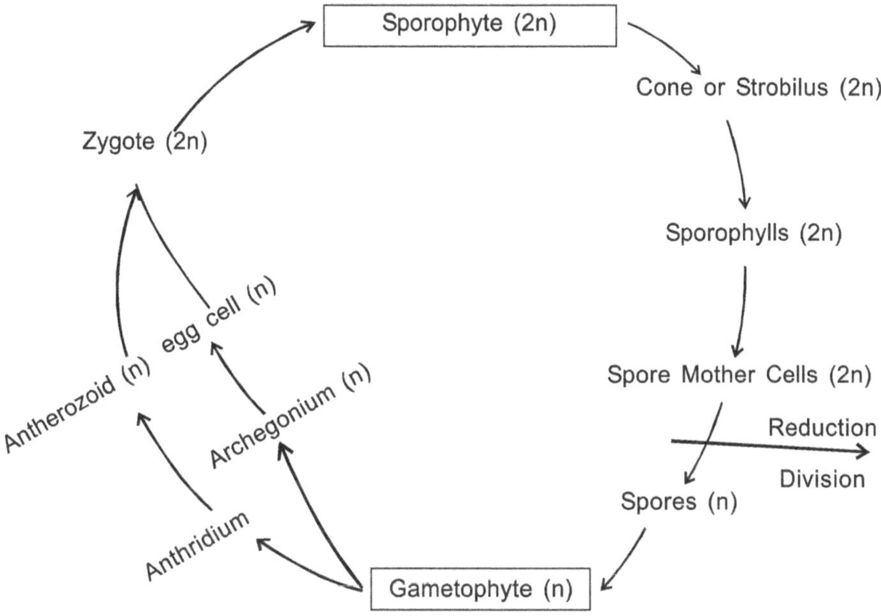

Fig. 1.19(c) Alternation of generation in a fern

1.11 Life History of Gymnosperms

The term Gymnosperms was first used by Theophrastus, in his book '*Enquiry into Plants*' to describe plants whose seeds are naked or unprotected. These are the simple primitive group of seed plants and form, a bridge between the pteridophytes and angiosperms. They posses certain features peculiar to themselves and some of them exhibit distinctly fern features and others show close affinity to angiosperms. However, they are primitive seed bearing plants; the seeds occur on the surface of the sporophyll and not contained within an ovary as in angiosperms.

Fossil History

Cycads first appeared during the close of Paleozoic but were more diversified and dominant during Mesozoic period. *Bjuvia simplex* a cycad of mesozoic era from the Rhaetic of Sweden shows general resemblance to *Cycas*. Cycads are derived from Cycadofilicales. Conifers are an early off shoot of Cycadofilicales. The fossil conifers belong to orders Cordaitales and Volziales.

Gymnosperms constitute one of the two major groups of seed producing plants, the other being the angiosperms. Living gymnosperms are evergreen, abundant in Northern Hemisphere, 60% of them are trees.

Important Features of Gymnosperms

1. Absence of vessels from wood (except *Gnetum*)
2. Absence of companion cells in phloem tissue.
3. Presence of many celled and many nucleated female gametophyte.
4. Presence of archegonia in the female gametophyte (except *Ginkgo* and *Welwitschia*)
5. Absence of double fertilization, (double fertilizations is unique feature of angiosperms only)
6. Absence of true flowers (true flowers is a unique feature of angiosperms only)

General Features

The Cycads are ancient palm like seed plants of tropics and cultivated in glass houses and garden and are easily mistaken for palms in view of their short thick stem with a crown of large feathery leaves at the summit. Cycads are dioecious. *Cycas, Zamia* and *Macrozamia* are the rich source of starch. *Cycas circinalis* is the "sago palm" of commerce. There are 9 genera and 100 species of cycads with only one family Cycadaceae for the world. Some of the cycads are rare botanically. In India several species occur; some are wild and endemic and others grown in gardens.

The Conifers are ever green pyramidal to conical trees of cool temperate regions and include yews, spruces hemlocks, cedars, pines, firs and junipers representing 600 species. Together they are the major component of vegetation in many parts of the world.

The conifers are the tallest (365' feet) and long lived (4,000 years old) red wood trees (eg. *Sequoia sempervirens*) of California, USA. Conifers are the most valuable trees of the world, valued for high quality wood, fuel, pulp, paper, plywood, veneer, resins, terpentines, essential oil, varnishes, edible seeds (chilgoza) and medicines such as *Ephedra* and *Ginkgo*.

There are 67 genera, of which 35 are restricted to North America. 20 to South America; 39 are found in the Old world and 11 in the New World. There are seven families, Pinaceae, Cup ressaceae Arucariaceae, Podocarpaceae, Taxaceae Taxoidiaceae and Cephalotaxaceae. The Taxales (5 genera and 20 species) and Coniferales (50 genera and 570 species) are important orders, the largest genera of conifers are *Podocarpus* with 80 species and *Pinus* with 95 species. *Ginkgo biloba* and *Welwitschia mirabilis* are other interesting monotypic gymnospermous plants. *Welwitschia* is a desert plant of Western South Africa with only two large, strap shaped perennial leaves arising from a short stout axis.

Habit and Habitat

Gymnosperm are predominately arborescent, woody, evergreen, perennial trees or shrubs. *Gnetum ula* and *Ephedra foliata* are lianes. *Zamia pygmaea* being the shortest (3-8 cms), *Sequoia gigantia* is the tallest and oldest tree in the plant kingdom.

Gnetum trinerve and *Podocarpus ustus* are parasites. The plants are xerophytic and grow in physical or physiological dry conditions. They are distributed through out the tropical and temperate zones including even the arctic regions of the world and form dense forests.

They posses taproot system. The stem is columnar either unbranched or branched as in *Cycas*. Branching is lateral. The leaves are monomorphic or dimorphic or fern like as in *Cycas*, simple needle like in *Pinus* but the Australian and New Zealand conifers (*Juniperus*) are *bushy* and have broad leaves. Angiosperm like leaves are found in *Gnetum*. Scale leaves are also present.

Anatomy

Anatomically gymnosperms are complex Vascular bundles consists of xylem, phloem and cambium. Plant shows secondary growth. Xylem is devoid of vessels (except *Gnetum, Ephedra)*. Xylem consists of tracheids. Xylem paenchyms and xylem fibres are absent. Phloem consists of albuminous cells instead of companion cells. The transfusion tissue in leaves is common in cycads. Wood is monoxylic (soft) in cycads but in conifers it is pycnoxylic (hard).

Reproduction

Gymnosperms are diploid, the cycads are dioecious while the conifers are monoeions. The flowers are produced in the form of cones. The male (staminate) flowers are unisexual and consist of microsporophylls that bear microsporangia. Microspore mother cells show meiosis and produce microspores (pollen grains).

The female (ovulate) flower consists of megasporophylls. The ovules are naked, attached laterally the megasporophylls. The main body of the ovule is called nucellus which is surrounded by a single integument. It consists a sporogenous cell that forms a linear tetrad of megaspores by meiosis. Only one megaspore becomes functional and the rest abort.

Gametophytes

In gymnosperms, the gametophyte is reduced but the sporophytic generation is the dominant and highly developed.

Pollen grain is the first cell of male gametophyte. It produces one or two male prothallial cells representing the body of male gametophyte. Antheridl cell consist a smaller generative cell and a larger tube cell. Generative cell produces a stalk cell and a body cell. The latter produces ciliated male gametes (*Cycas*) or non motile sperms (rest of the gymnosperms). Thus, male gametophyte is reduced to few celled *stage*. These details are schematically shown below.

The megaspore is the first cell of the female gametophyte. It undergoes numerous free nuclear mitotic divisions, followed by cytokinesis to produce a multicellular female gametophyte which is permanently retained in the ovule or megarporangium. It develops highly reduced archegonia; number varies from species to species. The egg cell is present in the venter. *Gnetum*, is a highly developed gymnosperm and does not produce archegonia and resembles angiosperm.

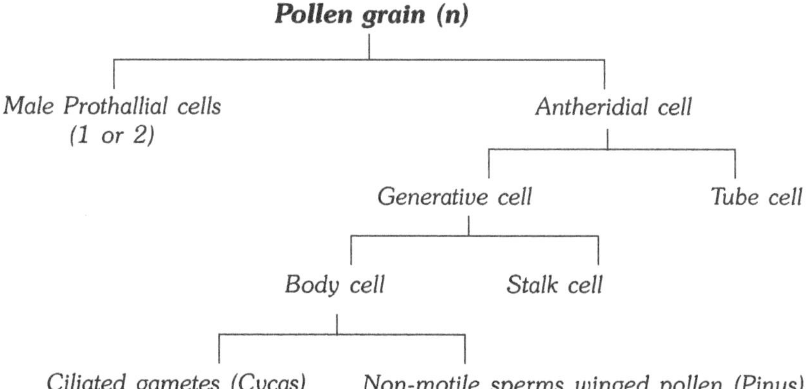

Schematic representation of the development of the male gametophyte in Pinus and Cycas

Pollination and Fertilisation

Pollination in cycads and conifers is anemophilous, and both are cross pollinated. Pollen grains germinate in the pollen chamber. Pollen tube acts as haustorium (*Cycas*) and also acts as carrier of male gametes to the vicinity of the egg cell. It is oogamous. Siphonogamy is accompanied by zooidogamy (*Cycas*) or Syngamy. As a result a diploid zygote is formed which later develops into embryo. Seed develops from ovule after single fertilisation in contrast to the angiosperms where seed is formed after double fertilisation. *Pinus* seeds show polyembryony.

Seed

Seeds are polycotyledonous and endospermic. Endosperm is formed from the female gametophyte before fertilisation.

The seed germinates with or without dormancy (*Cycas, Ginkgo*) and produces a new plant. Germination is hypogeal or epigeal.

Alternation of Generations

Gymnosperms show distinct heteromorphic alternation of generations. The plants are predominantly sporophytic which alternate with highly reduced dependant gametophytic generation. (Fig. 1.20 & 1.21)

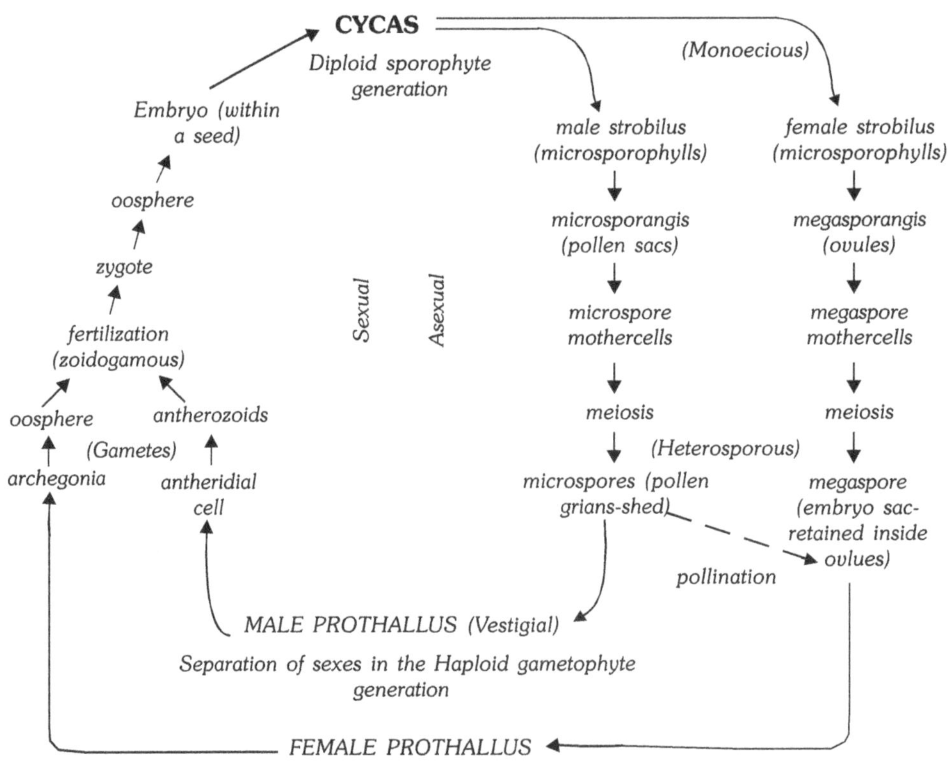

Fig. 1.20 Alternation of generation in *Cycas*

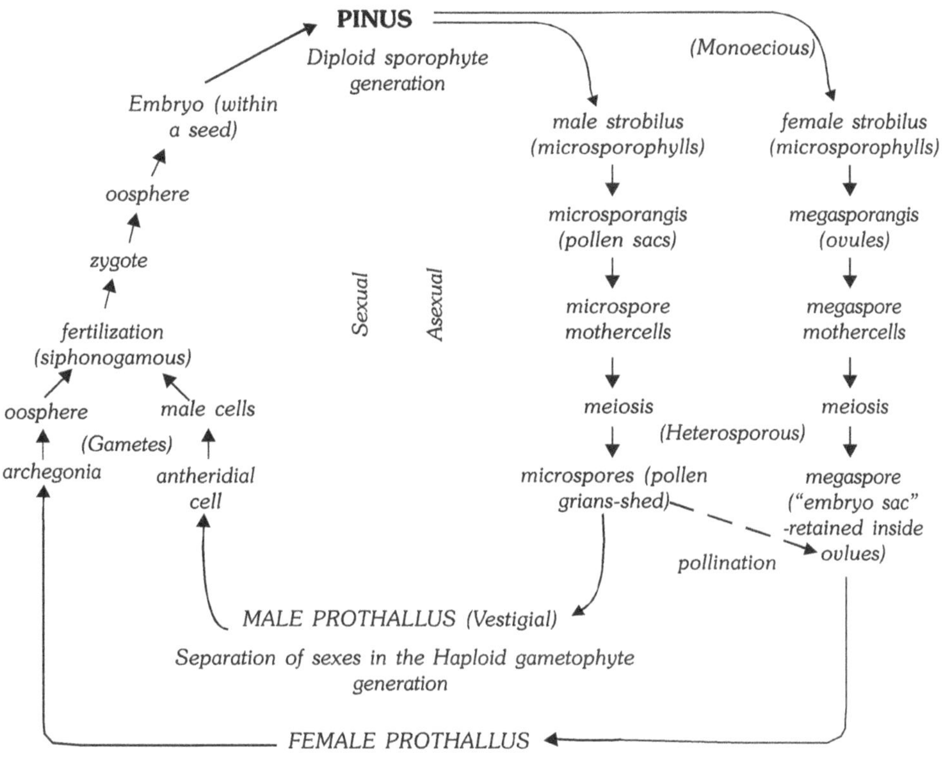

Fig. 1.21 Alternation of generation in *Pinus*

1.12 Life History of Angiosperms

The angiosperms (the flowering plants) are a dominant, highly advanced group of vascular plants which together with gymnosperms forms the dominant plant life of the geological era we live in. They are the products of long line of evolutionary development. The term angiosperm is derived from the Greek word (*Angios* = vessel, *sperma* = seeds) which is used to include all those plants that bear seeds enclosed within the ovary or fruit. They form the dominant vegetation on the earth represented by over 300 families, 12,500 genera and about 3,00,000 species. Yet there are hundreds of species to be identified from tropical forests, a feast for the upcoming taxonomists !

We are much more aquainted with flowering plants as we come across with them in daily life. They constitute the dominant vegetation of the earth and bear true roots, stems, leaves and of course flowers. They are more useful to us in food, clothing and shelter but are also useful to animals.

The angiosperms contains annual, biennial and perennial plants. The perennials may be monocarpic or hapaxanthic (e.g. *Agave*, bamboos) grow vegetatively for a number of years and then flower once and die or polycarpic (flowering repeatedly). Vegetative reproduction by the detachment of usually somewhat specialized plant parts may be pronounced. Trees and shrubs are numerous but herbaceous plants are in the majority forming the carpet vegetation.

Habit and Habitat

There are widely distributed and found in every nook and corner of the planet earth. They are mostly terrestrial but aquatic too. They can grow on the rocks, mountains, marshy areas, deserts, streams, ponds and perched upon their own kind as epiphytes. Thus they show maximum variation in their habitat.

Ecologically, the angiosperms range from desert inhabiting xerophytes to submerged aquatics (although they have been distinctly less successful in colonizing marine rather than fresh water habitats). They dominate plant communities from the tropical forests to the arctic tundras with their numerical superiority threatened only in cool-temperate coniferous forest areas.

The flowering plants display wide range of life forms may be herbs, shrubs, trees, lianes, twinners, stragglers and climbers. They range from (smallest 1-1.5mm) *Wolffia,* to tallest 350 feet in height *Eucalyptus*; to thallose Podostemaceae; to the huge Banyan tree that bears the greatest crown of foliage as seen in Calcutta Botanic Garden or at Mahboobnagar, A.P.

Nutrition

Angiosperms are mostly autotrophic yet show great variations; some plants develop symbiosis with fungi (eg. *Monotropa*). Some angiosperms are parasites which include root parasites, stem parasites partial as well as total. Some plants are insectivorous and saprophytes. A few examples are listed below.

1. *Rafflesia arnoldi* : Total root parasite of Java with largest flower (about a metre across).
2. *Cuscuta reflexa* : Green leafless, twining total stem parasite common in gardens.
3. *Striga densiflora* : Green plants that are partial root parasite can be seen in tobaco fields.
4. *Cassytha filiformis* : Green leafless twining partial stem parasite can be seen hanging on forest trees.
5. *Nepenthes, Utricularia, Drosera, Dionaea* are the various Insectivorous plants that grow in nitrogen deficient soils and utilize proteolytic enzymes to digest the captured insects.
6. *Microstylis*, *Neottia*, *Rhizanthella* are Saprophytes.

Many angiosperms (including all saprophytes) show an intimate association between their roots and soil-inhabiting fungi (mycorrhiza) which appears to be of nutritional assistance to them. Leguminous plants and some others develop root-nodules containing nitrogen fixing bacteria; a similar function may be performed by nodules developed in the leaves of tropical trees of the family Rubiaceae.

Fossil History

The angiospermic group is more modern than any other group of vascular plants. Their fossil remains though inadequate amply demonstrate the phylogeny of angiosperms. The fossilised wood and the impressions of leaf form and venation indicate that the angiosperms evolved by the middle Cretaceous period. Axelord (1952) suggested the existence of angiosperms during Triassic and Jurassic periods.

Axelord and Eames (1961) postulated that the angiosperms have been in existence at the end of the Paleozoic era. Lemsle (1946) regards Bennettitales as the probable ancestors of angiosperms. The earliest evidence of flowering plants is fossil pollen grain and fragments of flowers and fruit from lower Cretaceous rocks resembling living members of Chloranthaceae and date back to 110 million years.

Fig. 1.22 illustrates the various plant organs of a typical flowering plant which represent the sporophyte generation. As shown the plant is organised with a main axis; the root system below the soil meant for anchorage with the soil and absorption of water and mineral salts. The shoot system above the ground with leaves, branches and flowers adapted to perform exchange of gases, photosynthesis and reproduction.

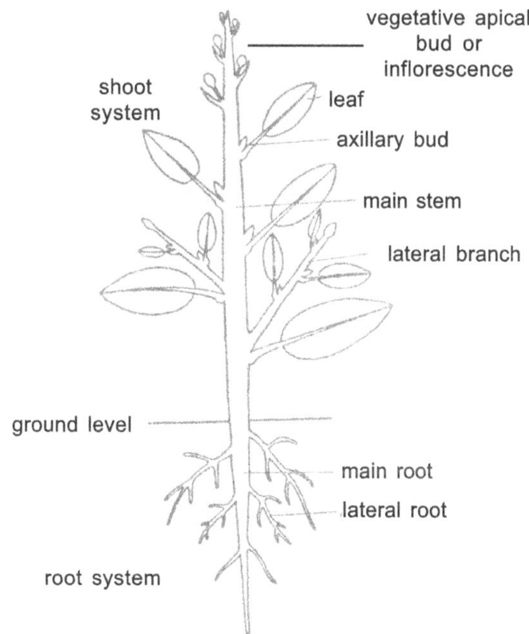

Fig. 1.22 Principal plant organs of a typical flowering plant

The Root

In the evolution of Tracheophytes, the root is apparently the last organ of the "big three" (roots, stems, and leaves) to develop. Stems in the form of rhizomes were the anchoring organs and rhizoids arising from the epidermal cells of the rhizomes provided the necessary absorptive function for the earliest primitive vascular plants, the Psilophytales and for a few species of Psilotales that are still living today. Development of the primary axis resulted in a differentiation of root and stem and both organs assumed an independent division of labour as well as conspicuous morphological differences. However, the root has remained simple structure. This is probably attributed to uniformity of the underground habitat in comparison with the aerial, where greater modification in tissue and organ development led to more specialized structurese in stems and leaves. (Fig. 1.23). The epiphytic roots of orchids are the velamenoid roots that are chlorophyllous. In *Taeniophyllum*, an orchid the roots are flat and green.

The Stem

A stem is an axis that bears leaves and buds and is characterized by the presence of *nodes*, the points of attachment of the leaves and *internodes*, the regions between two successive nodes. In the course of development as well as at maturity their relationship between a stem and its leaves the two structures are commonly treated as the shoot. The main apex is active (monopodial) or terminal bud inactive and growth is by uppermost lateral branch i.e., sympodial.

In addition to difference in the exent of internodal growth, stems exhibit other variations in form and behaviour; woody stems, characteristic of trees and shrubs, live for many years. In each growing season, they increase in length and become larger in diameter as a result of secondary growth. Herbaceous stems form little woody tissue and show little or no increase in diameter.

The form of a stem may be highly modified from the usual concept of an aerial axis bearing foliage leaves. The variations in structure are underground stem which may be thick and fleshy with major function of storage of food reserves as in ginger (rhizome), onion (bulb), colocasia (corm), potato (tubers).

The leaf

The leaf is a flat and lateral appendage borne upon an axis, or stem, and the stem and leaves together constitute a shoot. The diversity in size and shape of the leaves is enormous. A leaf is typically a flat and expanded organ, primarily concerned with photosynthesis; but frequent and drastic deviations in both morphologically and functionally. Although leaves are treated as distinct organs in classical morphology, the full significance of the leaf emerges only when it is considered as part of the shoot system, having its origin together with the stem in the shoot apical meristem.

The leaves of the Pteropsida (ferns and seed plants) are called megaphylls and have arisen through modification of parts of the shoot system so that they are equivalent to lateral shoots unlike the microphyll of primitive land plants which represent an enation (out growth) from the stem which gradually became vascularized in the course of its evolution.

The leaves may be simple or compound (with pulvinus), with leaflets and may be with stipules, spines, sheath or tendril. The venation may be parallel in monocots and reticulate in dicotyledons.

The Flower

A flower is a short branch appendages specialised for sexual reproduction of a typical angiosperm. The flowers may be borne singly or in grouped into inflorescence. Several flowers arranged in order. Thus, the "*flower*" of the Asteraceae is in reality a cluster of many, small flowers arranged on a common flat receptacle or head. The ray flowers are sterile, consisting of large attractive petals while the small disc flowers are usually incomplete but perfect and produce seeds. Similarly there are various types of inflorescences Cyathium (*Euphorbia*), Spadix (*colucasia*), hypanthodium (*big*), Catkin (*Morus*) etc.

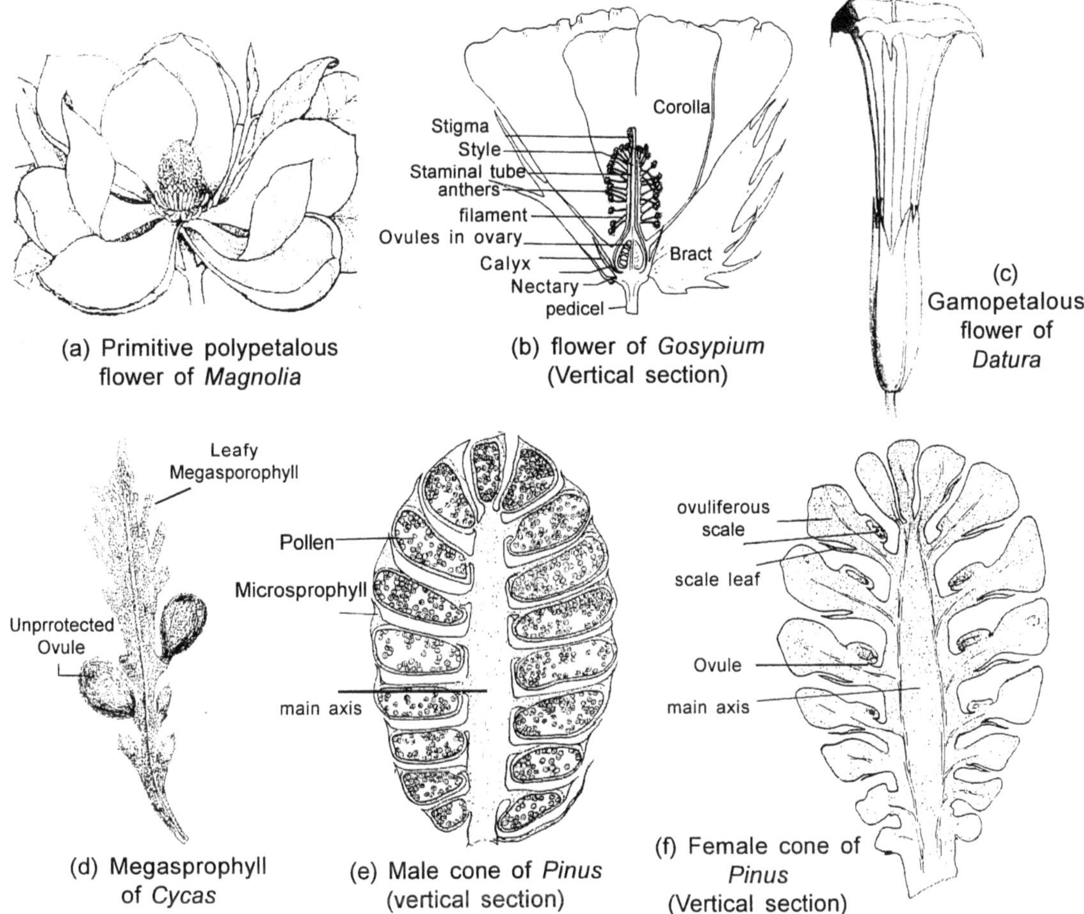

(a) Primitive polypetalous flower of *Magnolia*

(b) flower of *Gosypium* (Vertical section)

(c) Gamopetalous flower of *Datura*

(d) Megasprophyll of *Cycas*

(e) Male cone of *Pinus* (vertical section)

(f) Female cone of *Pinus* (Vertical section)

Fig. 1.23 Comparison of angiosperm flowers with pollen and ovule bearing structures of *Cycas* and *Pinus*

A typical flower consists of floral organs in (*cyclic*) pattern on a central axis the *receptacle*. The outer layer of green appendages is the calyx and consists of protective sepals. Next as the bud opens are one or more layers the corolla composed of brightly colored petals. The calyx and corolla are (not directly concerned with sexual reproduction) may be free or fused. Next to corolla are the male structures, the stamens (collectively, the androecium and the female parts of pistils together, the gynoecium). These are the essential or sexual parts of the flower. A stamen consists of a slender stalk (*filament*) and a terminal, saccular anther, the (*microsporangia*) and a connective. Pollen grains are produced in 2-4 sacs of anthers (Fig. 1.23 (a-c)).

A single ovule-bearing leaf is called a carpel. The ovule containing chamber is the locule. The pistil can be simple; composed of two or more fused together, (syncaprus). In such pistils the number of fused carpels is indicated in cross section by the number of locules. The arrangement of ovules on placenta is placentation which can be parietal (along the walls of a simple ovary; *Brassica*) : axile (along the central axis of a compound ovary; *Hibiscus*) or free central (on the axis of a simple ovary; *Primula*). There may a one (*Zea*) or many (*Antirrhnium*) ovules within the ovary and is supplied by vascular traces. The ovules are attached by a short stalk, the funiculus on the ventral suture with the *placenta*. The styles may be separated into the component number or the outer part of the ovary wall may be ridged; the stigma may be lobed, cleft or divided.

Botanists agree over the form of the primitive flower, most evidence points to a type similar to that of the present day *Magnolia*. The flower thus is homologous with the gymnospermous "cone" (Fig. 1.23 (d-f)) and is a highly modified shoot apex. Floral evolution from the basic strobilus like *Magnolia Michelia* (Champa) type has proceeded along many diverse lines as under :

(a) From complete flowers all four flower parts present (*Magnolia*) : to incomplete types one or more of the flower part missing (*Salix*). In some apetalous incomplete flowers (*Clematis*) the sepals are large and colored assuming the function of the missing petals.

(b) From *perfect, monoclinous*, or *bisexual* flowers, both of the essential cycles present (*Cornus*) : to *imperfect, diclinous* or *unisexual* only one of the essential series present (*Betula*). Imperfect flowers may be either *pistillate* (possessing pistils), and monoecious-pistillate and *staminate* flowers on the same plant as in castor or *dioecious* i.e., pistillate and staminate flowers on two separate plants as in papaya.

(c) From *hypogyny* perianth and androecium attachments below and free from the ovaries (*Pisum*) : through *perigyny* perianth and androecium partially fused into a floral tube surrounding, but not attached to the gynoecium (*Malus, Saxifraga*) to *epigyny* floral tube fused with the ovary wall so that the other floral parts appear to arise from the top of the ovary (*Cucurbita*). In hypogynous and perigynous flowers the ovary is said to be *superior* and the ovary is *inferior* in epigynous flower.

(d) From distinct floral parts individual floral appendages and cycles of appendages separate or free from one another (*Magnolia*) : to coalescent parts members of the same series united partly or completely into a tube, e.g., *synseplay* (*Nicotiana*), *synpetaly*, *synadelphy* or coalescence *Ipomoea* of staments (*Crotalaria*), and syncarpy or coalescence of carpels into a *compound* pistil (*Viola*); or *adnation* of parts union of members of different cycles e.g., stamens to pistils (*Calotropis*).

(e) From actinomorphic or radially symmetrical flowers (*Opuntia*) to zygomorphic flowers (*Salvia*).

The most effective and common agents of cross pollination are wind, and animals. Many angiosperms are wind pollinated (anemophilous). Most flowering plants are insect pollinated (entomophilous), and are strikingly adaped that the plant and animal pollination appear to have evolved in intimate correlation. Animals visit flowers for food in the form of pollen and nectar. Many of the floral structures may be interpreted in terms of their function in attracting pollinators. Bright colors and scent odors of flowers are definitely attractive to certain insects particularly bees. The foetid odour of *Sterulia foetida* suit the olfactory preferences of carrion fly pollinators, but not humans. Sugars exudates are produced by secretory cells called *nectaries* located in various parts of flowers. Nectar is familiarly collected by bees and other insects and is converted to honey. Flowers are usually constructed so that splashes of colour on the petals, the position of the nectary, or the symmetry of the flower serve to position the insect as it enters/alights so that cross pollination takes place.

Complexity in floral evolution reaches its zenith in orchids and asclepias which produce pollina and represent parallel evolution in mono and dicots respectively.

In flowering plants various breeding systems both morphological and physiological exist which control the relative frequency of cross and self pollination and fertilization (Table 1.6). Many types of adaptive mechanisms have evolved which promote out-breeding which is necessary in introducing genetic variability, so important in evolution. The advantages of cross pollination are reflected in species of the modifications developed by many flowers to favours effective cross pollination between flowers on the same plant or between flowers on different plants. Breeding systems can be tested simply by covering the flowers; if incompatible no seeds are set. If several such plants are grown together there will be seed set.

Table 1.6 *Plant Breeding systems*

I. *Sexual (Amphimixis)*
- (a) Inbreeding (autogamy)
  e.g. cleistogamy
- (b) Outbreeding (allogamy)
  1. Mechanical features of flower preventing self-pollination
  2. Temporal features having a smimilar effect
     e.g., protandry, protgyny
  3. Monoecism
  4. Gynodioecism or androdioecism
  5. Differential growth of pollen tubes
  6. Genetic incompatibility (Homomorphic)
  7. Genetic incompatibility (Heteromorphic)
  8. Dioecism

II. *Asexual (Apomixis)*
- (a) Vegetative reproduction - with or without sexual reproduction
- (b) Aganospermy - sporadic, facultative, obligate

However, some flowers are regularly self pollinated, the petals remaining closed (*Pisum*) while others have both open cross pollinated flowers and cleistogamus self-pollinated ones to ensure at least some successful fertilization and seed production (*Arachis*).

Flowering plants have adopted wide array of reproduction strategies that promote cross breeding. The most significient group of strategies is covered by the term self-incompatibility which is defined as the inability of fertilize bisexual seed plant to set seed upon self pollination or failure of functional pollen from the same plant to germinate on the stigma. Self incompatibility is wide spread in plants and in many species is controlled by a single multiple allele locus. 'S' phenotype is determined by its own genotype and sporophytic self incompatibility where the pollen S phenotype is that of the mother plant. The various types of incompatibility systems and their relevant details are summarized in Table 1.7. The incompatibility systems are broadly classified as gametophytic and sporophytic or may be homomorphic (flowers in the species of one type) or heteromorphic 2 or 3 types of flowers with reciprocal placement of anthers and stigmas within a species favouring cross pollination. At the opposite extreme, various mechanisms are known to promote self-fertilization or even the formation of viable seed in the absence of fertilization of the egg (apomixis) as in green sorghum.

Table 1.7 *Types of self-incompatibility systems in flowering plants*

Name	Site of recognition	Genetic basis	Features of stigma	Distribution
Gametophytic (normal type)	Pollen tube in the style	One locus with many alleles	Stigma with wet surface and cuticle with gaps	Widespread
Gametophytic (grass type)	Pollen tube on stigma surface	Two loci with several alleles	Stigma dry with continuous cuticle	Grasses (others ?)
Sporophytic	Pollen grain on stigma surface	One locus with many alleles	Stigma dry with continuous cuticle	Mainly Brassicaceae & Asteraceae
Heteromorphic (two forms)	Pollen grain on stigma surface or style	One locus with two alleles	Stigma normally dry with continuous cuticle	30-35 unrelated families
Heteromorphic (three forms)	Pollen grain on stigma surface or style	Two loci each with two alleles	Stigma normally dry with continuous cuticle	6-9 unrelated families
Late-acting	Mainly pollen tube at ovule	Unknown	Variable	Probably widespread

The Male Gametophyte

Pollen grain is the first cell of the male gametophyte. It is liberated from the anther at 2-celled stage (generative cell and tube cell) and ready for germination. Pollen grains may be single or in groups. The outer wall of pollen is the exine which may be variously sculptured and may be provided with various kinds of apertures through which pollen tube come out during germination. The pollen wall the sporopollenin is made of polymerised mixture of esters of fatty acids and carotinoids the cellular interactions between the pollen, stigma and embryo sac are summarised in Table 1.8.

Table 1.8 *Cellular interactions in angiosperm reproduction*

Fate of the male gametophyte		*Interaction with female → sporophyte involves :*
Capture of pollen	... ...	Stigma surface materials sugars and enzymes
Hydration } Germination of pollen	... ...	Stigma papillae surfae and underlying cells
Pollen Tube penetration	... ...	Stigma cuticle and underlying cells
Pollen Tube growth in the style	... ...	Transmitting tissue hollow canal of the style.
Pollen tube entry into the female gametophyte and gamete discharge.	... ...	None : interactions with the female gametophyte

Pollen grain germinates on the stigma surface which may be wet or dry may be papillate or non papillate. The pollen tube passes through the style (which may be solid or hollow) and then enters the embryo sac where the generative cell produces 2 non motile male gametes. The pollen tube thus acts as a carrier of male gametes. Male gametophyte reaches to a maximum 3-celled stage (a tube nucleus, 2 male nuclei). Thus, the male gametophyte in angiosperms is a highly reduced structure.

The Female Gametophyte

Unlike the male gametophyte, the female gametophyte is elaborate and represented by the embryo present in the ovules which may be straight (*orthotropous*) and reverse (*anatropous*) and may be unitegmic (with one integment) or bitegmic (with two integements). The main tissue of ovule is nucleus and the embryo sac is embedded in the centre. The ovules may be tenuinucellate (scanty chalazal tissue) or *crassinucellate* (bulky chalazal tissue). This has its origin from the megaspore mother cell which divides meiotically and forms a linear tetrad, the lower megaspore is generally functional and develops into *embryo* and the rest aborts. The surviving megaspore cell enlarges and divides mitotically to form two nuclei. These migrate to opposite poles where the nuclei divide twice forming four nuclei at each pole. The nuclei organise themselves at the pole; near the micropylar end, form an egg apparatus and antipodal cells at the *chlazal* end. Two haploid nuclei remain in the centre and are now called as *polar nuclei* and these fuse to form a large *central cell.* The mature embryo sac is eight nucleate and may be *monosporic*, *bisporic* or *tetrasporic* (Fig. 1.25). The egg apparatus consists of a central large *egg cell* with a conspicuous nucleus; two *synergid* cells which later *disintegrate* as the pollen tube enters. The *antipodal* cells also degenerate prior to fertilization. The two *polar nuclei* fuse to form a *secondary* nucleus which is diploid.

Fertilization

After pollination; the pollen tube traverses through the *nucellus* and reaches the *embryo sac* through the *synergids*; the pollen tube ruptures releasing the two sperms. The following important cvents occur during fertilization.

* The fusion of one male gamete with the female gamete (egg) is called *syngamy* and the diploid zygote results. This unique observation was first made by a famous botanist. Strasburger (1884). The fusion of the male and female gametes, the fertilization, is the succesful culmination of a series of recognition events leading to the formation of a zygote which finally develops into seed.
* The second male gamete fuses with 2 haploid polar nuclei and this is called "*Triple fusion*" and results in the development of a triploid nutritive tissue the endosperm. This unique observation was made by the Russian botanist Navoaschin (1898).
* Since two fertilizations are involved in a single embryo sac, this is called "*double fertilisation*", a feature unique to angiosperms only.

Immediately after ferilization many changes can be seen in the flower. Flower colour changes, stigma, style, stamans and petals wither. The ovules which have under gone fertilization develop into seeds and the ovary develops into a fruit. As shown in (Fig. 1.24) the fertilised zygote divides transversely to form a basal cell which under goes many divisions and develops into proembryo, heart shaped embryo etc. The basal cell also divides to form the suspensor. The entire post fertilisation changes leading to development of embryo and seed are called *embryogeny* and are shown in Fig. 1.25.

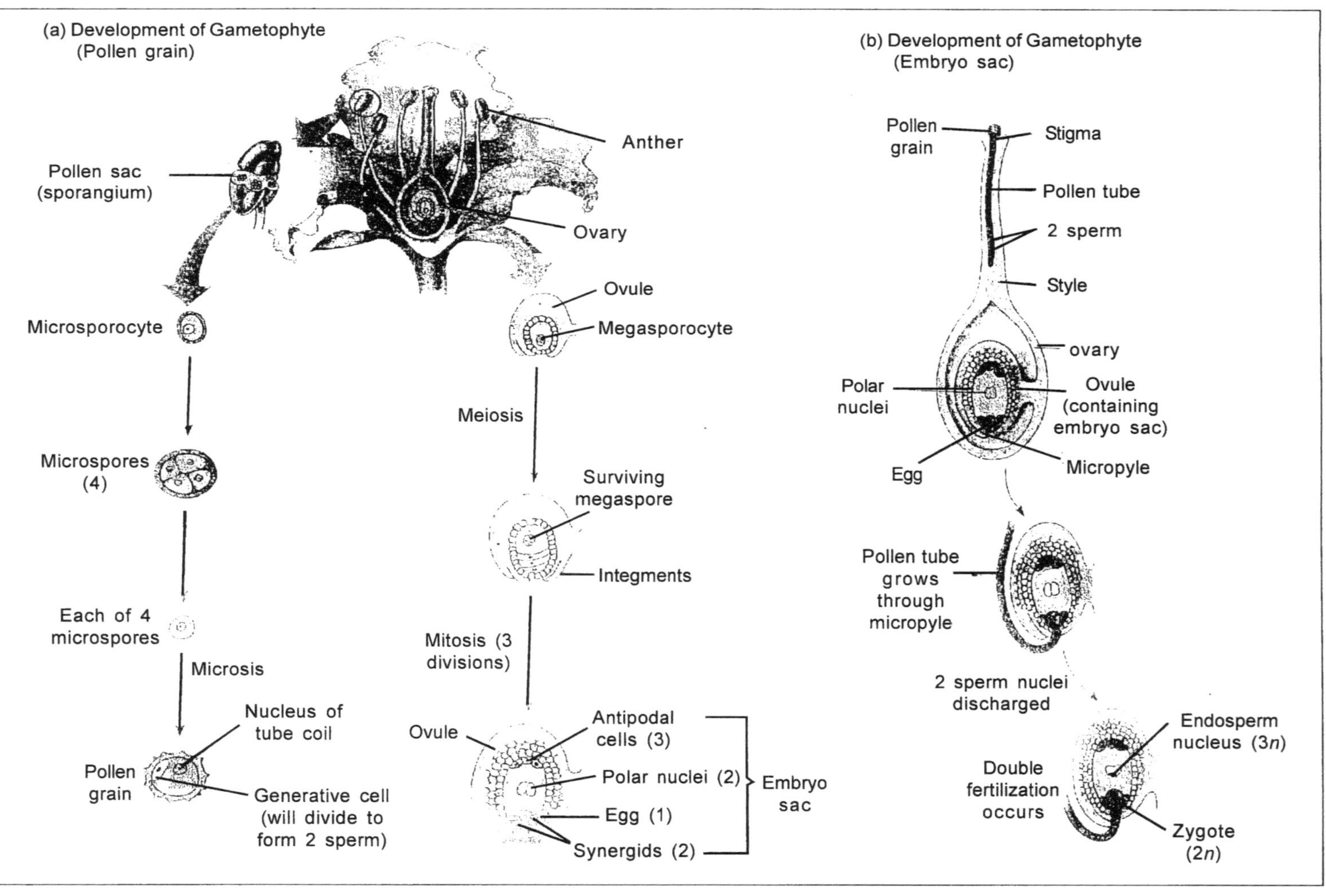

Fig. 1.24 The production of (a) male and (b) female gametes by the gametophyte of angiosperms

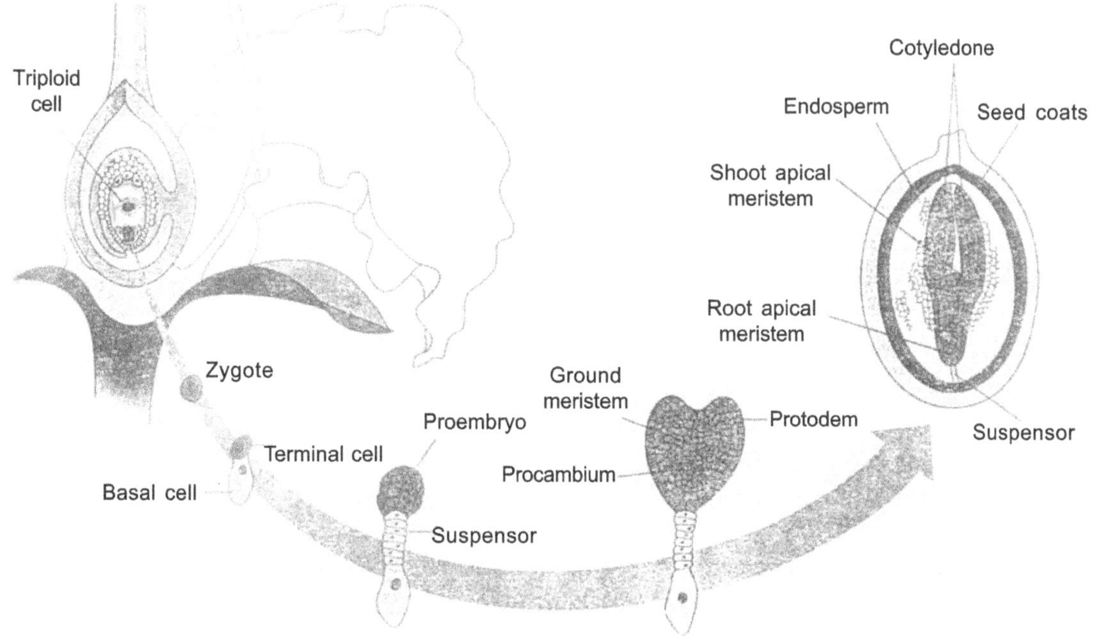

Fig. 1.25 Developmental steps in the formation of a angiosperm embryo

Fruit

A fruit is the whole mature gynaecial structure resulting from the fertilization of the ovules within it. In other words the fruit is a fertilized ripend ovary. There are numerous kinds of fruits. The fruits may be classified as simple, fleshy etc on the basis of their structure and development eg. Berry (tomato), pepo (pumpkin), pome (apple), hesperidium (citrus), capsule (okra), legume (peas), caryopsis (cereal). The fruit wall (pericarp) may be fleshy or dry.

Seed

An angiospermous seed is a fertilised ripe ovule with one or more protective seed coats (testa and tegmen), an embryo formed from the fertilised egg and remains dormant; the embryo has a plumule the shoot and radicle the root (Fig. 1.26(a). These may one cotyledon (monocotyledon) or two (dicotyledon) and may be endospermic as in castor, maize or non endospermic as in pea. The testa may be smooth or ornamental with a hilum near the micropyle, a scar, the site of attachment to the funiculus (placenta). Seed viability is the capacity of a seed to germinate. Some seeds readily germinate others show dormancy. Germination of a seed results in seedling with hypocotyle/epicotyl forming hypogeal (below the ground) and epigeal (above the ground) germination. The seeds may be tiny (0.25 mm long and about 3 μg in weight) as in orchids or tobacco and produced in vast numbers in a capsule or may be 45 cm long and weighs 3 kg as in *Lodoicea maldivica* the double coconut palm. Corresponding to the diversity of pollination systems there is great variety of dispersal mechanisms for fruits and seeds which show dispersal by water, wind and animals (by adhesion or passage through the gut). The testa surface is tough made of secondary thickend cells and variously ornamented and rarely provided with arils, elaisomes, wings or hairs etc.

Micropyle
Embryo
Outer integument
Inner integment
Perisperm
Endosperm

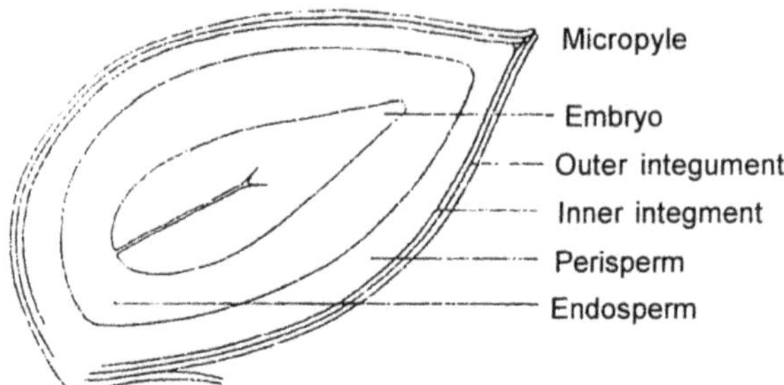

Fig. 1.26(a) An ideal dicotyledonous seed in longitudinal section

Alternation of Generations

In angiosperms alternation of diploid green sporophyte with haploid gametophyte exist. The important feature is the extreme reduction in the gametophytic stages (Fig. 1.26(b). The male gametophyte exist as pollen grain (2 or 3 celled) which germinates into a tube containing two gametes that affect fertilisation. The female gametophyte exist as a nucleate embryo sac which is dependent on sporophyte for nutrition and protection.

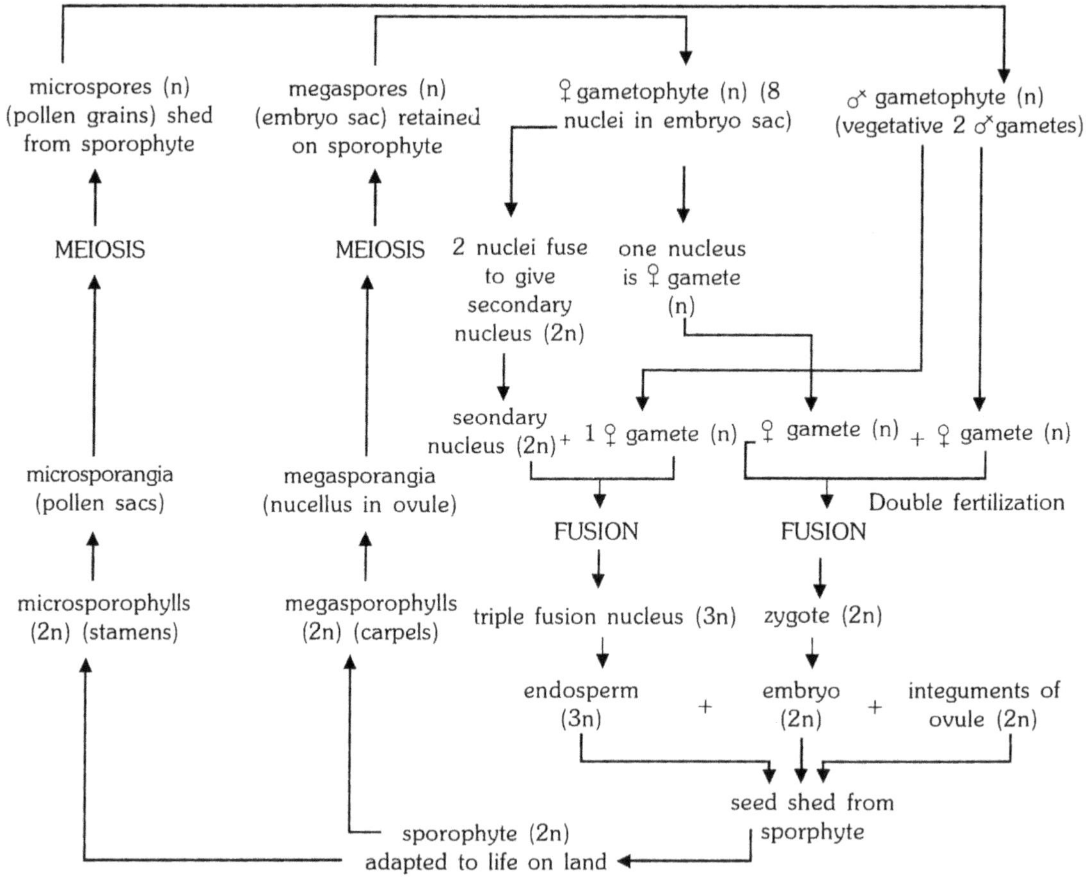

microspores (n) (pollen grains) shed from sporophyte
megaspores (n) (embryo sac) retained on sporophyte
♀ gametophyte (n) (8 nuclei in embryo sac)
♂ gametophyte (n) (vegetative 2 ♂ gametes)
MEIOSIS
MEIOSIS
2 nuclei fuse to give secondary nucleus (2n)
one nucleus is ♀ gamete (n)
seondary nucleus (2n) + 1 ♀ gamete (n)
♀ gamete (n) + ♀ gamete (n)
microsporangia (pollen sacs)
megasporangia (nucellus in ovule)
FUSION
FUSION
Double fertilization
microsporophylls (2n) (stamens)
megasporophylls (2n) (carpels)
triple fusion nucleus (3n)
zygote (2n)
endosperm (3n)
+
embryo (2n)
+
integuments of ovule (2n)
seed shed from sporphyte
sporophyte (2n) adapted to life on land

Fig. 1.26(b) Generalised life cycle showing alternation of generation in angiosperm

The fertilization of the egg is dependent upon adaptation to dry conditions.

1.13 Concepts of Growth

Growth is a visible, complex, fundamental, universally accepted characteristic feature of all living organisms. Can an organism develop without growing ? For example, when a bean seed, which has been soakded in water until it has reached a constant wet weight, is placed in a warm dark and humid place (to allow germination, but to prevent gain by carbon dioxide assimilation and loss by dehydration) and left, it will germinate and produce elongated root and shoot systems. In doing so it will have lost weight due to the oxidation of food reserves. It has undoubtedly developed, but has it grown. Growth, therefore, is the addition of mass resulting from the absorption and assimilation of matter from the environment or whenever anabolism exceeds catabolism. In periods of starvation, when catabolism exceeds anabolism, there is negative growth (degrowth). The organisms lose mass, but will regain it, at an increased rate of growth, when starvation ceases. Growth therefore is a permanant irreversible increase in mass, size, shape, structure, weight and volume of a cell, tissue, organ or organism which is accompanied by cell enlargement, cell division, i.e. cell differentiation new material and maturation. It is the net result of metabolic changes that occur in the cell or tissues and other related complex processes.

The criteria for the growth of any organism are :

1. Increase in cell mass
2. Duplication of genetic material in cells
3. Division process assuring the daughter cells that they receive equal and identical complement of genetic material again and again, generation after generation of cell cycles.

Cells life span is cell cycle and represents growth and increase in the cell following sequence of events. All cells of plant, animal and human show mitotic cell cycle and the following stages are distinguished :

1. Interphage, the resting nucleus; chromosomes are not visible.
2. G_1 : Cell growth and normal function (Pre DNA synthesis) Nobel Laureate J. Watson calls this phase as '*Programmed operations*'.
3. S : DNA synthesis & replication, chromosome duplication, formation of chromoatids.
4. G_2 : Cells grow and prepare for mitosis (Post DNA synthesis)
5. Mitosis, this is the final phase of cycle, the G_1 and G_2 are called the gap phases.

The time taken for G_2 interphase cell cycle varies from plant to plant. Onion root tip require 83.7 minutes and pea roots 110 minutes. Cell cycle is contolled by cell protein cyclines and cyclin dependent kinase (CDPKs). G_1 cyclins by cell in G_1 activate CDPKs which stimulates DNA synthesis at G_1/S control point. G_1 cyclins are degraded and produce M cyclins which activate CDPK at stimulate mitosis (G_2/M).

Growth Due to Just Enlargement

Growth in unicellular organisms like *Chlamydomonas*, *Chlorella*, yeast or bacteria is due to just enlargement. It occurs due to

1. The addition of protoplasm
2. Enlargement of *vacuole* and
3. Thickness in the cell wall.

Diffused Growth

In lower plants like algal *Spirogyra* and *Oedogonium*, every cell in the plant divides and produces new cell resulting in the linear growth of the individual. This is called diffused growth.

Meristems

In higher plants growth is restricted to definite regions in the plants, called *meristems*. In liverworts (Bryophytes) growth is due to the activity of one or more apical cells in the notches of the thallus. In mosses like *Funaria* growth is due to single pyramid shaped apical cell. In Pteridophytes like *Lycopodium*, growth is due to apical meristem that consists a group of cells. In *Isoetes* (Pteridophyte) growth occurs due to single apical cell. In gymnosperms and angiosperms growth occurs due to apical, lateral and intercalary meristems. The apical meristem increases the length of the stem and root, while the lateral meristem helps in increasing the width of the stem and root. Intercalary meristem helps in increasing the length of the internode and leaf.

Primary Growth and Secondary Growth

The increase in the length of the stem or root that occurs due to the apical meristem is called the primary growth. The lateral meristem such as cambium helps in increasing the width of the organ and the such growth is called as *secondary growth*; commonly seen in gymnosperms dicotyledons and woody monocotyledons plants.

Open Growth and Closed Growth

The stem and root of the plants can grow in any direction. This is called open growth where as in animals growth is of closed type. Though the plant as a whole shows open growth, its organs like leaves, flowers and fruits grow in a definite direction. Thus, the plants also show closed type of growth to some extent.

Plants keep track of the time, season length of the day and night, the direction of light gravity.

Indefinite Growth and Definite Growth

The plants in which the apex of the stem does not show any modification and grow continuously or indefinitely. eg. *Polyalthea, Eucalyptus, Pinus* etc. The plants like *Ficus* or *Datura* show definite growth because meristem ceases to function or undergoes modification. In all the plants leaves, flowers, fruits show growth of definite nature.

Growth and Development

Development is a parallel phenomenon to growth in plants and animals. Both are irreversible and permanent. Development is an ordered change or progress towards a more complex state. Development and growth may occur one without the other or one along with the other. Anabolism should exceed catabolism for growth to occur.

Stages of Growth

There are various stages of growth e.g. a zygote becomes embryo. A spore becomes a gametophyte. A seed becomes seedling after germination. A seedling becomes a mature plant. A bud becomes a flower and an ovary becomes a fruit. Thus there are several such cases where growth can be observed. The growth of the plants is slow in the beginning but becomes faster later and stops at some stage. Growth is generally accompanied by

1. Cell division
2. Cell elongation
3. Cell maturation and
4. Cell differentiation

Cell Division

Unlike animals where every cell divides and contribute to its growth, growth in plants is restricted to specific tissues of certain regions. These are called *meristems* and tissues are called *meristematic* and show continuous mitotic activity. The rate of cell division varies with age of the plant and various physiological status of the plant. Cell division in plants is linked to the functioning of phytohormones called cytokinins.

Mitosis (Greek : mitos, meaning thread) is the process which results in the equal division of the nucleus was first observed by Walter Fleming in 1879. Most of the cells in a sexually reproducing individual are derived by successive divisions from the single-celled zygote, The original cell divides into two, the products into four, they in turn into eight and so on. Under normal circumstances the nucleus of every cell contains exactly the same set of chromosomes as the zygote. The process of division whereby the chromosome complement of a cell comes to be repeated in the daughter cells is called mitosis; it is an essentially similar process in all plants and animals, from unicellular algae and protozoa to flowering plants and mammals. Mitosis although a continuous process, is conveniently divided, for descriptive purposes, into four stages : *prophase, metaphase, anaphase* and *telophase* (Fig. 1.27).

Prophase

In the interphase nucleus the chromosomes are invisible but with the beginning of prophase they become visible (in ordinary microscopical preparations), and each chromosome can be seen to be longitudinally divided into a pair of parallel *chromatids*. Just outside the nucleus as in most animals as well as in some algae and fungi is a small, darkly-staining centrosome. At the beginning of prophase the centrosome divides and the two halves travel round the nucleus in opposite directions until they are 180^0 apart. At the same time a system of rays, the aster, forms round each of the daughter centrosomes. The nuclear membrane breaks down and disappears at the end of prophae and may be partially incorporated into the spindle, which appears at this time and consists of fibres radiating from the two opposite poles of the cells (marked by the centrosomes in animals).

Metaphase

Throughout prophase the chromosomes condense, become shorter and thicker. At the beginning of metaphase they arrange themselves on the equator of the spindle by microtubules midway between the poles, metaphase plate.

Anaphase

The centromere splits longitudinally and the daughter centromers move up the spindle towards opposite poles, dragging the chromatids (which may now be appropriately called daughter chromosomes) to which they are attached, and so the members of each pair are separated. As the daughter chromosomes move towards the poles, the equatorial part of the spindle elongates and narrows.

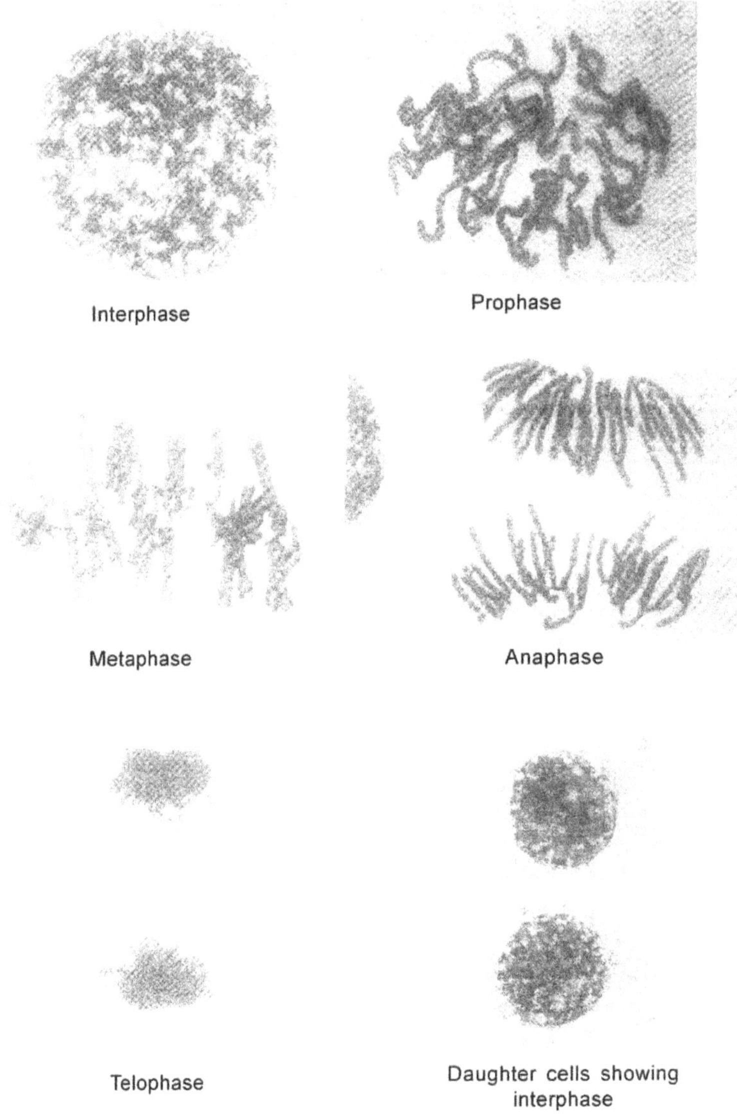

Fig. 1.27 Mitosis in root tip of lily

Telophase

The two groups of daughter chromosomes lose their smooth outline and come to appear as a tangled mass (as at the beginning of prophase), and a nuclear membrane is formed round each group. Thus two daughter nuclei are formed, each containing one half of the results of longitudinal splitting of the original chromosomes, i.e. the daughter cells are genetically identical. (Fig. 1.27)

A cycle consisting of cell growth and cell division is first established; cell growth leads to increase in cytoplasmic mass, chromosome mass and cell surface; is followed by cell division in which the cytoplasmic mass and the chromosomes get equally divided into daughter cells.

Mitosis is quickly followed by cytokinesis (cytoplasmic division) which involves the formation of a cell plate in plants and cleavage furrow in animals.

Cell Elongation

Newly formed cells show cell elongation and cell enlargement. The region of elongation in stems and roots can be easily demonstrated by making indian ink marks from 1 to 8 on young and stems at intervals and observing the position of the marks at various growth intervals afterwards (Fig. 1.28). The growth zone is more extensive in the stem. It manifests itself in the visible signs of growth and results in increase in size, shape, volume and weight of the plant. Gibberellins are chiefly concerned with cell elongation. (Fig. 1.29)

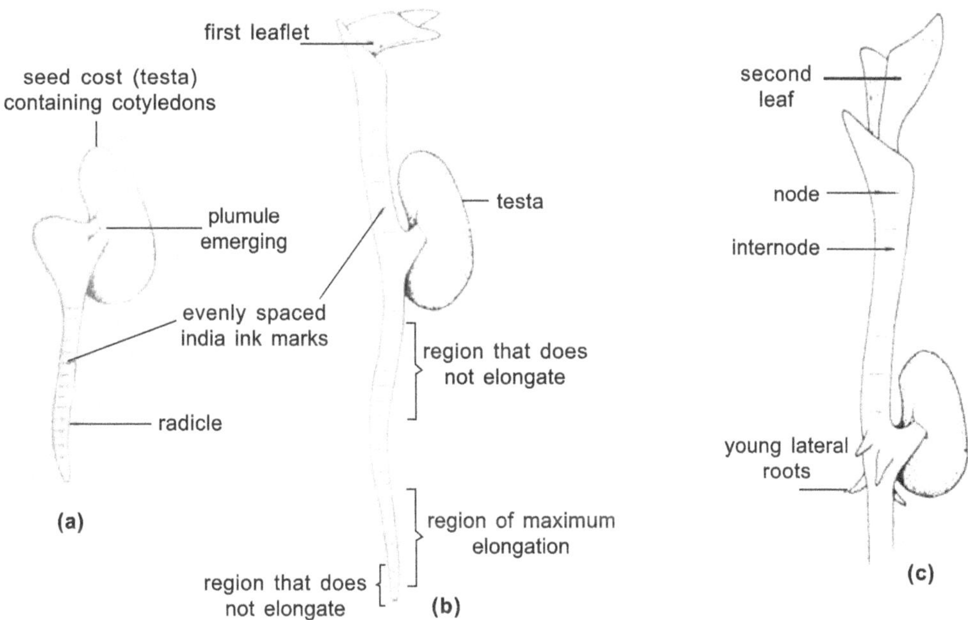

Fig. 1.28(a-c) Growth zones in bean seedling

Cell Differentiation

After the meristematic cells have increased in size, they differentiate into the parenchyma, collenchyma, sclerenchyma, xylem and phloem which make up the epidermis, cortex and the vascular cylinder, the three characteristic internal regions in the higher plant body. Roots also have an outer cap of parenchymatous cells which protects the meristematic apical region, as the growing root forces its way through the soil. Differentiation occurs according to the hereditary pattern and results in structural complexity which includes the organisation of cells into tissues and organs. The effect of cell differention is also externally seen and this is the visible stage in growth.

The total growing period is known as the Grand Period of growth and can be divided into three phases namely.

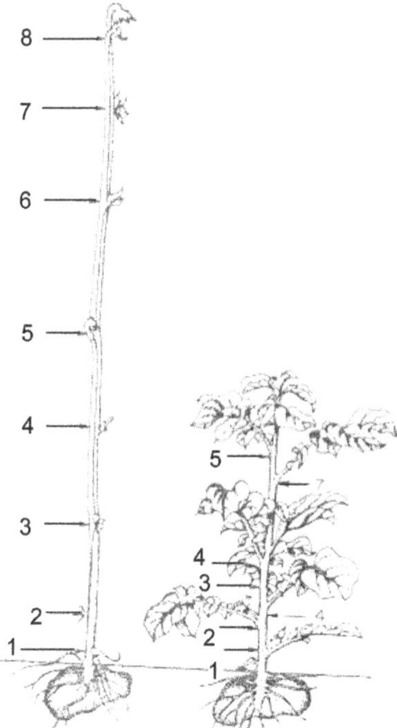

Fig. 1.29 Two potato plants grown in dark (a) and light (b) showing difference in growth, the numbers 1-8 indicates notes

1. The formation phase, characterised by multiplication of cells and slow growth.
2. The enlargement phase, characterised by enlargement of cells, cells begin to divide; most rapid growth phase.
3. The maturation phase is characterised by mature cells both in shape and size; the growth slows down.

Cell Maturation

The differentiated cells begin to mature. The organs or tissues take definite, permanent shape and form. For example, the vessels of the xylem show pitted thickenings and sieve cells show sieve pores in the sieve plate. A mature cell may be living or dead.

Growth Curve

Growth of the unicellular organisms occurs at constant rate. Organisms multiply at fixed regular intervals of time. The growth of any parameter (weight, volume, size, shape) is plotted against time. The growth curves takes many forms and growth symmetrical is called sigmoid curve.

The growth is slow in initial stages (Lag phase). Later the growth is at its maximum. Number of cells increases due to rapid cell division. This phase is called the Log phase or exponential phase. Later the growth rate decreases due to ageing of the cells or limiting of nutrients. This is the senescence phase.

Plant growth is greatly influenced by environmental conditions. As would be expected, it increases with increase in temperature. It is also more rapid in darkness, which under natural conditions is at night, though in practice this is often obscured by the retarding effect of lower night temperatures. The effect of darkness in accelerating growth is well demonstrated when seeds are allowed to germinate in the dark. The increase in length at the expense of increase in the size of the leaves is known as etiolation. This ability of plant is obviously of an adaptive value for it ensures that a seed buried in the soil rapidly grows into the light where photosynthesis can begin.

Growth in plants is markedly seasonal. Unlike the tropics in temperate climates growth slows down or ceases with the coming of the cold weather. In many cases, the leaves and flowers are shed, to be renewed with the return of the warmer weather. This characteristic seasonal growth however, is not entirely controlled by the environmental temperature because plants are also sensitive to day length. This is known as a *photoperiodic response*. In equatorial regions where variations in day length are less than at higher latitude, the main environmental factor in determing growth rates is the occurrence of the rainy season. This is very striking in hot desert regions which come life as the dormant seeds germinate after the seasonal rains. Annual plants may live for only one growing season, and do not show much increase in size but those living for more than one season continue growing year after year until they reach an enormous size as most trees. Generally the size of a plant is a measure of its age because at each growing season new cells are added in each of the growing regions, resulting in growth rings.

Measurement of Growth

Growth is generally slow and cannot be easily seen. Growth in general is more during the night time. Bamboos show fast growth rate among the flowering plants. Growth is measured variously by :

1. Horizontal microscope which measures the growth of an organ or plant at a particular time.
2. Arc Auxanometer is used to find out the increase in length of the plant.
3. Pfeffers auxanometer gives a graph that shows the growth in a fixed period of time.

Distribution of Growth

Growth is not uniformly distributed along the apex of stem and root. The region of cell division in the roots extends over a millimeter length. The elongating region in roots is restricted to about 10mm. The entire remaining root which may run to any length covers both the region of differentiation and maturation.

The growth of higher plants ie. those aspects involving both the patterns of stems, leaves and roots and the increase in bulk results primarily from cell division at the meristem followed by a secondary increase in size because of water uptake. These activities occur throughout the period of plant growth. The rates of both growth and cell division vary widely in different parts. This differential increase in size is a vital factor in controlling the shape of plant organ or organs.

Correlation

Overall control of growth is brought about by a series of growth correlations in the plants. Lateral buds may remain in a dormant state unless the shoot tip is removed, removal of a newly mature leaf may accelerate the development of the next, removal of the cotyledons may inhibit root and stem development but not that of the leaves. These examples of correlations suggests the pervasiveness of the phenomenon. Considerable evidence shows that many of the correlated phenomena are dependent on auxin distribution in the plant. In tissue cultures of tobacco pith little growth occured if indole-3-acetic acid alone was added to the sugar-inorganic element nutrient medium, but when Kinetin (KN) was supplied in varying amounts in addition to indole-3-acetic acid results in callus growth, but growth, or root growth could be independently stimulated. It was postulated that a balance exists between auxin and the nucleic acid constituents in the plant and that the state of the balance determines the kind and amount of growth which occurs.

1.14 Meristems

Plants, unlike most animals, possess so called '*Open*' system of growth. This means that plant possess perennially embryonic regions, from which new tissues and organs are periodically produced. Such embryonic regions in multicellular plants are referred to as meristems. Meristems are variously defined as perpetually young tissue concerned primarily with growth as 'regions' of more or less continuous cell and tissue initiation. In vascular plants, some meristems arise during the formation of the embryo while other types appear later in the development of the plant. One kind of meristem, present in all plants are the apical meristems at all growing points, whether they may be shoots, roots or primordia of leaves and reproductive organs (Fig. 1.30). In the simpler plants, algae, mosses and ferns the apical meristems consists of a single tetrahedron like cell which rise to cells on the three inner faces. This results in the formation of a surface layer of cells which then, by their own division, add cells to the central region of the apex (Fig. 1.30).

In seed plants the apical meristem is more complex, consisting of a region of actively dividing cells (Fig. 1.30). In some plants, the meristem is seen to be divisible into an outer layer and a central region, but often this distinction is not clear because the outer layer of cells donates cells to the central region. Whatever the details of the morphology of the apical meristems, their functions are to permit increase in the length (height) of the plant and to supply cells which can differentiate into the several specialised tissues as shown in below :

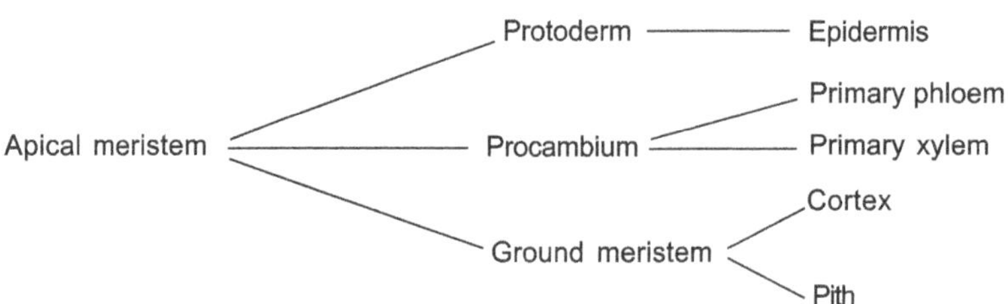

Meristems are specialised areas of cells in the plant body which have the potential for producing new cells, tissues, or organs in characteristically morphogenetical ways. Thus meristems are growing points that produce new cells during periods of growth. Cells are thus called meristematic which show active cell division from which permanent tissue is derived.

Meristematic cells are approximately isodiametric, thin walled, elongated, densely cytoplasmic to vacuolate, not well developed, plastids, poorly developed endoplasmic reticulum, the walls may be symmetrically or asymmetrically thickend, and are devoid of intercellular air spaces and show a higher metabolic rate than other mature living tissues. In gymnosperms the initials of shoot apical meristem are large, vacuolate with thickend cell walls. The plane of division of a given cell or cells comprosing a meristem is of importance in determining growth patterns in a tissue or an organ.

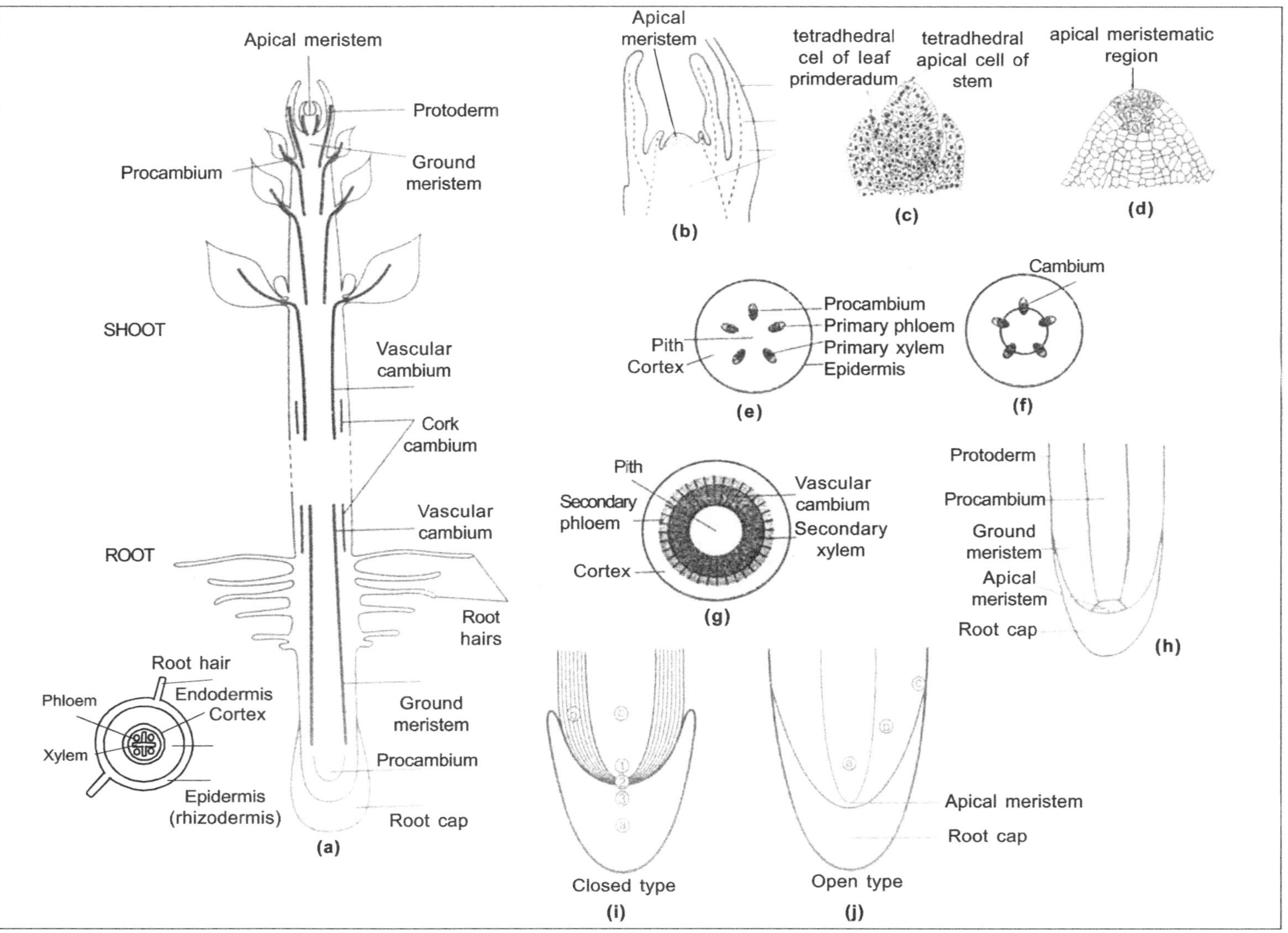

Fig. 1.30 The location of meristem in a dicot

A division plane parallel to the surface is called as *periclinal*, whereas at right angles to the long axis of a cylindrical organ is referred to as *transverse*. In addition, two types of divisions that are parallel with the long axis are possible : one, called *radial* occurs parallel with a radius of the circle making up the short axis and the other, called *tangential*, occurs at right angles to the radius. If cell division in a given meristem is limited to a single plane, such a meristem is called *rib or file* meristem, and longitudinal rows of cells are produced. If cell division is limited to two planes, a *plate* meristem is involved and a essentially flat tissue/organ like a leaf is produced. Cell division occuring in all planes is a feature of the mass meristem, resulting in the formation of a massive plant body or organ. (Fig. 1.31(a,b)).

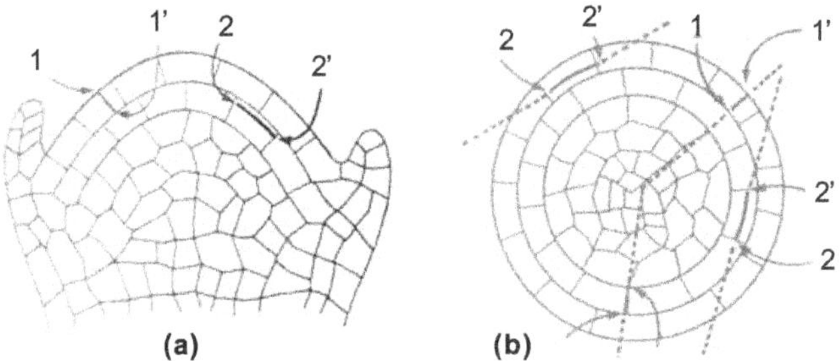

Fig. 1.31 (a, b) (a) Diagram of a median longitudinal section of a shoot apical meristem showing anticlinal (1-1') and perclinal (2-2') planes of division. (b) Cross-sectional diagram of a stem or root showing radial (1-1') and tangential (2-2') planes of division.

Classification

Meristems may be classified variously :

Based on Origin

1. Primary meristem that originate from the embryonic root or shoot cells eg. Apical and fasicular cambium
2. Secondary meristems that are formed from permanent tissue eg. cork and interfasicular cambia.

Based on Position

1. Apical meristem are found in vegetative spices of stem and root, may have one or more initials.
2. Intercalary meristem are present between permanent tissues, and
3. Lateral meristem are present laterally in axis and divide periodically (eg. vascular and cork cambium).

Apical Meristems

The cells of shoot and root apical meristems are organised in a precise fashion. A knowledge of the arrangement and organisation of cells at the apex is useful. Located at the terminus of the stem axis is a group of cells which are meristamatic and are ultimately responsible for the formation of all cells in the shoot. These cells constitute the apical meristem. The apical cells are thin-walled, of similar size and polygonal in shape. This is due to their being closely packed together. Intercellular spaces are usually absent. They have dense cytoplasm with relatively large nuclei. Generally, vacuoles are absent, or, if present, are of small size. In other words, they are typically undifferentiated plant cells.

Farther back from the apex intercellular spaces appear and the cells assume a more spherical form. Farther back still they are much enlarged and elongated. This is largely due to the formation of the central vacuole so characterstic of plant cells. The cells at the growing regions of plants are not firmly attached to each other. On the contrary, they easily slide over each other as they elongate because the lamellae between the cells are mucilaginous and the cell walls are saturated with water. In angiosperms, there are no individualised apical initials and further there is stratification into an outer mantle of cells called "*tunica*" one or more peripheral layers and an inner tissue called the "*corpus*" enclosed by the tunica. These two structures together constitute the shoot apex as per the *Tunica corpus* theory of Schmidt. Tunica-Corpus organisation is a feature of angiosperm apices. Some gymnosperms like *Ephedra* also show this pattern. The tunica is differentiated from the corpus in that cells of the tunica divide only in an anticlinical plane while the cells of the corpus divide anticlinally, periclinally and obliquely. Each tunica layer has its own initial, and the corpus and one layer of initials.

Topographically the shoot apex is conical in shape. The immediate derivations of the individualised apical initial(s) or upper most cells are in turn meristematic, constituting the primary meristems, protoderm, ground meristem and procambium. Through rapid cell division these meristems give rise to cells which will differentiate into elements of the dermal, ground and vascular tissues respectively. Leaves have their origin in cells of the flank of the shoot apex and the three primary meristems also are present in the developing shoot apex.

Root Apical meristem

The apical meristem of the root is different because of the presence of a fully differentiated sructure, the root cap. In addition, lateral appendages (lateral roots) arise endogenously from tissues some distance from the root tip. In others there may be a set of initials for root cap and future epidermis and other stratified initials for the remainder of the root. Like the shoot primary meristems continue meristematic growth proximal to the apical meristem. Following the *Histogen theory* of Hanstein (1868), the root apex is differentiated into three histogens or tissues, *dermatogen* the outer meristematic layer produces epidermis, the *periblem* produces the cortex and *plerome* giving rise to the central vascular zone. Further, each histogen has its own initials in addition calyptrogen forms the protective root cap. In monocots two types of root apical meristems called *closed* type as is in maize is characterised by three layers of initial cells in a single region in the apical meristem which form all the tissues of the root i.e., (a) gives epidermis, (b) the cortex and (c) gives the vascular cylinder and the *open* type as in onion, three distinct meristems near the apical meristems, the procambiam (a) forms vascular cylinder, the ground tissue (b) forms the cortex and the protoderum (c) gives rise to epidermis. (Fig. 1.32)

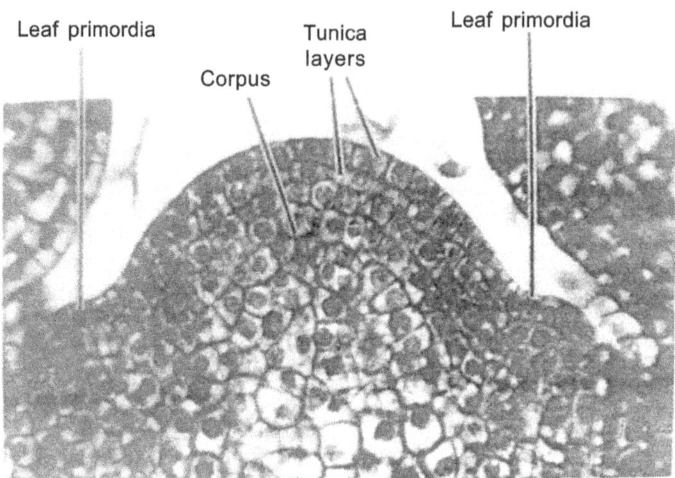

Fig. 1.32 Median longitudinal section through the shoot tip of coleus, a dicotyledon, showing tunica-corpus organization and leaf primordia.

Functions of Apical Meristems

The functions of all apical meristems are :

1. Produce cells that differentiate into primary tissue meristems, which in turn produce the primary tissues of the plant.
2. Intiate organs, primary leaves and branches in a co-ordinated manner and in a genetically determined pattern, and
3. Contribute to the increase in length of the plant.

Lateral Meristems

Usually two types of lateral meristems are present in plants; the vascular cambium which produces xylem and phloem cells, that originates from procambium that has not fully differentiated during the formation of primary xylem and primary phloem. The cambium which is present as a single row of cells produces, secondary phloem externally and secondary xylem (wood) internally. The production of these tissues is attributed to the lateral meristem that contribute to growth in stem thickness. Tissue meristems called the *vascular* cambium and the cork cambium (phellogen) is a meristematic tissue found in all woody plants. Cork cambium is a cylinder of cells located beneath the bark. It forms all internally and externally, the latter forms cork, of suberised cells. Cells formed internally are parenchymatous and close to phellogen and develop into phelloderm. Periderm refers to complex tissue comprising phellogen, cork and phelloderm.

The cambium thus is yet another meristematic region of general occurrence in gymnosperms and dicotyledonous angiosperms. Whereas the apical and intercalary meristems serve to increase the length of the plant, the cambium adds cells to the plant body in such a way as to increase its girth. This is because the plane of division of the cambium is in the plane of the axis of the plant.

Intercalary

Another type of meristem found in most higher plants occurs in the growing regions which are seperated from the apical meristems by differentiated tissues. They are known as intercalary meristems. A good example is the region at the base of the internode of grasses-the region where the leaves are attached to the stem. The intercalary meristem is regarded as part of the original apical meristematic region which has become separated from the apex during development by the differentiation of more distantly situated cells.

Intercalary meristems are usually located at certain positions on the stem, where leaves arise (ie. nodes) and are largely responsible for elongation as in grass and bamboo shoots and leaves.

Development of Different Plant Organs

Plant organs include structures such as leaves, stems, roots, flowers (parts of flowers including male and reproductive structures the anther and pistil), fruits and its units the seeds. Organ systems are groups of mutually functioning organs, the aerial shoot system with leaves and the underground system of roots. The shoot and root system are interrelated systems by which plants exploit the environment they live and hence are the products of evolutionary advances over geological time.

Evolution of Plant Organs and Organ Systems

The diversity of organs and organ systems among the many species of plants is the result of evolution by natural selection. The various structures enable the plant to be better adapted to its environment. The origin and evolution of diverse and complex organs and organ systems have been subject of great interest. The fossil record suggest that complex plants originated from simple plants. This is consistent with the fact that selective advantage was conferred in becoming more complex during the course of evolution. The multicellular condition enables cell and tissue specialisation, a division of labour among the parts of the whole plant body.

Some cells, tissues (organs) thus become specialised for capturing sun light and synthesising food (leaf), others anchor (root) the plant body while others play the role of reproduction (flower) resulting in corresponding increase in the ability of the plant to exploit its niche and enhance its chances for survival.

Development of Plant Organs

Very early in its development, the body of a plant exhibits a polarised (i.e. directional growth with definite shoot and root ends) differentiation into an axis composed of three organs stem, leaf and root. This differentiation occurs as the fertilised egg develops into embryo. Although each of these organs differ significantly in structure and function : are similar in important ways and share a common origin in young embryo. The tissues of each organ merge imperceptibly, an arrangement facilitating the performance of common functions namely, the movement of water dissolved minerals and food longitudinally from one part of the body to another. In addition, the stem and root actually constitute one continuous structure, exhibiting basically a cylindrical form. Both these organs also posses similarities in their method of growth in shoot and root tips mentioned earlier in the study of meristems.

The cotyledons (first leaves) are the first organs formed. As they begin to develop, the shoot apical meristem forms between them. At the end of the embryo opposite the cotyledons, is the root apical meristem. The root and shoot apical meristems continuously contribute cells to the primary tissue meristems which following differentiation and maturation give rise to the primary tissues and the primary body. The primary tissues are : primary xylem, primary phloem, cortex, pith and epidermis. In a young stem, the arrangement of tissues as seen in a cross section is as follow : outermost tissue the epidermis; followed by cortex, primary phloem and primary xylem. The pith tissue is in the centre (Fig. 1.32(a,b)).

The primary tissue meristems are designated protoderm, procambium and ground meristem. The protoderm functions to produce the dermal (skin) tissue system. The procambium gives rise to vascular (conducting) tissue system and the ground meristem the fundamental tissue system.

The plant body is made of various types of cells organised to form tissues. Tissues have been classified as simple and complex. Simple tissues have a common origin and common function Parenchyma, collenchyma and sclerenchyma are simple tissue often associated as components of complex tissues while complex tissues are heterogeneous : consisting of several cell types and may perform more than one function (vascular tissue). There is considerable diversity among plants in the pattern exhibited by various tissues. Tissues are also classified on the basis of their position in the plant, the types of cells they contain and the functions of these cells. Tissues give rise to the mature tissue systems of the plant of which three can be distinguished : 1. the *dermal tissue* system which make up the outer protective tissue of the plant. 2. the *vascular tissue* system, comprising the water and food conducting tissues, the basic units xylem and phloem (this system has evolved as early as the Silurian period about 430 million years ago) and 3. the *ground tissue system*, includes all tissues excluding of dermal and vascular tissue.

The fundamental tissue of various plant organs is as under :

1. Leaves are primarily made of parenchymatous mesophyll tissue which is largely photosynthetic.
2. Leaves are generally provided with epidermis which has stomata and hairs and forms the dermal tissue system which is much simpler
3. Root epidermis is provided with root hairs, whose function is water absorption.
4. Roots are made of parenchymatous cortex
5. Stem generally has epidermis with stomata and hairs in young condition. Stems are made of parenchymatous cortex and pith while cork, bark with lenicels are seen in older stem.

All vascular plants posses a well developed vascular system that conducts water and food material through the plant axis to the various organs in addition to the function of support. The vascular system consists basically of two complex tissues, the xylem and phloem.

Xylem (wood) are conducting tracheary elements perforated at both ends, forms hollow column, the vessles. The non conducting elements are xylem parenchyma cells and xylary fibres. Functionally xylem transports and stores water and nutrients, transports plant hormones and also provides mechanical support to the plant body.

Phloem is the conducting sieve elements, connected with each other through sieve areas forming sieve tubes. Sieve elements are sieve tube members that are connected through sieve plates to form continuous column. Non conducting cells of phloem are parenchyma cells and phloem fibres. Functionally the phloem provides path for distribution of the photosynthetic product sucrose and for translocation of protein and mRNA in growth and development of plant.

Each strand of the primary vascular tissue (derived from apical meristem) is a vascular bundle. The xylem and phloem of each vascular bundle is arranged in various pattern, phloem is outside of xylem (collateral); phloem on both sides of xylem (bicollateral); the phloem surrounds xylem (concentric bundle). The vascular bundle and the tissue associated with it is called stele, the simplest and primitive is protostele without pith. Siphonostele is with a column of pith in the centre. The siphonostele with several modifications is present in most vascular plants called the eustele. In monocots, the scattered vascular strands is called atatostele.

The development of various plant organs is beyond the scope of this book.

1.15 Economic Importance of Plants

Botany has been source for other arts and sciences such as medicine and chemistry as it still provides the base upon which depend various applied plant sciences, agriculture, agronomy, forestry, horticulture, pharmacology, ethnobotany and other related fields. In early civilizations, as also among primitive people today, knowledge of plants, especially those of medicinal importance and other practioners who often cloaked their knowledge with magic superstition, secrecy, religious rituals and other manifestions of professional mystery.

Gardens have been an integral part of practically every civilization from time immemorial. Gardening is a more specialised pursuit that lends itself to both vocation and avocation. The Gardens of Eden to the Hanging Gardens of Babylon are the fore runners of modern botanical gardens and arboreta, devoted to ornamental plants, fruit trees, timber trees, culinary herbs and interesting plants. Many kinds of trees, shrubs, and herbs are used as ornamentals in gardens and in landscaping. Plants have exerted enormous influence upon the social and economic life of peoples from earliest times.

The study of direct relationship between people and plants is ethnobotany. The agricultural system has supported human development. From the ethnobotanical view point, there is no comprehensive account of the seeds and nuts and by tribals and aborigines inspite of the fact that the tropics is inhabited by about 250 million tribal people. There are a multitude of ways in which the seeds are utilized. Fig. 1.33 summarise the produciton and uses of plants and the annual agricultural production is shown in Table 1.9.

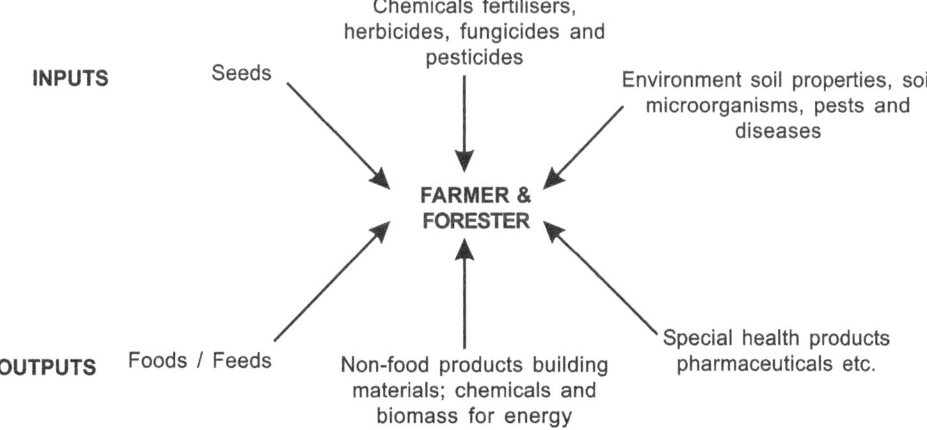

Fig. 1.33 Production and uses of plants

Table 1.9 *Annual world production of some important agricultural products*

S.No.	Sector	Product	Tonnes
1.	Foods and feeds	Cereals	1.8 billion
		Sugar	120.0 million
		Crude starch	1.0 billion
2.	Materials	Harvested wood	1.6 billion
		Paper	200.0 million
		Cotton	16.0 million
3.	Chemicals	Plant oils	43.0 million
		Natural rubber	4.0 milliion
4.	Others	Tobacco	4.0 million

The organs and organ systems of plants are most essential to the well-being of man. In fact, he is dependent upon plants for the essentials of his existence. The food of man comes from plants, either directly or via the flesh of animals that feed upon plants. Much of man's clothing and shelter are made from plant products. Paper is made from wood, and many drugs used to cure man's diseases are prepared from plant materials. The coal used to generate electric power is simply fossilized plants that lived millions of years ago. Plants generate oxygen, modify the climate, and aid immensly in the control of soil erosion, floods and even pollution.

Economically, seeds (and the fruits that contain them) are important primarily because they provide man with his most important foods; for example, the cereal grains, such as wheat, rice, and maize; the seeds of beans, peas, peanuts, soybeans, almonds, sunflowers, hazelnuts, walnuts, and various other nuts; the fruits of date palm, olive, banana, avocado, apple, and orange. Other useful products provided by seeds and fruits are abundant and too many to be listed for ornaments (Job's tears from the grass *Coix*, used for curtains; *Abrus*, *Adenanthera*, and *Rhynchosia* seeds for necklaces; others for rosaries) are also provided by seeds and fruits.

The reproductive system of plants, including fruits and seeds, provides numerous products of great economic importance to mankind. The entire inflorescence of *Artocapus* and *Ananas sp.* are edible. The fruits of many species throughout the world are eaten raw or cooked, including such plants as breadfruit, cucumber, eggplant, pumpkins, squash, tomato, apple, pear, watermelon, grape, cranberry, strawberry, orange, date, olive, pineapple, and many others.

There are a multitude of ways in which seeds are utilized. Nuts such as pecan, hickory, walnut, pine, almond, and chestnut provide a variable diet for many people. Betel nuts and kola nuts are chewed and enjoyed as mild narcotics.

Centres of Origin of Cultivated Plants

Agriculture became established about 20,000 years ago as art. Primary agriculture was merely the successful domestication of certain food plants in certain parts of the world. Actual agriculture of plants preceded domestication of animals. He then commenced tilling the soil. This removed him from the nomadic class and provided the basis on which higher civilization developed (Table 1.10).

Table 1.10 *Domestication sequence of plants cultivated by man. (from Burkill)*

Stage	System	Order of adoption
1.	Man static in one locality	Cereals, pulses, short lived green oil seeds (*Brassica sp.*)
2.	Man static and mobile	Root crops, herbaceous fruits, fibres, and dye plants.
3.	Urbanization	Woody plants, mostly fruit trees, various industrial plants

Agriculture is not the invention of any one man but numerous persons among diverse people in various part of the world. It is strange but true that the modern man has not added any thing new to the list of domesticated plants and animals, than those exist, selected and domesticated by pre-historic man.

Studies on the origin of cultivated plants were made during the 19th Century by A.de Candole and Charles Darwin and in the last century by the Russian botanist Vavilov (1887-1942). Vavilov concluded that most species of cultivated plants were domesticated in a relatively small number of countries. These countries are called the '*Centre of origin*' and he distinguished nine such centres as shown in Fig. 1.34. It is strange coincidence that the centres of origin of cultiated plants largely coincide with the ancient centres of human civilisation. About 300 plants and 30 animals are estimated to have been domesticated. The various centres are as follows with the list of some important crop plants that are characteristics to the centres.

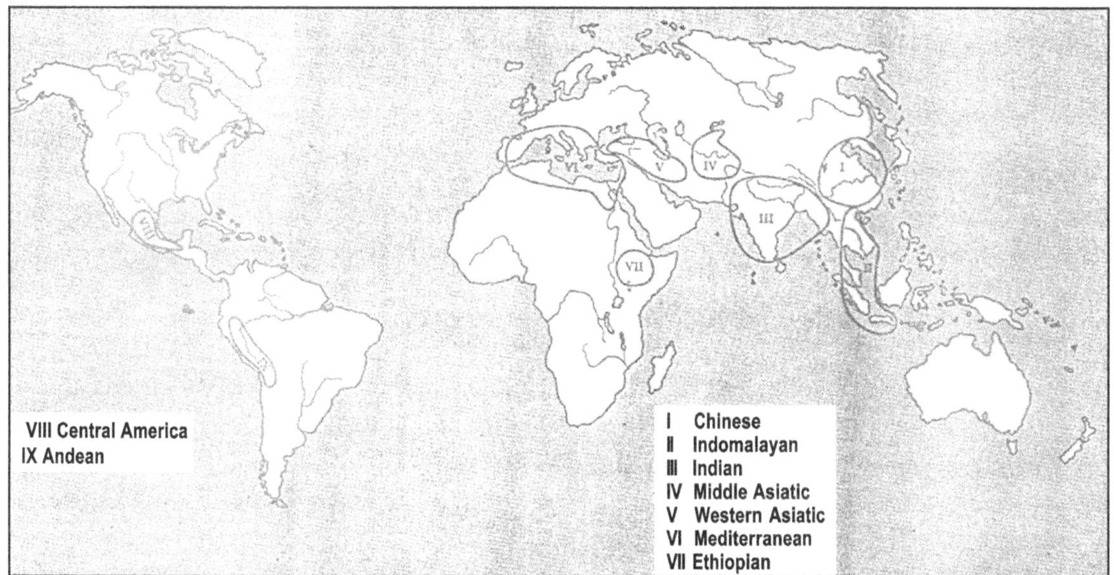

Fig. 1.34 World Map Showing the nine centres of origin of cultivated plants

I Chinese : China and Korea (Oats, Barley, Millets, Soya, Bamboo, Sugarcane, Oranges, Lemon, Peaches, Apricot, Pears, Cherries, Tea, some varieties of beans.

II & III Indomalayan & Indian; India, Indonesia, China, Burma (Mango, banana, pepper, bringal, sugarcane, cotton diploid varieties.

IV	Middle Asia : Kashmir (India), Pakistan, Afghanistan, South Russia, North Turkey (Almonds, Walnuts, Pears, Onions, Melons, Wheat (softbread)
V	Western Asia : Asia Minor, North Iran, Transaucasia (wheat, soft and hard, pats, pear, flax, poppycarrots, pears, cherries, grapes, almonds, figs etc.)
VI	Mediterranean countries around mediteranean sea. (wheat varieties less used, sugarbeet, peas, flax, olives, asparagus).
VII	Ethiopian : Ethopia, Eriteria, (Gingley, coffee, jowar, barley, flax)
VIII	Central America : S. Mexico, Guatemala, West Indies (corn, papaya, sweet potato, pepper, upland cotton).
IX	S. America and Andean : Peru, Bolivia, Equador (potato, tomato, tobacco, cacao, rubber, pumpkin, quinine, cotton).

From the above information it is obvioius that no one can claim that this or that plant belongs to his or her country. It is difficult to estimate the importance to the mankind of the achievements of the primitive agriculturists in the above mentioned nine centres. The people in the ancient times managed the first steps of the evolution under domestication of immensely difficult yet beneficial work without the benefit of modern knowledge or modern technology.

As with many other aspects of the plant sciences, the origins of agriculture precede recorded history. Most of today's cultivated plants were domesticated in pre-historic times. In fact, it is remarkable that so few important agricultural plants have been domesticated within historical times; on the other hand, most crop plants have been greatly improved through hybridization, selection, and other technological advances in breeding. The questions of where and how plants were domesticated has been studied by Russian Vavilov botanist and according to him there are several centres of origin of crop plants.

Beverage Plants

Man's earliest beverage in all probability was fruit juice. With advent of civilization and domestication of plants various types of soft drinks from variety of fruits were developed. Present day beverages are either alcohalic or non-alcohalic. Among the former are fermented from sugar and the other distilled from fermented liquors. Wines and beers are the oldest fermented beverages. Wine is the fermented juice of grapes with 7-16% alcohol, as against 3-8% alcohol in beers. Distilled alcoholic beverages are whisky, brandy, rum etc which have high caloric content than fermented drinks.

Non alcohalic beverages are tea, (*Camellia sinessis)* Coffee (*Coffea arabica*) and (*C. canephora*) and Cocoa *(Theobroma cacao)* all are tropical crops. The refreshing and stimulating properties of alkaloid caffeine (in coffee and tea) and cocoa these are enjoyed by many people all over the world.

Tea has a long history and the Chinese first discovered this around 2700 BC. There are two types of teas, the Chinese and the Assamese tea differing in their growth; the former slow growing and dwarf and the latter tall and fast growing. In view of the antioxidant properties tea is good for health and consumed by millions all over the world. Tea industry in India, China, Myanmar, Thailand and Srilanka is a flourishing business and several brands are available in market differing in taste, growing area and are accordingly priced. Indian tea export last year was 100 million kilogram worth Rs. 876 crore.

Coffee cultivation and consumption is nearly 500 years old. Brazil, Colombia is the largest producers and exporters. Indian's coffee mostly comes from Karnataka (75%) and Tamilnadu, Kerala and Orissa share the rest. The coffee of commerce are the beans (seeds) which are roasted and powdered. Instant coffee, chicory coffee, filtered coffee is consumed by millions and there is a good national and International market.

Cocoa and chocolate are two major products obtained from roasted ripe seeds of *Theobroma coca*. Its history is less than 500 year old. Cocoa butter, milk chocolates, cocoa powder are commonly used by millions all over the world and this is also a flourishing industry. Cocoa Research Institute, Ghana in Africa is a major centre of research.

Food Plants

The historical record, both written and archaeological, shows that at least 3,000-8,000 species of plants have been used as food. Most present knowledge of edible wild plants has been gained from aboriginal people around the world. Of the 1,000 or so species that truly have been domesticated and cultivated intensively, only about 200 have become productive enough to enter into international commerce as food plants. The rapid growth of world population has inevitably set the trend through the centuries to reduce or eliminate the cultivation of less efficient crops and to concentrate on the more productive ones. As a result, the majority of the world's population is now nourished by only 12 plant species, which include three cereals, rice, wheat, and corn (maize); two sugar plants, sugarcane and the sugar beet; three "root" crops, the potato, the sweet potato, and the cassava; two legumes, the common bean and the soybean; and two "tree" crops, the coconut and the banana. The next three most highly utilized species are the peanut, sorghum, and barley. These few crop plants stand between most men and starvation.

Mention must be made of food of grazing animals the fodder. It comprises of fresh and dry grasses, legume, leaves and pods of many species.

Cereals Plants

Cereals are the most important source of food. Cereal grains such as rice, wheat, and corn (maize) are staples in the diet of hundreds of millions of people throughout the world. Other cereals such as oats, barley, rye, and millet are important food plants in various regions. Wild rice and buckwheat are delights fancied by some. Other dietary mainstays of millions of people are the legumes, including beans, peas, soybeans, peanuts, and lentils. They have been grown since the dawn of history, and every major civilization has been founded on cereals as the principal source of food. Cereals in Latin is '*cerealia munera*' gift of Ceres the Roman Goddesses of agriculture. Today their production is calculated in millions of tons all over the world. Some ancient cereals are : Rice (*Oryza sativa*), wheat (*Tritium monococcum*) domesticated wheat *T. aestivum* is the bread wheat, Barley (*Hordeum vulgare*) and corn (*Zea mays*).

Secale cereale (Rye) and *Triticum* wheat have been crossed and a man made cereal called *Triticosecale* was produced and this is abundantly grown and used in Europe and North India. Hybrid varieties of various cereals are also grown. Cereals are food plants rich in proteins, starch, fats, vitamins and minerals. Variety of preparations are made depending upon the region from the various cereals or in combination with pulses. There are number of rice byproducts, wheat biprodut‌s, corn byproducts and so on.

The psendocereals are plants that look like cereal grains, are amaranthus, chenopods, buckwheat and are fancied by many gourmets.

Millet Plants

Millets are small seeded edible grass and known as '*poor mans*' cereal as they are low in protein but are rich in fat.

The important millet plants are : *Setarica italica* (Foxtail millet), *Eleusine coracana* (Finger millet), *Pennisetum typhoideum* (Pearl millet) and *Panicum miliaceum, Sorghum bicolor* (Jowar).

World production of millets is over 55 million metric tonnes annually mostly from India, Africa and China. Millets are non glutenous and unsuitable for making bread, therefore, consumed as porridge, chapati and also used as animal food.

Fibre Plants

In addition to food and shelter man needs clothing to call himself civilised. The minor bark of many trees was once beaten into a natural cloth and this antedated the weaving of plant fibre into cloth. Fibres plant species are long cells or cluster of cells that give strength to stem, leaves etc. Thousands of plant botanically yield useful fibre and have been used by primitive people all over the world. Fibres still in use were discovered in pre-historic times and were already highly refined by the earliest historic times. Fibres of vegetable origin are composd chiefly of cellulose, a carbohydrate substance manufactured by the plant. Although cellulose fibres are available from numerous plant sources, only a limited number of them are suitbale for textile use because of the difficulties involved in separating them from other substances. Unlike animal fibres (wool) natural plant fibres absorb moisture and hence ideal for clothing. Fibre plants are classified variously, in which they originate *Bast fibres*: (Jute, hemp) *Surface fibre*; (cotton, kapok) Fruit fibre (coconut). Cotton is the world's oldest and chief firbre plant. Two different cotton types originated in India and America. Flax is 10,000 years old and was cultivated in pre-historic times. Filling fibres for pillows are furnished from the silk cotton tree (Kapok) and other plants. Manila hemp is made from the leafstalks of banana a close relative *Musa textilis*. Coir, the important fibre produced from the coconut has long history of uses and is multipurpose. Hemp, the fibre from *Canabis sativa*, has been used since ancient times for the production of rope and cordage. In addition to hemp fibre, *Cannabis* also contains hallucinogenic compounds, the hashish and marijuana. Other important fibres of great economic importance are jute, and ramie (Bohemeria).

Fruit Plants

Fruit botanically is a fertilized ovary. There are so many kinds of fruits such as dry or fleshy, simple or multiple, dehiscent or indehiscent, legumes, capsules, berries and drupes. Important fleshy fruits are pome, pepo, berries etc. Fruits are consumed all over the world by young and old for its taste and value added nutrition. Many fruits, especially those of the citrus family; limes, lemons, oranges and grape fruits are rich in vitamin C (ascorbic acid), unpolished cereal grains in vitamin B_1 (thiamine); and wheat germ is vitamin E. The important fruits are listed below :

List of important fruits of India

Family		Scientific Name	Common name
Rosaceae	. . .	*Malus pumila*	Apple
		Pyrus communis	Pear
		Prunus persica	Peach
		Prunus domestica	Plum
		Prunus avium	Cherry
		Prunus armenaica	Apricot
		Rubus spp.	Raspberry
		Fragaria spp.	Strawberry
Rutaceae	. . .	*Citrus sinesis*	Orange
		Citrus limon	Lemon/lime
		Citrus paradisi	Pummelo
Moraceae	. . .	*Ficus carica*	Fig
		Artocarpus heterophyllus	Jackfruit
Musaceae	. . .	*Musa sapientum*	Banana
Anacardiaceae	. . .	*Mangifera indica*	Mango
Myrtaceae	. . .	*Syzgyium cumini*	Jamun
		Psidium guajava	Guava
Arecaceae	. . .	*Cocos nucifera*	Coconut
		Phoenix dactylifera	Dates
Vitaceae	. . .	*Vitis vinifera*	Grapes
Bromeliaceae	. . .	*Ananas cosmosus*	Pineapple
Sapotaceae	. . .	*Achras sapota*	Sapota
Sapindaceae	. . .	*Litchi sinensis*	Litchi
Clusiceae	. . .	*Garcinia mangostana*	mangosteen
Cucurbitaceae	. . .	*Cucumis melo*	kharbuz
		Citrulus lanatus	Tarbuz
Caricaceae	. . .	*Carica papaya*	Papaya
Annonaceae	. . .	*Annona squamosa*	Sitaphal
		A. cherimola	Hanumanphal

Fumitory and Masticatory Plants

Narcotics are closely related to medicinal plants for their physiological and biological effects. Most aboriginal peoples discovered some such plant that eased the chores of everyday life. These drugs were often under the control of chiefs or medicine men and were sometimes of religious importance. Examples are coca (from the Andes), opium (from the Orient), betel (from Indonesia) cola (from tropical Africa), cannabis (from Asia), datura (from Asia) and tobacco (from the New World).

Tobacco, betel leaf, areca nut, cola nut and several related products come under this category as these are used either for smoking or chewing.

Tobacco is chiefly derived from *Nicotiana tabaccum* and *N. rustica*. The leaves contain several alkaloids eg. Nicotine, anabasine, nicotyrine etc. Hundreds of brands of cigarettes are available in global market. In addition, cigars, beedis, chewing tobacco, gutka and hookah are widely used and all are habit forming and are certain source of cancers of lung, mouth etc.

India ranks 3rd in the world and AP and Gujarat are the top tobacco producing states. *Areca catechu* (nut) and *Piper betel* (pan) both contain alkaloid and are used in pan. Various types of single and combined products are sold in market and are habit forming and known to cause oral cancer.

The nuts of *Cola nitida* are rich in caffeine, theobromine, kolamin and are widely chewed.

Erythroxylum coca contains cocaine, the crude drug is rich in tropococaine and cinnamylcocaine. The boiled latex of *Manilkara achras*, a relative of sapota, yields the gutta percha, the chewing gum, widely used all over the world by young and old.

Spice and Condiments Plants

There are 70 species of spices cultivated all over the world. Broadly speaking spices are aromatic used for seasoning or garnishing foods and beverages. These are pungent with strong odour sweet or bitter eg. cloves, ginger, cardamom, nutmeg, vanilla, turmeric. The important spices are :

Zingiber officinale	:	Zingiberaceae
Curcuma longa	:	Zingiberaceae
Elettaria cardamom	:	Zingiberaceae
Cinnamomum verum	:	Lauraceae
Eugenia caryophyllus	:	Myrtaceae
Crocus sativus	:	Iridaceae
Piper nigrum	:	Piperaceae
Myristica fragrans	:	Myrsticaceae
Capsium anuum	:	Solanaceae
Vanilla planifolia	:	Orchidaceae
Foeniculum vulgare	:	Apiaceae
Coriandrum sativum	:	Apiaceae
Cuminum cyminum	:	Apiaceae

Condiments are flavouring agents with a sharp taste and are added to food after cooking. Various and curry powders come under this category. There is flourishing trade of spices at the national and international level.

Vegetable Plants

Vegetables are a rich source of vitamins, minerals, starch and above all roughage in human nutrition. Human diet include both rice eating and non rice eating area is made up from tuber crops which are high in carbohydrate but low in proteins, both root and stem crops are used.

The important root crops are :

Potato *Solanum tuberosum,* Sweet potato (*Ipomoea batatas*), Cassava (*Manihot esculenta*), Colocasia *Colocasia esilenta, Amorphophallus*, Yam *Dioscorea* species (*D. alata, D. esculenta*), Arrow root (*Maranta arundinacea*)

Potatoes are underground stems rich in starch and are consumed all over the world. Potato Research Institute in Simla, H.P has evolved several varieties including the one resistant to late blight.

Sweet potato are tuberous root rich in starch and consumed variously.

Underground stems have many food uses. Potato tubers feed millions of people in many parts of the world. In Southest Asia and Polynesia, the tubers of taro and dasheen are an important starchy food. Onion bulbs also are consumed by millions of people in Asia, Africa, USA etc.

Throughout the world, the fleshy storage roots of amerous plants are eaten and radishes add variety to meals. The roots of beets, turnips, parsnips, and rutabages are cooked and eaten as a starchy food by people in temperate regions. The roots of yams, cassavas, and sweet potatoes provide starch for many people, especially in tropical areas. Gourmets relish the falvours imparted to foods by ginger, turmeric, and horseradish.

A favourite vegetable of many people is Brussels sprouts, the axillary (stem) buds of a relative of cabbage.

The shoot system, or portions of it, is used in numerous ways. The leaves of cabbage, lettuce, and watercress are eaten in green salads; those of spinach, kale, collards, and turnips are consumed as a cooked vegetable.

Stems are also utilized in many ways. Gourmets relish the young, juicy, succulent shoots of asparagus; many people enjoy the large, spherical, juicy stems of kohlrabi. Both the spice cinnamon and the medicine quinine come from the bark of certain woody plants.

Inflorescences (flower clusters) consisting of immature flowers and flower stalks comprise the vegetables cauliflower, and broccoli. Important leaf and crops and fruit are listed below :

1. Leafy crops

Brassica oleracea		-	Cabbage
Brassica oleracea -	var. capitata.		cauliflower
Brassica oleracea -	var. *acephala*		knol kool
Brassica oleracea -	var. *gemmifera*		brussels sprout
Lactuca sativa		-	Letticeve
Spinacia oleracea.		-	Spinach (Pallakura)
Rheum rhapouticum		-	Chuka kura
Basella rubra		-	Batchella kura
Amaranths tricolor		-	Thota kura

2. Root crops

Solanm tuberosum (Potato)	-	Solanaceae
Ipomoea batatas (Sweet potato)	-	Convolvulaceae
Manihot esculenta (Casava)	-	Euphorpiaceae
Dioscorea alata D. esculenta (Yams)	-	Dioscoreaceae
Colocasia esanlenta (Arvi)	-	Araceae
Allium cepa (Onion)	-	Liliaceae
A. sativm (Garlic)	-	Liliaceae
Brassica capitata (Turnip)	-	Brassicaceae
Daucus carota (Carrot)	-	Apiaceae
Raphanus sativus (Radisk)	-	Brassicaceae

3. Fruit crops

Cucurbits :

Cucumis melo var. *reticulata*	-	Kharbuz
Momordica charantia	-	Bitter gourd
Citrulus lanatus	-	water melon
Luffa cylinderica	-	Sponge gourd
L. angulata	-	Torai
Curcurbita moschata	-	Pumpin
Cucurbita pepo	-	Kadu
Benicasa hispida	-	Petha
Trichosanthes dioica	-	Parwal

Non - cucurbits :

1. *Lycopersicon esculentum*	-	Tomato
2. *Solanum melongena*	-	Brinjal
3. *Capsicum annum* *C. frutescens*	-	Chillies
4. *Abelmoschus esculentus*	-	Bhendi

Forests Plants

Forest trees provide man with a number of useful products including wood, food, fibres, gums, and resins. Some aboriginal groups in tropical America, Africa, and Asia still depend entirely on the forest for all their needs and do no attempt any agriculture. Forest plants may be timber yielding, non timber and minor forest produce.

Wood Plants

Civilizations now and in the past depend on wood and wood products at great cost to the world natural resource. Forest trees provide man with a number of useful products including wood, food, fibres, gums and resins etc.

Wood from trees is used for building material, paper making, fuel, and furniture. Phloem fibres of flax are used to weave linen cloth; those of hemp and jute are woven into ropes and bags. Both stems and wood strips are made into baskets and matting. The large, hollow stems of bamboo are extensively used in constructing furniture, fishing rods, and many other articles.

The important timber yielding plants are :

Acacia, *Albizzia*, *Ailanthus*, *Bombax*, *Chloroxylon*, *Casurina*, *Dalbergia*, *Diospyros*, *Eucalyptus*, *Gmelina*, *Grevilla*, *Santialum*, *Swietenia*, *Terminalia*, *Tectona*, *Pterocarpus*, *Xylia*. In addition, 'bamboos' belonging to *Dendrocalamus*, *Calamus* and *Bambusa* are abundantly used.

Modern technology has increased the efficiency of wood distillation as well as the number of products that can be recovered during that process; among them, in addition to charcoal, are acetic acid, acetone, methanol, wood gas, and wood tar and oils. Softwood from conifers, especially old pine stumps, is distilled to produce turpentine, rosin, pine oil, and creosote, as well as charcoal and wood gas. Many modern techniques have been developed to give wood greater strength or to enhance other properties. Several chemical processes remove lignin and other materials from wood fibres, leaving almost pure cellulose ready for manufacture into paper. Other processes utilize cellulose to manufacture fibres such as rayon, films such as cellophane, and many different types of wood and plastic products.

Modern forestry today is concerned with the management of wooded lands and grasslands watershed production, grazing, protection of wild life, scenic values, recreation and even game etc.

Medicinal Plants

The earliest records of medicinal plants are those of the ancient Chinese, Hindu and Mediterranean civilizations. In 2735 BC, the Chinese emperor Shen Nung wrote an herbal in which he described the antifever capabilities of a substance known as Chang Shang, since shown to contain antimalarial alkaloids.

The number of plants that have been used for medicinal purposes is nearly as large as the list of edible plants. Root drugs that date back to 5000 BC include aconite, colchicum, gentian, squill, and valerian. Other root drugs in use are goldenseal, ginseng, ipecac, licorice, jalap, podophyllum, rhubarb, and senega; several have their origin in the New World and were well known in aboriginal medicine. All bark drugs in common use are from the New World; they include cascara, sagrada, curare, and quinine. Drugs extracted from stems and wood, ephedrine is of Chinese origin, and guaiacum and quassia are from tropical America. Many drugs occur in leaves, which are the primary source of chemical compounds in plants; such are aloe, belladonna, cocaine, digitalis, eucalyptus, witch hazel, henbane, horehound, lobelia, pennyroyal, senna, stramonium, and wormwood. A few drugs occur in flowers, of which the best known are chamomile and hops. Many seeds and fruits that are poisonous contain ingredients that, in moderation, may be of medicinal importance; examples are croton oil, nux vomica, opium, psyllium, strychnine, strophanthus, and wormseed.

Crude drugs

The term crude drug is usually applied to plant or animal organs or whole organisms or exudations of these, either in the fresh state or dried and either ground or unground. Some crude drugs such as acacia, belladonna, and starch are official; for these, rigid specifications are available. Plant-derived crude drugs may come from cultivated soures or may be collected in the wild. Cultivation and collection are carefully planned to obtain drugs with the highest possible therapeutic activity or content of active constituents.

Many crude drugs still used now, such as ipecacuanha root (ipecac), were known and employed by the ancients. The Egyptians treated constipation with senna pods and castor oil and employed caraway and peppermint to relieve indigestion. The Greek physician Galen (c. AD 130-c 200) included hyoscyamus, opium, squill (a plant drug used as an expectorant, cardiac stimulant, and diuretic), and viper toxin among other drugs in his apothecary shop. He also insisted on purity in drugs; i.e., the right variety and age of the botanical specimens.

Alkaloid compounds were first isolated from plant. Narcotine was obtained from opium by a French pharmacist in 1803 and was followed by Morpheus in 1806, emetine and strychicine and quinine (1820), nicotine (1828), atropine (1833), cocaine (1860), and physostigmine (1867). Glycosides such as digoxin are another important group of drugs obtained from plants. Isolation of these potent compounds was a milestone in pharmaceutical progress.

Pharmaceutical practice improved markedly in the 16th and 17th centuries. In 1546 the first pharmacopeia, (selected list of drugs and medicinal chemicals with directions for making preparations from them) appeared in Nurnberg, West Germany. Number of pharmacopoeia's are now available, British pharmacopoeia (BP), Indian Pharmoepia (IP), US Pharmacopoeia (USP) etc.

For further information on medicinal plants the reader should consult the chapters on medicinal and drug yielding plants.

Narcotic Plants

Narcotics are closely related to medicinal plants in their physiological and biological effects. Most aboriginal peoples discovered some such plant that eased the cares of everyday life. These drugs were often under the control of chiefs or medicine men and were sometimes of religious importance. Examples are coca (from the Andes), opium (from the Orient), betel (from Indonesia), cola (from tropical Africa), Cannabis (from Asia), datura (from Asia) and tobacco (from the New World). Narcotic plants are sleep inducing plants. They may be sedatives, hypnotics or halluinogens and illusions and cause a stage of estasy causing cerebral excitation.

1. *Papaver somniferum,* (the poppy plant) the dried latex is the morphine used as an excellent sedative in post surgery.
2. *Cannabis sativa,* the cannabis is known to mankind since 2737 BC. There are various names to various plant parts as listed :
   * Bhang are the dried leaves, stem and male and female flowers used in drinking.
   * Ganja is the dried unfertilized female flowers used for smoking.
   * Charas/hashish is the fresh yellow milky exudate from leaves and inflorescence used for smoking.

The chief component alkaloids are scopolamine, tetrahydrocannabinol. Marihuana/marijuana is used in USA and corresponds to Bhang in India.

Oil Plants

Vegetable edible oils are derived from seeds, endosperm and cotyledons of wide variety of plants. Depending on its ability to absorb oxygen from atmosphere the oils are variously classified.

1. *Non drying oils :* Groundnut (*Arachis hypogea*), palm (*Elaeis guineensis*), Castor (*Ricinus communis*), olive (*Olea europea*), rape (*Brassica napus*) and almond (*Pyrus sp.*).
2. *Drying oils :* Linseed (*Linum usitatissimum*) soybean (*Glycine max*) and Safflower (*Carthamus tinctorius*).
3. *Semidrying oils :* Cotton (*Gossypium herbaceum*), gingley (*Sesamum indian*), sunflower (*Helianthus annus*), corn (*Zea mays*).

Oils and fats are primarly glycerides. Most of the oils under the three categories listed above are used as cooking medium and used universally in various food stuffs; manufacture of margarine, cooking fat, soaps and for painting.

Vegetable oils differ from volatile oils in several respects :

1. They do not volatilise at room temperature
2. Glycerids are therefore form soaps with alkalies
3. They lack taste and odour
4. Leave a greasy stain on paper/cloth

Vegetable oil are used in food and in industrial applications and are rich in vitamins A D E & K. The oils are used in the manufacture of vanishes, candles, plastics, polishes, lubricant. The oil cake is used as feed.

Ornamental Plants

Ornamental plants whether cultivated for flowers, fruits or leaves are very numerous and cannot be listed. They may be trees, shrubs, herbs, climbers etc. They may be hardy that withstand the winter, summer, outdoor; half hardy which require protection during winter. Succulent plants are cactic etc. A visit to a garden or nursary is a rich experience to get close to various plants of different colours. In India about 100 plants are used for ornamentation. There are number of hardy trees with various flower colours to beautify the roads as avenue trees. There are number of bulb plants especially Irisis, gladioli, amaryllids, lilies etc.

Pulse Plants

The seeds of pods or legumes called beans or pulses are mankind's most important protein foods next to cereals. There are two types; the large seeded (*Phaseolus acontifolius, P. vulgaris*) and the small seeded *Glycine max* (soybeans) and *Vigna unguiulata* are ancient and important legumes of India. Annual acreage of pulses in India is 24 million hectares amounting to 12 million tonnes and 20% of acrage under food crops. Important pulses are :

Cajanus cajan
Vigna mungo
Vigna radiata
Vigna aconitifolia
Lens enlinaris
Dolichos biflorus
Cyamopsis tetragonoloba
Glycine max
Cicer arietinum
Pisum arvense, P. sativum
Lablah purpureus

The top six grain legume/pulse crops are soybean, peanut, dry bean, chick pea, pigeon pea and cow pea. There are numerous other minor legumes which have received less attention among them in the winged bean (*Psophocarpus tetragonolobus*) which has been under cultivation in tropics as a poor mans pulse. Its seeds have created the greatest interest internationally for its mutritional value but also for oil, vitamin and minerals. Today it is as significant as soybeans in world agriculture. Every part of the winged bean is edible.

Perfume Plants

Perfumes and scents have been integral part of man's quest for the life beautiful since time immemorial. Perfumes have an exotic and fascinating heritage. According to a legend, the goddess Venus was the first to use perfume. Cleopatra, Helen of Troy said to have used perfumes.

In olden times perfumes were made from flowers like rose, jasmine, juhi, chameli and keora. It is estimated that 12,25,000 roses are needed to make just one pound weight of pure rose oil.

The sense of smell is one of the keenest of man's five senses, no wonder most cosmetics are delicately scented to enhance their appeal to ones olfactory senses.

Rose oil (flowers of *Rosa species*), Geranium oil (leaves of *Pelargonium*) species, khus, vetiver oil (roots of *Vetiveria*), sandal wood oil from heart wood of *Santalum album*. Jasmine and *lavander* oil from (flowers of *Jasminum* and *Lavendula*) are some important perfumes.

Perfumes industry is a flourishing one and French perfumes and Indian Itars are famous. With Aroma therapy catching up, there is a boost to perfume industry all over the world. Southern France is famous for its perfumes.

Rose water, rose sherbat and gulkhand are commonly used in India.

Jalaluddin (1977) has reported the isolation and identification of a fungus which attacks trees of *Aquilaria agallocha*, the diseased wood yields a fragrant perfume upon steam distillation and it is used to make incense sticks at the level of an important cottage industry.

Petro Plants

Petroplants are plants that yield various forms of hydrocarbons. To date 50 species mostly Euphorbiaceae and Asdepiadaceae have been identified as potential source of hydrocarbons/biocrude, biodiesel and certain blending components which can improve the quality of fuels. There are two approaches

1. Plants which have high flow of milky fluid (latex) which is collected, preserved and coagulated and processed into liquid fuel. *e.g. Euphorbia caducifolia, E. nivulia, E. royelana, E. neriifolia.* The biocrude content varies from 26-30% by weight.
2. Plants which have less flow of milky latex, the whole plant mass is processed and the biocrude converted into liquid fuel by a process known as fluidised catalytic cracking (FCC). *e.g. Euphorbia antisyphitica, Pedilanthus, Calotropis, Crytostegia.* The biocrude content varies from 2.5–8.8% by weight.

E. antisyphilitica is the potential energy crop as it provides 160 tonnes of biomass and 3-37 tonnes of biocrude per hectare every year yielding 1.01 tonne gases, 0.76 tonne naphtha, 0.61 tonne kerosine, 1.05 tonne diesel and 0.12 tonne coke. In terms of properties and molecular weight petroplant fuel compares well with fuels currently in use.

In addition, *Jojoba curcas* (Simmondsiaceae) native of Mexico and Western USA has been found to equally good biofuel and the Rajasthan government is growing the petro plant as a commercial crop. The Indian Institute of Petroleum, Dehradun, has developed the technology for Jojoba oil for lubricants and related products.

Jatropha oil *Jatropha curcas* is used as a diesel substitute in several countries including Austria and Nicaragua. The IIT, Delhi and IIP Dehradun have tested and assessed its efficacy under Indian conditions. Accordingly the Central Govt. has drawn up a plan of about Rs. 1500 crores to make use of the jatropha oil as a bio-alternative to diesel. The plan once implemented is expected to produce 6 lakh tonnes of diesel quality oil valued at Rs. 1020 crores per annum. The Agriculture Ministry would raise Jatropha plantations in an area of 4 lakh hectares in eight states, AP, UP, MP, Karnataka, Gujarat, Rajasthan, Maharashtra, and Chhatisgarh. These plantations would have a life span of 45 years and an investment of Rs. 1.430 crores is expected to get a return of Rs. 46, 000 crores.

Fuel

Two major types of fossil fuel, (coal including lignite) and petroleum (including natural gas) both are derived from plant deposits accumulated millions of years ago and transformed into their present fuel state by heat and pressure.

Rubber Plants

Rubber has been derived from number of plants belonging to 10 families but only one, *Hevea brasiliensis* has remained consistently outstanding. Other plants that yield rubber are *Ficus elastica, F. vogelli, Castella elastica and Parthenium argentatum.*

Hevea brasiliensis was introduced in 1876 by British in their colonies in Far East and its cultivation soon became common. By 1950 *Hevea* became the sole source of rubber. Dunlop, Rubber industry played a key role. Good year (1888) invented pneumatic bicycle tyre which later gave into motor car tyre and this gave automobile and aviation transport a big filip to fast and superfast transport.

Rubber has many applications, rubberised apparel, atheletic goods, belt conveyor materials and in escalators. It is also used in plastics, civil and electrical engineering toys and upholstery etc.

Sugar Plants

It's difficult to imagine the world without sweet preparations. In ancient times honey was used as a sweetner. Sugar was sold in apothecaries as a rare commodity during the Middle Ages and it is only after the middle of 15th century sugar replaced honey as a sweetner.

World's major supply of sugar at present comes from stems of *Saccharum officinarum* and *Beta vulgaris*; the beetroot. There are several species of *Saccharum* and their hybrids grown in different parts of India. Sugar beet is extensively grown in Europe, USSR and USA. Sugar the white crystaline substance is the important source of energy in human diet largely used in soft drinks, ice creams, chocolates, confectionary, tonics, medicines etc. Jaggery and other related products are commonly used in India and several other countries.

Molasses is a by product of cane sugar and is used as a livestock feed and for the making of alcohol, beer, rum, vinegar, glycerol and several industrial products.

Sugar palm (*Arenga pinnata*) and palmyra palm (*Borassus flabellifer*) are other plants that yield sugar.

Stevia rebauldiana (Asteraceae) is a small delicate herb from Paraguay, South America and is a safe and natural sweetner commonly used in Japan. The leaves of the plant are very sweet; 30 times sweeter than sugar cane and are non caloric. It can be used as a subsitute for cane sugar as well as to treat many ailments like diabetes, high blood pressure, skin defects etc.

Stevia can be grown in pots or as a commercial field crop in well drain red soil and sandyloam. The crop can be harvested in 4-5 months of planting, with subsequent harvests once every three months. The cost of cultivation is Rs. 10 lakhs/hectare and gross income of Rs. 200/kg dry leaves.

Dioscorephyllum cumminsii (Merispermaceae) of W. Africa is another non caloric harmless, acceptable sweetner. The berries produce a protein which is regarded as the sweetest known naturally occurring substance (upto 3000 times sweeter than sucrose). It has not been exploited.

Thaumatococcus danielli : The berries of this plant contain a protein called Thaumatin said to be most sweetest this is being made to produce in genetically engineered organisms and is marketed and the trade name Nutra-sweet in Japan and Europe.

Dye Plants

Plant types are of two types, those produced from plants especially cultivated for the purse and those obtained from plants growing wild.

The use of plant dyes is deeply interowven with the use of fabric and is as old as civilization. The famous Kalamkari painting from Vizag are made with plant dyes only. There are so many dyes plants and for the ease of understanding they are arranged as follow :

I From stem/bark

Acacia catechu - Red dye
Symplocos sp - yellow dye
Alnus nepalense - red dye

II From roots/rhizome

Curcuma longa - yellow
Rubia cordifolia - red
Oldenlandia umbellata - Alizarin red

III Leaves

Indigofera tinctoria - Blue
Lawsonia inermis - Henna/Mehndi

IV Flowers

Butea monosperma - yellow
Nyctanthes arbor-tristis - orange
Tagetus erecta - yellow

V Fruit/seeds

Mallotus phillipiensis - Kumkum
Bixa orellana - Orange, yellow

VI Wood

Haematoxylon campechianum - Blue-black

There are over dozen plants which are used in tanning industry; *Acacia catechu* is used in the preparation of Katha used in making '*pan*' & '*pan masala*' etc.

Insecticidal Plants

In the past many narcotic plants and their products such as nicotine have been used to control pests. In ancient India, neem leaf was extensively used to preserve cereals and to ward off pests. Likewise the use of pyrethrum dried powder of flower heads of *Pyrethrum cinerariaefolium* were used and are still being used.

Thanks to chemical screening of number of plants 'rotenone' derived from the roots of several legumes especially

Derris elliptica, *D. ferruginea*, *Tephrosia, Mundulea* and *Lonehocarpus.*

In addition there are number of minor insecticides obtained from various plants belonging to atleast dozen plant families.

Sacred plants / Plants as symbols

Plants or their parts have been in use by man since paleolithic times and even in ancient India. This practice has led to the preservation of some plants. Plants thus have symbolic association. The use of rice as *'akshatalu'* is as symbolic and important ritual as the use of '*henna*' in any marriage. The worship of sacred tulasi, banyan tree, the use '*bel leaves*' and '*sacred lotus*' is typically associated with many sacred hindu functions. Similarly, the kadamba tree (*Nauclea kadamba*) and harshringar (*Nyctanthes arbor-tristis*) are associated with Lord Krishna. The Baobab tree (*Adansonia digitata*) with its swollen base is worshipped in Africa. It can be seen as a huge, 200 years old tree, in Golkonda Fort, in Hyderabad. The olive branch (*Olea europea*) is a symbol of global peace and its oil was used as currency in ancient times. *Commiphora* (myrrh) and *Boswellia* are used as incense or sticks. Liberal use of various flowers and leaves is a common practice not only in India but several Asian, African and Latin American countries on festivals, marriages, births and deaths. Shiva rudraksha (*Peganum hermala*) beads garland originally used by Rishis is now commonly used by young and old for its curative properties.

This account is not a complete survey of plants used in various walks of life. To conclude with a Oriya saying "*plant a tree to witness a birth, to commeorate a marriage and the the tree remains a friend to the dead also*".

ESSAY QUESTIONS

1. What is alternation of generatiaon ? Explain with suitable example.
2. Describe the various life cycles in algae.
3. Give an account of sexual reproduction in algae.
4. Describe the types of spores in algae.
5. Give a brief account of fungi.
6. Write an account of Actinomycetes
7. Describe the alternation of generation in a bryophyte
8. Describe the various pollination mechanisms in flowers
9. What is double fertilisation. Describe the details with illustrations
10. Describe the structure of flower and their importance
11. Describe the various theories of apical meristem
12. What are meristem ? Where do you find them ? Describe their structure and functions.
13. What is tissue culture ? What is its importance and applications ?
14. Give a list of T.C raised plants.
15. Differentiate between hybrid and cybrid. Describe their importance
16. Write an essay on lichens and their importance
17. How are angiosperms different from gymosperms. List their differences.
18. Enumerate the major differences between monocot and dicot plants.
19. What is micropropagation. Describe its advantages and applications.
20. Describe briefly the production of haploid plants
21. Write an essay on importance of plants in day to day life
22. What an account of vegetable plants ?
23. Describe the important fruits plants.
24. What is growth ? Describe.
25. Give brief account of plant breeding systems.
26. Describe the sporophyte of a bryophyte
27. Economic importance of gymnosperms
28. Compare ovule with a seed.
29. List important features of angiosperms.
30. Describe the ultra structure of plant cell.
31. Enumerate briefly the characters of various algal groups.
32. Describe mitosis in a Plant Cells.
33. What are the inclusions found in plant cells ? Give brief account and their importance.
34. Describe the structure of chloroplast and its functions.
35. Describe the structure of mitochondrian and its functions.

Short Questions

Explain Following :

1. Thallus organisation in algae.
2. Chloroplast in algae.
3. What are lichens.
4. Apothecism.
5. Soredium.
6. Thallus of lichen.
7. Isogamy and anisogamy.
8. Fruiting bodies in fungi.
9. Economic importance of algae.
10. Economic importance of fungi.
11. Economic importance of actinomycetes.
12. Important features of gymnosperms.
13. Heterospory.
14. Ascolichens.
15. Uses of lichens.
16. Protonema.
17. Megasporophyll of *Cycas.*
18. Male strobilus of *Cycas.*
19. Male cone of *Pinus.*
20. Female cone of *Pinus.*
21. Microphyll and megaphyll.
22. Winged pollen grain of *Pinus.*
23. Ovule in *Cycas.*
24. Ovules in angiosperm.
25. Syngamy.
26. Microsporogenesis.
27. Megasporogenesis.
28. Embryogeny.
29. Structure of gynaecium.
30. Testa and tegmen.
31. Endospermic seeds.
32. Embryosac.
33. Double fertilisation.
34. Endosperm.
35. Fruit.
36. Seed.
37. Incompatibility in plants.
38. Cell cycle.
39. Metaphase.
40. Protoplasm.
41. Mitosis
42. Cytokinesis
43. Chromosomes
44. Plant cell wall
45. Plastids
46. Mitochondrion
47. Totipotency
48. Callus
49. Meristem
50. Root meristem
51. Shoot meristem
52. Tissues
53. Plerome
54. Dermatogen
55. Periblem
56. Vascular cambium
57. Xylem
58. Phloem
59. Auxin
60. Cytokinin
61. Meristematic tissue
62. Rhizogenesis
63. Organogenesis
64. Somatic embryogenesis
65. Synthetic seeds/synseeds
66. Hybrids
67. Cybrids
68. Protoplast fusion
69. Auther culture
70. Laminar flow bench

2 Plant Biotechnology

2.0 Plant Tissue Culture

Every part of the plant has the innate capacity to produce a new plant, from roots in *Muraya*, from leaves in *Bryophyllum* and from stem in rose and sugarcane and this is called *vegetative propagation*. This unique property of the plants prompted the German biologists, Theodor Schwann and M. J. Schleiden (1838) to put foreward the '*totipotency theory*', which means that every cell of the plant, under appropriate conditions has the capacity to produce a new plant. Since all the genetic information needed to specify the whole plant is coded in its DNA. Vochting (1878) reported callus induction in several species followed by Sach's (1880-1882) work on callusing and wound healing. Rechinger (1893) described the formation of callus on isolated fragments of stems and leaves. Gottlieb Haberlandt, (1902) plant physiologist from Germany made first attempt to maintain the isolated mesophyll cells from leaf for 20-27 days in simple nutrient medium and noted cell wall thickening and cell enlargement but no cell division. However, within a few years, the zoologists employing the methods suggested by Haberlandt were successful in growing animal tissues in culture.

Haberlandt (1913), using pieces of potato noted cell division *in vitro* when the piece contained a fragment of vascular bundle and concluded that the "*leptome would secrete one or more stimuli which diffuse through the adjacent storage tissue*". The major break through was made by Robins and Kotte, when they could grow the isolated root tips. However, the first real success was achieved by Nobecourt, Gautheret and White (1937) were able to culture cambium tissue capable of unlimited growth and maintained the same for one year. These three are generally credited for laying the foundation for further work in plant tissue culture. However, plant tissue cultures made progress in another direction, the growth of excised organs. In 1934 tomato roots were grown through successive subcultures without any diminution in the rate of growth and by 1937 a chemically defined medium, which would support their continuous growth, had been developed. Skoog and Tsui (1948) demonstrated shoot initiation in segments and callus from the tobacco stem by chemically regulated by manipulating the medium. In 1956 single isolated plant cells were successfully grown in culture.

Skoog and Miller (1957) demonstrated organogenesis (shoot and root initiation) in tobacco callus cultures by regulating the auxin and cytokinin ratio in the medium. Murashige and Skoog (1962) proposed a medium for tobacco tissue culture and this medium the *MS medium* and its severel modifications are currently used all over the world.

The discoveries on enzymatically isolated proplasts and their culture by Cocking (1960), Maheswari and Guha (1964) on haploids from anther culture of *Datura innoxia*; production of somatic hybrids from fusion of protoplasts isolated from potato and tomato by Melchers (1978) opened up new avenues for plant hybridization by unconventional methods. Since that time a large number of different organs and plant parts have been grown in culture. Among these are stem tips, leaves, flowers, stamens, ovaries, ovules, gametophytes, endosperm and embryos in addition to tissues from several kinds of normal organs, pathological tissues of crown-gall, insect, or viral origin and spontaneous, genetic, and chemically induced tumors and galls.

Plant tissue culture provides a method by which living tissues may be maintained for prolonged periods outside the organism. The excised tissue, the explant is sustained by a nutrient medium which may also support additional growth and proliferation. When all the nutritional requirements of the tissue are met by the culture medium it is possible by the use of appropriate sub-culturc techniques, to maintain continuous growth of tissues through potentially unlimited periods of time.

Although the definition emphasizes the culture of plant tissues, the techniques have proven equally applicable not only to the culture of excised organ fragments, whole organs and intact organisms but to the culture of cell populations and single isolated cells. Sterile culture, *in vitro* culture callus culture, organ culture, or cell culture are other alternate terms used by some workers.

The two volume Bibliography of Plant Tissue Culture by S. S. Bhojwani (Elsever Publication, Holland) provides a comprehensive and complete list of world wide work on plant tissue culture and Plant Biotechnology upto 1998.

Set up of Tissue Culture Laboratory

A tissue culture laboratory basically requires aseptic conditions at every stage i.e. inoculation, callusing, sub-culture and regeneration until transfer of plant lets to pots/fields is a pre-requisite for a successful tissue culture programme.

To carry out cell and tissue culture studies the following minimum facilities are required :

1. Media preparation rooms with autoclaves of different capacities.
2. Wash rooms for glassware with ovens.
3. Laminar flow (sterile air flow) bench.
4. Air conditioned room with racks and desired lighting facilities.
5. Shakers for suspension cultures.

6. Growth chamber/mist chambers/glass house to grow regenerated plantlets under controlled or special conditions.
7. Deionizer
8. Single glass and Double water distillation units.
9. Sufficient glassware and chemicals & growth regulators.

In addition, a pH meter; Digital Balance, Autoclave, Inverted Microscope, Centrifuge and a stand bye generator etc. are also necessary. The equipment is arranged in an order from one facility to another, wash area for glassware, media preparation room with autoclave, laminar flow bench for aseptic inoculation to incubation of cultures in sterile AC rooms with temperature controls around 25 ± 1 °C etc. Modern high quality clean labs are provided with environmental control system, air handling technology, vibration protection. Wearing gown and head gear is also mandatory in such high tech labs.

Media Preparation

Tissues of different plants or different parts of the same plant may require different nutritional requirements for their satisfactory growth. Thus no single medium that can satisfy all the *in vitro* culture requirements in an experiment.

The composition of the nutrient medium is of great importance and in several laboratories considerable effort has been devoted to obtaining chemically defined media which will support continuous tissue growth. Such media usually consist of the following fractions:

1. A carbon-energy source,
2. A mixture of major inorganic salts,
3. A trace element mixture,
4. Various growth supplements *e.g.* B-vitamins hormones etc.

Basic culture media are prepared by using concentrated stock solutions (usually 20 or 40 times the original concentration) of macro and micro nutrients, vitamins and growth regulators. The stock solutions should be kept free from contamination in refrigerators. Some constituents like inositol, sucrose, yeast extract, casein hydrolysate, should be freshly prepared separately for each experiment. Agar 0.6% is generally used to solid by the media.

Various media formulations are now available, and several types of media are being used by more than 10,0000 tissue culturists throughout the world for specific purposes. Most of the media are composed of major elements, minor elements, organic constituents and optional constituents like various auxins and cytokinins etc. The commonly used media are *Murashige* and *Skoog* (MS), *Linsmaier* and *Skoog* (LS), *B5, Blaydes, SH, N6, Whites*. Although, a particular media is suggested for a specific purpose, the researcher must always screen and develop the proper media for efficient growth of callus cultures and regeneration as per the experimental requirements of the plant.

Phytohormones like auxins, cytokinins or gibberellins in ultra dilutions are added to the medium to stimulate growth and differentiation. Coconut milk, yeast extract, or fruit juices can also be used.

Four different procedures are commonly used for growing plant parts in culture.

1. Agar solidified medium for growth of many tissues and organs (leaf, shoot apex etc.).
2. Stationary liquid medium for growth of root cultures.
3. Shaking liquid medium for growth of cell populations.
4. Single drop medium is used in single cell isolations.

Asceptic Conditions

Tissue culture experiments are carried out under 100% germ free or asceptic conditions. Maintaining cleanliness in the working area is mandatory as also personal hygine.

Tissue cultures can be initiated from any part of the plant, the excised plant parts (tissues) are called as *explants*. Plants grown in an external environment are often contaminated with micro organisms (bacteria and fungi) which compete for growth with plant tissue in the medium. It is, therefore, essential that these "*explants*" be freed from contamination. Different methods of sterilization of explants, may have to be tested for evolving efficient sterilization protocol. Depending upon requirement disinfection the plant material is a pre-requisite for any *in vitro* culture experiment.

The plant parts, are from which explants, excised are washed under running tap water (and also with detergent solution like, teepol) for few hours. This is followed by surface sterilization with different reagents like sodium hypochlorite (5-10% w/v solution), mercuric chloride (0.005-0.25 w/v solution) etc. Seeds are often sterilized with 0.1% mercuric chloride for a desirable time period.

The sterilised explants are washed in sterile distilled water and are directly implanted onto the nutrient. Transfer of the sterilised explant into the test tube containing the nutrient medium is called inoculation. This is carried out using '*Laminar flow chamber bench*'. It maintains air flow free of micro organisms and provide aseptic conditions.

Nutrient Medium

The nutritional requirements supplied to the explant for its optimal growth during the in vitro culture experiments is called nutrient medium.

After inoculation and plugging, tubes or conical flasks are kept undisturbed in culture room. The room temperature should be 25 $\pm$ 2 °C and humidity above 50%. The incubation is done for 3-4 weeks and observations are made at regular interval for callusing and regeneration response.

Sterilisation of the Medium

The medium is sterilised in conical flasks or culture tubes in an autoclave (an improvised pressure - cooker type) for 15 minutes at 121 °C under 15 1b pressure and equally distributed to test tubes and plugged with cotton wool or plastic caps.

Type of in vitro Culture Experiments

There are three kinds of *in vitro* culture experiments.

1. Cell culture
2. Tissue culture, and
3. Organ culture.

If the explant is a cell, tissue, organs (like anther, ovule, ovary etc.) the *in vitro* culture experiments are called cell culture, tissue culture and organs culture respectively. Any part of the plant, be it a cell, tissue or an organ used in the *vitro* culture to produce a new plant is called explant.

Establishment and Maintenance of Cultures

Cultures are established by placing a suitable nutrient medium in a closed container under aseptic conditions. Plant parts root tips and stem tips from newly germinated seeds in an active state of division at the time of excision from the plant are used. Similarly, fragments from the dividing cambial zone of woody plants or the immature elongating zone of the stem usually grow well in culture. Mature parenchyma from fleshy tuber (potato) the root (carrot) may undergo dedifferentiation and resume meristematic growth when planted on an agar medium. Explanted tissues generally begin to proliferate within a few days, few weeks and give rise to voluminous parenchymatous callus out growths. Once the callus is established in the medium, the callus proceeds as a series of growth cycles. Each cycle is characterised by its sigmoid growth nature comprising of a lag phase, cell division phase to a stationary phase. The explant may not proliferate uniformly. Stem pieces, which are usually implanted in an inverted position, proliferate most at the exposed morphological base, and cambial explants proliferate most extensively from the face consisting of the youngest cells.

Tightly compacted buds may be surface sterilized and used to obtain leaf primordia and floral organs which usually do not require further sterilization. Root and stem tip sub-cultures may be made from the main axis tip or from lateral tips in species where branching is prevalent.

Techniques have been standardized with various explants like cotyledon, epicotyl, hypocotyl, shoot base, seedling root, internodes, glumes, anthers, ovaries and embryos of different plant, specific media in efficient callus initiation, plantlet regeneration and subsequent transfer to pots/field.

Differentiation of desired explant with suitable media supplemented with 2, 4D or required hormones results into callus in two to four weeks. Healthy callus, is processed for cell suspension cultures or sub-cultures (through flask/petridishes/test tubes) or redifferentiated through embryogenesis/organogenesis with the manipulation of different concentrations of plant hormones like, 2, 4D, IAA, NAA, BAP, Zeatin etc.

Morphology

Plant tissue cultures possess a characteristic morphology which is different from that of tissues of a plant. On agar medium the tissue forms a more or less coherent mass of undifferentiated tissue which may be firm e.g. lilac, or friable, breaking up into small nodules. Colour is characteristic pale. Chlorophyll is absent or present in reduced amounts. White yellow-green becoming scarlet in later stages of a sub-culture. Cultured tissues from different varieties of the same species may appear different, but in many plant tissues derived from different organs resemble each other in appearance.

Histology

Occasionally cultured plant tissues consist of a single cell type. In a few cases cultures are entirely parenchymatous, but a degree of heterogeneity may exist in that certain cells contain tannins and may be large.

Most cultured tissues are heterogenous consisting, in addition to the parenchyma ground tissue, of xylem and phloem elements, and occasionally of periderm elements and laticifers also. Vascular elements are usually produced in nodules which are formed by the activity of isolated cambia. There is no vascular continuity through the tissue. Some parenchyma cells may also become lignified forming occasional xylem elements. The degree of internal differentiation may undergo change in successive sub-cultures.

Tissue or organ cultures of several species are now known to regenerate into entire plants by a developmental sequence that more or less resembles the steps in normal embryogeny.

The callus can be made to differentiate into plantlets by transferring on to regenerating medium containing suitable combination of auxin and cytokinin. Regenerated plantlets are transferred to wick culture (filter boat) with liquid media and the 1-2 week old plantlets are transferred to small pots (supplemented with liquid nutrient medium) in a growth chamber, mist chamber which may help in better growth and establishment. The plantlets may require about 16 weeks to mature, depending on the species, and the seeds are harvested from tissue culture raised plants for further studies.

Plants regenerated from tissue culture involving a differentiated callus phase show variations in a wide range of characteristics. These are called somaclonal variations. Processing of this variability so generated for commercially releasable material is relatively simple.

In general plant tissue culture can be studied under two broad categories : (i) *culture of unorganized tissues* and (ii) *culture of organized structures*. Under the first category four different types are recognized : (a) *Callus cultures*, (b) *Cell (Suspension) cultures*, (c) *Protoplast cultures*, and (d) *Anther cultures*. Under the second category five different types are recognized : (a) *Shoot tip cultures*, (b) *Meristem cultures*, (c) *Node cultures*, (d) *Embryo cultures*, and (e) *root cultures*.

Progress of tissue culture technology of various crops in India is presented in Table 2.1

Some important plant tissue culture techniques are briefly described.

Callus Culture

The first major breakthrough in plant tissue culture, in the sense of achieving long-term cultures of unorganized tissue masses called callus, was with the explants of cambium tissue isolated from tobacco (White, 1939), potato and carrot (Nobecourt, 1939).

Under *in vitro* conditions callusing is a response to endogenous or exogenous growth regulators. Callus is an amorphous mass produced by in an unorganized manner due to cell proliferation.

Callus formation is dependent on the tissue (explant) type. Meristematic tissues are more suitable for callus induction than mature tissues. Callus cultures can be maintained for long period by sub-culturing the primary callus (callus established originally from the explant), at periodic intervals. Long term callus cultures generally suffer from spontaneously arising genetic variations, which are reflected in the phenotype of plants regenerated from such calli, and such variations are known as somaclonal variations, and are known in many species.

Organogenesis

The formation or induction of organ(s) is known as *Organogenesis* or *Morphogenesis*. Two types of organogenesis are distinguished based on their origin : *Direct organogenesis*-direct from the explant and (ii) *Indirect organogenesis*.

Table 2.1 *Tissue Culture Technology of different Crops in India*

S.No	Crop	Name of the plant	Technology available at
1.	Fruits	Citrus	NBRI, Lucknow; University of Jodhpur, Jodhpur
		Banana	IIHR, Bangalore; NCL, Poona, NBRC, Trichirapally
		Pineapple	BARC, Bombay
		Pomegranate	NCL, Poona
		Papaya	IARI, New Delhi, NCL, Poona
2.	Spices	Cardamom	NCL, Poona, Spices Board, Cochin
		Ginger	NCL, Poona, Spices Board, Cochin
		Turmeric etc.	NCL, Poona; CPCRI, Kasargod
3.	Ornamentals	Orchids	IIHR, Bangalore; NBRI, Lucknow
		Carnation	Delhi University, Delhi
		Gladiolus	NBRI, Lucknow; NCL, Poona; IIHR, Bangalore
		Bouginvillea	NBRI, Lucknow; IIHR, Bangalore
		Chrysanthemum	NBRI, Lucknow; NCL Poona
		Ferns	Baroda University, Baroda and Panjabi Univ. Patiala
		Rose	RFL : Rose Floriculture Ltd., Amritsar
4.	Forest trees	Teak	NCL, Poona
		Eucalyptus	NCL, Poona; IISc, Bangalore
		Sandalwood	BARC, Bombay; IISc; Bangalore
5.	Commercial crops	Sugarcane	NCL, Poona; SBRI, Coimbatore
		Tobacco	CTRI, Rajamundary
6.	Medicinal plants	Glycyrrhiza	GAU, Anand
		Jojoba	NBRI, Lucknow
		Dioscorea	NBRI, Lucknow; RRL, Jammu, NBRI, Lucknow
7.	Plantation crops	Various spp	CPCRI, Thiruvantapuram

Key : NBRI : National Botanical Research Institute
IIHR : Indian Institute of Horticulture Research
BARC : Bhabha Atomic Research Centre
NCL : National Chemical Laboratory
IARI : Indian Agricultural Research Institute
GAU : Gujarat Agricultural University
IISC : Indian Institute of Science
SBI : Sugarcane Breeding Research Institute
CPCRI : Central Plantation Crop Research Institute
RRL : Regional Research Laboratory
NBRC : National Banana Research Centre
CTRI : Central Tobacco Research Institute
RFL : Rose Floriculture Ltd., Amritsar

Shoot and root morphogenesis (*caulogenesis* and *rhizogenesis* respectively), embryogenesis and reproductive development (development of flowers) are possible. Micropropagation through somatic embryogenesis is the most widely used method. Somatic (or sexual) embryogenesis is the formation of embryo-like structures from somatic cells. Somatic embryos are structures similar to zygotic (sexually formed) embryos and they can be induced to germinate in tissue culture media. In somatic embryogenesis, embryos develop directly on the excised plant tissues without an intervening callus phase. Asparagus, celery, carrot, banana, cardamom and cassava are common examples of plants in which micropropagation is already a commercial success.

Cell (Suspension) Cultures

Such cultures may be continued indefinitely by sub-culturing a few drops suspension mostly of single cells by breaking cells of callus which can grow and divide as a result of shaking liquid culture. The simplest suspension culture system is a conical flask with agitation on the platform of an orbital incubator. The swirling motion generates oxygen in the medium. Cells in suspension culture show logarithmic growth with increase in fresh weight, DNA etc. and there after the growth declines. It can be sub-cultured with auxins and cytokinins to stimulate cell division. Cell cultures are more easily amenable for various experiments than those grown on semi-solid medium. This has led to the development of chemostats and turbidostats.

Muir *et al* (1954) first reported culture of single isolated cells. They placed *Nicotiana* cells on the upper surface of a square filter paper raft whose lower surface was in contact with an actively growing nurse cells (i.e. original tissue). Numerous cells divided and developed into callus masses.

Suspension cultures of single cells are useful in plant propagation *via* somatic embryogenesis, regeneration of plants *via* callus on solid media from single cells and for the production of secondary metabolites.

In Fig. 2.1 the flow chart of tissue culture is shown.

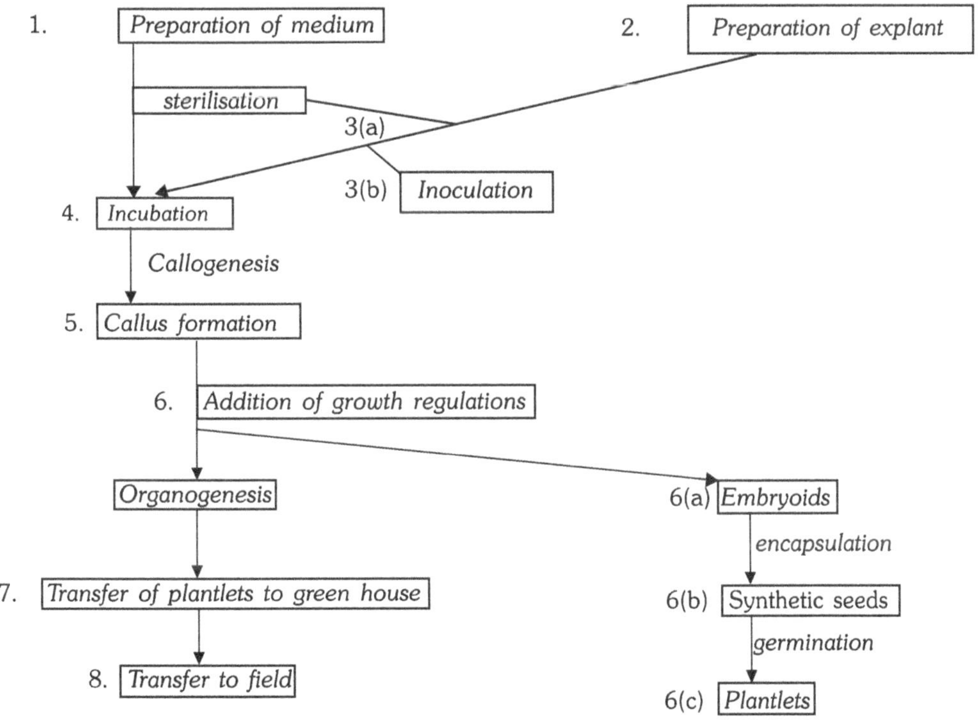

Fig. 2.1 Flow chart of Plant Tissue Culture

Applications of Tissue culture

1. *Micropropagation :* It is employed for rapid multiplication of plants in a short period. A shoot tip of *Eucalyptus*, if used as explant, millions of plants can be produced in a single flask and thus it is useful in silviculture, horticulture, in the propagation of forest or flower yielding plants like chrysanthemum, orchids etc. It also helps in improving the social forestry and in afforestation.
2. *Protection of Rare Plants :* Rare and endangered plants like Red sanders, pitcher plant etc can be protected and multiplied by tissue culture.
3. *Disease Resistant Plants :* Stem tip is the virus free part of the plant hence useful in raising disease resistant plants.
4. *Quick Growth :* The plants obtained through tissue culture show uniform and synchronous growth.
5. *Parthenocarpic Fruits* : Parthenocarpic fruits (seedless fruits) are obtained in ovary culture.
6. *Synthetic seeds / Synseeds :* Artificial seeds or synthetic seeds are produced by enclosing embryoids with sodium alginate. These can be preserved for prolonged periods and can be transported to any distance without any damage.
7. *Haploid Plants :* Rare haploids can be obtained from anther or ovule culture which can be used to produce homozygous diploids, essential for hybridisation experiments.
8. *Somaclones :* Clones obtained from the vegetative parts called somaclones show wide range of variation which can be exploited commercially.
9. *Mutant selection :* Selection of mutants is economically performed in single flask. Selection results in a population which consists of 100% mutant individuals. Each cell can potentially be regenerated into a whole mutant plant. Further single alleles can be rapidly added to a specific cultivar thus further enhancing agricultural usefulness.

   Mutations not found in nature in whole plant can be obtained. In addition many types of mutations can be selected.
10. It is possible to produce uniform plants in a small area.
11. Ther is no dependence on agronomic condition since growth occurs in a controlled environment throughout the year.

Importance of Tissue Culture

The tissue and organ culture technology has emerged as a general but very effective tool of research which is increasingly employed as science and industry. Consequent upon pioneering research during the last 50 years, plant tissue culture as a tool in plant biology has found wide applications with economic benefits in agriculture, horticulture, forestry and plant breeding.

Plant tissue culture has vast potential to resolve problems of experimental biology which otherwise through conventional means cannot be tackled. This technique has already played a very prominent role in genetic engineering, afforestation programmes and germplasm conservation and product biosynthesis. In addition it provides information on the following :

1. To gain knowledge of plant growth, metabolism and auxin physiology.
2. Viral specificity of tissues from species to species.
3. Transformation of normal cells to cancerous cells to understand the causal mechanisms.

4. Morphogenesis and histogenesis.
5. Pathology, cytology and radiation effects.
6. It is possible to select and grow cells with biochemical properties eg. nutritional variants, antigenic variants etc.
7. Biochemical mechanisms of mitosis may be studied with synchronised plant cell culture.
8. Growth and development can also be studied in cultures.

Current and Future Applications of Tissue Culture

These are summarised under the following heads. Further details on some of these are given in the next chapters to follow :

1. ***Applied Technology***

1. Micropropogation
2. Somatic embryogenesis
3. Virus elimination
4. Embryo rescue
5. Haploid production
6. Ploidy manipulation
7. Germplasm storage and transport / germ plasm exchange
8. Production of chemicals by cultured cells.

2. ***Developing Technology***

1. Modification by somoclonal variation
2. Plant gene transfer by protoplast fusion
3. Gene introduction by *Agrobacterium* or other means (consequences include introduction of disease, pest and herbicide resistance, regulation of plant growth and control of development).

Marketing of Tissue Cultured Plants

This is the end protocol of tissue culture and biotech products. The perishable and delicate nature of the plants is the first marketing hurdle. In other words the shelf life of the plants or other related products is limited. Therefore good infrastructure and user/farmer is necessary. Most of the production has to be customer tailored apart from packaging affordable/competitive prices, high air fright, air connections, export customs, sanitary conditions etc., makes this industry profitable, and competitive.

2.1 Plant Growth Regulators (*Phytohormones*)

The common observation that some parts of an organism grow at faster or slower rates than the whole organism suggests that these parts are synthesising new molecules and are producing new cells at above or below the average rate. This leads to the study of control of the growth process.

The Control of Growth

Growth therefore depends on the availability of the raw materials which can be synthesised into the molecules of which cells are composed. In plants amino acids transported to the growing regions are synthesised into proteins under the control of DNA in the growing cells. Likewise glucose gets polymerised into the cellulose of their cell walls.

Growth in plants is regulated by several categories of phytohormones namely, *auxins*, *gibberellins*, *cytokinins, brassinosteroids*, *polyamines* and *inhibitors*.

Phytohormones are organic substances produced in ultra-minute quantities in one part of the plant and transferred to other plant parts where it exerts profound effect. Since plants lack nervous system, the hormonal control system is essentially for communication between cells and tissues. In other words, the growth regulators are the only communicators in plants. Indeed the plants are special as the hormones (*auxins*) are produced in growing points (*meristems*) of a plant. From the meristem, the growth regulators move to other plant parts where they are used up or destroyed but never circulate in the plant body.

Plant Growth Regulators/Phytohormones

Plants have no nervous system and there is consequently little coordination of activities within the organism. Indeed, the different parts of the plant body are much more autonomous in their activities than those of animals. We can cut leaves, twigs or even whole branches off a tree with little or no effect on the plant. The phytosynthetic activities taking place in a leaf in one branch are quite independent of those occurring in a leaf or another. Perhaps the only activities common to all parts of a plant are donating to and receiving from a common pool the products of their individual activities. Nevertheless, within each part, there is a high degree of coordination between the various activities.

In responding to stimuli, which may be internal as well as external in origin, an organism spends energy. The maintenance of the organised stage of the individual, a necessary conditions for its continued survival, depends upon the organism making correct responses to these stimuli. Unless the individual is in an organised state it will not be able to provide the energy it needs to respond to stimuli. We thus have the situation in which there is interdependence between an organism's ability to respond to stresses (stimuli) and its steady-state.

Coordination in plants is by diffusion and by the transportation of chemical substances by the xylem and phloem. It is a slow mechanism even when the vascular system is used because it contains no pumping mechanism, like the heart to facilitate the circulation of these chemical coordinating substances around the body.

The general term for such chemically coordinating substances is a *hormone*. A hormone can be defined as a *naturally occuring organic sustance produced in low concentration by a specific group of cells which exerts a specific effect on a distant group of cells and coordinate growth processes.* Plant hormones are different from the animals hormones.

Organisms, (and parts) do not normally increase in size in an uncontrolled manner. In plants, cell division is stimulated to occur at the growing points (meristems) by a chemical substance produced by the cells at the tip of the meristem which diffuses away from the apex to regions of lower concentration. Some of these substances (Phyto hormones) have been identified. They may be natural or synthetic. One is indole acetic acid a plant hormone called an *auxin*. (Fig. 2.2)

Auxin secretion in the shoot meristems is partially inhibited by light. This is why growth of a plant is greater in the dark. If the illumination is unequal on the two sides of a plant, the dark side is stimulated to grow, mainly by cell enlargement, more than the illuminated side and the plant that grows towards the light.

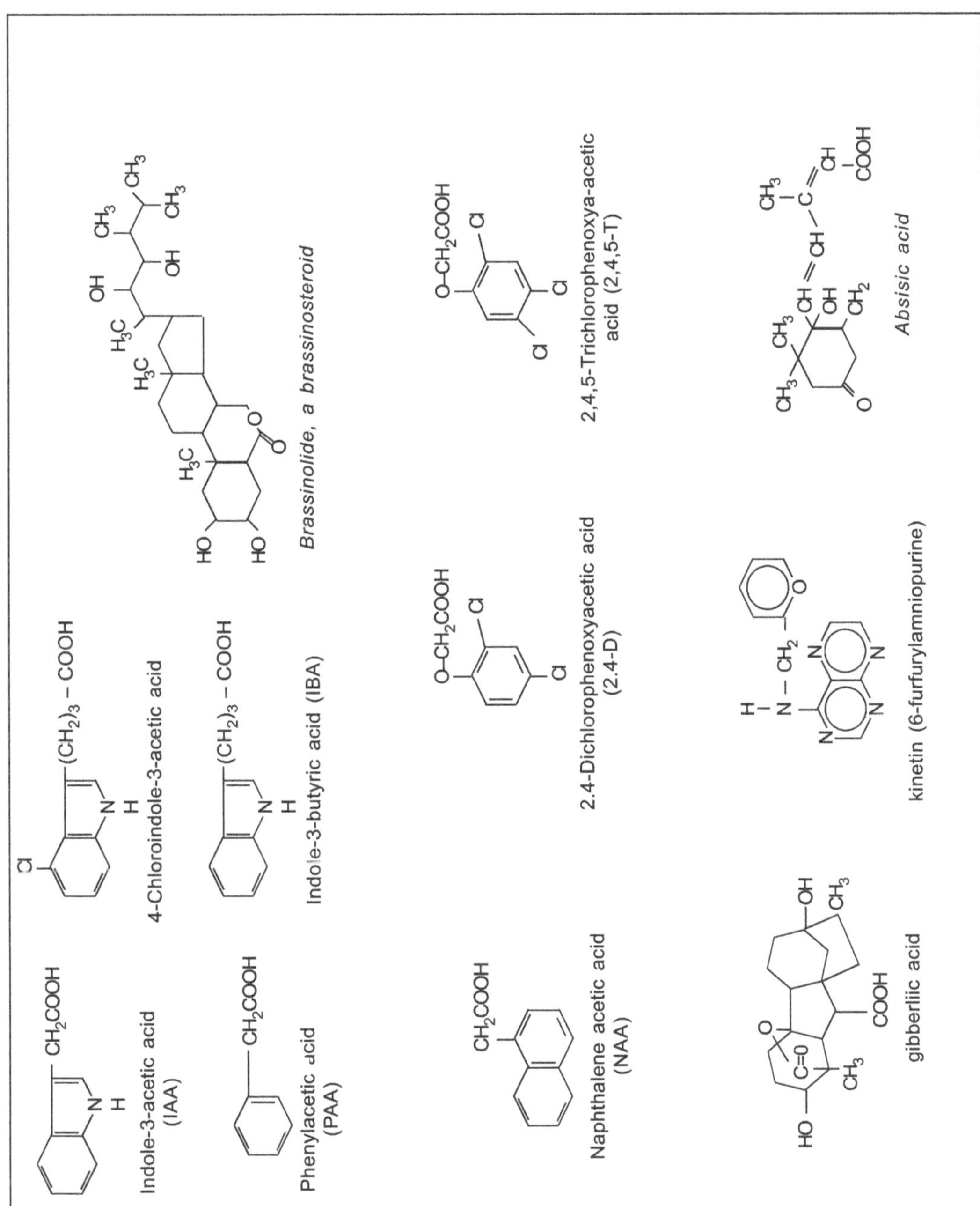

Fig. 2.2 The structure of Plant Hormones

The hormone secreted by the apical meristem inhibits the secretion of auxin by the lateral meristems and consequently these remain dormant, but if the apex of the shoot is destroyed, the inhibition of the lateral meristems is removed and these then grow. Gardeners make use of this phenomenon when they nip out the apical bud of such plants as in roses in order to promote the development of a bushy plant. Conversely, to remove the lateral buds of chrysanthemums, which are escaping the inhibitory influence of the apical meristem, when they wish to ensure that the available nutrients in that stem are assimilated into a single large flower.

Different concentrations of growth hormone are required by different plants. Use is made of this fact in the application of hormone weed-killers. These cause excessive and abnormal growth in the more sensitive plants and thereby kill them.

The respondes of green plants to the stimuli of light and gravity have been well studied. It has been found that in both cases the responses are in the form of growth and cellular differentiation. These responses are collectively known as tropisms. What form the response takes and whether it is positive or negative depends on the nature of the stimulus and the competence of the cells to respond.

Growth of plants towards a source of illumination (positive phototropism) is controlled by the indole hormones collectively known as auxins. The chief and most important of these is indole-acetic (IAA).

This plant hormone is synthesised in the apical meristematic cells from where it diffuses causing cellular elongation and initiation of cell division in the cells behind the apex. Seedlings of the grass family have been used in these studies.

Auxins are also responsible for the positive geotropism of roots and the negative geotropism of shoots. When a seed is buried in the soil the chances of its being orientated with the embryonic shoot pointing upwards and the root downwards is small, yet shoots invariably grow upwards through the dark soil into the light and the roots grow downwards deeper into the soil. We can easily demonstrate that it is not a phototropic response by excluding all light in our experiments. When a seed germinates the apical meristems of both shoot and root begin to synthesise auxin. If these are orientated sideways they soon bend up and down respectively. It is believed that, in both, the auxin synthesised diffuses downwards, but by what mechanism is not known. We can understand why the shoot should grow upwards from the experiments described previously but why should the root grow downwards? The answer seems to be that the auxin inhibits cell elongation in roots, the cells least stimulated undergo elongation and therefore bend the root in a downward direction. Since the same auxin (indole-acetic acid) is involved in both root and shoot, the difference in response is presumably one of sensitivity of the target cells (shoot and root cells) to the auxin. Root cells, it is suggested, have a much lower threshold than shoot cells, but exposure to excess hormone inhibits their growth. However, shoot cells require as much hormone as inhibits most cell elongation. It is these same auxins that are responsible for the phenomenon of apical dominance. The hormone secreted by the apical meristem inhibits the activity of the lateral shoots.

The auxins from the apical meristem should not be confused with those hormones formed in leaves by the action of light and whose function is to control the onset of flowering. It appears that the leaves of flowering plants contain a pigment called the *phytochrome*, which absorbs orange-red light of wave-length 660 nm. Orange-red light converts this phytochrome into a physiologically active form which absorbs light of wave-length 735 nm (infra-red). This physiologically active form regulates the production of a hormone, the florigen that controls flowering. Florigen diffuses from the leaves and is transported to the other parts of the plant. When it reaches those meristems capable of responding to its action, flowers differentiate.

While the conversion of phytochrome 660 into phytochrome 737 is a quick photochemical reaction, the reverse reaction is slow. It is catalysed by an enzyme and requires an energy source for its completion. As long as a plant, (rather its leaves) are illuminated during the daylight, the photochemical reaction dominates and the reaction proceeds far to the right. It is during the night

$$\text{phytochrome 660} \underset{\substack{\text{enzyme} \\ + \\ \text{energy}}}{\overset{\text{light}}{\rightleftharpoons}} \text{phytochrome 735}$$

that the reverse reaction takes place. In the summer, when the nights are short, less phytochrome 735 is converted back than in the spring and autumn when the nights are longer. There is presumably more florigen in plants in the spring and autumn than in the summer.

Variations in the amount of florigen control the onset of flowering in both short and long-day plants. It is observed that if the nights are artificially lengthened, summer-flowering plants fail to come into flower. Too much phytochrome 735 is apparently converted back in the 660 form. Conversely, if spring-and autumn-flowering plants are illuminated at night they fail to flower. Presumably not enough phytochrome 735 has been converted back in this case. Thus summer-flowering plants are referred to as short-night plants and spring and autumn-flowering plants as long-night plants, rather than as long-and short-day plants respectivley.

The paradox is that in summer-flowering plants phytochrome 735 is needed for florigen action while in spring-and autumn-flowering plants it inhibits florigen action. It is suggested that, long-night and short-night plants have different thresholds of response to the same flowering hormone. The former require a lot of hormone, too little not initiating a response in the undifferentiated meristematic cells. The latter require only a little hormone, too much inhibiting flowering it assumes, that there is a positive relationship between the amount of phytochrome 735 and production of florigen.

The Auxins

The auxins promote irreversible elongation of plant cells and this is the most critical property of auxins. Auxins inhibit bud elongation, decreases the growth of roots, initiate roots on plant cuttings, control geotropic and phototropic curvatures and are involved in activation of cambium. Auxins are widely used as weed killers in agriculture and as defoliating agents in chemical warfare; Vietnam war for example.

There are many auxins. Indole-3-acetic acid Indole 3-butyric acid, phenylacetic acid etc. are natural auxins. Many synthetic chemicals show auxin like activity such as 2, 4-D (2, 4 dichlorophenoxy acetic acid), 2, 6-trichlorobenzoic acid and napthlene-1-acetic acid etc.

Natural auxins are found in fruits, seeds, pollen, growing points, young leaves, developing buds. The polarised transport system in one direction i.e from tip of shoot to root is specific property. Further, growth and differentiation in the plant can be regulated by the production of auxin.

The Gibberellins

Gibberellins are endogenous plant growth substances with novel properties. About 100 naturally occurring gibberellins are known but only Gibberellic Acid (GA_3) is most significant commercially produced. GA_1 and GA_9 are involved in elongation and flowering. This is derived from tetracyclic diterpene named "Kaurene" and is present in higher plants and fungi. Gibberellins are named after the fungus *Gibberella fujikuroi*, which produces excessive growth and poor yield in rice called *Foolish seedling disease.*

Gibberellins are abundant in seeds, young leaves and roots. They move upward from the roots in the xylem and thus show reverse polarity of auxin. In experimental conditions gibberellins tend like auxins, to retard senescence. Gibberellins stimulate flowering resettle morphology. They inhibit rather than stimulate rooting of cuttings. Gibberellins promote the growth of main stems especially when applied to the whole plant.

The Morphactins

Morphactins are growth regulators and act as dwarfing agents by a direct competition with gibberellins at their site of action.

The Cytokinins

Cytokinins are found naturally in maturing seeds. Cytokinins were discovered when plant cells in culture enlarged but refused to divide in the presence of auxin alone. Cell division was found to be stimulated by various natural extracts whose activity is now known to be due to various derivatives of purine and adenine. Like auxin they show a directional polar movement within plant tissues but counteract the effect of auxin by promoting bud growth. Cytokinins contain isoprene subunits.

Cytokinins are found in coconut milk and is the only natural plant product which stimulates growth in fragments of plant tissues in vitro. Leaf expansion, seed germination is promoted by cytokinins. They also preserve the green colour of detached leaves by controlling protein synthesis. Kinetin is the most important and interesting cytokinin as it activates cell division. Best natural cytokinin is Zeatin isolated from liquid endosperm of tender coconut is an N6 hydroxyisopentenyl amino purine. Synthetic analog have similar activity. Cytokinins are derivatives of adenine or amino purine. Cytokinin and auxin levels control organogenesis in vascular plants.

Brassinosteroids

Brassins are a recently discovered, complex group of lipids synthesized from the sterol campesterol found mostly in Mustard family Brassicaceae. They are present at low level but have strong growth promoting effects, stimulating cell division and cell elongation. Brassinosteroids act with the other plant hormones to regulate growth and differentiation.

Polyamines are compounds that contain two are more amine groups. Three types are recognised as follows :

(i) Putrescine : $H_2N - (CH_2)_4 - NH_2$ – isolated from putrifying flesh

(ii) Sperimidine : $H_2N - (CH_2)_3 - NH - (CH_2)_3 - NH_2$ - isolated from sperms

(iii) Spermine : $H_2N - (CH_2)_3 - NH - (CH_2)_3 - NH - (CH_2)_3 - NH_2$ - isolated from sperms.

Their biosynthesis originates from the amino acids lysine and arginine. Levels of polyamines increase where rapid cell division occurs putrescine level increase in response to some forms of stress and they may be involved in some aspects of embryo and fruit development.

Growth Inhibitors

Various types of growth inhibitors have been identified in plants. Dormins are hormones and show inhibitory action on plant development and cause the abscission layer of the stalk to form when the leaves are shed.

Absisic Acid (AA) is the only naturally occurring dormin and is chemically related to cytokinins. It is universally present in higher plants and has variety of actions. The active form of ABA is + cis ABA. It promotes leaf fall, the development of dormancy in buds and the formation of tubers in potato. It is a regulator of dormancy, seed germination and responds to stress. It is synthesised in roots and shoots at high rates in stressed tissues and transported in xylem and phloem.

Etheylene ($C_2 H_4$) this is a remarkable natural hormone and was discovered in 1900 as a gas that regulates fruit repening, affects plant growth in as low as 0.06 ppm concentration and affects every aspect of plant growth, germination, elongation of cells, growing shoots. It promotes abscision in senescent leaves by destroying the auxins. Employed for onset of flowering and commercial ripening of banana fruits. Therefore, ethylene is regarded as a ripening hormone. Although it is a simple molecule, its affects are highly specific.

The chemical structure of various plant hormones is shown in Fig. 2.2.

2.2 Tissue Culture for Mass Propagation of Plants

The propagation of various plant species through the techniques of cell, tissue or organ culture is called micropropagation or invitro propagation or clonal propagation. It is one of the most important aspects of plant tissue culture technology. It is relatively a new technology and involves the application of innovative methods that have served to overcome barriers to progress in the multiplication of rare, endangered or elite species. It plays a key role in the mass production of horticultural, forestry, foliage, crop and medicinal plants.

The term micropropagation is derived from the miniature shoots/plantlets initially produced in this mode of plant propagation.

Micropropagation allows the benefit of maximum genetic gain from the genetic variability existing in natural populations. When suitable selection criteria and methods are adopted, micropropagation of economically important plants can be beneficial. The desire for modifying the crop plants for economic gain has been there with man for a long time. In the beginning it was restricted to selection of seeds from healthy and high yielding plants for raising the next crop. Then it was breeding better varieties. The quest for designing crop plants, by new methods, as per his needs is ever-continuing. The realization that the naturally occurring variations in crop plants, and the breeding barriers among distantly related crop plants and their relatives, are limiting factors for transfer of desired characters, has now led to some spectacular developments modern gene transfer techniques or Genetic Engineering-the engineering of genetic make up of plants so as to meet the present and future human needs.

Clonal propagation is a method by which identical plants of superior genotype can be multiplied within a short span of time. The cloning principle is based on the fact that every cell of the plant has all the information necessary to reproduce the whole plant if suitable conditions are provided. The most commonly used tissues for culture are the apical-or the axillary growing points. Morel (1960) on culturing the orchid meristems observed a large number of shoots which could be rooted to form whole plants. Following this method for obtaining plantlets from meristems were developed for several herbaceous and woody plants. (Murashige, 1974).

Micropropagation is one of the important method of Plant Tissue Culture to commercialise plant propagation and has vast economic significance. The chief advantages of micropropagation are cultures can be initiated from tiny segments of the mother plant in the laboratory under asceptic conditions.

Murashige (1974) first recognized different stages in the in vitro multiplication of plants. Stage-I, Stage-II and Stage-III. Treatment and preparation of stock plants treated as a separate stage (Stage-0) before Stage-I. The details of these stages are therefore briefly given :

(i) Stage-I involves establishment of tissue *in vitro*, (ii) Stage-II involves multiplication of shoots (often media are not changed between stage I and stage II), (iii) Stage-III concerns root formation and conditioning of propagules prior to transfer to the green houses; this stage requires high intensity and alteration of media for promotion of root formation. (iv) Stage-IV involves growth in pots followed by field trials. The number of steps in micropropagation may sometimes be reduced to two or three.

As discussed earlier propagation that involves an unorganized callus phase may produce variant plants. Thus, in practice, most micropropagation is achieved by maintaining organized tissues by the multiplication of meristems and axillary buds. The stages involved in micropropagation of a given plant are as follows :

1. *Selection and preparation of mother plants*
   - choose typical, healthy, disease-free mother plants
   - virus testing/elimination may be required
2. *Establishment of aseptic cultures*
   - surface sterilize and transfer explants to culture
3. *Multiplication*
   - multiplication of cultures to give rise to intact plants by appropriate procedures
   - cultures may be recycled many times to obtain required multiplication rates
4. *Preparation for transfer*
   - rooting of shoots (perhaps after shoot elongation; some shoots can be transferred to the next stage without roots)
5. *Transfer to natural environment*
   - wash agar from roots, transfer to sterile rooting medium or 'artificial soil'
   - initially maintain at high humidity
   - gradually harden off

There are several approaches to micropropagation : (a) Propagation from axillary buds (b) Shoot Tip Culture (c) Propagation from nodal pieces. (d) Somatic embryogenesis both direct and indirect (e) Haploid plants.

Micropropagation by organogenesis and somatic embryogenesis. Tissues from leaves, stems, roots and inflorescences can be cultured to promote direct morphogenesis (without going through a callus or suspension stage) or indirect morphogenesis (after going through callus and/or suspension cultures) to obtain either direct shoot generation (organogenesis) or somatic embryos (somatic embryogenesis).

(a) ***Propagation from Axillary Buds :*** The most common method is to excise apical or lateral shoots containing meristems, and to culture these on media which suppress apical dominance to encourage precocious development of axillary buds. To accomplish this cytokinin is added to the medium and this results in highly branched system that lacks root. After 3-5 weeks of culture, individual shoots can be separated and sub-cultured on the same medium for multiplication. This may be repeated times many of the shoots can be transferred to rooting.

(b) ***Shoot Tip Culture :*** Ball (1946) first demonstrated the possibility of regenerating plants *in vitro* from the shoot-apices of certain angiospermous plants. Morel's (1960) technique of shoot-apex culture of orchids has since led to clonal propagation of a large number of orchids. The technique is also used to commercially propagate large number of marketable ornamental plants like *Dracena, Cordyline*, and *Hemerocallis* etc. Clonal propagation of crop plants like potato also become successful.

In most cases, clonal propagation is achieved by placing sterilized shoot tips or auxiliary buds onto a culture medium that is sufficient to induce formation of multiple buds. This method has already been used to propagate a large number of marketable ornamentals.

Most micropropagation methods employ apical or lateral shoot tips containing meristems. Each of these shoot tips has the shoot apex and several leaf primordia, at the axils of which axillary shoots could potentially develop. Excised shoots are cultured on an agar-solidfied culture medium with high concentration of a cytokinin hormone. This induces a large number of axillary shoots to proliferate. Each of these axillary shoots are separated and the same process repeated to rapidly multiply the shoots in an exponential manner. The shoots are then planted in an agar culture medium containing root-inducing hormone so that a complete plantlet is obtained from each. (Fig. 2.3).

When shoots, meristems are cultured they continue their organized growth. The shoots/ multiple shoots produced can be rooted to produce plantlets. This capacity has practical application and economic significance for plant propagation of the meristematic zones of the shoot tip.

Eucalyptus and sugarcane propagation is also worth like micropropagation has been successful through shoot apex culture only.

As early as 1934, it was known that meristems of virus infected roots are free of the pathogen (White, 1934 a; 1934 b). Limasset and Cornuet (1949) found that shoot tips of virus infected plants were also virus free. Ball 1946, demonstrated the possibility of regenerating plants *in vitro* by culturing shoot apices of some angiosperms. Morel and Martin (1950; 1955) could produce healthy plants from virus infected *Dahlia* and potato plants. The production of virus-free plants through shoot-tip culture from infected mother plants is possible because the pathogen concentration is not uniform in the infected plants, and the apex of rapidly growing shoots are often not invaded by the virus. Morel (1960) used shoot apices of orchids for their rapid clonal multiplication. Shoot tip culture has two important practical applications : (i) *virus eradication* and (ii) *micro-propagation*. These developments were followed by *in vitro* propagation of plants from shoot tip culture. Methods for the micropropagation of a large number of both woody and herbaceous plants.

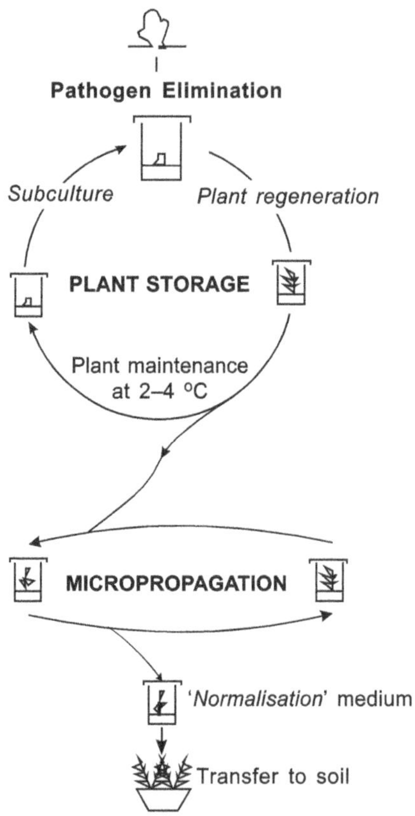

Fig. 2.3 Pathogen elimination, plant storage and micropropagation

(c) ***Propagation from Nodal Piece :*** In some plants, the addition of cytokinin does not readily induce multiple shoots, and multiplication is achieved more easily by serial sub-culture of stem nodes. Individual stem nodes comprising axillary buds are cut from the choice plant and the axillary buds are allowed to grow up to several centimeters in length on agar solidified medium and the axillary buds allowed to grow into shoots. Each stem is then cut to separate the individual nodes, and the process is repeated. This serial subculturing can produce few thousand plants from a single node in less than 3-4 months. This approach is normally used, for potato.

Various other explants (e.g. young inflorescences, stems, leaves) can also be used, and multiplication can also be achieved by direct organogenesis or embryogenesis on such explants. Because sub-culturing is labour-intensive, the use of automated systems has potential advantages, and is the subject of much research. Alternately the synchronous production of somatic embryos in suspension cultures can eliminate most of the labour of sub-culturing, but results in the production of naked propagules that must be coated in some way to protect them if they are to be used directly in the field as '*artificial seeds*'.

Production of virus and other pathogens free plants

Virus infection is a big problem with vegetatively propagated species like sugarcane, potato, etc. Conventional methods of virus elimination by heat treatment (30 °C for 10-20 days) are useful only with some species. The recovery of pathogen free plants through shoot-tip culture is based on the assumption that pathogen concentration is not uniform throughout the infected plant. In addition pack of vascular differentiation in the meristem prevents intercellular movement of viruses. Therefore if stem tip is excised and used as explant, it is possible to achieve virus free plants. These are disease resistant and also show increased vigour. Various chemical agents, including purine and pyrimidine analogues (*e.g.* 8 - azaguanine, 2 - thiouracil), amino acids, plant growth regulators and antibiotics (*e.g.* actinomycin D, glutamycin) can be applied either before excision or in culture media supplying the meristems, and in some cases such treatments aid virus eradication.

Callus is similar to apical meristem, therefore virus free plants can also be obtained via repeated sub-culture. This was successfully achieved in gladiolus, geranium, tobacco, potato etc. Fig. 2.3 shows pathogen elimination, plant storage and micropropagation upto transfer of plants to soil.

The production of disease free mother stocks has led to rejuvenation of old cultivars of both crop and horticultural plants for rapid growth and higher yields. Pathogen free plants have also been produced from stocks infected with mycoplasmas, fungi and bacteria.

Virus free plants from 65 species and 136 cultivars have been reported world wide. Its application in India however is limited and reported for potato, banana and sugarcane etc.

(d) ***Somatic Embryogenesis :*** Steward's work (1958) on cell suspension cultures of carrot led to the discovery of embryogenesis i.e. differentiation of bipolar somatic embryos. Somatic embryos can be derived from several tissues like hypotyle cotyledons and any other vegetative cells.

Individual somatic embryos can be encapsulated in an alginate matrix by dripping embryos, coated in sodium alginate and nutrient solution, into 50 ml $CaCl_2$ solution. The calcium ions induce rapid cross-linking of the alginate, to form firm beads, of diameter about 5mm, containing single embryo. Additives, such as fungicides and nutrients, can be included in the solution before cross-linking, as for encapsulated seeds. Somatic embryos can be germinated readily on culture medium without hormones until plantlets reach a suitable size for transfer to soil or vermiculite. For bulk handling of cloned somatic embryos of agricultural species it is important to generate a population of embryos that are as synchronized, uniform and have as high a proportion of viable individuals as possible.

Somatic embryogenesis involves the formation of somatic embryos or embryoids from an explant or via callus. These embryos are similar to normal zygotic embryos found in the seeds, but their initials are the somatic cells and not embryos. The plantlets produced are free from diseases particularly viral infections. This technique has been used to produce disease free strawberries, potatoes, sugarcane, citrus and many ornamental plants like *Gerbera*, *Cordyline* etc. Large scale propagation of somatic embryos is possible through cell suspension cultures which can be manipulated to yield large numbers of somatic embryos that float freely on the culture medium and can be harvested mechanically. Millions of embryos of plant can be grown in bioreactors. Banana, Cardamom, Carrot, Cassava etc are examples of commercial success of micropropagation. (Fig. 2.4).

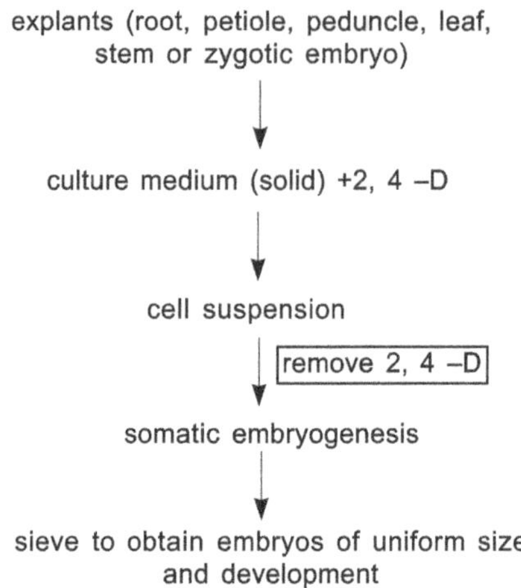

Fig. 2.4 Somatic embryogenesis from cell suspension culture

The growth of plant cells in suspension cultures offers immense potential for the large scale propagation of plants. Somatic embryogenesis in particular offers tremendous promise in view of their bipolar nature bearing both root and shoot apices. In one step both the meristems necessary for complete plant growth are initiated. Proliferating cells and proembryos typically separate and suspension cultures can optimally produce embryonic units. Suspension cultures of somatic embryos therefore are amenable to mechanical handling.

Somatic embryogenesis can be induced by auxins such as 2, 4-D. Somatic embryos are allowed to mature by with drawing 2, 4-D from the medium and the embryos are allowed to germinate in reduced concentrations of macro and micro nutrients but without hormones.

Somatic embryogensis is a preferred method of plant propagation *in vitro*. It allows rapid production of large number of "true to type" plants, which are more suitable for commercial multiplication and even for mutation research. Regeneration via somatic embryogenesis is a preferred method for cloning of stable mutants/hybrids and genetically engineered plants.

In addition somatic embryogenesis also offers the following advantages :

(i) Single cell origin,
(ii) Separate units,
(iii) Suitable for genetic manipulation,
(iv) Rapid multiplication of plants by large scale production of synchronous population,
(v) Synthetic seed production,
(vi) Regenerated plants are "*true to type*".

Induction of somatic embryogenesis is the most important mode of micropropagation. It has been achieved using shoot apex or axillary buds of plants with premium characters like more yield, increased vigour, disease resistance, earlyness acclaimatisation are selected and placed on to shooting medium. Multiple shoots does obtained or transferred to rooting medium. Plantlets with roots are obtained. The *in vitro* environment considerably reduces the ability of such plants to transfer to soil because,

1. These plants do not develop sufficient cuticle on leaves and stems epidermis.
2. Cutine (wax) coating is discontinous, ie. cuticle not fully developed as a layer.
3. Stomata in such plants are imperfectly formed therefore not fully functional
4. Hypolignification is found in these plants.

The net result of these abnormalities is that plants suffer from excess loss of water, when removed from culture vessel and require extremely difficult hardening procedures. These problems (collectively called as vibrifration) are due to high humidity and high ammonium content of the culture medium. The use of higher percentage of gelling agent results in more normal plants.

Synthetic Seeds/Artificial Synseed

The concept of "Synthetic seeds" is recent and refers to the somatic embryos (SE) enclosed in an artificial coat. This method was first advocated by Bryan (1978). Such seeds can be stored, transported and preserved as germplasm. A synthetic seed is an analogue to true (zygotic) seed is primarily based on the similarity of somatic embryo with zygotic embryos in morphology, physiology and biochemistry. The synthetic seed must be able to grow directly in greenhouse/field like true seedlings. However, there are various aspects which need to be considered for the successful development of synthetic seed.

(a) Production and application of synthetic seed coat of Sodium Alginate gel,

(b) Production of improved quality of somatic embryo.

Large scale propagation by somatic embryogensis/shoot tip cultures combined with the technique of synthetic seeds has opened up a new field of low cost mass propagation of plants.

Crops which need the production of synthetic seeds come under two categories;

1. Those having strong commercial basis, such as high seed costs due to low fertility, genetic instability, labour intensive hybrid seed formation etc.
2. Those having strong technological basis such as plants that are newly created by the recombinant DNA techniques in which foreign genes that are introduced into plant genome are generally unstable and may be lost upon seed set (Liu, 1988). Even, if the genes for useful characters can go through meiosis in a stable manner they get scattered among the progeny and this can be prevented by artificial seed production.

These embryos are maintained in arrested state of development by either cold storage or by using mitotic inhibitors. They could be incorporated into artificial seeds either by coating or encapsulation. These artificial seeds could then be handled like normal seeds and stored and transported.

In addition, somatic embryogenesis offers an excellent *in vitro* system to study the fundamental questions of growth, differentiation and development.

(e) ***Haploid Plants (Androgenesis) :*** Guha and Maheshwari (1964) first obtained haploid plants by culturing anthers of *Datura innoxia*. This basic discovery has applications in plant breeding and plant biotechnology. Since the plants are diploid screening of mutations is not possible and secondly haploid plants offer scope to develop homozygous diploid following chromosome doubling, which is done by colchicine treatment. Homozygous diploids are very useful as seed producing parent. Haploid plants from anther or ovule culture are recorded in 250 species belonging to about 40 families and widely used in crop improvement programe especially wheat and rice. Haploid rice varieties are under cultivation on over 100,000 hectares in China. As shown in Fig. 2.5 haploids are induced via somatic embryos. Callus from haploid is used to diplodise and homozygous diploid plants are obtained.

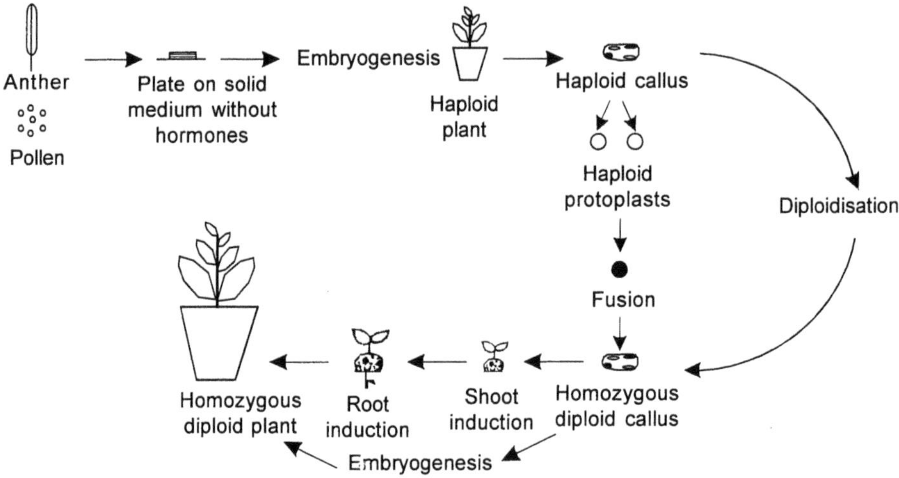

Fig. 2.5 The production of haploid plants

Embryo Rescue

In sexual crosses where the parents are taxonomically distant (unrelated) barriers that prevent successful hybrid production may be overcome by in vitro culture techniques. Specifically, in the cases where fertilization is successful, but the embryo fails to develop, immature zygotic embryos can be excised and cultured, and hybrid plants can be regenerated and multiplied by this novel technique.

Commercial Micropropagation

There are new many commercial companies producing millions of plantlets by micropropagation. For example, in the Netherlands alone where there is an efficient pot-plant and glasshouse industry there are more than 10 companies producing up to 10 million plants per year, and over 50 companies serving smaller markets down to individual glasshouse complexes propagating plants only for their own requirements. Similar micropropagation companies, biotech labs exist in the U.K., France, Germany, Israel, the United States, South America, Africa, Middle and South East Asian Countries, Australia, India. The world wide application of micropropagation techniques in that there are several hundred companies globally, each capable of producing more than a million plants per year through micropropagation. A select list of micropropagated plants is given in Table 2.2.

In a commercial micropropagation laboratory it is important to carryout sub-cultures in laminar flow benches in rooms that are as clean as possible. Thus the whole laboratory may be supplied with filtered air. The micropropagation must be able to work rapidly but maintain sterility, and also be able to recognize and eliminate contaminated cultures. Culture media, perhaps developed specifically

Table 2.2 *Select list of some important in vitro micropropagated plants*

Ornamental species	Vegetable and crop species	Woody species
Authurium	*Allium*	*Betula*
Chrysanthemum	*Apium*	*Coffea*
Cymbidium	Arachis	Eucalyptus
Dianthus	*Asparagus*	*Grevilea*
Ferns (many species)	*Beta*	
Cordyline	*Brassica*	*Hevea*
Fuchsia	*Cicer*	*Pinus*
Gladiolus	*Glycine*	*Pyrus*
Gloxinia	*Phaseolus*	*Rosa*
Hemerocallis	*Solanum*	*Salix*
Hippeastrum	*Vigna*	*Santalum*
Iris	*Zea*	*Sequoia*
Orchids *(many species)*	*Trifolium*	*Tectona*
Pelargonium	*Musa*	*Vitis*
Phlox	*Tomato*	*Rubus*
Rhododendron		*Thuja*
Saxifraga		
Prunus		
Tulipa		
Saintpaulia		
Narcissus		

for individual plant genotypes, must be produced in media 'kitchens' with careful control of the medium components. The propagation rates must be closely monitored in order to produce the required number of plants at the appropriate time. The importance of this control is evident if, for example, 200 000 plantlets of an individual fern must be produced each week, or alternatively, if large numbers of plants must be produced in time for Easter (e.g. lilies) or Christmas (e.g. chrysanthemums or poinsettias). Since each species or even specific genotypes may have different requirements for culture media, methods of propagation, rooting, or transfer to the environment, it is not surprising that companies specialize more in producing a limited range of plants as the numbers of individual plants required increases.

Tissue culture industry in India has potential to grab a sizable chunk of the growing international market for flowers and other plants and earn valuable foreign exchange. Government has given license to several units to export tissue cultured plantlets of flowering and ornamental plants. India's share and its performance in the world market is commendable. Netherlands stand first in the world market of tissue culture plant business. A.V.Thomos and Co(AVT) is first to commercialise plant tissue culture in Kerala. They created a superior variety of Cardamom through tissue culture and exported it. Another success Indian story is that of ITC Agro Tech which developed good variety of sunflower hybrid (PAC3425). In many cases the companies specialized, and propagate, for example, only ferns, or specific plants for the cut flower industry (e.g. Gerberas). Work is usually carried out on a contract basis, with the production of small rooted plantlets that are passed on to other companies specializing in glasshouse growth upto marketable size.

They are also involved in propagation of ornamental plants for export. The Bangalore based Indo American Hybrid seeds (IAHS), a pioneer in hybrid vegetables and seeds has a capacity to produce over 10 million plants per annum. A tissue culture bank is also being maintained for preservation of breeding lines of vegetables such as tomato, cabbage, onion, chilli etc. This company proposes to introduce ready to use diagnostic kits and also to takeup disease indexing and screening resistance against major viral, fungal, bacterial diseases mainly for vegetable crops. The list of success stories is too long.

The most viable technology in plant tissue culture today is the clonal/micropropagation of elite plants. Several industries and nurseries have standardized protocols for the multiplication of ornamentals like orchids gerbera, carnation, roses, chrysanthemum and plantation crops like banana and cardamom. In India, there are more than 20 private industries engaged in commercial production of horticultural plants, forests and fruit crops such as apples, peaches, pears, strawberries, plums; and forest trees, radiata pine. The major benefits of this method include the following : (i) rapid multiplication of superior clones and maintenance of uniformity, (ii) multiplication of disease free plants, and (iii) multiplication of sexually derived sterile hybrids.

However, no serious attempts have been carried out on micropropagation medicinal plants for use in industry, although laboratory scale cloning has been demonstrated for *Dioscorea, Solanum, Simmondsia, Tylophora, Rauvolfia* and *Coleus* etc. Phytochemical variability is a common event in the medicinal and aromatic plants. In tree species the variability could lead to major problems of commercial viability. By standardizing protocols for clonal propagation of selected elite plants, it is possible to achieve eight to ten fold increase in the products per unit area of cultivation. Clonally propagated plants would also have identical phytochemical profiles. For large scale cloning, standardization of protocol is a pre-requisite which is a laborious process and requires considerable inputs.

The micropropagated plants produced in the laboratory are acclimatized to the greenhouse conditions before they are transplanted to the field (Fig. 2.6).

Advantages of Micropropagation

1. *Small space requirements :* Shoot multiplication can be achieved in small space because small shoots are multiplied. Cultures can be initiated with very small pieces of plants (explants) and thereafter small shoots are propagated (hence the term 'Micropropagation' to describe the in vitro methods). This means that only a small amount of space is required to maintain and multiply large number of plants.
2. *Sterility :* Since the propagation is carried out under sterile conditions, the plantlets produced are free from bacteria, fungi and nematodes. Once cultures have been started there is therefore no loss of plants through any disease and the plantlets finally produced will be free from pathogens.
3. *Certified virus free :* If virus free material was used to initiate cultures, large number of microbiologically sterile (virus, fungi, bacteria) free plants can be produced. In other words guaranteed disease free plants exempt from quarantine regulations that may apply to export and import plants. Thus, micropropagation via tissue culture technology has been a fruitful business for floral crops, foliage plants, agricultural crops, plantation crops, fruit trees, forest trees, rare endangered species by small or big company.

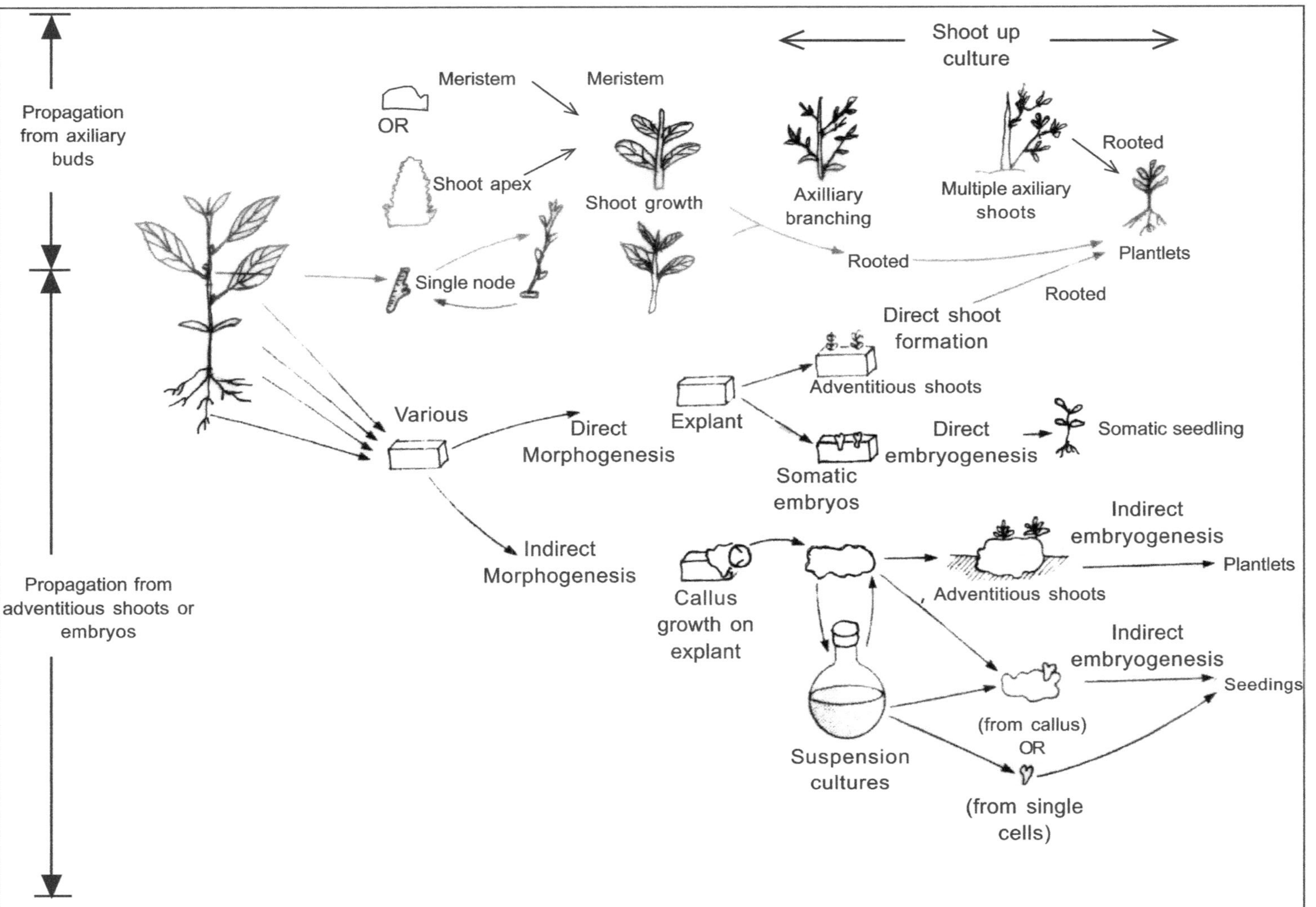

Fig. 2.6 Diagrammatic representation of different source material and methods that can be used for micropopagation. (Adapted from George and Sherrington, 1984)

4. *Control conditions :* The conditions for *in vitro* multiplication can be strictly controlled light, medium composition, growth regulatory levels, temperature. Reproducibly high rates of propagation can be achieved.
5. *Labour :* Little or no attention is required between sub-cultures (e.g. watering, weeding, spraying etc.) cost of labour is drastically reduced, therefore cheaper.
6. *Glass house space :* Less expensive glass house is required.
7. *Continous production :* Micropropagation can be done continuously without break without seasonal variation all around the year.

   The rate of propagation is therefore much greater than in macropropagation and many more plants can be produced in a given time. Large number of plants can be produced in a short time.
8. *Stock plants :* Stock plants can be stored *in vitro*.
9. *Range of species :* Some plants that can be propagated *in vitro* are difficult or impossible to propagate
10. *Mechanisation :* It may be possible to automate propagation procedures for some species and this leads to reduced labour requirements. Modern green houses meant for raising or hardening tissue culture plants use high technology and have provision for automatic control of shading, light intensity, temperature and fogging. The green house atmosphere is computerised controlled system. Thus the plants can be multiplied all around the year within the laboratory. Unlimited number of superior plants are produced. Since only one parent plant is involved in giving rise to copies, each plantlet is genetically identical with the mother plant. Production can be continued all the year round and is independent of seasonal changes.

Disadvantages of Micropropagations

1. *Specialsed facilities :* Fairly expensive and specialised facilities are required (e.g. Culture rooms, laminar flow hoods etc.)
2. *Skilled operations :* Skilled staff requirements, must be able to work under sterile conditions and decide where to divide different cultures at the appropriate time.
3. *Contamination :* If bacterial or fungal contamination occurs during early multiplication many potential propagules may be lost.
4. *Micropropagation conditions :* Specific methods for efficient micro propagation for each species need to be developed including the conditions for rooting and plantlet establishments.
5. *Plantlet size :* Initial plantlet produced are small, therefore great care is necessary.
6. *Genetic stability :* Propagation techniques must not introduce genetic instability.
7. *Cost :* Because of the sophisticated facilities required and labour intensive procedures, the cost of plantlets produced is relatively high and must therefore be offset by a large scale production by high added value of the plantlets produced.

Conclusion

Micro/Mass propagation is useful in the following :

1. A million plants can be produced from a small piece of plant tissue within a year.
2. Plants can be produced through out the year irrespective of season.
3. Rapid multiplication of superior plants is possible.

4. Multiplication of disease free plants (free from bacteria, viruses and fungi) is possible.
5. The problem of deforestation can be answered.
6. Enhanced production of secondary metabolites
7. Uniformity among the plants can be maintained.
8. More useful to propagate trees that lack vegetative reproduction.

Risk Involved

Capital investment in the commercial tissue culture laboratories are very high. So for the risks involved due to contamination and imperfect hardening procedures could result in heavy loss. It is extremely labour intensive and most of the expenditures (90%) goes to labour costs. Foundation and working of tissue culture laboratories though good still it needs further research since the protocol varies from one plant species to other.

2.3 Genetic Manipulations in Plants

Biotechnology is as old as civilization and as new as todays break through. Egyptians were the first to bake yeast raised bread as early as 2000 BC; Sumerians and Babylonians drank beer/wine by 2000 BC. The first wheel of cheese, the age old Indian Idly, the Indonesian Temph, the African Porridges and '*Koji*' method of Japanese enzyme preparation were all dramatic steps forward for the science of biotechnology. The prospects of biotechnology as the last major industrial technology of the 20th century are sufficiently exciting to attract the attention of biologist, biochemist, biotechnologist, government, investors, public, industries and even social scientists.

Genetic engineering has rightly an image of science fiction. The time when new species with any wanted combination of genetic properties can be ordered from an animal or plant breeding seems for away. The layman's view that the science fiction of today is the reality of tomorrow is certainly an insufficient argument to justify optimism. If this were so, we should by now be able to produce hybrids between members of the animal and plant kingdom as was foreseen by 19th century equivalent of Fred Hoyle (Fig. 2.7). Despite the sceptics expressed by Sir Macfarlane Burnet in his book '*Genes, Dreams and Realities*' (1971), recent advances in molecular genetics have raised new enthusiasm but uneasiness which make people speak of genetic engineering as something to just correct inborn errors of metabolism. However, genetic engineering of plants not only as a future method to improve species, but also as a fundamental approach to the study of gene expression, especially with respect to cell differentiation. Literally, the definition of genetic engineering might be any intentional genetic manipulation to alter species or to make new ones. In this sense genetic engineering was practised long before Mendel (1865) proposed his laws heredity. In 17th century the bulb growers of the Low Countries in Holland produced new varieties of Tulips costing a thousand Florins each. This is indication of strong impact of genetic engineering on society; the so-called crazy tulip trade caused a financial disaster on the Amsterdam stock-exchange comparable to the Wall Street crash in the 1930s.

Biotechnology Variously Defined

Given below are some definitions of biotechnology :

The integrated use of biochemistry, microbiology and engineering sciences in order to achieve technological (industrial) application capabilities of microorganisms, cultured tissue cells and parts thereof. The controlled use of biological agents, such as micro-organisms or cellular components, for beneficial use. The science of the production processes based on the action of microorganisms

Fig. 2.7 Tigerlillia terribilis, hypothetical tigerlily flower said to be found in the valley of Verrikwie, near Lake Odd-grow and the Hill Orfeltugg by Prof. Bosh

and their active components and of production processes involving the use of cells and tissues from higher organisms. Medical technology, agriculture and traditional crop breeding are not generally regarded as biotechnology.

The *European Federation of Biotechnology* (EFB) considers biotechnology as '*the integration of natural sciences and organisms, cells, parts thereof, and molecular analogues for products and services*'. The main aims of the federation are as follow :

1. To advance biotechnology for the public benefit.
2. To promote awareness, communication and collaboration in all fields of biotechnology.
3. To provide governmental and supranational bodies with information and informed opinions on biotechnology.
4. To promote public understanding of biotechnology.

The EFB definition is applicable to both '*traditional*' and '*modern*' biotechnology. Traditional biotechnology refers to the conventional techniques that have been used for many centuries to produce beer, wine, cheese and many other foods, while '*new*' biotechnology embraces all methods of genetic modification by recombinant DNA and cell fusion techniques, together with the modern developments of 'traditional' biotechnological processes. According to the Netherlands Directorate General for International Co-operation, "*Biotechnology is the integrated use of Genetics, Biochemistry, Cytology, Microbiology, Molecular Biology and Process Technology to supply goods and services, employing micro-organisms, parts of micro-organism or cells and tissues of plants and animals*".

Unlike other scientific disciplines, biotechnology is multidisplinary and draws upon a wide range of relevant fields, such as microbiology, biochemistry, natural sciences, genetics, morphogenesis, molecular biology, cell biology, immunology, protein engineering, enzymology, breeding techniques, and bioprocess technologies (Fig. 2.8). Biotechnology is not itself a product or range of products like microelectronics; rather it should be regarded as a range of enabling technologies that will find significant application in many industrial sectors.

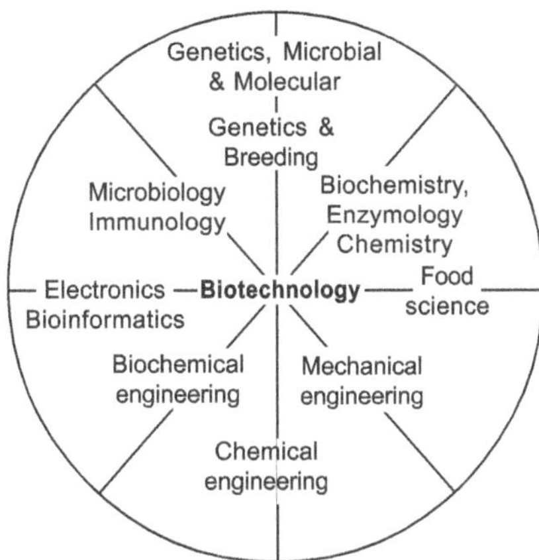

Fig. 2.8 The Multi disciplinary nature of biotechnology

The increasing diversity of modern biology has been derived primarily from the post-war introduction into biology of other scientific disciplines such as physics, mathematics and computer science which has made possible the understanding of life processes at the cellular and molecular level. No wonder, in the last 25 years well over 26 Nobel prizes have been awarded for discoveries in biotechnology and allied fields especially genetics.

The new biotechnology revolution that began in the 1950s and later, when genetic recombination in micro organisms were discovered. This 'genetic engineering' has had a profound impact on almost all areas of traditional biotechnology and further permitted break throughs in medicine and agriculture.

Biotechnology in India

The Department of Biotechnology in the Ministry of Science and Technology was established by Govt. of India in 1986. Several Biotechnology centres in India were established at Indian Agricultural Research Institute, now called as the Lal Bahadur Shastri Centre for Advanced Research in Biotechnology; New Delhi, Indian Veterinary Research Centre; Izatnagar, U.P and National Dairy Research Centre; Karnal, Bangalore and Thiruvanthapuram were established.

In addition, several plant molecular biology centre were also established viz, National Botanical Research Institute, Lucknow, Bose Research Institute, Calcutta, Tamilnadu Agricultural University Coimbatore, JNTU, New Delhi and Osmania University, Hyderabad. In addition, the DBT has also set up service oriented facilities to strengthen research and development in the following areas.

1. Germplasm banks for plants, animals, algae and microbes.
2. Animal houses.
3. Oligonucleotide synthesis (DNA synthesizers)
4. Production and distribution of enzymes, and radio-labelled compounds etc.
5. Bioprocess optimization and pilot plants.
6. Genetic engineering R and D units.
7. Reproductive biology and molecular endocrinology.
8. Carbohydrate cell surface and cellular transport.

9. Protein and peptide sequencing.
10. NMR facility
11. Marine cyanobacterial germ plasm bank.
12. Antibody development consortium.

Plant Genetic Engineering Techniques

Plant genetic engineering technology is an off-shoot of many remarkable scientific developments in the late twentieth century has witnessed in the fields of plant science; namely genetics, physiology, pathology, biochemistry, molecular biology, cell and tissue culture etc.

Plant genetic engineering is an area in which tremendous progress has been made during the last twenty years. A number of useful foreign genes specifying pest resistance, herbicide resistance and virus resistance have been successfully transferred to crop plants. The first practical system for genetically manipulating plants occurred in 1983 with the discovery of the ability of the bacterium *Agrobacterium tumifaciens* to transfer part of its Ti plasmid (T-DNA) into the host plant genome.

To introduce a desired gene or set of genes by conventional methods requires a sexual cross between two lines, and then repeated back-crossing between the hybrid offspring and one of the parents until a plant with the desired characteristics is obtained. This process, however, is restricted to plants that can sexually hybridize, and genes in addition to the desired gene will be transferred. Traditional breeding programmes involving sexual crosses will continue to dominate the approach to improve the agronomic characteristics of food crops but will be increasingly used.

Genetic manipulation of plants has been practiced to improve the quality and productivity of agricultural important plants for many hundreds of years with great success by plant breeders in designing new genotypes primarily by shuffling the available genetic variablity in various crops and plant breeding has now become a sophisticated branch of applied genetics. Plant breeders have developed elegant schemes for cross fertilizing plants to introduce and maintain desirable traits in inbred lines, and the yields of crops like maize and wheat have steadily increased over the past 50 years. However, the methods of classical plant breeding are painstaking slow and uncertain.

However, the results of classical methods should not be underestimated. Great advances have been made in the adaptation of North Americal cereals to the climate conditions in developing third world countries; resulting in '*green revolution*'. Norman E. Borlaug was awarded the Nobel prize for Peace in 1970 for his extensive genetic work.

Plants offers greater opportunities and are more amenable for genetic engineering than animals, owing to their ease of growing complete plants from cell cultures coupled with their totipotent nature. Using restriction enzymes it is possible to transfer genes freely between one plant to another with obvious potential benefits in the development of improved food crops and for improving and increasing the supply of human food.

The biotechnological revolution (the Second Green Revolution in plant sciences) relies on novel strategies of cell/tissue culture techniques, microbial and molecular genetics techniques for crop production, protection and improvement is the most powerful and revolutionary technology of modern genetics. All these technologies rely on cell and tissue culture as the pivot.

Among the several biotechnologies available today, areas of special interest to various crops are as follows : (Fig. 2.9).

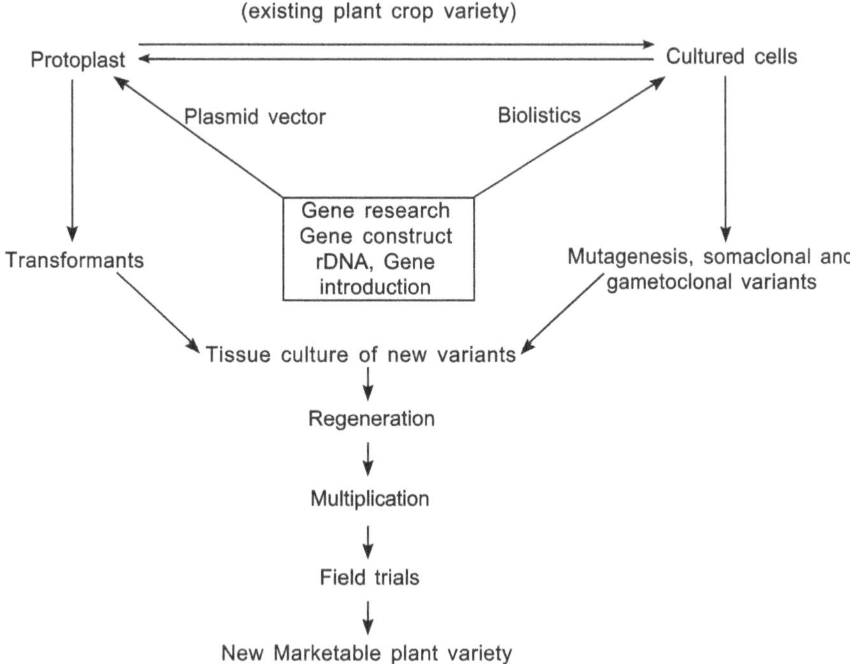

Fig. 2.9 Flow chart of experimental approaches currently used to create new plant varieties

1. Plant cell, cell culture, regeneration and micropropogation
2. Somaclonal variation
3. Androgenesis and ovulogenesis
4. Somatic embryogenesis and organogenesis
5. Protoplast fusion, somatic hybridization
6. Transformation (Transgenics)
   (a) *Agrobacterium* mediated
   (b) Microinjection into cells and zygotes etc.
   (c) Direct DNA uptake
   (d) Virus mediated
   (e) Electroporation
   (f) Shotgun etc.
7. Pollen Biotechnology
8. Embryo rescue, overcoming incompatibility barriers in plants.
9. Molecular markers etc.

Genetic manipulation essentially involves insertion of foreign genes or altering the activity of existing genes. Methods to insert foreign genes coupled with new molecular techniques of gene mapping, RFLP, RAPD, AFLP, PCR etc. in combination with plant tissue culture to regenerate transformed plants has set a new trend in plant molecular biology.

The various genetic technologies or cloning strategies currently in use are :

1. Production of defined DNA fragments
2. Production of recombinant DNA molecule
3. Vectors (Plasmids - R factors), cosmids, and viral vectors
4. Insertion of DNA molecule into vector
5. Stiching DNA fragments
   (a) Restriction enzymes (Restriction endonucleases)
   (b) DNA ligase
6. The use of cDNA
7. Determination of recombinant DNA molecules
8. Screening for particular recombinant

A brief account of some of the above strategies are described below.

Isolation of a Desirable or Selected or Foreign or Target Gene

Genes are intrinsic parts of chromosomes and are present along the length of chromosomes like beads in a necklace. But we do not know where one ends and another one starts. Gene is a length of DNA of any length. The isolated DNA is cut into a number of fragments by RE. Different RE recognise different nucleotide sequences and cuts DNA into several fragments.

The fragments of DNA are separated from each other by using a technique called Electrophoresis.

Restriction Endonuclease (RE)

The enzymes which break the DNA backbone (sugar-phosphate) at general or specific sites are called *Restriction Enzymes* or *Nucleases.* A nuclease that breaks the backbone at the terminal part of DNA, is called exonuclease. The other category of enzymes which nick the backbone at the inner sites (between the two ends) are called Endonuclease. The endonuclease enzyme which cuts the DNA at a specific site is called Restriction Endonuclease Enzyme (or RE) are site specific. The site where RE cuts is species specific. Different types of RE cut the DNA at different sites. Arber, Nathans and Smith were awarded Nobel prize for their work on RE in 1972. Paul Berg called them as molecular scalpels and sutures. Discovery of RE revolutionised the conventional breeding methods. It helped as the best food in r - DNA technology. These are mostly found in Bacteria.

Palindrome

The site where RE recognises a sequence of nucleotides in DNA is called Palindromes. It is a kind of nucleotide sequence which reads the same in both the left and right directions. It is a symmetrical repetitions inverted sequence like '*Madam*' in English. It usually consists of 4 - 6 nucleotides.

Types of Restriction Enzymes :

These are two types of restrictions enzymes on the basis of how they cut the DNA.

1. ***At opposite ends :***
   RE breaks the two strands of DNA precisely at opposite points and yields blunt ended fragments of DNA (ds DNA). eg. Hae III, Sma I, Alu I
2. ***At staggered ends :***
   RE breaks the two strands of DNA at staggered points and yields short, complimentary, single stranded *sticky* ends or *Cohesive ends*. eg. Eco RI, Hind III.

Nomenclature of RE

The name of RE has three or four letter abbreviation. First letter indicates the name of genus from which RE is obtained. The second and third letters indicate the species and variety of the source (organism) respectively.

The RE is given a Roman number at the end. It indicates serial number of that enzyme among the enzymes derived from the same source. eg. E co R I.

E = *Escherichia* (Genus)
co = *coli* (species)
R = RY (variety)
I = First enzyme derived from *E.coli* RY

Some of the important Restriction Enzymes are listed in Table 2.3.

Table 2.3

S.No	Name of RE	Source
1.	Eco RI	*Escherichia coli* RY 13
2.	Hind II	*Haemophilus influenza* Rd
3.	Pvul	*Proteus vulgais*
4.	Bam HI	*Bacillus amyloliquifaciens* H
5.	Sal I	*Streptomyces albus* G
6.	Bgl III	*Bacillus globigi*
7.	Bal I	*Brevibacterium albidum*
8.	X ba I	*Xanthomonas badril*
9.	Sfa I	*Streptococcus faecalis*
10.	EcoR II	*E.coli* R 245

Vectors

A biological structure unit that carries a desired gene from one organism to the other is called *vector*. The common vectors used in r DNA technology or gene manipulation are plasmids phages and cosmids etc.

Plasmids

The term plasmid was coined by the Nobel Laureate Lederberg. Plasmids are small circular, complete, naked, closed, double stranded, extra nuclear; cytoplasmic, self replicating, independent, free DNA molecules. These are mostly found in bacteria and in some fungi and yeast.

A plasmid may contain 25-30 genes. A bacterium may contain one or many types of plasmids. Plasmids may be one or many copies in a bacterium. They can be easily isolated cloned, inserted. They can be multiplied to any number. They confer resistance to the bacteria against the chemicals, minerals, environment etc. They serve as ideal cloning vectors. The cloning of DNA in a plasmid is shown in Fig. 2.10.

PBR 322 is a standard cloning vector used in gene manipulation of plants. It is a hybrid synthetic best studied vector and obtained from col E, plasmid and PSC 101 plasmid and is named after Bolivar and Rodriquez. It is acted by several RE.

Cosmids

A brief account of plasmids and cosmids is given in Unit VI.

Transfer of a gene from one cell into another is the basic procedure in genetic engineering. There are four stages.

1. The gene is cut out from its location in the DNA.
2. The DNA fragment containing the gene is linked to another piece of DNA in the vector, which allows it to be replicated in an appropriate cell. The fused DNA molecule is the recominant DNA or r-DNA.
3. The r-DNA is introduced into the host cell where it is either incorporated into the host chromosome or maintained in the cytoplasm.
4. The cells which have successfully taken up the DNA are isolated, cloned in large number of genetically identical individual are produced from the original cell and only those of interest are identified (Fig. 2.10).

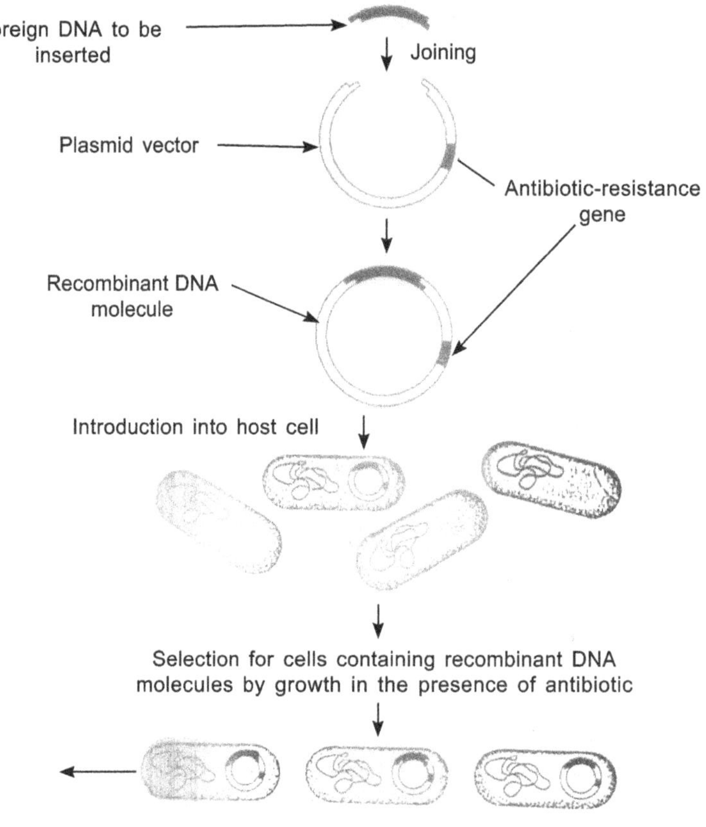

Fig. 2.10 The Cloning of DNA in a plasmid

Plant Vectors

The only and the best studied plant vector system of introducing foreign genes into plants is the naturally occurring crown gall tumor system, caused by the soil borne agro bacteria. *Agrobacterium tumefaciens* also known as '*Crown gall bacterium*'. *A.rhizogens*, similarly infects wounded plant tissue but induces production of roots and is the causative agent of 'hairy root disease' because they contain Ri plasmids.

A. tumefaciens can infect susceptible plants of thousand of dicot species but not cereals and causes tumors (uncontrolled growth) at the wound site. Taking advantage of this property Hosch *et al* (1985) first reported *A. tumefaciens* mediated transformation in plants and since then considerable work has been done on engineering Ti-plasmids to contain other genes that are transferred to host plants during infection (transfection) creating transgenic plants.

A. tumefaciens (Rhizobiaceae) is a gram negative soil bacterium that has an in built natural genetic engineering system, as it contains a segment of T-DNA (Ti-plasmid) that is transferred from the bacterium to plant cells, since they are independent replicating genetic unit. In Fig. 2.11 the structure of Ti-plasmid is shown and the use of T-DNA in cell transformation is also illustrated. During infection the bacteria present in the soil. 1. infect a plant cell the plant cells begin to proliferate and form galls (tumors) at the base of plant (i.e crown). The gall infact is made of transformed plant cells. 2. As a result of infection, the plant cells begin to synthesise an amino acid an arginine derivative called an 'opine' which may be either "nopaline" or "octopine" depending upon the strain of *A. tumefaciens*. Thus, there are octopine type and nopaline types of Ti plasmids. These opines are catabolised and used as energy source by the infecting bacterium. Strains of *A. tumefaciens* can be grown on nopaline but not an octopine and the vice versa. This interesting relationship between *tumefaciens* strains and their respective host plants evolved during the course of evolution.

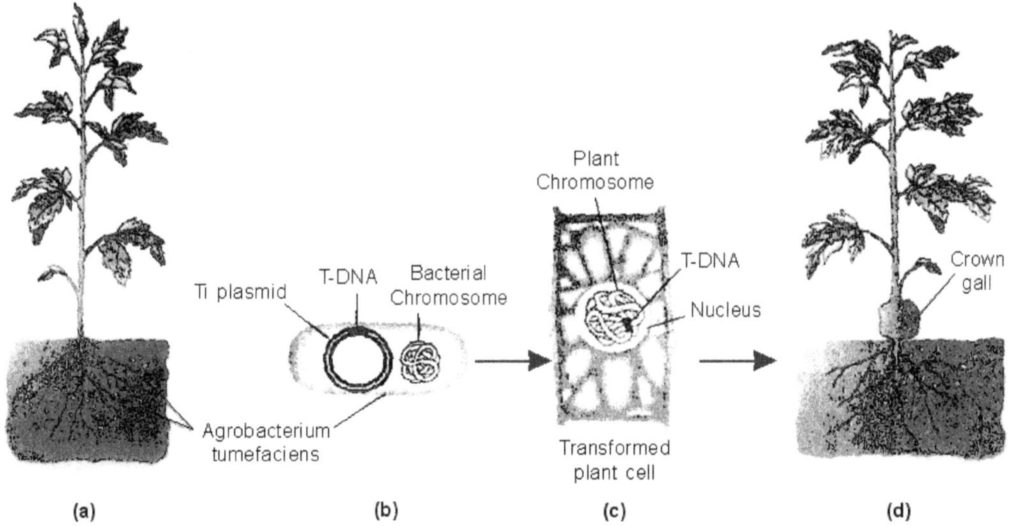

Fig. 2.11 Stages in crown gall formation and cell transformation

When the bacterium infects a plant cell, a part of the Ti-plasmid a region called T-DNA is transferred and gets integrated covalently at random into the genome of the host plant cell as shown in Fig. 2.11. Rarely multiple copies of T-DNA integration occurs in the same cell.

The ability of *A. tumefaciens* to induce crown galls is controlled by genetic information carried on a large plasmid called the Ti-plasmid is a long circular DNA about 200 kb. This includes a small transposable element of 30 kb flanked by 25 nucleotide base pair inverted repeats which are required for T-DNA excision and transformation (Fig. 2.12).

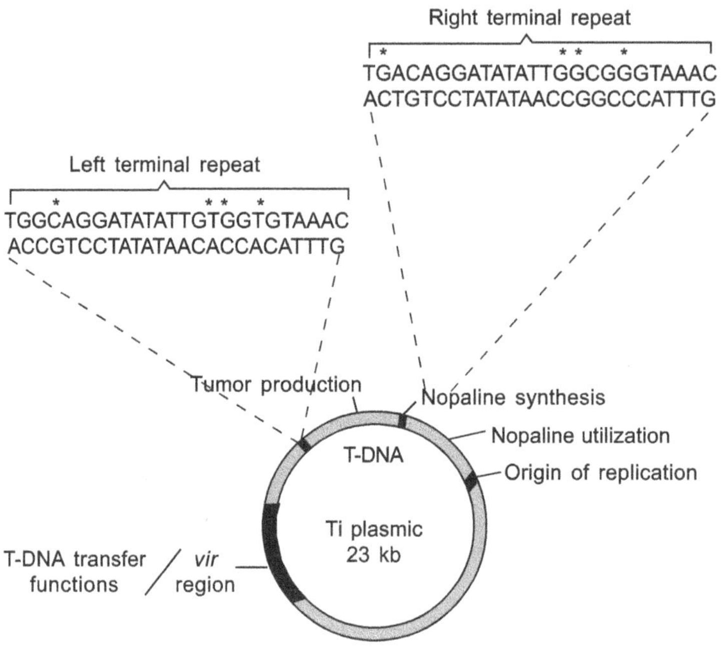

Fig. 2.12 Simplified representation of the major regions of Ti plasmid of *A. tumefaciens*. Structure of the nopaline Ti plasmid pTi C58. The nucleotide-pair sequences of the left and right terminal repeats are shown at the top; the asterisks mark the four base pairs that differ in the two *border* sequences.

The Ti-plasmid has two functional components.

1. The T-DNA (transferred DNA) is flanked by left and right borders involved in excision of T-DNA.
2. Vir region (T-DNA transfer function), virulence region for infection that facilitates transfer of T-DNA.

The T-DNA region encodes enzyme that catalize the synthesis of auxin (IAA) and cytokinin (isopentyle adenosine) synthesis as well as for amino acid and phospage derivative that the bacterium needs for tumor growth.

The Vir genes (is 40 kb with 24 genes and 8 operons) can supply the functions needed for T-DNA (single strands) transfer when located in trans position to the host DNA. In soil their expression is low (takes 10-15 hrs for tumor formation) but at wound site in plant the expression is high. Further, phenolic compounds such as acetosyringone and over drive segment to right border of

T-DNA act as inducers of the vir-genes and enhance the rate of transformation by the addition of acetosyringone. This way, the T-DNA region of Ti-plasmid of *A.tumefaciens* is transferred to the plant chromosome and bring about cell transformation. Therefore, T-DNA system is good for transfer of foreign genes (from fungi, animal, bacteria) into plants but the problem is that the transformed plant cells carry wild type T-DNA that lose their normal control of cell division and tumor formation. This property of T-DNA renders wild type Ti-plasmid undesirable with the goals of most gene transfer experiments. By deleting one or more gene from T-DNA produces a disarmed T-plasmid with wild type Ti-plasmids the recipient plant cells form tumors and are easily identified by tumor formation with disarmed Ti-plasmids, the recipient plant cells continue to grow like other cells that lack T-DNA. This led to development of selectable marker located within the T-DNA region of the disarmed Ti-plasmid.

The Ti-plasmids are too large to be easily manipulated and cannot be made smaller as they contain few unique restriction sites. The transposae functions are not present in the T-DNA but in the plasmid region outside the T-DNA. In infected cells, the transposae functions mobilise the T-DNA and cause its insertion into the plant genome.

A good selectable marker gene is resistance to antibiotic that arrests or kills the growth of normal plant cells. There are three such selectable markers now commonly used; viz.

1. Kanamycin and related aminoglycoside G 418
2. Hygromycin
3. Methotrexate

Of these Kan^R gene from *E. coli* transposon (Tn 5) has been used as a selectable marker in plants, it encodes an enzyme called NPT II (neomycin phosphotransferase type II) is one of the prokaryotic enzyme that detoxify kanamycin family of amino glycoside antibiotics by phosphorytaking them. Since the promotor sequence of transcription, termination signals are different in plants and bacteria, the native Tn 5 Kan^R gene cannot be used in plants. Therefore the NPT II coding sequence has to be provided with a plant promotor 5' to the coding sequence and plant termination and polyadenylation signals (3' to coding sequence). Such construction of prokaryotic coding sequence flanked by eukaryotic regulatory sequences are called chimeric selectable markers.

Development of efficient plant regeneration protocol is a pre-requiste to carry out effective transformation. For successful development of transgenic plants, identification of a suitable target tissue (with acetosyringone) equally suitable and efficient gene transfer protocol is necessary.

Agrobacterium tumefaciens mediated transformation is most frequently used method for generating transgenic crop plants (Fig. 2.13 & 2.14). Transgenic plants are tested for stable integration, expression of genes by PCR or southern hybridisation and such transgenic plants are designated as T_0 generation and seeds obtained from them are designated as T_0 seeds. Plants obtained from T_0 seeds are designated as T_1 generation.

Transgenic tobacco plants that have been transfected by Ti plasmids containing the luciferase gene from fireflies. This enzyme catalyzes the reaction of a chemical called luciferin with ATP, in this process, light is emitted which explains why fireflies glow in the dark. The value of these experiments is not the production of glowing plants, but rather the use of the glow to "report" the action of specific genes. In other experiments, the promoters and enhancers of certain genes were attached to the luciferase gene. Luciferase thus would only be produced when these promotors were activated thus, the glowing areas of the plant shows the region where the transfected gene is active. The luciferase gene is useful as a "*report*" or announcer system used to monitor the activity of the gene of interest and thus this system holds much promise.

Other genes used as reporters in plants are the bacterial GUS (β-glucuronidase) gene, which turns the compound X-glue blue, and the bacterial lac (β-glucurosidase) gene, which turns the X-Gal blue, cells in which these reporters are expressed turn blue and the blue colour can be easily seen either by the naked cye or under the microscope.

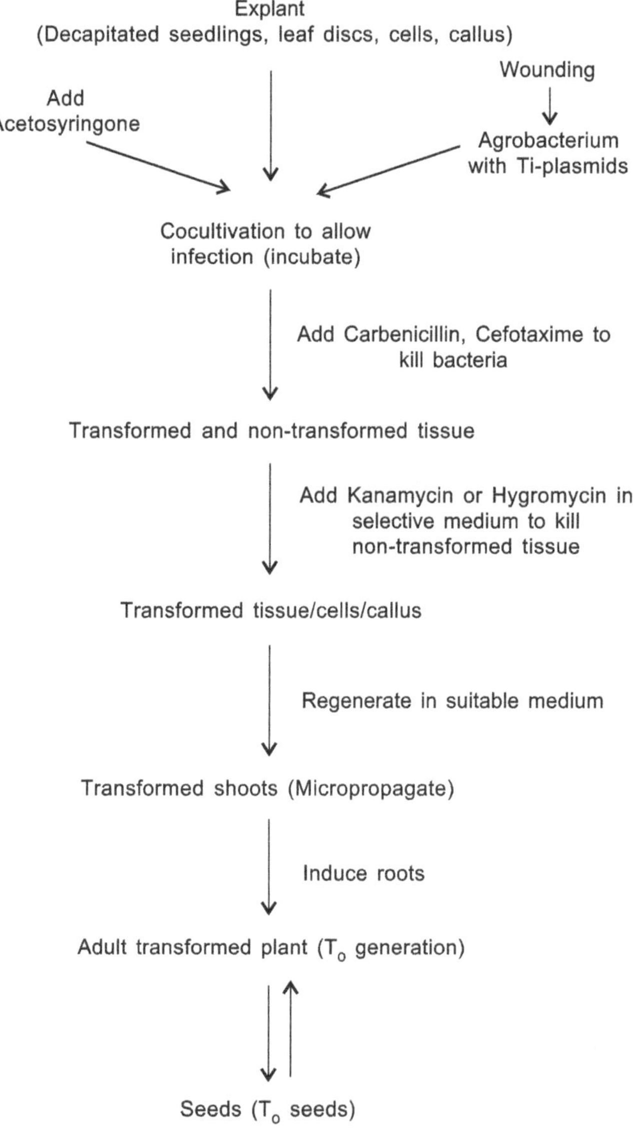

Fig. 2.13 Simplified representation of *Agrobacterium tumefaciens* mediated transformation

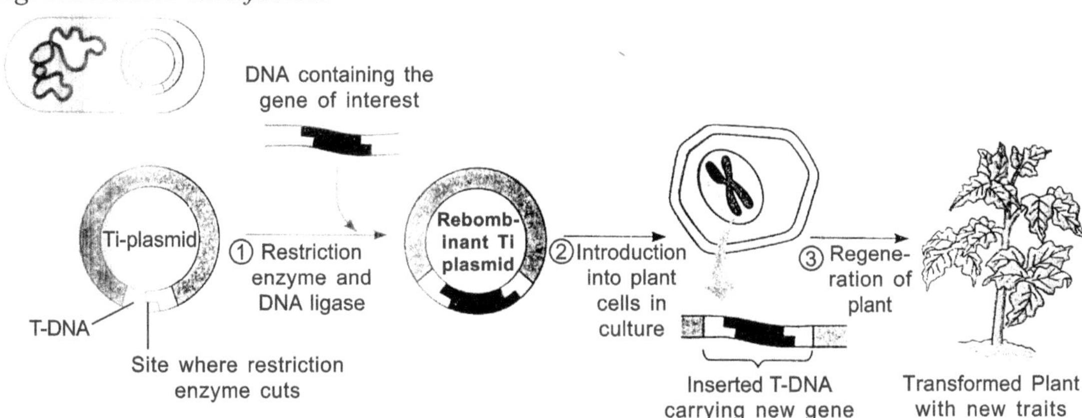

Fig. 2.14 The transfer of genes into plants using the Ti plasmid from A. tumefaciens as a vector

Tomato plants were made resistant to tobacco mosaic virus (TMV) infection by introducing the TMV coat protein gene into plant using the Ti-plasmid containing *A.tumefaciens* system. Plants that express the coat protein gene are resistant to attack by the TMV virus presumably by a mechanism that involves preventing the native viral DNA from being released from its coat protein. The tomato yield from the engineered plant, was more than three times that of the controls. Transgenic plants carrying any one of a variety of foreign genes are in current use and many more in development.

DNA Shotgun

'*Shotgun*' transformation is a novel approach that uses gun powder to deliver DNA into plant cells. (Klein *et al* 1987). A thin coat of DNA is coated onto the surface of 1mm diameter tungston or gold particles by precipitation with $Cacl_2$. The beads are placed on the end of a plastic bullet; the macroprojectile in a barrel as shown in Fig. 2.15. The target plant tissue/cell suspension are placed next to a small opening at the end of the barrel. Using an explosive charge the macroprojectile is propelled towards the cells, liberating the DNA particles to pass through the aperture and hitting the cells. Following DNA bombardment, the cells are transferred to a nurse cell culture plate and the plants are regenerated. Microscopic examination of bombarded cells show microparticles in the cytoplasm, the DNA entering the nucleus causing transformation.

The approach though innovative, it is unclear if this method can be used in routine basis for gene transfer in morocots. However, fertile transgenic plants in cereals and soybean have been obtained.

Other methods presently available for introducing genes into plant cells are :

Chemically mediated DNA uptake

The most successful approach has been to use polyethylene glycol as such an agent. Successful introduction of foreign selectable DNA, followed by recovery of transformed cells, has been achieved for a number of cereals and grasses. In this and the following methods, excess 'carrier' DNA is usually added with the selectable gene to reduce degradation by nucleases and perhaps also to facilitate integration of the marker gene. Treatment of protoplasts with the chemical polybrene is also known to enhance transformation (Fig. 2.16).

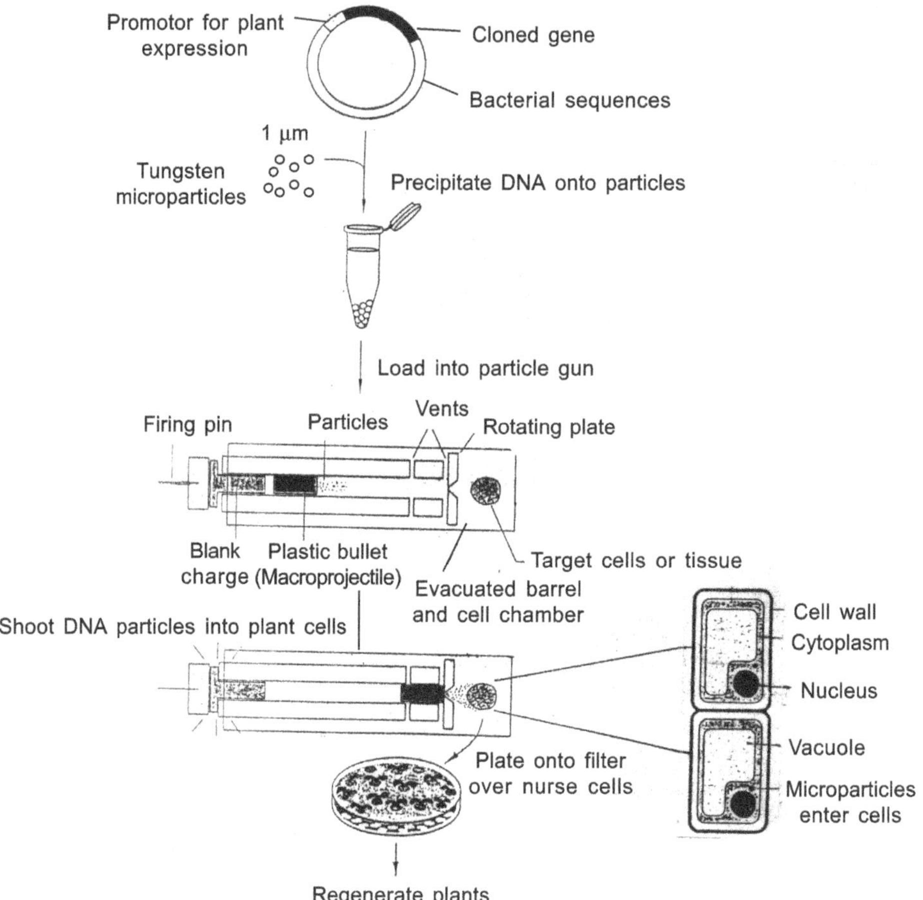

Fig. 2.15 Shotgun transformation technique

Electroporation

The protoplasts are subjected to electric current that creats transient pores in the plasma membrane through which foreign DNA can enter a cell. Transgenic rice, maize and tobacco plants have been obtained.

Macroinjection

There is a report that a simple technique has enabled the recovery of three rye plants containing a foreign gene conferring resistance to the antibiotic kanamycin. The procedure involved the injection of an aqueous solution of the kanamycin resistance gene (in a plasmid) into the space surrounding the developing inflorescences about 14 days before meiosis. The injected plants were allowed to produce seeds, which were screened for resistance to kanamycin. Stable integration of the resistance gene was demonstrated in resistant seedlings by Southern analysis.

Microinjetion

Microinjection of DNA using fine glass micropipettes DNA is directly introduced into the cell where direct DNA transfer is not possible or isolation of protoplast difficult or impossible as in cereals. Injecting DNA copies into the cell or directly into the nucleus the probability of integration increases with 10-14% success. Microinjection is a promising tool for DNA transfer into microspores germinating pollen grains and other organelles. In rape, transformants were obtained upto 50% after microinjection of embryoids. This is labour intensive procecdure and requires services of skilled worker. By this method cell organelles or isolated chromosomes can be transferred.

Direct DNA uptake

Certain chemical or electrical treatments allow direct uptake and incorporation of foreign DNA into plant protoplast allowing transformation. Since a large number of cells can be treated, transformation is relatively simple. Cells expressing the desired trait can be regenerated and tested further. Cereal dehydrated embryos have been transformed by this method.

Silicon carbide fibers mediated DNA delivery

This method involves vortexing a mixture of plant cells, plasmid DNA (which encodes selectable marker genes) and silicon carbide fibers. Silicon carbide fibers are single crystals (0.6 m average diameter and 10-80 m long). This method is rapid, inexpensive and requires little preparation of the DNA before its use.

Somatic hybrids and cybrids

Protoplast (Parasexual hybridisation) fusion is a equally novel and important promising method for producing hybrid plants that otherwise cannot be obtained by sexual means. Of the several somatic

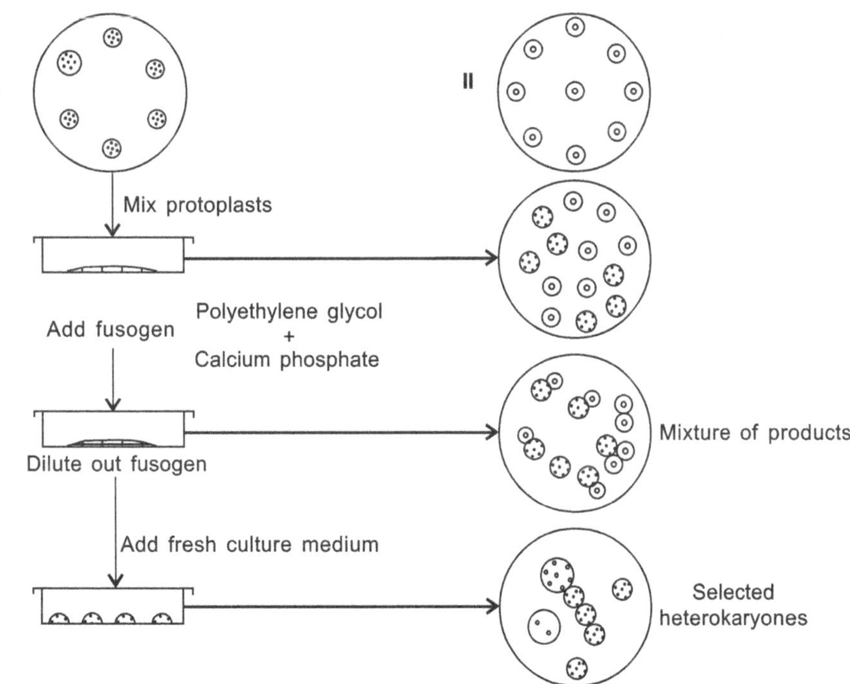

Fig. 2.16 Flow chart for achieving protoplast fusions using chemical fusogens

hybrids, the hybrid *Brassia naponigra* resistant to *Phoma lingam* is the result of fusion of protoplast of *B. napus* and *B. nigra.* Somatic hybrids of *Solanum tuberosum* + *S. brevidens* is resistant to certain virus diseases. This method is useful for transferring multigenic traits. Successful gene transfer by this method depends on the ability to regenerate mature plant from the fusion product. Intergeneric fusion of protoplasts of potato and tomato is "*topato*" worked out by Melchers *et al* (1978). In addition to somatic hybrids, cybrids, i.e fusion of nuclear and cytoplasmic genome of one parent and cytoplasmic genome of another parent. These can be of particular advantage in the transfer of cytoplasmic male sterility in crop plants, an important trait in plant breeding.

Plant protoplast represent the finest single cell system and offer exciting possibilities in the fields of somatic cell genetics, and crop improvement. They serve as an excellent starting material for cell cloning and development of mutant lines. They also provide experimental material for many other fundamental and applied studies for the production of hybrid plants by protoplast fusion and production of hybrids. Thus, the genetic diversity and potential of plants can be altered without involving sexual process of fertilisation : by production of haploid, triploid and tetraploid cells; by the use of protoplast fusion between different species and even unrelated genera; and by transformation, i.e. transferring DNA from one plant (or even another type of organism) into the cells of another plant. The flow chart for protoplast fusion and for obtaining hybrids and cybrids is shown in Fig. 2.17 & 2.18.

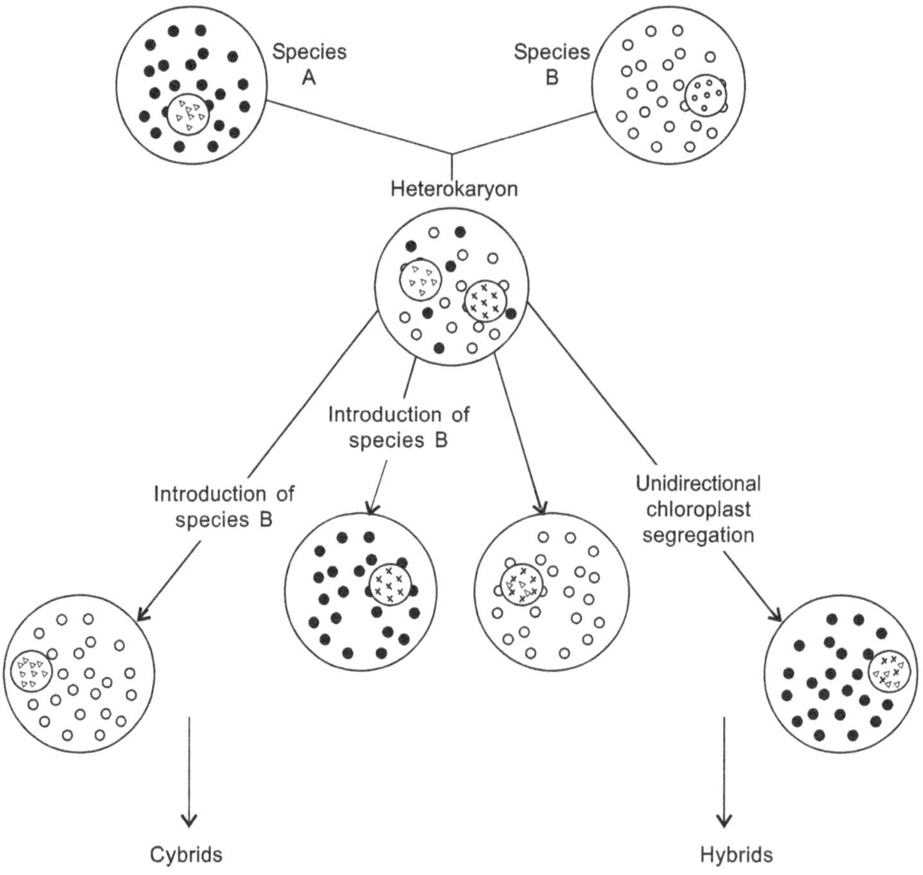

Fig. 2.17 Flow chart of cell products obtained through protoplast fusions

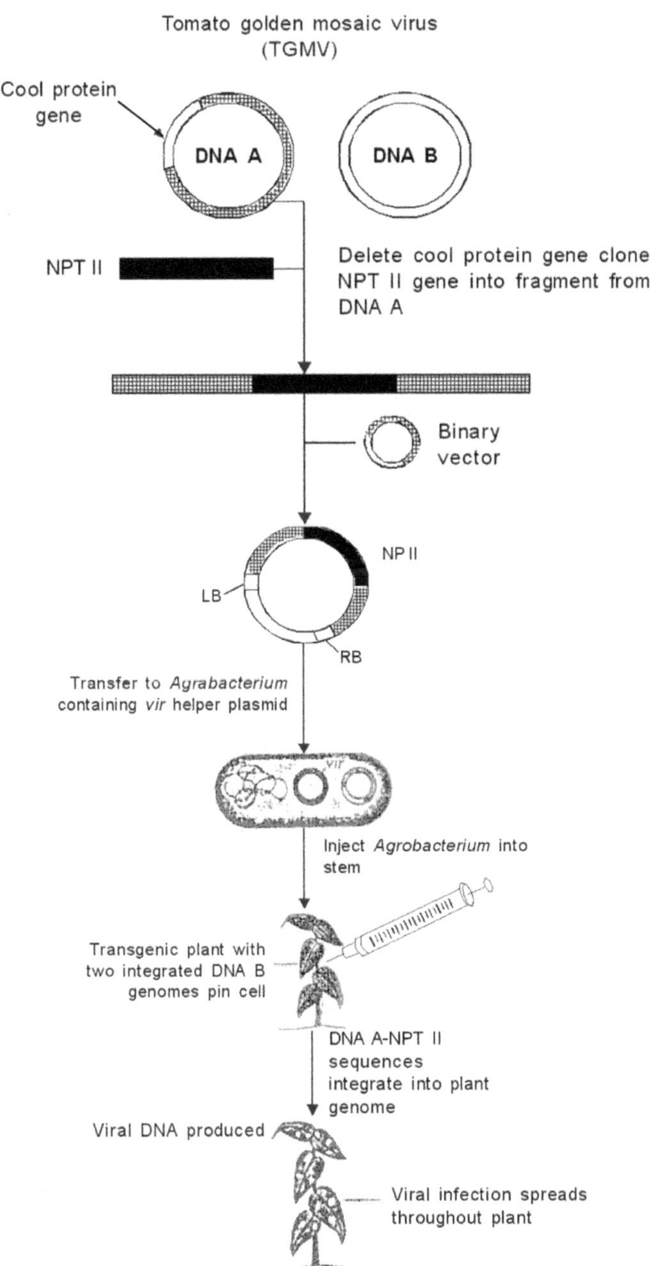

Fig. 2.18 Systematic infection of a plant with recombinant geminivirus vectors

Embryo culture

Embryo culture implies excision of the embryo from the seed, to be grown on an artificial medium. This technique is used to circumvent seed dormancy to shorten life cycle to prevent embryo abortion. Embry abortion is very useful in several interspecific crosses and also in obtaining triploids by crossing tetraploid X diploid parents. Since regeneration of plants from somatic cells in cereals is nearly impossible, regeneration of plants from embryo deduced protoplast has been achieved in *Pernisetum* american (*pearl millet*) and in wheat.

Manipulation of Sex cells

The manipulation of fertilization is a fundamental concern of the plant breeder, and can be studied at three stages : 1. the development of the pollen grains in the anther and the female gametes (the egg cell and central cell) in the ovule; 2. the fate of the pollen grain once it has alighted on the surface of the stigma (incompatibility) and 3. the transmission of the male nuclei to the egg and central cells. Thus, the sexual cells (gametes or gametophytes) have considerable potential to act as natural vectors of foreign gene material. Being natural vehicles for the delivery of DNA from one to another, they are preferred targets for the insertion of foreign gene/DNA. Through these agents, tissue culture manipulations are circumvented entirely. Therefore, the sexual cells furnish a via media for plant gene manipulations since most higher plants reproduce *via* the sexual route.

Pollen biotechnology

This is an upcoming area and during the last 10 years some advances have been made to bring about genetic transformation through pollen sperm cells. The sperm cells are isolated by osmotic shock or by degrading pollen wall by enzymes.

Microinjection of exogenoms DNA into pollen grains/sperms cells/eggs is considered more effective to achieve in vitro fertilization and development of zygote plants. Microinjection of sperm nucleaus into egg cell of maize of a desired plant to bring about wide hybridization can be done. In the direct DNA uptake method, pollen and DNA containing a selectable gene are mixed and applied to stigma. The exogenous DNA carried by pollen fuses with the egg and brings transformation.

Virus mediated transfer

Virus mediated transfer of DNA has played pivotal role in animal biotechnology. But viral vectors for plant systems have not been exploited due to underdeveloped knowledge. Cauliflower mosaic virus (CaMV) has been used with some success in turnips and brome mosaic virus in barley. Developing suitable generic virus-mediated transfer systems could perhaps accelerate progress in plant biotechnology.

Plant viruses appear ideal for transfer of DNA into host cells as they replicate fast to as also their infection. Despite these advantages most viruses contain DNA instead of DNA and RNA does not get integrated into plant genome but replicates autonomonsly with the virus. This implies that the germ cells and hence the seeds are free from virus. Therefore, vegetative crops like potato, sugar cane and bulb flowers are suitable for viral transformation. Two plant DNA viruses have been extensively studied. Cauliflower mosaic virus (Ca MV) and gemini viruses. Ca MV and other caulimo viruses have only a limited host range, however, gemini viruses exhibit a wide host range and therefore more suitable for general application. It is claimed that Km^r and few other bacterial genes have been successfully transmitted to cultured wheat and maize cells by using wheat dwarf virus, a member of the gemini virus group.

Tomato golden mosaic virus (TGMV) is composed of 2.5 kb single stranded DNA molecules packed into a protein coat. DNA-A encodes the seed protein and replication while DNA-B is responsible for cell to cell infection. Naked DNA cause infections in plants and these have been engineered as vectors that transfer cloned DNA of interest into plant cells via infection. This simple technique eliminates the seed to regenerate plants from transformed cells. To test the validity of the vector system, the NPT II gene for resistance to kanamycin was ligated into a cloned fragment (Fig. 2.18) of DNA-A molecule, replacing the gene for coat protein. This DNA fragment as shown is figure, positioned between T-DNA border sequences (LB and RB) in a binary vector, was transferred to *Agrobacterium*. The DNA was then transferred to tobacco plants by injecting the *Agrobacterium* directly into the stem. The tobacco plants had previously transformed with T-DNA containing two copies of TGMV - DNA-B. The infection spreads all over the plant in three weeks. Single stranded DNA-B and DNA - NPT II recombinant molecules were detected in leaves from infected plants by Southern blot DNA hybridization. Extracts from the leaves showed high levels of NPT II activity. An alternate binary vector containing both DNA-A and DNA-B was developed and any plant capable of participating in *Agrobacterium* mediated transfer can be infected thereby eliminating the transformation with DNA-B.

In 1983 the first GM transgenic plant was announced in a scientific meeting at Miami. Florida, USA. During the past 20 years this technology has picked up very fast specially in the field of herbicide resistance, plant protection, quality aspects and production of new biomolecules etc. Initially dicots and of late monocots plants have been successfully manipulated by trans technology to introduece/regulate the expression of desired traits. More than 4000 field trials of more than 50 different plant species have been conducted all over the world especially in North America and Europe. Some of the commercially important plants which have been transferred by various agronomic and/or disease tolerance or quality traits include alfalfa, apple, barley, cotton, maize, pine, potato, rice, sorghum, soybean, sugarbeet, sunflower, tobacco, tomato and wheat. More details are given under GM food.

Table 2.4 is a list of some crop plants which have been genetically manipulated. Production of GM/transgenic plants has begun on a large scale. Genetic engineering is creating a revolution in agriculture allowing an ever-increasing range of plants. This hopefully will lead to increased stability in the market and much less wastage. Agricultural biotechnology will allow higher quality standards with lower costs of production.

Table 2.4 Summarises applications of GM plants that have great potential in world agriculture.

Table 2.4 Transgenic crops

Cereals	Beverage plants	Fibre	Legumes oil seeds	Vetables fruits	Flowers	Pasture grass	Trees
Rice	Coffee	Cotton	Linseed	Carrot	Rose	Lucerne	White spruce
Maize		Flax	Soybean	Tomato	Chrysanthemum		Walnut
Wheat			Rape	Strawberry	Carnation		Apple
Rye			Sunflower	Celery	Geranium		
			Canola	Cauliflower	Orchid		
			Palm	Tobacco			
				Sugar beet			
				Cucumber			
				Potato			

Note : *This list is incomplete (see G. M. Crops for more details)*

The future prospects of GM crops are also summarised (Table 2.5). Achieving the goals may be difficult but not impossible. Considerable success has already been achieved in the recent past.

Table 2.5 *Applications of genetically engineered crops*

AIM	Application	Method	Crops transformed
Herbicide tolerance	Use of herbicides post emergence of seedlings at lower doses than required before seedling emergence	Introduce bacterial gene for enzyme which degrades the herbicide or which bypasses the point of plant metablism inhibited	Soybean, canola (oil-seed rape), corn, cotton
Insect resistance	Reduce losses pesticide spraying not required	Insertion of gene from the bacterium *Bacillus thuringiensis* confers resistance to a range of insect pests	Corn (against European corn borer); cotton (against boll worm, tobacco budworm etc); potato (against Colorado beetle)
Post-harvest quality	Increasing shelf-life reduces losses in transport and harvest	Modified activity of polygalacturonase or other repening enzymes	Tomato
Virus resistance	Reduced losses due to viral diseases	Insertion of viral coat protein gene into plant	Tobacco (tobacco mosaic virus); potato (potato viruses X and Y)

Risks of Genetic Manipulation

Like any technology GM technology has its own pros and cons. There are three major risk factors

1. Environmental risk cross pollimation of GM crops with native flora damage of fauna
2. Food safety
3. Socio-economic factors.

The discovery that plants can synthesise functional human antibody fragments opens up a whole new area for plant biotechnology. This is described in detail under plantibodies.

Biotechnology is developing at a phenomenal pace and will increasingly be seen as a necessary part of modern life and not simply a way to make money. Biotechnology will increasingly affect the efficiency of all fields involving life sciences and it is now realistically accepted that in another 10-20 years, it will contribute many trillions of dollars to world markets.

The impact of biotechnology should be felt through solving the basic problems of food, fibre, fuel and medicine. Current forecast predict that by the year 2030 the world population will probably be doubled. This means that between now and the year 2030, farmers will have to produce more food more calories than they have done from the beginning of agriculture until now. To achieve such a level of food and agriculture production, serious and sustained investments and developments in technologies that can raise yield level further without damage to the planet, have to be made. Biotechnology will be one of such technologies for the developed world, it can help ensure environmentally sustainable supplies of safe, nutritious, affordable food. For developing countries, it can provide readily accessible, economically viable means of addressing primary food production needs. The future possibilities of GE crops are summarised in Table 2.6.

Table 2.6 *Future possibilities of genetically engineered crops*

S. No.	AIM	Application
1.	Salinity tolerance	Increased crop yield in areas affected by saliniy (e.g. in long-term irrigation).
2.	Drought tolerance	Increased crop yield in marginal, semi-arid zones
3.	Waterlogging tolerance	Improved survival in temporary flooding
4.	Enhanced flavor, storage and properties	Improved consumer acceptance; decreased losses; decreased energy inputs to processing or storage; enhanced product value or usefulness
5.	Enhanced amino acid content	Dietary improvement and hence health
6.	Antibody and pharmaceutical production	Less energy input and cost than use of animal cell culture
7.	Improved disease resistance	Reduced pesticide inputs; increased yields mean population can be fed using smaller land area

In conclusion, genetic engineering has revolutionized the gene manipulation, gene structure, gene functon and gene expression "*at will*".

Gene manipulation through biotechnological approaches is not only the hope but a reality of the future and will provide unlimited opportunities for meeting the expectation of the Third World people for solving the problems of hunger, disease, environmental pollution, toxic wastes, energy shortage and above all over population.

2.4 GM Crops

Plant breeding has played a major role in modern agriculture by providing plants with superior traits for disease resistance and increased yields. Genetic engineering promises to have a significant impact on crop improvement by accelerating breeding efforts and by dramatically expanding the available gene pool. Identification and transfer of genes of agronomic importance into plants have begun and some degree of success has already been achieved in engineering selective tolerance to herbicides, viral diseases and insect pests. Plant biotechnological techniques, in conjunction with conventional breeding methods, have begun to yield significant contribution in the areas of agriculture.

Plants which have been manipulated to carry genes not normally present in them are termed transgenic. Biotechnologists have been fairly successful in developing transgenic plants in the recent past.

Methodology of Transformation in Development of Genetically Modified Plants

Natural (Biological Vectors) Plasmids Method

Agrobacterim mediated gene transfer; Agro-infection. These vectors are more efficient natural vectors therefore used for genetic transformation.

Transformation Using Ti Plasmid

Agrobacterium is a gram negative bacterium and harbours a Ti (tumor-inducing) plasmid. *Agrobacterium* is attracted by hormones released from an injured plant; infects plant and transfers part of Ti plasmid (T-DNA) to plant cell; T-DNA (tumor-DNA) gets integrated into plant nuclear DNA resulting in crown gall tumor. T-DNA contains genes that direct the synthesis of plant hormones auxin (makes plant cells bigger) and cytokinin (makes plant cells divide); the result of rapid growth and division is a tumor. T-DNA also contains gene for synthesis of opines, nutrient molecules that can only be utilized by bacteria that posses a Ti plasmid.

The Ti plasmid of wild type *Agrobacterium* is relatively large (about 200 kilobases) and difficult to manipulate experimentally. It was found that a region of the Ti plasmid outside the T-DNA, referred to as the virulent region, carries genes (vir genes) that are involved in the transfer of T-DNA into plant cells. Although the precise mechanism of T-DNA transfer is not fully understood, molecular analysis of T-DNA integrated in host chromosomes has shown the presence of 25 nucleotide directly-repeated sequences flanking the boundaries of the integration sites. Genes that are to be introduced into plant cells must be inserted between these 'left' and 'right' border sequences or next to at least one border of T-DNA. Thus by cloning foreign genes into the T-DNA of the Ti plasmid, it is possible to exploit the natural ability of *Agrobacterium* to transfer new DNA into the plant genome.

The production of transgenic plants has become a routine practice in many laboratories around the world. GM crops from being laboratory curiosities have caught the attention of agriculturist, farmers and biotechnologist the world one and despite opposition to GM technology in some countries the economic benefits that accrue to farming community has ensured rapid acceptance of transgenics globally and in particular USA.

In vitro (Direct Gene Transfer) Methods

Various techniques like Pollen Mediated transformation; Gene transfer by Imbibition; Macro Injection; Micro Injection; Particle Bombardment (Biolistics), Electroporation; Ultrasound-induced DNA uptake; Silicon Carbide Fibre-mediated DNA uptake; Laser-mediated DNA uptake; Liposome-mediated DNA uptake; PEG-Mediated Transformation; Agrolistics are being used for efficient gene transfer.

The selected genes are isolated, cloned and transferred to plants using the techniques described above. The plant into which the gene is transferred is called transgenic plants or genetically modified plant and the technology is called *Transgenic Technology*.

More effort is focussed directed toward making plants not only higher yielding but resistant to: herbicides, insects, fungal, and viral infections. Between 1996 - 2000, 40 plant varieties and 6 'microorganisms' have been deregulated. The field release in 2000 were 5012 as against only 5 in 1987. Total number of field released transgenics and their percentages upto April 2000 are : herbicide tolerant 1853 (28.4%), insect resistant 1595 (24.5%), product quality 1195 (18.3%), viral resistant 650 (10.0%), agronomic 385 (5.9%), fungal 319 (4.9%) and others 529 (8.1%) comprise bacterial and nematode resistant crops. In terms of global area, under GM crops in million hectares, upto1999, the USA dominates (72.0%) followed by Argentina (17%), Canada (10%), China (1%) and Australia, South Africa, Mexico, Spain, France, Portugal, Romania, Ukraine constitute 1% each. Between 1996-1998 eight countries (5 industrial and 3 developing) have contributed to more than 15 fold increase in the global area of transgenics.

The world population around 1900 was 1.5 billion and likely to reach 8 billions. At the time of Independence, India's population was 3.5 million as against one billion now and likely to be 1.6 billion by 2050. Population growth at the global (8 billion) and national level (1 billion) demands more food. The food and nutritional security is the major challenge now and transgenics are expected to meet this challenge.

Transgenic Technology

GM crops are now being grown widely and its products are available in international markets. Five principal transgenic crops are grown in 8 countries in 1998 namely Soybean, Corn, Cotton, Canola, rapeseed and potato.

Transgenic plants dicots or monocots ?

Though transgenic plants were first obtained and mostly restricted to dicotyledons, but now the list is extended to monocotyledons plants also. The list of transgenic plants is as follows. This includes cereals, beverage, fibre, legumes, oil seeds, vegetables, fruits, flowers, pasture and trees as well.

I **Dicotyledons**	*II* **Monocotyledons**
1. Tomato	1. Rice
2. Tobacco	2. Wheat
3. Carrot	3. Oat
4. *Petunia*	4. Maize
5. Rose	5. Asparagus
6. Papaya	6. Bamboo
7. Pea	7. Sugar beet
8. Alfalfa	8. Lotus
9. Lettuce	9. Rye
10. Sunflower	10. Orchard grass
11. Rape	11. *Festuca arundinacea*
12. Flax	12. Palm
13. Cotton	13. Orchids
14. Pear	**III Gymnosperms**
15. Cabbage	Lablolly pine.
16. Cauliflower	
17. White clover	
18. *Arabidiopsis*	
19. Mustard	
20. Brinjal	
21. Chicory	
22. *Chrysanthemum*	
23. Apple	
24. Celery	
25. Walnut	
26. Neem	
27. *Populus*	
28. *Ipomoea batatas*	
29. *Vigna*	
30. Cucumber	
31. *Digitalis*	
32. Annoraiea	
33. Licorice	
34. Morning glory	
35. Strawberry	
36. Grapes	
37. *Dianthus*	
38. Soybean	
39. Coffee	
40. Potato	
41. Geranium	

Why the GM Crops ?

The availability of transgenic crop plants would especially benefit farmers in the developing countries where farmers can often not afford the luxury of using expensive herbicides to protect their crops. There is no denying the fact that more than new technologies and products, it is eduction and awareness that is crucial for progress.

Following are select examples of the success story of transgenic technology in the area of environment ie. use of agrochemicals in agriculture.

Herbicide Resistance

Herbicides are used to control weeds. The environment is contaminated by herbicides due to their excess use, some of them are not biodegradable. Therefore, new herbicides that are safer and biodegradable are now being developed. These herbicides inhibit biosynthesis of essential amino acids. Cells deprived of these amino acids, can not synthesise proteins and die.

Glyphosate is a widely used broad spectrum herbicide. It is the active ingradient of Roundup herbicide. It is used to control annual and biennial sedges, grasses and broad leaved weed species. The herbicide is sprayed on the foliage. which rapidly gets absorbed by foliar tissue and translocated to the other plant organs. It inhibits EPSPS (5-enol pyruvic-shickimate-3-phosphate synthase encoded by aro A gene) which is involved in aromatic amino acid biosynthesis. The amino acids, deadly to plants and bacteria are safe for human and animals because human beings can not synthesise aromatic amino acids. Identified bacteria that were resistant to glyphosphate, cloned the mutant '*aro a*' gene, used glyphospate, cloned the mutant '*aro A*' gene, used it to transform plant cells in culture and make resistant, tomatoes, potatoes, petunias etc. Shah (1986) utilised high level expression of a plant EPSPS gene to engineer glyphosate-resistant plants. The EPSPS gene was isolated from ghyphosate-resistant *Petunia hybrida* that over produces the herbicide sensitive form of EPSPS. A chimeric gene was constructed and introduced into *Petunia* cells. The transgenic *Petunia* plants were found to survive glyphosate spraying.

Similarly plants have been engineered for selective tolerance to other herbicides like Sulphonylurea (active ingredient of Glean and Qust herbicide), Imidazoline, Phosphinothricin, Triazine, Gluphosinate, Biolaphos, 2, 4-D, Chlorsulphron and Bromoxynil.

Transgenic tobacco plants resistant to phosphinothricin (PPT), an active ingradient of herbicide called Basta are also produced by the scientists at Hoechst Laboratories (Germany).

Apart from the above method in which target gene is modified, another approach is also followed where the herbicide is detoxified or degraded. Such that the herbicide kills the weed but not the crop plants. The herbicides are detoxified by enzymes such as Glutathione - S transferase (GST), Nitrilase Phosphinothricin acetyl transferase (PAT) which detoxify the herbicides Atrazine, Bromoxynil and PPT respectively. These enzymes are found in microbes and higher plants as well. Nitrilase is coded by ***bxn*** gene in *Klebsiella pneumoniae* and bar gene codes for PAT in *Streptomyces* spp.

The genes ***bxn*** and bar were isolated from their respective owner cells and introduced into Tomato plant which was found to be herbicide resistant. Similarly, Potato, Rape, and Beet root with introduced bar gene proved to be herbicide resistant and thus transgenics were obtained.

It seems clear that within next 5-10 years, commercial plants with selective tolerance toward major potent broad-spectrum herbicides will be available.

Virus Resistance

Plants, suspectible to viruses, can be protected by a phenomenon called cross protection. This was first discovered by Mc Kinnery (1929) who observed that tobacco plants infected with one strain of virus resisted infection by a second related strain. This phenomenon was used to reduce losses in tomato, potato, citrus against Tomato Mosaic Virus, Potato Spindle tuber Viroid and Citrus Tristeza virus respectively. Hamilton (1980) first suggested that protection from viral disease can be achieved with the help of gene transfer technology.

Powell Abel (1986) synthesised a Chimeric gene consisting DNA of the coat protein (CP) gene of TMV inserted to Ti plasmid of *Agrobacterium tumefaciens* from which Ti genes had been removed. The chimeric gene is transferred to Petunia cells from which Petunia transgenic plants are regenerated. The plants showed resistance to virulent TMV. Using the same TMV-CP gene, virus resistant tomato and tobacco plants were also obtained. These plants either did not develop infection or disease symptoms developed slowly than in the control plants after inoculation with a severe strain of TMV.

This phenomenon of genetically engineered cross-protection was extended to another virus, alfalfa mosaic virus (AmV). Recently, cross-protection against PVX and cytomegalovirus has also been obtained.

Transgenic tobacco plants have been produced by using sat RNA technique. Scientists at Cornell University and the University of Hawaii, USA have developed independently two lines of genetically engineered Papaya that are resistant to Papaya Ring Spot Virus (PRSV). This research protected the Hawaiian Papaya Industry which otherwise would have been destroyed by the viral disease.

Insect Tolerance (Self Protecting Plants)

Bacillus thuringiensis produces a natural insecticide that kills a number of caterpillars, but when cloned into plant, not much toxin was produced. Thus plants have been engineered to protect themselves by expressing proteins which are detrimental to insect feeding hence works as a very sucessful biocontrol agent.

Another practical application of genetic engineering with important practical aplications for crop improvement has been the production of insect resistant plants. Generally synthetic chemicals are used as insecticides and pesticides to protect the crop plants from insects and pests. Insect resistance transgenic plant have been achieved through the expression in plants of the insect toxin gene of *Bacillus thuringiensis*. It is an entomocidal bacterium. It produces a parasporal protein (delta endotoxin) crystal during sporulation. This protein in fatally toxic to larvae of various insect pests but specific in their activities orders namely Lepidopteran. Genes encoding the lepidopteran-specific toxins from several strains of *B. turingiensis* have been isolated, cloned and used for *Agrobacterium* Ti plasmid mediated transformation of Tobacco, Tomato, and Cotton plants. The plants become resistant to insect pest. Plants of a species suspectible to insect attack can thus be made resistant by adding chosen Bt genes into it. The transgenic form of insects would be of great use to agriculture because this new resisance would be an integral 'built-in' one and would also be inherited by the following generations and the use of insecticides would be made unnecessary.

Transgenic crops containing insect resistant genes from *Bacillus thuringiensis* have made it possible to reduce significantly the amount of insecticide applied to cotton plants in the USA. Bt cotton is grown in many states in India but because of less awareness among farmers there is still opposition for Bt cotton. When inheritance of the insect resistance was studied using crosses with normal control plants, F_1 plants showed resistance and F_2 generation exhibited expected segregation. So the farmers will have to procure new seeds every year.

Monsanto, a US based multinational company has genetically engineered a cotton which has Bollgard transgenic genes toxic to the Cotton bollworm, tobacco budworm and pin boll worm. Other crops incude maize and tomato. Popularity of such crops is increasing and is being accepted the world one. The insecticidal proteins are specific to target pests and thereby do not affect other insects.

Bacteria Resistant Plants

Rice plants are commonly attacked by a bacterium *Xanthomonas oryzae pv. oryzae* (Xoo) that causes a devastating pestilence Blight of Rice. The bacteria spreads like fire from plant to plant, field to field through water droplets. The world's first transgenic disease resistant rice plants were produced at the University of California (USA). John C. Sanford (1987) and Pamela at International Laboratory for Tropical Agricultural Biotechnology (ILTAB) found Xa 21 gene resistant to Xoo. They introduced Xa 21 gene into rice cells (of Taipei 309 variety). Plants were regenerated from transformed cells that could resist bacterial blight.

Salt-Tolerant Plants

Considerable region of the earth's surface is much too salty to support agriculture. Researchers at the University of Toronto, Canada, have genetically engineered plants that manage to flourish even when watered with concentrated salt solution. This resulted in boosting world crop production of wheat, corn and other commercial crops.

The salt-tolerant plants are *Arabidopsis*. Although most plants are salt-sensitive and deteriorate under conditions of high salinity, some plants manage to live by directing sodium ions away from regions of plant cells where salt can disrupt plant metabolism. These salt-tolerant plants deal with extra pinch of salt by sequestering sodium ions in vacuoles. To bestow salt tolerance upon plants that ordinarily stick to a low sodium diet, the Toronto botanists identified the gene that holds the instructions for producing the antiport protein, cloned it and have grown modified plants that over produce the ion-directing protein.

Stress Resistant Plants

Genes responsible for providing resistance against stresses such as heat, cold, etc., have also been identified. A gene that codes for glycerol-phosphated acetyl transferase has been isolated from *Arabidopsis* (resistant to chilling) and introduced into Tobacco. Thus, transgenic Tobacco plants resistant to chill have been developed.

Desert plants reduce trienoic fatty acid at high temperature. This condition was mimicked in transgenic plant by co-expressing *desaturase gene*, wherein plants were able to photosynthesize at 40° C. On the other hand by expressing the gene *trehalose* sugar, which acts as osmoprotectant, it was possible to overcome freezing temperatures. These approaches have made it possible to grow food plants under adverse conditions.

GM Crops in India

Mahyco-Monsanto's insect resistant Bt cotton is India's first approved GM crop in India. A number of companies and research institutes are engaged in transgenic work on a variety of GM crops like Mustard, Tomato, Brinjal, Cabbage, Cauliflower etc. The first company to research GM crops in India is Proagro PGS, a Delhi based company and a part of Aventis (MNC, USA) has produced transgenic Mustard. Much such work is also carried out by universities and research institutes in Delhi, Hyderabad, Coimbatore, Shillong, Mumbai, Shimla, Kolkota, etc.

Metal Toxicity

Aluminium toxicity affects a very large percentage of arable Indian land limiting the use of acidic soils for cultivation. This problem has been overcome by over expressing a bacterial gene for Citrate synthase. The transgenic plants so produced are able to produce about ten times more Citric acid than normal plants and this helps them to grow and produce normally in these acidic soils. Statement of glycine betanine gene from *Escherichia coli* bacterium helps to tolerate salt concentrations three times more than a normal plant can tolerate. The above approaches will reduce pressure to bring more forest under food production. Additionally, it helps in utilizing degraded soils, which are not suitable for plant growth.

Pharming

Presently, several of the pharmaceuticals are produced via fermentation or animal cell cultures, which needs an industrial setup. It has been shown that, some of the peptides could be produced in transgenic plants. Gamma interferon, blood coagulant Factor VIII, Factor XIII and Thrombin clotting enzyme have been produced in transgenic plants. It is estimated that, yield from one plant may be equal to a 1000 L fermentation product of mammalian cells. The added advantage from these plants is that, there is no danger of passing of viruses like HIV, Hepatitis B and oncogenes. Overall, pharmaceutical peptide production from plant source would be much simpler and less energy consuming as compared to the fermentation route and cheaper as well.

Edible Vaccines

Lack of refrigeration and cold chain facilities in many developing countries of Asia, Africa, South America is one of the main constraints for effective immunization programme. It has been possible to express surface antigen (protein responsible to trigger immune reaction) in plants. Clinically acceptable level of immune reaction and protection was demonstrated by oral feeding of these transformed plants. Hepatitis B surface antigen in tobacco, cholera B sub-unit in potato and banana, have been successfully expressed. These transgenic plants would eliminate the need for refrigeration and make it available easily in even the most remote areas, without the medical personnel to administer the vaccine. (This is also discussed under GM food).

Nutritious Food (Golden Rice)

GM technology has helped to make food more nutritious. 'Golden Rice', a genetically modified rice which has genes for increased betacarotene (precursor for Vitamin A) and iron has been produced. This golden rice can prevent blindness and iron-deficiency anemia, in several Asian countries especially school children. Repeseed has been engineered to produce high beta-carotene, a table spoon of oil is sufficient to meet the daily requirement of Vitamin A. The 'golden rice' is a practical application for augmenting nutritional deficiency. A gene for higher content of protein isolated from *Amaranthus* (AMA-1) has been transfered into potato, which makes potatoes rich in proteins too. Another gene, Aspartokinase is also known to help in making the plants to produce more protein.

But a record of the major Indian developments on the transgenic front suggests that apart from Mahyco's Bt cotton and Proagros hybrid Mustard, we may be still many years behind from the commericalisation of other crops. Although India has developed a regulatory frame work for administering, GM crops, it has yet to pass legislation relating to plant variety protection, biodiversity and intellectual property rights. If India is going to benefit from GM crops, the authorities must lose no time in putting such legislation in place.

GM Crops - Challenges Ahead

According to Dr. M.S. Swaminathan, the following are important agricultural challenges and their solutions have to be sought on priority basis.

1. Genes for Leguminous lectin and haemoglobin have to be moved into cereal plants. i.e, conversion of cereal plants to nitrogen fixing plants like legumes.
2. Weed plants synthesise chemicals that render them resistant to various kinds of pests. If the concerned genes can be transferred to crop plants, the latter can be made inherently pest resistant.
3. Transgenic C_3 plants to carry on C_4 pathway to be achieved.

Product Based Assessment

GM plants with particular gene in question are regulated. In order for a transgenic field experiment to be carried and a number of questions like, origin of gene, function of gene in donor organisms, effect of GM plant phenotype, evidence of toxins, allergenicity, non target effects on range of organisms, gene transfer by pollination to other crops, weeds etc. have to be taken into account and duly approved by Advisory Committee on Release to the Environment (ACRE).

The use of transgenics in field trials in keeping the environmental and other regulations are very important. The issue of "terminating seeds" came to light only in March 1997 when Delta and Pine Land Company of Scott, Mississippi USA received as US patent on a technique to block seed germination so that the embryos turn sterile. The patent covered both transgenic and conventional varieties, though tried for cotton and tobacco only. The issue became a major concern when biotech giant MONSANTO which now operates in India announced its plan to acquire this technology.

GM Plants : Prospects Controversies and Risk

This can be studied under the following heads, better health, better products, better for environment and better for commercialisation

Better Health

As a cheap source of human therapeutics, the GM plants offers immense potential. In 1996 US researchers genetically engineered banana plants to produce an antigen found in the outer coat of hepatitis B-virus. Such GM banana plants could be used to immunsie children of developing countries which cost few US cents/dose, as against the hepatitis B vaccines which costs Rs 100-200/ dose. Likewise, banana or any fruit eaten raw could be genetically engineered to vaccinate against various dreaded diseases.

Better Products

This attribute is possible by providing the plant with pests and disease resistance or increasing crop tolerance to a wide range of climatic conditions or making the food more attractive to the consumers. Research is under way to introduce antifreeze gene into fruits and vegetables to make them frost resistance. Likewise, by introducing Bt genes crops resistant to insect/pest attack have been developed.

Product Qualities

GM plants have been developed to improve the quality of their products. For example alien genes have been inserted into tomato which delays ripening and softening, resulting in longer shelf life thereby reducing spoilage. Canola and soybean plants have been engineered to contain higher concentration of fats and amino acids. GM plants can be used for plastic production also. Recently, scientists working at Agracetus, Wisconsin, USA, have developed cotton plants which produces cotton fibres whose middle lamella is filled with polyester (polyhydroxybutrate), such cotton-polyester retains 8.6% more heat that cotton alone and hence ideal for dressing in cooler climates.

2.5 Genetically Modified Food

In November 1996, several British Newspapers published front page photographs of inflatable rafts belonging to the Green Peace Organisation blocking the entry of a freighter into Liverpool harbour, U.K. and displayed a huge banner on a barge, showing the words "*Floodgate : Genetic Pollution*" The protest was staged against the first shipment of genetically engineering sobybeans from the United States to Britain. In December 1996 Greenpeace was in Hamburg harbor, Germany, using a powerful slide projector to beam the words "Genetic experiment - don't buy it" onto the side of a similar freighter. The soybeans contained a bacterial transgene that conferred resistance to a weed killer called glyphosate, inserted so that fields of crops could be sprayed and kept weed free. The advantages to the farmer are obvious, in as much as weeds are a considerable problem in agriculture. The protestor's objection to such transgenic plants was that the transgene might be harmful to human health, the concern is possible allergic reactions to the transgene's protein product. The company that developed the GM syobeans claimed that there was no danger, and this view was supported by the regulatory agencies. In addition, as a result of the crop being resistant, consumers might be exposed to higher levels of herbicides. Further, the engineered plants might escape and may introduce the transgene into related species.

Whether these objections were justified or not, the reference to a "*floodgate*" in the Greenpeace banner was provably right because transgenic crops are flooding into farms and hence into the super markets. The popular term for transgenic crop plant is "*GM foods*" (for "*genetically modified foods*") or the food obtained from transgenic crops is called *GM Food.*

The application of Plant Genetic Engineering to food production is primarily intended to enhance the utility and desirability of characteristics of the plant and to eliminate the undesirable. The main aim of the food industry with respect to genetic engineering is to improve the quality and quantity and properties of existing food productions, to produce new food productions and good financial returns. The consumer is willing to pay more for better products. Plant genetic engineering thus offers plenty of opportunities to tailor foods to public taste and demand.

Regarding the applications of genetic engineering in the application of food organisms both plants and animal though the scientific case is persuasive but the public perception as per the Euro barometer is ambivalent. Though the public has readily accepted the various medical products like insulin etc. all over the world, they are less willing to accept the "*GM food*" as unnatural. However, there is change in public opinion and the concern factor is greatly reduced "*GM food is catching up fast in many American and European supermarkets*".

Biotechnology contributes to the food industry by improved practices, greater sophistication in the use of microorganisms and enzymes. Biotechnology offers good for the malnourished as well as for the well-fed. The aims of the GM food are

1. Food in more abundance
2. Food at lower cost production
3. Food of higher nutritive value (tailored products with improved qualities)
4. Food of greater variety (Novel products)
5. Food preservation for longer periods (Enhanced shelf life)
6. Improved food micro organisms

The final aim of GM food is to provide good food as good health without good food is next to impossible. The GM food achieved can be of animal or plant origin.

Table 2.7 summarise the global area of GM crops in 12 countries. A perusal of the figures show that in USA alone there was increase in area of GM crops upto 72%. Tables 2.7 & 2.9 show the global area of GM crops during 1998 and 1999 and the GM crops already commercialized in USA (Table 2.10).

Bioengineered plants have been in the news for a decade now (Table 2.8), the Flavor Savr tomato which has raised the hackles of the antibioengineered food group led by Jeremy Rifkin has been cleared by USDA in 1992, because it possesses no environmental risk. In 1994, FDA gave this

Table 2.7 *Global area of genetically modified crops (millions of hectares)*

Country	1998	%	1999	%
U.S.A.	20.5	74	28.7	72
Argentina	4.3	15	6.7	17
Canada	2.8	10	1.0	10
China	<0.1	<1	0.3	1
Australia	0.1	1	0.1	<1
South Africa	<0.1	<1	<0.1	<1
Mexico	<0.1	<1	<0.1	<1
Spain	<0.1	<1	<0.1	<1
France	<0.1	<1	<0.1	<1
Portugal	0.0	0	<0.1	<1
Rumania	0.0	0	<0.1	<1
Ukraine	0.0	0	<0.1	<1
Total	27.8	100	39.9	100

Table 2.8 *Some examples of horticulturally important genes and potential genetically modified plants*

Crop	Gene	Organization/Company
Apple	ACC synthase	Cornell University, USA
	Attain esterase	Cornell University, USA
	PG	Cornell University, USA
Banana	ACC synthase	NRCPB, IARI
Brassica oleracea	ACC oxidase antisense	Seminis-Veg Seed
		Seminis-Veg Seed
Melon	ACC oxidase	INRA, France
	SAM hydrolase	Agritope
	SAM transferase	Agritop
Pepper	Acetolactase synthase	DNAP
	Glacanase	DNAP
	Hericellulase	DNAP
Strawberry	Acetolactase synthase	DNAP
	Chitinase	DNAP
	Glucanase	DNAP
	PG inhibition protein	DNAP
	SAM transgerase	Agritope
Tomato	ACC deaminase	Calgene
	ACC oxidase	Peto Seed
	ACC synthase	Cornell University (USA)
	Acetolactase	DNAP
	Etr-1-1	University of Florida (USA)
	Glactanase	Zeneoa
	Lpt	Calgene
	Pectin esterase	Peto Seed
	PME	Purdue University (USA)
	PG	Calgene
	SAM ase	Agritope
	SAM transferase	Agritope
	SPS	Colgene
	Transcriptional activation	DNAP
	SAM decarboxylase	USDA

DNAP : DNA Plant Technology; USDA : US Department of Agriculture (1999)

transgenic tomato the green signal saying that it is satisfied that the antibiotic resistance selective marker will not poison the consumer or trigger allergies or interfere with people taking antibiotic or add to the spread of antibiotic resistant pathogens.

Tomato is one of the best examples of producing tasty and long lived transgenic fruits and is the first GM food crop released in US Market under the name '*Flavr Savor*'. The calgene biotechnologist USA cloned C-DNA to tomato PG (Polygalaturonase) and inserted into tomato, the

DNA in the antisense orientation by inhibiting the polygalaturonase by antisense genes the tomato can remain on the plant until mature and be transported in a firm solid state. The Flavr Savr tomato with improved flavour and shelf life is now marketed in the USA. The current US market for fresh tomatoes is $US 3.5 billion. Such techniques will now be used in a wide variety of soft fruits. Other studies are considering ethylene synthesis/inhibition as a means of controlling fruit and flavour maturation. The transgenic tomato had 99% PG level, thus enhancing shelf life and long distance safe transport.

Table 2.9 *Global area of genetically modified crops in 1998 and 1999 : By crop (millions of hectares)*

Crop	1998	%	1999	%
Soybean	14.5	52	21.6	54
Corn	8.3	30	11.1	28
Cotton	2.5	9	3.7	9
Canola	2.4	9	3.4	9
Potato	<0.1	<1	<0.1	<1
Squash	0.0	0	<0.1	<1
Papaya	0.0	0	<0.1	<1
Total	27.8	100	39.9	100

Table 2.10 *Genetically Modified Crops Already Commercialized In USA.*

Crop	Trait Improved
Canola	Herbicide tolerance; Hybrid technology; High lauric acid
Maize	Control of corn – borer; Herbicide tolerance; Hybrid technology
Cotton	Bollworm control; Herbicide resistance
Potato	Resistance to Colorado beetle
Soybean	Herbicide tolerance; High oleic acid
Tomato	Delayed / improved ripening
Papaya	Virus resistance

Biotechnology has helped to make food more nutritious. 'Golden Rice' a genetically modified rice which has genes for increased betacarotene (a precursor for Vitamin A) and iron is produced. This rice can prevent blindness and iron deficiency anemia, in several Asian countries. A gene for higher content of protein isolated from *Amaranthus* (AMA -1) has been put into potato, which makes potatoes rich in proteins too. Another gene Aspartokinase is also known to help in making the plants to produce more proteins.

Genetic manipulation of flower and leaf colour, abundance of flowers, perfume, and shape are now targets for the decorative plant industries. In the Netherlands there is an annual market for cut flowers in excess of $US 1.5 billion.

Edible Vaccines

One of the main constraints for effective immunization programme is lack of refrigeration and cold chain facilities in developing countries. It has been possible to express surface antigen (protein responsible to trigger immune reaction) in plants. Clinically acceptable levels of immune reaction and protection was demonstrated by oral feeding of these transformed transgenic plants. Hepatitis B-surface antigen in tobacco, cholera B sub-unit in potato and banana have been successfully expressed. Such transgenic plants would eliminate the need for refrigeration of vaccine and make it available easily, even by the most interior and remote areas. In addition such vaccines do not require the need of medical personnel to administer the vaccine.

Production of vaccines by transgenic plants opens up exciting possibilities for bio-pharming of vaccines. Transgenic plant technology makes it possible to identify and remove known allergenic agents from foods. Research has reached a major milestone where problem of cyanide in cassava (*Manihot*) is being examined. The other results could be allergen free peanuts and rice etc.

Medically useful proteins have been secreted in honey by expressing these proteins with nector specific proteins.

2.6 Plantibodies (Plant Antibodies)

Discovery and development of biotechnology based drugs, known as biopharmaceuticals, has revolutionized the medical world. According to one survey, by the end of year 2000, 84 therapeutic biomolecules were marketted in US and European Union. In subsequent years, more than 60 biomolecules have been approved for therapeutic use and over 500 are in various phases of clinical evaluation. The present annual global market of biotechnology medicine is more than $ 30 billion, compared to $ 12 billion in the year 2000 (Walsh, 2003). It is estimated that molecular farming of monoclonal anitbodies (Mabs) will emerge as 100 billion dollars industry by 2020.

Plantibodies (antibodies derived from plants) are bioactive molecules that owing to their individual and specific binding properties, allow a large diversity of potential applications. These include medical diagnosis and therapy, the sensitive detection and removal of environmental contaminants, control of pathogens and industrial purification processes. Plantibodies provide an invaluable tool in fundamental research because of their ability to interfere with metabolic processes within an organism.

The concept of using plants as heterologous expression hosts for recombinant antibodies (plantibodies) or rAbs is hardly 15 year old (Hiatt *et al* 1989). The combination of antibodies and plant genetic engineering : two rapidly advancing technologies has resulted in the expression of a diversity of molecular forms in different plant species (Fisher and Emans, 2000).

Apart from various antibodies like polyclonal antibodies, monoclonal antibodies, murine monoclonal antibodies, chimeric and humanized monoclonal antibodies produced by r-DNA technology are gaining importance.

Antibodies are produced using cell fusion and genetic engineering (rDNA) technology. In the cell fusion technique, two combinations-mouse/human and human/ human are used, but this technique has the limitations of karyotypically unstable mouse/ human hybridoma cell line and availability of human myeloma cell. Recombinant DNA technique on the other hand proved to be more suitable for the production of mAbs. The main advantages of this technology are : 1. stable antibodies can be produed on a large scale, 2. antibody function can be changed by exchanging the C-regions 3. antibody affinity can be engineered by mutagenesis of variable regions and 4. *Fab* fragments of antibodies can be easily engineered.

Contemporary applications in agronomic research include immunomodulation of physiological processes and engineering of antibody mediated resistance to pathogen infection. The most advanced application apart from animals, is the utilization of plants as bioreactors to produce antibodies required for medical use or industrial processes.

Agrobacterium mediated transformation and particle bombardment have been used to introduce antibody genes into plants (Giddeings *et al*, 2000). However, particle bombardment enables the simultaneous introduction of multiple constructs, thereby expediting the recovery of transgenic lines expressing multimeric antibodies such as secretory immunoglobulin (slg A) (Larrick *et al* 2001).

The recombinant proteins are deposited throughout the plant or in specific organs (seeds). The deposition and storage of antibody molecules in seeds of various crop plants has been demonstrated by several researchers (Table 2.11).

Table 2.11 *Expression is stable through generation but gets diluted with lower gene dose*

Seed generation	Location	Gene dose	Total soluble protein (%)
R_2 inbred	Greenhouse	3	3.0
R_3 inbred	Field A	3	4.5
Hybrid	Field A	0 - 3	1.9
R_3 inbred	Field B	3	3.3
Hybrid	Field B	0 - 3	1.0

For example, high accumulation of a single chain Fr fragment (scFv) antibody in pea seeds was reported using the seed specific USP promoter by Saalbach *et al* (2001) while the high stability of an scFr antibody was confirmed in tobacco seeds over one year period by Ramirez *et al* (2001). Plant cell or organ culture in bioreactors is comparatively more expensive than agricultural production but offers advantages as rAbs can be produced in containment and under controlled conditions. Several researchers have recently reported the expression of a murine immunoglobulin (Ig) GI in hairy root cultures leading to secretion of rAb into the culture medium. (Sharp and Doran 2001) Apart from nuclear gene transfer in the transformation of organelles like cloroplast resulting in plastid expression of multimeric vaccine. Expression of recombinant proteins via the chloroplast genome has some advantages compared to nuclear gene transfer in view of high levels of expression and containment.

As shown in Fig. 2.20 there is considerable similarity in the aspargine linked glycan structure in mammals and plants as also the initial glycan transfered to the antibody although minor differences exist. This has raised much concern regarding the potential immunogencity of plant specific complex N-glycans present in plantibodies.

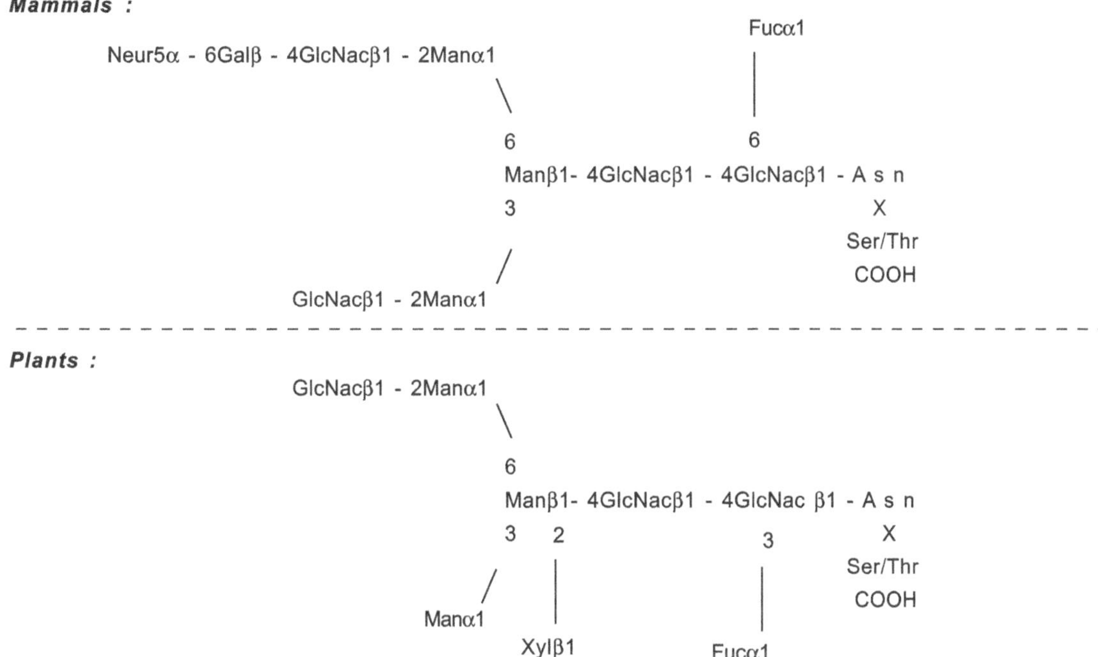

Fig. 2.20 Comparison of asparagine-linked glycan structure from mammals and plants. *key :* Fuc, Fucose, Gal, galaclose, *Glenac*-N-acetyl glucosamine : Man, mannose; Neur-neuraminic acid: *Xyl*, xylose.

Recently human β-1-4 galactosyltransferase was expressed in tobacco plants (Bakker *et al*, 2001), upon crossing these plants with those expressing a murine antibody, a plantibody with partially galactosylated N-glycan was obtained.

Plants do not have the capacity to synthesise or link neurominic acid to a glycan chain. The chain tend to end in mannose. Thus glycan and Mabs produced in plants are altered in their CMC and ADCC activities. Plants also have the capacity to add novel sugars during triming process. As shown in Fig. 2.20 fucose is added as an a-1, 3 linkage and a novel b-1, 2 xylose is added to the branch point mannose. These residues are antigenic when endogenous plant glycoproteins are used to immunise rabbits. Mabs produced in plants indicate the glycosylation is consistent with that of endogenous plant glycoproteins. The initial glycan transferred to the antibody is identical in mammalian cells and transgenic plants.

Therapeutic antibody production system should not only meet volume and cost requirement but time factor also. In corn from gene transformation to initial seed expression require 8 months and for sustainable production from a master seed bank (MSB) harvest require 18 months. Part of scaling is dependent on seed amplification (production) rate by the plant expression level and total harvest mass. An inbred corn plant yields 100-500 seeds every 4-6 month generation of 0.2g each. The variability is due to genetic line differences as well as greenhouse or field grown conditions, Tobacco yields many logs more seeds per plant, but at 11,000 seeds/g. Corn, therefore is a better production system but tobacco offers faster scaling times and is useful for a leaf derived product.

Seeds are the best source of therapeutic proteins of greater than 10kg volume, since seeds have the potential to capture the full value of plant production system, flexibility in scaling, culturing and processing. Storage and shipping of large quantities of seeds allows flexibililty not seen with other systems. Inbred lines differ in their transformation efficiency. As shown in Table 2.12 in inbred corn expression is stable across generatons and field locations but the gene dose may lower expression. The transgenic plant must be transgenic pure for only one drug product.

The details of purification are summarised in Table 2.12. The detail required at every step of this process is called good manufacturing practice (GMP). Once purified the product needs to conform to standards both at molecular and functional characterization. As shown in table using number of parameters, the purified antibody is effectively pure and homogenous. The sequence homogenety is used with the antibodies from tobacco seeds.

In view of efficiency of agricultural production systems coupled with commercialisation transgenic crops; The use of plants to replace more traditional mammalian of cell production is based on similarity of protein synthesis in both systems including post transtational steps.

Biopharmaceuticals are being increasingly developed and marketted as diagnostics and therapeutic agents. Monoclonal antibodies constitute a specific class of biomolecules, with around 210 mAbs being clinically evaluated worldwide. An update of on applications of these molecules only plantibodies in health care is summarised in Table 2.12.

Table 2.12 *End points and results of Mab purification from corn seeds*

Parameter	Initial	Final
Total gram Mab	38	23
Bio burden cfu/g Mab	15,000	0
Endotoxin units/mg Mab	220,000	< 0.1
HPLC purity	NA	> 99%

In Table 2.13 examples of pharmaceutical antibodies produced in transgenic plants is given. A perusal of data shows that several companies have chosen various seed crops and in particular corn (*Zea mays*), as a vehicle for the commercial production of pharmaceutical antibodies. The details of various antibodies, their application in medicine and company are also shown. There are four types of plantibodies for therapeutic, diagnostic, contraceptive topical oral and even vaccine. Although various plantibodies have successfully completed early phase clinical trials, regulatory guidelines and public acceptance remains to be resolved. Plants are highly competitive in terms productivity, safety, cost and timeliness.

Indeed, plantibodies represent a viable alternative to mammalian and prokaryotic expression system. Easy scale up is a major advantage of transgenic plants. Long term targets for plant bioreactors may therefore encompass high volume, low cost antibodies which do not require extensive purification.

The plantibodies have revolutionized the world of immunology as well as health care. Once perfected, patented and commercialised plantibodies will be the top class biomolecules based medicines.

Table 2.13 *Examples of Pharmaceutical antibodies produced in transgenic plants*

S. No.	Antigen	Plant	Antibody form	Application	Company/reference
1.	*Streptoccus* surface antigen SAI/II	Tobacco	I_gA/G (CaroRX)	Therapeutic (topical)	Planet Biotechnolog y. CA (Lawrick *et al* 2001 Ma *et al* 1998
2.	Herpes simplex virs	Soybean, rice	I_gG	Therapeutic (topical)	EPIcyte (Zerttin *et al.* 1998) (Briggs *et al* 2001)
3.	Non-Hodgkins lymphoma idiotypes	Tobacco, virus vector	sccFv	Vaccine	Large Scale BIology Corporation. CA (Ziegler *et al* 2000)
4.	Human IgG	Alfalfa	IgG	Diagnostic	(Khoudi *et al* 1999)
5.	CEA (carcinoembryonic antigen)	Tobacco, rice, wheat tomato	scFv	Therapeutic/diagnostic	(Vaquero *et al* 1999) (Stoger *et al* 2001)
6.	CEA (carcinoembryonic antigen)	Tobacco (transient and stable	Diabody	Therapeutic/diagnostic	(Vaquero *et al* 2000)
7.	Respiratory syncytial virus,	Corn	IgG	Therapeutic (inhaled)	EPIcyte (www.epicyte.com)
8.	*Clostridium difficile*	Corn	IgG	Therapeutic (oral)	EPIcyte (www.epicyte.com)
9.	Sperm	Corn	IgG	Therapeutic (topical)	EPIcyte (www.epicyte.com)
10.	Various	Corn	IgG	Therapeutic/diagnostic	Integrated Protein (Technologies c/o Monsanto, MO (Baez *et al* 2000)
11.	Colon cancer antigen	Tobacco	IgG	Therapeutic/diagnostic	(Verch *et al* 1998)

2.7 Phytoremediation

Biotechnology is transforming world agriculture, adding new traits to crop plants at a greatly accelerated rate. Plants are becoming more efficient producers of food, fiber, medicines, and construction materials.

Alternative agriculture, which expands the use of plants well beyond food and fiber, is beginning to change plant biology. Two plant-based biotechnologies were recently developed that take advantage of the ability of plant roots to absorb or secrete various substances. They are (i) **phytoextraction**, the use of plants to remove pollutants from the environment and (ii) **rhizosecretion**, a subset of molecular farming, designed to produce and secrete valuable natural products and recombinant proteins from roots.

With the exception of root crops, plant roots in general are less utilized and less studied than shoots. However, this situation may be changing because of the emerging biotechnologies described below that exploit the ability of plant to transport valuable molecules into and out of their roots.

Throughout, arthropomorphic activities, such as mining, agricultural activities, oil and explosives have been the sources of contamination of much of the world's soil and water. With increase in public opinion about the state of the environment, the process of remediating these areas has increased during the last few decades. Methods have now been established to handle many of the pollutants, but due to the unique nature of metal in soil and water, few techniques have been adopted to clean-up these areas. One possible method to remove the metal contaminant from a soil or water system is with the use of the treatment process known as phytoremediation.

Definition

Phytoremediation combines the Greek word "phyton", "plant," with Latin word "remediare", "to remedy," to describe a system whereby certain plants, working together with soil organisms, can transform contaminants into harmless and often, valuable forms. This practice of phytoremediation technology is increasingly used to organic compounds.

Phytoremediation is an intriguing process for the biologically promoted removal of heavy-metal ions from wastewaters. *Phytoremediation is broadly defined as the use of living green plants to remove, contain, or render harmless environmental contaminants.* Phytoremediation processes are capable of removing heavy metals directly from contaminated soil or from wastewater. Phytoextraction is the use of plants to remove heavy metals from soils, whereas rhizofiltration is the use of plants to remove heavy-metal ions from aqueous streams. A general scheme for the application of phytoremediation technology to metal contaminated soils is illustrated in Fig. 2.21.

Phytoremediation, the process of using green plants based systems to remove or stabilize a contaminant in soil or water, provides a low cost, effective, and non-invasive method to deal with a contaminated area.

Compared to most traditional remediation and engineering techniques, it is a fledging technology that addresses a wide variety of surfacial contaminants.

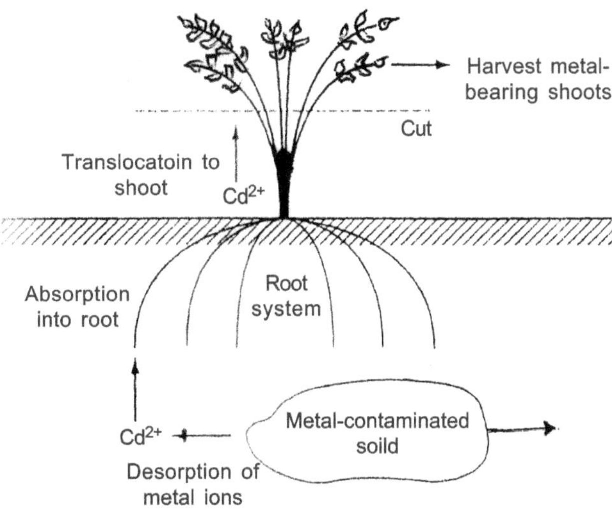

Fig. 2.21 General concept of phytoremediation for removal of heavy metals from contaminated soil and groundwater.

Remediation can imply : (a) clean-up where either the contaminant is removed from the matrix (leaching, bioremediation, etc.) or the entire contaminated matrix is removed from the site (excavation and landfilling), or (b) stabilization where the physical or chemical form of the matrix or contaminant is transformed to a more inert condition. Stabilization and biological activity are the basis of plant-based approaches for the cleanup of inorganic and organic environmental contaminants. Phytoremediation uses a plant's natural ability to take in water and nutrients through roots, evaporate water through leaves, and act as a natural reactor to metabolize organic compounds like oil and pesticides or to absorb and concentrate toxic trace elements such as lead and selenium.

A new and rapidly developing form of bioremediation is *phytoremediation*, which uses green plants and their associated biota to destroy, remove, contain, or otherwise detoxify environmental contaminants (Cunningham, et al 1996). Five types of destruction and removal reactions can occur in phytoremediation. Definitions of the five pathways are :

1. *Phytovolatilization* is an enhancement of the volatilization process from the soil or through the plant's roots or shoots. Enhanced volatilization can occur via plant transpiration of volatile compounds or transformation of contaminants to more volatile forms.
2. *Phytodegradation* involves uptake of organic contaminants by the plant and subsequent metabolism by plant enzymes to form benign (non toxic) products.
3. *Phytoaccumulation* phytoextraction involves uptake of organic contaminants by the plant and absorption of the contaminant into plant tissue (bio mass) which subsequently is harvested and disposed or metal recovery through smetting and organic substances. Hydrophobic contaminants are most susceptible to phytoextraction.
4. *Microorganism Stimulation Rhizo (sphere) degradation* to benign products is catalyzed by plant enzymes excreted by the roots or by the microorganisms found in the rhizosphere.

5. *Phytostabilization :* Plants prevent contaminants from migrating by reducing runoff, surface erosion, and ground-water flow rates. "Hydraulic pumping" can occur when tree roots reach ground water, take up large amounts of water, control the hydraulic gradient, and prevent lateral migration of contaminants within a ground water zone. The benefit of phytostabilization is that it makes the contaminants less bioavailable to humans and other receptors. Three forms of phytostabilization can be identified :
   (i) Humification incorporates contaminants into the soil organic matter, or humus, through binding (polymerization) reactions of plant and microbial enzymes.
   (ii) Lignification incorporates the contaminant into plant cell-wall constituents.
   (iii) Aging slowly binds the compounds into the soil mineral fraction.

The distinctions among the various stabilization, destruction, and removal mechanisms often are not sharp. For example, rhizosphere peroxidases can catalyze oxidation reactions that lead to contaminant mineralization (degradation) or to polymerization (humificaction).

The basic mechanism of phytoremediation for removal of heavy metals involves two steps : (i) active uptake of heavy metal ions by the root system, and (ii) translocation of the heavy-metal ions to green shoots. Plants which can sequester high concentrations of heavy-metal ions are called hyperaccumulator metallophytes. For example, the hyperaccumulator ions, including Ni, Zn, and Cd. In this plant, over 18,000 mg of Zinc and 1000 mg of Cd per kg of shoot biomass can be accumulated without inhibition of growth. The exact mechanisms for active uptake for heavy-metal ions are under debate, but could involve the use of biologically expressed phytochelatins that complex with heavy-metal ions and thus project the plant's biochemical machinery from the toxic heavy-metal ions. Genetic engineering offers a means to improve growth rates of hyperaccumulators or introduce heavy-metal remediation activity into plants. For example, the bacterial gene for mercuric reductase, which reduces Hg^{2+} to Hg^0, was recently cloned into the *Arabidopsis* plant.

Configuration of engineered reef-bed systems and constructed wetland systems for removal of heavy-metal ions from wastewaters using phytoremediation is illustrated in Fig. 2.22. The root system of the hyperaccumulator plant penetrates a permeable rock bed. The waste water is introduced into one end of the bed and flows through the permeable rock layer. The rock layer should be inert to heavy-metal ion binding so that it does not unwittingly serve as sink for heavy metals. Heavy-metal ions are sequestered by the root system and translocated to the shoots. The treated effluent is discharged. Periodically, the metal-containing shoots are harvested. A scheme for the post harvesting of the metal-laden biomass for disposal or recovery of metals is shown in Fig. 2.23. The biomass can be burnt off or composed to yield a low volume of metal-rich ash.

In the last 20 years much progress has been made in the clean up of contaminated soils, sediments and ground waters but existing methods have proven to be either costly or limited in various ways.

The unexpected pollutant threat pose new problems as such as MTBE problem (Methyl Tertiary Butyl Ether) of ground water. In the early 1990 the use of phytoremediation i.e. the use of plants to remediate environmental toxicity emerged with promises of significant economies. Drawing upon the contaminated with metal (Nickel) the use of plants to extract and accmulate toxic heavy metals was proposed. It was proposed that toxic organic compounds might be degraded by the action of microorganisms peculiar to the rhizosphere of plants. This technique is potentially economically beneficial and equally impressive. Plants are delicate to robust and solar powered. Their root systems

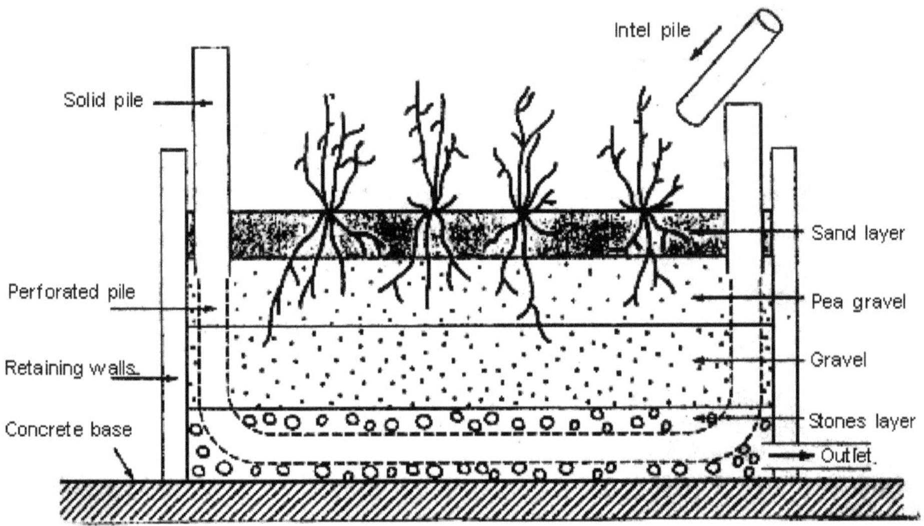

Fig. 2.22 Sample configuration of an engineered reed-bed system for phytoremediation of heavy-metal ions from wastewater.

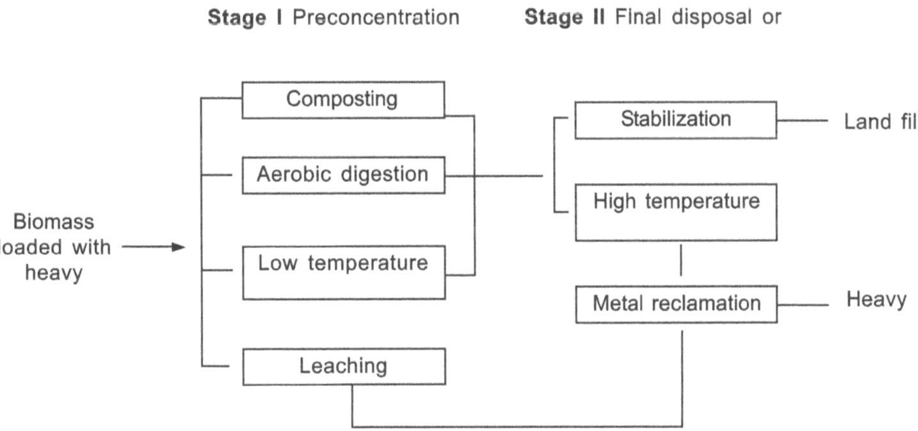

Fig. 2.23 Strategies for concentrating biomass or carbonaceous adsorbents loaded with heavy metals prior to metals reclamation or disposal.

permeate the soil and sediment environment with an extensive and active membrane system. The soil near the roots has microbial populations orders of magnitude greater than non root soil. These benefits are provided with little or no maintenance requirements. Furthermore, plant based systems are welcomed by the public due to their inherent and superior aesthetics and the societal and environmental benefits that their presence provides.

Valuable metal-resistance traits can be found in metal hyperaccumulating plants that are endemic to soils naturally enriched with heavy metals. These plants can accumulate exceedingly high amounts of essential and nonessential heavy metals in their foliage, to levels that are highly toxic to most other plants. For example, several *Thlaspi species* can accumulate Ni and Zn, to 1-5% of its dry biomass. This is an order of magnitude greater than concentrations of these metals in the nonaccumulating plants growing nearby. The prevention of herbivory and disease is thought to be the main function of this unique phenomenon. It has recently been established that the ability of *T. goesingenese* to hyperaccumulate metals is the result of high resistance to the metals rather than the greater rates of metal uptake. Unfortunately, most hyperaccumulating species are not suitable for phytoextraction for several reasons : (i) metals that are primarily accumulated (Ni, Zn, and Cu) are not among the most important environmental pollutants; (ii) most plants have very low biomass and carpricious growth habits therefore unsuitable for monoculture; and (iii) agronomic practices and crop protection measures for their cultivation have not been developed. However, many metal-hyperaccumulating species belong to Brassicaceae (mustard) family, and are related to *Brassica juncea*, the preferred plant for phytoextraction of lead. Unfortunately, *Brassica juncea*, while exhibiting a high capacity for metal uptake and translocation, is not very resistant to high levels of lead or other heavy metals in its foliage. Therefore, chelate-assisted phytoextraction is toxic to *B. juncea*, requiring harvesting several days after chelate application.

In the last 8 years it has become clear that phytoremediation using conventional plants has some potential for use in remediation but at the same time encountered by limitations that prevent widespredad benefits from its significant economies. Plants that accumulate metals to level high enough to make harvesting and metal recovery economical are hampered by very small biomasses. The most important heavy metals are immobilised in soils and taken up poorly by known metal accumulating plants. Experiments with enhanced rhizosphere microbial degradation of toxic organism such as polyaromatic hydrocarbons have yielded only small increases in degradation rate and marginal improvements in the ultimate residual pollution levels that can be achieved. Studies using deep root trees have shown that Chlorinated Hydrocarbons (CHCs) such as premier ground water pollutant, Tri Chloro Ethylene (TCE) are taken up by trees and are also degraded in tree plantations but degradation rates are low and large areas are required for such plantations. Therefore phytoremediation is largely a major niche technology and cannot play a major role in providing the order of magnitude cut in remediation costs needed.

The potential of genetic engineering for enhancing the biodegradation of xeonbiotic toxic compounds has been recognised since the early 1980's. Initially it was hoped that transgenic bacteria would provide powerful methods for remediation of environmental problems. However this approach has failed due to two factors. 1. inconsistent survival of genetically modified microorganism in the solid and water environments and 2. reluctance by regulatory authorities to approve the release of genetically modified microbes in the environment. This is due to the fact that due to high growth rate and small size of bacteria and the horizontal gene transfer among dissimilar prokaryotes. These problems can be overcome by using plants rather than microorganism as the delivery system for genetically engineered environmental cleanup biosystems. Primarily by virtue of their size, the vascular plants are more easily controlled than microorganism. By selecting sterile uncontrolled genetic releases from phytoremediation plantations can be prevented. Thus, by using plants, the promise of genetically engineered environmental clean up systems can be realised. During the beginning of this century progress has been made towards this goal and three types of genetic modifications of plants for phytoremediation have been reported.

1. Transfer of Mercuric reductase genes from bacteria into plants results in plants that can transfer the toxic mercuric ions to the much less toxic elemental mercury which can then be volatilsed and diluted into the atmosphere, avoiding the more dangerous potential of exposure by soil contact.
2. Bacterial genes for removal of nitroso groups from explosive compounds have been transferred into the plants and used to detoxify Tri Nitro Toluene (TNT).
3. Mammalian genes for Cytochrome P4502E1, the liver enzyme responsible for activating TCE, have been expressed in plants resulting in an increase in TCE oxidation by two orders of magnitude.

These above examples provide ample evidence of the potential of genetically modified plant for bringing the promise of phytoremediation to fruition. However, much work remains to be done. The problem of heavy metals uptake and accumulation in useful levels for phytoremediation remains for organic phytoremediation to be useful a wide variety of trees must be transformed and activities optimised.

How much time required for Phytoremediation

The times it takes to clean up a site using phytoremediation depends on several factors :

* Type and number of plants being used
* Type and amounts of harmful chemicals present
* Size and depth of the polluted area
* Type of soil and conditions present.

These factors vary from site to site. Plants may have to be replaced if they are destroyed by bad weather or animals. This adds time to the cleanup. Often it takes many years to clean up a site with phytoremediation.

Advantages of Phytoremediation

Phytoremediation processes have two major advantages. First, phytoremediation is an environmentally friendly process for on-site remediation of heavy metals. The only potential drawback is that the species of hyperaccumulator plant introduced to the contaminant site may not be endogenous to the local ecosystem. Second, phytoremediation processes are simple in configuration, require minimal chemical or energy input, and are inexpensive to operate. Further, like many biologically based processes, they are difficult to control.

Phytoremediation is an affordable technology that is most useful when contaminants are within the root zone of the plants (top 3-6 ft.). For sites with contamination spread over a large area, phytoremediation may be the only economically feasible technology. The process is relatively inexpensive because it uses the same equipment and supplies routinely used in agriculture.

When poplars trees are used, the idea is to move as much water through them as possible, so that they take up as much of the contaminants as possible. Once the heavy metals are absorbed, they are sequestered in the tree's roots. Any organic compounds that are absorbed are metabolized.

Absorption of large amounts of nutrients by plants and only a small amount of plant toxins that might be harmful to them, is the key factor. Plants generally absorb large amounts of elements they need for growth and only small amounts of toxic elements that could harm them. Therefore, phytoremediation is a cost-effective alternative to conventional remediation methods. Cleaning the top 6"(15 cm) of contaminated soil with phytoremediation costs an estimated \$2,500-\$15,000 per hectare (2.5 acres), compared to \$7,5000-\$20,000 per hectare for on-site microbial remediation. If the soil is moved, the costs escalate, but phytoremediation costs are still far below those of traditional remediation methods, such as stripping the contaminants from the soil using physical, chemical, or thermal processes, according to Dr. Scott Cunningham, of Dupont Central Research for Environmental Biotechnology, USA.

Plants are effective at remediating soils contaminated with organic chemical wastes, such as solvents, petrochemicals, wood preservatives, explosives, and pesticides. The conventional technology for soil clean-up is to remove the soil and isolate it in a *hazardous waste* landfill or incinerate it.

Phytoremediation, according, to Dr. Ray Hinchman, of Argonne National Laboratory, USA is "an *in situ* approach," not reliant on the transport of contaminated material to other sites. Organic contaminants are, in many cases, completely destroyed (converted to CO_2 and H_2O) rather than simply immobilized or stored.

Phytoremediation takes advantage of plants nutrient utilization processes to take in water and nutrients through roots, transpire water through leaves, and act as a transformation system to metabolize organic compounds, such as oil and *pesticides*. Or they may absorb and bioaccumulate toxic trace elements, such as the heavy *metals, Lead, Cadmium*, and Selenium. In some cases, plants contain 1,000 times more metal than the soil in which they grow. Heavy metals are closely related to the elements plants use for growth. In many cases, the plants cannot tell the difference, according to Dr. Ilya Raskin, Professor of Plant Sciences, Center for Agricultural Molecular Biology at Rutgers University, USA.

Another aspect of phytoremediation is hydraulic control. During their growing season, plants transpire large volumes of water to the atmosphere. This transpiration can stop the downward flow of precipitation from the vadose zone to the saturated zone. It also can "pump" water from the top of the saturated zone to the atmosphere. Transpiration is being exploited as a means to stop or slow the migration of contaminants that are in the vadose zone or the boundary between the vadose and saturated zones. Of course, this pumping action only occurs when the plants are active in photosynthesis. For example, deciduous trees are ineffective in the winter.

Phytoremediation is still in its early stages of research and development. Nonetheless, it has promise for the decontamination of petroleum hydrocarbons, a range of chlorinated aliphatics and aromatics (including PCBs), and pesticides (Cunningham *et al.*, 1996). Phytoextraction of heavy metal is another promising application (Moffat, 1995).

Dr. Ilya Raskin and his interdisciplinary team at Rutgers AgBiotech Centre were the first to demonstrate the utility of certain species of Brassicaceae in removing such metals as chromium, lead, cadmium, and zinc from contaminated soil. Related technology developed by Raskin's group used hydroponic plant cultures to remove toxic metals from aqueous waste streams.

Plants can accelerate bioremediation in surface soils by their ability to stimulate soil microorganisms through the release of nutrients from and the transport of oxygen to their roots. The zone of soil closely associated with the plant root, the rhizosphere, has much higher numbers of metabolically active microorganisms than unplanted soil. The rhizosphere is a zone of increased microbial activity and biomass at the root-soil interface that is under the interface of the plant roots. It is this symbiotic relationship between soil microbes that is responsible for the accelerated degradation of soil contaminants.

The interaction between plants and microbial communities in the rhizosphere is complex and has evolved to the mutual benefit of both organisms. Plants sustain large microbial populations in the rhizosphere by secreting substances such as carbohydrates and amino acids through root cells and by sloughing root epidermal cells. Also, root cells secrete mucilage, a gelatinous substance that is a lubricant for root penetration through the soil during growth. Using this supply of nutrients, soil microorganisms proliferate to form the plant rhizosphere. (Table 2.14)

Table 2.14 *Media, contaminants and typical plants for the types of phytoremediation.*

Application	Media	Contaminants	Typical Plants
Phytovolatization	Soil, groundwater Landfill leachate, land application of wastewater	Herbicides (atrazine, alachlor); Aromatics (BTEX); Chlorinated aliphatics (TCE); Nutrients; Ammunition wastes (TNT, RDX)	Phreatophyte trees (poplar, willow, cottonwood, aspen); Grasses (rye, Bermuda, sorghum, fescue); Legumes (clover, alfalfa, cowpeas)
Microorganism stimulation	Soil, sediments Land application of waste water	Organic contaminants (pesticides aromatic, and polynuclear aromatic hydrocarbons	Phenolics releasers (malberry, apple, osage) Grasses with fribous roots (rye, fescue, bermuda); Aquatic plants for sediments
Phytostabilization	Soil, sediments	Metals (Pb, Cd, Zn, As, Cu, Cr, Se, U), Hydro-phobic Organics (PAH, PCB, DDT, dieldrin)	Phreatophyte trees to transpire large amounts of water (hydraulic Grasses to stabilize soil erosion; Dense root systems are needed to sorb/bind contaminants
Phytoaccumulation /extraction	Soil, Brownfields, sediments	Metals (Pb, Cd, Zn, As, Cu, Cr, Se, U) with EDTA addition for Pb, Selenium	Sunflowers; Indian Mustard; Rape seed plants; Barley, Hops; Crucifers; Serpentine plants; Nettles; dandelions
Degradation	Soil, groundwater, Landfill leachate, land application of wastewater	Herbicides (atrazine, alachlor); Aromatics (BTEX); Chlorinated aliphatics (TCE); Nutrients Ammunition waste (TNT)	Phreatophyte trees (poplar, willow, cotton-wood, aspen); Grasses (rye, Bermuda, sorghum, fescue); Legumes (clover, alfalfa, cowpeas)

In addition to the rhizosphere effect, plants themselves are able to passively take up a wide range of organic wastes from soil through their roots. One of the more important roles of soil microorganisms is the decomposition of organic residues with the release of plant nutrient elements such as *carbon, nitrogen, potassium*, *phosphate*, and *sulfur*. A significant amount of the *carbon dioxide* (CO_2) in the atmosphere is utilized for organic matter synthesis primarily through *photosynthesis*. This transformation of carbon dioxide and the subsequent sequestering of the carbon as root biomass contributes to balancing the effect of burning *fossil fuels* on global warming or cooling.

Table 2.15 *Summary of the main advantages and disadvantages of phytoremediation.*

S.No.	Advantages	Disadvantages
1.	Works on a variety of organic and inorganic compounds	May take several years to remediate
2.	Can be either In situ/ ex situ	May depend on climatic conditions
3.	Easy to implement and maintain contamination within rooting zone	Restricted to sites with shallow
4.	Low-cost compared to other treatment methods	Harvested biomass from phytoextraction may be classified as a RCRA hazardous waste
5.	Environmentally friendly and aesthetically pleasing to the public	Consumption of contaminated plant tissue is also a concern.
6.	Reduces the amount wastes to be landfilled	Possible effect on the food chain

Environmental Protection Agency, (EPA), uses phytoremediation because it takes advantage of natural plant processes. It requires less equipment and labour than other methods since plants do most of the work. Trees and plants can make a site more attractive as well. The site can be cleaned up without removing polluted soil or pumping polluted groundwater. This allows workers to avoid contact with harmful chemicals. Phytoremediation has been successfully tested in many locations, and is being used at several Superfund sites in USA. EPA uses many methods to clean up pollution at Superfund and other sites since phytoremediation, is considered new and equally *innovative* method.

One major concern with phytoremediation is the possible affects on the food chain. For example vegetation is used that absorbs toxic or heavy metals and moles or voles eat the metal contaminated plants. The predators of the moles or voles then become victims of intoxication. Although the possibilities of such scenarios are being looked at, more fieldwork and analysis is necessary to understand the possible effects phytoremediation can have.

Dr. David E Salt, Associate Professor of plant molecular physiology and principal investigator, says, "This is really one of the first tools that we have got to manipulate this process of metal hyper-accumulation," he says. "So what we are going to do now is to start expressing these genes in non-accumulating plants to see if we can turn them into metal-accumulating plants". Besides mustard greens, pumpkin vines were used to clean up an old Magic Marker factory site in Trenton, New Jersey. Hydroponically grown sunflowers were used to absorb radioactive metals near the *Chernobyl* nuclear site in the Ukraine, Russia and a Uranium plant in Ohio, USA. The mustard's hyper-accumulation results much less material for disposal. The composting of plant material can be another highly efficient stage in the breakdown of contaminants removed from the soil. (Table 2.15)

Regulatory Issues

As of now phytoremediation is too new a technology to be approved by regulatory agencies such as the (Environmental Protection Agency, USA). Eventually the main question that regulators will focus on is; will phytoremediation remediate the site to the standards and reduce the risk to human health and the environment. In developing regulations for phytoremediation the following questions will need to be answered.

* Can it clean up the site below action levels if so on what scale ?
* Does it create any toxic intermediate or products ?
* Is it cost effective as alternative methods ?
* Does the public accept this rather simple technology ?

Conclusions

Neither phytoextraction nor rhizosecretion will directly contribute to feeding world population in the next century. However, these technologies will improve the quality of life for many people if their development continues. The future challenge for metal phytoextraction is to further reduce the cost and increase the spectrum of metals amenable to this technology. This goal can be achieved by creating superior plant varieties for phytoextraction by using genetic engineering to introduce valuable traits into plants, developing better agronomic protocols for their cultivation, and designing safer and more effective soil amendments. A recent, and probably the only, example of the successful use of genetic engineering applied to metal phytoremediation is the use of bacterial mercuric reductase (merA) gene to achieve mercuric ion reduction in transgenic *Arabidopsis* and yellow poplar plants. Elemental mercury produced in transgenic plants is much less toxic than ionic mercury and can be volatilized from transgenic plants by the process of phytovolatilization.

Phytoextraction and rhizosecretion are begining to change the scenario, while allowing scientists to take a radically new look at the dark corners of plant biology because roots are underground. These technologies will open the doors to the value-added, nonagricultural uses of plants in future.

2.8 Medicinal Plants

The cultivation of medicinal plants for food and cure date back to the late Mesolithic to early Neolithic age roughly about 10,000 BC. Assyrians recorded 250 medicinal plants and Sumerians 1000 plants. Civilizations of ancient Egypt, Greece, Arabia, Europe, Africa, North America and Latin America recorded the use of medicinal plants in their traditional medicines. The Chinese were the pioneers in using plants as valuable source of home medicines. No wonder etheno-botanical medicine has been in use in China for over 5000 years. Emperor Shen Nung studied medicinal plants and verified their pharmaceutical properties and over 11,000 herbal remedies were developed and used in China for thousands of years.

The people of India from time immemorial have been using various medicinal plants for curing various diseases and to ward off pests. Charaka Samhita (900 BC) mentions 395 medicinal plants and Susruta Samhita added invaluable knowledge to the science of medicinal plants. But the folk knowledge system, as well as the more formal Ayurvedic system, is said to be not standardised as the modern system of biological nomenclature.

In India, the earliest reference to the medicinal value of plants appears in Rigveda (4500-1500 BC) which lists 67 plants of therapeutic values; with brief references to the healing property of plants like simal, palas, peepal etc. Likewise, Yajurveda lists 81 medicinal plants followed by Atharvanaveda which mentions 290 medicinal plants. However, it is in the Ayurveda that definite properties and uses of drugs and drug yielding plants have been discussed in detail. There is no plant, our sages say, devoid of medicinal value. So when Rama and Lakshmana were badly wounded during the war with Ravana; Jambuvanta, the chief medical adviser to Rama's army, ordered '*sanjeevani*' and three other herbs from the Oushadhagiri, the medicine mountain. Hanuman offered to fetch them, as he was not a botanist, instead of risking choosing the wrong herbs, he carried back the entire mountain. Oushadhagiri probably was the same size as Nilgiri, of South India. If so, it might have supported about 4,000 types of flowering plants, in addition to algae, fungi, lichens, and ferns.

Ancient Indian medical heritage is based on 7000 plant species and about 8000 herbal remedis have been codified in the Ayurveda. Plants in Ayurveda have Sanskrit names and synonyms and each plant may have synonyms ranging from five to hundred. There is no good publication today which has listed out all the synonyms that are mentioned in the Ayurvedic texts and grouped them according to the plants they refer to. The exact number of plants mentioned in Ayurvedic literature is not correctly known. But the three great works of Ayurveda – Charaka Samhita, Sushrut Samhita and Ashtanga Hrdaya; mention between them 19,000 plant names referring to around 600 distinct species of 4% of the total number of species in folk use.

Aristotle (350 BC) described number of drug plants. Pedanius Dioscorides a Greek botanist of 1st century AD in his famous book '*De Materia Medica*' written in Greek language described 600 kinds of plants with authoritative comments not only on their habitat but their medicinal properties. He first classified plants under the heading as aromatic, culinary and medicinal plants. This illustrated herbal of medicinal plants remained the last word on Medical Botany upto 1500 AD.

In later centuries, the doctrine of signature was advocated, which reasoned that plants medicinally useful to man, possessed certain forms and shapes which had close resemblance to the shapes of human body organs and could be used for their treatment. For example walnut seeds show superficial resemblence to human brain and, therefore, considered to be a brain tonic etc.

The period of Renaissance (16th and 17th centuries) is called the era of Herbalists; notably Brunfels, Fuchs and Bobel who have described various medicinal plants in their herbals. It is common to see medical gardens in several British Universities (Cambridge, Oxford and Birmingham) even now.

People in every society depend on plants for staying healthy and improving their quality of life. According to WHO 3.5 billion people in developing countries use plant based medicines for their primary health care. WHO further indicates that over 30% of world's plant species have at one time or another been used for medicinal purpose.

The only well-documented, somewhat limited, listing of medicinal plants is available in modern systems of medicine like allopathy and homeopathy: the allopathy list of Indian medicinal plants is less than 50. Homeopathy uses 571 medicinal plants but these include many exotic species.

India occupies the top position in the use of herbal drugs and also enjoys unique position in exporting plant drugs and their derivatives largely because of the following factors :

1. Its huge diverse flora spread over the entire Indian sub-continent
2. Ayurveda, Unani and Homeo system of medicines dwell heavily on medicinal plants.
3. Scientific cultivation of medicinal plants and phytochemical research
4. Well developed pharmaceutical industry.
5. Rapidly growing phytochemical and herbal drug industry.

With the advent of organic chemistry, several synthetic drugs flooded the market resulting in decline of natural medicines from plants. Thanks to the efforts of several NGO; and other research organisations there is welcome revival of interest in native herbal medicines and medicinal plant at large.

During the last 15-20 years medicinal plants have attracted a lot of attention and number of conferences, seminars, workshops and books have been published focusing the various aspects of medicinal plants.

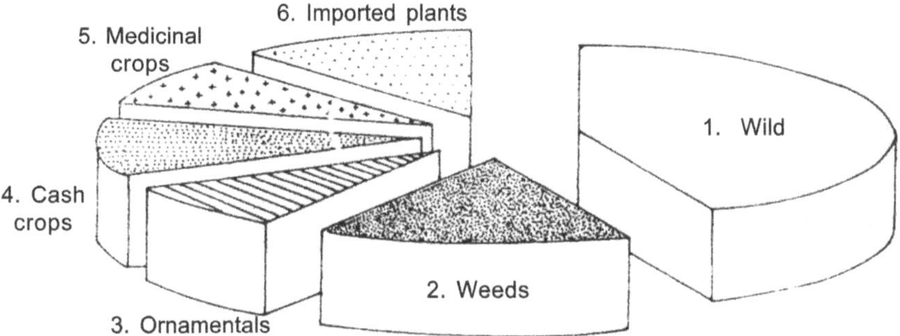

Fig. 2.24 Plant species providing medicinal raw materials to Indian Drug and Pharmaceutical Industry (by origin)

Characteristics of medicinal plants

Medicinal plants represents a wide variety of plant groups ranging from algae, fungi, lichens, and flowering plants and represents various life forms, herbs (50%), shrubs (20%), climbers (10%) and trees (35%).

The drug-yielding medicinal plants are generally classified on the basis of the morphology of the plant part from which the drug is extracted. These are as follows :

(i) Drugs obtained from roots
(ii) Drug obtained from rhizome
(iii) Drugs obtained from bark
(iv) Drugs obtained from stem and wood
(v) Drugs obtained from leaves
(vi) Drugs obtained from flowers
(vii) Drugs obtained from fruits
(viii) Drugs obtained from seed parts
(ix) Drugs obtained from gum-resins and
(x) Drugs obtained from the entire plant

Majority of the drug plants found in the tropics grow in wild condition and are mainly used by herb doctors and ayurvedic vaids, who refer to them as *jaributis*. The medicinal value of the drug plants is due to the presence of specific chemical substanes present in them, which when consumed in prescribed doses, produces physiological action and provide relief to the human body.

Some of these important compounds include alkaloids, glucosides, resins, gums, mucilages, tannins, essential and fatty oils, compounds of carbon, hydrogen, oxygen and nitrogen etc. compounds which are generally present in the roots, stems, bark, leaves, seeds etc.

The plants used for medicinal properties range from leaves, stem, rhizomes, bark, flowers, seed etc. (Fig. 2.25) and depending upon the requirement, the various plant parts are harvested safely to enable the plant to regenerate. Plant based forest management requires skills. The number of plants with confirmed therapeutic properties with a clinically useful chemical compound is around 700 species of these 50% plants are regularly and largely used as raw materials by Indian drug and pharmaceuticals industry. This also includes the raw materials imported from other countries such as ephedra, liquories, henbane, galangal, cassia bark and long pepper.

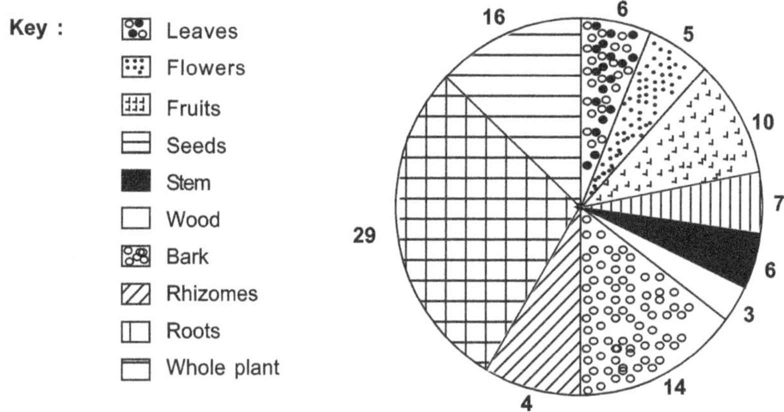

Fig. 2.25 Plant part used for medicinal properties (%)

Table 2.16 *Cardiovascular plants (Cardiac glycosides)*

S.No	Name	Family	Important chemicals
1.	*Digitalis purpurea*	Scrophulariaceae	Digoxin, Digitoxin, Lanatoside
2.	*Thevetia nerifolia*	Apocynaceae	Thevetin
3.	*Coleus forskolii*	Labiatae	Colenol
4.	*Podophyllum sp.*	Berberidaceae	Podophyllotoxin, Podophyllin
5.	*Urgenia scilla*	Liliaceae	Scillaren
6.	*Convallaria majalis*	Liliaceae	Convallotoxin
7.	*Acokanthera schimperi*	Apocyanaceae	Onabain
8.	*Ginkgo biloba*	Ginkgoaceae	

Table 2.17 *Antidiabetic plants*

S.No	Name	Family
1.	*Salacia oblonga*	Hippocrateaceae
2.	*Gymnema sylvestre*	Asclepiadaceae
3.	*Syzygium jambolana*	Myrtaceae
4.	*Pterocarpus marsupium*	Fabaceae
5.	*Momordica charantia*	Cucurbitaceae
6.	*Trigonella foenum-graceum*	Fabaceae
7.	*Acokanthera schimperi*	Apocyanaceae

Table 2.18 *Select list of Medicinal plants based on plant part used*

S.No	Plant part		Common name	Scientific name	Family	Cure
1.	I	Roots	Sarpagandha	*Rauvolfia serpentina*	Apocynaceae	Blood pressure, schizophrenia
2.		Roots	Ashwagandha	*Withania somnifera*	Solanaceae	Sedative, Transquiliser
3.		Roots	Liquorice	*Glycirrhiza glabra*	Fabaceae	Cough, Bronchitis
4.		Roots	Ginseng	*Panax ginseng*	Araliaceae	Antistress,antifatigue antianging
5.	II	Rhizome	Yam	*Dioscorea florifunda*	Diocoreaceae	Contraceptive
6.		Rhizome	Sweet flag	*Acorus calamus*	Zingiberaceae	Cold cough, dyspepsia
7.		Rhizome	Turmeric	*Curcuma longa*	Zingiberaceae	Antimicrobial, antiseptic
8.		Rhizome	Ginger	*Zingiber officinale*	Zingiberaceae	Antimicrobial, antiseptic
9.	III	Bulb	Autumn crocus	*Colchicum autumnale*	Liliaceae	Anticancerous
10.		Bulb	Garlic	*Allium sativum*	Liliaceae	Antitumor, hypolipidimic
11.		Bulb	Onion	*Allium cepa*	Liliaceae	Blood purifier
12.	IV	Bark	Cinchona	*Cinchona officinalis*	Rubiaceae	Antimal arial
13.		Bark	Arjuna	*Terminalia arjuna*	Combretaceae	Cardiotonic
14.		Bark	Yew	*Taxus brevifolia*	Taxaceae	Anticancerous
15.		Bark	Dalchini	*Cinnamomum zeylanium*	Lauraceae	Cold, asthma
16.	V	Stem	Ephedra	*Ephedra gerardiana*	Ephedraceae	Carminative, asthma
17.		Stem	Pomegranate	*Punica granatum*	Punicaceae	Anthelmentic
18.		Stem	Miswak	*Salvodora persica*	Salvodoraceae	Dental diseases
19.	VI	Leaves	Foxglove	*Digitalis purpurea*	Scrophulariaceae	Cardiac stimulant
20.		Leaves	Brahmi	*Centella asiatica*	Apiaeceae	Memory enhancer
21.		Leaves	Brahmi	*Bacopa monnieri*	Scrophulariaceae	Memory enhancer
22.		Leaves	Ginkgo	*Gingko biloba*	Ginkgoaceae	Cardiovascular
23.		Leaves	Madhumashini	*Gymnema sylvstre*	Asclipiadaceae	Antidiatbetic
24.		Leaves	Antamul	*Tylophora asthamatica*	Asclipiadaeae	Anti Asthamatic
25.	VII	Flowers	Pyrethrum	*Chrysanthemum cinerarifolium*	Asteraceae	Antimalarial

Table 2.18 Contd...

S.No	Plant part	Common name	Scientific name	Family	Cure
26.	Flowers	Santonine	*Artemesia annua*	Asteraceae	Antimalarial
27.	Flowers	Clove, (lavang)	*Syzygium aromaticum*	Apiaceae	Cough, brouchitis, gum diseases
28.	VIII Fruits	Senna	*Cassia angustifolia*	Coesalpinaceae	Antihelmenthetic
29.	Fruits	Poppy	*Papaver somniferum*	Papaveraceae	Sedative
30.	Fruits	Carawy	*Trachysperum ammi*	Apiceae	Expectorant
31.	Fruits	Amla	*Phyllanthus emblica*	Euphorbiaceae	antidiabetic, antidysentric.
32.	IX Seeds	Vishamushti	*Strychnos nux-vomica*	Loganiaceae	
33.	Seeds	Jimson weed	*Datura stramonium*	Solanaceae	Indigestion, flatulence
34.	Seeds		*Solanum viarum*	Solanaceae	
35.	Seeds	Methi	*Trigonella foenum graceum*	Fabaceae	Antidiabetic
36.	X Seed husk	Isabgol/psyllium	*Plantago ovata*	Plantaginaeae	Constipation
37.	XI Seed aril	Nutmeg	*Myristica fragrens*	Lauraceae	Digestive, thermogenic
38.	XII Seed hairs		*Mucuna prureata*	Fabaceae	Parkinson's disease cure
39.	XII Stigme/style	Saffron	*Crocus sativus*	Iridaceae	Bronchitis, diuretic
40.	XIV whole plant	Chirayata	*Swertia chirata*	Gentianaceae	
41.	whole plant	Kalmegh	*Andropgraphis paniculata*	Acanthaceae	
42.	XV Resin	Gogul	*Commiphora wightii*	Bursuraceae	Nervous and bone diseases
43.	Resin	Hing	*Ferula asafoetida*	Apiaceae	Carminative, expectorant
44.	Resin		*Agathis sp.*	Pinaceae	Medicinally useful
45.	Resin		*Guaicum officinale*	Zygophyllaceae	Medicinally useful
46.	Resin	Roal	*Shorea robusta*	Dipterocarpeae	
47.	XVI Gum	Gum arabic	*Acacia nilotica*	Mimosae	Medicinally useful
48.	Gum	Karaya gum	*Sterculia urens*	Sterculiaceae	Medicinally useful
49.	Gum	Tragacanth gum	*Astragalus gummifer*	Fabaceae	Medicinally useful
50.	Gum-Resin	Rumi Mastagi	*Pistacea lentiscus*	Anacardiaceae	

Table 2.19 *Antimalarial plants*

S.No	Name	Family	Important chemicals
1.	*Simarouba amara*	Simaroubaceae	Glaucarubinone
2.	*Eurycoma longifolia*	Simaroubaceae	Eurycomalatore, Eurycomanol
3.	*Melia azadirachta*	Meliaceae	Limonoid, gedunin
4.	*Dichoroea febrifuga*	Saxifragaceae	Febrifugine
5.	*Artemisia annua*	Asteraceae	Artemisinin

Table 2.20 *Anticancerous plants*

S.No	Name	Family	Important chemicals
1.	*Catharanthus roseus*	Apocynaceae	Uinblastin, Uinorelbine, Uindesine
2.	*Ochrosia sp*	Apocynaceae	
3.	*Phyllanthus acuminatus*	Euphorbiaceae	Phyllanthoside
4.	*Ostodes paniculata*	Euphorbiaceae	
5.	*Podophyllum peltatum*	Podophyllaceae	Podophyllotoxin, Podophyllinic acid
6.	*Camptotheca acuminta*	Nyssaceae	Camtothecin
7.	*Taxus brevifolia*	Taxaceae	Palitaxol (Taxol)
8.	*Heliotropium indicum*	Boraginaceae	
9.	*Dirca occidentalis*	Thymeleaceae	
10.	*Trichosanthes kirilowii*	Cucurbitaceae	

Table 2.21 *Anti HIV plants*

S.No	Name	Family	Important chemicals
1.	*Asplanthus linearis*	Leguminosae	Polysaccharide
2.	*Erythrina glauca*	Leguminaceae	Pterocarpans, Isoflavonoids
3.	*Calophyllum sp*	Guttiferae	Autoflavonoids
4.	*Garcinia multiflora*	Guttiferae	Biflavonoids
5.	*Terminalia bellerica*	Combretaceae	Lignans
6.	*Buchenavia capitata*	Combretaceae	Flavonoids
7.	*Hedera taurica*	Araliaceae	Saponins
8.	*Maprounea africana*	Euphorbiaceae	Triterpene
9.	*Syzigium claviflorum*	Myrtaceae	Betulinic acid platanic acid
10.	*Dianthus caryophyllus*	Caryophyllaceae	DAP 30, 32
11.	*Momordica charantia*	Cucurbitaceae	MAP 30

Table 2.22 *Important aromatic plants*

S.No	Botanical Name	Family	Economic parts	Active principle
1.	*Astemissia pallens*	Asteraceae	whole plant	Davanone, davona furans
2.	*Acorus calamus*	Araceae	rhizome	Eugenol, calamine, p-asarone
3.	*Anethum graveolens*	Apiaceae	seed	Carvone
4.	*Cymbopogon winterianum*	Gramineae	whole plant	Methyl eugenol, eugenol
5.	*Busrera pencillata*	Burseraceae	fruit husk	Linalool, geraniol, narol
6.	*Eucalyphus citriodora*	Myrtaceae	leaves	Citroneellol, gernaniol
7.	*Jasminum grandiflorm*	Oleaceae	flowers	Jasmone
8.	*Lavendula latifolia*	Lamiaceae	leaves, flowers	Lavendin, comphene
9.	*Mentha piperata, M.arvensis*	Lamiaceae	whole plant	Menthol, Menthone
10.	*Marjorana hortensis*	Lamiaceae	whole plant	Lindool, Linalylatate
11.	*Pogostemon patchouli*	Lamiaceae	leaves	Patchoulene, patchouli alwhol, linalool
12.	*Rosa damascena*	Rosaceae	flowers	Geranial, linalook, nerol
13.	*Rosemarinus officinalis*	Lamiaceae	whole plants	Pinine, cineol
14.	*Polyanthus tuberosa*	Liliaceae	flowers	Generoil, forvesol
15.	*Vanilla fragrans*	Orchidaceae	pod	Vanillin
16.	*Vetiveria zizanioides*	Gramineae	whole plants	Vetiverone, Veticerol
17.	*Thymes vulgaris*	Lamiaceae	whole plant	Thynol, carvacrol
18.	*Ocimum camphora*	Lamiaceae	whole plant	Camphene

The occurrence and availability of raw materials derived from plants is as under.

I. plants occurring in forests, grasslands, aquatic and desert ecosystems constitute about 150 species, some very important plants are :

Ephedra geradiana, Saussurea costus, Podophyllum hexandrum, Taxus baccata ssp. wallichiana, Aquillaria malaccensis, Valeriana jatamansi, Strychnos nuxvomica, Terminalia chebula, Rauvolfia serpentina and various species of *Berberis*. The raw material of more than 80 plants consists of root, rhizome or other underground parts: heart wood or whole herb which are non-renewable. The plants where the raw material is leaf, flower, fruit, seed, exudates or other renewable part also suffer if the collection method is destructive.

Plants growing as weed constitute 54 species in this category, provide the raw materials such as *Adhatoda zeylanica, Andrographis paniculata, Phyllanthus amarus, Boerhavia diffusa, Convolvulus microphyllus, Gymnema sylvestre* and various species of *Datura* have large demand. These plants usually occur in fallow agricultural land, along roads and railway tracts, in gardens and orchards, and other waste places. Some of these are escape from cultivation or colonizers in cleared or degraded forest land. The medicinal plants are traditionally derived from forests by local communities who depend on them for their subsistance needs.

Plants cultivated as ornamentals or as cereal, fruit, vegetable, spice, oil seed etc. constitute 68 plants and in this category are also source of medicinal raw material, such as clove, cinnamom, turmeric, cumin jamun, papaya, castor seeds (for oil) etc.

Table 2.18 summarises the select list of plants whose plant parts are used in treatment of various ailments. Tables 2.16 to 2.21 is a select list of specific ailment plants such as heart, diabetes, malaria, cancer and AIDS while Table 2.22 sumarise important aromatic plants which have wide applications in medicine and various industries especially perfumery.

Plants cultivated as medical crop, 30 species under this category are under regular or large scale cultivation. The major crops are: *Plantago ovata* (psillim) *Cassia angustifolia* (senna), *Withania somnifera* (ashvagandha),several *Cinchona* spp, *Saussurea costus*.

As per an estimate 80, 000 species out of 2.5 lakhs species are globally known to be medicinal and are of aromatic value.

Indian medicinal flora

Medicinal and aromatic plants flora is widely distributed in different parts of the Indian states with specific phytogeographical entites. There are some pockets in far off and inacessible areas of Himalayas, and N.E. India where several plants are still unexplored while many plants are endemic.

Medicinal plants in the Indian forest ecosystem range from Kerala and Andaman islands to hot desert of Kutch to cold desert of Ladak and Himalaya.

The Indian sub-continent is rich in various types of medicinal plants and 8000 plants are used by various traditional systems of medicines. A total of 18000 plants belonging to 387 families comprising 2220 genera constitute the Indian medicinal flora. The following families are the major components.

1. Asteraceae . . . 424 species
2. Euphorbiaceae . . . 219 species
3. Lamiaceae . . . 218 species
4. Fabaceae . . . 217 species
5. Rubiaceae . . . 210 species
6. Apiaceae . . . 118 species
7. Poaceae . . . 269 species
8. Acanthaceae . . . 142 species
9. Rosaceae . . . 30 species

Indian medicinal plants are both indigenous and introduced and about 2500 are used in ethno medicine. According to a report, Ministry of Environment and Forest, Government of India about 7500 are used as ethnomedicine. It is relevant to state that Ayurvedic Materia Medica contains 260 (Dhanvantri's nighantu) and 560 plants (Bhavprakash's nighantu) while the Dept. of Indian system of medicine list 387 plants and the Unani system list 440 to 750 plants. 360 plants are of exotic origin. According to a report ayurvedic and other traditional healers in South Asia use at least 1800 different plants.

Sixteen medicinal plants are of exotic origin and part of Indian flora. Most important are: senna, psyllium, Belladona, *Cinchona*, *Eucalyptus*, Ipeac, *Digitalis*, Mexican *Dioscorea*, *Artemesia annua*. It is estimated that 30000 effective and safe medicines are prepared from plants all over the world. A total of 627 plant drugs and adulterates are used and traded in India and only 70 species are under pure cultivation, the remaining species are partly cultivated or wild, occur in wild state only and thirdly of purely exotic origin.

One third of prescription drugs in US have a plant component and over 120 important drugs are derived from plants (Table 2.24).

Table 2.23 *Raw materials imported entirely or partially to supplement indigenous production from other countries (except Nepal and Bhutan).*

Plant Name	Plant Part	Official/Trade Name	Demand
Acacia Senegal	Gum	Gum Arabic; Gond Babul	High
Aloe (various sps.)	Dried leaf juice	Aloe, eluva, Mussabar	High
Anacyclus pyrethrum	Whole herb	Akarkara	Marginal
Aquillaria agallocha	Infected wood	Agarwood, agaru	Medium
Artemisia absynthium	Dried herb	Absynth; Afsanatin	Marginal
Astragalus gummifer	Gum	Tragacanth gum;katira gond	Medium
Bambusa arundinacea	Manna	Bamboo manna,Bansalochan	High
Boswellia carteri	Oleo-gum resin	Olibanum, Kundyr	High
Carum carvi	Seed	Caraway; Siah zira	High
Cinnamoum verum	Stem bark	Cinnamon; Dalchini	Medium
Commiphora molomol	Oleo-gum resin	Heerabole	Low
Crocus sativus	Dried pistils	Saffron; Kesar	Low
Cuscuta epithymum	Whole plant	Aftimoon Vilayati	Marginal
Ephedra geradiana	Stem	Ephedra	V. high
Ferula asafoetida	Gum resin	Asafetida; Heeng	Low
Gaultheria fragrantissima	Leaf oil	Winter green; Gandhapura	Low
Gentiana lutea	Root	Gentian, traiman	Medium
Glycyrrhiza glabra	Root	Liquorice; muletthi	Medium
Hyscyamus muticus	Seed, leaf	Hyoscamus, Ajwain Khurasam	Low
Hyssopus officinalis	Dry herb	Jufa	Low
Illicuim verum	Fruit	Star anis; Badian Khatai	Medium
Juniperus communis	Fruit	Juniper berry; Hauber	Medium
Lavandula stoechas	Whole herb	Uste-khuddus	Low
Myristica fragrans	Seed/aril	Nutmeg/mace; Jaipal/Javitari	Medium
Matricaria chamomilla	Flower	Gulbabunah	Marginal
Operculina turpethum	Root	Ud-salab	High
Piper cubeba	Fruit	Cubebs; Kabab chini	High
Paeonia officinalis	Root	Ud-salab	Low
Piper sylvaticum	Fruiting spike	Long pepper; Barapipal	High
Pistacea integerrima	Insect gall	Kakarsinghi	High
Pistacea lentiscus	Gum resin	Rumi mastagi	Medium
Quercus infectoria	Insect galls	Gallnut; majuphal	Medium
Rheum palmatum	Root	Rhi-chin; Revand chin	Medium
Salix capera	Whole plant	Bedmushk	Marginal

Table 2.24 *Some Natural products in the prescriptions and over-the-counter preparations in the USA*

S.No	Name of the natural product	Use	Source plant family	Parts used
1.	Quindine	Suppressor of out of sequence heart beats	*Cinchona* sp (Rubiaceae)	Bark
2.	Quinine	Antimalarial	*Cinchona* sp (Rubiaceae)	Bark
3.	Pilocarpine	Glaucoma treatment	*Pilocarpus* sp. (Rutaceae)	Leaves
4.	Picrotoxin	Nervous stimulant	*Anamitra* sp. (Menispermaceae)	Fruits
5.	L-dopa	Treating parkinsonism	*Mucuna sp* (Fabaceae)	Fruits (seeds disease and hairs)
6.	Bromelain	Anti-inflamatory	*Ananas* sp. (Bromeliceae)	Fruit
7.	Scopolamine	Sedative	*Datura* sp (Solanaceae)	Leaves, flowers and seeds.
8.	Digitalin and digoxin	Heart drug	*Digitalis* sp. (Scrophulariaceae)	Leaves
9.	Atropine	Powerfull pupildilator	*Atropa* sp.	Roots and leaves
10.	Curare	Muscle relaxant	*Chondrodendron* sp. (Menispermaceae)	Leaves
11.	Ephedrine	Decongesatant	*Ephedra* sp(Gnetaceae)	Fruit
12.	Ipecac	Emetic and anti dysentric	*Cephalis* sp.(Rubiaceae)	Root
13.	Sennosides	Laxative	*Cassia* sp(Caesalpiniaceae)	leaves and fruits

Table 2.25 *Threatened medicinal plants of India*

Sl. No	Name	Family	Present status
1.	*Acontium deinorrthizum*	Ranunculacae	Almost extinct
2.	*Aconitum heterophyllum*	Ranunculacae	Greatly threatened
3.	*Angelica gluca*	Apiaceae	Threatened
4.	*Amebia benthemii*	Boraginaceae	Threatened
5.	*Artemisia brevifolia*	Asteraceae	Likely threatened
6.	*Artme`w`sia martima*	Asteraceae	Threatened
7.	*Atropa acuminata*	Solanaceae	Threatened
8.	*Berberis aristata*	Berberidaceae	Greatly Threatened
9.	*Bunium persicum*	Apiaceae	Threatened
10.	*Colchicum luteum*	Liliaceae	Threatened
11.	*Corydalis govaniana*	Papaveraceae	Likely to be Threatened

Table 2.25 *Contd...*

Sl. No	Name	Family	Present status
12.	*Dactylorthiza hatagirea*		Threatened
13.	*Dioscorea deltoidea*	Dioscoreaceae	Threatened
14.	*Ephedra gerardiana*	Gnetaceae	Likely to be Threatened
15.	*Ferula jaeschkeana*	Apiaceae	Threatened
16.	*Gentiana kurroa*	Gentianaceae	Threatened
17.	*Hedychium spicatum*	Zingiberaceae	Likely to be Threatened
18.	*Jurinea dolomiaea*	Asteraceae	Likely to be Threatened
19.	*Nardostachys jatamansi*	Valerianceae	Threatened
20.	*Orchis latifolia*	Orchidaceae	Threatened
21.	*Picrorrhiza kurroa*	Scrophulariaceae	Likely to be Threatened
22.	*Podophyllum emodi*	Berberidaceae	Threatened
23.	*Rheum emodi*	Polygonaceae	Threatened
24.	*Swertia chirata*	Gentianceae	Threatened
25.	*Valeriana wallichii*	Valerianceae	Likely to be Threatened
26.	*Xanthoxylum alatum*	Polygonaceae	Likely to be Threatened

Threatened Medicinal Plants

1. According to International Union for Conservation of Nature and Natural resources, (IUCN), 10% of world's vascular plants are under varying degrees of threat (20000-25000 plants). Likewise, India which is home for 15000 species, 10% are said to threatened. This is because most of the medicinal plants grow wild in forests and remain at the mercy of the collectors. As a result, the list of threatened and endangered species are gradually increasing. In India, 26 plant species have reached the stage of almost extinction greatly threatened and likely to be threatened (Table 2.25). The centre for environmental research, University of Rajshahi, Bangladesh estimated that 37 plant species in that country are lost/rare/endangered/threatened due to habitat destruction and over extraction. According to Bhattarai, 14 plant species are threatened in Nepal.
2. The Botanical Survey of India has published in 1980 booklets on threatened plants of India. Nayar and Shastri (1987) in their 'Red Data Book' mention 235 plants to be endemic to India and state that they are under threat of endangered or vulnerable. Incidentelly over 70% of the threatened plants are in active trade involving harvesting of plant parts and products.

Trade of Medicinal Plants

Medicinal plants are important to the global economy. The WHO estimated that the world trade at US $500 million in 1980. In both Europe and North America, the demand being fueled by an outburst of consumer interest in products that are 'all-natural' as well as by aggressive marketing of herbal medicine sales in Eastern Europe, Asia, Japan and North America are US $6000,2300, 2100 and 1500 millions respectively with a derived from leafy plants amounted to be US $11 billion.

It is further estimated that the total economic value to the United States from plant derived pharmaceuticals to be total sale of US $12400 million. (Srivastava *et al*, 1995). The value of traditional medicines in China as US $571 million and the country-wide sale of crude plant drugs.

The trade of medicinal and aromatic plants has two components; the export of plant/plant products/ plant extracts to other countries and the imports of plants from other countries to meet the national demand.

The internatioinal trade of medicinal plants is estimated at US $70 billion and is growing at a rate of 7% annually. The Planning Commission (Govt of India) plans to increase the trade of medicinal plants extract to Rs.3000 crores by 2005 and Rs.10000 crores by 2010.

The domestic demand of medicinal plants in 1999-2000 was about Rs.1100 crores and is likely to go up to Rs.2000 crores by 2005.

About 40 raw materials are imported regularly and/or in substantially large quantities from other countries (Table 2.23). Some of these are as follow liquorices, asafoetida, gall nut and a number of materials used in unani drug formulation, ephedra, long pepper etc.

Table 2.26 *Medicinal plant export from India*

Sl. No	Species	Part Exported
1	*Plantago ovata*	Seed,husk
2	*Cassia angustifolia*	Leaf, pod
3	*Rheum australe*	Rhizome
4	*Inula racemosa*	Rhizome
5	*Rauvolfia serpentina*	Root
6	*Hedychium spicatum*	Rhizome,
7	*Zingiber officinale*	Rhizome
8	*Colchicum luteum*	Rhizome,seed
9	*Valeriana wallichi*	Rhizome
10	*Acorus calamus*	Rhizome
11	*Adhatoda vasica*	Whole plant
12	*Juglans regia*	Bark
13	*Punica granatum*	Flower, root bark
14	*Barberis aristata*	Root
15	*Juniperus communis*	Fruit
16	*Juniperus macropoda*	Fruit
17	*Heracleum candicans*	Rhizome
18	*Picorhiza kurroa*	Root
19	*Actinium species*	Root
20	*Saussurea lappa*	Rhizome
21	*Swertia chirata*	Whole plant
22	*Podophyllum emodi*	Rhizome

Table 2.27 *Medicinal plants imported by India*

Sl. No	Species	Family	Parts used
1.	*Glycyrrhiza glabra*	Fabaceae	Under ground stem and root
2.	*Pimpinella anisum*	Apiaceae	Fruits
3.	*Thymus vulgaris*	Lamiaceae	Whole plant.
4.	*Operculina turpethum*	Convolvulaceae	Root and stem
5.	*Custcuta epithymum*	Convolvulaceae	Whole plant
6.	*Smilax omata*	Liliaceae	Roots
7.	*Lavendula stoechas*	Lamiaceae	Flowers and leaves.

India exports parts of 22 plant species (Table 2.26) and imports parts of 7 plant species (Table 2.27) but the quantitative information on these exports and imports are not available. The annual international trade of *Ginkgo biloba* leaf extract is approximately US $500 million. It is the only plant which is cultivated in France and U.S.A for the supply of 100 tonnes of dried leaves per month (Hamberger et al, 1991). There are several artificially synthesizable drugs which are also extracted from the wild collections because of the lower cost of production. India and Africa export *Rauvolfia* sp. to various countries for its drug.

As stated, India is a major exporter of medical plant raw materials and their extracts, to various countries. Majority of export include senna leaf and od, psyllium husk and seed, galangal roots, bael fruits, karaya gum; kamila powder, nux-vomica seeds, chebulic,belleric and emblic myrobalan and atleast 100 other materials. India exported 42000 tonnes of medical raw materials to other countries upto 2000-01 of this psyllium husk and seeds, senna leaf and pod, periwinkle and some other account for 32209 tonnes. The export of material employed in Indian system of medicine was about 9740 tonnes.

Hongkong has the largest herbal market in the world and import drugs with about US $190 million per year. The herbal market consists chiefly of 1. Gising products 2. crude drugs other than ginsing and 3. over the counter drugs and medicated tinctures etc.

In Japan pharmaceutical export for Kampo medicine in 1983 was US $150 million. The herbal product market in Germany in 1989 was about US $ 1.7 billion, while in USA the sale of pharmaceuticals was $68 billions per year.

India is next to China by exporting 32600 tonnes of medicinal raw material worth US $ 46 million. With increasing demand, the market for medicinal plants has expanded , thus opening new avenues to improve the economic status of local commodities.

With globalization of Ayurveda, there is high demand for herbal and medicinal plants. The Asian countries account for 16% of global market of the total US $ 62 billion/year, Chinese have a major export share followed by India. Global demand of medicinal and aromatic plants is growing @ 7% per annum.

Indian Drug and Pharmaceutical Industry

India has well established and rapidly growing drug and pharmaceutical industry utilizing plant based raw materials. There are around 7000 small and big pharmacies, manufacturing medicines and over the counter (OTC) products like digestives, carminatives. The popularity of herbal medicines in traditional health care can be gauged by the following statistics. There are 55, 000 licenced pharmacies, 13, 770 dispensaries, 7000 licenced manufacturing units 16, 900 hospitals, 98 Ayurvedic colleges, 400, 000 registered practioners.

While much of the plants are gathered by herb collectors and small traders for use in Ayurvedic and Unani systems of medicine, there is shortage of these materials for maintaing the sustained supply to the plant-based durg industries. There is thus need for domestication and mass farming of these plants as well as producing the durg products of uniformly high potency and quality.

Protection, Conservation and Biodiversity

As on today, the rising global interest in medicinal plants is burgenoning legitimate and "underground" trades in plant materials. Most samples are collected from the forests in developing countries in a completely unregulated manner. The collection, processing and marketing of medicinal plants are seldom handled within an organized and regulated sector. There is also no international pricing policy for export or import of these plants or plant parts, which remain as uncontrolled exploitation with little or no regard to the future. Escalation of consumers demand, results in more indiscriminate harvest damaging both to ecosystems and the biodiversity. As the demands for roots, rhizomes, whole plants, flowers and seeds are high so are also the losses of the reproductive potentials of these plants. In view of the increasing demand and decreasing reproduction, the protection of medicinal plants becomes imperative.

The Government of India therefore has taken up special strategies to conserve and to cultivate medicinal plants. The Council of Scientific and Industrial Research (CSIR) has a specialized institute, Central Institute for Medicinal and Aromatic Plants (CIMAP) in Lucknow, that deals with conservation, cultivation, and processing technologies of endangered and commercial medicinal plants. This institute has also established Gene Bank under G-15 GEBMAP programme. There are also several regional research stations of the CSIR with similar working objectives.

The Indian Council of Agricultural Research (ICAR) initiated an all India coordinated improvement project on medicinal plants with number of centers in different agricultural universities with the objective to develop cultivation technology and to incorporate it into the farming systems. The Ministry of Agriculture, Government of India also supports projects universities and institutes to establish conservatories and nurseries of medicinal plants with the objective to conserve them and bring them under cultivation. The Department of Biotechnology and the Department of Science, the Central Indian Medicinal Plants Organization (CIMPO) was established by CSIR, new Delhi in 1957 which was later shifted to National Botanical Research Institute campus, Lucknow and renamed as Central Institute of Medical and Aromatic Plants (CIMAP) in 1978. It is full fledged CSIR institute in Lucknow and has 7 sub-centers all over the country.

Domestication of plants involves extensive exploration and collection of the germplasm of a particular species from the whole range of it's distribution, quantitative and qualitative analysis of it's active principle from different sources of collection, multilocational trial of the species, recurrent selection and breeding, establishment of seed orchard or clonal multiplication area, standardization of propagation techniques, planting techniques and cultural practices, streamlining the production system and devlopement of proper harvesting schedule.

Genetic Resources

National Gene Bank at National Bureau of Plant Genetic Resources (NBPGR) New Delhi was established for safe conservation of germplasm of agri-horticultural crops followed by three exclusive gene banks for medicinal and aromatic plants have been established at NBPGR, New Delhi, CMAP, Lucknow and TBGRI, Thiruvananthapuram; under G-15 group of countries programmes for safe and effective conservations of medicinal and aromatic plants (M & AP) genetic material for posterity.

From the Andhra Pradesh state point of view, Bhadrachalam,Seshachalam, Nagaram, Tirumala hills, Paderu, Srisailam and Godavari forests are a home for numerous medicinal plants and traditionally the local tribes depended on these plants for curing common ailments and the knowledge has been passed on from generations to generations. Due to various reasons, especially increasing biotic pressure and mass exportation, the forest wealth is reduced. The AP forest department has given a flip for scientific management and conservation of the medicinal plants along lines of regular forest species and several research and development councils at Tirupati, Rajahmundary, Achutapuram, Warangal and Hyderabad (Mulugu) have medicinal gardens with about 600 species each.

Micropropagation

Micropropagation has become one of the most convenient and beneficial ways to propagate economic plants which has many advantages such as rapid clonal propagation, establishment maintenance and distribution of pathogen free clones and maintain true-to-type plant species. Commercial micropropagation of ornamental, and horticultural species, have become a profitable industry. During the last 10 years interset in *in-vitro* propagation of medicinal plants has considerably increased for many reasons. In India many laboratories are engaged in tissue culture research and developed protocols for micropropagation of number of medicinal plants.

With the increased general awareness on micropropagation technique and increasing demand for herbal drugs many pharmaceutical firms are showing interest to exploit the benefit of the technique. However, to harvest rich dividends the next step which needs attention to evolve cost effective mass propagation methods for commercial exploitation. It is hoped that micropropagation of medicinal plants is expected to become multi dollar industry in the years to come. Despite impressive account of number of *in vitro* protocols for rapid regeneration of number of medicinal plants, little attention has been paid towards commercial exploitation of the technique for cultivation of medicinal plants.

Conservation of Medicinal Plants

Conservation of biodiversity has assumed considerable significance in view of over increasing human population as well as due to depleting natural resources, due to ruthless exploitation of the plant wealth. Efforts are being made at global level to preserve biodiversity to achieve genetic conservation. In vitro technology and cryopreservation (freeze preservation) i.e. storage in liquid nitrogen (-196° C) have received attention and complement significantly to existing methods such as seed storage in gene bank (-20°C), maintenance under field conditions, orchards, plantations and botanic gardens. Protocols for in vitro storage of several medicinal plants are also now available.

National Medicinal Plants Board (NMPB)

This was set up under a Government resolution on 24th November 2000 under the Chairpersonship of the Union Minister for Health and Family Welfare, Government of India.

The following are the objectives. To promote conservation and sustaniable use of medicinal plants from shrinking wild source. NMPB has been established with similar Boards at State and Regional levels. This has identified 60 plants on priority basis (Table 2.28).

Transgenic Medicinal Plants

Just as several crop plants have been genetically engineered (see Unit II.) and are called Transgenic crops in a similar manner some very important medicinal plants have been genetically transformed mostly by *Agrobacterium tumafaciens* (using Ti plasmid) and some by particle bombordment technique. A list of transgenic medicinal crops, the important gene transferred for the high alkaloid content are summerised in Table 2.29.

With ever increasing population and decrease in forest land area of medicinal plants and ever increasing demand at the national and international market, the future appears to be safe and sound *in vitro* technology as the alternative. The possibilities of using elicitors, bio transformations and gene transfer to enhance the manurability of the technology and adding it to more spice and flavour. The future developments in this up coming and fertile area will be paralleled by better understanding of the metabolic pathways and control mechanisms that govern the biosynthesis of various secondary metabolities. This is dealt under plant drugs.

Departmetnt of Indian Systems of Medicine and Homoeopathy of the Ministry of Health and Family Welfare, Govt. of India has a Central Scheme for the Cultivation and Development of Medicinal plants in which 46 species of Medicinal plants, (Table 2.28) have been identified for promoting their cultivation in order to reduce pressure on their natural habitat and to meet the shortage in their demand by the industry involved in producing the ISM and Homoeopathy medicines etc.

Emerging Plant Medicines

Artemisinin is the most recent antimalarial drug developed from plant based traditional medicine. It is isolated from the leaves and flowers of *Artemisia annua* (Asteraceae) and is indigenous to China. The extract of this plant has been used to treat malaria in China for over 2000 years. Its active component, artemisinin, was first isolated in the 1970's by Chinese scientists. Artemisinin is an endoperoxide of the sesquiterpene lactone whose chemical structure is too complex to be synthesized effectively. Unlike quinine and chloroquine, this compound is non-toxic, rapid in effect, even safe for pregnant women and it is effective against chloroquine - resistant *Plasmodium falciparum* malaria and in patients with cerebral malaria. It kills the parasites directly so parasitaemia is quickly controlled. This work was confirmed by the WHO in Africa and other parts of South East Asia.

The WHO and USA are jointly engaged in the cultivation of Chinese *Artemisia* for worldwide use. This recent development offers renewed hope for using traditional medicine to provide new drugs for future medicines.

Taxol is an antileukaemic and antitumour agent isolated from the bark of the *Taxus breviofolia*. This plant, a slow-growing runt, is found mainly in the US national forests in Oregon and Washington. Taxol is alkaloid with a complex ring structure and its anticancer activity, was first observed 20 years ago. Recently there has been renewed interest in this compound; approved for use against ovarian cancer. Taxol is currently under going phase II clinical trials for breast cancer and lung cancer.

Table 2.28 *Priority species identified by the Ministry of Health and Family Welfare, Government of India*

Sl. No	Botanical Name	Common name
1.	*Aconitum chasmanthum*	Vatsnabh
2.	*Aconitum heterophyllum*	Ativsha
3.	*Aconitum napellus*	Monk's Hood
4.	*Aegle marmelos*	Bilva
5.	*Anacycius pyrethrum*	Akarkartha
6.	*Arnica montena*	Leopard's Bane
7.	*Asparagus recemosus*	Shatavari
8.	*Asparagus adscandens*	Safed Musali
9.	*Carum carvi*	Krishnajirak
10.	*Celastrus paniculatus*	Jyotishmati
11.	*Commiphora mukul*	Guggulu
12.	*Convolvulus scammonia*	Sagmonia
13.	*Coptis teeta*	Mamiran
14.	*Crocus sativus*	Kesara
15.	*Desmodium gangeticum*	Salparni
16.	*Glycyrrhiza glabra*	Yastimadhu
17.	*Gmelina aroborea*	Gambhari
18.	*Hemidesmus indicus*	Anantmool
19.	*Hydrastis canadensis*	Golden Seal
20.	*Illicium verum*	Badiyan Khatai
21.	*Mesua ferrea*	Nagkesar
22.	*Myristica fragrans*	Jatiphyala
23.	*Nardostachys grandiflora*	Jatamansi
24.	*Onosma bracteatum*	Gaozaban
25.	*Oroxylum indicum*	Syonak
26.	*Picrohiza kurroa*	Kukti
27.	*Pilocarpus jaborandi*	Jaborandi
28.	*Piper longum*	Pippali
29.	*Pistacia lentiscus*	Mastagi Roomi
30.	*Pluchea lanceolata*	Rasna
31.	*Plumbago zeylanica*	Chitrak
32.	(a) *Premna integrifollia* (b) Premna micronata	Agnimanth Agnimanth

Table 2.28 Contd...

Sl. No	Botanical Name	Common name
33.	*Saraca indica*	Ashoka
34.	*Saussurea lappa*	Kushtha
35.	*Senecio cienerraria*	Dusti Miller
36.	*Smilax china*	Chobchini
37.	*Stereospermum suaveolens*	Patola
38.	*Swertia chirayita*	Chirayatha
39.	*Terminalia arjuna*	Arjuna
40.	*Tinospora sinensis*	Guduchi
41.	*Uraria picta*	Prishnipami
42.	*Valeriana jatamansi*	Tagar
43.	*Viburnum opulus*	Gradberry Highbush
44.	*Viola serpens*	Banafsha
45.	*Zizphus sativa*	Unnab

Following medicinal plants are also included in the above mentioned priority list of the Ministry of Health and Family Welfare, for developing their agronomic techniques.

1.	*Atropa acuminata*	Indian Belladona
2.	*Angelica glauca*	Chora
3.	*Colchicum luteum*	Hirantutiya
4.	*Ephedera gerardiana*	Khander
5.	*Jurinea macrocephalla*	Dhup
6.	*Podophyllum hexandrum*	Bankakri
7.	*Taxus baccata*	Thuner
8.	*Cholorphytum tuberosum*	Safed musli
9.	*Cholorphytum tuberosum*	Kali musli
10.	*Curcuma caesia*	Kali musli
11.	*Curcuma aromatica*	Jangli haldi
12.	*Curcuma angustifolia*	Tikhur
13.	*Dioscorea daemona*	Bacchandi
14.	*Cyperus scariosus*	Nagarmotha
15.	*Pimpinella anisum*	Anise

Clinical development of taxol is limited by a shortage of *Taxus* bark and low yield of taxol. About 60 Ib of bark, the amount produced by three fully grown trees, are needed to produce taxol for the treatment of a patient for one year. Now suddenly, it has become the most precious medicinal plant, in high demand all over the world.

Over two million people in the US, and five to ten million people world wide, are infected with the human immunodeficiency virus (HIV) that causes AIDS. Recent projections indicate that the majority of those now infected will develop AIDS have been reported in the US and more than half of these patients have died. Clearly, AIDS represents an unprecedented threat to national as well as global health, and the search for effective therapies is of paramount importance.

Table 2.29 *Transgenic medicinal plants*

	Species	Genes transferred	Alkaloids
1.	*Catharanthus roseus*	Strictosidine synthase (GUS-A)	Strictosidine
2.	*Tabernaemontana pandacaqui*	Strictosidine synthase	Strictosidine
3.	*Atropa belladonna*	Hyoscyamine hydroxylase Hyoscyamine 6-hydroxylase	Anabasine, Scopolamine Scopolamine,
4.	*Nicotiana rustica*	Ornithine decarboxylase	Putrescine,Nicotine
5.	*N. tabacum*	Lysine decarboxylase	Cadaverine
6.	*Solanum lacinatum*	Ti, Ri plasmid	Steroid alkaloid
7.	*Papaver somniferum*	Ri plasmid	Sanguinarine
8.	*Peganun harmala*	Tryptophan decarboxylase	Serotonin

In the past few years, many traditional medicinal plants have been studied for their antiviral, anticancer and other drug activities. Their isolation, purification and characterisation of a new class of anti-HIV proteins from several unrelated species of plant, *MAP 30, TAP 29, GAP 31, DAP 32* has been reported. These proteins are unique in that they not only inhibit acute infection of new cells by the AIDS virus, but also block the replication of the virus in already infected cells. In addition, they are not toxic to normal human cells. In contrast, currently available anti-AIDS drugs such as *AZT* and ddI are effective only against the acute infection of HIV-I but are not able to eliminate the virus in already infected cells.

All of these anti-HIV agents are proteins with molecular weights ranging from 29,000 - 32,000 daltons, consisting of 260 - 280 amino acids. The amino acid sequences have been partially determined some of these agents are glycoproteins that contain varying amounts of glucose, mannose, xylose, fucose and glycosamine. A brief summary of the physicochemical properties of these proteins is shown Table 2.30.

Plant Medicines Future Prospects

Recent advances in molecular biology and biotechnology have significantly improved the production of many protein and polypeptide drugs by genetic engineering. However, in the case of most plant-derived medicines, the application of these techniques remains limited because many of these plant components are not proteins and cannot be cloned. They are secondary metabolites, whose biosynthesis involves complex pathways using multiple enzymes, many of which have yet to be identified and characterised. Genetic engineering of these secondary metabolic drugs is currently a difficult task. Thus, plants or plant cell cultures will most likely continue to be the primary source for the production of these drugs. (More information under the chapter *Plant Drugs*).

In contrast, plant-derived drugs that are proteins, for example the anti-HIV plant proteins, are primary metabolites of plants. Molecular cloning of their genes and recombinant production of these proteins is feasible. The application of these biotechnologies could ensure sufficient supply of these anti-HIV agents for clinical trials as well as for the design and development of more effective second generation drugs.

It is thus obvious that there is a wealth of potential medicines in our forests and plants that we have only just begun to realise. The study of plants used in traditional medicine in various cultures has yielded important drugs that are critical to modern medicine. While new compounds are being rediscovered in medicinal plants, one can only wonder about potential drugs that have not yet been discovered. The plant kingdom thus holds the key that will unlock the secrets to many other drugs. We must use our wisdom and foresight to preserve and protect these precious resources for the present and future generations.

Table 2.30 *Physicochemical properties of anti-HIV proteins*

Plant source	Anti-HIV agents	Carbohydrates	Mol. wt.
Momardica charantia	MAP 30	–	30, 000
Trichosanthes kirilowii	TAP 29	–	29, 000
Gelonium multiflorum	GAP 31	+	31, 000
Dianthus caryophyllus	DAP 32	+	32, 000
Dianthus caryophyllus	DAP 30	+	30, 000

As stated by Adjanohoun, (1996). "The biodiversity of the tropical regions of the world at large may be well exploited more particularly a wide range of pharmaceutical industries".

2.9 Drugs and Chemicals from Plants

Drug is a material that can alter the structure or function of the living organism. Organs or occasionally plants or animals, and dried frozen for preservation and are termed crude drugs. They may be used medicinally in essentially this form, as in the case of cardiac drug, digitalis, or as source of mixtures or of chemicals obtained by processes of extraction. Pepermint oil (steam distillation), podophyllum resin (percolation) and others can be aqous solutions. These are not to be confused with the exudations, like opium from poppy fruits, turpentine from conifers etc.

Drug and pharmaceutical are interchangably used. The former means pharmacologically active ingredient in a medical formulation. A pharmaceutical is a thereapeutically active drug, that is, a chemical substance that alters the structure or function of a living organism and cure or alleviate disease. Most pharmaceuticals are purchased by public in the form of prescription drugs which are produced by the pharmaceutical industry under strict regulatory procedures by methods closer to those of conventional laboratory chemistry. Since 1935 the pharmaceutical industry has become a large and profitable sector of the chemicals and allied products industry. It is research intensive and competition is frequently between patent protected products by "molecular roulette". This industry offers most prized possessions; good health, issues of quality control and testing and safety of new drugs are of paramount importance. Fifty important drugs for seven main illnesses are shown in Table 2.31 while Table 2.32 summarise of drugs. By value of products, fermentation is the most important since almost all antibiotics of fungal origin are produced by this method. Even the chemically modified antibiotics require a fermentation step to provide their precursors and modern biotechnology processes may also be included under this category. By tonnage, chemical synthesis is the most important. It provides majority of heart drugs, anti-histamines and analgesics.

The term plant chemistry and pharmaceutical chemistry were used interchangably and this field is active even todays and promises to continue for long time to come because it is estimated that less than 10% of the known flora (plants) have been examined.

Table 2.31 *Categories of drugs in the top 50 (U.S. 1983)*

Category	Number of drugs	Number of prescriptions (millions)
Cardiovascular drugs	13	316
Antibiotics and antibacterials	8	182
Drugs affecting the central nervous system	9	86
Analgesics (excluding asirin and acetaminophen)	3	65
Antihistamines	5	52
Steroid drugs	4	45
Non-steroid antiinflammatory drugs	4	41
Miscellaneous	4	32

Table 2.32 *Sources of drugs*

Source	Examples
Chemcial synthesis	Heart drugs, psychotropic drugs, antihistamines, analgesics. Also chemical modification of materials from other sources.
Fermentation	All antibodies except cloramphenicol, steroids from diosgenin, vitamins B12 and C. Biotechnology
Animal extracts	Heparin, thyroid, insulin, steroid hormones.
Vegetable extracts	Alkaloids (e.g., digoxin, papaverine, atropine, codeine, and quinidine), steroids.
Biological sources	Vaccines and serums (e.g., against diphtheria, polio, and whooping cough).

The food and drug legislation of 1906 led to the broadening of research activities, development of analytical procedures for the maintenance of the quality and potency of drugs and this has resulted in revisions of various pharmacopeias the world over.

Pharmcognosy literally means knowledge of drugs, their mode of action particularly on humans. Pharmcognosy is restricted to natural products, drugs of plant and animal origin, their biosynthesis and identification of the pharmacodynamic constituents.

During the thousand of years of early human existence many natural material were used for combating human ailments either by instinct or intuition or by trial and error. In other words there was a pill for every ill. The earliest mention of the medicinal use of plants is found in “Rig Veda”, which were written between 4000 and 1600 BC. In the “AtharvaVeda,” where one can find use of drugs. It is in the “Ayurveda” which is considered as an “UpaVeda,” that definite properties of drugs and their uses are given in considerable detail. “Charaka Samhita” is another earliest treatise on “Ayurveda” (600 BC) which provides list of 341 plants and their products for use in health management. “Susruta Samhita” also dealt with plants related to medicine. Subsequent authors of later treatises have extended the list of Ayurvedic single plant drugs to the 600 species of plants. Biksu Atreya a well known scholar of the University of Taxila, his student Jivaka, who was the physician to king Bimsisara of Magadha; Dhanvantari and Nagarjuna were the well known authorities with an intimate knowledge of the characteristics of many medicinal plants.

Initially the plants were the main part of folk medicine practiced by the ancient people in different parts of the world especially India, China, Middle East, Africa and South America. The folk medicine in various countries gave rise to the traditional systems of medicine. Some of the important systems that are still practiced in the third world countries are : Unani is practiced in India, Pakistan, Afghanistan, Bangladesh and the Middle East, Ayurveda and Siddha are practiced in India, Nepal, Bangladesh and Srilanka, and the Chinese system of medicine is practised in China and Far east and derive more than 80% of therapeutic agents from plants. The Tibetan system is becoming popular and this traditional systems make use of about 1000 genera and 2500 species in Western medical therapy.

Isolation and Characterisation of Active Principles

Isolation of active principles from medicinal plants and their characterization, can be traced to the beginning of 19th Century. An apothecary from Einbeck, Germany was primarily interested in the isolation of the active constituents of plants; this resulted in the discovery and description of many alkaloids, glycosides, carbohydrates, volatile oils and fixed oils. From the crude drug Ma Huang (*Ephedra spp)* of China, ephedrine was isolated and later introduced as drug. Likewise, from opium *(Papaver somniferum),* morphine was isolated F.W.A Serturner in 1804 and introduced as drug in 1818. From *Cinchona* spp. of Peru, quinine was isolated in 1820 and introduced as drug in 1825. Ipecacuanha (*Cephaelis ipecacuanha*) of Brazil, emetine was isolated in 1894 and introduced as drug in 1912.

Large number of drugs from medicinal plants were discovered and introduced in modern pharmacopoeias during 1850 – 1950. Some of the important crude plant drugs are belladonna (*Atropa belladona*), cascara (*Rhamnus purshina*), digitalis (*Digitalis purpurea*), ipecae (*Cephaelis ipecacuanha*), opium (*Papaver somniferum*), rouwolfia (*Rauvolfia sepentina and* veraturm (*Veratrum viride*). More than hundred pure compounds derived from higher plants find their place in modern medicine and some of the traditional active plant principles are atropine (*Hyocyasmus muticus*), codeine *(Papaver somniferum*), digoxin (*Digitalis lanata*), digitoxin (*Digitalis purpurea*) *Colchicicum autumnale*, hyoscine (*Datura metal*), pilocarpine (*Pilocarpus jaborndi),* quinidine and quinine (*Cinchona spp*). From 1950 onwards, there was examination of ethenotherapeutics constitute in search of new drugs from higher plants and marine flora. During this period there was outstanding contribution of five drugs from higher plants like reserpine, deserpidine and rescinnamine from *Rauvolfia serpentina*, and vinblastine and vincristine from *Catharanthus roseus*.

Transformation of Medicinal System

Plants continue to constitute one of the major raw materials for drugs for treating various ailments although there has been significant development in the field of synthetic drug chemistry. During the last two decades considerable advances have been made in the medicinal system all over the world. Because of the general awareness of the widespread toxicity and harmful after effects associated with the prolong use of synthetic drugs and antibiotics, the Western Society of late prefer the drugs from natural sources rather than the synthetics. In USA and UK, the plant based drugs are being used substantially, while Russia and China have adopted an integrated system of allopathic, traditional and folk system of medicine.

Commercialization of Plants and Modern Medicine

Over 100 medicinal plants are used in modern medicine and in traditional systems of medicine more than 500 plants are used. Most of the raw material for traditional medicine of pharmaceutical houses is collected from wild sources but many of the medicinal plants are cultivated commercially for extraction of important active constituents. For steroidal hormones which are synthesized from diosgenin, hecogenin or solasodine different species of *Dioscorea, Solanum* and *Agave sp.* are cultivated. *Costus speciosus* is also a source of diosgenin. *Papaver somniferum* is a source of different types of alkaloids, out of which morphine, codeine, papaverine and nascopine are used routinely. Various species of *Cinchona* also contain large number of alkaloids in the bark, and most important are quinine and quinidine. Different species of *Datura, Hyoscyamus, Atropa, Duboisia, Scoipolia and Physochalaina* contain *tropane alkaloids* which include hyposcyamine, hyooscine, atropine and are routinely used. *Rauvolfia serpentinan, R.canescens* and *R.vomitoria* contain reserpine, *resinamine and deserpidine*, and the species *R.vomitoria* also contain ajamaline. *Catharanthus roseus* has become important plant modern medicine as it contains more than 100 alkaloids of which Ajmaliciune is of great importance. Leaves of this species also contain two more anticancer alkaloids viz., vincristine and vinblasline. *Cephaelis ipecacuanha* and *C. acuminata* yield number of alkaloids of which emetine and psychotrine are most important. Various drug are obtained from the dried sclerotia frinking body of the fungus *Claviceps purpurea* that grows on rye plant-(*Secale cereale*), contains over 20 alkaloids of which ergotamine, ergometrine and ergotoxine are used in treatment of migraine and others diseases not to mention the production of so many antibiotics from various fungi and lichens.

Table 2.33 is a select list of plants, their active principle and curative properties are given. In addition secondary metabolites previously unknown as constituents in intact plants are briefly summarised below (Table 2.34).

Phytochemicals from plant cell cultures

Number of phytochemicals (plant metabolities) such as alkaloids, terpenoids, steroids, anthocyanins, anthraquinones and polyphenols are used as drugs, food flavours, pigments, perfumes and agrochemicals and could potentially be obtained from cell culture. A unique feature of most of the phytochemicals is that they are secondary metabolitics of plants origin and unique to specific groups of plants. In many cases they are highly tissue specific, for example crocin and picrocrocin from saffron are synthesized only in the stigmas of the plant *Crocus sativus*. Many of these chemicals have a complex structure and stereospecificity and are difficult to synthesise chemically. Efforts have been made in the recent past to explore the possibility of using cultured cells as the source of these valuable phytochemicals. Plant cell and tissue culture techniques are of promise to agument the commercial production of different phytochemicals used for human welfare.

Which may be essential for biological activity plant cells are totipotent, all the necessary genetic and physiological potential for natural product formation present in an isolated cell (Zenk 1978). According to this theory, cultured cells obtained from any part of a plant might be expected to yield secondary compounds similar to those of the plant grown in vivo. Therefore the exploration of tissue culture techniques in medicinal plants for the extraction of important chemical compounds is indeed desirable (Tabata 1977). The root, stem and leaves of regenerated plants or the induced callus may be used fresh or dried as raw drugs, or different secondary metabolitics may be extracted from them. The plant tissue selected for culture may be critical for regeneration or callus culture of desired plants may be maintained in tissue culture.

Table 2.33 *List of important drug plants*

Plant Name	Active Principle	Therapeutic property
Acorus calamus	Asarone and oil	Tranquilizer
Andrographis paniculata	Andrographolide	Hepatoprotector
Antemisia annua	Artemisinin	Antimalarial
Atropa belladona	Atro pine	Antispasmodic, sedative
Azadirachta indica	Nimbin, gedunin	Antifungal, antimalarial
Asparagus racemoses	Shatavarin I	Antiabortifacient
Adhatoda zeylanica	Vasicine	Branchodilator, oxytocine
Bupleurum falcatum	Saikosaponin	Antiallergic, analgesic
Bacopa monneri		Memory enhancer
Butea frondosa	Palasonin	Antihelminthic
Boerhaavia diffusa	Puranavine	Divretic, antienflamatory
Coptis japonica	Berberine	Stomachic
Coleus forskohlii	Forskolin	Cough, asthma
Catharanthus roseus	Ajamalcine	Cancer
Centella asciatica	Asiaticosides	Psychotropic, memory enhancer
Curcuma longa	Curcumin	Antiinflammatory
Datura sp.	Hyoscyamine, hyoscine	Antispasmodic, sedative anticholinergic
D. stramonium	Hyoscyamine	Narcotic, antispasmodic
D. metal	Hyoscyamine	Narcotic, antispasmodic
D. mexicana	Hyoscyamine	Narcotic, antispasmodic
Digitalis lanata	Digoxine, lanatoside-C	Cardiotonic
D. purpurea	Digitalin, Digitoxin, Gitalin	Cardiotonic
D. deltiodea	Corticosteroids	Antiinflammatory, hormonal, antifertility
D. composita	Diosgenin	Antiinflammatory, hormonal, antifertility
D. floribunda	Diosgenin	Antiinflammatory, hormonal, antifertility
Ephedra sp.	Ephedrine, pseudoephedrine	Bronchodilator Vasoconstrictor
Echinacea sp.	Polysaccharides	Immunomodulator
Erythroxylum	Cocaine	Local anaesthetic
Eucommia ulmoides	Pinoresinol	Anti-hypertensive
Foeniculum vulgare	Ancthole	Stimulant, Carminative
Glycyrrhiza glabra	Glycyrrhizin, glycyrrhetic acid	Antiinflammatory, expectorant, demulcent
Ginkgo biloba	Bilobalide, Ginkgolides, Flavonoids	Anti-ischemic, Antihypoxidotic, PAF-antagonistic, Memory enhancer

Table 2.33 *Contd...*

Plant Name	Active Principle	Therapeutic property
Hydnocarpus weightiana		Leprosy
Hyoscyamus niger	Hyoscyamine, hyoscine	Anticholinergic
H.muticus		antispasmodic
Harpagophytum procumbens	Harpagoside	Anti-inflammatory and Antiexudative
Holarrhena antidysenterica	Conessine	Anti-dysenteric
Hypericum perforatum	Hypericins	Antideppressant
Hyperzia serrata	Huperzine A and other alkaloids	Active cognition chnancer, facilities memory and motor activity in the aged
Lithospermum erythrorhizon	Shikonin	Diuretic, sedative
Panax ginseng	Ginsenosides	Adaptogen
P. quinquefolius	Ginsenosides	Adaptogen
Phyllanthus amarus	?	Hepatoprotector
Picrorrhiza kurroa	Picroside, Kutkoside	Hepatoprotector
Piper methysticum	Methysticine and related pyrones	Anesthetic, anticonvulsant central muscle relaxant
Prunus africana	?	Benign prostratic hyperplasia
Psoralea corylifolia	Psoralen, bakuchiol	Antileucoderma, antibacterial
Physostigma venenosum	Physostigmine	Cholinergic
Papaver somniferum	Morphine, codeine, nascopine, Papaverine	Analgesic, antitussive, smooth muscle, relaxant
Pilocarpus jaborandi	Pilocarpine	Parasymthomimetic
Plantago ovata	Mucilage and hemi celluloses	Laxative
Podophyllum peltatum	Podophyllotoxine (Etoposide)	Anticancer
P.emodi	Etoposide, teniposide	Anticancer
Rauvolfia serpentina	Reserpine rescinnamine	Hypotensive, tranquillizer
Ruta gravolens	Rutacridone	Anthelmintic, antispasmodic
Rubia peregrina	Anthroquinone	Carminative diuretic
Silybum marianam	Silymarin	Antihepatotoxic, promotes ribosome formation and protein synthesis
Solanum laciniatum	Solasodine	
Strophanthus gratis	Quabain	Cardiotonic
Serenoa repens	?	Benign prostratic hyperplasia

Table 2.33 *Contd...*

Plant Name	Active Principle	Therapeutic property
Swertia chirayita	Xanthons citratin	Febrifuge, asthma
Urginea indica	Scillaren A & B	Cardiotonic
U.maritima	Scillaren A & B	Cardiotonic
Valeriana officinalis	Valpotriaes	Tranquillizer, Hypnotic
V. wallichi	Valpotriaes	Tranquillizer, Hypnotic
Vinca minor	Vincamine	Vasodilator
Vitis vinifera	Stilbane glucosides	Free radical scavenger
Woacanga africana	Tuberosonine used for synthesis of vincamine	Vasodilator

Table 2.34

Culture	Metabolite
Stephania cepharantha	Armorine
Plagiorhegma dubium	Dehydrodiconiferylalcohol-D-glucoside
Morinda citrifolia	5,6-Dihydroxylucidin
Ochrosia elliptica	Epohrosine
Thuja occidentalis	Hinokiol
Andrograhis paniculata	Paniculid A
Picralima nitida	Pericine
Ruta graveolens	Rutacultin
Gardenia iasminoides	Tarennosid
Voacanga africana	Voafrine A and B

1. ***Biotransformation*** **:** One of the recent techniques for the commercial exploitation of secondary metabolites from cell cultures is biotransformation. Production of valuable plant products by biotransformation from inexpensive precursors that cannot be transformed effectively by chemical or microbial methods is an interesting area for commercial application of plant tissue culture. The process of biotransformation may be simple, where one or more enzymes with many steps mediate the process. Single step biotransformation is comparatively efficient, as the yield decreases with increase in steps. Natural or synthetic substrates are used for biotransformation. Example of such biotransformation is the conversion of digoxin by C-12 hydroxylation to digitoxin, a cardiac drug from cultured cells of *Digtalis lanata.*
2. ***Cell permeabilization :*** One of the problems of the production of herbal drugs is that these compounds are stored intracellularly which renders the products recovery expensive. Permeabilization of cells result in the release of the product. When the cells of *Catharanthus roseus* were subjected to a lone current for 18 hours, the rate of release of alkaloids almost doubled. Pretreatment of cells with permeation agent, dimethylsulfoxide (DMSO) further enhanced the secretion.

Table 2.35 *Medicinal compounds produced through in vitro biotransformation*

S.No	Species	Precursor	Culture type	Final products
1.	*Catharanthus roseus*	Accumulated alkaloid	CS	Indole alkaloid
		Mevalonic acid Secologanin	CS	Monoterpene alkaloids
		Loganin / Loganic acid		
		Absciscic acid	CS	Indole alkaloids
	Vanadyl sulphate			
2.	*Cephaelis ipecaluanha*	Shikimic acid	CS	Cephelin
		Phenyl alanine		
3.	*Cinchona ledgeriana*	Cinchonine	CS	Quinoline
4.	*Digitalis lanata*	Methyl digitoxin o-Methyl digitoxin	IC CS	Cardiac glycosides Methyl dogoxin
5.	*Dioscorea balanica*	Cholesterol Norflourazon	C	Diosgenin, phytosteroids
6.	*D.pentaphylla*	Cholesterol	C	Diosgenin
7.	*N.tabacum*	Secologanin, Tryptamine	CS	Strictosidine
8.	*Rauvolfia serpentina*	Ajmaline	CS	Unknown compounds like raumachine
9.	*Sanguinaria canadensis*	Calcium chelate Calimoduline	CS	Benzophenathridine
		Quercetin	CS	Benzophenathridine
10.	*Valerina wallichii*	Colchicine	CS	Valepotriate & 14 new compounds

3. ***Cell immobilization :*** Capsaisin, an alkaloid used in pharmaceutical preparations as a digestive stimulant and for rheumatic disorders has been produced in *vitro* using immobilized cells freely suspended cell cultures of *Capsicum frutescens*. Kim and Chaing (1990) showed the enhanced production of shikonin by the immobilized cell culture of *Lithospermum erythrorhizon.* Furuya *et al.* (1984) reported transformation of codeninone to codeine using immobilized cell culture of *Papaver somniferum.*

4. ***Elicitation :*** Fungal extracts and certain chemicals have been used as elicitors of cells of production by feeding tryptophan to *Cinchona succirubra* suspension culture. Fungal extract was used as elicitor in *Papaver somniferum* cell culture to induce enhanced secretion of sanguinarine and 5-adenosyl methonine from *Ruta graveolans.* Addition of precursor secologanin increased ajamaline production tenfolds in *Rauvolfia serpentina.*

5. ***Hairy root cultures :*** Roots are good source of variety of natural products like tropane alkaloids, cathranthus alkaloids, atropine, hyoscyamine, scopolamine and steroidal precursors. Considerable attention has been given in recent years for obtaining secondary plant products by hairy root culture technique. In summary, plants belonging to 15 families are known to produce major secondary products in root cultures. The ability to exploit plant root cultures as a commercial source of natural products depends on the development of a suitable bioreactor, which increases the ability of plant root culture as a commercial source.

Two main types of root cultures are employed, viz. :

1. Untransformed excised root cultures and
2. Transformed hairy root cultures are recognized. Even though the former type provide good source system for the accumulation of secondary compounds, the exploitation of the latter seems to have greater potential for secondary metabolite production.

Recently, attempts have been made to induce root tumours in plants using the bacteria *Agrobacterium rhizogenes* that carries Ri plasmid which cause hairy root disease. Root tumours obtained in this way grow rapidly in culture and have been found to be very useful for the production of phytochemicals.

The transformed roots are genetically stable and retain their biosynthetic activity for longer periods. The growth in these roots is comparable to that of cell suspensions and is faster than that of normal roots. Genes required for the expression of biosynthetic pathways can be transferred and expressed in hair root cultures.

A wide variety of compounds have been synthesized through the transformed or hairy root cultures. The secondary metabolites produced from hairy root cultures exhibit several advantages over the cell cultures such as 1. the relative ease of culturing, 2. simple media they require, 3. the genetic stability 4. rapid growth with greater accumulation of biomass in short period, 5. steady level of product synthesis over a relatively long period and 6. the production of new compounds which are absent in cell cultures but present in roots of intact plants. Genetic transformation of excised roots of *Atropa belladonna* with *Agrobacterium rhizogens* caused increased production of tropanes in hairy roots. Shimomura *et al* (1991) a hairy roots have been produced in the antitumor drug producing plant. (*Euphorbia lathyris :* Follansbee *et al* 1995) anti malarial drug producing plant. *Artemisia annua* (Weathers *et al* 1997), parasympathicolotics drug producing *Datura stramonium* Sikuli and Demeyer, 1997). The hairy roots of *Rubia peregrine* contained double the amount of anthraquinones as compared to one year old field grown plants (Lodhi et al 1996).

Bioreactors

Large scale cultivation of plant cells is carried out in bioreactors. Bioreactor is a large culture vessel made up of glass for use at laboratory scale (up to 10L) but large scale reactors are made up of stainless steel. These are fitted with microprocessor control unit for the control of pH, dissolved oxygen, gas flow rate, agitation, speed nutrient factors, temperatures inside the vessel and cell density for optimal growth and/or production. Small reactors are autoclaved while commercial scale reactors are sterilized in *situ* by passing jet steam at suitable pressure. Various types of bioreactors have been designed depending upon the requirements.

Types of Bioreactors

Depending upon the mode of agitation, bioreactors can be classified into following types.

(a) Mechanically agitated bioreactors (employs Rush tone turbine)

(b) Pneumatically agitated bioreactors

(c) Surface-immobilized plant cell bioreactor

Pharmaceuticals

During the past three decades several phytochemicals with biological activities have been identified from higher plants. The potential market for plant secondary products (Table 2.36). Vincristine, vinblastine, artemisinin, taxol and podophyllotoxin are in the list of approved drugs : some products under clinical trials are Forskolin, Castanospermine, Ginkgolides and Hypericin. The products listed are of very high value ranging from US $ 0.5 million to US $ 9 million per kilogram and these are excellent candidates for production through biotechnological approach.

Table 2.36 The potential market for plant secondary products

Compound	Use	Estimated retail market ($US million)
Vinblastine/vicristine	Leukaemia	18-20 (USA)
Ajmalicine	Circulatory problems	5-25 (World)
Digitalin	Heart disorders	20-55 (USA)
Qyuinine	Malaria, flavour	5-10 (USA)
Codeine	Sedative	50 (USA)
Jasmine	Fragrance	0.5 (World)
Pyrethrins	Insecticide	20 (World)
Spearmint	Flavour, fragrance	85-90 (World)

Production of food additives such as pigments, flavours and fragrances is also an important application of Plant Biotechnology. Cell cultures are being investigated in different laboratories all over the world for the production of crocin, Betaxanthines, Bixin and Anthocyanins. Amongst the flavours and fragrances capsaicin and vanilla are commercially important. The global vanillin requirement is around 2000 MT of which only 20 MT is from the natural source. The consumer preference for natural vanilla costing US $ 2500/kg as against synthetic vanillin US $ 50/kg makes it a potential plant for tissue cultures. Escagenetics Corpn, USA have a patent for production of vanilla by plant cell cultures (Angela stafford, 1991). The aromas of cocao and coffee have also been produced by cell cultures.

At present there are 35 biotechnology derived therapeutic and vaccines approved by the US FDA for medical use and over 500 additional drugs and vaccines by 150 companies are in various phases of clinical trials. Biotechnology has also spurred growth in diagnostics and over 600 biotechnology based diognostics are now available in clinical practice.

Drugs/Medicine derived from fungi and lichens need mention. At least 80% of antibiotics in use are derived from actinomycetes. Pencillin (*Pencillium*) Ergot alkaloids from *Claviceps* are widely used. According to Richardson (1988) lichens have a long history of usage of herbal medicine. Several antifungal and antibacterial compounds and about 350 secondary metabolites contain pharmacologically active compounds.

ESSAY QUESTIONS

1. Given an account of plant tissue culture and its application.
2. Write an account of plant growth regulators.
3. Differentiate between auxins and gibberellins.
4. Differentiate between auxins and cytokinins.
5. Write an essay on micropropagation in plants.
6. Write an account of important achievements in plant biotechnology.
7. What is somatic embryogenesis and its applications ?
8. What are the various important approaches to create new plant varieties ?
9. Describe the mechanism of Ti mediated transformation.
10. Differentiate between hybrids and cybrids.
11. What are transgenic crops.
12. Comment on the applications of GM crops.
13. Write an essay on plantibodies.
13. What are plantibodies ? What are their applications.
14. What is Virus and insect resistance ? Elaborate the role of genetic engineering.
15. What is herbicide resistance ? How this has been achieved by genetic engineering ?
16. Write an account of genetically modified food.
17. What is phytoremediation ? Discuss its advantages and disadvantages.
18. What are bioreactors ? Describe their importance and discuss their in genetic engineering.
19. What are Restriction Enzymes ? Describe their role in genetic engineering.
20. What are Haploids ? Describe their importance.
21. Write an account of medicinal plants.
22. What are the threatened Indian medicinal plant ?
23. What are drugs ? Categorize and list their sources.
24. Describe the various categories of drugs
25. Give an account of phytochemicals derived from plant tissue culture.

SHORT QUESTIONS

Explain the Following :

1. Define Tissue Culture.
2. Totipotency.
3. Callus.
4. Haploids.
5. Somatic Embryogenesis.
6. Name some Micropropagated Plants.
7. Name some famous Tissue Culture Laboratories of India
8. Define Biotechnology.
9. What are Vectors.
10. Restriction Enzymes.
11. What are Plasmids.
12. What are Ti-plasmids.
13. Name some Transgenic Crops.
14. Edible Vaccines.
15. Phytoremediation.
16. What are Antidiabetic Plants ?
17. What are Antimalarial Plants ?
18. Name Anticancerous Plants.
19. Anti HIV Plants.
20. Define Gibberllins.
21. Name some auxins.
22. Name some Cytokinins.
23. What are Polyamines ?
24. What are Brassinosteroids ?
25. *Agrobacterium*.
26. List GM medicinal plants.
27. Name some transgenic vegetables.
28. Differentiate between plasmids and cosmids
29. Aromatic plants
30. Suspension culutres

3 Animal Science

3.1 Introduction to Animal Kingdom

Zoology (Gr. *Zoon* = animal, *logos* = discourse) deals with the study of animals. The aim of Zoology is to understand the structure and body functions, relationship to one another and to the surroundings in which they live, their classification, development, origin, past history (evolution) and economic importance.

Animals are found everywhere on the earth : in water, on land and in the air. They are found underground, in deserts, in thermal and sulphur springs, on high mountains, in deep wells, and even within our own bodies. There is in fact, not a square inch of the earth's surface which is not inhabited by some kind of animal or other. Over a million different kinds of species of animals living at the present time have been named and described by zoologists.

The animals first appeared on the earth nearly a thousand million years ago. They all first lived in the sea and were mostly minute and very simple in organization, After a long geological time, they gradually became more and more complex and developed various organs for performing different, often complex functions. Many of them left the sea in search of food or to escape from enemies and gradually occupied the land or even invaded the air. Man, is also an animal, came last of all, hardly a million years ago.

Zoology has many branches. They are the following ones.

1. *Morphology* : The study of external features of organisms.
2. *Physiology* : The study of functional mechanisms of animals.
3. *Taxonomy* : The study of classification of animals.
4. *Endocrinology* : The study of endocrine glands and the hormone actions in animals.
5. *Embryology* : The study of development of animals from its beginning till they attain the adult stage.
6. *Ecology* : The study of organisms in relation to their environment.
7. *Genetics* : The study of heredity and variations of living organisms.
8. *Evolution* : The study of origin and differentiation of animals over a long period of time.
9. *Palaeontology* : The study of fossil remains of extinct animals.
10. *Zoogeography* : The study of distribution of animals on the surface of earth.
11. *Cytology* : The study of the structure and functions of individual cells and their components.
12. *Pathology* : The study of nature of diseases, causes and symptoms of pathogenic organisms.

The following branches are restricted to the special study of a particular group of animals.

1. *Osteology* : The study of bones
2. *Protozoology* : The study of unicellular organisms i.e.. protozoans.
3. *Helminthology* : The study of helminthes.
4. *Parasitology* : The study of parasitic animals.
5. *Entomology* : The study of insects.
6. *Malacology* : The study of molluscs.
7. *Ichthyology* : The study of fishes.
8. *Herpetology* : The study of amphibians and reptiles
9. *Ornithology* : The study of birds.
10. *Mammology* : The study of mammals.

Besides these, based on the practical application, there are certain other branches, such as :

1. *Animal Husbandry* : Breeding of domestic animals.
2. *Poultry* : Fowl rearing for egg production.
3. *Apiculture* : Bee-keeping for honey production
4. *Sericulture* : Silk worm culture for raw silk production.
5. *Aquaculture* : Cultivation of fish and other aquatic animals.
6. *Piggery* : Pig rearing for pork production.
7. *Dairy Science* : The science of milk and milk products.
8. *Vermiculture* : Rearing earthworms for manure production.

3.2 Numerical Taxonomy

Taxonomy is the science of classification of organisms which includes principles, procedures and rules for nomenclature. The term taxonomy was proposed by the French botanist De Candolle.

The principal objectives of taxonomy are :

1. To discriminate among organisms and to provide means for the subsequent identification of the discriminated entities.
2. To develop a suitable procedure (nomenclature) for designating them for reference purposes and
3. To device and perfect a scheme of classification in which the named species or taxa can be arranged.

The functions of the classification are :

1. To provide a means for the communication and retrieval of information concerning each of the 1.9 million different animal species that are identified so far.
2. To facilitate the gathering of new information by permitting the prediction of characters in unfamiliar organisms and
3. To demonstrate at once the unity and diversity of organic life by expressing the evolutionary origins of various taxa.

The basic raw materials of taxonomy are the individual organisms which make up the species, sub species and other population units which arc found in nature. It is estimated that 1.9 million different animal species have been described in the zoological literature and that each year, about 10,000 new ones are identified and added. Taxonomy's necessity is evident when one imagines trying to gather information on a particular unfamiliar organism.

The first person to collect and organize knowledge of animal classification was Aristotle (384-322 BC). He classified the animals by their way of living, their actions, their habits and their bodily parts.

Later, English Biologist John Ray (1627 - 1705) devised a system of classification utilizing structural similarity as the basis.

Carolus Linnaeus (1707 - 1778), a Swedish botanist was the first to truely systematize taxonomy. He identified 236 different animal species and his classification scheme was based primarily on morphology. He also proposed the *binomial system of nomenclature* which is universally used.

Then, Charles Darwin (1809 - 1882) published a book "*The Origin of Species by means of Natural Selection*" gave a new dimension ie, evolutionary considerations (phylogenetic relationships) also along with morphological differentiations.

Ernst Haeckel (1834 - 1919) further refined the taxonomic system in 1886 by representing phylogeny by means of a branching diagram, tree like phylogenetic diagram, called dendrogram.

With further sophistication, genetic, ecological, geographical and morphological characteristics were taken into account giving taxonomy a greater biological perspective.

In view of several functions of a classification ie. information retrieval, prediction of unobserved characters and expression of evolutionary relationships, much interest is now surrounded on the methods by which higher classification is based. The construction of the higher classification involves the evaluation of two distinct sources of evidence 1. the similarity among different organisms and 2. the clues to the relative recency of their common ancestry. For this, two separate approaches to classification are recognized. 1. The phenetic and 2. the phyletic.

The phenetic approach places emphasis on the similarities and differences existing among taxa as judged by their phenotypes. The phyletic approach emphasizes the reconstruction of ancestral lineages.

Degree of similarity is determined by the number of taxonomic characters which species have in common. For this purpose, characters may be drawn from the same sources (comparative morphology, comparative physiology, comparative ecology) as those utilized for taxonomic discrimination. A rigorous form of phenetic classification is called Numerical Taxonomy.

Numerical Taxonomy

It is the numerical evaluation of the affinity or similarity between taxonomic units and arranging these units into taxa on the basis of their affinities. Numerical taxonomy concept is based on the use of maximum number of characters, and that too, all are given equal weight. The characters may not be necessarily derived only from the external or internal morphology but may now include any attributes of the operational taxonomic units (OTUs) ie, bio chemical, behavioural, cytological, ecological, developmental, etc. Distinct taxa can be constructed due to the diverse character correlation in various groups under study. Following this method. ie., numerical methods on mites, Moss (1967) designated a number of new families and genera. In the mite, *Dermanyssus,* he used 135 characters belonging to 15 OTUS from all areas of the body. Other important works on numerical taxonomy are those of Michner and Sokal (1957), Crowson (1970), Lerman (1970), Jardine and Sibson (1971) and others. Dr. J. A. Peters devised a computer programme for the identification of genera of Central South American snakes.

A character may exist in two or more conditions or states eg. different number of teeth. Each of the initial taxa or *OTUS* is examined for each character and the state of the character recorded. Numerical methods are used to evaluate the degree to which each OTU is similar to each of the other OTUS under study. The extensive computations usually require the services of an electronic computer. The result of this analysis called the *similarity matrix* contains the most information about how similar or dissimilar an OTU to each of the other OTUS. Using numerical methods the OTUS may be grouped or clustered together, beginning with those having the highest index of similarity and progressively grouping those less similar into a hierarchic scheme. The classification may be graphically represented as a tree like diagram called a dendrogram or phenogram. However phenogram should not be confused with a phylogenetic diagram because the phenogram does not imply ancestral lineages. The strictly phenetic classification is well suited for prediction because of its statistical basis but may fail to reflect the evolutionary origins of the taxa.

The orthodox taxonomists do not accept the claims of numerical taxonomists regarding their ability to establish real blood relationships between organisms. Blackwelder (1967) and many others also doubt the usefulness of numerical methods. The approach is exposed to the great risk of reaching unsound classification because by giving equal weight to all characters, it does not allow for mosaic evolution, special adaptations, convergence, parallelism, development and genetic homeostasis in addition to evolutionary, genetic and developmental phenomena that disturb the expected close relation between phenetic similarity and phylogeny. Moreover, the use of complex mathematical and statistical methods by numerical taxonomists has further gone against its due recognition because of great difficulties by the biological taxonomists to follow them.

Yet today all systematics to some extent are numerical. Computers and numerical taxonomic programmes are now standard resources in every museum and systematics library all over the world.

3.3 Phylogeny of Invertebrate and Vertebrate Phyla

The animal kingdom is the most diverse kingdom with more than one million species. Animals are usually multicellular heterotrophs which ingest food and are generally mobile. A few animals are sedimentary and are attached to the substratum.

Primitive animals originated in the sea. Later fresh water and terrestrial animals evolved from these organisms. Only echinoderms are exclusively marine and have no fresh water or terrestrial forms. The aquatic animals may be benthic, pelagic or planktonic. Some terrestrial animals have evolved the capacity to fly. Several groups such as protozoans, helminths, Arthropoda etc., have evolved parasitic forms also.

The animal body construction is characterized by a body plan (i.e., either made up of clusters of cells and no tissues or organs as seen in sponges or like a blind sac i.e., the digestive cavity with only one opening for ingestion as well as egestion or a tube within a tube plan i.e., the digestive system is open at both ends for ingestion and egestion and also with a body cavity, the body symmetry (asymmetrical, with no definite shape; radial symmetry - body can be divided into two equal halves in any plane; bilateral symmetry - body can be divided into two equal halves in only one plane), the body cavity or coelom i.e., a cavity between the digestive tract and the body wall situated between two mesodermal layers (acoelomate with no body cavity, Pseudo coelomate - false body cavity and coelomate with a true body cavity), body segmentation and appendages (locomotory, sensory and feeding) and a body supporting system (endoskeleton) and for protection by way of exoskeletal structures.

The number of germ layers formed during the development is important. It may be diploblastic (i.e., ectoderm and endoderm) or triploblastic (ectoderm, mesoderm and endoderm).

Based on the presence or absence of notochord, the animals are grouped into non-chordates or invertebrates (with no notochord or vertebral column) and chordates or *vertebrates* (with a true notochord or vertebral column).

The major non chordate phyla are : *Protozoa, Porifera, Cnidaria* (*Coelenterata*), *Platyhelminthes, Nemathelminthes, Annelida, Arthropoda, Mollusca* and *Echinodermata.*

The major groups of phylum Chordata are Cyclostomes, Fishes, Amphibians, Reptiles, Birds and Mammals.

Protozoans are unicellular or acellular organisms. All the life activities are carried on by a single cell. These are free living, fresh water or moist terrestrial habitats. Reproduction is by asexual or sexual means. Examples of Protozoa are *Amoeba*, *Plasmodium*, *Euglena*, *Paramoecium* etc. (Fig. 3.1)

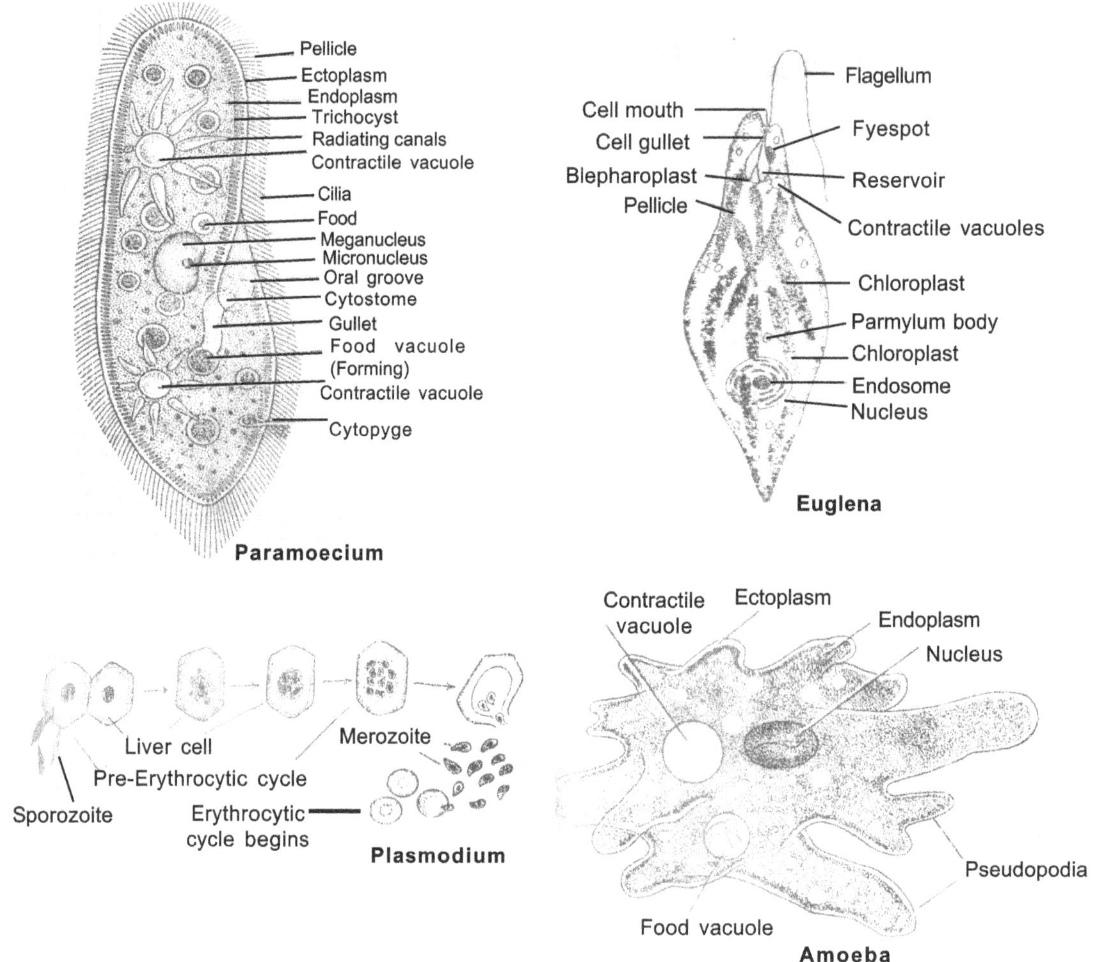

Fig. 3.1 Protozoans

Phylum *Porifera* comprises sponges. These are primitive, sessile, often asymmetric, mostly marine forms and a few fresh water forms are present. These are multicellular organisms with the cell aggregate body plan. There are numerous pores on their body surface connected by way of canals to a central hollow chamber. Life activities are distributed to several specialized cells i.e., division of labour is seen. Reproduction by asexual and sexual means. They have a high power of regeneration. Examples of Porifera are *Sycon*, *Grantia*, *Spongilla* (fresh water sponge), *Euplectella*, *Hyalonema* etc. It is generally believed that sponges were derived from the choano flagellate ancestors of protozoa as these highly specialized cells are found in choanoflagellates and sponges. Sponges depend on these cells for existence. (Fig. 3.2)

Coelenterates or Cnidarians (Phylum *Cnidaria*) are aquatic, mostly marine, radially symmetrical animals with a blind sac type digestive system i.e., ingestion and egestion take place from the same opening. These are diploblastic with ectoderm and endoderm layers. Low grade of tissue formation

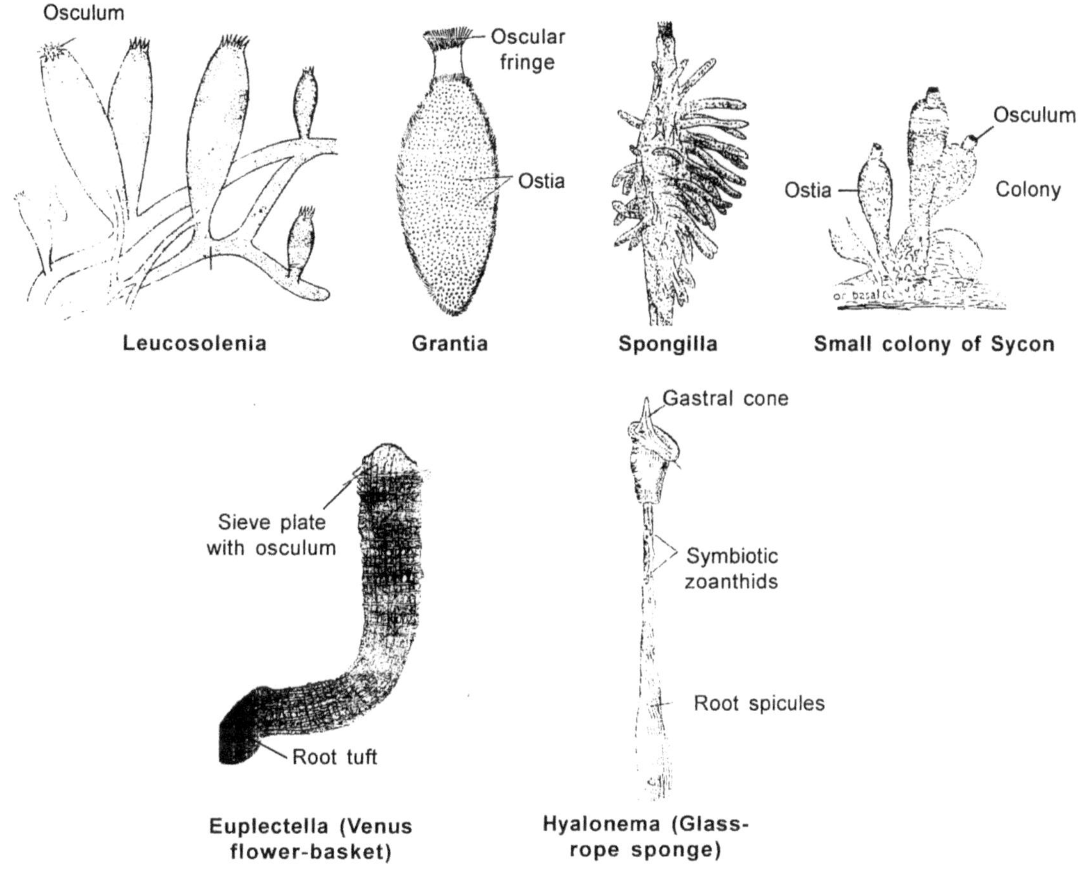

Fig. 3.2 Poriferans

is seen. They have nerve cells forming a nerve net and stinging cells. Most Cnidarians exhibit two body forms, sessile hydra like polyp form and jelly fish like, free living, swimming medusa. Corals are polyp colonies which secrete a protective calcareous skeleton. Body generally sensitive, some have sensilla such as eye spots, statocysts etc. (Fig. 3.3)

Tentacles are present encircling the mouth serving for food intaking, capturing the prey and defence. Reproduction takes place by asexual and sexual methods. Larval forms such as Planula, Scyphistoma, Ephyra etc. are there in the development of certain Cnidarians like *Aurelia.*

Examples of this Phylum are *Hydra, Obelia, Halistemma, Porpita, Physalia* (Hydro zoans), *Aurelia* (Scyphozoan) *Adamsia* (Sea Anemone), and Corals such as *Corallium, Tubipora, Gorgonia, Pennatula* etc (Anthozoans).

The evolutionary relationship of Cnidarians is much controversial. Some are of the opinion that Anthozoans (corals) are derived from polyp (asexual form) of Scyphozoans (*Aurelia, Rhizostoma* etc). A few others are of the opinion that actinula larva of hydrozoans (*Hydra, Obelia* etc) is a recapitulation of ancestral coelenterates from which medusoid form (sexual form) appeared. General

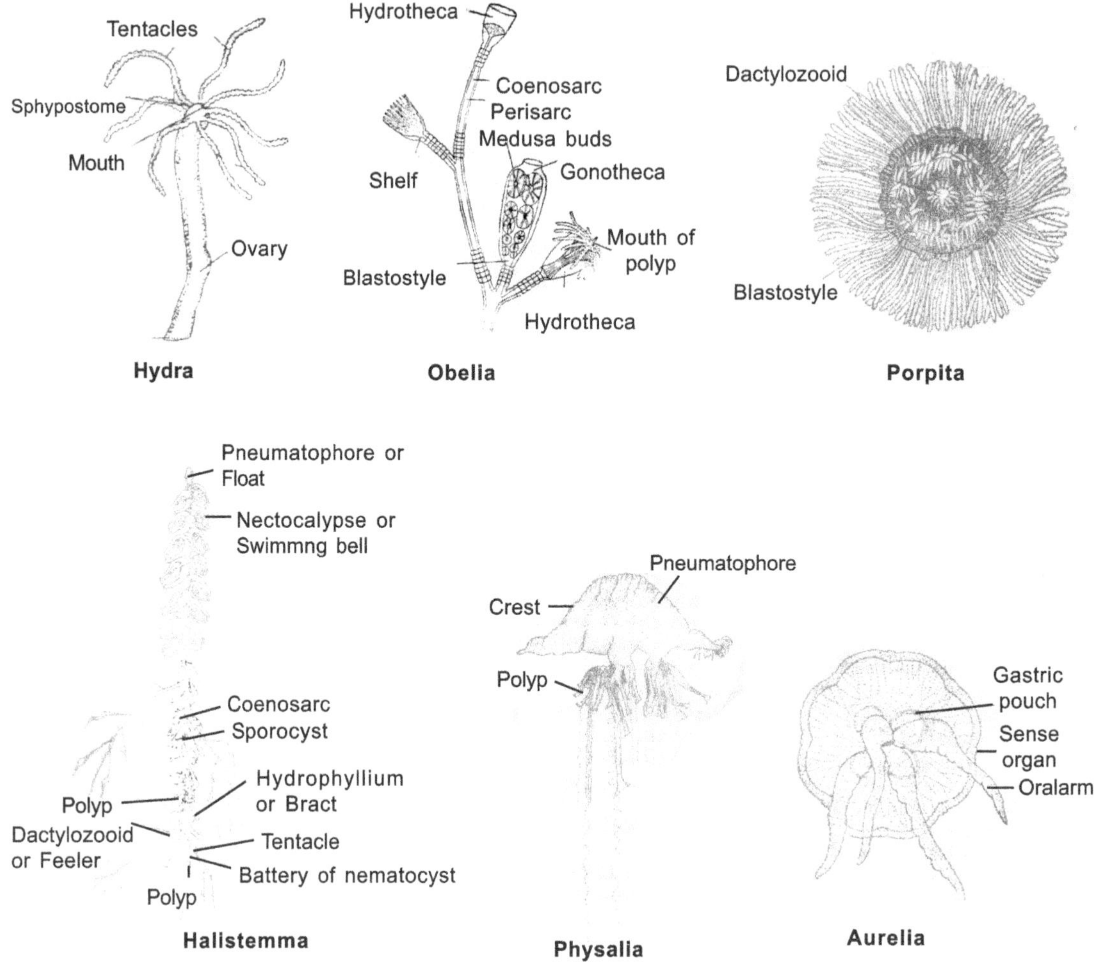

Fig. 3.3 Coelenterates

and most accepted opinion says that cnidarians arose from a polypoid ancestral form. The Scyphozoans and anthozoans evolved from a common polypoid ancestor and the medusae arose independently as a means of dispersal.

Flat worms (Phylum *Platyhelminthes*) are either free living, primarily aquatic animals like planaria or parasites such as flukes and tape worms. These animals are acoelomate, triploblastic, bilaterally symmetrical animals and they are devoid of anus (i.e. blind sac body plan).

Tissues organized into definite organs. Excretory organs are peculiar flame cells. Nervous system is primitive and consists of a pair of ganglia with longitudinal nerve cords. These are hermaphrodites (i.e., both the sexes are present in the same organism). Reproduction takes place by asexual and sexual methods. Development includes several larval stages such miracidia, sporocysts, redia, cercaria, metacercaria, hexacanths, cysticercus etc., in Trematodes and Cestodes. (Fig. 3.4)

Examples of *Platyhelminthes* are *Planaria* (*Dugesia*), Liver fluke (*Fasciola hepatica*), tape worm (*Taenia solium*) etc.

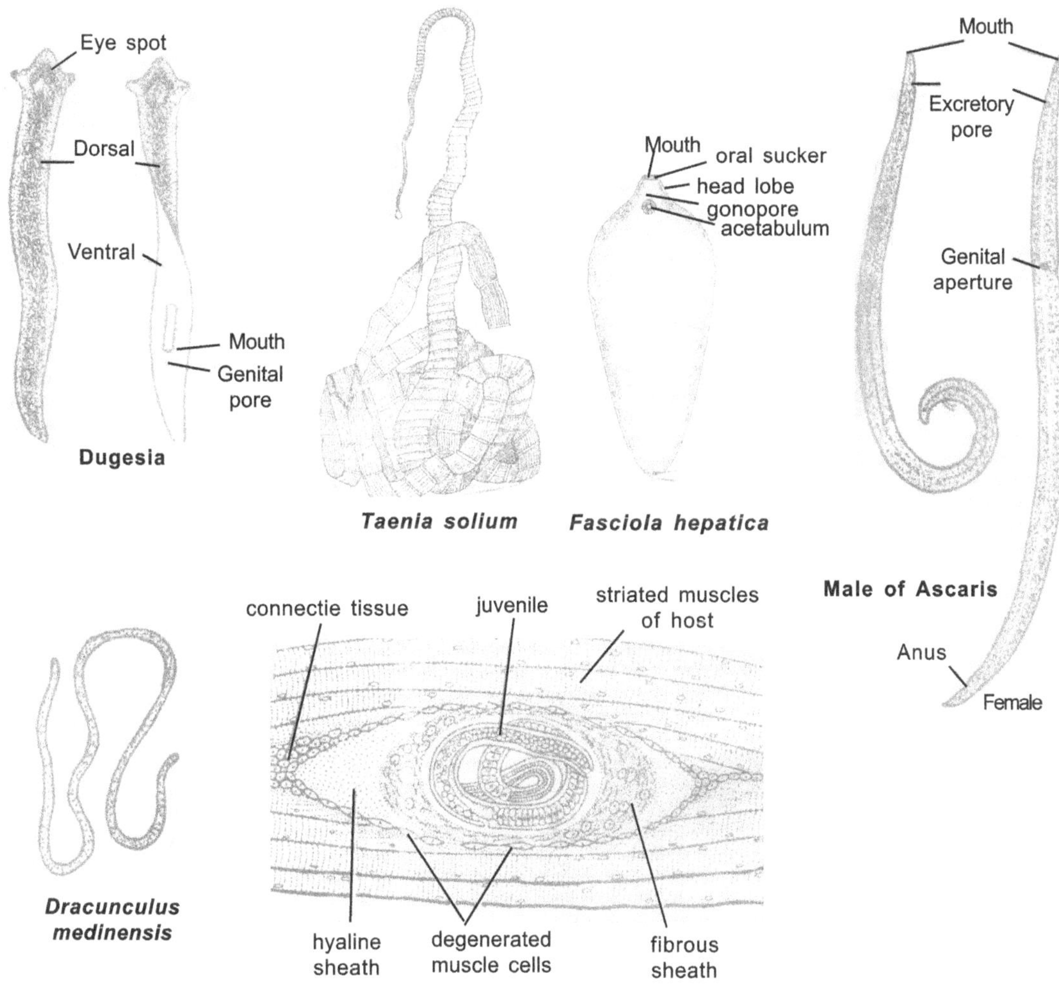

Fig. 3.4 Helminthes

Round worms (Phylum *Nemathelminthes*) live in the soil or as parasites in animals and plants. They are triploblastic, bilaterally symmetrical with a tube within a tube body plan i.e., mouth and anus are present in the digestive system. Their false body cavity or pseudocoelom allows the body wall and the digestive tract muscles to act independently. Sexes are usually separate. Fertilized eggs have a thick wall and survive adverse conditions. Parasitic round worms in human beings include hook worm, guinea worm, filaria worm, pin worm, etc.

The flat worms must have originated from a radial ancestor. This ancestor was a planuloid form resembling the planula larva of Cnidarians. It had an elongated, radially symmetrical, ciliated body having endodermal cells and nerve net having neurosensory cells concentrated towards the anterior end. Nutrition in this ancestral form continued to be of the protozoan type and hence mouth and enteron were not developed, being not required. This planuloid ancestor may be supposed to have given rise to both the coelenterate and helminth (Bilateral) lines. It gave rise to the Coelenterate line by the formation of mouth at the place where the blastopore closed, hollowing of the endodermal mass to form an enteron. It gave rise to the bilateria by assuming creeping mode of locomotion, flattening of the body, developing bilateral symmetry and forward shifting of nervous centre.

The planuloid ancestor resembled the Acoela. Acoela are the most primitive Platyhelminthes. The acoeloid form later produced the higher turbellarian by hollowing out of the central mass to form an enteron and greater differentiation of anterior region as the head. The Trematodes and Cestodes arose independently from the parasitic Rhabdocoel ancestors (a group of *Turbellaria*).

Nematodes resemble platyhelminthes in the absence of segmentation, lack of circulatory system and nature of nervous system. However they differ from them in shape, in the absence of cilia, circular muscles and suckers, presence of complete digestive system and a pseudocoel, separate sexes etc.

Some workers hold that the flame cells have secondarily disappeared in the nematodes. This view brings the excretory system of nematodes close to that of the flat worms.

Among the nematodes themselves, the marine forms are the most primitive. The terrestrial and fresh water nematodes have evolved from marine ancestors and themselves gave rise to the parasitic forms.

Annelids or *segmented worms* (Phylum *Annelida*) include polychaetes, earthworms and leeches. They are bilaterally symmetrical, eucoelomate animals. A true, closed circulatory system with 'hearts' and oxygen carrying blood is noted for the first time in annelids in the evolutionary history of animals. Excretory organs are nephridia. The nervous system consists of a brain ganglion and a double ventral nerve cord. Annelids are either hermaphrodites or the sexes may be separate. (Fig. 3.5)

A fairly close relationship exists between Annelida, Mollusca and Arthropoda than helminthes. The Annelida and Mollusca resemble with each other in having spiral cleavage, mesoderm derived from special embryonic cells and a ciliated free swimming trochophore larva. The Annelida resemble the Arthropoda in having a segmented body covered by a cuticle secreted by epidermis, in the plan of the nervous system and in forming mesoderm from special embryonic cells. This resemblance suggests that all these phyla have descended from a common hypothetical ancestor called the Trochozoon.

The Trochozoon gave rise to the ancestral annelid from which several independent lines of evolution diverged developing into various families of polychaetes. The present archiannelids are specialized survivors of the ancestral annelid. The aquatic oligochaetes were evolved from the polychaetes and gave rise to the terrestrial Oligochaeta. Leeches are derived from some Oligochaete ancestor. *Peripatus* belonging to Onychophora is the connecting link between Annelida and Arthropoda. (Fig. 3.5)

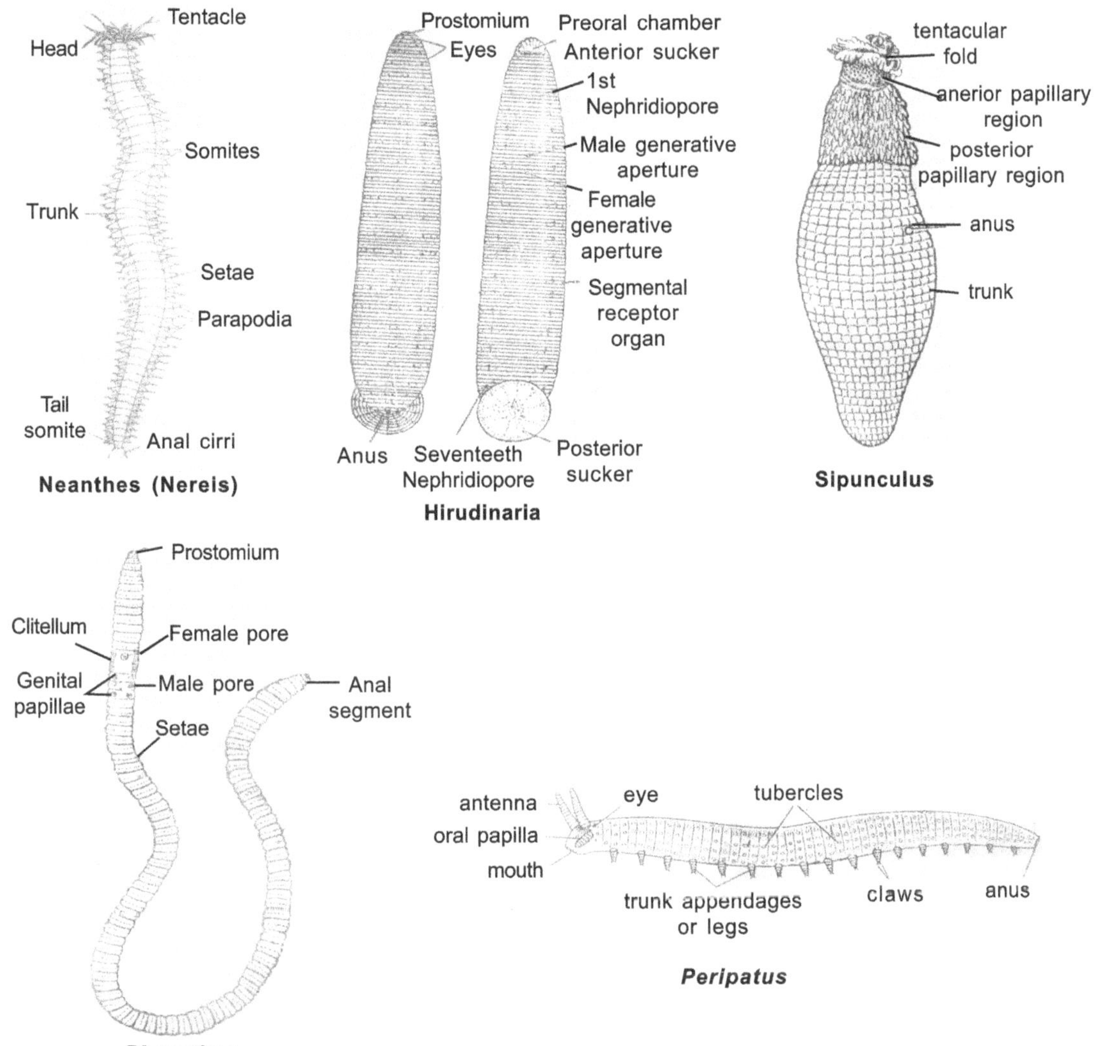

Fig. 3.5 Annelidans

Molluscs (Phylum *Mollusca*) are soft bodied animals. Many possess a protective hard shell which constitute the various sea shells. The body consists of a ventral foot with a dorsal hump containing the visceral organs. The three major molluscan groups are the Gastropods, Bivalves and Cephalopods. Gastropods have a single coiled shell, includes snails and the shell-less slugs. Their larvae are bilaterally symmetrical, (eg. Glochidium larva) but this symmetry is lost in adults because of the twisting of the viscera as the shell formed.

Bivalves have two hinged shells and include oysters and clams. They are filter feeders. The Cephalopods include shell-less octopuses and squids. Some of these are the largest invertebrates. Their head and foot are fused. They have 8-10 tentacles with suction cups, a well developed brain with learning capacity and image forming eyes. They are at the top of invertebrate evolution in terms of the learned behaviour they exhibit. (Fig. 3.6)

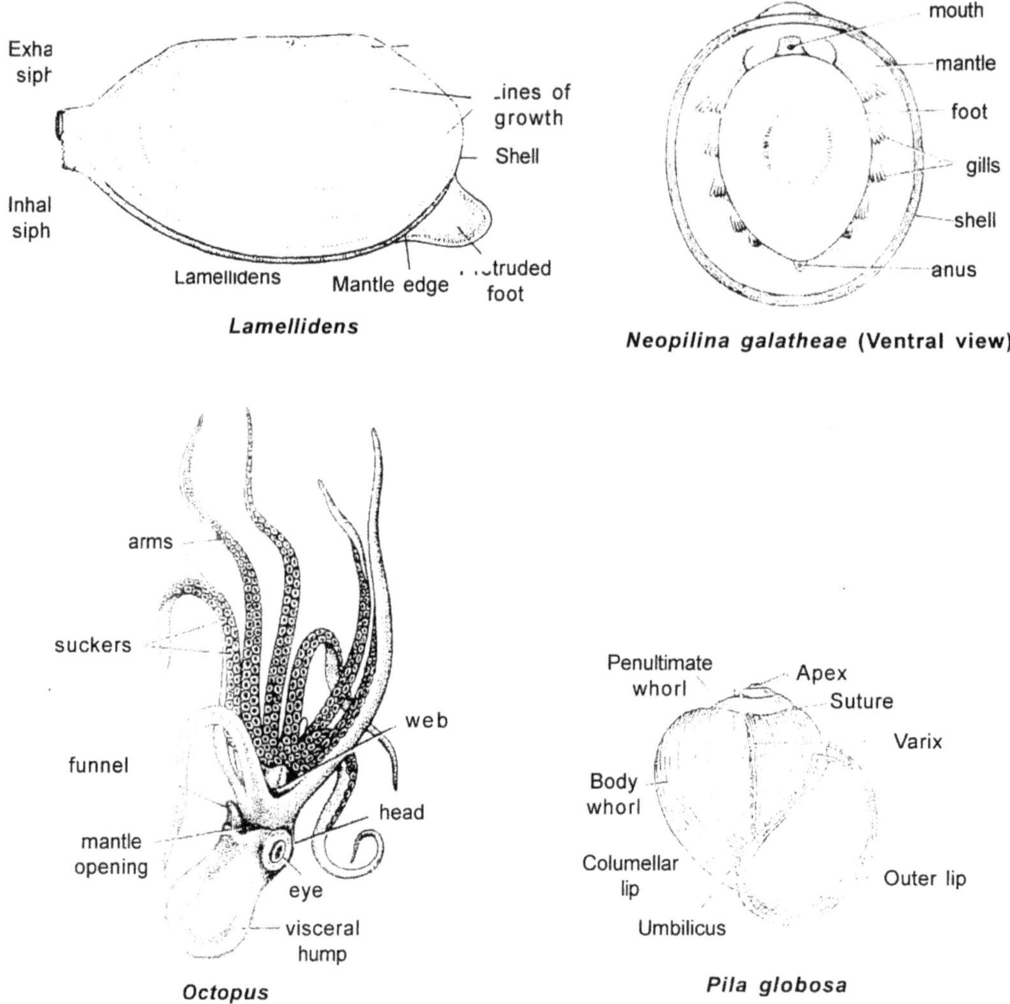

Fig. 3.6 Molluscans

The molluscs show affinities with Turbellarians and marine annelids. They resemble the turbellarians in the ladder like nervous system of their primitive representatives like *Chiton* and others. This resemblance suggests the origin of molluscs from the turbellarian like ancestor.

They also resemble the marine annelids in having spiral cleavage and a trochophore larva. Recently discovered mollusc *Neopilina* have the metamerism similar to annelids which further strengthens the relationship. Presence of a few pairs of ganglia and 1 or 2 pairs of nephridia are possible relics of segmentation in molluscs. All this indicates that the Mollusca and Annelida have descended from a common hypothetical ancestor called the Trochozoon.

The phylum *Arthropoda* is the largest in the animal kingdom. It includes the classes Arachnida, Crustacea, Myriapoda and Insecta. The success of Arthropoda on land is due to their water impermeable and light exoskeleton of chitinous cuticle. Arthropoda are bilaterally symmetrical,

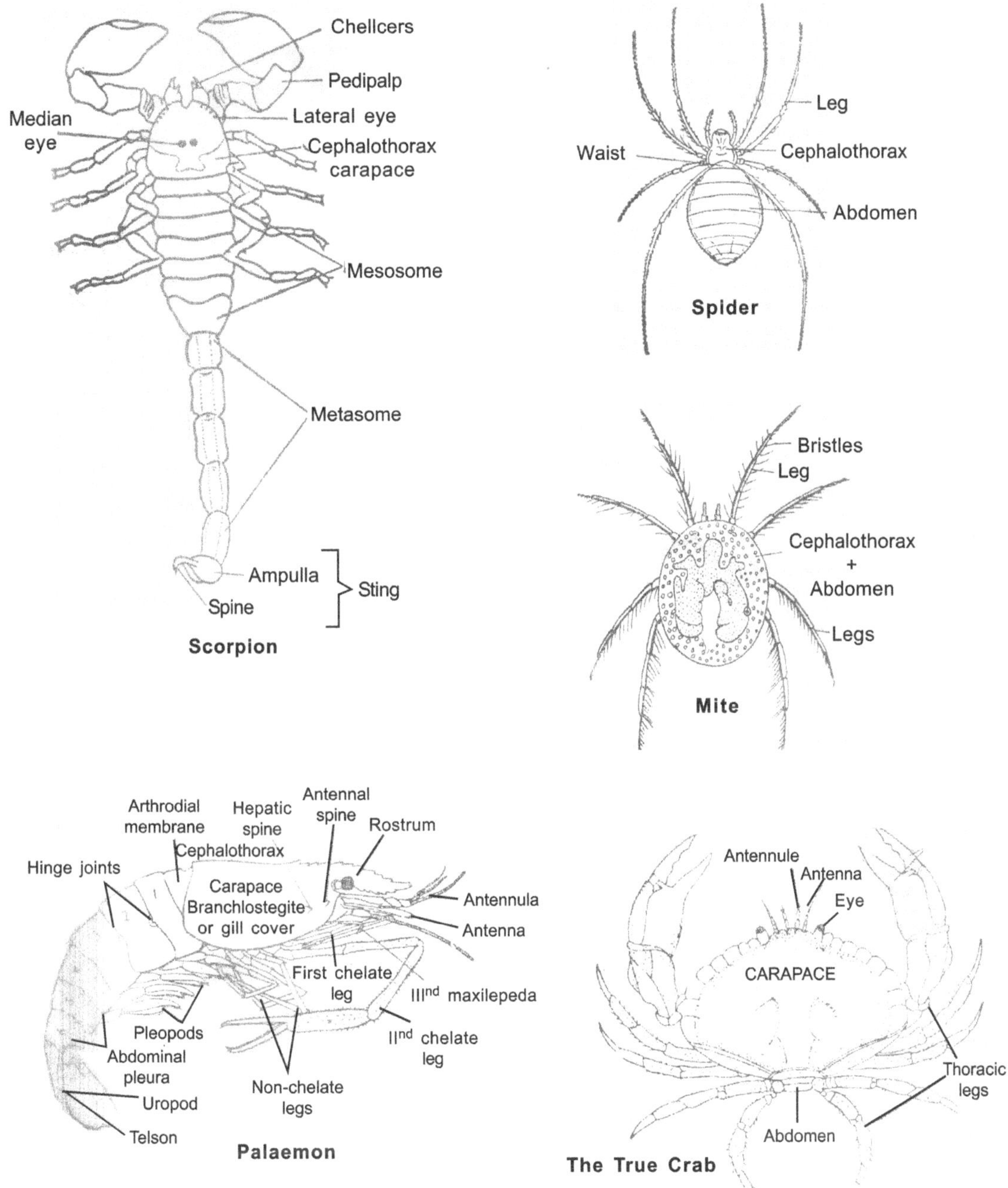

Fig. 3.7 Arthropods

segmented animals with appendages serving as antennae, mouth parts, pincers or legs. Sensory structures in Arthropoda include odour - perceiving antennae, simple and compound eyes, statocysts, receptors for taste or sound. The sexes are separate and fertilization is internal in land Arthropoda.

(Fig. 3.7)

Arthropoda are generally oviparous and a few are viviparous. Development may be direct, or indirect exhibiting metamorphosis as in many insects. The young grow by moulting.

The Arachnids include scorpions, spiders, ticks and mites. These animals have cephalo thorax, abdomen and 6 pairs of appendages. The first 2 pairs help in feeding and the last four pairs are legs. They have no antennae.

Crustaceans are predominantly aquatic Arthropods with a protective carapace and compound eyes. Prawns, lobsters and crabs are economically important crustaceans.

Myriapods include herbivorous millipedes and carnivorous centipedes. Millipedes have 2 pairs of legs per segment while centipedes have one pair. They have simple eyes. Centipedes have poison claws. (Fig. 3.8)

Insects form the most important and the largest group of Arthropoda. They have one pair of antennae and compound eyes. The thorax of 3 segments bears 3 pairs of legs. The abdomen has no appendages. Most insects have 2 pairs of wings. Respiratory gas exchange is through tracheae and excretion is by Malpighian tubules. Insects affect human life in many ways. Common examples

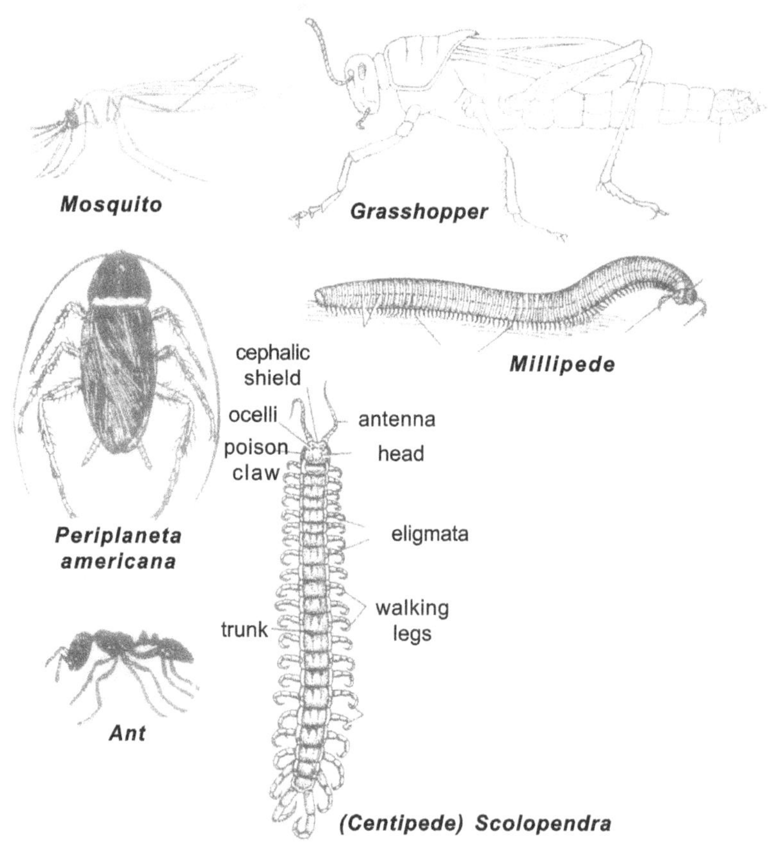

Fig. 3.8 Insects and Myriapodes

are Mosquitoes, houseflies, bed bugs, lice, fleas, cockroaches, grasshoppers, ants, beetles, moths, butterflies, dragon flies and so on.

Arthropoda evolved from the same ancestor as the annelids. This is indicated by a close resemblance of the onychophores with the annelids and arthropods, perhaps, the most important factor that made annelids and arthropods evolve along separate lines is the development of a hard exoskeleton in the arthropod stem in place of a soft cuticle of the worm-like ancestor. The hard exoskeleton was developed obviously after the divergence of the ancient onychophorans. Formation of hard exoskeleton created certain problems which necessitated several other changes. The problem of growth was solved by the process of moulting. To coordinate the activities involved in moulting, an endocrine system was elaborated. With rigid exoskeleton, the body could not bend or change its shape and therefore the annelid type locomotion by lateral undulations or by stretching and shortening of the body became impossible. Hence flexible inter segmental joints were developed and this caused a change in body musculature. The continuous muscular sheet was replaced by bundles of flexors and extensors to bend one part of the body in relation to another. To make locomotion more efficient, segmental appendages with joints and muscle bundles were acquired. Originally, perhaps all segments and appendages were alike. Later, specialization occurred, producing body division, each with a special set of appendages. Regional specialization was followed by reduction in the number of segments. This apparently was of survival value as it made the body compact and hence more efficient. Reduction took place by fusion of segments and their ganglia and by giving up anamorphic growth.

External changes were accompanied by internal changes. The latter included reduction of coelom and segmental excretory organs and formation of haemocoel.

All these changes produced aquatic arthropods (crustaceans). The crustaceans had two main requirements to evolve into land arthropods i.e., mechanism to conserve water and ability to breathe air. Malpighian tubules and tracheae fulfilled these requirements and this paved the way for aquatic forms to come on to the land. Insects added another feather to their cap i.e., power of flight, and this has made them the dominant animals.

Echinoderms (Phylum *Echinodermata*) are marine, spiny skinned animals. Adult echinoderms such as Starfish and sea urchin are radially symmetrical. Larval forms are bilaterally symmetrical. The echinoderms are characterized by a water vascular system and tube feet used for locomotion and capture of prey. The echinoderms are evolutionarily closest to the chordates. The examples are starfishes, sea urchins, sea cucumbers etc. (Fig. 3.9)

The echinoderms were at one time combined with coelenterates in the Radiata on the basis of their radial symmetry. As the coelenterates lack coelom and have tissue grade of organization, and the echinoderms have a coelom and complex organ system grade organization, these differences reject the relationship between these two.

The relationship with annelida was also suggested on the basis of resemblances of the larval stages i.e., trochophore larva of Annelida with Auricularia and Bipinnaria larvae of echinoderms, presence of coelom etc. Sipunculida, a group of Annelida shows many characters such as spiral cleavage, tentacles around the mouth etc. The basic adult anatomy of sipunculida shows that they are in close resemblance with the primitive adult echinoderm.

However, there are certain differences also. Sipunculida is protostomous i.e., blastopore is near the mouth end while echinoderms are deuterostomous i.e., blastopore becomes the adult anus. This difference in mode of development is probably a divergence at an early stage of evolutionary history.

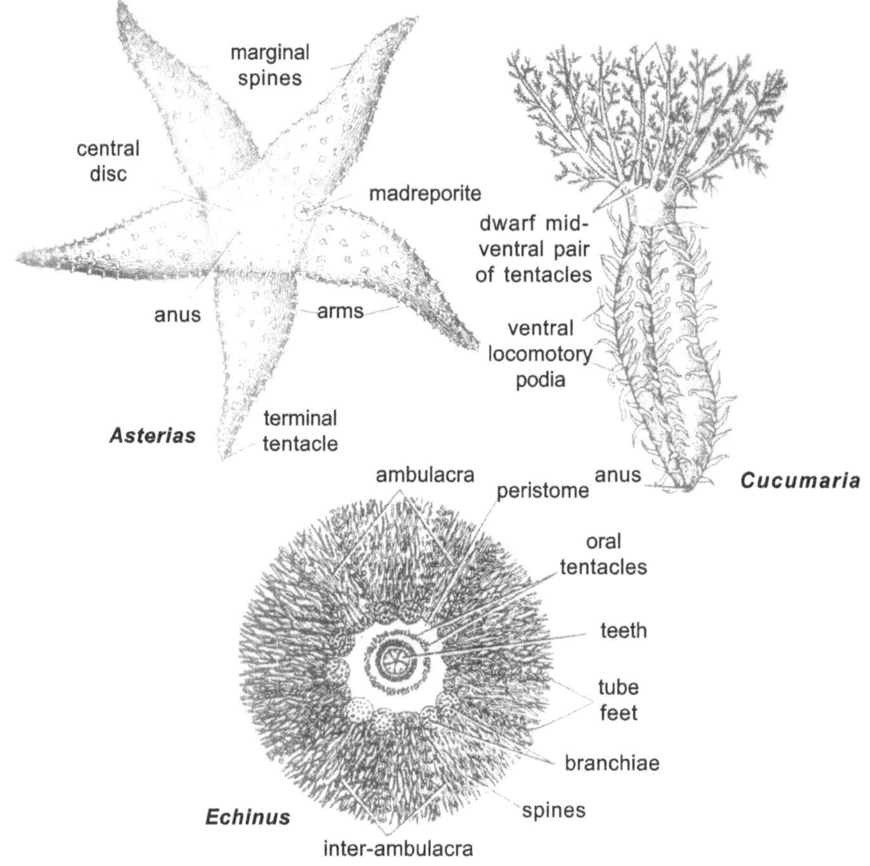

Fig. 3.9 Echinoderms

The protostomous condition is seen in Sipunculida, Phoronida, Ectoprocta and Brachiopoda. All these four are collectively called *Lophophorata* due to the presence of lophophore, which is a hollow, tentacled feeding funnel around the mouth.

The examples of Deuterostomous condition are Chaetognatha, Pogonophora, Hemichordata and Echinodermata. Out of these Chaetognatha is regarded at the base of the line towards the echinoderms and chordates. The remaining three phyla such as Pogonophora, Hemichordatea and Echinodermata have many common features such as enterocoelic retention of blastopore, indirect development, a dipleurula type larva. The larvae of hemichordates and echinoderms are closely related. It is regarded that the dipleurula concept represents a possible larval stage common to echinoderms and Hemichordates.

The Echinoderms resemble the chordates in having :

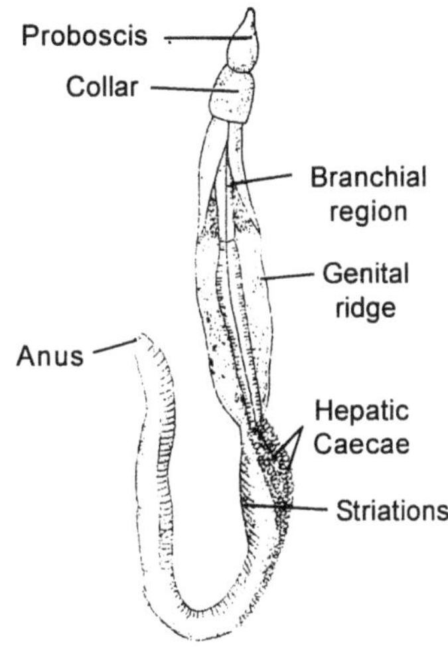

Fig. 3.10 Hemichordates (Balanoglossus)

1. Mesodermal endoskeleton
2. Mouth derived as an ectodermal invagination (Stomodaeum)
3. Enterocoelous coelom.
4. Radial cleavage
5. Adult anus derived from embryonic blastopore
6. Larvae similar to those of hemichordates.

From this resemblance, it is generally held that the Chordates at least the Hemichordates (eg : *Balanoglossus*), have been derived from the echinoderms. This evolution is considered to have taken place through forms produced by shortening of the life cycle of an echinoderm ancestor by omission of the radially symmetrical cleavage. (Fig. 3.10)

The distinctive features of chordates are the notochord, dorsal hollow nerve cord and gill slits on the pharynx. Vertebrates which represent the largest group of chordates have 2, 3 or 4 chambered heart and 2 pairs of lateral appendages i.e., fins or limbs. Vertebrate classes are Agnatha *(jawless*

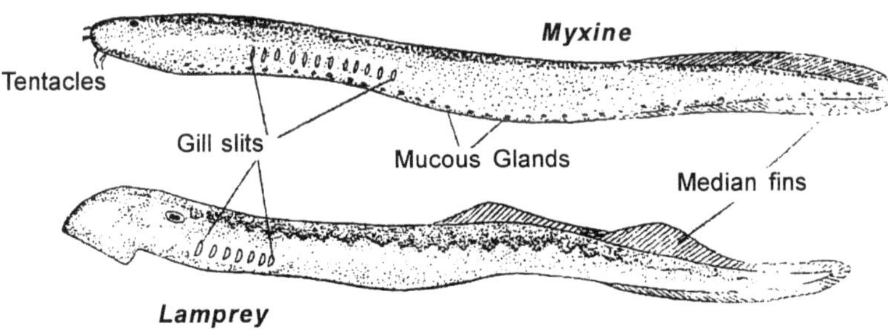

Fig. 3.11 Cyclostomes

vertebrates or *Cyclostomes*), Chondrichthyes (Cartilaginous fishes), Osteichthyes (bony fishes), Amphibia, Reptilia, Aves and Mammals.

Jawless Vertebrates (Agnatha) like Lampreys are ectoparasitic on fishes (Fig. 3.11). Cartilaginous fishes include sharks and rays. They have an endoskeleton of only cartilage and five pairs of gill slits but no gill cover. Electric rays generate electric current to stun prey. The bony fishes (*Osteichthyes*)

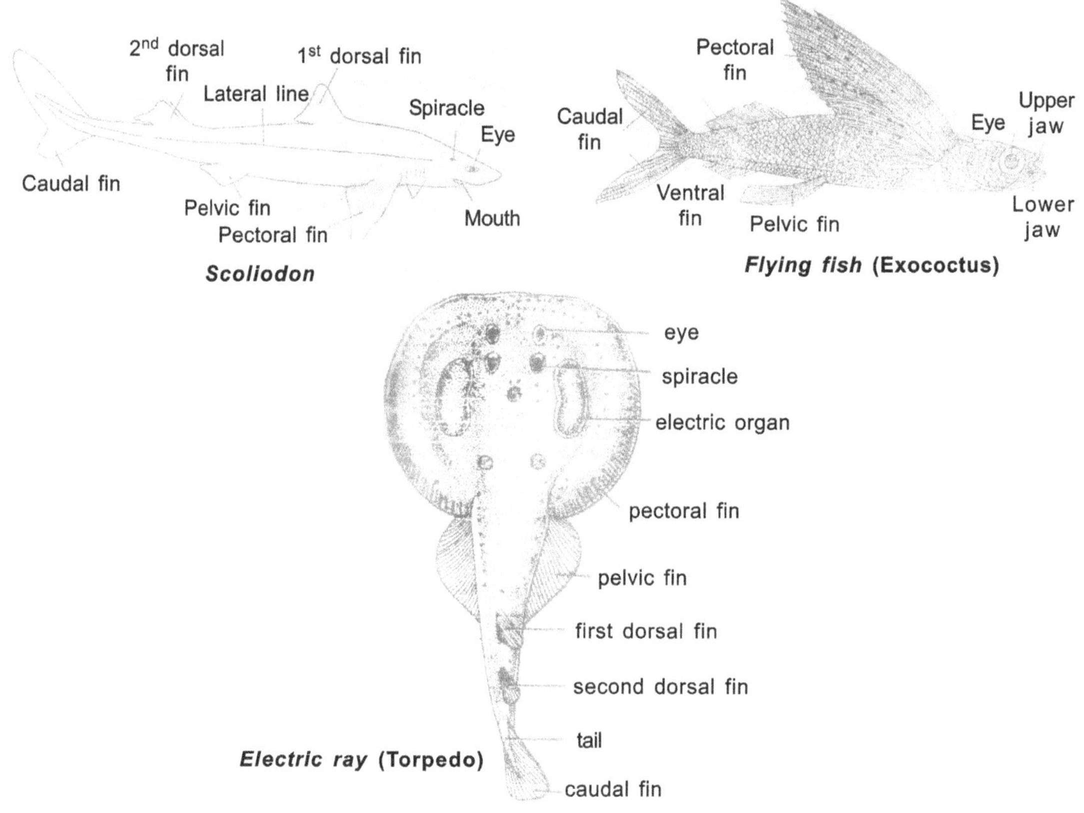

Fig. 3.12 Fishes

include diverse kinds of marine and fresh water forms. Their gills are covered by gill cover and their eyes lack eyelids they have lateral line sense organs to detect vibrations and electric fields. Fishes are oviparous and fertilization is external. Some species exhibit parental care (Fig. 3.12).

Amphibians include tail less frogs and toads (Anurans), tailed newts and Salamanders (Urodelans) and limbless forms (Apodans). They exhibit metamorphosis. (Fig. 3.13)

As larvae they are generally gill breathers. Adult amphibians have four limbs and lungs. The skin is not covered by scales and is often kept moist for gaseous exchange. The heart is three chambered. Fertilization is generally external and the eggs are laid in water or in a moist environment.

The present day reptiles (*Reptilia*) include lizards, (Lacertilia), Snakes (Ophidia), turtles and tortoises (Chelonia) and crocodiles and alligators (Crocodilia). In the past, giant reptiles, the dinosaurs, dominated

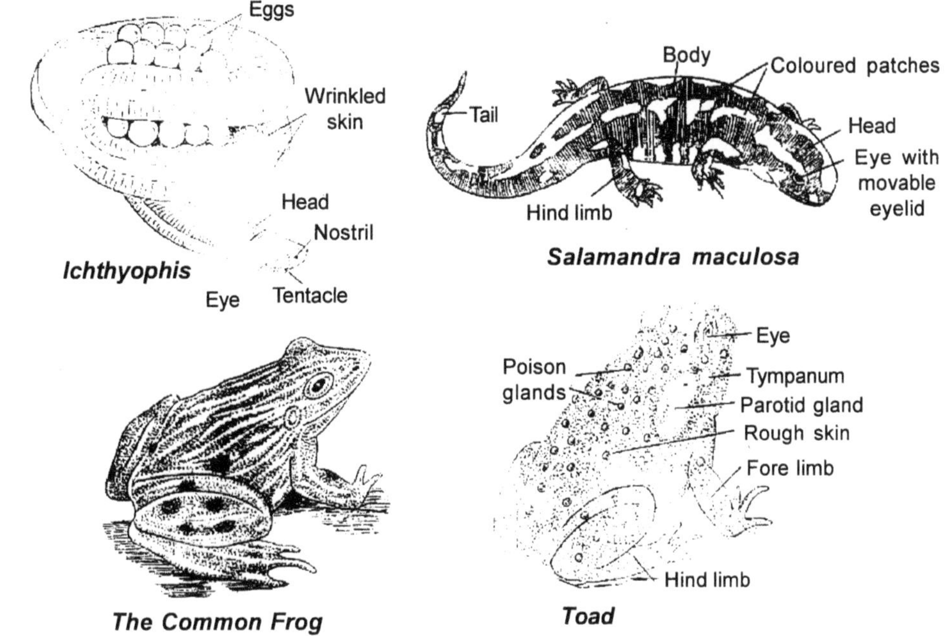

Fig. 3.13 Amphibians

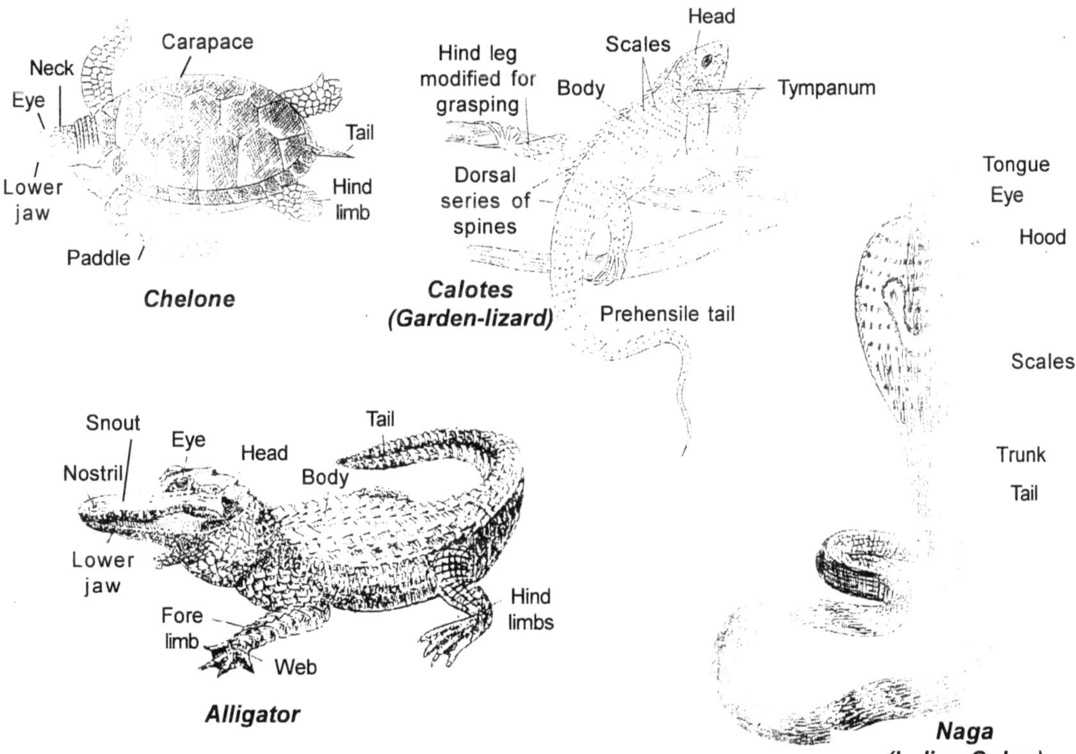

Fig. 3.14 Reptiles

Lizards are insectivores whereas snakes feed on small animals. Snakes are limbless, lack ears but have a keen sense of smell. Some possess poison fangs. Turtles and Tortoises have a characteristic bony shell into which they can retract their head and limbs. Turtles are among the longest lived animals. Crocodiles and alligators live in water or close to it and are generally fish eaters. The Gharial is a unique Indian species with a thin snout. *Archaeopteryx*, a fossil bird is the connecting link between reptiles and birds.

Birds are distinguished by feathers and a capacity to fly. A few birds are flightless. Birds have descended from reptiles. Their feathers are modified scales and their eggs resemble reptilian eggs. The birds show several features adapted for an aerial mode of life. Birds maintain a high metabolic rate and a constant body temperature. The heart is four chambered with separation of oxygenated and deoxygenated blood. The beaks are adapted to many ways of feeding. The hind limbs are variously modified for walking, hopping, grasping, perching, wading or swimming. (Fig. 3.15)

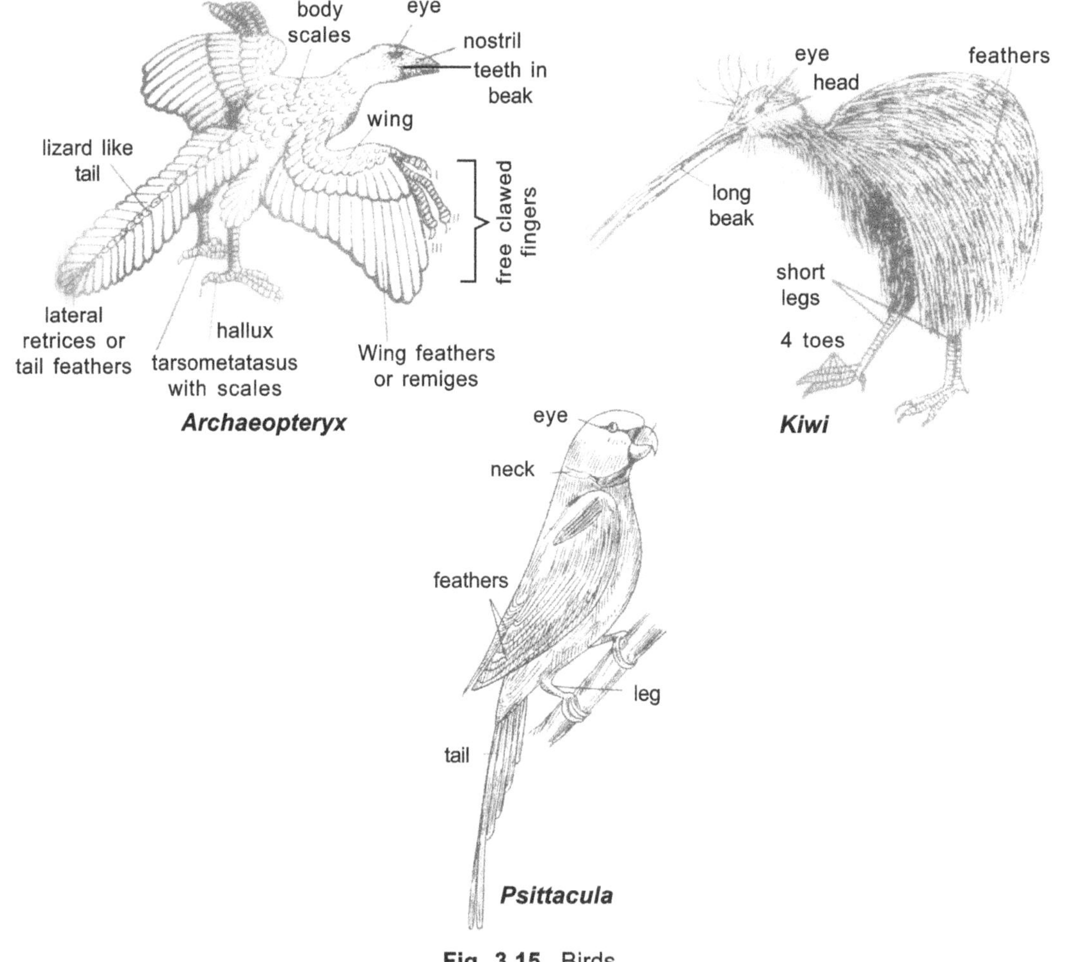

Fig. 3.15 Birds

Birds have a keen sense of sight and hearing and are able to coordinate movement and balance. They exhibit courtship, nest building, parental care and migratory behaviour.

Mammals (*Mammalia*) are dominant animals that inhabit diverse environments. Their distinctive features are milk-producing mammary glands, hair on the body, sweat glands, external ear and a diaphragm. They produce two sets of teeth. They maintain a constant body temperature and exhibit elaborate parental care. The heart is four chambered.

The Mammals are classified into 3 groups 1. egg laying mammals (Monotremes-Duck Bill *Platypus*) 2. Pouched mammals (Marsupials-Kangaroo) and 3. True placental mammals which nourish the young in the womb through the placenta (dog, elephant, monkey, man, etc.). (Fig. 3.16)

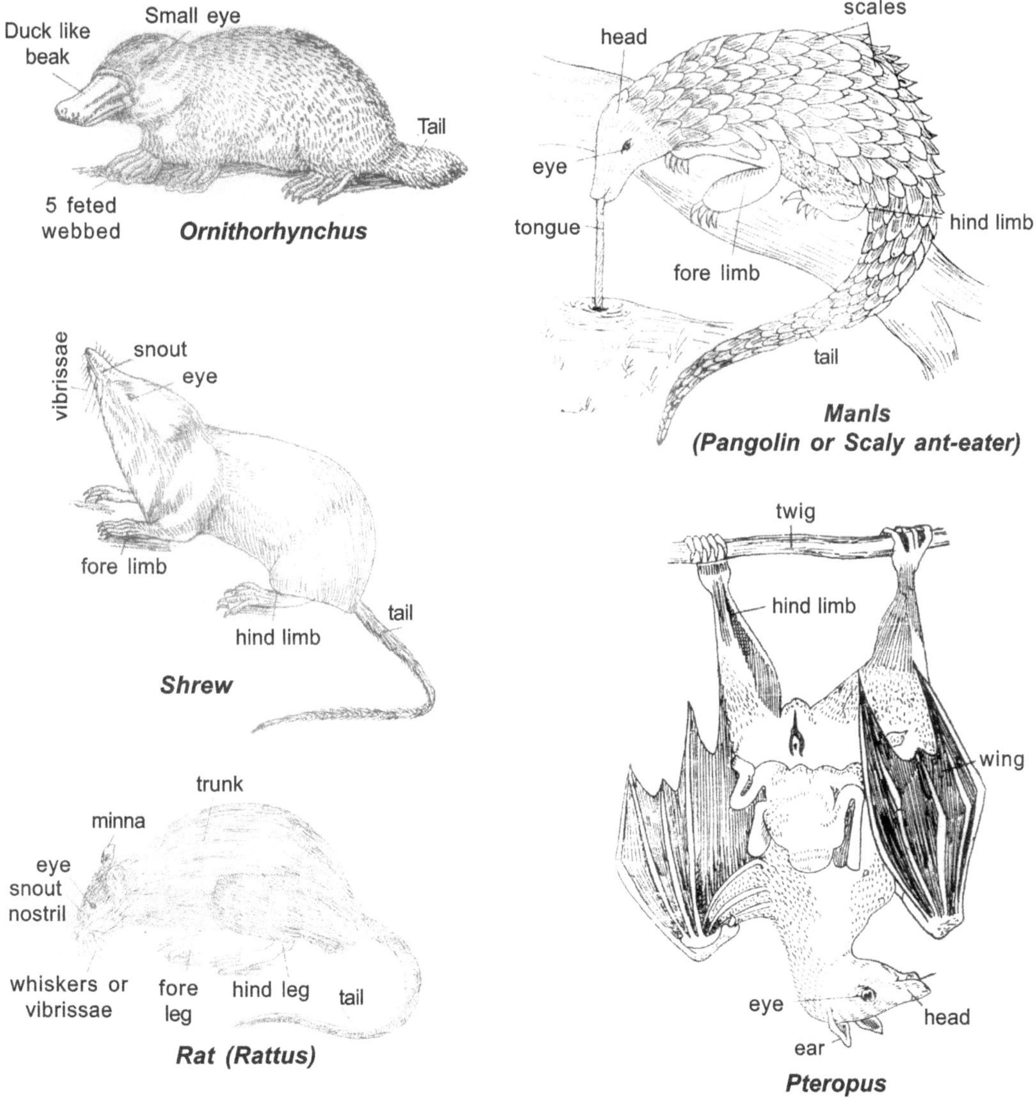

Fig. 3.16 Mammals

Contd. Fig. 3.16

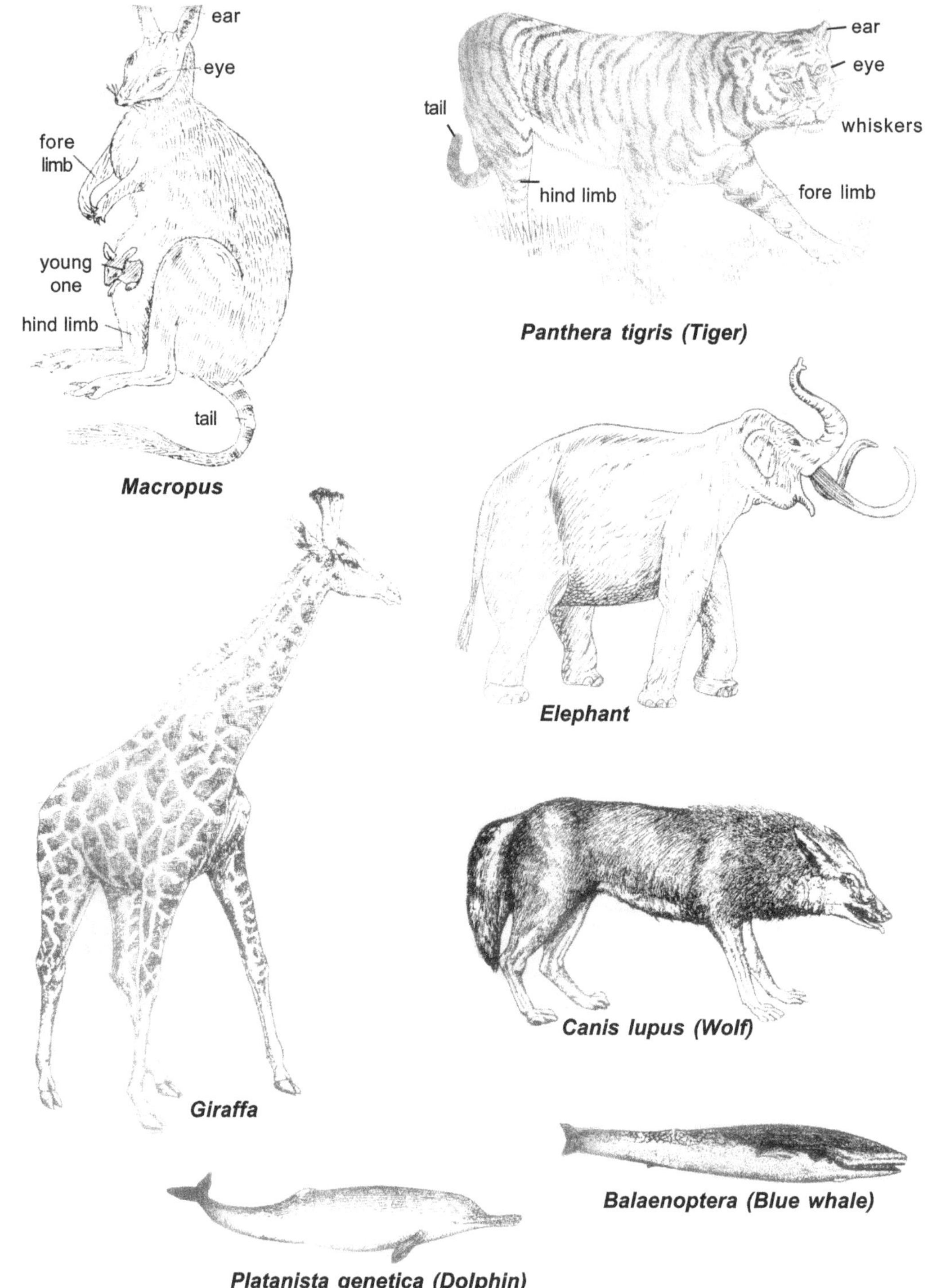

Fig. 3.16 Mammals

Fig. 3.16 *Contd...*

Fig. 3.16 Mammals

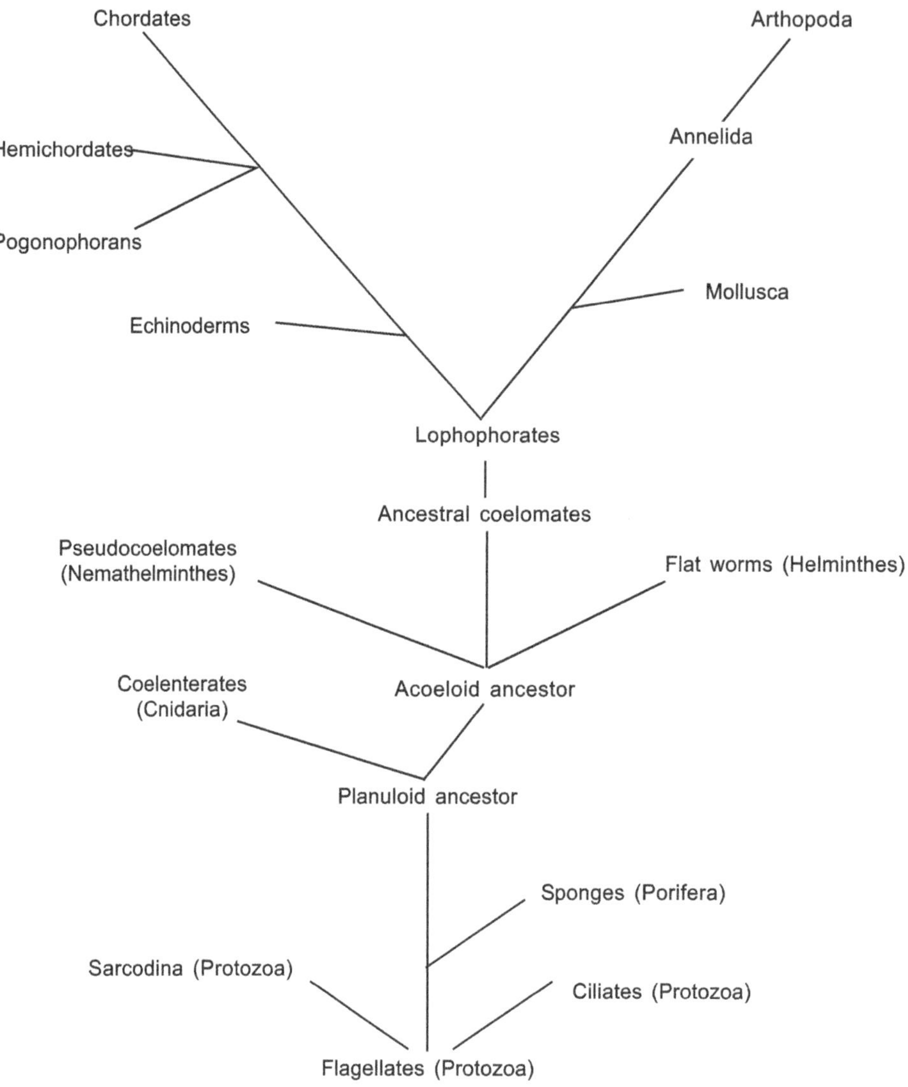

***Phylogeny of the Animal Kingdom* (Marcus, 1958).**

Placental mammals include several lines, such as rodents, rabbits, whales and dolphins, carnivores, the ungulates (hoofed animals) and elephants. The primates include monkeys, apes and man have evolved from insectivores.

Shrews, moles and hedgehogs are examples of insectivores. Many of them burrow and live underground. Bats are flying mammals and echolocate their prey in darkness. Some bats feed on insects and others live on fruits. The Indian pangolin or Scaly ant eater is a toothless mammal (edentate).

Whales and dolphins are aquatic mammals with fore limbs modified as flippers. They lack hind limbs. As lung - breathers, they have to surface to breathe. Blue whales are the largest animals. Dolphins are highly intelligent mammals.

Rodents include rats, mice and squirrels. They have chisel like incisors for gnawing and can be very destructive.

Carnivores are flesh eating mammals with large canines for tearing flesh. They include the tiger, lion, leopard, wolf, jackal, fox and bear. Seals are aquatic carnivores.

The ungulates are herbivores. They may be even-toed or odd toed. Some examples of ungulates are cow, buffalo, goat, deer and giraffe. Cow, buffalo and camel are examples of ruminants. Elephants are related mammals with padded feet instead of hooves.

The primates include the prosimians i.e, lemurs, tarsiers and lorises and the simians (Old world and new world monkeys), apes and man. The distinctive feature of primates is the grasping limb with apposable thumb. Indian monkeys include the rhesus and langurs. The great apes include the chimpanzee, gorilla and Orang-Utan. They can walk erect and swing on trees. A large brain, bipedal gait, dietary habits, power of speech, unique memory and ability to make use of tools distinguish man from apes.

3.4 Functional Morphology

The single cell of a unicellular organism performs all life activities. In multicellular organisms different cells share different biological activities of the organism. Cells of multicellular organisms undergo differentiation and each type of cells is specialized for a limited number of specific functions. The cells coordinate their activities, support one another, exchange nutrients, metabolites and provide information. One or more types of specialized cells are set in specific extra cellular materials to constitute a tissue. Animal tissues are divided into six major classes on the basis of their functions (1) Epithelial tissue covers the free surfaces of other tissues. (2) Connective tissue joins, supports, holds other tissues together (3) Muscle tissue causes the movement of skeleton and the internal organs by contraction (4) Nerve tissue transmits messages in the form of impulses (5) Blood and (6) Reproductive tissues.

An organ is made up of different tissues. Each organ (such as stomach, liver, pancreas, urinary bladder etc) performs specific functions, which depend upon the collective and integrated activities of its tissues.

Several organs constitute an organ system. Organs of a system function in a coordinated manner to carry out a major life process. For example, kidneys form urine, ureters conduct it to the urinary bladder and the latter stores it and periodically expels it through the urethra. Thus, these organs of the excretory system are involved in the elimination of waste products. The digestive system carries out digestion and absorption of food. The respiratory system is concerned with the exchange of gases between the organism and its environment. The circulatory system is responsible for the transport of nutrients, respiratory gases, metabolites, hormones and waste products between different parts of the body. The endocrine system consists of glands which secrete hormones for regulating the functions of other organs and tissues.

The nervous system receives information of external and internal changes to give rise to sensations, transmits information among different parts of the body in the form of propagated potential changes, and regulates other systems. The reproductive system is responsible for the production of offsprings. The skeletal system supports the body and protects the softer internal organs from external injury.

The muscular system participates in movements and locomotion. Likewise blood is composed of Red cells (RBC) and white cells (WBC) performing the function of transport of O_2 and CO_2 and produce antibodies respectively. Thus, the life of an organism is maintained by the cooperative activities of its organ systems.

Morphological Basis of Classification

The feature that distinguish closely related species of animals are usually superficial differences such as colour, size, and proportion. In contrast, the major divisions or phyla of the animal kingdom are distinguished by characteristics that, although usually not unique to a single division or phylum, occur in unique combinations in each.

One morphological feature useful in classifying animals and in determining their evolutionary relationships is the presence or absence of cellular differentiation i.e., animals may be either single celled or composed of many kinds of cells specialized to perform particular functions. Some multicellular animals have only two embryonic cells, or germ layers ie., an outer ectoderm and an inner endoderm. Other animals have three, in addition a mesoderm, which lies between two layers ie., ectoderm and endoderm. The bodies of the Coelenterata (eg : the jellyfish) and other primitive many-celled animals consist of a double-walled sac surrounding a single cavity with a mouth. Higher animals have two cavities, and their bodies are constructed on a so called tube-within-a-tube plan. An inner tube, or digestive tract, is lined with endoderm and opens at each end to form the mouth and the anus. An outer tube, or body wall, is covered with ectoderm. Between the two tubes a second cavity, or coelom, lies within the mesoderm and is lined by it. Another major distinguishing morphological feature of animal phyla is the presence or absence of segmentation. The members of several phyla have bodies characterized by the presence of a row of segments, or body units, of the same fundamental structure. Segmented animals include the annelids (eg : the earthworm), and the arthropods (eg : the insects). In some segmented animals such as man and most vertebrates, however, the segmental character of the body is obscured. An evolutionary tendency in many animal phyla has been the progressive differentiation of the anterior end to form a head with conspicuous sense organs and an accumulation of nervous tissues into a brain; the tendency is termed cephalization. Some morphological structures are specific to a particular phylum; for example, only the Coelenterata have stinging cells (nematocysts); the Echinodermata (eg : star fishes) have a peculiar water vascular system, and only the Chordates have dorsally located, hollow nerve cord.

The Scope of Contemporary Morphology

Fundamental Concepts : Homology and analogy

Homologous structures develop from similar embryonic tissues and show similar basic structural and developmental patterns, reflecting common genetic endowments and evolutionary relationships. In contrast, analogous structures are superficially similar and serve similar functions but have quite different structural and developmental patterns. eg. The arm of a man, the wing of a bird, and the pectoral fins of a whale are homologous structures in that all have similar patterns of bones, muscles, nerves, blood vessels, and similar embryonic origins; each, however, has a different function. The wings of birds and butterflies, in contrast, are analogous structures i.e., both allow flight but have different developmental processes.

Homology and analogy are also applied to the molecular structures of cellular constituents. The haemoglobin molecules from different vertebrate species contain remarkably similar sequences of amino acids, they may be termed homologous molecules. In contrast, haemocyanin is present in crab blood. The haemoglobin and haemocyanin are termed analogous molecules because they have a similar function (oxygen transport) but differ considerably in molecular structure. Corresponding, similarities occur in the structures of other proteins from different species e.g., cytochrome C and other enzymes (biological catalysts) such as the lactic dehydrogenases etc.

Body Plan and Symmetry

The bodies of most animals and *plants are organized according to* one of three types of symmetry: *spherical, radial, or bilateral.* A spherically symmetrical body can be cut in any plane through the centre to yield two equal halves. A few of the simplest plants and animals are spherically symmetrical e.g., protozoans such as radiolaria and heliozoa. *Radially symmetrical* bodies, such as those of starfishes have a distinguishable top and bottom and usually have a cylindrical shape, with the body parts radiating from the central axis. A starfish can be cut into two equal halves by any plane that includes the line, or axis, running through its centre from top to bottom. The anterior, or oral, end contains the mouth; a posterior, or aboral, end may have an anus. In the *bilaterally symmetrical* body of higher animals including man, only a cut from head to foot exactly in the centre divides the body into equivalent halves. An anterior, or head end and a posterior, or tail end can be distinguished; and the dorsal or back side can be distinguished from the ventral, or belly side. But because some internal organs of man are not symmetrical (e.g., the heart), even the right and left halves of the human body are not exactly equivalent. A few organisms-amoebas, slime molds and certain sponges-with an irregular form, or one that changes as the organism moves, have no plane or symmetry.

Animal structure

Detailed comparisons of the morphological features of different animals, termed comparative anatomy, provide strong arguments for the evolutionary relationships among different species. In the course of evolution, animals tend to undergo adaptive morphological changes that enable them to survive under certain environmental conditions. As a result, animals unrelated evolutionarily, may come to resemble each other superficially because of common adaptations to similar environments, a phenomenon known as convergent evolution. Structural similarities ie., streamlined shape, dorsal fins, tail tins, and flipper-like forelimbs and hindlimbs for example, have evolved in such varied animal groups as the dolphins and propoises the extinct Icthyosaurs. (reptiles) and both the bony and cartilaginous fishes. Similarly, the mole, an insectivore, and the gopher, a rodent, have both evolved shovel like forelimbs, an adaptation for digging.

Divergent evolution, an opposite phenomenon, occurs when animals originally closely related adapt to different environments and come to be superficially quite different. Sea lions and seals, are carnivores and closely related to bears, cats, and dogs, their adaptations to an aquatic existence have resulted in morphological characteristics distinct from those of the terrestrial carnivores. In the course of mammalian evolution, many features have changed to permit specific animal groups to adapt to particular environments e.g., the number and shape of the teeth, the length and number of bones in the limbs, the number and attachment sites of muscles, the thickness and colour of the hair and the length and shape of the tail etc.

Careful study of adaptive morphological aspects permit inferences about the course of the evolutionary history of various animals and of their successive adaptations to changing environments. The present-day Australian tree-climbing kangaroos, are the descendents of a ground-dwelling marsupial, from whom evolved forms that began to live in trees and eventually developed limbs adapted to tree climbing. But the events may have occurred in the reverse sequence; that is, specialized limbs may have evolved before the animal adopted an arboreal mode of life. In any event, some of the tree-dwelling kangaroos subsequently left the trees, became readapted to life on the ground (i.e., their hindlegs became adapted for leaping), and then went back to the trees but with leg so highly specialized for leaping as to be useless in grasping a tree trunk; consequently, present day tree Kangaroos climb by bracing their feet against a tree trunk, as do bears. Careful comparisons of the feet of many kinds of living Australian marsupials reveal the stages in this complicated process of adaptation and re-adaptation.

Changes in gene or mutations constantly occur and may result in a decrease in size and function of an organ; on the other hand, a change in the environment or in the mode of life of a species may make an organ unnecessary for survival. As a result, many animals contain organs or parts that are useless, degenerate, undersized, or lacking some essential part when compared to homologous structures in related organisms. The human body, for instance, has more than 100 such organs e.g., the appendix, the fused tail vertebrae (coccyx), the wisdom teeth, the muscles that wiggle the ears, and the hair on the body.

3.5 Adaptive Biology

Living organisms are equipped to cope up with their environment in a number of ways. For example, the aquatic animals have paddle like structures for swimming, aerial animals like birds have wings to fly and terrestrial animals have strong appendages to run or walk. In the same way, burrowing animals are provided with digging organs. In this way animals are well adapted to live successfully in different environments by acquiring special structures to suit those environments. This phenomenon is known as adaptation. Adaptation can be defined as the adjustments made by individuals in response to specific environmental conditions. The organisms that are well adjusted or well adapted are successful in their survival in a particular environment and those which can not adjust to a particular environment are said to be ill adapted or poorly adapted and they perish in a course of time.

Animals in general are adapted variously and the following types can be recognized.

1. Aquatic adaptation
2. Cursorial adaptation
3. Desert adaptation
4. Cave adaptation
5. Fossorial (burrowing) adaptation
6. Arboreal adaptation
7. Flight or volant adaptation

1. *Aquatic Adaptations*

The aquatic adaptations are of two types,

(i) Basic or Primary Adaptation and

(ii) Secondary Adaptation.

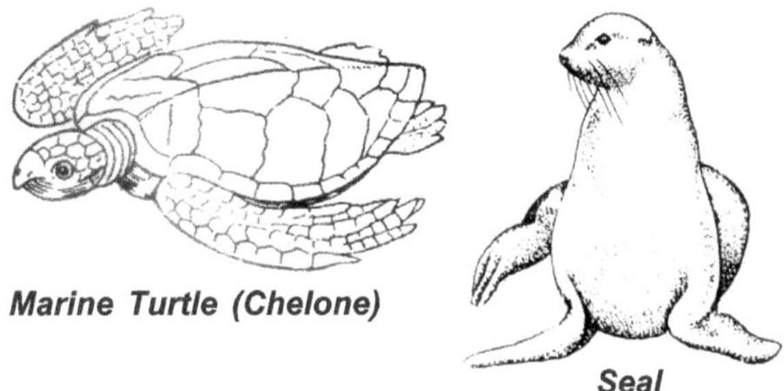

Fig. 3.17 Aquatic Adaptation

The primary aquatic adaptations are seen in fishes. They are

(i) Stream-lined body for little resistance in swimming

(ii) fins acting as paddles help in swimming and changing directions

(iii) gills for gaseous exchange

(iv) lateral line sense organs to detect temperature, pressure and water current etc.,

(v) eyes with transparent nictitating mucous membrane to facilitate vision and for protection

(vi) Mucous glands that secrete mucus for prevention of diffusion of water through skin

(vii) scales on the body for protection, etc.

Aquatic animals have developed different methods to maintain the internal body fluids in the same concentration for survival.

For example, the body fluids of fresh water fishes are concentrated to the surrounding fresh water. Hence endosmosis occurs. The excess of water that enters the body is removed by the kidney. The salt lost through skin is regained by the chloride cells that are present in gills. They absorb salts from fresh water and add them to the body fluids.

In marine fishes exosmosis occurs because sea water is hypertonic to the body fluids. The loss of water is regained by drinking sea water. As the sea water contains a large amount of salt, the salt thus entered the fish is again removed by the chloride secretory cells that are present in the gills.

Apart from these characteristic features, presence of cilia, flagella etc. also help in locomotion, light weight body helps the organisms to float on water, adhesive structures for attachment on other objects or organisms for locomotion or for parasitic mode of life (feeding on the host).

The secondary aquatic adaptations are seen in those animals which were terrestrial but secondarily adapted to aquatic life *eg :* turtles, dolphins, porpoises, sea lions, whales etc. (Fig. 3.17)

These animals also have a stream lined body and fins like those of fish. However there are certain special structures. They are (i) flippers : they are paddle like and help in swimming as seen in turtles (ii) Webbed feet for swimming as seen in turtles and ducks (iii) absence of neck to reduce resistance as seen in whales and dolphins (iv) Skin lacking scales and hairs as in dolphins, whales etc, (v) Blubber a fatty layer beneath the skin preventing loss of heat and reduces specific gravity as found in whales, dolphins and porpoises (vi) absence of pinna for reduction of resistance in water (dolphins, whales, sea lions) (vii) Oxygen carrying capacity of haemoglobin is high (viii) The bones are light and spongy (ix) Instead of a nictitating membrane a fatty substance secreted by lacrimal glands protects the cornea and conjunctiva and help in vision.

A lot of specializations are there for respiration in secondary aquatic organisms. Networks of blood vessels called Ritia mirabilia are found especially in thoracic region. They store oxygen to be used when the animal is under water. The rate of heart beat is reduced during diving of the animal. This phenomenon is known as *Bradycardia.* This is an oxygen saving mechanism. When the aquatic mammal remains underwater, it can accumulate high *oxygen debt*, which can be repaid when the animal comes to the surface.

2. *Cursorial Adaptations*

Cursorial means fast running. The animals living on land are adapted for fast running exs : flightless birds, mammals (bandicoot, tiger, horses, deer) etc. The following are the cursorial adaptations. (Fig. 3.18)

1. *Body Shape :* The body is stream lined for reducing resistance against air while running.
2. *Digitigrade Gait :* Maximum speed is obtained when walking on digits (dogs, cats etc).
3. *Unguligrade Gait :* Walking on hooves is called unguligrade gait (horses, deers, etc).
4. *Plantigrade Gait :* Walking on foot is called plantigrade gait (Man, bear etc).
5. *Elongated Distal Limb Segments :* The distal limb segments like carpals, metacarpals, tarsal and metatarsal segments are elongated to facilitate to cover a long distance in each step.
6. *Loss of Digits :* The fast running animals have reduced number of digits. For example, planti grade animals like man has 5 digits, digitigrade animals like dog have 4 while in unguligrade animals the number is reduced variously ranging from four to one.

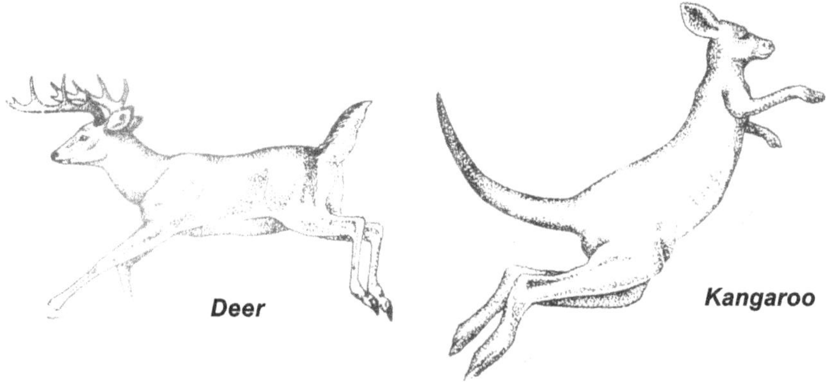

Fig. 3.18 Cursorial Adaptation

7. *Reduction of Ulna and Fibula :* As an adaptation for fast running, the Ulna from fore limbs and fibula from hind limbs are reduced to vestiges.
8. *Movement of Limbs in One Plane :* The limbs move only in one plane at the joints. This avoids interference between fore and hind limbs during running.
9. *Limb Muscles :* Limb muscles are powerful. The rate of contraction of limb muscles is rapid, helping the animal to run faster.

3. *Desert Adaptations*

A desert is a waterless, treeless, large wasteland often covered with sand. It is characterized by low or no rainfall, scarcity of water, extremes of temperatures and dust storms. The important deserts are Rajasthan (India), Gobi (China), Sahara and Kalahari (Africa).

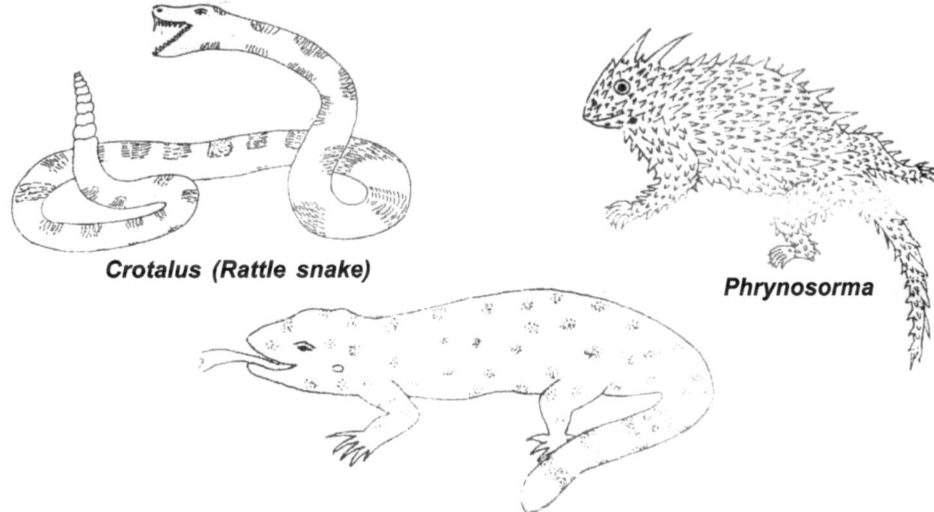

Fig. 3.19 Desert Adaptations

Desert Fauna

The important desert animals are uromastix, heloderma, Mabuia, rattle snake, Phrynosoma, birds like quails, Indian bustard, grey partridge and mammals like bats, hedgehog, hare, ant eater, porcupine, mouse, wild boar, sambar deer, spotted deer, wolf, red foxes, jungle cats, panther etc. (Fig. 3.19)

Desert animals are adapted for different purposes such as water intake and conservation, Protection, and Heat tolerance.

As the desert is characterized by dry atmosphere, desert animals tend to lose water from their body through evaporation. To conserve water, the desert animals have special adaptations.

1. The skin is thick and hard and may be covered by scales, spines, scutes, shields etc.
2. To avoid high temperatures of the day, they are adapted to nocturnal habit or take rest or shelter under shady places or live in the burrows.

3. Certain animals aestivate to prevent desiccation like Pocket mice, Kangaroo rat, etc.
4. Their excretory product is chiefly uric acid, as it does not require water. Water is resorbed from urine and faeces in desert animals hence they are in the form of a paste or pellets. Eg. Desert insects, reptiles and birds.
5. The desert rodents and ruminants have lost the sweat glands. As a result no sweat is produced and no water is lost through sweat.

Certain animals like African antelope reduce their metabolic rate and conserve water.

For acquisition of water, they go to different places or obtain it in different forms. Some of the animals go to water holes or migrate short or long distances to get water. Some of the animals acquire water through food like leaves, tubers, seeds, other animals or depend on a small amount of water produced as a bye product of metabolism. Certain animals have a hygroscopic body to absorb any moisture that is present in the surrounding. Certain animals store water in their body such as intestine (*Uromastix*) water cells in the rumen of stomach (camel) etc. Certain animals have learnt to live without water drinking such as desert birds like black throated sparrow, lark, and kangaroo rat (mammal) etc.

Desert animals are well adapted to heat tolerance. For example, camel permits its body temp upto 40.6 °C without sweating during day time and drop to 33.8 °C at nights. African antelope stores heat in its body during day time resulting in a substantial rise in body temperature (hyperthermia).

They are well adapted to protect themselves from enemies or environmental factors like temperature (heat, cold), moisture etc. The nocturnal habit, living in burrows etc. help them to avoid high temperatures, to conserve water and escape from enemies. They have adaptations for swift running such as long legs, well developed claws. Their nostrils, eyes and ears are protected from dust storms. The skin is spiny so that predators have a problem in feeding, thus not preferring them. There are also poisonous animals such as rattle snakes, *Heloderma* (lizard) to kill the enemy.

4. *Cave Adaptations*

Caves are natural hollows on the earth. They may be present on the surface of the earth or in the hills, mountains, rocks or underground or under bodies of water.

Caves might have been formed by any one of the following ways such as (i) Abandoned channels through which rivers must have flowed in the past (ii) Over a period of time lime stone might have dissolved completely or partially to form caves (iii) Caves in sand are formed by wind erosion and (iv) Caves in the sea are formed by wave erosion.

Based on the penetration of light, a cave is divided in to a light zone (dysphotic or twilight zone), a middle dim zone and an inner dark zone.

The cave flora includes fungi such as mould and mushrooms etc. and the cave fauna includes turbellarians, chaetopod worms, leeches, springtails, mites, spiders, silverfish, cockroaches, crickets, snails, fishes like Amblyopsis, Typhlicthyes, Gronias etc, amphibians like protius, Typhlotriton, birds like owls, mammals like white footed mouse, bats etc. (Fig. 3.20)

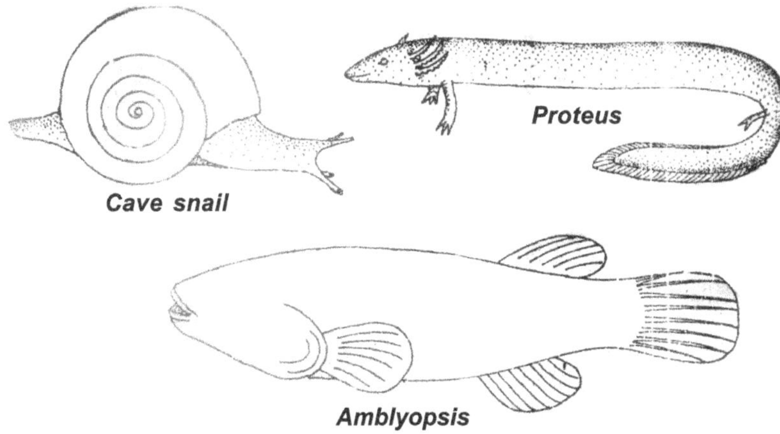

Fig. 3.20 Cave Adaptations

The animals living in caves are called '*Cavernicolous*' animals and their adaptations are called cavernicolous adaptations. They have following adaptations.

1. *Loss of pigmentation :* As there is no light in the caves, pigments are absent in cave animals. When these animals are brought to day light, they develop pigmentation.
2. *Absence of Eyes :* As there is no light in the caves, the eyes are either non functional or rudimentary or completely lost.
3. *Weak Exoskeleton :* The exoskeleton of cave arthropods is weak and fragile, the shell of cave snails is thin, the cave fishes have lost the scales completely.
4. *Scarcity of Food :* Caves are characterized by scarcity of food due to lack of green vegetation. Hence cave animals have adaptations to live without food for many days.

As there is little or no vegetation, most of the cave animals are carnivorous in habit.

5. *Fossorial or Burrowing Adaptations*

Animals living inside the burrows or under the surface of earth are called fossorial animals. They have adaptations for digging and burrowing. They are characterized by cylindrical or spindle shaped body to reduce resistance, a tapering head, a short tail, reduced eyes, reduced pinna, a short or reduced neck, a strong vertebral column, a strong shoulder to withstand pressure while digging, protected skin with mucus (annelids), Chitin (arthropods), scales (reptiles), fur or spines (mammals) etc and special organs for digging purpose such as snout, teeth, claws on forelimbs and hind limbs, fossorial legs etc.

Burrowing animals are seen in both invertebrates and vertebrates. Nereis, Chaetopterus, Crabs, Arenicola, Lamellidens, Dentalium, Sea Cucumber are some of the examples of invertebrates and *Balanoglossus*, *Amphioxus*, *Ichthyophis*, *Platypus*, *Echidna*, *Moles* etc are some of the examples of vertebrates.

6. *Arboreal Adaptations*

The animals which can climb on trees, rocks, stones and other vertical surfaces are called arboreal animals. Wall lizards, varanus etc can climb on vertical surfaces like walls of houses. Some animals like Squirrel can climb the trees well but they prefer to live on land. Such forms are known as terrestrio arboreal forms. There are true arboreal forms like lemurs, chamaeleons etc. which move on the branches freely or some animals like bats, sloths and flying lemurs etc. hang from the branches of trees while some others like monkeys swing from branch to branch or tree to tree.

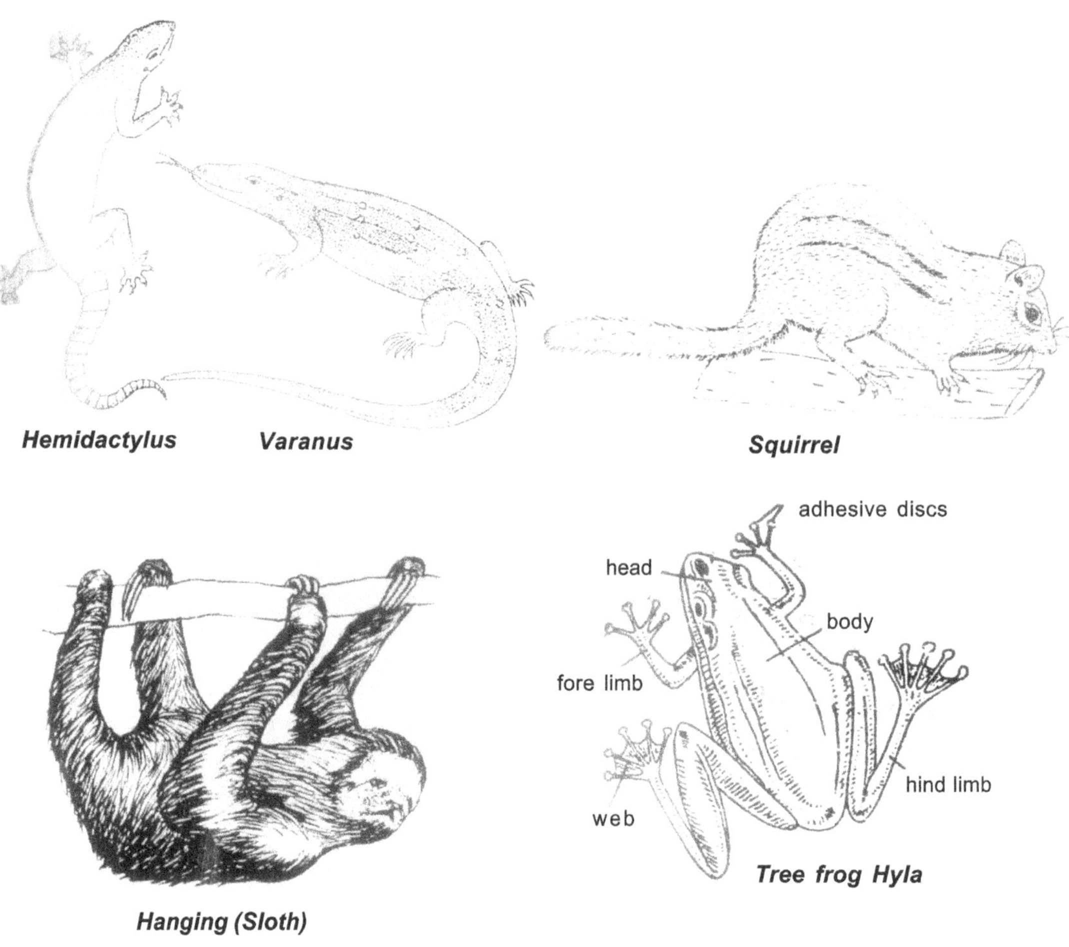

Fig. 3.21 Arboreal Adaptations

For such arboreal mode of life, these animals have acquired certain adaptations. They have a stout body, long limbs, reduced number of digits, (syndactyly i.e., Union of two or more digits by a membrane; Zygodactyly ie., two toes directing forwards and two toes backwards to hold to the branches), clawed limbs, Adhesive discs on the digits of the limbs, a prehensile tail etc., These adaptations are seen in various animals to lead successfully an arboreal life. (Fig. 3.21)

7. *Flight Adaptations*

The animals which can move in the air are called flying forms or aerial forms or volant forms and the adaptations are called flight or volant adaptations.

These animals can not live forever in air. They have to return to the ground for rest. Thus they possess adaptations for aerial and terrestrial or aquatic modes of life. (Fig. 3.22)

Flight is of two types,

(i) Active flight or true flight and

(ii) Passive flight or gliding flight.

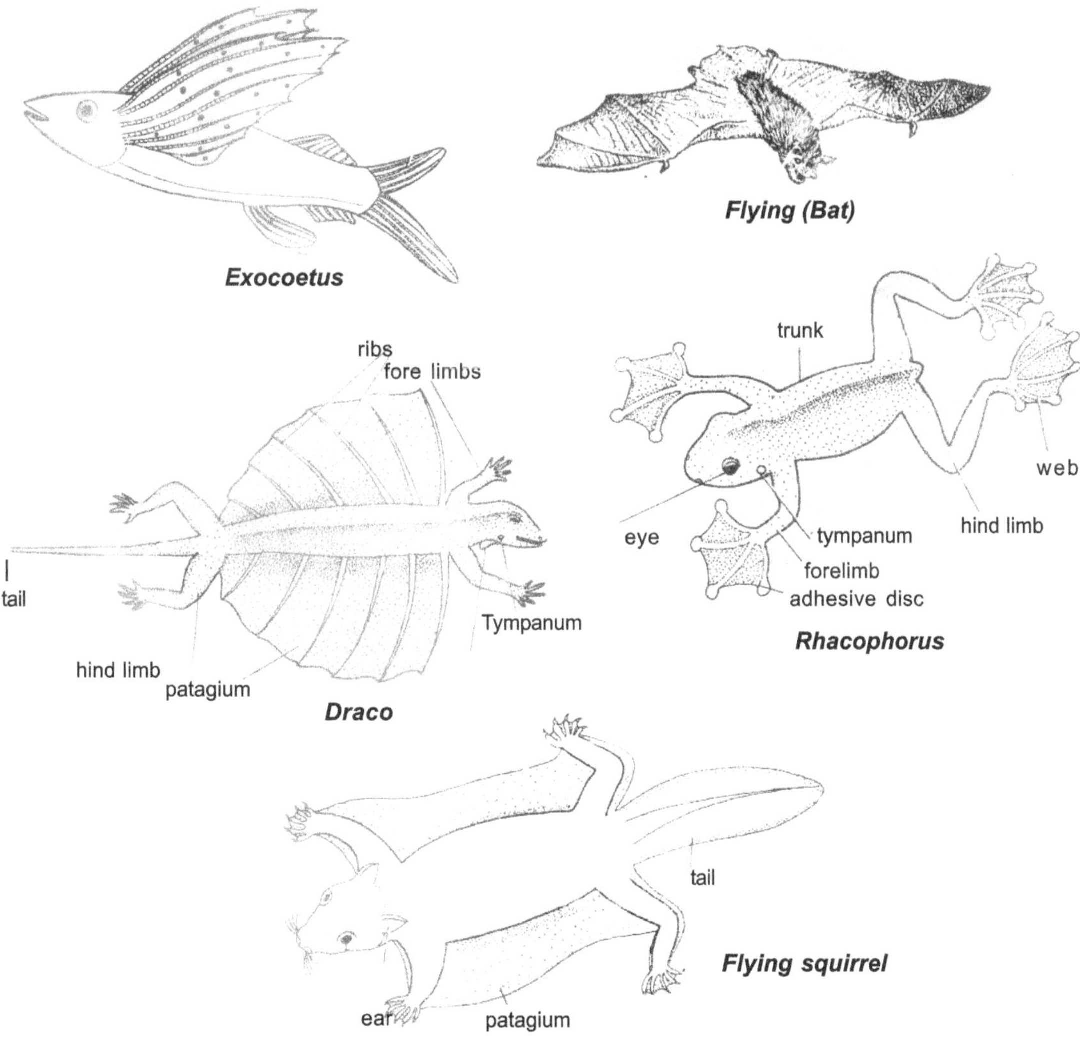

Fig. 3.22 Flight Adaptations

1. *Active Flight :* In birds, this is facilitated by the wings. The general structure and functions of the birds are highly modified for movement in the air. The shape of the body is streamlined ie., boat shaped or spindle shaped to afford little or no resistance in flight. Forelimbs are modified into wings while hind limbs are used for standing on the ground or sitting on a branch or swimming in the water. Wing (flight) muscles are well developed. The body is covered by feathers for reducing weight, friction, insulation, buoyancy etc. Light body weight due to pneumatic bones, fusion of bones, absence of teeth, loss of parts such as urinary bladder, one side ovary and oviduct etc, Presence of air sacs, efficient circulatory system, perching ability on branches, a very high power of sight etc, also facilitate in efficient aerial life.

   In mammals (bats), the fore limbs are modified into wing like patagia for flight. The flight muscles are well developed. The hind limbs are provided with claws which help to hang from the trees. The bones of the body are light in weight, the bones of fore limbs are elongated. Bats are also provided with Ultrasonic sound system which helps the bats to locate the obstacles lying ahead and used in way-finding as they are nocturnal in habit.

2. *Passive Flight or Gliding Flight :* This is seen in different vertebrates like *Exocoetus* (flying fish), *Rhacophorus* (flying frog) *Draco* (flying lizard), flying squirrel etc.

   In *Exocoetus* the pectoral fins are well developed into wing like structures. The fish can glide above water for a distance of upto 400 yards. *Rhacophorus* has webbed digits, which function as parachute. It can leap upto 6-10 metres in the air. The digits are also provided with adhesive pads for attachment to the trunks of trees. *Draco* has a wing like patagium developed between the limbs supported by 6 ribs. With these structures, it can jump from branch to branch. The flying squirrels too have patagia between the limbs, with the help of which they glide from tree to tree or tree to ground Exs; African flying squirrel, *Anomalurus*, Travancore flying squirrel, *Petinomys* etc.

3.6 Concept of Species

The different kinds of animals that we see in nature are actually species. The existence of species has been recognized by man since the dawn of civilization. Ernst Mayr (1957) reviewed the works of others on species and its definitions and he broadly grouped them into three main concepts.

1. Typological or essentialist species concept.
2. Nominalistic species concept and
3. Biological species concept.

   Later, two more concepts were added by other workers.

4. Evolutionary species concept and
5. Recognition species concept.

1. *Typological or Essentialist Species Concept :* If two individuals or groups of individuals appear sufficiently different, they are considered as different species. The species can be recognized by their essential features and these are expressed in their morphology like physical characteristics, colour, size, habitat etc, segregate them from all other organisms. It is also called as the Morphological species concept.

2. *Nominalistic Species Concept :* According to this concept, nature produces individuals and nothing more. Species have no actual existence in nature. They are mental concepts and nothing more. These have been invented in order that we may refer to great numbers of individuals collectively.
3. *Biological Species Concept :* It says that species are groups of interbreeding natural populations that are reproductively isolated from other such groups.

The species has 3 separate functions.

(a) *It forms a reproductive unit.* The species specific genetic programme of every individual ensures intra specific reproduction.

(b) *It is an ecological unit.* It interacts as a unit with other species with which it shares the environment and

(c) *It is a genetic unit.* It consists of a large inter communicating gene pool. The individual members of a population possess some cytological, genetical or ecological isolating mechanisms which maintain the gene pool of the species.

4. *Evolutionary Species Concept :* An evolutionary species can be defined as a lineage (an ancestral-descendent sequence or populations) evolving separately from others and with its own unitary evolutionary role and tendencies.
5. *Recognition Species Concept :* According to this concept, a species is that most inclusive population of individual biparental organisms which share a common fertilization system.

It is not still settled as to which species concept should be accepted and it cannot be decided until a way is found to correlate the species composed of populations supposedly phylogenetically related.

Recently Florkin (1964) gave a biochemical definition of a species as groups of individuals with more or less similar combinations of sequences of purine and pyrimidine bases in their macro molecules of DNA, and with a system of operators and repressors leading to the biosynthesis of similar amino acid sequences.

3.7 Speciation

Before discussing speciation as a process, one should know the definition of a species, so that, when the differentiation in a population is progressing, one may recognize the stage where a new species due to differentiation has originated.

A species may be defined as an inter breeding population which is reproductively isolated from other similar but morphologically distinguishable populations.

New characters are evolved in homogenous populations due to organism-environment inter- action. Since environment keeps on changing in space and time, such new characters are operated upon by natural selection. These will be selected, if they have any adaptive value. In this process of origin of new variation and the operation of natural selection pressure, differentiation of an originally homogenous population will be brought about some of this variability will enable a part of the population to adopt itself to changed environment. The entire process thus brings about adaptive radiations and the species which formed the original homogenous population now becomes a *Polytypic Species.* The sub-populations of this polytypic species can be easily recognized and are called sub species.

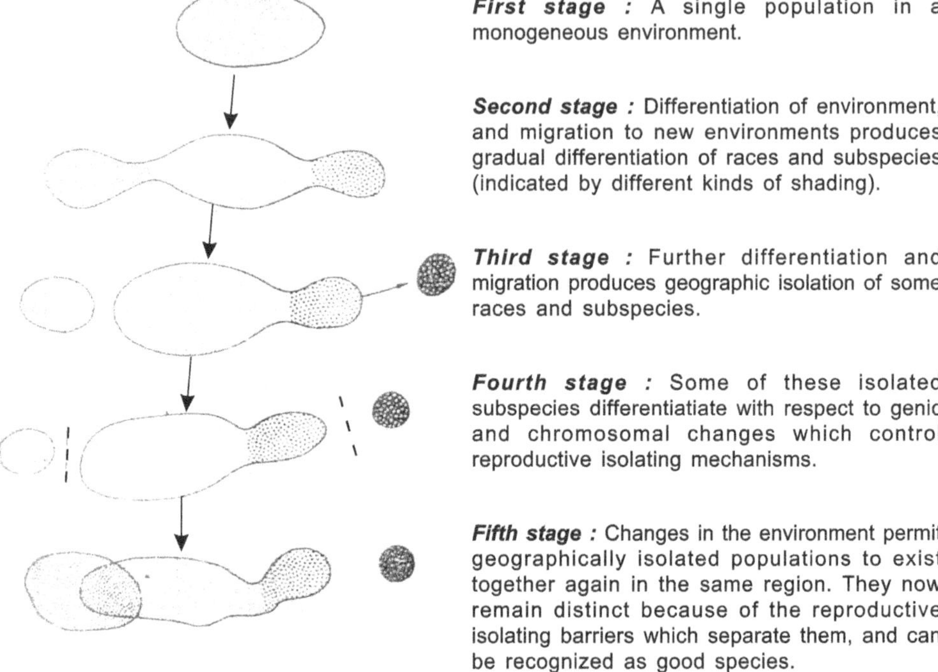

Fig. 3.23 Diagrammatic representation of the process of speciation

This is followed by development of barriers to the free exchange of genes. Consequently, the mating between different sub species or races of a polytypic species will not take place or the fertility of the hybrids will be greatly reduced. These barriers which would check the free exchange of genes between populations are called *Isolation mechanisms.* (Fig. 3.23)

3.8 Allopatric and Sympatric Speciation

The concept of allopatric and sympatric populations is important for understanding the process of speciation and the development of isolation mechanisms.

Allopatric populations are those populations of species or races which inhabit different areas and *Sympatric* populations are those which inhabit the same geographical area. The idea of allopatry and sympatry has been introduced because one may come across populations, which have been isolated geographically for millions of years and thus developed morphological distinction, but still have the ability to exchange genes freely if brought together. On the other hand, there are populations inhabiting the same area but still maintaining their distinct morphological entity. These distinct species are valid species. It should be realized that populations recognized as separate species should pass the test of sympatry. In other words, they should be able to maintain their identity when they live or made to live sympatrically so that the opportunities for cross pollination exist, but interbreeding does not take place.

Such a situation results due to varied factors which are collectively called isolating mechanisms. These may be purely spatial or may involve specific mechanisms covering reproductive isolation which will operate in sympatric populations. From the point of view of speciation, spatial isolation is not enough, since unless these populations have passed the test of sympatry, they cannot be called separate species, although they do maintain their distinctness due to spatial isolation.

The reproductive isolation is of two types,

(a) Pre-zygotic mechanisms and

(b) Post-zygotic mechanisms.

The prezygotic mechanisms either prevent contact between species at the time of active reproductive period or else the fertilization is checked even if mating may be effectively brought about. These may be ecological isolation, seasonal or temporal isolation, ethological isolation (different courtship patterns) or mechanical isolation (incompatible reproductive structures) etc.

In the post-zygotic mechanisms, the fertilization is brought about but the hybrids are either inviable or weak, or the hybrids are sterile or in still other cases, the F_1 may be fertile but F_2 progeny may be sterile.

As soon as these isolation mechanisms become strongly developed, the populations will be said to be reproductively isolated from each other. Then it would be said that these populations have attained the status of an independent species. In other words, the process of speciation through the building of isolating mechanisms has now been completed. This will then be followed by further differentiation and thus further speciation.

3.9 Multivariate Models in Species Composition

Each species or each biological discipline or model has its own set of variables which includes morphological measurements, concentration of chemicals in body fluids, frequencies of certain events etc. A variable can be defined as a property with respect to which individuals in a sample differ in some ascertainable way. Length, height, weight, number of teeth, genotypes etc are examples of variables in ordinary, genetically and phenotypically diverse groups of organisms.

Types of biological variables in a species or a model

The biological variables have the following types.

1. *Measurement variables :* Measurement variables are all those whose differing states can be expressed in a numerically ordered fashion. They are of two types (a) the continuous variables which can assume an infinite number of values between any two fixed points. For example, between two length measurements 2.5 and 2.7cm, there is an infinite number of lengths that could be measured if one were so inclined and had a precise method of calibration to obtain such measurements. (b) the discontinuous variables or meristic or discrete variables : These are variables that have only certain fixed numerical values with no intermediate values possible in between. Thus, the number of segments in a certain insect appendage may be 4 or 5 or 6 but never 5.5 or 4.5. Examples of discontinuous variables are numbers of certain structures such as segments, bristles, teeth or glands etc.

2. *Ranked variables :* Some variables cannot be measured but atleast can be ranked by their magnitude. For example, in an experiment one might record the rank order of emergence of ten pupae without specifying the exact time at which each pupa emerged. In such cases the data is coded as a ranked variable i.e. order of emergence. Thus, by expressing a variable as a series of ranks such as 1, 2, 3, 4, 5, we do not imply that the difference in magnitude between ranks 1 and 2 is identical to or proportional to the difference between 2 and 3.
3. *Attributes :* Variables that can not be measured but must be expressed qualitatively are called attributes. These are all properties such as black or white, pregnant or not pregnant, dead or alive, male or female etc. When such attributes are combined with frequencies they can be treated statistically. For example, of 80 mice, 4 were black, two agouti and the rest gray. When attributes are combined with frequencies into tables suitable for statistical analysis, they are referred as *enumeration data.*

Variate

Variate is a single reading, score or observation of a given variable. Thus, if we have measurements of the length of the tails of 5 mice, tail length will be continuous variable and each of the 5 readings of length will be a variate.

Derived variables

There is an other important class of variables in biological research that we may call the derived or computed variables which are generally based on 2 or more independently measured variables whose relations are expressed in a certain way. Here the ratios, percentages, indices, rates etc are referred.

A ratio expresses as a single value i.e, the relation that two variables have one to the other. For example, 64:24 which may represent the number of males versus females. The above example implied ratios based on counts. Ratio may also be expressed as fractions such as 1.8 : 1.2.

Percentages are also a type of ratio. Ratios and percentages are basic quantities in biological research.

An index is the ratio of one anatomic variable divided by a larger, so called standard one. An index could be the average of two measurements such as 1/2 (length of A and length of B). Another example of index is the *cephalic index* in physical anthropology.

Rates

The amount of a substance liberated per unit weight or volume of biological material, weight gain per unit time, reproductive rates per unit population size and time (birth rates) and death rates etc. are grouped in this category.

The models may be constructed for a variety of reasons. The mathematical models are developed for prediction of dynamic change over time. The failure of a model to predict change is in itself useful, because it may point out flaws in the conceptual frame work from which the model was developed. Models can be evaluated in terms of three basic properties ie, realism, precision and generality. Realism refers to the degree to which the mathematical statements of the model when translated into words, correspond to the biological concepts that they are intended to represent. Precision is the ability of the model to predict numerical change and to mimic the data on which it is based.

Generality refers to the breadth of applicability of the model (the number of different situations in which it can be applied).

There are innumerable examples where the individual members of the same species have considerable differences in structure and function depending on their stages of development, sex or specialization of the members. For example, in moths, butterflies and many other insects, the adult, larval and pupal stages are considerably different. In frogs, toads and salamanders, the adult and larval (tadpole or axolotl) stages have little resemblance in structure and function.

In many organisms, the males and females may be different in size, form and colour. The occurrence of two forms among the organisms of the same kind is known as dimorphism. Ex :- man and woman etc. They show sexual dimorphism. The individuals in a colony of the social insects such as the bees and termites differ considerably in structure and function. The occurrence of many forms of individuals within the same kind of organism is called polymorphism. Various types of individuals are seen in termites i.e, queen, king, soldiers and workers. In honey bees there are queen, drone and workers.

In the same way, *Physalia* colony consists of hundreds of individuals of different kinds called zooids performing different functions such as feeding, defence, reproduction etc.

There are considerable physical differences among the people of different human races such as the Negroids, Caucasoids, Mongoloids etc.

3.10 Concepts of Ecosystem

The living organisms cannot live in isolation. They live in an environment which supplies all the essential things that the organisms need such as food, shelter, water and so on. Hence the biotic community together with the physical environment forms on interacting system called ecosystem. Any structural and functional unit of the environment that can be identified and studied is called ecosystem for eg : forest, grassland, pond, lake etc.

Thus an ecosystem has two main groups (i) Biotic component or living organisms and (ii) Abiotic component or non living things. The biotic component consists of different kinds, numbers and distribution of living organisms.

The abiotic component consists of the kinds, quantity and distribution of physical and chemical factors such as light, temperature, water, oxygen, carbon, nitrogen and minerals etc.

Biotic Components

All organisms require energy for their life processes and materials for the formation and maintenance of body structures. Food supplies both energy and materials for the sustenance of life. Green plants produce carbohydrates by photosynthesis and also synthesize proteins and fats. Thus, the green plants are called *producers*. The animals that consume this energy or food are called *consumers*. The fungi and some bacteria which are incapable of producing their own food and hence live on the dead and decaying plants or animal parts and are called decomposers. In this way the plants (producers), herbivores (primary consumers), carnivores which feed on the herbivores (secondary consumers) and bacteria etc. that feed on dead and decaying organic matter (decomposers), all put together, constitute a *food chain*. Each step in the food chain represents a trophic level. Depending upon the availability and choice of food, different organisms at each level have food relationship with more than one organism at the lower levels. For example a frog feeds on many types of insects, rats feed on many types of food grains, stems, roots, fruits etc. Thus a network of food chains exists and this is called *food web*. The food web becomes more complicated because of variability in taste and preference, availability and compulsion and several other factors at each level. Another common feature in food chain is that at the successive higher level, the numbers will be smaller. For example insects are fed by insectivorous birds which in turn are fed by preying birds. Thus the number of organisms in one trophic level is higher than that at the next higher level. When the number of organisms at successive levels is plotted, it assumes the shape of a pyramid. This is called a *Pyramid of numbers*. In the same way, the amount of food or biomass or energy that is transferred from lower level to the higher level also becomes less and less. This can be measured and graphically represented. This also assumes the shape of a pyramid, called the *pyramid of energy* or the *Pyramid of Biomass*. For example if we see the pyramid of energy, at the lowest level the plants take solar energy to produce carbohydrates by photosyntheiss (the producers). At the same time, they also utilize some energy for their own requirements. Herbivores (primary consumers) which feed the plant material consume only a portion of energy that is trapped by the producers. Animals being more active than plants also use much of the energy acquired before they are consumed by the next trophic level. So, at each stage of energy transfer, a considerable amount of energy is lost from the food chain. The amount of energy transferred to the next higher level gradually decreases. The decomposition of dead organisms also releases chemical energy. Eventually, all the solar energy that entered the living system through the producers goes back into nature not as light but as heat. Chemical energy is transferred in the living organisms through food. Food means material containing energy that organisms can use. (Fig. 3.24 to 3.29)

The number of organisms at any trophic level depends upon the availability of organisms which serve as food at the lower level. For example, in the rainy season, lot of vegetation is seen. Simultaneously, the number of phytophagous insects increase in number. This results in the increase of insectivorous animals like frogs, lizards, birds etc. In this way preying birds or carnivorous animals also increase. Thus availability of food is the main factor which maintains the balance of nature. In the absence of vegetation, likewise, decrease in the number of primary consumers, secondary, tertiary consumers etc also takes place.

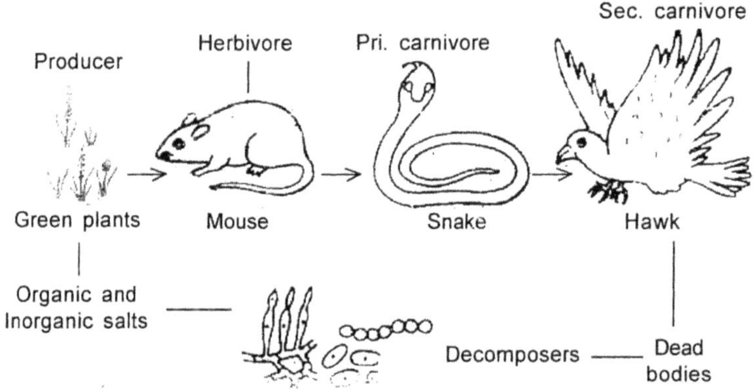

Fig. 3.24 Food Chain in a Grassland

Fig. 3.25 Food Chain in a Forest

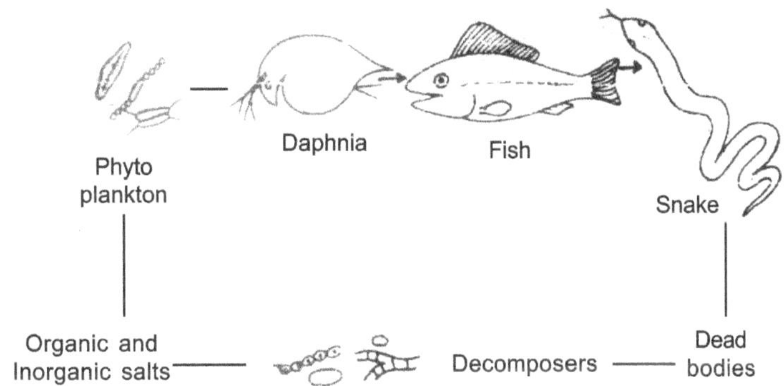

Fig. 3.26 Food Chain in a Pond

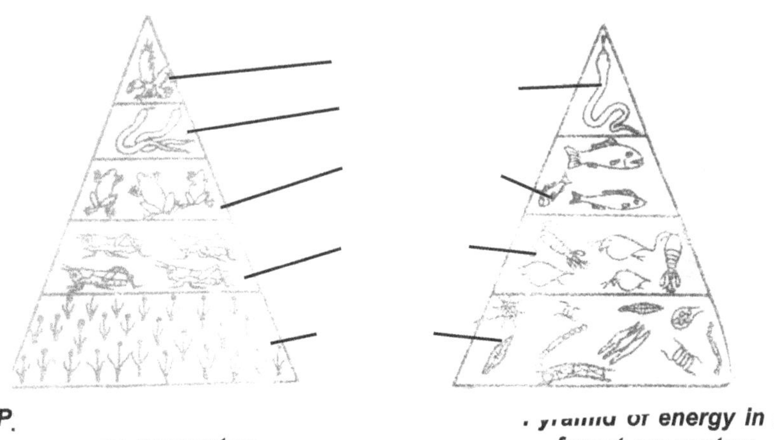

P
ecosystem

of energy in a forest ecosystem

Fig. 3.27

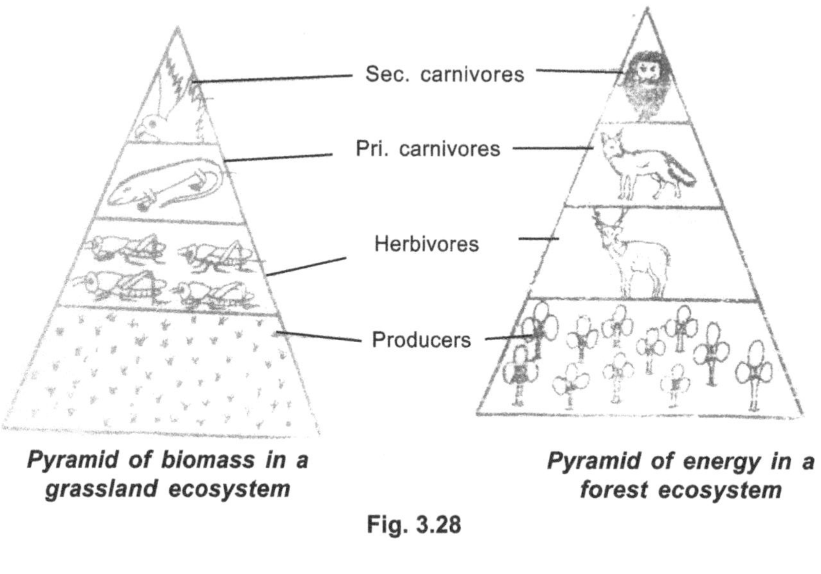

Pyramid of biomass in a grassland ecosystem

Pyramid of energy in a forest ecosystem

Fig. 3.28

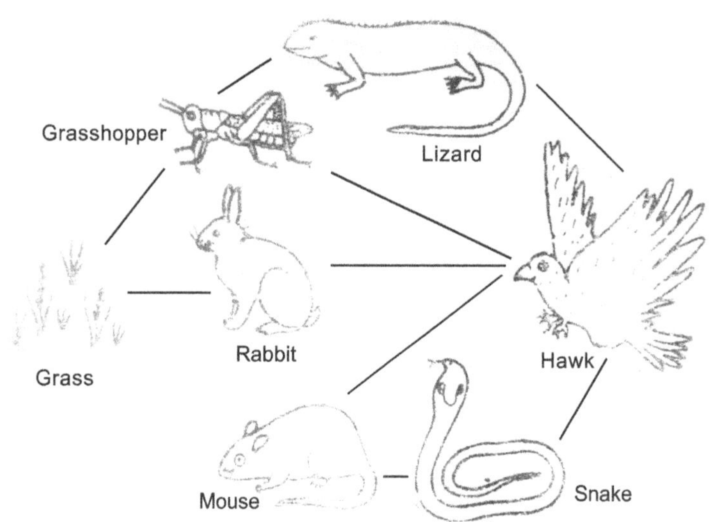

Fig. 3.29 Food web in a grassland ecosystem

Abiotic Components

Among the abioitc factors, the most important are climatic and edaphic factors. The climatic factors include temperature, humidity, light, wind, pH, topography, background etc. The edaphic factors include the soil and substratum.

The ability of organisms to utilize, tolerate, or combat the various abiotic factors are different and it may limit their distribution, behaviour and relationship with other organisms. Some of the most important abiotic factors are described below.

1. *Temperature :* Living organisms can live only in a narrow range of temperature which allows their metabolism. Wherever they live, the organisms must develop physiological and behavioural adaptations to withstand extremes of temperature. The polar bear can live in very cold regions. Some birds migrate to warmer places in winter to avoid extreme cold. Some desert animals live inside burrows to avoid intense heat.
2. *Water :* Water is an essential requirement of life. No life can exist without water. A large number of organisms live only in water. The requirement of water varies from organism to organism. The distribution of organisms depends upon the extent of the need and special adaptations for conserving water. Fishes live in water, frogs live both in water and on land, snakes, lizards, etc live on land.
3. *Light :* Light is essential for Photosynthetic organisms for the preparation of food on whch rest of the living organisms are dependent. Many animals have special photo receptors. The rhythm of functioning of a large number of organisms is modulated by light. Some are active in the absence of light (Nocturnal organisms like cockroaches, bats, snakes etc). Some are active in the presence of (day) light (diurnal organisms eg : butterflies, birds, cattle, man etc.)
4. *Humidity :* The humidity in the atmosphere directly regulates the rate at which water evaporates from the body surface of terrestrial organisms by transpiration, perspiration and other means. Different organisms show different adaptations to withstand dry conditions.
5. *Wind :* Wind determines the rate of transpiration in the plants. Where wind velocity is very high, only the plants with strong root system and strong stems can survive. Wind also helps in the dispersal of fruits and the seeds. The seeds are either minute or with appendages.
6. *pH :* Most organisms thrive in an optimal pH range. Some plants and aquatic animals require acidic conditions, others need neutral or alkaline conditions. pH of the soil and water has a strong influence on the distribution of organisms.
7. *Mineral Elements :* Availability and concentration of essential mineral elements control the distribution of microbes, plants and animals. Plants living in soil deficient in nitrogen have developed special adaptations for obtaining it, such as, harbouring nitrogen fixing bacteria and the carnivorous habit. High concentration of minerals can limit the distribution of organisms. Excessive salts (salinity) are eliminated by plants and animals through special salt secreting glands. Fresh water animals usually can not live a marine habitat. Animals inhabiting an estuary have special physiological or behavioural adaptations for withstanding fluctuations in salinity accompanying tidal rhythms.
8. *Topography :* Topography of an area (physical features such as hills, plains, slopes etc) influences distribution of organisms as much as wide geographic separation. Topographical effect may be direct. It also indirectly affects other factors like wind, water current, light or wave action. The centre and edge of a pond or stream, topside and underside of a rock, north and south face of a ridge or a wall are generally inhabited by different species of organisms.
9. *Background :* The background of the habitat also determines the distribution of animals by enabling them to camouflage against the colour, general texture and pattern. Desert animals like lion and the camel are sand coloured. Most of the jelly fishes, sea cucumbers are glassy. The chamaeleon can change colours according to the background.

Ecosystems

A habitat is a specific place where an organism lives. It is a physical entity comprising the sum total of abiotic factors to which a species or a group of species is exposed. Habitat usually refers to a relatively large area such as a pond, a forest, an estuary or an ocean. However, they can be further sub divided into Microhabitats which show distinct properties and harbour different types of organisms. For example, muddy bottom, surface water of a pond, tree canopy and forest floor are microhabitats.

Each species of a community lives in a very specific part of a habitat and performs certain functions. The habitat together with the functions form the Niche of the species. Niche includes factors such as food relations, predators, tolerance levels and other habits as well as abiotic factors. A population of each species within a community has a separate ecological niche.

There are distinct major ecosystems or life supporting zones also called biomes in marine and fresh water environments. The distinctive features of the aquatic systems are different from the terrestrial biomes.

The major biomes are 1. Tundra 2. Northern conifer forests 3. Deciduous forest 4. Tropical rain forest 5. Chapparel 6. Trophical savannah 7. Grassland and 8. Desert

1. *Tundra :* Tundra refers to the treeless plain of the arctic region which remains marshy during summer and frozen hard in winter. About 5 million acres of tundra stretch across North America, Northern Europe and Siberia, generally above 60^0 N latitude. Tundra is characterized by the following features like low rain fall, cold climate, continuous darkness in winter and continuous light in summer. The frozen deeper soil is called permafrost. The vegetation grows only for a brief perod of 60 days in summer. Flora that are present in Tundra are Lichen (reindeer moss), mosses, grasses, sedges and shrubs. The Tundra fauna is arctic fox, hare, reindeer, lemmings, oxen, polar bear, caribou, walruses, seals, penguins, migratory birds and insects.
2. *Northern Coniferous Forests or Taiga :* This biome is thickly populated with pine trees characterized by long needle like leaves. These trees belong to the class Gymnosperms. egs : cedar, pine, fir, hemlock, spruce etc. The leaves of the trees are ever green. It is also called taiga because it gives thick shade and prevents the growth of herbs and shrubs. This biome stretches as a belt south of tundra in North America, Europe and Asia. The rainfall ranges from 10 to 35 cm and the average temperatures are 6^0 C in winter to 20^0 C in summer. The annual rate of production is great because of the ever green leaves and continuous photosynthesis throughout the year. It forms the major supplier of timber. The seeds of the pine trees form the major source of food for certain animals. The Taiga fauna is Hyla, Rana, Wood peckers, Cross bills, Squirrels, Hares, Beavers, Deers, Canadian Porcupine, Wolves, Bears, Foxes, Caribou, Lynx, elk, insects like Bark Beetles, Saw flies, Geometrid moths etc.
3. *Deciduous Forests :* These are found in Australia, Europe, Japan and Eastern United States. These forests are characterized by tall trees with broad thin leaves which fall during winter. The climate is warm during summer and cold during winter. The annual precipitation ranges from 60 to 150 cm per year. With the onset of autumn, most of the trees and shrubs shed their leaves. The characterisitc flora are oak, elm, birch, maple and hickory. The

fauna includes deer, fox, bear, beaver, squirrels, flying squirrel, raccoon, wild boar, rattle snake, song birds, birds of prey such as horned owl, hawks, crested fly catcher, thrushes, wood pecker, amphibians, insects, etc.

4. *Tropical Rain Forests :* Rain forests are found in tropical countries like India (Western ghats), Malaya, East Indies, Africa, Northern Australia, Central America along the Amazon and Orinoco rivers, South America etc. The forest is characterized by thick growth of tall trees. The leaves of these trees are broad and ever green. In these forests high precipitation (ie, more than 200 cm / year), high humidity, low illumination because of thick vegetation, etc are seen. Here vertical stratification is seen. It is formed of 3 layers namely an upper layer of leaves and branches, a middle layer of shrubs and small trees and a lower layer of forest floor.

   The forest fauna are worms, snails, millipedes, centipedes, scorpions, spiders, insects, planarians, leeches, tree frog, chamaeleon, geckoes, snakes, birds of paradise, insectivores, leopords, jungle cats, ant eaters, flying squirrels, monkeys, sloths, arboreal kangaroos, tiger, elephant, deer, gaur, etc.

5. *Chapparel :* This is also known as mediterranean scrub forest. The chapparel extends along the mediterranean, pacific coast of North America, Chile, South Africa and South Australia. It is marked by very limited winter rain followed by drought in the rest of the year. The biome is characterized by broad leaved evergreen vegetation. It is composed of fire resistant resinous plants and drought adapted animals. Bush fires are very common in this biome.

6. *Tropical Savannah :* Savannahs are warm climate plains chracterized by coarse grass and scattered trees on the margins of tropics where rain fall is seasonal. They are located primarily in South America, Africa and Australia. The average total rain fall is 100 - 150 cm. Wet seasons alternate with dry seasons. Plants and animals are drought tolerant and do not show much diversity. This biome supports the greatest variety of hoofed herbivore species including Zebra, Giraffe, Elephant and several kinds of antelopes. Kangaroos are found in the Savannahs of Australia.

7. *Grassland :* The main grasslands are the Prairies of Canada and USA, the pampas of South America, the steppes of Europe and Asia and the veldts of Africa. The annual rainfall is between 25 - 75 cm. Winter blizzards and summer drought can be severe, with periodic fire accidents. The dominant plant species comprise short and tall grasses. Animal species include larks, the burrowing owl, prong-horned antelopes, badgers, coyote, jack-rabbit and bison.

8. *Desert :* A desert is waterless, treeless, large waste land often covered with sand. It is characterized by low or no rainfall, scarcity of water, extremes of temperature and dust storms. The important deserts in the world are Rajasthan (India) Gobi (China) Sahara and Kalahari (Africa). The organisms having specialized structural, physiological and behavioural adaptations only can survive in this biome. Desert plants include different kinds of Acacias, cacti, euphorbias, and other succulents and hardy grasses. Ants, locusts, wasps, scorpions, spiders, lizards, rattle snakes, a large number of insect eating birds such as swifts and swallows, seed eating quails, doves, desert rats, rabbits, foxes, jackals and various cats are the common desert animals.

Aquatic Ecosystems

There are many kinds of aquatic ecosystmes. They range from open ocean to small temporary puddles, have conditions of salinity, depth, fluctuations of temperature, etc. Even the so called marine and fresh water ecosystems vary very much. The adaptations of organisms to aquatic environments also vary in degree and diversity. Some are exclusively aquatic (fishes), some amphibians (frog, crocodiles, hippopotamus, aquatic birds), some totally marine (echinoderms, coelenterates) while others are totally fresh water (*Clarias, Ophiocephalus, Catla, Tilapia*) some fish like Hilsa can live both in fresh water and sea.

Marine Environments

The marine environment is chracterized by its high concentration of salt and mineral ions. The salt concentration is roughly 3.5 % while the mineral ions that are present are soidum, chloride, sulphur, magnesium and calcium. The important determinants of marine ecosystems are depth of water, distance from the shore and drainage of glaciers and rivers. From the coast line a gradually sloping region extends to an average of 160 km into the sea. This zone with a depth of 8 - 200 metres is called *continental shelf.* The angle of the slope then suddenly steepens to form the continental slope. The continental slopes often have ridges, basins and trenches formed of mud and sand, deposited by slow flowing rivers. The contimental slope levels off into a more or less horizontal expanse forming the ocean floor. The depth of the ocean floor is in the range of several thousand metres.

The vertical zones of the ocean are determined by the availability of light for photosynthesis. Light can be reached upto 200 metres. This zone is called *photic or Euphotic zone.* The next zone from 200 - 2000 metres gets less light which is insufficient for photosynthesis. This zone is called *Aphotic zone.* Below 2000 metres there is complete darkness. This zone is called *abyssal zone.*

Three major environments can be recognized in the ocean basin 1. The littoral zone comprising the sea floor from the shore to the edge of the continental shelf 2. The benthonic zone formed of the sea floor along the continental slope and the aphotic and abyssal zones 3. The pelagic zone constituting the water of the ocean basin.

1. *Plankton :* It consists of passively drifting or floating organisms. These are minute organisms with very weak or no locomotory organs and are confined to the euphotic zone. The plankton includes autotrophs such as diatoms and heterotrophs such as small crustaceans (zooplankton). (Fig. 3.30).
2. *Nekton :* These consist of actively moving organisms with well developed locomotory organs. They include actively swimming organisms found in the surface or deep waters. They feed mainly on the planktons or smaller nektons. The nektons include small and large fishes, squids and whales.
3. *Benthonic Organisms :* These are found along the floor of the sea bed and include creeping, crawling or sessile organisms. Many of them are scavengers.

The photic zone contains planktons and nektons. The aphotic zone contains nektons while abyssal zone contains predators, some of which have enormous jaws and sharp teeth, while a few have luminiscent organs to attract other organisms.

The benthonic fauna comprises predators, scavengers and decomposers.

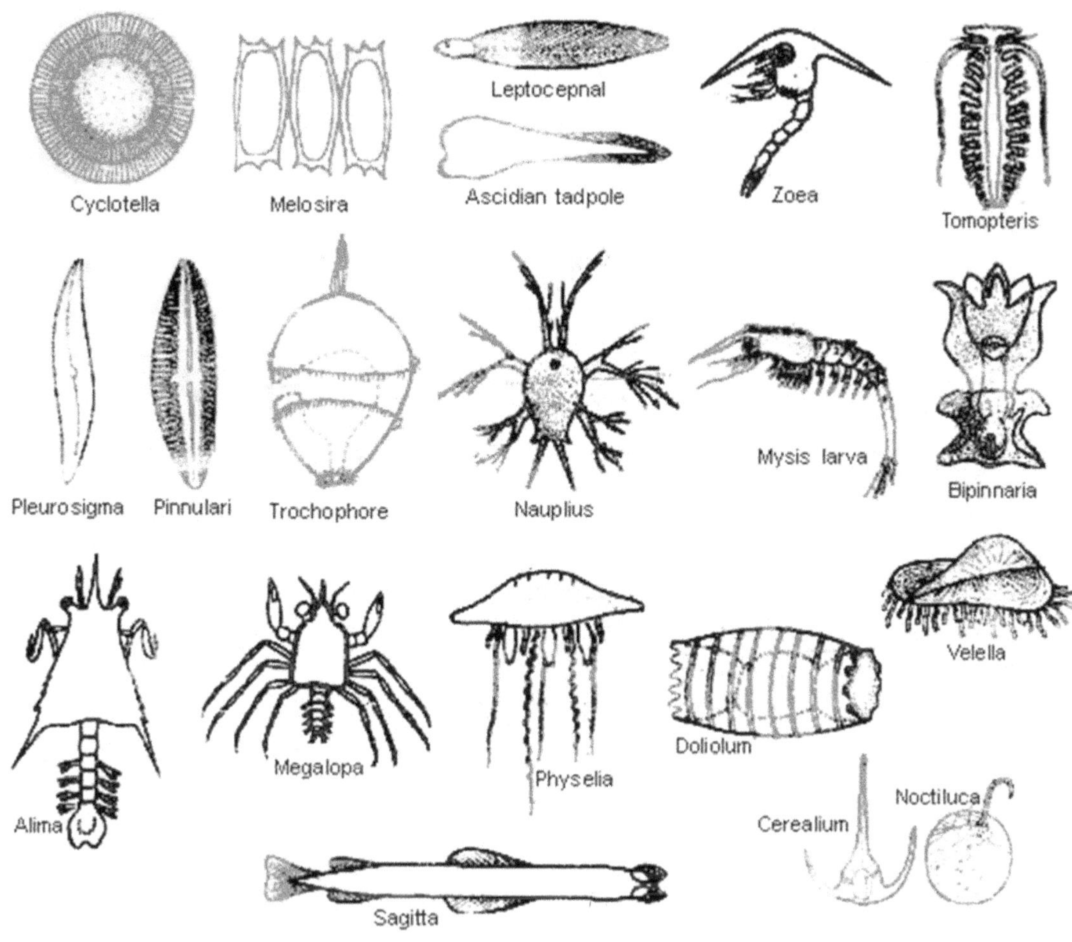

Fig. 3.30 Phytoplankton and Zooplankton

Fresh Water Environment

Lakes and Ponds

These are stagnant fresh water bodies. They vary in size from less than a hectare to hundreds of hectares. Their depth varies from a few centimetres to a few hundred metres. Small ponds dry up within a few months (one season) whereas some lakes exist for many centuries. The relatively shallow lakes called eutrophic lakes have a rich accumulation of organic matter. Deep lakes with steep and rocky sides have poor phosphate nutrients and are called oligotrophic lakes.

Ponds have autotrophs and heterotrophs. The autotrophs comprise phytoplankton and other floating plants such as *Azolla*, *Pistia*, *Lemna*, *Ceratophyllum* and *Utricularia*, *Vallisneria*, etc.

The heterotrophs include ciliates, flagellates, rotifers, small crustaceans like Daphnia, Cyclops, Cypris and larval stages of mayflies, dragonflies, damsel flies, water bugs, beetles etc.

Streams and Rivers

Rivers, streams and brooks are bodies of flowing fresh water. They differ widely in volume of water, speed of flow, O_2 content, temperature and many other physical and chemical conditions. The nature and composition of flora and fauna also depend on the source of the river and the land environment they pass through. Rivers with rapid currents have no floating planktons. In the middle reaches, the water current becomes slower and the river body becomes wider. The sediments are deposited in the river bed. The autotrophs and heterotrophs are seen in this region. In the lower reaches the current diminishes still further. The water is often muddy, cutting down light at the river bed. Both phytoplankton and zooplankton are quite common. The nature and composition of the fauna resemble that of the lakes. Many reptiles, mammals and birds obtain their food from this area.

Man-Made Ecosystems

Man has changed the environment to a great extent for his benefit. The examples of man made ecosystems are villages, cities, orchards, plantations, gardens, parks, large dams and reservoirs, lakes, canals, fish tanks, aquaria etc.

Most important man made modifications in the biotic community came with the use of fire, cultivation of plants and domestication of animals. Man has converted very large areas of forest and grasslands into fields for growing very selected plants such as cereals, pulses, oil seeds and forages, etc.

The study of the relation between agricultural crops and environment is called agroecology.

The physical condition of the farm depends upon the latitude, overall climatic and edaphic factors. The physical deficiencies are overcome by artificial irrigation and the use of fertilizers and manures. The type of crop depends upon the climate, season and choice of the farmer. The composition of the biotic community also varies accordingly. It is richest when the crop is in the field. In a field along with the main crop, weeds, grasses etc (autotrophs) are there. The animal population comprises earthworms, nematodes, domestic animals, pollinators, pests, parasites, predators, rodents and many birds.

All man made agro ecosystems lack the diversity of natural ecosystems. The consequence of diversity is stability. A single crop may be totally destroyed by drought, floods, diseases, pests or pathogens. A diversified system has scope for many adjustments and substitutions.

Boundaries between one ecosystem and another are indistinct and overlapping. Some exchange of materials and energy always occurs between ecosystems through biological, climatic or geological processes.

3.11 Remote Sensing

Remote sensing is a multi-disciplinary activity that combines aerial and space photography and electronic devices and deals with the inventory, monitoring and assessment of natural resource through the analysis of data obtained from a remote platform. Remote sensing is a science and art of acquiring information about an object or surface feature of earth from measurement made at a distance without coming in physical contact with the object of feature of interest. This information is acquired by non contact methods usually from aeroplanes or satellites in any portion of the electromagnetic spectrum.

Since remote sensing satellites reappear and reobserve the same areas at periodical intervals, there is added advantage of monitoring changes that occur in the area of coverage, to plan for remedial measures for any adverse happening, much sooner than what we were able to do in the past, when the facility of remote sensing from space altitudes was not available. It is in this context that we could view forest cover or vegetation cover of a large area and the changes that occur to the area and the data is quantifiable. If the observation is made due to the electromagnetic radiation from the sun or due to the self emitted radiance it is called "*Passive Remote Sensing*". It is also possible to produce electromagnetic radiation of a specific wave length or a band of wavelengths to illuminate the terrain. The interaction of this radiance can then be studied by sensing the scattered radiance from the target. This is called "*Active remote sensing*". Remote sensing system consists of a sensor to collect the radiation and a platform-ballon, rocket, aircraft, satellite or *even a ground based*. The information received by the sensor is suitably manipulated and transferred back to the earth may be telemetered as in the case of unmanned space craft or brought back through films, magnetic tapes etc. as in aircraft or manned space craft systems. The data are reformatted and processed on the ground to produce either photographs or computer compatible magnetic tapes (CCTS). The photographs/CCTS are interpreted visually/digitally to produce thematic maps and other resources information.

Photography from the ground aerial platforms is a conventional form of remote sensing used by foresters, geologists and geographers. These days topographic maps are made by stereoscopic photography. It is one of the areas of rapidly advancing technologies that makes it possible to deal more effectively with very large ecosystems. The exciting prospect is that remote sensing will be a logical bridge between intensive ecological research on small areas and the application of principles thus revealed to planning and management of large political units such as townships, countries or states, or whole natural units such as watersheds, tropical rain forests, or ocean basins.

Today, the remote sensing technology has been put into operation to cover diverse areas such as forestry, agriculture crop coverage and yield estimations, drought and flood monitoring and damage assessment, land-use and cover studies, waste land identification and reclamation, water resources development and management, ground water survey, marine resources survey, urban planning, mineral prospecting, environmental impact assessment and so on; thus covering almost every facet of sustainable resource development and management (Kasturirangan et. al. 1996).

Such automatic sensing, automatic interpretation, classification and mapping is an important technological break-through to mankind.

Infrared photography is a comparatively new and very useful technique in the study of vegetation because the molecules of pigments in plants do not absorb infrared wavelengths. Instead, the infrared is either transmitted through the leaf or is reflected by the cellulose cell walls. Cells of one species of plant have a different reflectivety than those of another and normal cells have a different reflectivety than abnormal ones. Therefore, infrared photography has been used to distinguish species and to detect differences in environmental temperatures. Such pictures or thermographs in which differences of temperature appear as contrasting bands of colour, permit the detection of thermal pollution.

Technological advances in space research created a diversity of new airborne sensors that capture energy from various portions of spectrum, thus adding much to information gathering potential of aerial overflights. Electronic systems analyse and display the data in pictorial format or on tape for input to a computer. In order to utilize the information given by airborne sensors must have basic concepts of energy and matter relationships implicated in the understanding of the images. Energy absorption, emission, scattering and reflection by any kind of matter are selective with regard to wavelength and are specific for that matter, depending upon its atomic and molecular structure. Thus each kind of biomass or physical substrate emits a spectrum of characteristic frequency and intensity. Since most of the earth's land surface is covered by some kind of vegetation, this is primarily the plant leaf surfaces which are perceived by remote sensors. Remote detection of individual animals can be much facilitated by attaching to the main animal an energy source that emits an easily detectable signal. Such a procedure of attaching transmitters to animals is known as biotelemetry.

The recent introduction of a laser profilometer which measures very small differences in elevation by the time versus distance relationship of a highly focused light pulse, promises to replace expensive ground surveying for vertical control of these maps. In forestry, tree height, stem diameter, and therefore, timber volume can be estimated from panchromatic aerial photographs with as little as 20 percent error. Since the ratio of dbh (diameter at breast height) of tree stems to crown diameter is consistent, the total dbh or basal area of overstory stems can be estimated by measuring the crown diameters as revealed on the photographs.

One of the most sophisticated remote sensing techniques is called "*Multispectral Scanning*". i.e., several or many spectral bands from Ultraviolet to Infrared are recorded. The visual data are gathered by telemetry or are recorded on film magnetic tape. The image data are fed into a multi channeled sensor, which sorts out the various displays or images. Because all the channels are recorded simultaneously, each image is in perfect register. This enables the interpreter to pick out and study an object of interest with different types of images taken at the same time.

Appropriate electronic systems analyze and display the data in pictorial format or on tape for input to a computer. Some of these systems are electronically sophisticated and require specially trained operators. It should be emphasized that the meaning and validity of remote sensing data, especially that obtained in non-visual portions of the spectrum, require familiarity with ecosystem recorded. Pertinent measurements and observations obtained on the ground at the time of overflights are called "*ground truths*" and are essential to the success of most remote sensing operations. Once the indicator value of remotely obtained data is established, ground measurements may no longer be required. Thus, a multidisciplinary approach, including the work of computer engineers and ecologists and biologists is required for meaningful interpretation of data.

Remote Sensing in India

In India, the first Landsat Series of Satellites with a multispectral scanner on board was launched in 1972, and the remote sensed data of our entire country was recorded in a near real time mode, although one could not realise the real impact of such information immediately. The first three series of Landsat satellites had on board a multispectral scanner giving a ground resolution of approximately 80 meters and in four spectral bands.

The National Remote Sensing Agency (NRSA) was established at Hyderabad in A.P. (India) in 1975, which took up a study to establish credibility of using satellite remote sensing for forest surveys to compile gross forest vegetation resources information of the entire country in a single format and monitor the changes. The study was conceived by the Chairman of the Department of Space, Prof. Satish Dhavan and carried out by scientists of NRSA. By the early eighties, remote sensing scientists, representing various disciplines came forward to undertake near operational investigation projects using satellite remote sensed data products. Scientists also began to delineate forest cover types, to conceptualise the ecological status of the forest. Essentially, NRSA made rapid assessment of the forest/vegetation cover as existed in India during 1972-75 and again in 1980-82 to demonstrate the deteriorating condition of the forest and also use of remote sensing technology for making rapid assessment. India is covered by about 190 scenes of the multispectral scanner imagery of landsat, each covering an area of about 185×185 kilometers. So 1972-75 had to be taken, although theoretically data should be got every 18 days in view of the satellites periodicity of reappearance at the same place. Similarly data pertaining to 1980-82 was also used, covering the whole country. Both these sets of data in the form of false colour imagery were visually interpreted to delineate gross open and dense forest and maps on 1:1 million scale were prepared covering all the States and Union Territories of India. Thus the use of satellite remote sensing has been found to be a reliable and useful tool for creating forest information.

The IRS satellites have been proposed to last for three years. But the IRS IA which was successfully launched in 1988, was working till 1994. Subsequently, IRS-IB and IRS-P2 satellites were launched in 1991 and 1994 respectively. The IRS-P3 just like its predecessors, ceases to be effective after a few years when it sinks in orbit and dies.

The IRS-P3, the country's fifth remote sensing satellite, which is now revolving around the earth 14 times day (once every 104 minutes) at an altitude of 817 km after being launched on March 21, 1996, has started sending data to the designated earth station at Shadnagar in Mahabubnagar district of Andhra Pradesh about 60 km from Hyderabad. The data signals received by mammoth antennae and introduced into magnetic tapes were brought to Hyderabad and decodified and multicoloured imageries were made out of them at the NRSA the only such facility in the country.

The photographs sent by the satellite of excellent quality, and could indicate in the seas the exact location of even dynamic variables like chlorophylls, fish food, by virtue of which scientists can issue fishing forecasts. The satellite's Indian cameras, wide field sensors, have proved that they could see and photograph even objects measuring a few square meters on the earth. The indigenously built satellite now journeying at a speed of 8 km per second is also taking pictures of the atmosphere over its head with the help of "X-ray astronomy" sensor, a job not done by the earlier IRS series. For the purpose of oceanic studies, the three modular opto-electronic scanner cameras with Germany kept in the satellite as its payload under a collaboration have been operating.

Apart from taking pictures of the usual land scapes like vegetation, forest coverage, earth strata and watershed area, as is being done by the IRS series, the IRS-P3 can do not only mapping of chlorophyll in the seas, but also specify percentage of sedimentation/transportation at mouths of rivers which indicates whether the water at that spot is disturbed (and so not favoured by fish), and monitor even the biomass in oceans consisting of living organisms like plants, a totally uncovered area. Owing to this, the fishing forecasts are going to be more accurate hereafter and they will be transmitted to the fishermen community through a network of fax centres opened all along the coastline–Bombay, Goa, Thiruvananthapuram, Madras, Machilipatnam, Visakhapatnam, etc.

For successful organisational implementation on a national basis, the coordination of various ministries, state governments, organisations is necessary. The National Remote Sensing Agency (NRSA) in India with its Head quarters and Analysis complex in Kukatpally, Hyderabad and its aerial photo interpretation facilities at the Indian Photo-interpretation Institute, Dehradun and Research Flight Facility at Bangalore are playing an important role in compilation of data.

Applications

This technique is now being increasingly used for over all development of India through rapid exploration, assessment, inventorying, exploitation, management and conservation of natural resources.

1. Forest cover mapping and surveillance : The increasing population is exerting pressure on forests for fuel, fodder and timber. Hence India's forest is undergoing rapid changes leading to forest degradation. The problem is further activated due to increased activities of mining, industrialisation and construction of multi-purpose river valley projects. This has made necessary for nationwide monitoring of forest cover to make strategic planning and to maintain adequate forest cover as prescribed by the Forest Act (1956). Towards this remote sensing technology has been employed to map the entire country's forest cover. The forest cover mapping provides the information on forest density classes with respect to dense, open and degraded forests. Satellite remote sensing is ideally suited for monitoring of vegetation cover.
2. Forest ground fire damage assessment : Out of the 16 forest types in India, tropical moist deciduous forest forms the major percentage accounting for 37% of the total forest cover followed by tropical dry deciduous forest which forms 28.6% of the total forest cover. Generally during summer periods these forests are threatened due to extensive man made ground fires for regeneration of forest grasses for cattle. The fires occur primarily during the course of burning dry matter/grasses of the understory leading to damage of plantations, forest trees and prevent regeneration of saplings. The use of satellite data permits to account the ground fire damages periodically during the dry periods. The use of multi temporal satellite data in the vulnerable areas would greatly help in making fire protection plans.
3. Ecological Hotspots : The 13 biosphere reserves are selected in India considering them as virgin areas for preserving the already existing flora and fauna without making further drain in the country's germ plasm. However, the conservation strategy need also to be concentrated on the hot spot areas where the ecological degradation is rapidly occurring. It is necessary to identify the hot spot areas in all the agro-climatic zones of India and declare them to be ecologically sensitive zones and needs top priority basis for their conservation/reclamation strategies. Remote sensing can play a vital role in ameliorating these areas as ecologically sustainable.
4. Biosphere Reseves : There are 54 national parks and 372 sancturies in India with a total area of nearly 1,09,650 sq.km. India has done well in identifying the zones for environmental protection based on biospheres. However, there is a need to document and inventory the resources with respect to flora and fauna periodically to account for the changes that are taking place. Studying such large biogeographical zones obviously is made possible only by use of remote sensing.
5. Wetland environment : India has immense potential on the vast extent of wetlands. The unique characteristic of wetland is the presence of water and a water saturated soil. This may be either persistent feature or occurring partially for a part of the year. This ecosystem is a zone between the terrestrial and the aquatic systems characterising an unique environment. Diversity of character, size, type and mode of occurrence of wetlands have given rise to diversified biota. These ecosystems are useful for nutrient recovery and recycling, releasing excess nitrogen and toxins absorption of heavy metal residues through tail end of waste water channels. The fragile and the unique wetland environment need to be intensively studied and the areawise conservation plans to be chalked out to preserve the less attended but most important fragile systems. This is possible through remote sensing data.

6. Coastal vegetation and coral reef mapping : The coastal zones provide an ideal environment for growth and survival of coastal resources such as mangroves and coral reefs under certain environmental conditions. The mangroves are salt tolerant forest ecosystems which grow in intertidal regions. Several studies have been carried out using remote sensing in identification and mapping of mangroves in middle Andamans, Sunderbans, Mahanandi delta and around gulf of Rann of Kutch. The mangroves protection is important as they protect the coasts and harbours and a good breeding ground for fishes. Coral reefs are shallow water, tropical marine eco-systems known for the high rates of biological productivity. They are exceptionally diverse in fauna and flora and have complex food web and trophic organisation. Satellite remote sensing can facilitate in knowing the extent and distribution of mangroves and coral reefs of India and this will help in timely and cost effective management of these ecosystems.
7. Wild-life management assessment : The monitoring of wild life habitat is an important environmental aspect to conserve the wildlife habitability. The assessment of vegetation cover and water resources of an area provide necessary information to plan the carrying capacity of game animals. The satellite remote sensing provides information on areas vulnerable for wildlife and identification of appropriate locations as their habitats. Satellite data permits in identifying suitable areas as corridors connecting the core of the national parks which are under pressure for migrating wildlife to safer areas in situations of flooding, forest fires etc.
8. Waste land and land degradation : Land degradation can lead to formation of wasteland. One of the naval applications of remote sensing relates to mapping and monitoring of wastelands. Monitoring the extent of waste lands and effective implementation of reclamation measures has become a national need. Satellite data has been successfully used to identify and map the extent and type of waste lands at village level.
9. Environmental impact assessment : The construction of major hydroelectric dams in the critical catchments would result in inundation of large tracts of thick forest, agricultural land, habitation areas etc. Likewise construction of major industries, mining activities etc. also result in degradation of habitats. Remote sensing data will help in assessing the environmental impact assessment of major activities.
10. Flood damage assessment : The damages caused by floods are manifold and include damages to agricultural crops. In order to assess the damage due to floods and to take appropriate flood management and control measures, it is essential to map flooded areas with minimum delay and to identify flood prone areas for future management and control measures. Satellite remote sensing data (flood prone area maps) provide this information in time and help to avoid recurrence of floods and to reduce damage caused by floods.
11. Monitoring of areas : Mapping of suspended sediments in the major reservoirs, mapping of river migration and loss of land area, monitoring of coastline and estuarine environment mapping of desert locust breeding centres and their identification are also possible through the remote sensing data.
12. Desertification : Desertification is a worldwide phenomenon affecting all the continents of the earth and gradually decreasing the natural resources of land. Desertification is process of destruction of the biological potential of the land, which leads ultimately to desert like conditions and is an aspect of the wide spread deterioration of ecosystem under the combined pressure of adverse and fluctuating climate and excessive exploitation. The application of space based images for monitoring the spread of deserts and the measures to arrest them is being attempted.

13. Geosphere-Biosphere studies : Taking in account the concern shown in the environmental degradation in recent times and its possible impact on the global habitability, several countries have launched specific programmes to study the geosphere-biosphere interactions. In the Indian context the major concerns are changes in monsoon strength and flow, possible sea level rise, greater snow melt, changes in hydrological cycles with recurrent droughts and floods etc. It will call for studying of the various air-land-ocean interactions. One such major impact could be due to the increase in minor constituents such as CO_2, CFC, N_2O, CH_4 etc. which could lead to global warming and associated effects on climate, sea level and vegetation. The space technology has developed several tools which could be utilized towards studies related to atmosphere, land and ocean aspects. The satellite based remote sensing could be useful in monitoring of various parameters related to land and ocean.

14. *Environmental Planning :* Cultural expansion, is changing the environmental conditions also. Pollution of water (by industrial effluents) is already seen in many parts of India as a result of heavy industrialisation. Cultural incursions with mass explorations have caused the forests to be depleted, deserts to expand, soil to be eroded, wild life to disappear, and fish to abandon their usual spawning water. In short, the natural ecological balances have been disturbed with consequences that would assume grave proportions in the future. In tackling such problems, detection of the sources of pollution, and continuous monitoring over wide areas for disturbances of ecologial equilibrium is the first requisite. Remote sensing is one of the most powerful tools for this purpose.

15. *Land use :* Land use is one of the most common applications of remote sensing. In a changing landscape and continuously increasing and competing demands for land for various purposes (agriculture, forestry, grasslands, housing, industry etc), it is necessary to have an inventory of actual land use. Many village records need to be up-dated. A land use map can be used to monitor urban growth effectively. A land use map coupled with hydrological information and a soil map can be converted to a land-capability map on which important decision can be taken by the government.

16. *Meteorology :* The Indian economy is mainly agricultural, and very much dependant upon the meteorological conditions. The outbreak of monsoon, its march across the country, the spacing of the showers, the amount of precipitation, all determine whether that year for India would be a deficit one for food or not. The pre-monsoon meteorological activity in the Arabian Sea is therefore of utmost importance. India is using APT stations for weather data collection, and also use the international operational weather satellites but still, the information is far from being complete. Remote sensing information through LANDSAT satellites coupled with NOAA and other meteorological satellites will augment the weather information to a considerable extent. It will also help to map the progress of the monsoon on the ground because, surface water bodies are easy targets to detect through remote sensing methods.

17. *Agriculture :* Agriculture is the chief occupation of majority of the Indians. With the pressure of increasing population, food has to be imported in those years where the monsoon has been insufficient or failed monsoon. There is also a loss of food crops due to diseases and pests not detected and diagnosed in time. Forest areas are being cleared to provide living space for people and cultivating space for their food. Deforestation is also adversely affecting natural environment. As a result of unregulated deforestation and cultural development, the area of these forests have decreased considerably during the last 35 years. This has caused

the march of the desert in western India, and erosion of land in several parts of India and ecological imbalance–not to speak of threat to the survival of several species of wild life, flora and fauna. Another problem arises out of misoptimisation of land use amongst commercial crops (like cotton, spices, coffee etc), food crops, grass and fodder and fuel energy crops. The demand and supply of all these types have to be properly measured and allocation of land optimally distributed. Thus, a good and quick method of producing classified thematic land use maps is very essential. A good prior estimate of the yield is a very important factor to decide whether to import food or not, to fix the procurement prices in advance, to fix the quantum of grains in buffer stocks and to plan the mix of crops for the next year. It is a known scientific fact that in the infrared region of the electro magnetic spectrum, diseases manifest themselves even before they are seen in the visible (which is the basis of conventional methods) parts of the spectrum. It is these areas that remote sensing can afford several benefits by giving good land use maps, classification of crop types and forests, early warning of diseases and yield prediction.

18. *Hydrology :* Water is another important natural resource which need to be managed very carefully in India. Droughts and floods in the same year in different parts of the country are not uncommon. Both cause severe damage resulting in loss of considerable revenue every year. Study of all river basins in an integrated manner, snow-melting pattern of the Himalayas, study of ground water resources, reservoirs and movements, study of salinity, carriage of salt, are all very important aspects of the hydrological problems. Our river basin studies form the basis of estimation of available water for power or irrigation during different parts of the year. Some of the rivers are flood-prone due to highly variable flows during different seasons. We need to study the river basins in an integrated manner using reliable data in order to design the management system for the regulation of the water they carry. The ground water sources to be studied in association with regional geological structures. Costal water studies are very useful in fisheries, water navigation, keeping the estuaries and the harbours free for navigation. The surface water situation is highly dynamic. Only a space-based platform with a large field of view and a repeatability of observations can follow the dynamic changes.

19. In addition, the remote sensing has also been found to be a very useful tool in mineral development, exploration of energy resources, geology, seismology, oceanography, hydrology and water resources urban and rural planning etc.

20. The determination of the presence of life on other planets is very difficult. Various techniques are designed to detect life forms, artifacts produced by intelligent life, waste products of metabolic reactions, remnants of former life, pre-biological molecules that may represent early evolutionary stages or substances such as carbon that are necessary for life as it is experienced on earth. The remote sensing is being used to find out about the existence of life on other planets. In this the light reflected or emitted by a particular planet can be examined. This method is directed towards finding some thermodynamic disequilibrium on the planet. This may be a chemical disequilibrium, a mechanical or a spectral disequilibrium. For example, it is easy to determine spectroscopically from Mars that the Earth's atmosphere contains large amount of O_2 and 1 ppm of methane. From Mars, this thermodynamic discrepancy would be considered as a significant hint of life on earth. In fact, the methane abundance on the earth is the result of bacteria. If an extra terrestrial observer knows how to interpret the methane disequilibrium accurately, it would be possible for him to deduce cows on earth by spectroscopical analysis.

The existence of relatively large amounts of Methane in the presence of excess of O_2 would remain a tantalizing but enigmatic hint of life on earth. Similarly the large amount of O_2 might itself be a sign of life one could reliably exclude the possibility that photodissociation of water was the source of O_2. Also, such relatively complex reduced organic molecules as terpenes, a hydro carbon given off by plants, might conceivably be detected spectroscopically in orbit about the earth. Thus, chemical disequilibrium of terpenes in an excess of O_2 would be suggestive of life. The terpenes are much more abundant over forested areas than over deserts.

21. Recently Tran *et al* (2002) have developed a method using remotely sensed data to map the incidence of Q fever in the vicinity of Cayenne, French Guiena. A satellite image was processed to map land cover and generate a population density index, which was used to determine areas of high disease incidence during the 1996 - 2000 period. A comparison with results obtained using population census data showed strong concordance between the two incidence maps. All high-incidence areas identified using population census data were also detected using remotely sensed data. This demonstrates the potential of remote sensing as a new tool for rapid mapping or disease incidence in epidemiologic surveys.

3.12 Environmental Laws

In India there have been several legislative measures eversince pre-independence period, both at State and Central Government levels to control different components of the environment. Some of the laws or Acts are, the Indian Ports Act 1907, the Motor Vehicles Act, 1938, The Orissa River Pollution and Prevention Act 1954, the Prevention of Food Adulteration Act 1954, the Atomic energy Act, 1962, Indian Fisheries Act 1987 and so on.

The real awareness about environmental protection at global level was recognized at the United Nations Conference on the human environment in 1972. Keeping in view the need for environmental protection, the Indian Parliament passed the water (prevention and control of pollution) Act, 1974, the Air (Prevention and Control of Pollution) Act 1981 and the Environment (Protection) Act, 1986. At present Central Pollution Control Board (CPCB) and the State Pollution Control Boards (SPCB) are actively engaged is dealing with environmental problems.

Presently, there are four Acts being enacted for the control of air, water and noise pollution. They are 1. The Water (prevention and control of pollution) Act, 1974. 2. The Air (Prevention and Control of Pollution) Act, 1981. 3. The Environment (protection) Act, 1986 and 4. The motor vehicles Act, 1988. The Water and Air Acts are implemented through CPCB and the SPCBs.

The Environmental (protection) Act, 1986

The aim of this Act is to provide protection and improvement of environment and related matters. It particularly lays down procedures and safeguards for manufacturing and handling hazardous substances. In this Act there are clarifications on various terms. For example 1. Environment includes water, air and land and the interrelationship which exists among and between water, air and land and human beings, other living creatures, plants, microorganisms etc. 2. environmental pollutant means any solid, liquid or gaseous substance present in such concentration as may be or tend to be injurious to environment 3. environmental pollution means the presence of any environmental pollutant in the environment. 4. hazardous substance means any substance or preparation which by reason of its chemical or physico chemical properties or handling, is liable to cause harm to human beings, other living creatures, plants, micro-organisms, property or the environment.

Some of the significant features of the Act are given below :

1. Along with the air and water pollution hazardous substances have been brought under the scope of the environmental regulation
2. If any offence is committed by any firm, every responsible person of that firm will be liable for prosecution under this Act.
3. The Ministry of Environment and Forestry (MOEF) can initiate action to close down any firm, which has violated the provisions of the Act.
4. It gives rights to citizens to file cases in appropriate courts against polluting industries, after giving 60 days notice to the concerned State Pollution Control Board.
5. It makes the top management responsible for environmental protection and provides for severe punishment and penalties.

The Act also provides the guidelines to safeguard the environment by laying down standards for 1. the quality of environment in its various aspects 2. the emission or discharge of environmental pollutants from various sources 3. prevention of accidents which may cause environmental pollution 4. handling of hazardous substances and so on.

This is achieved through carrying out and sponsoring investigations and research relating to problems of environmental pollution, wide publicity of information on environmental pollution by way of manuals, guides, charts, posters. News papers, Radio, TV, etc. relating to the prevention control and abatement of environmental pollution.

In accordance with the provisions of the Environment Protection Act (EPA), 1986, the MOEF has served closure notices to many firms violating the pollution control laws. Under EPA 1986, if any firm receives a closure notice for violating the laws, it has either to initiate a time bound programme to control pollution or close down.

Under 1992 amendment to the Environment Protection Rules, Rule 14 mentions that all the industries are required to submit 'Annual Environmental Statements' to the concerned SPCBs. In those statements, the entire information regarding the quantities of pollutants discharged, hazardous and solid wastes generated, their characteristics and pollution abatement measures taken should be incorporated.

Amendments in Air and Water Acts

The Air (prevention and control of pollution) Act 1981 was amended in 1987 to remove the difficulties encountered during implementation. It confers more powers on the implementing agencies and also to impose more heavy penalties for violation of the provisions of the Act. The Amended Act is known as the Air (prevention and control of pollution) Amendment Act 1987.

The water (prevention and control of pollution) Act, 1974 was also amended in 1988 through the water (prevention and control of pollution) Amendment Act, 1988. In this Amendment, CPCB and SPCBs have been given powers to close or stop supply of water and electricity to offending establishments. The citizens may file criminal complaints against offenders after 60 days notice to boards. At the time of establishment of any industry, the concerned person will have to take consent of the board.

The Motor Vehicles Act 1988

The salient features of the Act are as follows :

1. For registration of a new vehicle, the vehicle has to be produced before the registering authority for inspection to satisfy that the vehicle is fit for registration.
2. Every motor vehicle shall comply with the following emission standards (i) idling emission limit for 2 wheeler and 3 wheeler petrol driven vehicles should be not more than 4.5% by volume and for 4 wheeler petrol driven vehicles, it is 3% by volume.
3. The vehicle carrying goods of dangerous or hazardous nature should have prescribed labels indicating their nature. The packages should also bear such labels. The driver should know the fundamentals in case the vehicle is involved in accident.
4. The horn to be used is to be in accordance with the approved specifications of Bureau of Indian standards.

To help the owners of different vehicles to maintain the prescribed level of exhaust emissions, the directorate of transport or Road Transport Authority (RTA) has established pollution checking centres.

The Government of India has issued a notification in 1994 imposing prohibition on the expansion of any activity and undertaking of any new projects in any part of India, unless environmental clearance has been accorded. The managements of all the major projects have to submit an Environmental Impact Assessment Report as well as an Environmental Management plan before seeking permission.

Forest Conservation

Forest conservation through law was enforced by way of The National Forest Policy 1952 and Forest (conservation) Act 1980. The Act was enacted with a view to check indiscriminate dereservation and diversion of forest land to nonforest purposes. Under this Act, prior approval of Central Government is required before any reserved forest is declared as dereserved or forest land is diverted to nonforest purposes. If diversion is permitted, compensatory afforestation should be raised over equivalent area of non forest land.

The forest (conservation) Act 1980 was ammended in 1988 to incorporate stricter panel provisions against violators.

Wild Life Protection

To protect wild life, the wild life (protection) Act 1972 was enforced. The wild life (protection). Act covers International trade in Endangered species, export and imports of various animals etc. The wild life (protection) Amendment Bill 1991 looks after the protection of rare and endangered species.

Indian Board of Wild Life (IBWL) the main advisory body to Government of India was first constituted in 1952 and then reconstituted in 1991. The functions of the IBWL are as follows.

1. To advise central and State Governments for promotion of conservation and effective control of poaching of wild life.
2. To advise on setting of national parks, sanctuaries and zoological gardens.
3. To advise the Government on policy regarding export of living animals, trophies, skins, furs, feathers and other wild life products.

4. To review the progress in the field of wild life conservation in the country and suggest measures for improvement.
5. To promote public interest in wild life and on need of its preservation in harmony with natural and human environment.
6. To assist and encourage the formation of wild life societies and to act as Central Coordinating Agency for such bodies.

3.13 Biology of Pests of Rice, Cotton, Sugarcane & Groundnut

A pest is an organism whose population increases above a certain level so as to cause economic losses to crops or a nuisance and health hazard to man and his livestock. Insects form a major part in causing huge losses to the cultivated crops, the stored products, the household articles, the forests, the domestic and wild animals and human health as well.

Many insect pests that cause damage to the crops are known as agricultural pests and the damage they cause annually runs into thousands of crores of rupees. These phytophagous (feeding on plants) insects can be categorized into *monophagous* (restricted to a single species of plant), *Oligophagous* (feeding on plants of same family) and polyphagous (species from different families) types.

The injury or the damage caused by these insects is almost entirely due to their feeding activity. Principally, the insects having chewing and biting type of mouth parts feed upon the plant parts such as leaves, stems, flowers, fruits or roots externally or by boring into the internal tissues. The insects with piercing and sucking type of mouth parts suck the plant sap from the soft tissue parts of the plant and sometimes the feeding of such insects involve the transmission of plant pathogens like bacteria, viruses, fungi, etc.

The following is an account of certain important insect pests that attack paddy, cotton, sugar cane and groundnut crops.

Pests of Rice

The paddy (*Oryza sativa L.*) is an important crop in India, Japan, China, Korea etc., and is cultivated throughout the world. In India, Andhra Pradesh, Tamilnadu, Kerala, West Bengal, Orissa, Bihar, Madhya Pradesh etc are among the chief paddy growing states. Paddy is grown twice a year ie., Kharif season (July to October) and Rabi season (October to March). At some places like Tamilnadu and Kerala, third crop is also grown between March and June and the season is known as Zaid season.

The paddy crop is affected by many insect pests like moths (order Lepidoptera), beetles and weevils (order Coleoptera), bugs (order Heteroptera), grasshoppers (order Orthoptera), plant hoppers or leaf hoppers (order Homoptera), flies (order Diptera) etc.

1. *The Yellow Stemborer Scirpophaga (=Tryporyza) incertulas (Order Lepidoptera) :* This pest is a borer and has chewing and biting type of mouth parts in its larval stage. The female moth has bright yellow fore wings. It lays its eggs in masses of 40-100 upto a total of 600 on the lower surface of the leaves and covers the eggs with buff coloured hairs. The eggs hatch in about 5 days. The larvae or caterpillars bore into the stem and feed on the internal tissues and make tunnels. Because of their attack, the upper parts of the plant die and form

dead hearts or white heads with immature grains. The larval period varies from 20-29 days depending upon the temperature. In order to pupate, the caterpillars descend inside the stems to the lowest part of the plant, often well below the soil surface. They make a hole known as exit hole on the stem at the basal region so that the adult, on emergence, can escape out. They make silken tube and then pupate there. The pupal period is completed within 7 - 8 days. The adult after emergence comes out from the exit hole and starts a fresh cycle. The adult stage lasts for 4 - 7 days. On the whole, the life cycle is completed within 45-55 days and there are 2 - 3 generations in a year. (Fig. 3.31)

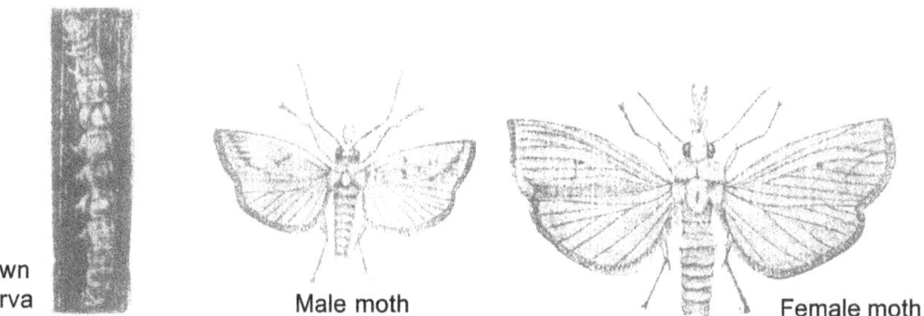

Fig. 3.31 Yellow stem borer, *Tryporyza incertulas*

2. *Brown Plant Hopper Nilaparvatha lugens (Order : Homoptera)* : This is a sucking pest. As a result of the feeding by both the young and adult stages of the insect, the leaves turn yellow and wither. Brown circular patches in the field are characteristic symptoms of this insect. These patches are also referred as hopper burn. The insect is also known to transmit a viral disease known as *grassy stunt*. (Fig. 3.32)

   The adult female lays upto 200 eggs in the leaf sheath. The eggs are white and elongate. They hatch in about 5 - 8 days. The larvae or nymphs are brown in the last instar and the larval period is completed in about 11 - 18 days. Adult males live for 14 - 21 days while females live for 19 - 23 days. The life cycle is completed in about 20 - 25 days and there are 3 - 4 generations in a year.

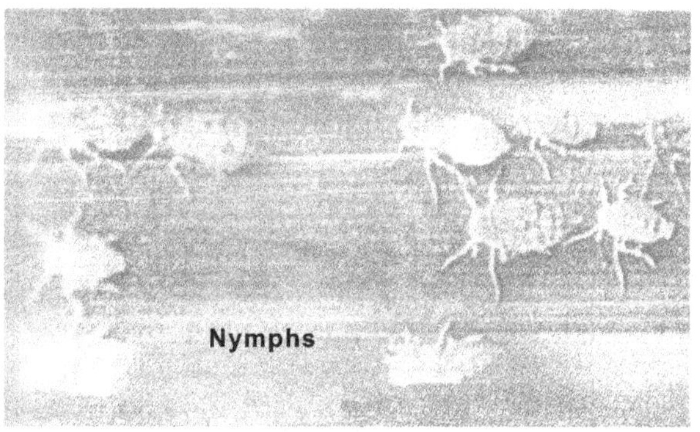

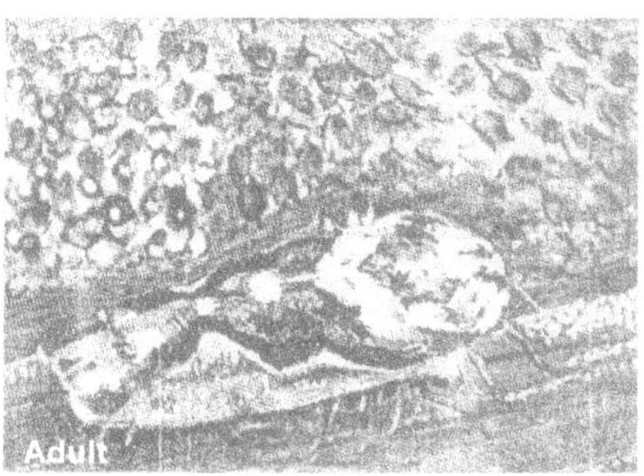

Fig. 3.32 Brown plant hopper, *Nilaparvatha lugens*

3. *Rice Gall Midge Orseolia (=Pachydiplosis) oryzae (Order : Diptera)* : The insect has rasping type of mouth parts. The female fly has a bright red stout abdomen while the male is darker. The female lays reddish eggs singly or in groups on the leaves. It can lay about

250 eggs in its life time of 5 days. The eggs hatch in 3 - 4 days. The hatched maggots (larvae) crawl to the growing points of the tillers and reach the interior of the buds where they lacerate and feed till pupation. Their feeding stimulates the tillers to grow into a tubular gall or silver shoot which resembles an onion leaf. Such tillers do not bear panicles. The larval period is completed in about 6 - 15 days. The pupal period is 4 - 7 days. The life cycle is completed within 13 - 32 days. (Fig. 3.33)

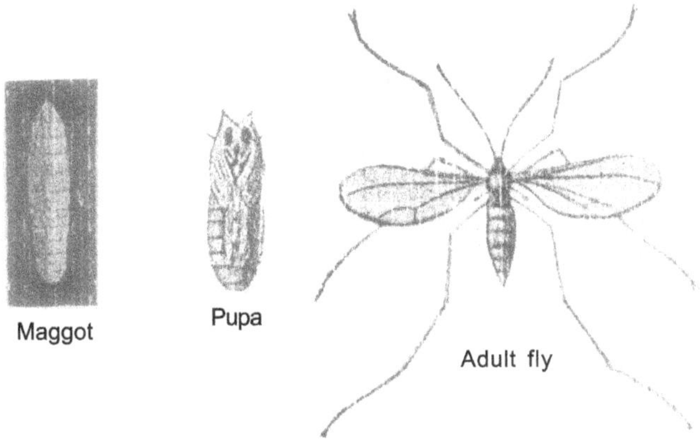

Fig. 3.33 Rice gall midge, *Pachydiplosis oryzae*

4. *Green Paddy Leaf Hopper Nephotettix apicalis (Order : Homoptera) :* This is a sucking pest. The adult males have a median black wedge shaped mark on the fore wings and the apical third portion is black while in females, this area is brown. The female leaf hopper lays about 22 - 37 eggs under the epidermis of young leaf sheaths. The incubation period is 5 - 6 days. The nymphs on hatching, in groups, feed on the leaves by sucking the cell sap which results in browning of leaves or hopper burn. Nymphal period takes about 16-18 days. The adults are nocturnal ie. they hide during the day time and feed on leaf tissues during night. The adult stage lasts for 20 - 26 days. The insect is also known to transmit dwarf (tungro) and yellow dwarf viral diseases. (Fig. 3.34)

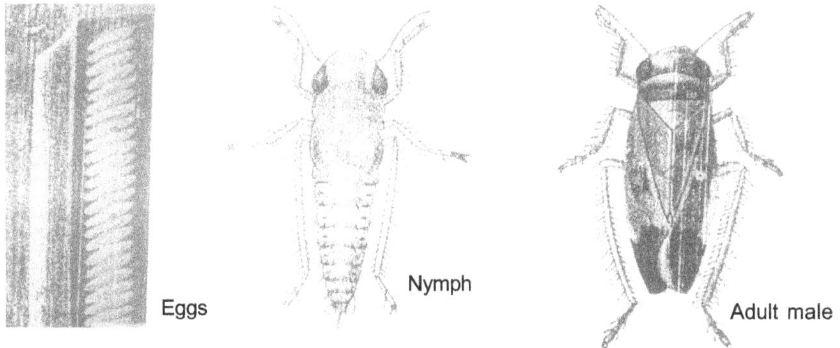

Fig. 3.34 Green paddy leaf hopper, *Nephotettix apicalis*

5. *Rice Hispa, Hispa (= Dicladispa) armigera (Order : Coleoptera) :* The beetles are square shaped, dark blue or blackish in colour with spines all over the wings. The adult female beetle lays the eggs in the tissue of the young leaf. About 55 eggs are laid by a female in its life time. The eggs hatch in about 3 - 5 days. The grubs (larvae) on hatching, start burrowing or mining into the leaf tissue, feeding on the green matter and produce characteristic blister patches on the leaf. There are 6 stages in the grub stage which is completed in 15 - 16 days. Full grown grubs pupate in the larval burrows and the pupal stage lasts for 3 - 6 days. On adult emergence, the beetles also feed on the green matter of the leaves in characteristic parallel white lines on the leaf surface. The adult longevity is 7 - 8 days. The total life cycle of the insect is completed in 28 - 35 days. Two to three generations are completed in a single paddy season. (Fig. 3.35)

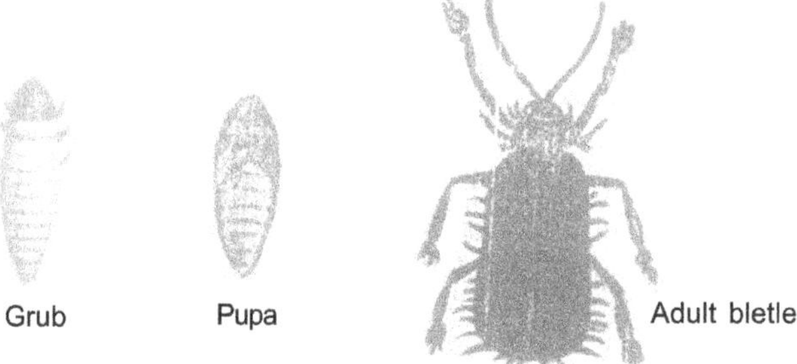

Fig. 3.35 Rice hispa, *Hispa armigera*

6. *Rice Gundhi Bug Leptocorisa varicornis (= acuta) (Order : Heteroptera) :* It is a sucking pest. The adult bugs are slender, long and greenish brown with long legs and emit a typical foetid smell hence the name Gundhi bug. Brownish seed like eggs, in clusters of 10 - 20 are laid in 2 - 3 straight rows along the midrib on the upper surface of the leaf blade 200 - 300 eggs are laid by a female. The incubation period is about 6 - 7 days. The slender greenish nymphs on hatching, start sucking the sap from the grains in the milk stage and tender shoots and leaves for about 25 - 30 days and later attain adult stage. The pest appears when the crop is developing earheads and is in the milky grain stage and usually disappears when the grains in the earheads become hard. The adult longevity is 14 - 15 days. The life cycle is completed in about 27 - 50 days. Usually the insect hibernates in its adult stage. (Fig. 3.36)

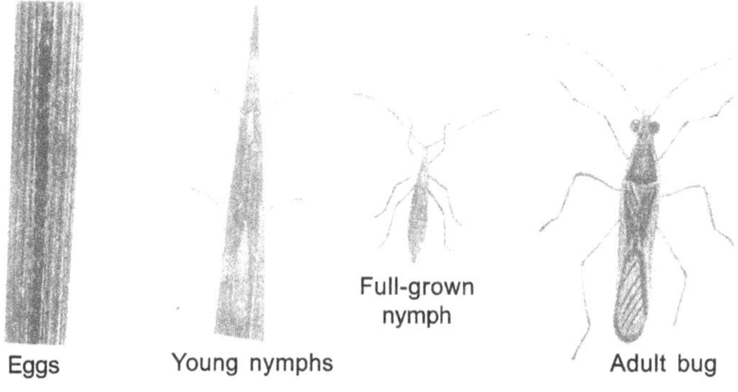

Fig. 3.36 Gundhi bug, *Leptocorisa varicornis*

7. *The Paddy Grasshopper Hieroglyphus banian (Order : Orthoptera) :* The insect has chewing and biting type of mouth parts. The adult grasshopper is greenish with tibia of the hind legs coloured blue. Both the nymphs and adults eat the foliage and developing ear heads of paddy crop. The eggs are laid in batches of 30 - 40 inside the soil, especially along the sides of bunds and other uncultivated areas. These egg masses remain in the soil during the hibernating stage till the next monsoon. After the first showers of rain of the following year, young grasshoppers emerge and start feeding on the grasses. Later, they migrate to the main crop and start feeding on leaves. They become full grown through 5 moults in about 70 - 80 days. There is only one generation in a year. (Fig. 3.37)

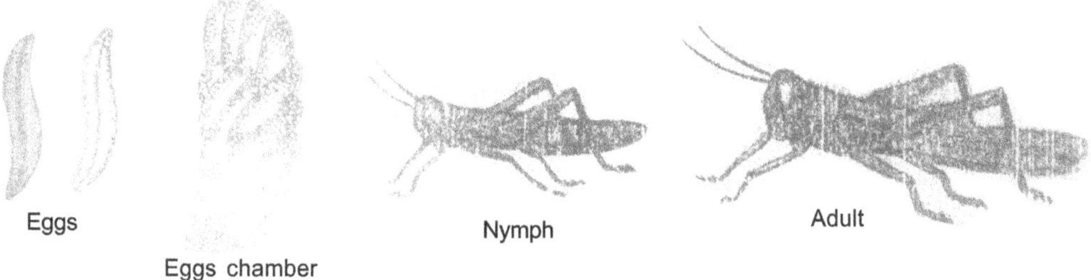

Fig. 3.37 Paddy grasshopper, *Hieroglyphus banian*

Pests of Cotton

In India, Cotton (*Gossypium spp*) is the most important fibre crop and contributes almost 75% of the total fibre consumed in the textile industry. The other uses of cotton products are as edible oil, cattle feed, flour, and ofcourse, the multiple uses of cotton are well known.

The cotton crop (also known as white gold) is attacked by over 135 types of insect pests. Of them, the important insects are moths (Lepidoptera), aphids, white flies and jassids (Homoptera), thrips (Thysanoptera), bugs (Heteroptera) etc.

The following is an account of some important cotton pests.

1. *Bollworms* : i.e, American boll worm, *Helicoverpa armigera ;* spotted boll worm, *Earias insulana* and pink boll worm, *Pectinophora gossypiella (Order : Lepidoptera).*

   The larvae are borers with chewing and biting type of mouth parts. Because of their attack, affected bolls fall or fetch low price to the low quality cotton. The moths are nocturnal and lay eggs singly on the leaves, flowers or immature bolls. The eggs hatch within 4 - 7 days. The larvae or caterpillars feed on leaves, tender shoots, flowers or bolls. They bore into the

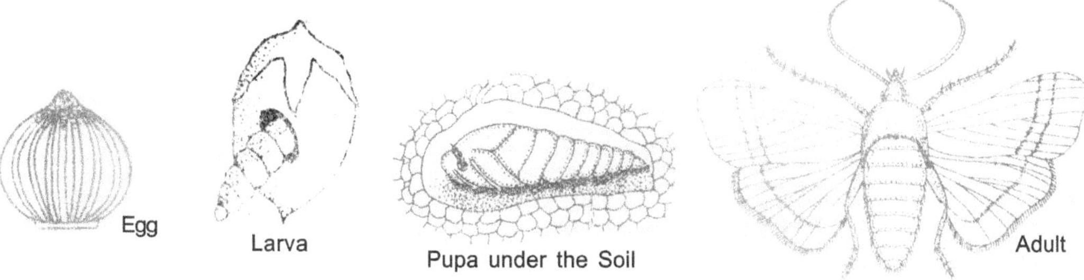

Fig. 3.38 Life cycle of *Helicoverpa armigera*

bolls and feed there on lint or seed hence the name boll worms. The larval period is completed in 9 - 20 days depending upon the climate. They pupate inside the bolls or in the soil. The pupal period is completed within 8 - 14 days. The adult longevity is 9 - 20 days. On the whole, there are 2 - 8 generations in a year. (Fig. 3.38, 3.39(a,b))

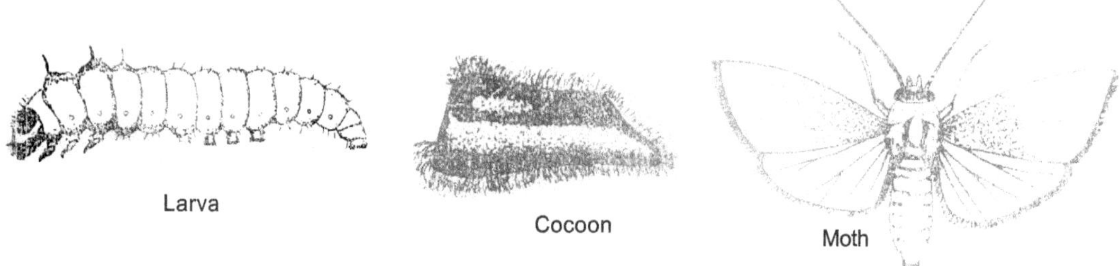

Fig. 3.39(a) The spotted boll-worms of cotton *Earias* insulana

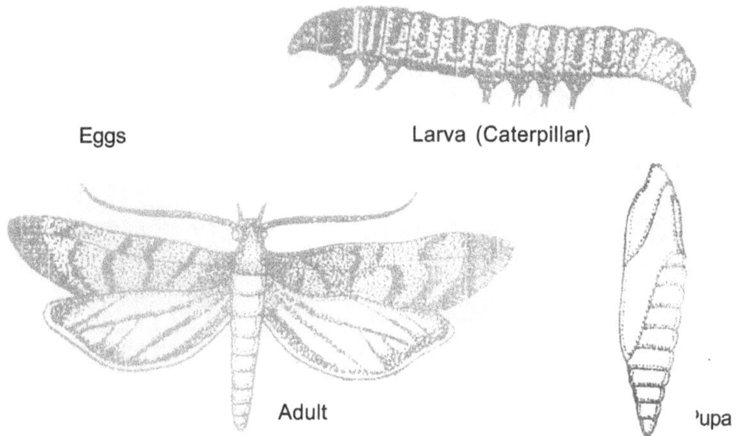

Fig. 3.39(b) Life cycle of Pink bollworm, *Pectinophora gossypiella*

2. *Tobacco Cutworm, Spodoptera litura (Order : Lepidoptera) :* The caterpillar damages leaves as well as bolls. It has chewing and biting type of mouth parts. The adult moths have pale grey to dark brown fore wings having wavy white markings and whitish hind wings. 150 - 200 eggs are laid in masses covered with brown hairs on tender leaves. The eggs hatch within 4 - 5 days. The caterpillars on hatching, start feeding on the soft green layers of leaves gregariously during the night. They become isolated in the later instars and full grown in 2 - 3 weeks. They descend to the ground and pupate in earthern cells in the soil. Their pupal period lasts from 9 - 14 days. The adult emerges out of the earthern cells and starts a fresh cycle. The adult longevity is 9 - 10 days. On the whole, the life cycle is completed within 20 - 35 days. There are 4 - 6 generations in a year. (Fig. 3.40)
3. *Cotton Aphid Aphis gossypii (Order :Homoptera) :* It is a sucking pest. The adult is oblong, dark, yellowish green or black in colour and has two projections called cornicles on the posterior side of the abdomen. Due to the feeding of these insects the leaves become yellow and subsequently dry and the plant growth is stunted. (Fig. 3.41)

The adults may be winged (alate) or wingless (apterous) forms. They reproduce parthenogenetically and viviparously. A single apterous female gives rise to 8 - 22 young ones per day called nymphs. They pass through 4 moults before reaching the adult stage. The duration of life cycle varies from 7 - 9 days. During the feeding activity, the aphids excrete drops of sticky honey dew which attracts sooty mould which affects the photosynthetic activity of the plant.

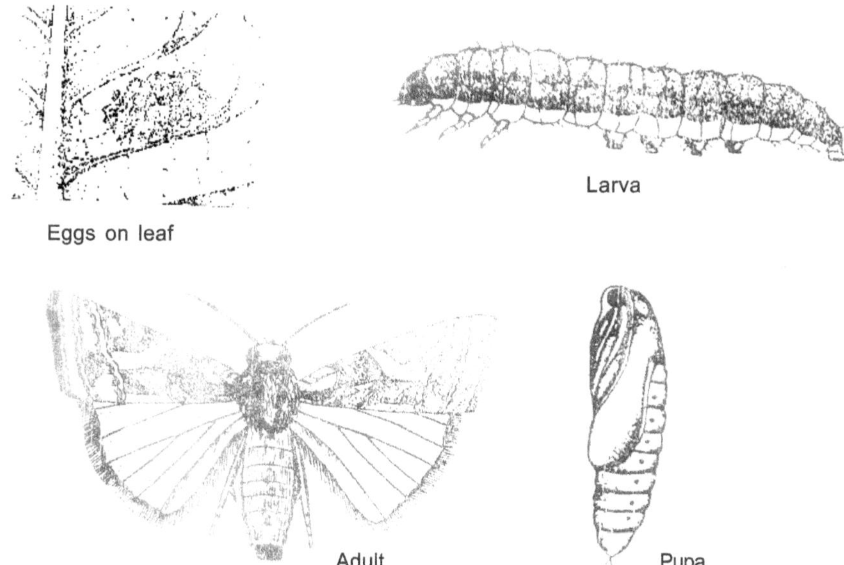

Fig. 3.40 Life cycle of the tobacco caterpillar, *Spodoptera litura*.

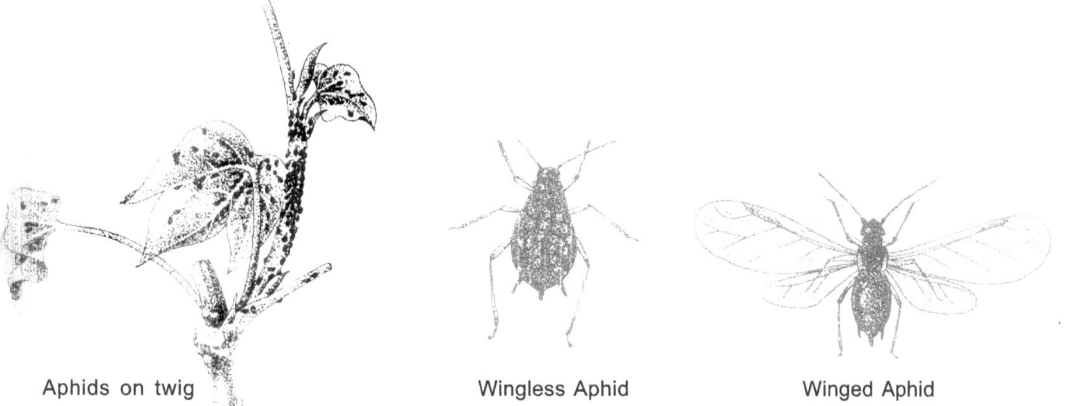

Fig. 3.41 Life cycle of the cotton Aphid *Aphis gossypii*

4. *Cotton Jassid, Amrasca (= Empoasca) devastans (Order : Homoptera) :* This is a sucking pest. The adult is wedge shaped and pale green in colour. The nymphs and adults pierce the plant tissues and suck the cell sap. Due to the feeding, curling, bronzening and sometimes drying of the leaves is observed. Plant growth is affected. The adult female lays

about 30 eggs in the tissue of the leaf vein and minute nymphs emerge from the eggs within 4-11 days. The nymphal period is 7 - 9 days and the adult longevity is about 3 months. Their entire life cycle is completed in 2 - 4 weeks. There are 7 - 8 generations in a year. (Fig. 3.42)

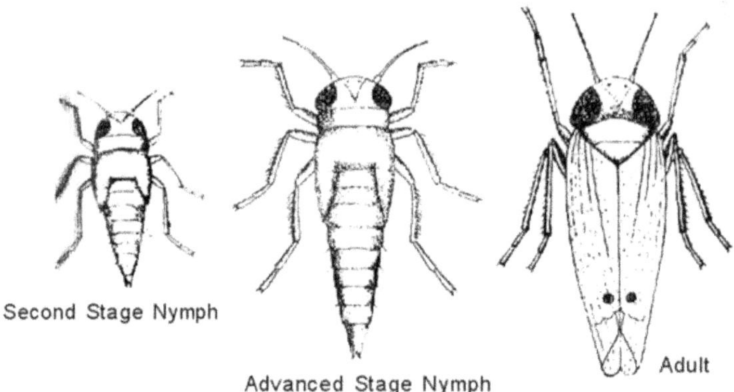

Fig. 3.42 The cotton jassid, *Amrasca devastans*

5. *Cotton thrips Thrips tabaci (Order : Thysanoptera) :* It is a sucking pest. The adults are minute, delicate and light yellow insects. Their wings have a fringe of hairs throughout. Both the nymhs and adults scrape the epidermis and suck the oozing cell sap. In such regions, the plant part dries up and initially becomes whitish but later often turns brown. The plant growth is stunted.

   The adult female insects lay 30 - 50 eggs in the leaf tissue. The incubation period is 4 - 5 days. The nymphal stage is very short and is completed in 2 - 5 days. When full grown, the nymphs drop to the soil and pupate. The quiescent pupal stage takes about 3 - 4 days. The adult longevity is 3 - 10 days. (Fig. 3.43)

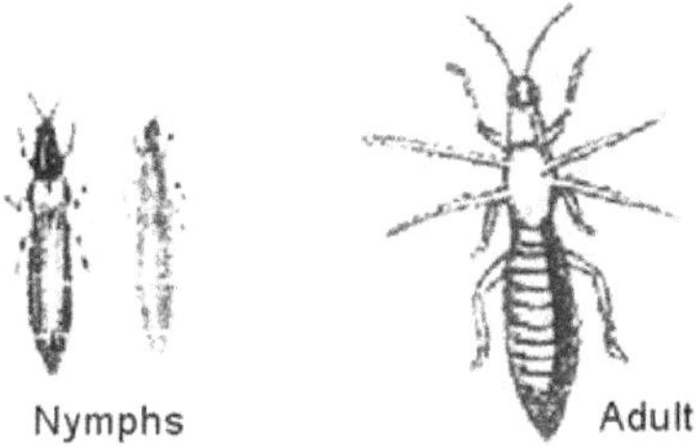

Fig. 3.43 Thrips (*Thrips tabaci*)

6. *Red Cotton Bug, Dysdercus koenigii (Order : Heteroptera) :*It is a sucking pest. The adult is bright red with eyes, scutellum and antennae coloured black. A series of white transverse bands is present on the ventral side of the abdomen.

   The adults and nymphs suck the plant sap and devitalize the plant. They also feed on the seed and lower their oil content. The infested seeds are useless for sowing. The female bug lays round and bright yellow eggs in a mass of 150 - 200 eggs on the soil near the plant. The eggs hatch in 6 - 7 days. The nymphs which are bright red pass through 5 instars in 25 - 30 days before reaching the adult stage. The adult longevity is 20 - 25 days. There are 3 - 4 generations in a year. (Fig. 3.44)

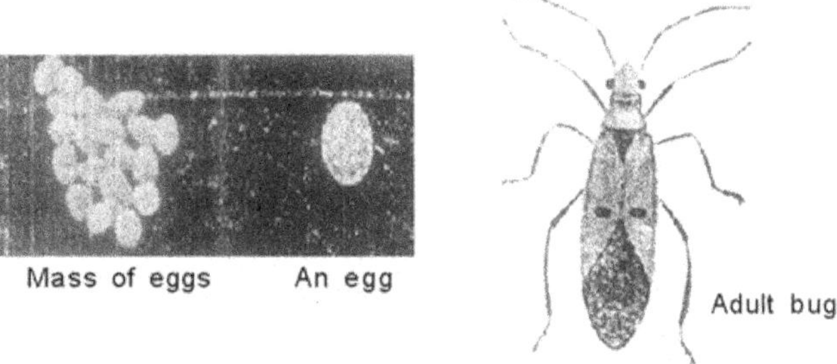

Fig. 3.44 Red cotton bug, *Dysdercus koenigii.*

7. *White Fly Bemisia tabaci (Order : Homoptera) :* It is a sucking pest. The adult is a minute insect having white or greyish wings, yellowish body and red eyes. Nymphs and pupae are oval, scale like with marginal bristle-like fringes. The nymphs and adults suck the sap from the leaves thus affecting the plant growth. The nymphs secrete a honey dew which attracts fungus (sooty mould) which gives the plant a blackish appearance and affects the photosynthesis.

   The female whitefly lays about 150 to 200 yellowish eggs on the underside of the leaves. These hatch in about 3 - 8 days. Freshly hatched nymphs crawl on the leaves till they find a suitable place to settle in. They become full fed in 9 - 25 days. They change into oval blackish pupae and the pupal period lasts from 2 - 8 days. It is actually the 4th stage immobile nymph and not a true pupa. The adult longevity is 10 - 20 days. There are 12 - 15 generations in a year. (Fig. 3.45)

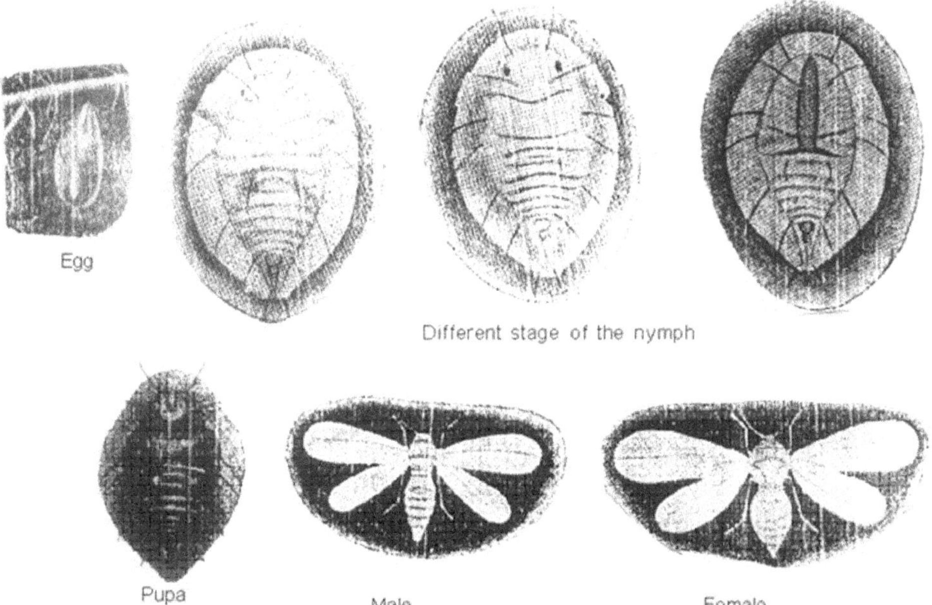

Fig. 3.45 White fly, *Bemisia tabaci*

Pests of Sugarcane

Sugarcane (*Saccharum officinarum* L) is the main source of sugar in India and holds a prominent position as a cash crop. More than 5 million acres of land is used for the cultivation of this crop. Sugarcane is vulnerable to several pests and diseases. More than 170 insect pests are known to feed upon sugarcane in India of which insect pests of major importance are moths (Lepidoptera), white flies, leaf hoppers, (*Order : Homoptera*), grass hoppers, (*Order : Orthoptera*) termites (*Order Isoptera*) etc. The following is an account of the serious pests of sugarcane.

1. *Borers :* Sugarcane Top shoot borer, *Scirpophaga* (= *Tryporyza*) *nivella*, Shoot borer, *Chilo* (= *Chilotraea*) *infuscatellus*; Root borer, *Emmalocera depressella (Order : Lepidoptera)*

   These moths (borers) attack the sugarcane plant at different regions feeding on the internal tissues (Fig. 3.46(a, b)). Because of the attack of the borers the central shoot dries up causing dead hearts.

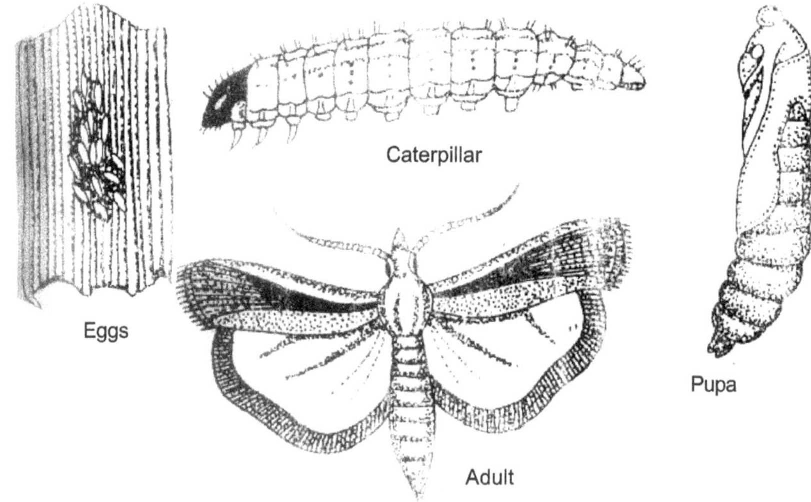

Fig. 3.46(a) Life cycle of Shoot borer, *Chilo infuscatellus*

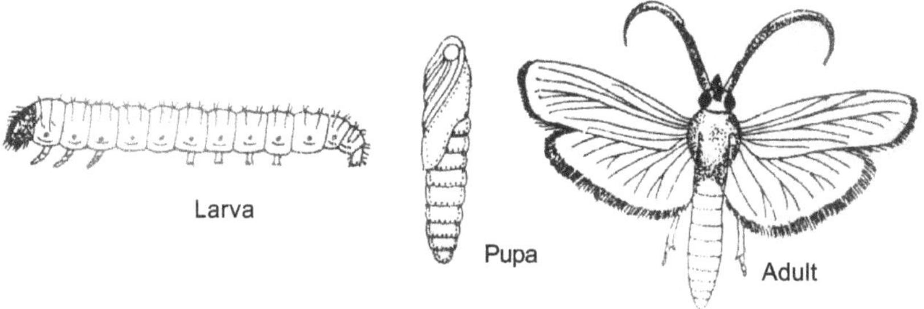

Fig. 3.46(b) Life cycle of Rootborer *Emmalocera depressella*

The eggs are laid singly or in batches on the underside of the leaves. The incubation period is 3 - 5 days. The first instar larva soon after hatching enters the shoot through the midrib of the leaf. The larva is full grown through 4 moults in about 20 - 30 days. Before pupation, the larva makes a round hole (exit hole) on the cane which is covered by a silken membrane from where the moth can escape. The pupa is found in the larval tunnel. Pupal stage is completed within 7 - 10 days. The adult longevity is about 2 - 5 days. The life cycle is completed within 30 - 40 days. There are 6 - 8 generations in a year.

2. *The Sugarcane Leaf Hopper Pyrilla perpusilla (Order : Homoptera)*: It is a sucking pest. It is a straw coloured small insect with a pointed beak. The nymphs and adults suck the sap of cane leaves, as a result of which the leaves dry up gradually. The hoppers secrete a honeydew on which black fungus develops that affects photosynthesis.

   The eggs are laid on the underside of the leaf and are covered with white, cottony, waxy filaments by the female. The eggs hatch within a week into tiny hoppers that start sucking the sap of the leaves. The nymphs become adult hoppers within 50 - 60 days. There are several generations in a year. (Fig. 3.47)

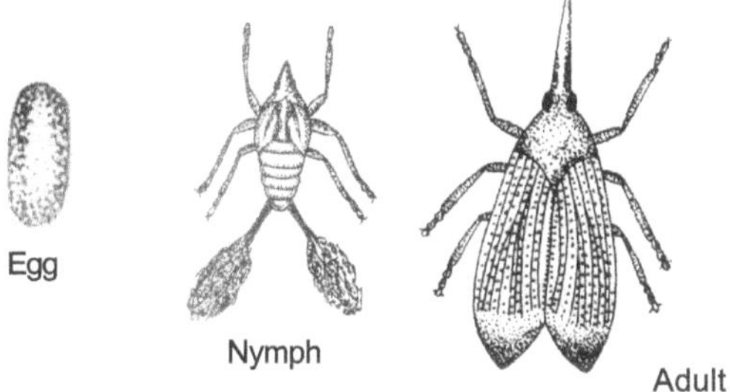

Fig. 3.47 Life cycle of *Pyrilla perpusilla*

3. *Sugarcane White Fly Aleurolobus barodensis (Order : Homoptera) :* It is a sucking pest, pale greenish insect with white wings having black dots. Because of the attack of the insect, the leaves wither and dry up and the blackening of leaf surface is due to the black mould which is attracted by the honey dew that is secreted by the insect. (Fig. 3.48)

   Life cycle is similar to that of cotton white fly, *Bemisia tabaci*.

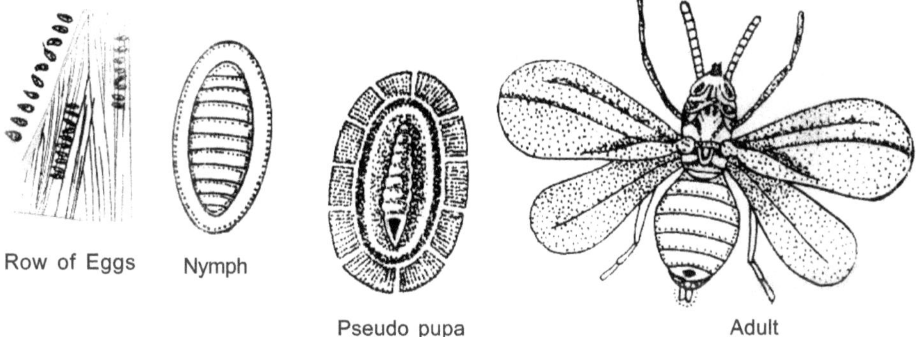

Fig. 3.48 Life cycle of whitefly, *Aleurolobus barodensis*

4. *The Sugarcane Grasshopper Hieroglyphus banian (Order : Orthoptera)* : It has chewing and biting type of mouth parts. The adult grasshopper is greenish in colour with blue tibia. The nymphs and adults feed on leaves and in case of a severe attack, completely defoliate the plants. Consequently, the growth of the cane is affected.

   The life cycle of the insect is same as Paddy grasshopper, *Hieroglyphus banian*.

5. *Termites or White Ants Odontotermes obesus (Order : Isoptera) :* The termites have chewing and biting type of mouth parts. These insects are medium sized, soft bodied, cream to light or dark brown insects. Due to their attack, the freshly planted sets (pieces with eyes near nodes) get damaged near the eye bud or from the cut end. Sometimes they attack the entire roots through which they ascend the cane killing the plant.

   Being social insects, termites are polymorphic ie., there are workers, soldiers, king and queen. The queen lives for 5-10 years and after mating with king, lays 70 - 80 thousands of eggs per day. The workers and soldiers are sterile and short lived (2-3 years). The workers do all the house hold work while the soldiers defend the colony from intruders. On the death of the queen, some workers may become replacement reproductives which are transformed into females. Only one female will be accepted and the rest will be killed by them. (Fig. 3.49)

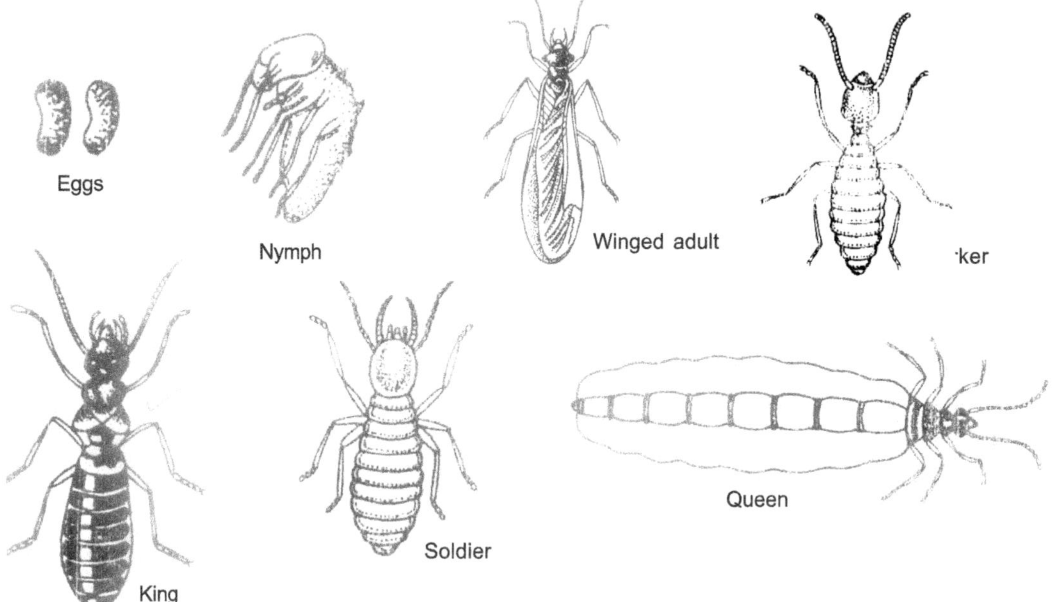

Fig. 3.49 Life cycle of termite *Odonotermes obesus*

Pests of Groundnut

Groundnut (*Arachis hypogaea*) is an important oil seed crop. India occupies the top position in ground nut production in the world. More than 6 million tons are produced every year. Though a hardy crop, it is vulnerable to attack from several insect pests. Among these, the important pests are aphids (Homoptera), moths (Lepidoptera), beetles (Coleoptera), termites (Isoptera) etc. The following is the account of some serious pests of groundnut crop.

1. *Groundnut Aphid Aphis craccivora (Order : Homoptera) :* It is a sucking pest. These aphids are small, black, soft bodied insects and are found on the lower surface of the leaves. They reduce the vitality and growth of the plant by their feeding. Also, they are the vectors of a viral disease commonly known as 'Rosette' of groundnut.

   The females are viviparous and produce young ones (nymphs) which undergo 4 moults to become adults within 5 - 8 days. The wingless females start producing the brood within 24 hours of attaining adult stage. Breeding occurs almost throughout the year. (Fig. 3.50)

Fig. 3.50 Ground nut Aphid, *Aphis craccivora*

2. *Groundnut Leaf Miner Stomopteryx nertaria (Order : Lepidoptera)* : The caterpillars of these moths eat the green matter of the leaves while mining, as a result of which skeletonizing of the leaves takes place. The crop gives a scorchy appearance. The insect is nocturnal. The female moth lays hundreds of eggs singly on the leaves and shoots. The eggs hatch within 3 - 4 days. The larva soon after hatching, mines into the leaves and later on bites its way out. It then webs together a number leaflets and feeds inside the chamber, so formed. The larvae are full grown in 9 - 17 days. They prepare silken cocoons to pupate. The pupa is reddish brown and the pupal period lasts for 4 days. There are several generations in a year.

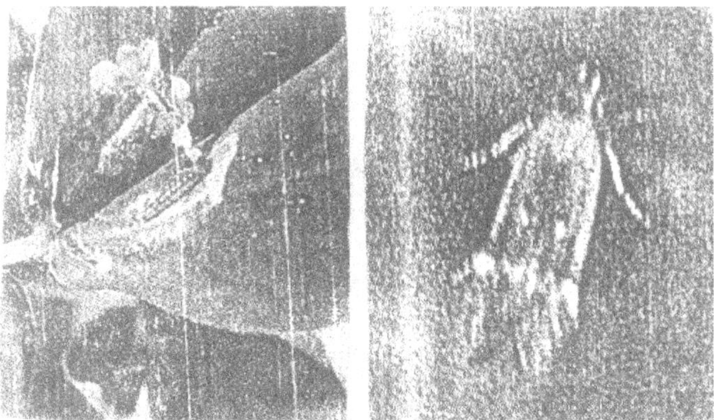

Fig. 3.51 Ground nut leaf miner, *Stomopteryx nertaria*

3. *The White Grub Holotrichia consanguinea (Order : Coleoptera)* : The insect has chewing and biting type of mouth parts. The adult is large sized, copper coloured and nocturnal while the grubs are white coloured. The female lays 60 - 70 eggs in loose moist sandy soil on the onset of monsoon. The eggs hatch in 7 - 10 days. The grubs attack the roots and rootlets, often killing the plants. The larval (grub) period is completed in 60 - 80 days. After the monsoon

the grubs migrate to a considerable depth in the soil for pupation. The pupa is creamy white. The pupal stage lasts for about 14 - 20 days. The adults are nocturnal hence come out in nights for feeding. The adult beetles feed on the foliage and are capable of defoliating the plants and even the trees. Diapause takes place in the adult stage. There is only one generation in year. (Fig. 3.52)

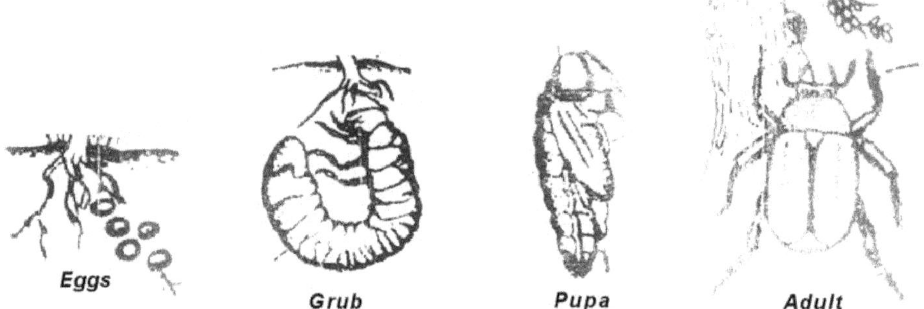

Fig. 3.52 Life cycle of white grub, *Holotrichia consanguinea Blanch*

4. *The Groundnut Stem Borer Sphenoptera peroletti (Order : Coleoptera)* : They have chewing and biting type of mouth parts. The grub of this beetle bores into the stem and the root, feeds on the contents often killing the plants. The adult is a small jewel like beetle with a striking metallic shine over a dark brown background. The beetles lay scale like eggs and glue them onto the branches. The incubation period is about 10 days. The slender pale white grubs called 'flat headed grubs' bore into the branches and travel down the main stem and ultimately reach the tap root. The larvae become full grown in about 2 months through 5 moults. They pupate inside larval tunnels. The pupal period is completed within 10 days. The adults emerge from the tunnels by cutting their way out and they start the next generation. There are 2 - 3 generations in a year. (Fig. 3.53)

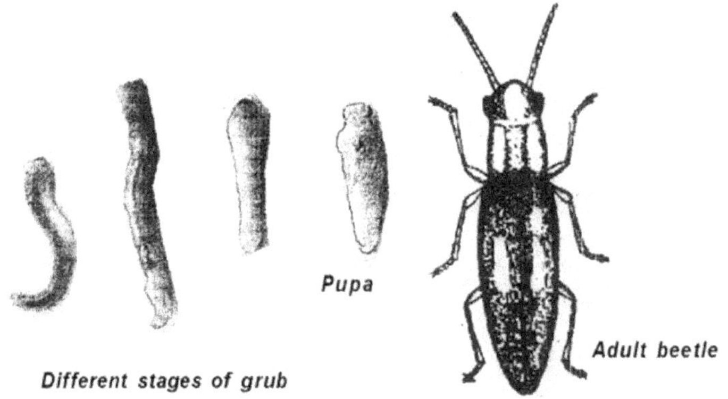

Fig. 3.53 Stem borer, *Sphenoptera peroletti*

5. *Termites, Odontotermes obesus (Order : Isoptera)* : The account of these termites has been described under sugar cane pests.
6. *Tobacco caterpiller, Spodoptera litura (Order : Lepidoptera) :* Groundnut fields are often attacked by this pest in post rainy season. The description of this pest is given under the pests of cotton.

3.14 Biology of Fin Fish and Shell Fish

Fishes are defined primarily as aquatic and gill breathing vertebrates equipped with fins. Fishes are the vertebrates with jaws. They are cold blooded animals and live in all the seas, rivers, lakes, reservoirs, canal and tanks. Fishes are economically a very important group of animals because they are used as food throughout the world. Fish liver contains vitamin A and D. It is the main source of liver oil. Body oils of fishes are extensively used in soap industry and tanneries. Capture and business of fishes is known as fisheries.

The fisheries can be categorised into two types ie., fin fisheries and shell fisheries. The fin fisheries are further divided into capture fisheries and culture fisheries. The capture fisheries is to capture the naturally available fishes from seas, rivers etc., whereas culture fisheries means introduction of fish seed, in confined areas like ponds, lakes etc., culture and capture them at a desired stage. Due to the culture of a variety of fishes including shell fish like prawn, lobsters, crabs, molluscs etc., culture fisheries has been termed as aquaculture.

Based on the type of water, aquaculture is classified into three types ie., fresh water aquaculture, brackish water aquaculture and marine water aquaculture or mariculture.

In India, six types of fresh water fishes are suited for culture. They are :

1. Major carps
2. Minor carps
3. Murrels
4. Cat fishes
5. trouts and
6. Exotic fishes.

The notable brackish water fishes are mullets, milk fish etc., while marine fishes are oil sardines, mackarel, Indian Shad, Bombayduck, Silver pomfret, Tuna, flat fish etc.

The following is the description of some important fishes.

Major carps

Catla catla, Cirrhina mrigala and *Labeo rohita* are commonly known as major carps. They can live even in low oxygen environments. They can tolerate high temperatures and can grow by feeding on aquatic plants, decaying organic matter, phyto and zooplankton etc. They can reproduce in artificial ponds.

Catla catla

This is the fastest growing species and the largest Indian carp. It has a deep body with a conspicuous head, large upturned mouth, a broad dorsal fin with 14 - 16 branched rays and devoid of barbels. The colour of the body is grey above and silvery on sides. It feeds on the zooplankton of the pond surface. The young ones feed only on zooplankton but also on phytoplankton. It breeds naturally in rivers during the rainy season. The female lays about 70 thousand to one lakh eggs. Its eggs are non-adhesive and float in water. In the first year it grows to a size of 35 - 45 cm and weighs 1.5 to 2 kg. It grows to a maximum size of 1.8 metres (upto 45 kg). It matures in the second year. (Fig. 3.54).

Cirrhina mrigala

This fish is commonly known as *mrigal.* This is a slender fish. It has a small head with blunt snout, a sub terminal mouth and the dorsal fin with 12 - 13 branched rays. The body is silvery but dark grey along its back. It is a deep water living fish and grows to a maximum size of 0.9 metres (6 to 7 kgs), it feeds on decaying organic and vegetable debris while the young ones feed on Zooplankton. The fish grows to about 30 cm (700g) in the first year. It matures in the second year. It breeds in the rainy season in rivers and tanks. The female lays about 1.2 to 1.4 lakh eggs. (Fig. 3.55)

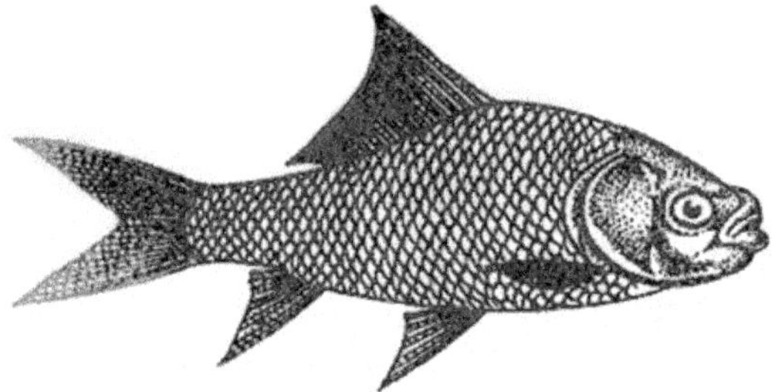

Fig. 3.54 *Catla catla* (Ham.)

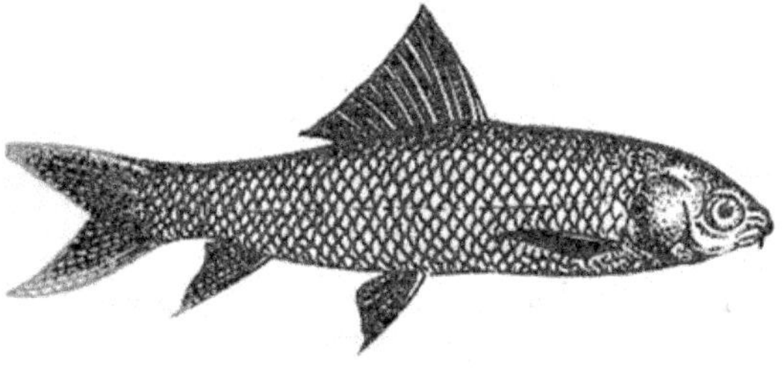

Fig. 3.55 Cirrhina mrigala (Ham.)

Labeo rohita

This fish is commonly known as *Rohu* and occurs in all rivers and canals. It has a small and pointed head, terminal small mouth and the dorsal fin is with 12-13 branched rays. The colour of the body is bluish or brownish grey above. The scales are dull reddish. It grows to a maximum size of 1 metre. It is a column feeder on phytoplankton, plant debris or decaying debris of aquatic plants white the young ones feed on zooplankton. It grows to a size of 35 - 40 cm (weighing about 900g) in the first year. It grows to a maximum size of 90 cm to 1 metre (weighing approximately 20 - 25 kg). It matures in the second year. The female lays about 0.70 to 1.0 lakh eggs. (Fig. 3.56)

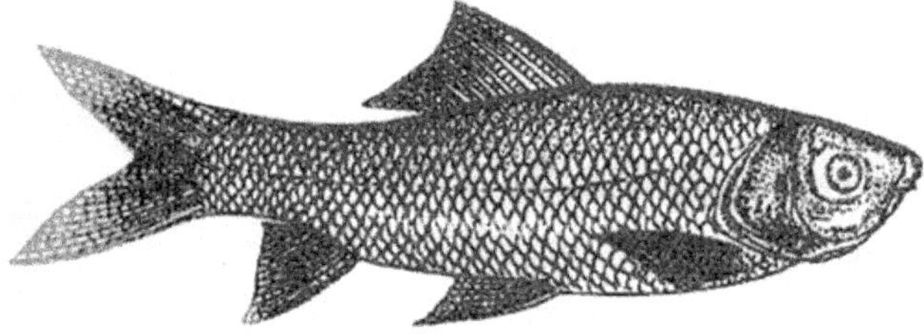

Fig. 3.56 *Labeo rohita* (Ham.)

Minor Carps

The minor carps, like major carps are found in fresh water bodies, rivers, lakes etc. The examples are *Labeo calbasu, Labeo bata, Cirrhinus cirrhosa* etc. They grow to a length of 30 - 100 cm and attain 1 to 1.5 kg of weight. (Fig. 3.57 to 3.59)

1. *Labeo calbasu :* It is commonly known as *calbasu* or *black rohu.* Its body is bluish green and has a small tapering head with subterminal fringe-lipped mouth with barbels. The dorsal fin is with 12-13 branched rays. It is an omnivorous bottom feeder and feeds

on detritus and animals. It attains a size of 30 to 35cm (about 450 g) in the first year and a maximum size of 1 metre. It matures in the second year. The female lays thousands of eggs and one gram contains about 1400 eggs.

2. *Labeo bata :* It is commonly called as *bata*. The body has greenish iridescence at the base of the scales and the dorsal fin with 9 - 10 branched rays is seen. It is an omnivorous fish and feeds on phytoplankton, zooplankton and filamentous algae. The fish attains a maximum size of 60 cm. The fish breeds in the rainy season and the female lays thousands of eggs and one gram contains about 290 eggs.
3. *Cirrhinus cirrhosa :* It is commonly known as *white carp*. It has a small head with blunt snout and thin lips. The dorsal fin is with 14-15 branched rays. The first few rays are much elongated. It has a silvery body and scales with reddish tinge except on the abdomen. It is a bottom feeder and feeds on detritus, algae and sometimes on the zooplankton. It grows to a maximum size of 45cm (about 1 kg). The fish breeds in the rainy season and lays thousands of eggs.

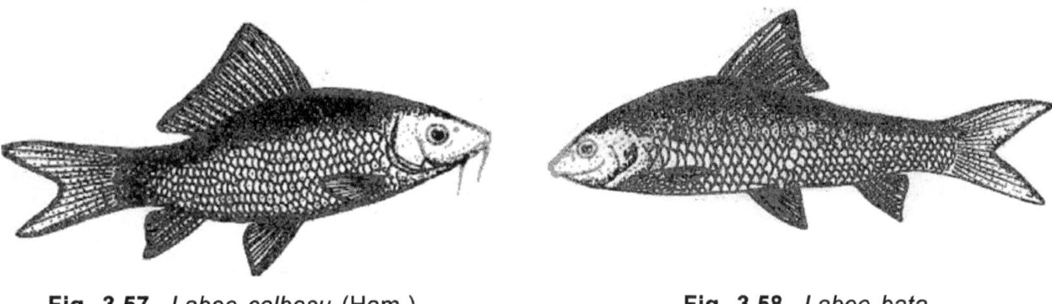

Fig. 3.57 *Labeo calbasu* (Ham.)

Fig. 3.58 *Labeo bata*

Fig. 3.59 *Cirrhinus cirrhosa*

Murrels

These are air breathing fishes and are capable of breathing atmospheric air directly. They have elongated and cylindrical body, depressed head and a large and protractile mouth. They can be cultured in irrigation wells also apart from lakes, ponds and rivers. *Channa striatus* is commonly known as *striped snake head* or *common murrel*. It attains a maximum size of 90 cm. Dorsal and anal fins are long and without spines. Stripes are present on the body. They mature within 12 months. The female lays 2000 - 4000 eggs. (Fig. 3.60)

Channa punctatus is small and grows to a size of 11 cm. Parental care is exhibited in murrels as they protect their spawn. However, cannibalism is also seen and when hungry, they feed on the young ones. The young ones grow to a size of 25 - 30 mm within 15 to 18 days. The murrels have a good demand. However, the murrels suffer from a dreadful disease called epizootic ulcerative syndrome. This disease wipes out the whole murrel population of natural and cultured populations. The culture had to be stopped for a few years because of this disease. (Fig. 3.61)

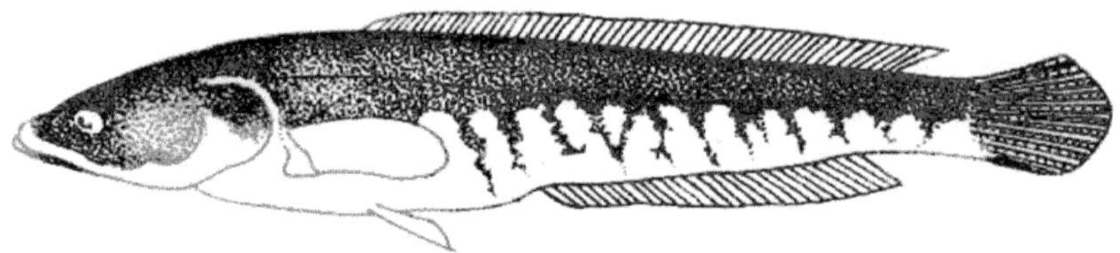

Fig. 3.60 Channa striatus (Bloch)

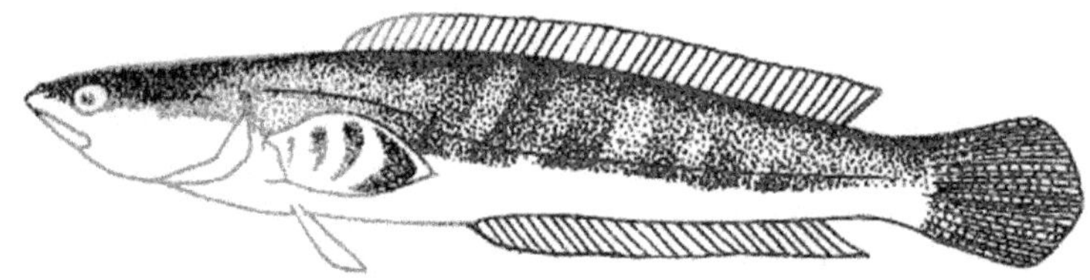

Fig. 3.61 Channa punctatus (Bloch)

Cat fishes

Cat fishes are also air breathing fishes. They are capable of breathing atmospheric air directly. These are generally used as food by middle class people. Examples of cat fishes are *Clarias batrachus, Wallago attu, Heteropneustes fossilis* etc. (Fig. 3.62 to 3.64)

Fig. 3.62 *Clarias batrachus*

Fig. 3.63 *Wallago attu* (Schneider)

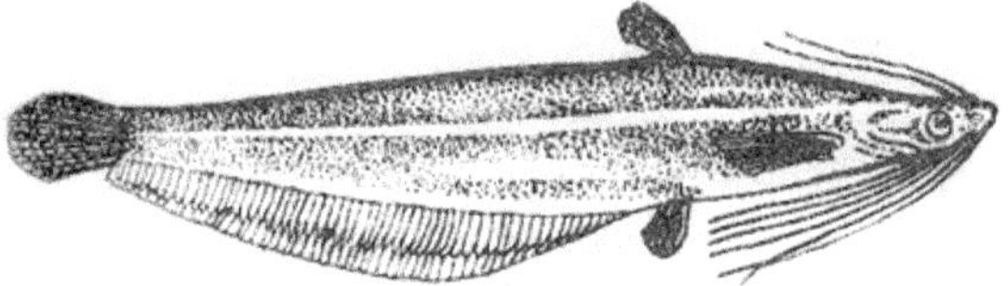

Fig. 3.64 *Heteropneustes fossilis* (Bloch)

1. *Clarias batrachus :* It is commonly known as *Magur*. It is identified by four pairs barbels present around the mouth. The mouth is non protractile having jaws with teeth. The body is without scales. Elongated first dorsal, long anal and round caudal fins are present. It

attains a maximum size of 40 cm. It is carnivorous, feeds on other animals, young fry and also on algae and higher plants of the bottom. The fish reproduces in the rainy season and 15000 eggs may be laid by a female. Each egg is 1.3 to 1.6 mm in diameter. The young ones (fry) reach 4cm within 2 weeks. Within 4 - 6 months the size may reach upto 25cm (weighing 200g). The fish attains maturity within a year.

2. *Wallago attu :* This is commonly known as *fresh water shark*. It is a game fish. The fish has a large mouth extending beyond the eyes, numerous teeth on the jaws, two pairs of barbels, laterally compressed body, long anal tail and a bent caudal fin. Adipose fin is absent. It is the largest cat fish growing to a maximum size of 2 m. It occurs in rivers and lakes. It is a predatory fish and feeds on fishes and crustaceans. It is not suitable for pond culture. The fish matures within a year.
3. *Heteropneustes fossilis :* It is commonly known as *Singhi*. A long laterally compressed body, flat head, four pairs of barbels on the jaws, a short dorsal fin without rays, devoid of adipose fin, rounded caudal fin and a long anal fin etc. are its characteristic features. Besides animals, it feeds on algae and higher plants of the bottom. It grows to a maximum size of 45 cm. It matures within one year and breeds in the rainy season. It is suitable for culture in swampy areas also.

Trouts

The fishes popularly known as trouts or game or sport fishes are cold water fishes. The examples are brown trout and rainbow trout etc. These fishes have been imported in India by Europeans. These fishes have elongated cylindrical body with scales in the skin and have an adipose fin posterior to the dorsal fin. These fishes feed mainly on insects, molluscans, small fishes and tadpole larvae. Cannibalism is also exhibited by them. (Fig. 3.65 to 3.66)

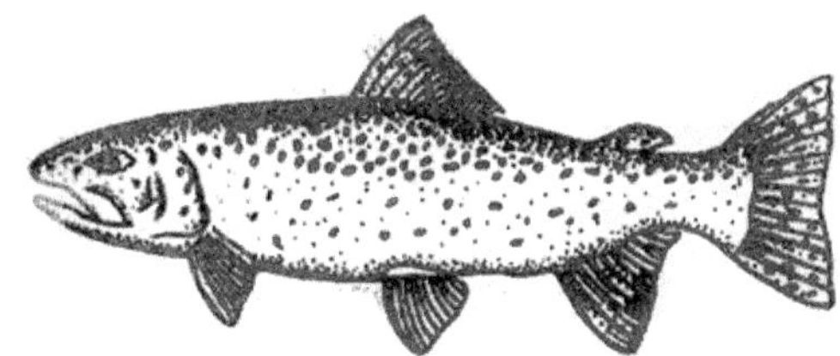

Fig. 3.65 *Salmo trutta*

Salmo gairdneri (female)

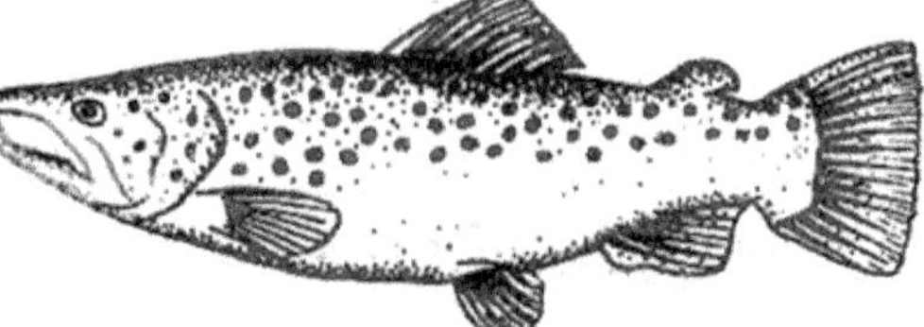

Salmo gairdneri (male)

Fig. 3.66

1. *Salmo trutta :* This is commonly called as the Brown trout. It is a river and lake fish of cold northern hemisphere and got introduced into Himalayan rivers. Its adipose fin is above the posterior part of the anal fin. The scales are minute. The back is deep brown coloured while the sides are somewhat lighter. The upper part of the head and body above the lateral line are marked with numerous dark spots. The fins are dark. It attains a maximum size of 35 cm and weighs 500 g. The breeding season is from October to January.
2. *Salmo gairdneri :* This is commonly known as Rainbow trout. It is a North American lake species which was introduced to Nilgiris and high ranges of Kerala. It is characterized by a small mouth, an adipose fin above the anal fin, emarginate caudal fin, head and dorsal side are steel blue coloured and a reddish band along the sides, light coloured belly and there are numerous dark spots above the lateral line. The dorsal and caudal fins are marked with dark spots. The sides have a rainbow iridescence. It attains a maximum size of 40 cm and weighs upto 1.3 kg. The breeding season is from January to May.

Exotic Fishes

These are the fishes of foreign origin. These fishes not only give nutritious food but also earn foreign exchange because of high export value. These fishes can be grown along with other carps. Examples of exotic fishes are *Cyprinus carpio* (Chinese carp), *Ctenopharyngodon idella* (grass carp), *Hypopthalmichthyes molitrix* (Silver carp). (Fig. 3.67 to 3.69)

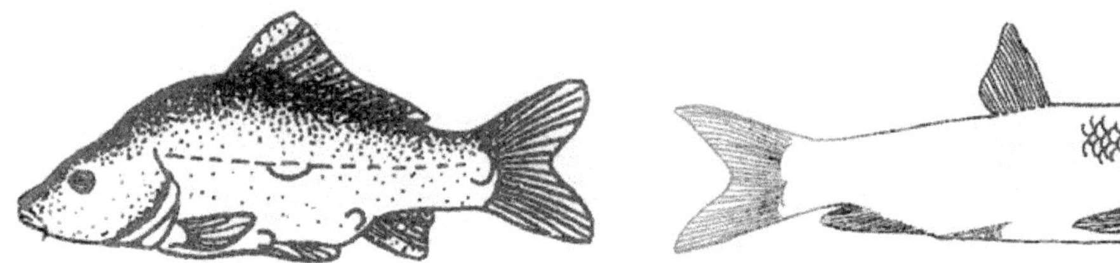

Fig. 3.67 *Cyprinus carpio*

Fig. 3.68 *Ctenopharyngodon idella*

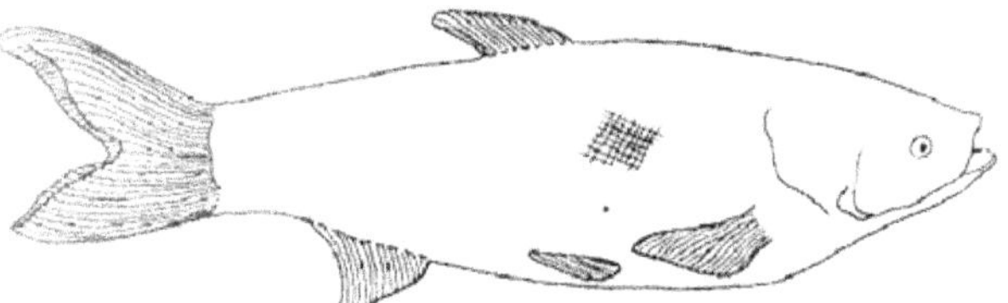

Fig. 3.69 *Hypophthalmichthys molitrix*

1. *Cyprinus carpio :* This is commonly known as Chinese carp or common carp. It is characterized by a deep body and short head. The fish has short round abdomen. The dorsal fin has three bony spines. The scales have golden shade, the fins are rose coloured. It has two pairs of barbels. It has a faster growth rate and grows upto one to 1.5 kgs within a year and upto 5 kgs in the second year. It attains maturity within 6 months and breeds 4 to 5 times and lays about 1.3 lakh eggs.

The eggs are adhesive, and are attached to the submerged aquatic weeds like *Chara, Hydrilla, Ceratophyllum, Myriophyllum, Eichhornia* etc., or tufts of nylon fibres that are placed for this purpose. The eggs hatch within two days. The fry grow at a faster rate and on an average grow at the rate of 1.5 to 2.5 cm per month (100 to 200 g by weight).

There are 3 types of Cyprinus carpio. They are

1. Scale carp, *C. carpio* var. *Communis*
2. Mirror carp, *C. carpio* var. *specularis* and
3. Leather carp, *C. carpio* var. *nudus*

2. *Ctenopharyngodon idella :* This fish is commonly known as Grass Carp. It is a Chinese carp introduced in India in 1959 by importing from Hongkong and Japan. It has an elongated body, broad head with rounded snout, slightly longer upper jaw, moderately sized light greenish scales etc are the characteristic features of this fish. It feeds on aquatic weeds and terrestrial grass (Napier) and in the absence of plants may even feed on fish fry. Newly hatched ones feed on zooplankton including mosquito larvae. Change of carnivorous to herbivorous habit is observed on attaining a length of 2 - 3 cm. The fish grows to a maximum size of 1.5 metres (approximately 30 kg by weight). The growth rate is fast and it grows upto 90 cm within one year (6.9 kg by weight). Sexual maturity takes place in the second or the third year. Breeding is seen in the rainy season.

3. *Hypopthalmichthys molitrix :* It is commonly known as silver carp. It was imported from Hongkong and Japan. An upturned mouth, larger lower jaw, small eyes, broad snout and laterally compressed body covered with small silvery scales are its characteristic features. It is a surface feeder, feeding mainly on phytoplankton. However it also feeds on small fish, worms, insects, crustaceans and also cattle feed etc. Young ones of less than 4 cm feed on zooplankton. Maximum size attained is 60 cm. It grows faster like other exotic fishes and weighs around 1.5 kg within one year. Sexual maturity is attained in the second or third year. The eggs are 3.9 to 4.3 mm in diameter. The largest specimen found was of 20 kg.

Mullets

Mullets are the most important of all the culturable brackish water species. Several species are found in the estuarine waters. These are *Mugil cephalus, Rhinomugil corsula* etc. The fry of mullets migrate into the brackish water areas and are caught for culture along the tidal creeks and pools. The mullets spawn in the sea and do not reproduce in fresh waters. (Fig. 3.70 and 3.71)

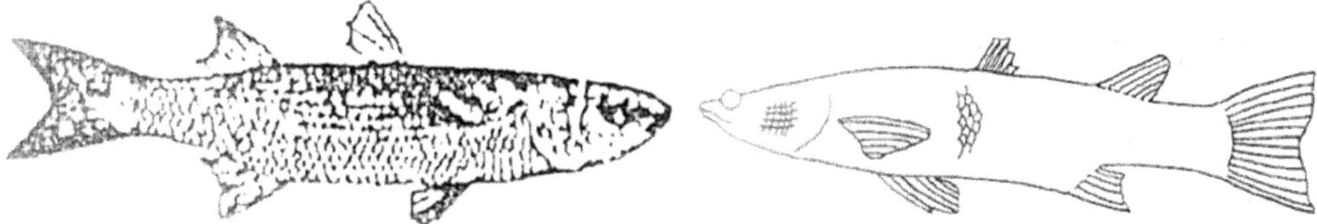

Fig. 3.70 *Mugil cephalus*

Fig. 3.71 *Rhinomugil corsula*

Mugil cephalus

It is commonly known as *Grey mullet* or *striped mullet.* It is the largest mullet. The body is covered with large cycloid scales. The snout is round. Pectoral fin is long. The dorsal and anal fins lack finlets. It grows upto 70 cm.

The maturity is attained by the male in the first year and by the female in the second year. Female lays about 1.2 to 3 million eggs. The spawning period is winter in the sea. The eggs are spherical, transparent, non adhesive. The eggs are 1 mm in diameter and float. The young are captured when they reach 15 to 18 mm in length and grown in brackish waters. They feed on phytoplankton and zooplankton. The growth is rapid. They weigh about 300 g in the first year and 550 to 700 g and sometimes upto 1kg under ideal conditions. The young ones are surface feeders and destroy mosquito larvae but they cannot be used extensively for malaria operations.

Chanos chanos

This is commonly known as milk fish. The body is laterally compressed. A small mouth, devoid of teeth on the jaws, small scales on the body, bright silvery shining of the body, green head are its characteristic features. The fish tolerates wide fluctuations in salinity. The fish grows upto 350 mm in the first year and weighs 250 g and upto 400 mm in the second year. (Fig. 3.72)

It feeds upon phytoplankton and the young feed on algae, diatoms, fish eggs etc. The breeding period is March to August and October to December. The sexual maturity is attained when the males attain a size of 750 - 900 mm and females 800 - 970 mm. The female lays 2 - 5 million eggs. The eggs are spherical and measure 1.2 mm in diameter and float.

Oil Sardines

The sardines constitute an important marine fishery along the west and south east coasts of India. The sardines are represented by nine species such as *Sardinella longiceps, Thrissocles* sp Dussumieria, Kowala etc.

Fig. 3.72 Milk fish (*chanos chanos*)

Fig. 3.73 Oil Sardine (*Sardinella longiceps*)

The oil sardine *Sardinella longiceps* is a very valuable commercial fish due to its food value and industrial uses. The fish is captured during the months of August to March. The body oil content of sardines is very high and is extracted for use in jute, leather and soap industry and as a base for insecticides. It is used as a manure in coconut, coffee and tea plantations. It reduces cholesterol percentage in blood.

The fish attains sexual maturity in the second year when it measures 16 cm in length. It feeds on phytoplankton of surface waters while the young feed on diatoms. The spawning occurs from June to August. The fish grows to a size of 15 - 16 cm within first year. Thereafter the growth rate is slow. Females grow faster than males and maximum size attained is 21 cm in the third year. (Fig. 3.73)

Indian Mackarel, Rastrelliger kanagurta

The Indian mackarel, *Rastrelliger* (*Scomber*) *kanagurta* is an important commercial fish of the West coast of India. The fishing season is from October to February or March. The fishery constitutes about 20% of the total marine catches. This fish attains about 25 cm in length and weighs approximately 4.5 kg. It is pelagic and feeds on phytoplankton and zooplankton. The fish measures 10 cm in the first year and about 18 cm at the end of the second year. The fish attains sexual maturity when they reach a size of 22.4 cm and no sexual dimorphism occurs. Spawning occurs from April to September. (Fig. 3.74)

Fig. 3.74 Indian Mackerel (*Rastrelliger kanagurta*)

Indian Shad, Hilsa ilisha

This fish is widely distributed throughout the Indopacific region. It is a pelagic fish and found in marine waters, brackish waters and fresh waters. Hilsa adult attains a length of 30 cm (males) and 35 cm (females). Sexual maturity is attained at the age of one and a half years. The fertilization is external. Spawning generally occurs from June to February. (Fig. 3.75)

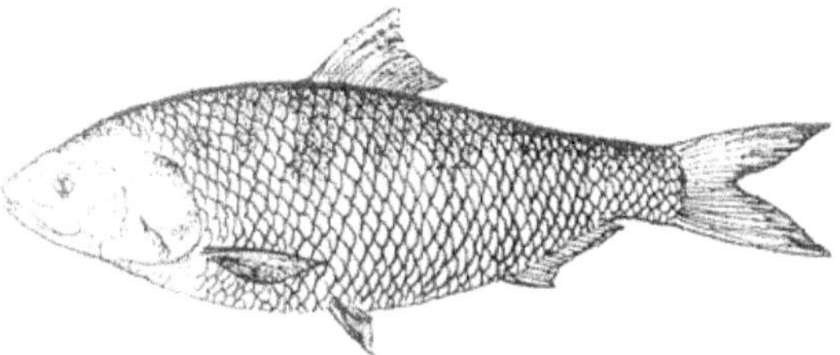

Fig. 3.75 *Hilsa* (*Tenualosa*) *ilisha* (Ham.)

The female lays upto 2.5 to 16 lakh eggs. The eggs float and are 2 mm in diameter. Hatching takes place within 20 hours at 28^0 C. Hilsa larvae feed on zooplankton (copepods and diatoms).

The fish is euryhaline and can tolerate high salinity variations. It is a strict plankton feeder. Hilsa is known to live for 5 - 7 years. It is an anadromous fish ie., it migrates from sea to fresh waters for breeding.

Bombay Duck, Harpodon nehereus

The Bombay duck fishery in India is more or less restricted to Bombay coast and occurs in abundance in Gujarat and Maharashtra coasts. The Bombay duck, *Harpodon nehereus* is a medium sized fish and attains a length of 33 cm. It is estuarine and carnivorous feeding chiefly on prawns. It breeds throughout the year with two peak spawning periods. The spawning ground seems to lie in off shore waters. Fishing is conducted from September to March. About 20% of the fish is marketed in fresh conditions while the remaining 80% is sun-dried and marketed. (Fig. 3.76)

Fig. 3.76 Bombay Duck (*Harpodon nehereus*)

Grey and Silver Pomfret, Stromateus argenteus

This fish is also called as Butter fish on account of their soft flesh and delicate flavour. It is ranked high on the basis of nutritious ingredients. About 300g of raw fish flesh is sufficient to supply about half of the total protein and fat requirement and one fourth of the calories needed in a balanced daily diet for an adult human being. These fishes occur along the east as well as the west coast of India. (Fig. 3.77)

Body ovate, compressed from side to side and covered with thin, small scales. Dorsal fin with a distinct spinous portion which is hidden in the skin. The upper parts are grey or silvery grey while head and body have fine black spots over a silvery background. A dark spot is present on the operculum.

Tunas or Thunnies

Tunas are world famous pelagic fishes of oceanic waters. They resemble mackarel in shape but are very large in size, sometimes weighing 80 - 82 kg. Dorsal and anal fins followed by 6-9 detached finlets. Scales minute or absent. The flesh of this fish is red in colour. About a dozen species are represented in the Indian ocean belonging to the genera *Thunnus*, *Euthynnus*, etc. *Thunnus albacares* is commonly called as yellow fin fish. The body is stout and elongated, the dorsal side is metallic blue or blue black. About 20 broken vertical lines. The fins are yellow in colour and the fish grows upto 2 metres. Tuna fishery is a deep sea fishery. It is not being exploited fully because of the lack of infrastructure. The tuna export will find immense market in Japan, America and Europe.

Flat Fish or Sole Fisheries

Sole fishes or flat fishes are deep sea fishes represented by a few genera like. *Psettodes*, Pseudorhombus, Solea, *Paraplagusia, Cynoglossus* etc. Of these, *Cynoglossus semifasciatus* forms an important fishery in Kerala. (Fig. 3.78)

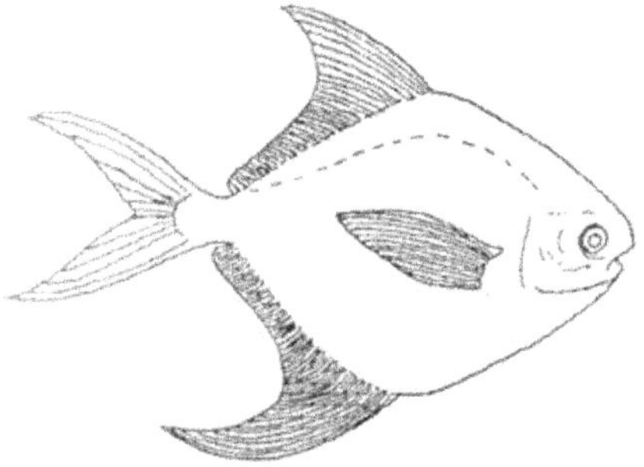

Fig. 3.77 *Stromateus argenteus*

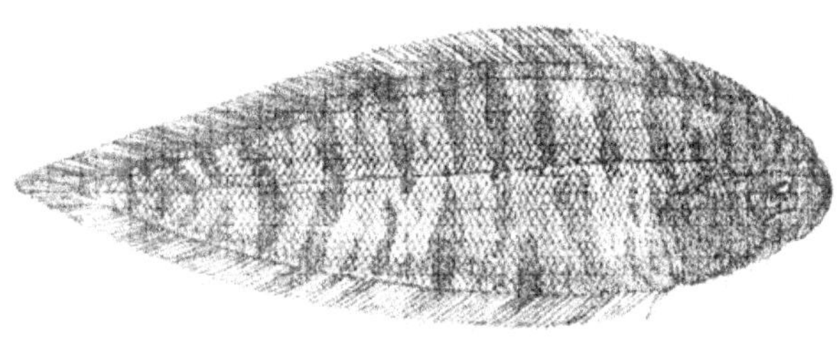

Fig. 3.78 Sole (*Cynoglossus semifasciatus*)

This fish attains a size of 17 cm. It is a bottom feeder consuming a variety of bottom organisms such as polychaete worms, crustaceans and bivalve molluscs. Spawning occurs during monsoon. Eggs are pelagic and start appearing from October. Larvae metamorphose to become adult like young ones which appear from January. The fishing is done between August and September when maximum catch is obtained. The catches during the peak periods are composed of individuals of 10 - 13 cm size group, the catches of October - November consist of spawners and those of January onwards until next monsoon are composed of small young ones.

The flat fishes are generally cured or sundried and marketed. They are also used as manure.

Shell Fisheries

Shell fisheries comprise prawns, crabs, lobsters, and molluscan species.

Crustacean fisheries constitute 10% of the total landings. The edible crustaceans include the prawns, shrimps, lobsters and crabs. The marine prawns and shrimps include several genera of the family Penaeidae. Lobsters include species belonging to the genus *Panulirus*. Crabs include *Scylla*, *Portunus*, *Varuna* etc. The prawns contribute to the major bulk (90%) of crustacean fisheries. Major prawn fisheries occur on the west coast.

Shrimps are smaller (about 2.5 cm in length) than prawns (5 cm to 30 cm) and include several species of the genus *Acetes* Prawns include the following species of commercial importance.

1. *Fresh water prawns :* About 30 species of prawns occur in fresh water. All belong to the genus *Macrobrachium*. Prawns have a very high export value. Indian prawn grounds are the most productive ones both for their high yielding as well as superior quality. *Macrobrachium rosenbergii* and *M. malcolmsonii* are the two important species of fresh water prawns suitable for culture. (Fig. 3.79)

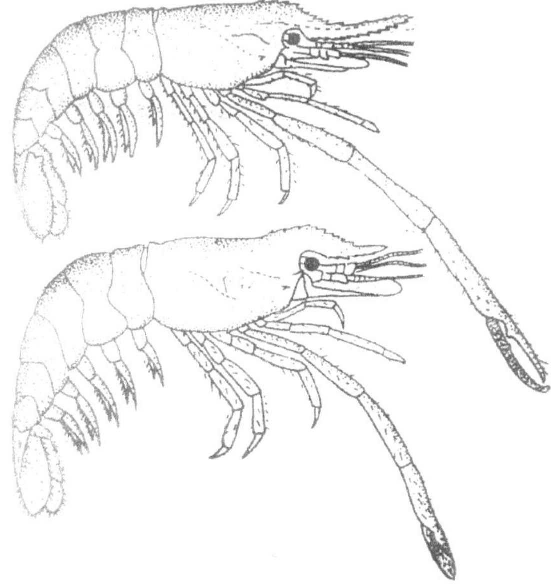

Fig. 3.79 *Macrobrachium rosenbergii, Macrobrachium malcolmsonii*

Macrobrachium rosenbergii

It is commonly known as the *Giant river prawn.* It is the largest among fresh water prawns. It occurs in all rivers of east and west coasts of India. It has a long sword shaped rostrum with equal number of teeth on both upper and lower edges. The second pair of legs (chelipeds) of male prawn are much longer and stouter than the body thus showing sexual dimorphism. Legs are large. It is a bottom feeder. Omnivorous. They attain a maximum size of 25 - 30 cm and a maximum weight of 450 g. It migrates to estuaries during the breeding season. Males are larger than females. The female lays around one lakh eggs. The breeding season is from August to December. The incubation period of the eggs is 15 to 19 days. Juveniles of various sizes (18 - 90 mm) show horizontal lines on the carapace ranging from one (in the smallest) to 7 - 8 (in the largest). After 90 mm stage these markings disappear and the dark abdominal belt like continuous patches develop on the junctions of the abdominal segments. The juveniles (young ones) feed on Crustaceans, annelids, small worms and also fish meal.

The eggs are extruded and attached to the pleopods of female. The planktonic larvae undergo 11 moults before metamorphosing into post larvae. Feeding may begin within a day or two of hatching and they feed on zooplankton. Post larval stage is reached within 35 - 40 days. The post larval shrimps become benthic and migrate upstream and reach their abode ie., rivers, ponds etc and thus the cycle is repeated.

Among brackish water prawns, there are about 7 species that can be cultured in brackish water of these, two species i.e. *Penaeus monodon* and *P. indicus* are extensively used for culture in India.

Penaeus monodon

This is commonly called as tiger prawn or jumbo prawn. This is the largest of marine prawns and grows to a maximum size of about 320 mm - 340 mm. This species completes its life cycle partly in sea and partly in the estuaries. It can tolerate low salinity. It is the highly preferred species for aquaculture because of large size and adaptability to varying conditions. (Fig. 3.80)

The body of the prawn is dark brown or reddish in colour. Antennae dark and light in colour. Pleopods are brown to blue with distinct yellow bands. It feeds on Phytoplankton, zooplankton, diatoms, algae, crustaceans and also detritus. The prawn is also capable of withstanding a wide range of salinity and temperatures of brackish waters.

Penaeus indicus

This is commonly known as *white prawn.* It occurs on the east and west coasts of India extending in distribution to estuaries, back water lakes etc. It grows to a maximum size of 200 - 230 mm and forms 10% of the total prawn production of the country. It completes its life cycle partly in sea and partly in estuaries. It feeds on detritus, small crustaceans, molluscans, polychaete larvae, phytoplankton, sea weeds etc. It attains sexual maturity when it is of 130 mm size. The spawning season is October to May. Female lays 68000 to 7.25 lakh eggs. The prawns are cultured for about 100 days and marketed when attain a size of 80 - 100 mm. (Fig. 3.81)

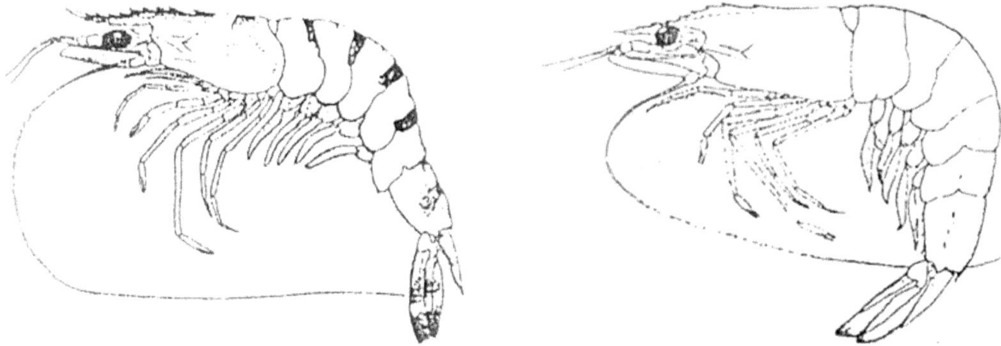

Fig. 3.80 *Penaeus monodon*   Fig. 3.81 *Penaeus indicus*

Crab Culture

The crabs have high export value. They stand second to shrimps among commercially important aquaculture species. The important species for culture are *Scylla serrata* and *S. oceanica*. These are known as mud crab and green crab. (Fig. 3.82)

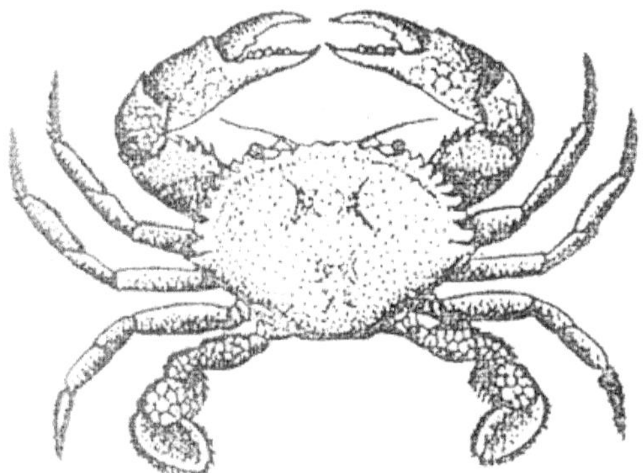

Fig. 3.82 *Scylla serrata*

Crabs attain 15- 2 0 cm (carapace width) size and weigh about 2 kg. Crab meat is sweet and tasty and it is not only a good food for man but also has medicinal value in chronic fevers and asthma.

Crabs are distributed along east and west coast line. A total of 11 species of crabs are of commercial importance of which *Scylla sp* are used as food crabs. Crabs are solitary swimmers and burrowers which inhabit estuarine or marine environments.

S. serrata inhabits in low saline waters in deep burrows dug in the mud. It migrates in the nights to the mangroves in search of food. The ventral side of the body is brick red and the carapace is metallic brown in colour. It grows to a size of 120 - 130 mm across carapace and weighs about 700 g - 1 kg. They breed almost throughout the year. *Scylla serrata* reaches a marketable size of 250 - 400 g in 7 to 8 months. The crab seed is the juveniles which are collected from estuaries for culture.

Lobsters

Commercially important lobsters are *Panulirus* (*Puerulus*) *polyphagus, P. ornatus, P. homarus, P. sewelli, Linuparus somniosus* and others. Of these, *P. homarus* is very common in landing of south west coast of Madras and Kerala. The lobsters are exported to USA and Japan and earn a significant foreign exchange. Lobster larvae are collected from inshore waters to be cultured in laboratory tanks. After a certain period of growth, the young lobsters are transferred into the pens of inshore water. Growth recorded is 7 - 9 mm per month.

P. sewelli is commonly known as *whip* or *spiny lobster*. The lobster attains sexual maturity when it attains a size of 120 - 129 mm. The breeding period is long with peak during January to April. It lays its eggs on its pleopods. It lays 10,170 to 36,700 eggs. The lobster grows to a maximum size of 196 mm. It is carnivorous feeding on sea prawns, fishes and molluscs. The sexual dimorphism is noticed. The males have greater carapace length when compared to the females.

Molluscan Fisheries

Edible molluscs are the mussels, clams and oysters apart from Pearl oysters comprise the molluscan fisheries. Edible shell fishes are more abundant on the west coast of India occurring in numerous Creeks, bays, rocky coastal regions, estuaries and back waters.

The important commercial edible species include the following :

1. *Mussels : Mytilus* (*Perna*) *viridis* (green mussel), *M. indica* (brown mussel) are important among mussels. The green mussel occurs on both coasts. It is liked as food rather widely. This has a high growth rate of about 11.3 cm in eight months. It attains sexual maturity only when it reaches the size of 25 mm. The annual yield from these mussels is about 1200 tonnes. The yield from rope culture in Kerala is about 200 tonnes/hectare/year. They are harvested when they attains a size of 60 mm size. The spawning starts from July and lasts till December. (Fig. 3.83)

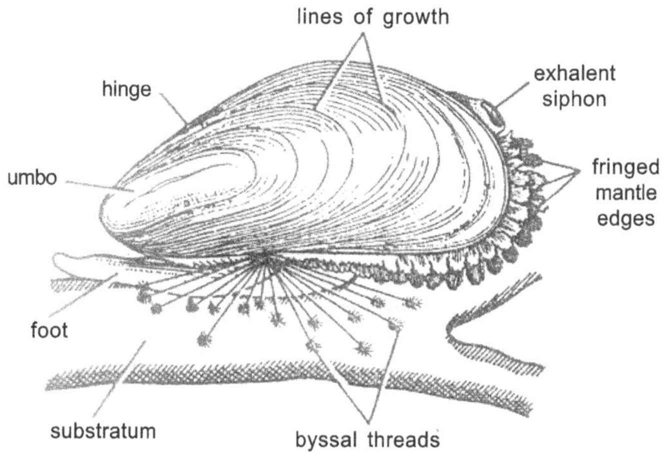

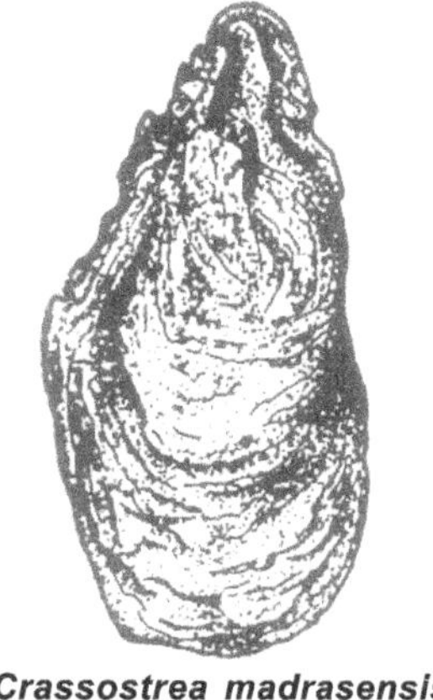

Crassostrea madrasensis

Fig. 3.83 Mytilus

2. *Clams : Meretrix meretrix (Bay clam) M. casta (Back water clam), Paphia malabarica* (false clam) etc., are commercially important clams of India. There are abundant in estuaries, back waters and bays along the coasts. They are relished as food. They are of common occurrence near estuaries and coastal regions and are of considerable food value.
3. *Oysters :* Eight species of edible oysters are known to occur along Indian coasts of which four species are abundant supporting fisheries of local importance. They are *Crassostrea madrasensis, C. gryphoides, C. discoides and C. cucullata.*

   The oyster *C. madrasensis* grows wildly in most of the tidal creeks and estuarine regions along the east coast of India. Oysters reach 85 - 90 mm size in 12 months time with an average weight of 120 g. Seed are collected from tidal pockets close to the natural habitat of oyster stock. It takes one year to grow to maximum size with tasty meat.

   The main larval feed is composed of flagellates and adults feed on detritus, diatoms etc. The larval growth is at a faster rate and grows to 38 mm in 3 months. By the end of year they grow to about 86.7 mm attaining a weight of 120 - 130 g. Oysters which are males in the beginning gradually change to females in about 1½ years. They appear to be bisexual ones in between 9-15 months when gametogenesis is totally arrested. These stages are thought to be the transitional ones during sex transformation. Eggs measure 48 - 60 mm and about 10 - 15 million eggs are released during spawning periods. Fertilization is external.

Pearl Oysters

The pearl oysters yielding pearls of high quality are *Pinctada vulgaris, P. margaritifera, P. chemmitzi, P. anamioides* and *P. atropuropurea*. The pearl oysters are gregarious and occur on rocks and corals at a depth of 10 - 12 fathoms about 20 km off the shore called pearl banks or parts. Pearl oysters become sexually mature when they measure about 16 mm. Spawning occurs more in summer months when water temperature is in the range of 28.5^0 - 35^0 C. Fertilization is external. The eggs are oval in shape. Through different stages such as trochophore, veliger, umbostage, eyed stages, pediveligers, plantigrade larvae within 20 days, grow to a size of 220 µm. (Fig. 3.84 and 3.85)

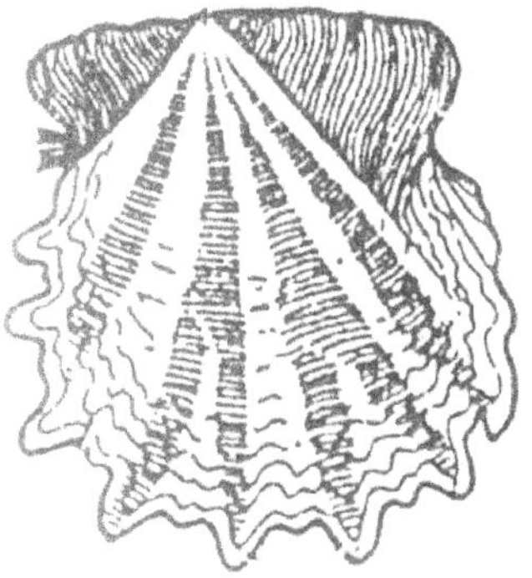

Fig. 3.84 *Pinctada*

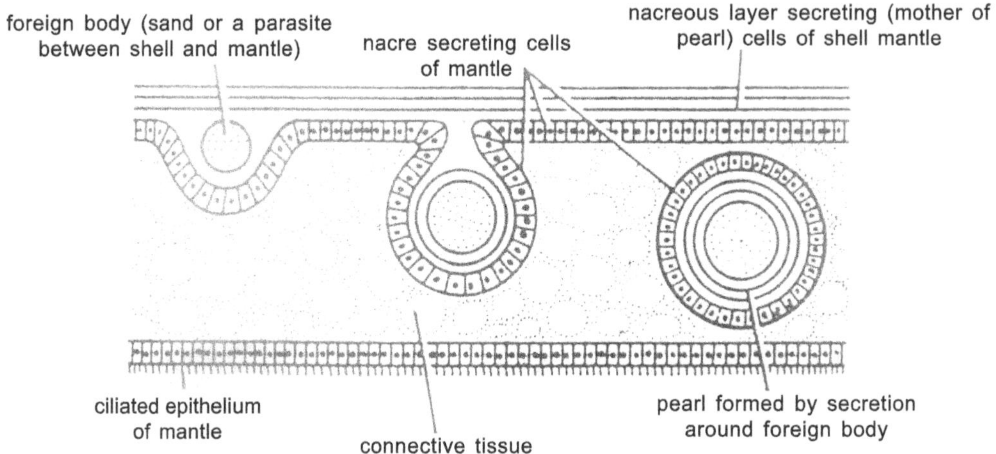

Fig. 3.85 Stages in pearl formation

3.15 World Aquaculture

Statistics on aquaculture production are compiled each year by a number of nations and by the Food and Agriculture Organisation (FAO) of the United Nations. In many instances, landing reports include data from both commercial capture fisheries and aquaculture. In addition, a considerable amount of fin fish and shell fish undoubtedly reaches the market without passing through a point where statistical information is collected. Thus obtaining reliable estimates of aquaculture production is difficult.

The data available shows that aquaculture production continues to increase. In year 1957 the total aquaculture production was 29 million tons, in 1989 it was 92 million tons, in 1995 it was 112.30 million tons, and it is predicted that by the year 2020, the aquaculture production would reach around 125 - 135 million tons.

On the whole, the aquaculture production is dominated by Asia, in which about 86% of the world's aquaculture commodities are grown. Table 3.1 shows the world aquaculture production by regional percentage. Table 3.2 shows the top 10 countries in terms of economic value of their aquaculture production in 1988.

Table 3.1 World aquaculture production by regional percentage contribution (FAO 1991)

Region	% of Total Production
Asia	77.8
Europe	11.4
North America	3.6
South America	3.5
Russia & Allies (USSR)	2.7
Pacific Island Nations	0.6
Africa	0.4

Table 3.2 Percentage of Total World Aquaculture production and primary commodities marketed by the leading producer nations in 1988 (FAO 1991)

Country	% of Total Production	Commodities
China	35.4	Molluscs, Shrimps, Carp
Japan	20.4	Algae, Molluscs, Yellow tail
Taiwan	5.4	Molluscs, Shrimp, eels
Philippines	3.2	Algae, Shrimp, Milk fish, Tilapia
USSR	2.7	Carp
United States	2.7	Molluscs, Cat fish, Salmon, Trout
Norway	2.6	Salmon, Trout
Ecuador	2.6	Shrimp
Republic of Korea	2.5	Algae, Molluscs
Indonesia	2.5	Algae, Shrimp, Carp, Milk fish, Tilapia

Substantial amount (about 40%) or the total marine catch is consumed immediately (unpreserved). The remaining catch is preserved and processed. Of them, 8% is marketed as frozen fish, 9% is canned, 24% is preserved by curing, 18% is used to make fish oil and fish meal.

There are certain problems in the way of achieving greater yields. Numbers of various species of fish have peaked and declined from over fishing. Lower catches per boat and a dominance of smaller fishes are indicative of the fact that certain fish species in certain areas are being overharvested, i.e., being harvested faster than the supply can be replenished. To overcome these problems, artificial production of fish through aquaculture, in both fresh and salt water has been undertaken.

Role of Aquaculture

The expanding human population on earth demanded maximization of the agricultural production. However because of the politics and economic constraints sometimes prevent nations with surplus food reserves to transfer to the needy. Due to the poverty, draught conditions, non productive lands etc. and added to these, no import or insufficient imports led such countries to search for alternate sources of food.

One logical environment that would seem to be available for additional exploitation is the world ocean, which covers about 70% of the surface of the globe, and this vast expanse of water provides a limitless source of animal protein for mankind. Overall harvests from the ocean have increased to some extent. The FAO (1991) reported that in 1989, the total marine catch of finfish, crustaceans, molluscs, and other animals, including species raised in aquaculture was 99.5 million tons. The demand for sea food for human consumption will double by the year 2005. Hence there is a need to enhance capture fishery and culture fishery.

Inland fisheries of the world

The total area of fresh water of the world is estimated at 2.5 million sq.km. The fresh waters are the most productive resources yielding an annual catch of over 12 kg/ha as against only 1.1 kg/ha value of marine catch.

Inland fisheries comprise both capture fishery from natural waters and fish culture. A considerable amount of catches are taken by individual fishing from local water bodies for subsistence of the family. Such fishing is dominant in south east Asia, USA, Africa and Sweden. In Philippines, it accounts for about 60% of the commercial catches.

Inland sources of fish consist of the great river systems, lakes, ponds and canals, other water bodies, paddy fields and back waters.

China tops in the output from freshwater sources, i.e., about half of the total world's inland catches. USSR comes next followed by other regions such as South East Asia, North Europe, USA, Canada, Africa and South America etc.

The fresh water capture fishery or fish culture has attracted world wide attention. Expansion of irrigation and industrialization etc posed problems for fish culture. However it is overcome by establishing artificial water bodies (irrigation canals, reservoir), restocking of natural water bodes with farm raised seed, transplantation of species of fish or its food organism from one region to the other where these do not occur.

The fresh water capture fishery covers a wide variety of fish and non fish organisms. A total of 350 to 400 species are used. The principal organisms other than fishes that are used are frogs, shrimps and prawns, lobsters, cray fishes, crabs, oysters, clams and cockles, mussels, squids etc.

The present status

The total global aquaculture production is estimated at 75 million tons in 1986. All modern devices for improvement in fishing and implementation of the exclusive economic zone (EEZ) concept, the world over not withstanding, the fishery gives depleting catch rates year by year. The depletion is further increased by rising fuel costs, trawler costs and other operational expenses.

Of the 75 million tons of global production, 66 million tons from capture fishery and 9 million tons from aquaculture. A total of 15 to 20 million fishermen are engaged in the fishing industry all over the world.

World Bank and other funding agencies are financing for aquaculture or fishery projects. Demand for more fish production has been recognised and discussed in various fora and a global network INFOFISH was started to impart regional information, services for the trade, strategies etc. The decisions that were taken at world fisheries conference, 1984 are as follows.

(i) Management of stocks of conventional fished species should be coupled with intensive culture practices for additional potential to raise production annually.

(ii) Wastage of fish to be avoided.

(ii) Pollution of marine resources and their upsetting due to man made activities to be prevented.

Realistically, aquaculture can be expected to provide an increasing share of the world's aquatic animal protein, but that protein will be expensive. As additional technology is required to meet increasingly tough effluent standards and ensure that incoming water is of proper quality for the animals or plants under culture, the costs of production can be expected to increase. It has only been through technological breakthrough and adoption of the results of scientific investigation that the price of aquaculture products has been kept as low as current levels. Additional technology development and scientific investigation are going to be needed to meet the challenges of the future if aquaculture is to remain viable.

Aquaculture in India

In olden days, Indians never faced the shortage of food as the supply was more than demand. So they led a peaceful and a happy life.

In 1947, the food production was only 50 million tonnes for a population of 33 crores which was never sufficient to the people. The Government of India in its five year plans, gave proper prominence to agriculture but there was only 2.7% increase in food production till 1974. Low percentage of increase is due to traditional practices in agriculture besides non utilization of water resources.

Our country gave priority to scientific methods practices in its five year plans since 1974. Agricultural produce increased from 120 metric tonnes (1977) to 210 metric tons (in 1994). All these food materials cannot provide necessary amount of proteins and fats. Hence man tried to invent supplementary food to provide nutritious food to the needy people. White revolution (Milk and milk products), poultry (eggs and chicken) and Blue revolution (culturing of aquatic organisms having nutritious value) are the attempts to provide balanced and nutritious food.

Our country can meet not only the demands of the people but also earn foreign exchange by utilizing aquatic resources (like canals, rivers, lakes, reservoirs, brackish and marine waters of east and west coasts) for culturing food organisms life fishes, prawns, oysters, lobsters and other molluscs.

Aquaculture is the farming of economically important aquatic animals and plants under controlled conditions. Aquatic animals include fish, crabs, crustaceans and molluscs. Aquatic plants include sea weeds. If an acre of water is farmed properly, it yields better than an acre of land. To achieve maximum productivity from the available water resources artificial culture of aquatic organisms has been takenup through blue revolution.

Today aquaculture industry is growing rapidly and over 15 million metric tonnes of aquatic products are harvested each year all over the world. Prawn culture in sea water is called as shrimp culture where as in fresh water is called as prawn culture. In America shrimp industry dominates aquaculture. The trouts, carps, oysters and mussels are cultured in large numbers in Europe.

In the world, there are about 300 species of fish that can be cultured in tanks. Only 10 - 15 suitable species of fish are available in India. India has a great potential of Inland and marine water resources for fishing.

Among the fish producing countries of the world, the top 7 fish producing countries along with their production in million metric tonnes are China (15), Japan (8.46), Peru (6.84), Chile (6.50), Russia (5.61), America (5.6) and India (4.17). China accounts for about 40% of the total fish production whereas India constitutes 38% of the total fish production.

Fisheries can be defined as capture of fish and other edible aquatic organisms from the fresh water or brackish water or sea.

When compared to other nations, India has freshwater, estuarine and marine water resources for utilization in fishery industry. Besides these sources, ponds, tanks, lakes, canals, reservoirs also contribute a lot to fishery industry.

India has about 0.338 m.ha of fresh water, 0.76 m.ha of estuarine and 0.58 m.ha of marine water area available for culturing aquatic organisms. Fishes, prawns, crabs, edible oysters, lobsters, snails and sea weeds are grown in these water areas.

Based *on the method of capture*, fisheries can be divided into capture fishery and culture fishery.

Capture fishery is the process of catching the fish from natural resources by using different craft and gear. It includes both fresh water and marine fisheries.

The catches include both desirable and undesirable varieties. Environmental factors and pollution influence the fish yield. Due to the indiscriminate catching of fish including juveniles and brooders, fish yield decreases gradually in capture fisheries.

Culture fishery can be defined as growing of fish or other edible organisms under controlled conditions inspite of its nutritional value. It is conducted in freshwater, brackish water and sea waters. The seed is stocked, nursed and reared in confined waters, then the crop is harvested. To get maximum yield, culture takes place in artificial ponds and supplementary food is provided to fish.

Farming of a wide variety of aquatic organisms like prawns, crabs, molluscs, oysters, mussels, frogs, sea weeds have come under culture fisheries because of the development and expansion of new culture systems. Considerable attention is being given to the culture fisheries to overcome the problems found in capture fisheries, so that the production can be increased.

Due to the culture of a variety of aquatic organisms, culture fishery is termed as aquaculture. It is under water agriculture. Aquaculture has been called as the rearing of aquatic organisms under controlled or semi controlled conditions. It provides inducement to establish other subsidiary industries in the country.

Classification of Aquaculture

Based on the type of water, aquaculture is of three types, 1. Fresh water aquaculture 2. Brackish water aquaculture 3. Mariculture or marine water aquaculture.

1. *Fresh water aquaculture :* Culture of fishes and prawns in fresh water bodies such as ponds and tanks is termed as fresh water aquaculture.

   Eg : Catla catla, Labeo rohita, Cirrhina mrigala (fishes), *Macrobrachium rosenbergii* (fresh water prawns)

2. *Brackish water aquaculture :* Estuaries are the areas where rivers open into seas. Water here is diverted to ponds established on shores which are called as brackish water areas. In these areas some fishes and particularly 'prawns' can be cultured.

   Eg : Penaeus monodon (prawn)

3. *Marine aquacultue :* The culture of mussels, clams, oysters and sea weeds in the sea is termed as marine aquaculture.

Fishing crafts

The vessels used for transport and collecting fish is termed as fish craft. Catamarans, Dugout Canoes are the fishing crafts used for the capture of organisms in reservoirs and rivers.

Mechanised boats, catamarans, satpati type of boats are used for fishing in sea water areas. Deep sea fishing vessels are introduced to catch tunas and other organisms from the deep waters of the sea.

Fishing gear

The various devices used for capturing fish are termed as fish gear. Dip nets or lift nets, Cast nets, purse nets, drag nets, stake nets, fish traps etc., are used for catching fresh water fish.

Boat seines, shore seines, trawlers, gillnets, linenets etc., are used for catching marine fish.

Field Culture, Culture Technology, Production, Utilization and Demand

Field culture

Field culture means promoting the fish production by growing fish in available water bodies.

Preparation of a Pond

The pond where fish is to be reared should be a fairly deep and clean pond. The water should not go down from a certain level. The quality of water as well as the percentage of Chlorine, Iodine, Fluorine, Silica, Arsenic, Manganese also play a vital role in fish culture. Carbon, nitrogen, phosphorus, calcium, potassium and sulphur are directly obtained from the soil and they are also necessary. So the quality of soil, its vegetation and environmental position is also an important factor in the preparation of a pond.

Fish farming requires various types of ponds which vary in depth, area and quality of the food. The ponds are made close to each other. Different types of ponds are used for different purposes. The ponds are classified as :

1. Nursery pond
2. Rearing pond
3. Stocking pond
4. Breeding pond
5. Marketing pond.
6. Hatching pond

The general ratio is nursery pond 5%, rearing pond 20%, stocking ponds 60%, breeding, hatching and marketing ponds are made in 5% areas each.

Selection of the Site

Generally the ponds should be made in areas where the water storage capacity is maximum. There should be proper inlet and outlet for water currents in a pond. Nature of soil, construction and layout plan of a pond are important factors in fish culture. Humus clay is best for pond as it retains water. In the absence of humus clay other soils like a mixture of sand and 60% clay are used for preparation of a pond. Ponds can be made in any size and shape. Rectangular pond is supposed to be the best. (Fig. 3.86)

The following physical and chemical factors must be taken into consideration in preparation of a pond.

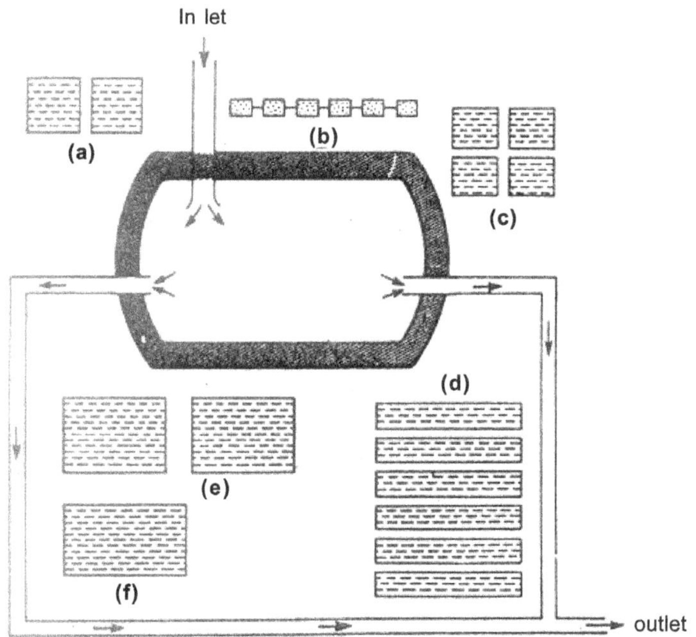

Fig. 3.86 A Fish farm
(a) Nursery pond (b) Rearing pond (c) Stocking pond (d) Breeding pond (e) Marketing pond (f) Hatching pond.

Physical factors

1. *Depth :* For maximum productivity 2m deep ponds are considered to be good.
2. *Pressure and movement :* Productivity can be favoured by the movement of water due to waves, currents and renewal.
3. *Temperature :* Temperature influences feeding, breeding, respiration and all other physiological activities.
4. *Turbidity :* Increase in Turbidity will create a serious problem of respiration mechanism. This is due to the deposition of clay mud and suspended material on gills. So care should be taken during floods and rainy season.
5. *Light :* Light directly affects the vegetation and O_2 concentration of water. Maximum amount of light should be allowed to the surface of water and to bottom by decreasing the turbidity and vegetation.

Chemical Conditions

1. *Oxygen and Carbon dioxide :* The pecentage of dissolved oxygen varies with temperature variations of water. High temperature reduces the oxygen. The algal growth on surface, over crowding of aquatic animals, pollution, etc., rottening leaves reduce O_2. During night when no photosynthesis takes place, carbondioxide increases.

2. *pH and alkalinity :* The alkaline or acidic water kills fishes. So the p^H range is supposed to be between 7.5 - 8.5. It is best for fish farming.
3. *Salts and Minerals :* Phosphorus, Potassium, Calcium and nitrogen are necessary for the development of fishes.

Construction of Fish Ponds

For rearing various stages of fish a fish culturist needs different types of ponds. The main considerations in pond construction are

1. Topography
2. Soil type
3. Water supply

    1. *Topography :* Topography is the surface feature of the area. The topography of a pond should be like a bowl shaped area or like a gently sloping terrain of a wide valley with high lands on three sides and a narrow outlet on fourth side.
    2. *Soil type :* Humus clay is best. Secondly a mixture of clay and sand is best.
    3. *Water supply :* we should maintain constant water level. To allow clear water even during the rainy season a very little inlet with mesh is required.

The size of farm depends on the purpose of the farm. Larger areas are needed for constructing the nursery, rearing, stocking and breeding ponds for commercial and experimental farms.

1. *Hatching ponds :* Hatching ponds are used for hatching the fertilized eggs. Hatching ponds are located near the nursery pond for easy transport. These ponds require continuous slow flowing water. (Fig. 3.87)

   Types of hatching ponds : Hatching ponds are of two types.

   (a) *Hatcharies* : These are small sized ponds in which fertilized eggs are transferred. After 2 to 15 hours the eggs are hatched. Some draw backs make the hatcharies unfit for advanced fish culture programme. These drawbacks are as under :

   (i) sudden rise and fall in temperature, (ii) entrance of predators in ponds, (iii) drying of water from ponds may cause mortality of eggs.

   To overcome all these drawbacks specially designed hatcharies and hatching hapas are constructed.

   (b) Hatching hapas : Hapas are rectangular trough shaped tanks made up of cloth supported by bamboo poles fixed in the river. In these hapas fish eggs are aerated by continuous flow of current. The size of hapa is about 3' × 1.5' × 1' and is made up of mosquito net cloth which is fixed into outer larger hapa made up of coarse cloth. Two types of hapas are designed.

   (i) Fixed type hapa : If possible to fix the perpendicular poles (arms of hapa) only then the fixed type of hapa is used for hatchings of eggs otherwise floating type of hapas should be used

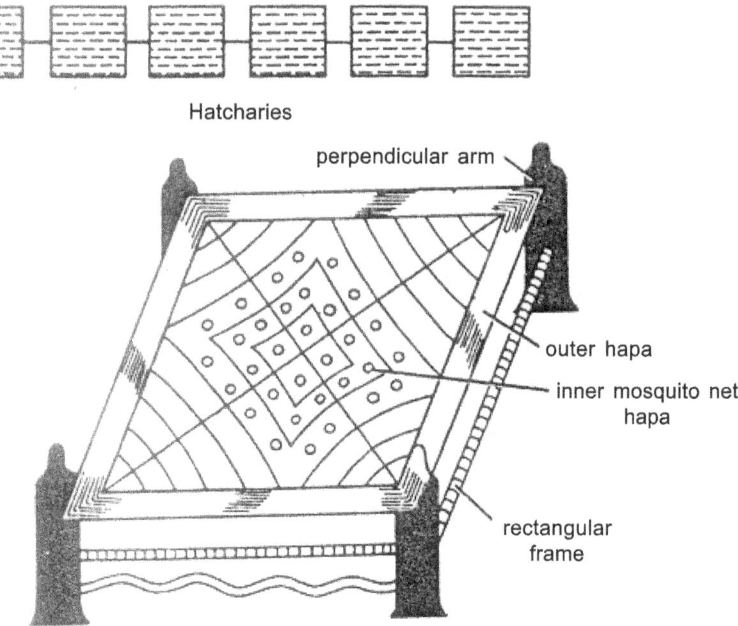

Fig. 3.87 Hatching hapa

(ii) Floating type hapa : In hard bottom areas where it is not possible to use fixed hapas, floating hapas are built. These hapas are arranged in large number in series attached with the bamboo and floated on water surface. To avoid overcrowding of eggs, only one layer of eggs is spread in each hapa. Hatching of eggs takes place in the outer hapas leaving the egg membrane in the inner one. The hatchings are kept for 36 to 48 hours in hapas and then transferred to the nurseries.

2. *Nursery ponds :* These ponds are used for rearing younger fish or fingerlings. It needs so much care for its construction because the fingerlings are very sensitive to external climatic conditions. The pond is allowed to stand unused atleast for 1 - 2 months. These ponds are shallow and rectangular.

3. *Rearing ponds :* These ponds are used for rearing advanced fry of 2 - 5 months. To facilitate netting, these ponds are made long and narrow and gradually sloping.

4. *Stocking ponds :* These are fairly large perennial tanks varying from 0.2 to 2 hectares in area with an average depth of 1.5 to 2.0 metres. The p^H of water is maintained between 8.5 to 9 by adding quick lime.

Besides the above, a few marketing ponds may also be provided for keeping the fish ready for market. For pond breeding species a couple of breeding ponds are also necessary. For breeding ponds one of the marketing ponds can be converted into a breeding pond.

Culture Technology

Culture technology means following various scientific methods the growing fish or rearing the fish basing on the food and water. Fish culture can be carried out in different methods. They are :

1. *Traditional system :* It is an age-old practice. Fish are grown with natural feed available in tanks i.e without fertilizers.
2. *Extensive fish farming :* In this method fishes and prawns are grown with the natural feed and the fertilizers are used to a limited extent.
3. *Semi-intensive system :* In this method, use of feed and water exchange are present. Oxygen is supplied by using aerators or sprinklers.
4. *Intensive system :* In this method, use of essential feeds, water exchange are used. Oxygen supply is increased by increasing the number of aerators. The production is very high in this method.
5. *Monoculture :* A single species of fish or prawn is cultured in this method.
6. *Polyculture or composite fish culture :* Introduction of selected varieties of compatable fish having different feeding habits to achieve high productivity is called composite fish culture. An important advantage of composiste fish culture is the maximum utilization of the available resources. When the density of available natural feeds is more, then different stages of the same fish species can be introduced and cultured. This facilitates high productivity. If the productivity and yield are more, profitability from such a culture is highly encouraging.

   Eg : The culture of Indian major carps (*Catla, Rohu, Mrigal*) and common carp, silver carp and grass carp.
7. *Integrated fish culture :* Integrated fish culture means culturing of fish in combination with agriculture, poultry, piggery, duckery and livestock. It is cheap and brings high results.
8. *Cage culture :* Cages are constructed with wood or bamboo frames covered inside by nylon cloth with holes. Fishes are released inside the cage and these cages are kept in natural flowing waters, particularly the drains or swamps. Through flowing water food comes into the cages and wastes are carried away from the cages. This method needs less management and capital investment. Harvesting is also easy.
9. *Fish feed :* Most of the fishes selected for pond culture are omnivorous in their habit. Different stages of fish such as fry, fingerlings, young fishes and adults depend upon both natural and artificial feed. Fish feed includes phytoplankton, zooplankton, crustaceans, their larvae, annelids etc.

Fishing Techniques :

Lift nets or Dip nets, Cast nets, stake nets, triangular nets, purse nets, drag nets etc are used to capture freshwater fishes. Boat seines and shore seines are used to capture organisms in the inshore waters of the sea.

Gill nets (both the stationary and floating) are used for migratory fishes. It requires no extra labour for fishing. (Fig. 3.88)

Eg : Sardines, Long lines are used for capture of tunas.

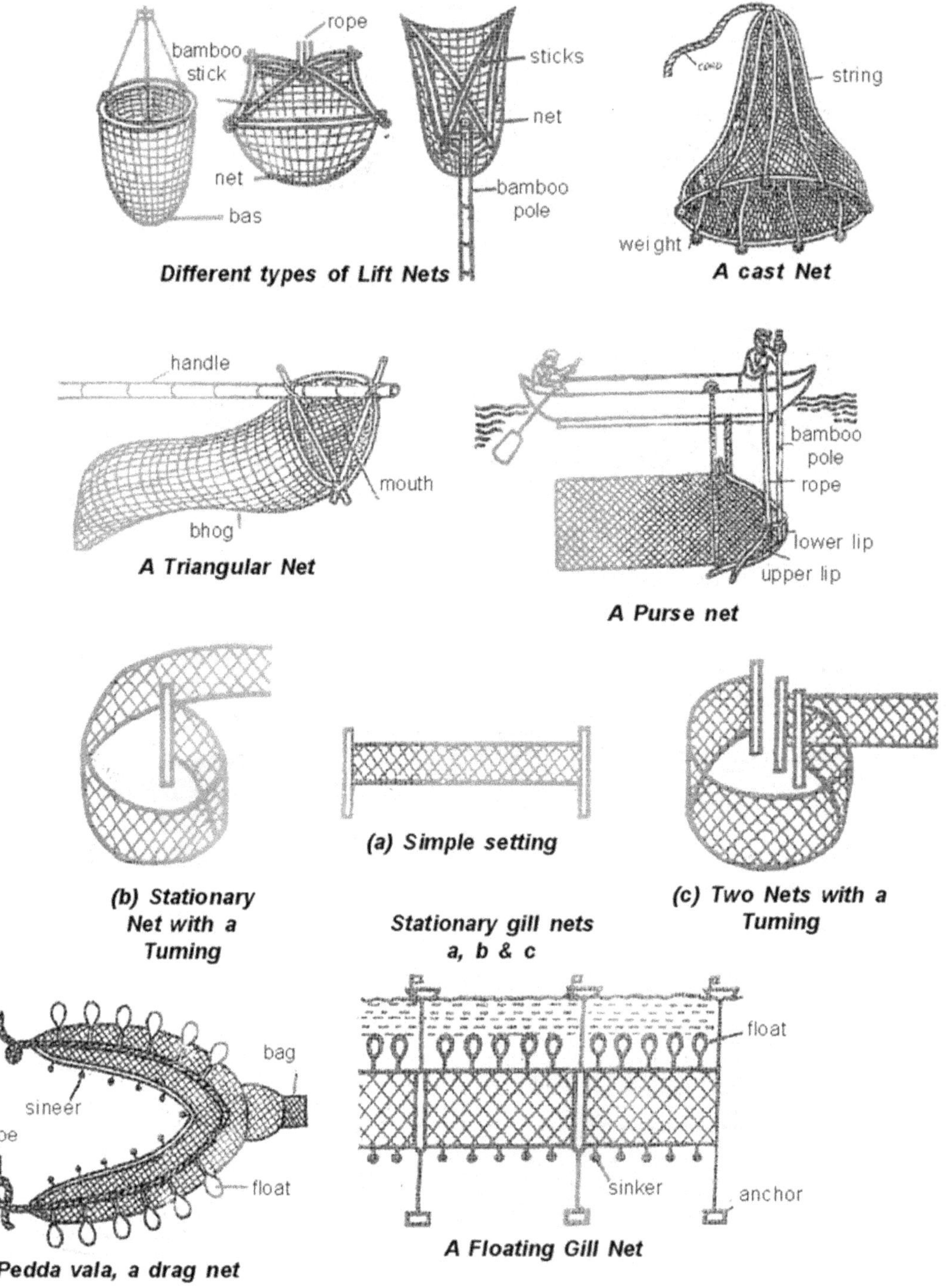

Fig. 3.88 Different types of Nets (Fishing devices)

Harvesting

During the course of harvesting well developed fishes are collected and they are sorted out on the basis of their weight. Harvesting is done generally in the second year of the farming.

Preservation

Preservation depends upon the environmental conditions, type of fishes, economics of the farmer and climatic conditions. Small fishes can be preserved by washing with salt water and mixed with salt in 1:3 ratio and dried in the open sun. Medium sized fish can be preserved by sun curing.

The most popular way is preserving in ice and salt. Besides this, deep freezing, freezed drying and smoking are various other methods used by big firms.

Breeding

The breeding is of two types 1. Natural breeding and 2. Induced breeding.

Natural breeding

It requires

1. Slow movement of water
2. Uniform temperature
3. Excess of dissolved oxygen
4. Shallow ponds
5. p^H = 7.5 to 8.5.

These factors along with many other factors play a vital role in the gonadial development and oviposition.

Induced breeding

Induced breeding is done by injecting required doses of gonadotropin hormones which stimulate the follicles and production of eggs as well as sperms also. One dose of this pituitary extract is enough for the male and two doses are necessary for a developed female to oviposit in desired space and time.

Management

The aim of fish management is to produce and make available surplus fish to the consumer at a cheaper rate.

The techniques of management include.

1. Control of fish mortality
2. Alteration of the physical habitat
   Eg : construction of suitable structures for the settlement of fingerlings, etc.
3. Control of bilogical habitat.

Production

The production of stocked fish mainly depends upon the suitability of species cultivated and the condition of pond where the cultivation is done. By means of correct choice of fish, artificial feeding, rate of stocking, maintenance as well as improvement of pond may control or increase the production.

Production is of three types ie., quantitative, qualitative and economic. Quantitative production mainly insists on the total weight of the fish harvested, irrespective of the grade and quality of fish. The main aim of qualitative production is producing high quality fish i.e of uniform size, weight and commercial value. The main aim of economic production is producing high quantity of fish of great market value.

Measures for increasing Fish Production

1. By means of control over the species cultivated
2. By means of control on the conditions the pond.
3. By limiting the pond.
4. Use of fertilizers
5. Use of artificial food and feeding.
6. By controlling weeds
7. By controlling fish enemies (Beetles, bugs, dragonflies, amphibians, birds, etc)

Fish Products and Bye-Products

When fish is sold in fresh condition, it fetches maximum price. Some times fish is preserved for long term as food for several reasons like non-availability of a ready market, lack of facilities for fast refrigerated transport. But there are yet other utilizations of fish in which protein, fat and other useful contents are processed into a number of valuable products and bye-products for a variety of usages. They are

1. *Fish liver oil :* Highly useful medicinally. It is obtained from the liver exclusively.
2. *Fish body oil :* It is the oil obtained from the entire body of fish. It may serve as a source of vitamin A and D for medicinal purpose.
3. *Fish meal :* Fish meal is ground cooked and dried preparation of the body of a fresh fish. It is an excellent poultry and animal food.
4. *Fish flour :* It is a hydrolyzed protein of muscle and other tissue. It is a superior quality of fish meal meant for human consumption.
5. *Fish manure and guano :* These are inferior quality of fish meal products. These are unfit as animal feed. The fish manure is particularly useful in raising tea, coffee and tobacco crops. The Nitrogen content of guano is very high, so it is several times more effective than any animal manure.
6. *Fish glue :* It is obtained from trimmings, bones and skins of fish. It is considered as a good adhesive.
7. *Isin glass :* It is a good substitute for gelatin. It's principal use is in clarification of beer, vinegar etc.

Fresh Water Fisheries of India

Fresh water area available for fishing industry in irrigation canals, rivers, rivulets etc., is around 2.5 million sq. km. whereas in reservoirs, lakes, tanks, fresh water ponds it is around 6.9 million hectares. Out of this total available area, only 1.6 lakh hectares of water is used for artificial culture.

By adopting modern methods of culture, proper management techniques and regularly monitoring the physico-chemical parameters high productivity can be achieved from the available waters.

Fresh water fishes cultured in India are Major Carps, Minor Carps, Cat fishes, Murrels, Exotic fishes and cold water trouts. These fishes are marketed in live state, hence they are also known as "Live fishes". They account for 15% country's total Inland fish market. They have high protein, high iron and low fat content.

In India, about 30 species of the fresh water prawn of the genus *Macrobrachium* are available. Among them, the giant fresh water prawn, *Macrobrachium rosenbergii*, *M. malcolmsonii* are the two most favoured species. Fresh water prawns are very popular and command good market price. They also have high export value.

Estuarine and Brackish Water Fisheries of India

Brackish water is a mixture of marine water and fresh water. Brackish water is found in estuaries, coastal lakes, tidal swamps, lagoons and back waters. Estuaries are the areas where rivers open into seas. In our coastal area, around 0.94 million hactares of brackish water resources are available for fisheries. Of this, only 2.91% of water is used for fisheries. Brackish water aquaculture received great attention during the recent years because of its high profits and export value. In addition to typical resident brackish water species, the fauna of brackish water includes marine and fresh water forms, which migrate to brackish water areas for purposes of feeding, breeding and larval development. Brackish water culture is of several types.

(a) *Tidal pond culture :* In this type of culture, ponds are constructed near the estuaries and are filled with water during high tidal action.

(b) *Pen culture :* Pens i.e specialy designed nylon or bamboo made enclosures are hung in water into which fish are released for culture.

(c) To culture mangrove crabs i.e. *Scylla serrata* bamboo cages are constructed in shallow waters.

(d) *Monoculture :* Culture of a single species is known as monoculture. In this type of culture, only one species of prawn or fish is stocked in the pond and reared till they attain marketable size.

(e) *Polyculture :* Culture of two or more species with compatible behaviour and feeding habits in the same pond is known as poly culture. In this type of culture, we must select suitable species that are non-predatory in feeding habit and should have a rapid growth rate, a capability to tolerate wide fluctuations in salinity and temperature, the ability to accept natural feed as well as supplimentary feed and a good market demand.

In brackish water aquaculture system the primary cultivated organims are prawns. Prawns are euryhaline forms i.e they can withstand the fluctuating salinity.

Out of 27 species of prawns that occur in the coastal waters of India, only 11 species are suitable for culture because of their attractive sizes, growth, high market price and other biological factors. The most preferred species for culture are the tiger prawn *Penaeus monodon* and white prawn *P.indicus*. Species of fin fish suitable for brackish water culture are Hilsa *Chanos chanos, Mugil cephalus* etc.

Brackish Water Prawn and Fish Farming

Suitable areas for brackish water aquaculture are mangroves, intertidal zones away from the estuaries, marshes, coastal swamps and tidal mudflats.

To decide the merits and demerits of brackish water pond site, it is essential to collect information about the assured source of water supply, tidal amplitude, duration of tidal influence, geographical conditions, climatic conditions, nature of soil and water. It is also important to collect information about the type and density of vegetation, fresh water run off, road and communication facilities and pollution free environment. It is essential to supply brackish water throughout the year to the ponds. This serves to replace water lost through seepage and evaporation and to exchange water from the confined pond.

Supply of water to the farm sould be through natural creeks or by pumping through electric and diesel pumps and wind mills. Important factors in the selection of site are soil type, its texture and properties. Sandy-clay soil or clay-loamy soil is good for prawn culture.

Brackish water farm will have stock pond and nursery ponds and they serve different purposes- stock pond is the production unit and nursery ponds are for seed rearing and production.

It is essential to eradicate the pests and predators, remove metabolites and excessive silt from pond before stocking juveniles in pond. To improve the natural productivity of the pond, it is necessary to apply fertilizers. Quality of water determines the growth, health and survival of prawn and fish stock under cultivation. So care should be taken to maintain the water quality.

The primary objectives of brackish water pond management are

1. Pond preparation to receive the stock for culture
2. Providing good quality water
3. Providing supplementary feed to the growing stock
4. Harvesting at appropriate time.

Marine Fisheries of India

Mariculture is the organised culture of marine organisms in sea water. Generally mariculture is conducted in the edge of sea, its inlets and in Bays. Bivalves are most suitable molluscs for culture. This is because they are almost sessile in habit, require no caging or impoundment to prevent escape. They mainly feed on phytoplankton and suspended particles of detritus through the filtering of seawater.

Molluscs which are cultivable in India are mussels, clams, edible oyster, window pane oysters, pearl oysters, etc. Molluscs are most ideal for culture since they give exceptionally high yield. Many species of fishes such as milk fish, perches and cels are also suitable for culture.

Mussel culture

The species used in our country for culture is the green mussel *Perna* (*Mytilus*) *viridis*. This has a high growth rate of about 11.3 cm in 8 months. It attains sexual maturity only when it attains 15.5 mm size. The annual yield from green and brown mussels is about 1,200 tonnes. The yield from rope culture in Kerala is about 200 tonnes per hectare per year. Spawning occurs from May to September. For mussel farming along the west coast, seed mussels are collected from natural beds from October to December. The spat are collected on tiles suspended from rafts or on ropes made up of either nylon or coir.

Clam culture

The most important bivalve resources of India are clams. *Meretrix meretrix, Meretrix costa, Katelysia opina, Katelysia marmorata, Villorita cyprinoides* are commercially important clams of India. These are abundant in estuaries, backwaters and bays along the coasts. Clam culture is mainly for lime, white and grey cement, calcium carbide bricks etc.

Oysters

Eight species of edible oysters are known to occur along our coasts, four species are abundant and are supporting fisheries of local importance.

They are

1. *Crassostrea gryphoides*
2. *C. madrasenses*
3. *C. discoides*
4. *C. cucullata.*

Oyster farming on a small scale has been carried out at Pulicat lake in Tamilnadu using *C. madrasensis* to Broach. Ribbon fish fishery is a major fishery of the coast of Tamilnadu. Sharks, rays and skates support a fishery of considerable importance throughout the year along the east and west coasts of India.

Window pane oyster, *Placuna placuna placenta* supports the local fisheries in the Gulf of Kutch and in the Kakinada Bay. Window pane oysters are used in the pharmaceutical industry and for preparation of lime.

Pinctada fucata is the pearl oyster of India. They are found in undisturbed areas of the sea, away from the coasts at a depth of 10 - 15 fathoms. All the pearl oysters are not able to form pearl, its rare occurance is due to its mode of formation. Rearing pearl oysters, seeding them with a nucleus and developing pearl is termed as Pearl culture. Pearl culture is done in natural climatic conditions in sea water under suitable depth and temperature.

Pearl formation is a rare and accidental phenomenon which takes about 2 - 6 years for attaining a good diameter. The longer the time the bigger would be the size of the pearl. Pearl is small about 10 mm in diameter, rounded, flawlessly white coloured, Pinkish, yellowish or bluish coloured calcium deposit. Pearls can't be produced in all the oceans. They are more common in Indian ocean, coast of China, Japan and Pakistan.

Sea weeds are identified as food because they contain more than 60 trace elements, minerals, proteins, iodine, bromine, vitamins and several other substances. Agar, agaroid and algin are manufactured from *Gracilaria, Sargassum, Laminaria, Ulva*, etc. which are used as gelling and thickening agents in food, pharmaceutical, dairy, textile, paper, paint and varnish industries. Sea weeds can be cultivated by means of vegetative propagation or by different kinds of spores such as zoospores, tetraspores, monospores and carpospores : In India Central Marine Fisheries Research Institute (CMFRI) and Central Salt and Marine Chemical Research Institute (CSMCRI) have undertaken experimental cultivation of sea weeds like *Gracilaria* and *Ulva* only.

Fish culture

Our country has North-West, South, East and North-East coastal line. It extends to 6,090 km in length. Water available for fishing is 5.8 million hectares. Oceans provide rich source of water for fishing. Oceanic area is divided into two zones. They are pelagic and benthic zones. Pelagic zone covers off shore fishing upto a depth of 200 meters. Oceanic zone covers off shore deep sea fishing upto a depth of 3000 - 4000 metres. There are about 1,865 fishing centres along the coastal line.

Principal fisheries of west coast and east coast are clupeids, sharks, rays, bombay duck, sardines, mackerels, cat fishes, flying fishes, perches, sharks, rays and skates.

Hilsa fishery is an important fishery of Chilka lake of the Hooghly Matlah estuary. Oil sardine fishery exists as a commercially important regular fishery all along the West coast. Kerala coast offers the richest waters for oil sardine fishery. Besides the sardine fishery, the Mackerel fishery is practiced. Bombay duck fishery and Ribbon fish fishery are fisheries of local importance. Bombay duck fishery is the local fishery of the Bombay coast from Ratnagiri.

3.15 Fouling Organisms

The exposure of hulls of ships and boats, pillars of piers buoys, underwater cables and equipment to the sea water results in the settlement of large number of organisms, called *foulers* or *fouling organisms*.

Many groups of animals and plants are recognised as foulers. Even though they are from entirely different groups, they show some common characteristics. They are chiefly intertidal and are mostly sedentary in the adult stage. These organisms together form a community, known as the fouling community. The most important members of the fouling community include algae, representatives from almost all invertebrate groups such as protozoa, porifera, tubeworms, coelenterates, polyzoa, annelida, crustacea (Arthropoda), Mollusca, the tunicates, bryozoa (either branching or flat or encrushing hydroids and encrusting sponges. (Fig. 3.89)

This succession starts with the primary film constituted by benthic diatoms, various filamentous algae and bacteria. The subsequent succession varies according to the various conditions prevailing in that area. Usually Bryozoa appear next, followed by mussels, whereas in certain other situations, hydroids, barnacles, tube worms attach at an early stage. Many other organisms join this community later to take advantage of the presence of the early settlers which provide shelter as well as food.

Those that settle in the beginning may not be represented at all in the final fouling community. The animals and plants that settle first may perish in large numbers, owing to the great struggle for space. Some, however succeed and form a pioneer community. This provides a new environment which may then shelter the young ones and other individuals. Thus, step by step, the fouling community develops in numbers and in the variety of organisms represented changing from one type of community to another.

There are some organisms in the marine environment that specifically target shellfish for food (predators) or as hosts (parasites). Parasites that can affect shellfish culture may be microcells (*Mikrocytos mackinii*) the *causative agent* of Denman disease and protozoa to flatworm and parasitic copepods. Predators have developed special adaptations to be able to penetrate the shells of their victims to feast on the meat inside or otherwise cause mortality. Crabs, starfish, moon snails are common shell fish predators.

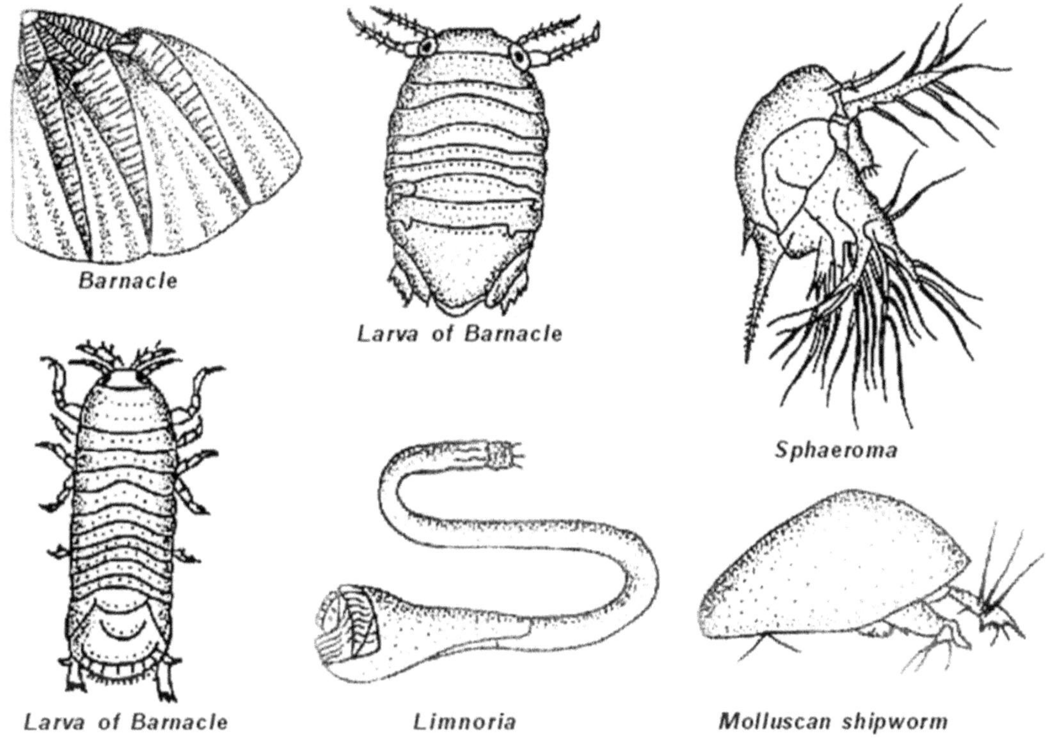

Fig. 3.89 Fouling Organisms

Nature of fouling

Among the fouling organisms (Fig. 3.89), the mussels (molluscs) are the most important foulers. They settle heavily especially in the pipe systems carrying sea water on board such as the fire mains and obstruct the flow of water. The constant flow of water is conducive to their large scale settling and rapid growth.

The larvae settle inside the pipes especially around the various isolating valves in the system, preventing their proper-functioning. This may some times lead to total failure of the water supply. Similarly, loose-mussel shells may sometimes enter condenser tubes, cooling systems etc., causing serious trouble.

Factors for the settlement of fouling organisms

Several factors influence the settlement of fouling organisms. These factors are primary film, illumination, colour of the substratum, the stage of the tide and seasons. The most important of these factors is recognised to be the formation of primary film. This is an essential requisite for many foulers since the settlement of the larvae of these organisms requires such a film.

(a) *Primary Film :* In the tropical waters, the primary film is composed chiefly of diatoms and algal spores with a relatively small proportion of bacteria. Even though the exact function of the film is not clearly understood, it seems that the filaments of algae may afford foothold on the substratum for the larvae. Further, the algae and diatoms probably serve as the food of these larvae.

(b) *Illumination and colour of the substratum :* Illumination at the site of attachment has a direct relation to the intessity of the settlement of these fouling organisms. The colour of the substrum also influences the settlement of larvae of fouling organisms. Among banacles, some species settle during day time on darkly coloured or shaded surfaces. These light preferences probably play an important part in leading the larvae to the different levels favoured by each species.

(c) *The stage of the tide :* The stage of the tide is believed to exercise an influence on attachment of certain species of fouling organisms. The rate of settlement in some species is highest during the low tide. In others an increasing tide i.e high tide favours the settlement while a third category is indifferent to the stage of the tide.

(d) *Seasons :* In tropical waters around India, fouling of underwater surfaces takes place at all times of the year. It is more pronounced during the warmer months in cold waters. Some of the fouling organisms show distinct seasonal maxima of frequency whereas in other seasons they are rare or entirely absent. These rhythms are mainly influenced by temperature and reproductive cycles of the concerned species.

Harmful effects of Fouling Organisms

Fouling organisms are of economic importance, as fouling of ships results in increased resistance to movement through water, reducing the efficiency, lowering of speed, increasing fuel consumption and leading to wear and tear of the engine.

The efficiency of underwater sound equipments fitted on to commercial and naval establishments is seriously affected as a result of accumulation of fouling organisms. Bacteria reduce the heat transfer efficiency of condensers. Foulers may jam the condenser tubes in the space between the turns of cooling coils, thereby causing corrosion.

Thus the problem of fouling organisms is really serious and therefore an understanding of the ecology of fouling is necessary for solving the problem of attachment of these organisms.

Prevention

Prevention of fouling routinely done by using copper sheathing in the case of wooden hulls of boats. A great disadvantage of this practice is that copper is subjected to corrosion when exposed to sea water. Later, attention was directed towards a suitable antifouling agent. Barnacles can be prevented from settling by painting the ships with Bis-N-Tributylin oxide (or) Neoprin commonly known as *Nofoul.*

Fouling on ships surface costs shipping industry approximately three billions US dollars a year. A significant cost is the increased fuel consumption needed to overcome hydrodynamic drag. Increased fuel consumption also contributes to pollution, global warming and acid rain.

Many anti fouling compounds used world wide are; Orgonotin antifoulants such as tributylin oxide (TBTO). These are effective in preventing fouling but have wide spread environmental problems.

Search is on for environmentally safe marine antifoulant which can be used as an alternative to Orgonotin. 3-Isothiazolone is a new compound chosen and work on its effectiveness is in progress.

Essay Questions

1. Write an essay on Numerical Taxonomy and its importance.
2. What is a Species ? Write in detail on different concepts of species.
3. Describe in Allopatric and Sympatric speciations.
4. Discuss the pests of Paddy or Cotton.
5. Discuss the pests of Sugarcane or Groundnut.
6. Write the biology of Alphids which attack different crops.
7. Write the biology of Jassids which attack different crops.
8. Write the nature of damage and biology of cotton boll worms.
9. Describe briefly the phylogeny of Invertebrate phyla.
10. Describe the phylogeny of Vertebrates.
11. Write the adaptations of Aquatic or Desert animals.
12. Describe Flight adaptations of animals.
13. Describe Biotic components of the ecosystem.
14. Describe the Abiotic components of the ecosystem.
15. Give an account of different kinds of biomes (ecosystems) you have studied.
16. Write an essay on aquatic ecosystems and its components.
17. Give an account of functional morphology and its significance.
18. Describe the Multivariate models in species composition.
19. Write an account of various environmental laws that are in force now.
20. What is remote sensing. Comment on its applications.
21. Write in detail on the fresh water fishes.
22. Write an account of shell fishes.
23. Give a detailed account on the preparation of a fish pond and its composition.
24. Write in detail on Fish products and Bye-products.
25. Write an essay on Fresh water fishes.
26. Write an essay on Brackish water fishes.
27. Write an account of Marine fishes.
28. Give a detailed account on the world Aquaculture.
29. Write an essay on aquaculture and add a note on its importance.

SHORT QUESTIONS

Explain the Following :

1. Speciation.
2. Alloptric population.
3. Sympatric population.
4. *Scirpophaga incertulas.*
5. Brown plant Hopper or Rice gall Midge
6. Green paddy leaf hopper.
7. Boll worms or Tobacco cut worm/White fly.
8. Sugarcane leaf hopper/White fly.
9. Termites or Aphids.
10. Trochozoon.
11. Origin and evolution in arthropods.
12. Similarities of Echinoderms with Chordates.
13. Agnatha
14. Present day reptiles.
15. Mammals.
16. Adaptations of Aquatic animals.
17. Cursorial/Desert/Cave/Fussorial/Arboreal
18. Flight adaptations of animals.
19. Pyramid of Energy/Biomass/Food chain.

18. Tundra
19. Taiga
20. Chapparel
21. Fauna of marine environment.
22. Body plan and Symmetry.
23. Measurement variables.
24. Derived variables.
25. The Environmental Protection Act 1986.
26. The Motor Vehicles Atc, 1988.
27. Wild Life Protection.
28. Major carps
29. Minor carps
30. Exotic fishes
31. Mullets
32. Prawn culture
33. Oyster culture
34. Composite.
35. Fish culture
36. Pearl culture
37. Mussel culture.
38. Fish breeding.
39. Fish Products and Bye- products.
40. Homology and analogy

4 Biology of Parasites and Immunology

4.0 Biology of Parasites

The animal species which thrives at the expense of another organism is called as a parasite. The organism which harbours the parasite is called a host.

There are different types based on the relationship that exists between the parasites and their hosts.

Mutualism

An association in which both the organisms derive benefit from one another.

Symbiosis

An association of dissimilar organisms in which both are so dependent upon each other that one can not live without the help of the other but without causing any harm. Such organisms are called symbionts.

Commensalism

An association of members of different species in which the parasite only derives benefit without causing injury to its host.

Parasitism

An association of two organisms, one living on another in which the parasite derives benefit from the host, and the lattar always suffers some injury.

The parasites can be classified as ectoparasites and endoparasites.

1. *Ectoparasites :* The parasites living on the surface of the body of the host.
2. *Endoparasites :* The parasites living inside the body of the host such as blood, tissues, body cavities, digestive tract and other organs.

Similarly the hosts can be classified into definitive hosts and intermediate hosts.

The definitive host

It harbours the adult stage of the parasite or where the parasite utilizes the sexual method of reproduction.

Intermediate host

In this, the host harbours the larval stages of the parasites. In some cases, larval stages are passed in two different intermediate hosts which are then referred to as first and second intermediate hosts respectively.

The parasitism can also be classified into following categories.

1. *Accidental :* Some free living animals are capable of leading a parasitic life for a certain period in a host, if swallowed accidentally.
2. *Facultative :* When the animals can live both as parasitic or free living, they are known as facultative parasites. Eg : Crab.
3. *Obligatory :* When the animals have lost the power of living a completely free life, and must live on or in some other suitable organism during all or part of their lives, without which they cannot live. Eg : Ascaris.

Adaptations

For leading a parasitic mode of life the organisms have acquired several adaptations. They are as follows :

1. *Thick cuticle :* All parasites (particularly endoparasites) are covered by a thick cuticle. It is formed of an insoluble protein. This protein can resist the digestive enzymes of the host.
2. *Organs of attachment :* Most of the parasites are provided with organs of attachment. These organs include hooks, suckers, rostellum, claws, adhesive secretions, etc.,
3. *Loss of sense organs :* The sense organs are helpful in understanding the environmental changes. As the parasite lives inside the host and there the environment is uniform, the sense organs are not necessary for the parasite. Thus, there is a loss of sense organs in the parasites.
4. *Loss of locomotory organs :* There are no locomotory organs since they are not needed, as the hosts transport the parasites along with them.
5. *Poor nervous system :* As the sense organs and locomotory organs are reduced, the nervous system is also poorly developed.
6. *Loss of alimentary canal :* Parasites feed on the digested food of the host. Hence, the alimentary canal is reduced or absent completely. Similarly the digestive glands are also absent completely.

7. *Antienzymes :* Intestinal parasites secrete certain enzymes called antienzymes to neutralize the digestive enzymes of the host.
8. *Anaerobic respiration :* Parasites live in an environment which is free from oxygen. Hence they have the ability to respire without oxygen.
9. *Complicated reproductive system :* There are many problems for a parasite to reach the adult stage. Therefore the parasites have a wide variety of adaptations for producing large number of eggs and young ones.

For this, they possess the following adaptations.

(a) Most of the parasites are hermaphrodites.

(b) The parasites have more than one set of reproductive system. For example, in *Taenia solium* there are 800 proglottides and each proglottis is provided with a set of male and female reproductive systems.

(c) The uterus is large and encloses a large number of eggs.

(d) The parasites have a tremendous capacity to produce thousands of eggs. For example *Ascaris* produces 20, 000 eggs at a time. Each proglottis of *Taenia* contains 30,000 to 50, 000 eggs.

(e) In some cases like *Schistosoma*, the females are permanently attached to the males.

10. *Parthenogenesis :* Some parasites exhibit parthenogenesis for rapid multiplication. For example the germ cells of Miracidium larva develop Rediae by parthenogenesis.
11. *Polyembryony :* The production of many embryos from a single germ cell is called polyembryony *eg : Taenia*. A single cysticercus larva produces many young *Taenia*.
12. Parasites find means of passing out fertilized eggs from the host into suitable places. The eggs have thick shells which are protective and withstand desiccation.
13. Most parasites have one or more intermediate hosts which act as transmitting agents to new definitive (final) hosts.
14. Parasites produce cysts which can be carried to different places by different methods such as wind, water or animals. In the same way, there may be motile larval stages for dispersal for eg : Miracidium larva, cercaria larva etc.
15. Certain parasites have more than one host to complete their life cycle. In the absence of one host, the parasite can survive on another host.

Effects of Parasitism

By parasitism the parasite multiplies in the host body and gradually kills it. Apart from this, it also causes diseases in the host. Along with these, the parasitism has other effects in different organisms.

1. *Plasmodium* parasite causes malaria in the host. Similarly *Ancylostoma* causes *Ancylostomiasis*.
2. In certain hymenopterous larvae, the diapause is averted when parasitised by certain insects.
3. The metamorphosis of Dipteran insect *Sciara* is influenced by a nematode parasite *Aproctonema*.

4. When a male crab is parasitised by *Sacculina* it changes its sex characters from maleness to femaleness.
5. The host growth is retarded because of the parasitism.
6. Because of the parasitism, the immunity of the host is reduced. This will lead to the secondary infections by bacteria and viruses.
7. Parasites also induce the production of antibodies and immunity in the host.

The parasites are commonly found in the *phyla* Protozoa, Helminthes and Arthropoda.

Table 4.1 *The protozoan parasites pathogenic to man.*

Class	Genus	Species	Habitat	Pathogenic effects
Rhizopoda	*Entamoeba*	*E. histolytica*	Large intestine	Amoebic dysentery, hepatitis, Liver Abscess
Mastigophora	*Giardia*	*G. intestinalis*	Small intestine	Diarrhoea
	Trichomonas	*T. vaginalis*	Vagina	Vaginitis
	Trypanosoma	*T. gambiense*	Blood, lymph node, C.N.S	African sleeping sickness
	Leishmania	*L. donovani*	R. E. system	Kala azar
Sporozoa	*Plasmodium*	*P. vivax* *P. falciparum* *P. malariae* *P. ovale*	RBC	Malaria
	Eimeria	*E. gubleri*	Liver	Hepatic coccidiosis
Ciliata	*Balantidium*	*B. coli*	Large Intestine	Dysentery.

Table 4.2 *The pathogenic helminths to man.*

Class	Genus	Species	Habitat	Pathogenic effects
Trematoda	*Schistosoma*	*S. haematobium*	Blood	Haematuria
		S. mansoni	Blood	Dysentery
		S. japonicum	Blood	Dysentery and cirrhosis of liver
	Gastrodiscoides	*G. hominis*	Large intestine	Mucous diarrhoea
	Watsonius	*W. watsoni*	Small intestine	Diarrhoea
	Fasciola	*F. hepatica*	Liver of sheep and man	Biliary colic
	Fasciolopsis	*F. buski*	Small intestine	Diarrhoea
	Clonorchis	*C. sinensis*	Liver	Jaundice
	Opisthorchis	*O. felineus*	Liver of cat, also man	Jaundice
	Heterophyes	*H. heterophyes*	Small intestine	Diarrhoea
	Metagonimus	*M. yokogawai*	Small intestine	Diarrhoea
	Paragonimus	*P. westermani*	Lungs	Haemoptysis.

Table 4.3 *The cestode parasites infecting man*

Genus	Species	Definitive host	Intermediate host	Stages found in man
Diphyllobothrium	*D. latum*	Man, dog, cat	Cyclops and fish	Adult worm
	D. mansoni	Dog, cat	Frog, snake, man	Larval stage
Taenia	*T. saginata*	Man	Cow	Adult worm
	T. solium	Man	Pig	Adult ; larval stage occasionally
Echinococcus	*E. granulosus*	Dog, wolf, jackal	Sheep, cattle, pig, man	Larval stage
	E. multilocularis	Fox, Dog, wolf	Field mouse, Tundra mole, man	Larval stage
Multiceps	*M. multiceps*	Dog	Sheep, goat, cattle, man	Larval stage.
Hymenolepis	*H. nana*	Man, rat, mice	Sheep, goat, cattle,	Adult and larval stages in intestine
	H. diminuta	Rat, man	Rat, flea	Adult worm

Table 4.4 *The nematode parasites infecting man.*

Common name	Scientific name	Site of infection
1. Trichina worm,	*Trichinella spiralis*	Small intestine and muscles
2. Whip worm	*Trichiuris trichiura*	Caecum, appendix and colon.
3. Old world hook worm	*Ancylostoma duodenale*	Intestine
4. Common Round worm	*Ascaris lumbricoides*	Small intestine
5. Bancrofts filaria	*Wuchereria bancrofti*	Lymphatic vessels, lymph nodes
6. African eye worm	*Loa loa*	Subcutaneous connective tissue of eyes
7. Guinea fire worm	*Dracunculus medinensis*	Subcutaneous tissue
8. –	*Srongyloides stercoralis*	*Small intestine*
9. New world hook worm	*Necator americanus*	intestine
10. Thread worm or pin worm or seat worm.	*Enterobius vermicularis (Oxyuris) Brugia malayi*	Large intestine, caecum and appendix
11. Blinding filaria	*Onchocerca volvulus*	*Subcutaneous connective tissues*
12. Ozzards filaria	*Mansonella ozzardi*	*Mesentery*

Table 4.5 The arthropod parasites pathogenic to man or vectors of disease causing organisms to man

Common name	Scientific name	Site of infection
Body louse	*Pediculus humanus humanus*	Skin reactions to bites, vectors to Rickettsiae and spirochaetes
Head louse	*P.humanus capitis*	Skin reactions to bites
Crab louse	*Pthirus pubis*	Skin reactions to bites
Bed bug	*Cimex lectularius*	Skin reactions to bites
Kissing bug	*Triatoma infestans*	Skin reactions to bites
Cone-nosed bug	*Rhodnius prolixus*	Vectors of *Trypanosoma cruzi*, the cause of chagas disease
Sand fly	*Phlebotomus* spp.	Skin reactions to bites, vectors of *Leishmania*, pappataci fever and carrions disease
Black fly	*Simulium* spp.	Serious biting, skin reaction to bites, vectors of *Onchocerca* and *Mansonella*
Mosquito	*Anopheles* sp. *Culex* sp. *Aedes* sp.	Serious biting pests, skin reactions to bites, vectors of viruses, protozoa and filaria
Horsefly	*Tabanus* sp.	Biting pests, skin reactions to bites
Deer fly	*Chrysops* sp.	biting pests, vectors of Tuleremia and loa loa
House fly	*Musca domestica*	Mechanical transmitter of pathogens
Stable fly	*Stomoxys calcitrans*	Serious biting pest
Tsetse fly	*Glossina palpalis*	Vectors of Trypanosomes of man and animals
Rat flea	*Xenopsylla cheopis*	Vector of bubonic plague
Cat flea	*Ctenocephalides felis*	biting pest
Dog flea	*C. canis*	biting pest
Human flea	*Pulex irritans*	biting pest, plague vector
Scorpions	*Palamnaeus* sp, *Buthus* sp, *scorpio* sp, etc.	painful bites followed by systemic reactions
Spiders	*Latrodectus mactans, Laxoceles reclusa, etc.*	bite usually painless. Initial blister at wound site
Soft ticks	Various genera and species	Skin reactions to bite, tick paralysis, vectors of relapsing fever
Hard ticks	Various genera and species	Skin reactions to bites, tick paralysis, vectors of Rickettsia viruses, bacteria and protozoa
Mites	Various genera and species	Occasional dermatitis from bites.
Human Itch mite	*Sarcoptes scabiei*	Burrows in skin causing severe itching
House mouse mite.	*Allodermanyssus sanguineus*	Vector of *Rickettsia akari*, the cause of Rickettsial pox
Hair follicle mite	*Demodex folliculorum*	Biting reactions

The biology of some of the important parasites is described in detail.

4.1 Biology of Protozoan Parasites

The following is the account of the biology of *Entamoeba histolytica, Trypanosoma gambiense, Giardia intestinalis, Leishmania donovani, Plasmodium.* sp. and *Balantidium coli.*

1. *Entamoeba histolytica*

Phylum : Protozoa

Class : Rhizopoda

Order : Amoebida

Genus : *Entamoeba*

Species : *histolytica*

This protozoan parasite lives in the mucous and submucous layers of the large intestine of man. It causes diarrhoea, dysentery, hepatitis and liver abscess in man.

Entamoeba histolytica is monogenetic i.e, its life cycle is completed in only one host. It exists in two stages 1. Trophozoite or Magna form and 2. Precystic or Minuta form.

1. *Trophozoite :* It is active, motile and feeding stage. It measures 20 - 30μ in diameter. It resembles amoeba in all structural details. It is covered by plasmalemma. The cytoplasm is divisible into outer clear ectoplasm and inner granular endoplasm. The endoplasm contains a vesicular nucleus and food vacuoles containing RBC. Contractile vacuoles are absent. The pseudopodia are short, broad and rounded. During movement, a single lobe like pseudopodium or lobopodium is seen. Trophozoite is pathogenic.
2. *Precystic form* or *Minuta form :* It is nonmotile, non feeding and round form. Pseudopodia and food vacuoles are absent. It is a non-pathogenic and non reproducing stage.

Life History

Entamoeba histolytica completes its life cycle in a single host i.e., man (Fig. 4.1).

The magna form or Trophozoite multiplies asexually by binary fission. Most of the daughter individuals become Trophozoites again. But some of them remain smaller in size and are known as minuta forms or precystic forms.

The minuta form enters into the lumen of the colon. Its nucleus divides twice mitotically to form four nuclei. It develops a cyst wall around. This stage is called *Tetra nucleate cyst.* It is the infective stage of man. The cyst goes out along with faecal matter. The infection occurs by contaminated food and water. The *tetra nucleate* cysts reach the intestine of man. In the small intestine of man excystment takes place. The enzyme trypsin dissolves the cyst liberating four nucleated amoeba called *metacyst.* This metacyst divides and produces *eight amoebulae.* These amoebulae attack colon walls and transform into trophozoites. They increase in their number and cause the disease to man.

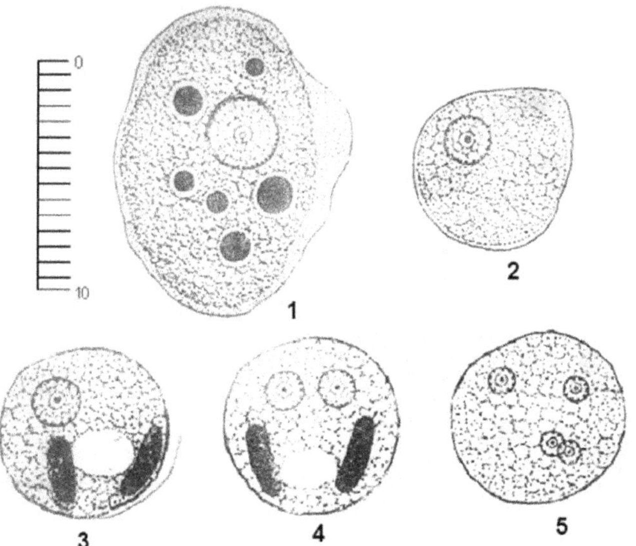

Stages of Entamoeba histolytica.
1. trophozoite; 2. pre-cystic stage; 3. 4. 5. uninucleate, binucleate and quadrinucleate cysts.

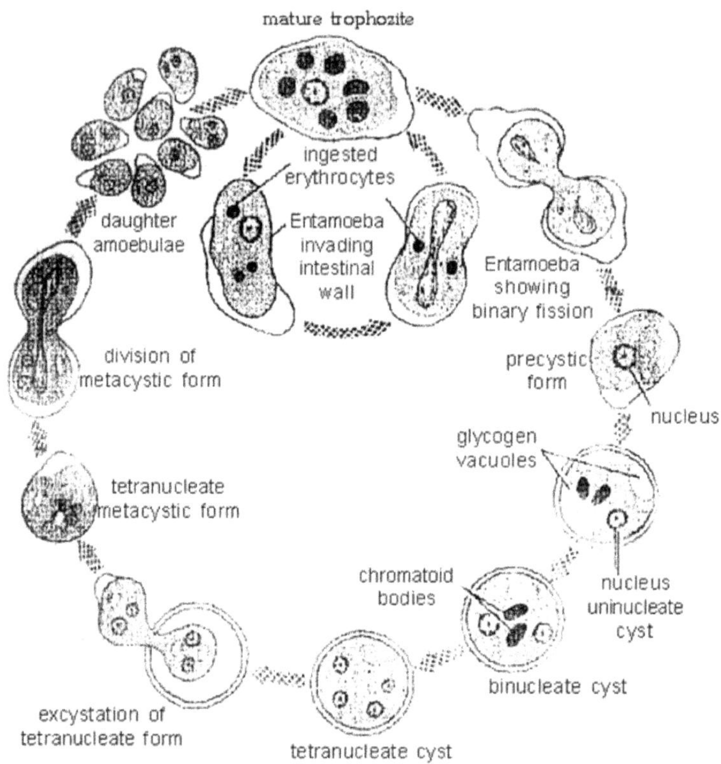

Fig. 4.1 Life cycle of *Entamoeba histolytica*

2. *Trypanosoma gambiense*

Phylum : *Protozoa*

Class : *Mastigophora*

Order : *Protomonadina*

Genus : *Trypanosoma*

Species : *gambiense*

This protozoan parasite lives in the blood and cerebrospinal fluid of man. It causes the disease Trypanosomiasis or African sleeping sickness in the affected person. It is a digenetic parasite and completes its life history in two hosts 1. Man and 2. Tse-tse fly.

Structure

The adult parasite is long and fusiform i.e., narrow and pointed at both ends. The organism is 20-40μ long and 2-10μ wide. The body is covered by a thin firm pellicle that gives fixed shape to the parasite. The cytoplasm consists of a large vesicular nucleus, green volutin granules or reserve food particles (glycogen bodies) and a kinetoplast in the posterior side. The kinetoplast comprises a basal granule or blepharoplast or kinetosome, a para basal body and a connecting filament rhizoplast. A flagellum arises from the basal granule, runs along the margin of the body and becomes free at the anterior end. Pellicle extends between the body and the flagellum and forms undulating membrane. Food vacuoles and contractile vacuoles are absent.

Life History

Trypanosoma completes its life cycle in two hosts i.e, man and Tse-tse fly (Fig. 4.2).

1. ***Life history in man :*** Infected tse tse fly bites a man and injects a few parasites in the blood of man. These are known as *metacyclic forms* which are the infective stages of the parasite. In the blood of man these parasites transform into long and slender forms. They multiply through longitudinal binary fission and produce three types of forms i.e., (i) long slender forms with a free flagellum, (ii) short and stumpy forms with a reduced flagellum and (iii) intermediate forms.

   These forms enter into cerebrospinal fluid of the nervous system. When the parasites are in the blood they cause gambia fever and when they are in the cerebrospinal fluid they cause sleeping sickness.

   It has been observed that the parasites periodically increase and decrease in number in the blood of man. During the period of decrease, the short and stumpy forms, which have great resistant power, survive and the rest die.

These short and stumpy forms are capable of development in tse tse fly.

Life cycle in Tse-tse fly

When a tse tse fly sucks the blood of an infected man, a number of parasites enter into the midgut of the fly along with the blood. These parasites remain in the midgut of the fly for a few days and start multiplying by longitudinal binary fission. After 10th to 15th day, long, slender forms appear

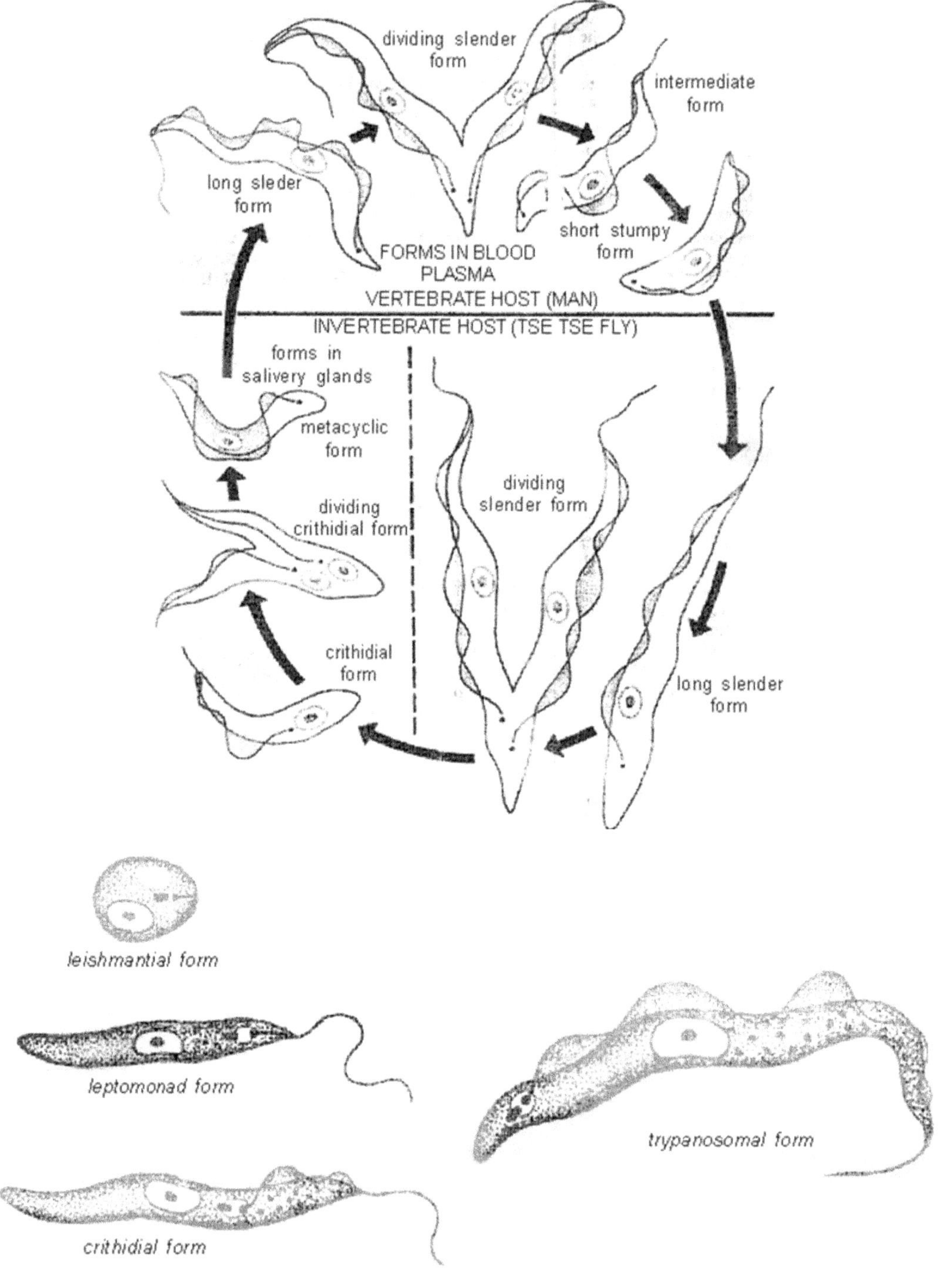

Fig. 4.2 Life cycle of *Trypanosoma gambiense*

in great numbers which move forward to the foregut. Afterwards, they reach the salivary glands. In the salivary glands, they transform into crithidial forms. These forms are characterized by a shorter flagellum and undulating membrane. They increase in their number by longitudinal binary fission and transform into short, stumpy individuals called metacyclic forms. They are the infective stage of man.

Trypanosoma takes about 20-30 days to complete its life cycle in Tse-tse fly.

3. *Giardia intestinalis*

Phylum : *Protozoa*

Class : *Mastigophora*

Order : *Diplomonadina*

Genus : *Giardia*

Species : *intestinalis*

This parasite occurs in the duodenum and the upper part of the ileum of man. It causes Giardiasis or diarrhoea in man (Fig. 4.3).

This parasite exists in two forms. 1. Trophozoite and 2. Cystic form.

1. *Trophozoite :* The trophozoite is pear shaped i.e., round and broad on the anterior side and pointed on the other side. The dorsal surface is convex and the ventral surface is concave with a sucking disc. The *sucking disc*, serves for the attachment of organisms. The size of the trophozoite is 14μ long and 7μ broad. The trophozoite possesses 8 flagellae and 2 nuclei. These flagellae are arranged in 4 pairs i.e, anterior, lateral, ventral and posterior. It also contains 2 rigid structures called blepharoplasts which occupy the entire length of the organism. The nuclei lie in the area of the sucking disc in the anterior part of the body.
2. *Cystic form :* The cyst is oval and measures 12μ long and 7μ in width. It has four nuclei, attenuated flagella and a blepharoplast. The cyst represents the resting stage of the organism. Its rigid outer wall protects the organism against unfavourable conditions. The cyst is the infective stage of the organism.

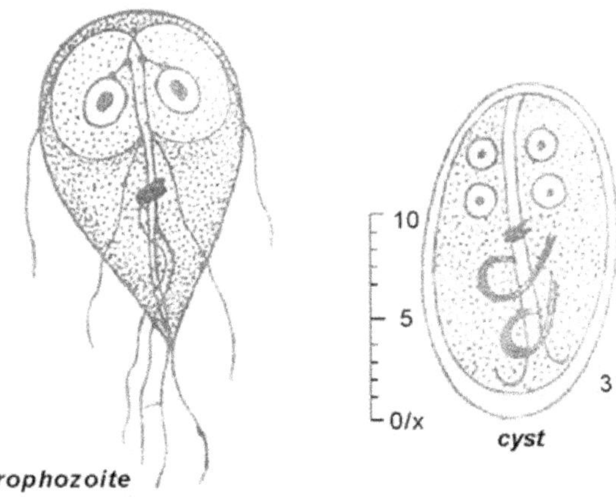

Fig. 4.3 *Giardia intestinalis*

Life Cycle

In the trophozoite stage, the parasite multiplies in the intestine of man by binary fission. When conditions in the duodenum are unfavourable, encystment occurs, usually in the large intestine. During encystment, a thick resistant wall is secreted by the parasite and the cell then divides twice within the cyst. Infection of man is brought about by ingestion of cysts through contamination of water and food. Within 30 minutes of ingestion, the cyst hatches out into two trophozoites, which then multiply in enormous numbers and colonise in the duodenum.

4. *Leishmania donovani*

Phylum : *Protozoa*
Class : *Mastigophora*
Order : *Protomonadina*
Genus : *Leishmania*
Species : *donovani*

This parasite causes visceral leishmaniasis or kala-azar. It is an intracellular parasite occurring in the reticulo endothelial system of liver, spleen and bone marrow.

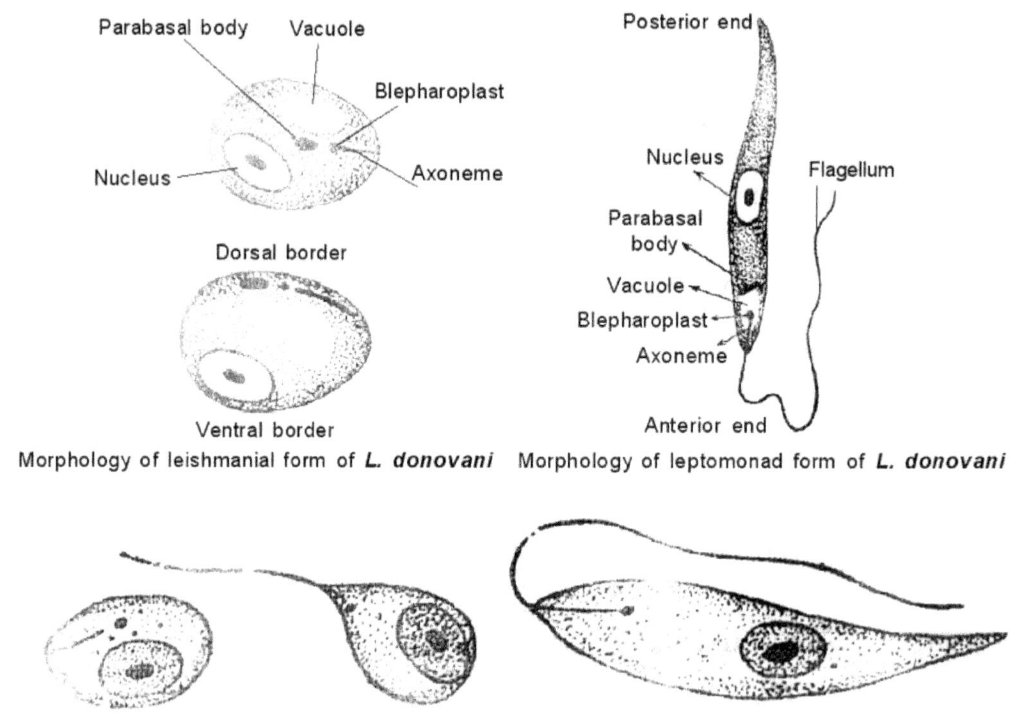

Morphology of leishmanial form of *L. donovani* Morphology of leptomonad form of *L. donovani*

Leishmania donovani, which causes Kala-azar in Asia. Left-Stage in man. Middle and right-stages in Phlebotomus (sand-fly)

Fig. 4.4 *Leishmania donovani*

Leishmania is dimorphic occurring in two forms.

1. Leptomonad form
2. Leishmanial form.

1. *Leptomonad form* or *flagellar form :* The body is long and slender and is 14-20μ in length. The pellicle covers the body. Cytoplasm shows vesicular nucleus. Towards the anterior end, it has a parabasal body, a rhizoplast and a basal granule. The flagellum arises from the basal granule at the antenior end. Contractile vacuole and food vacuoles are absent. This form appears in sandfly, the intermediate or secondary host.
2. *Leishmanial form or Aflagellar stage :* It is small, round or oval in shape and measures 2-4μ in diameter. The body is covered by a pellicle. A central vesicular nucleus is present in the cytoplasm. At the anterior end, the parabasal body, rhizoplast and a basal granule are present and the *flagellum* is absent. It is the adult stage and is seen in the primary host i.e, man.

Life History

The life history of *Leishmania donovani* is completed in two hosts :

1. Man, the primary host, and
2. Sand fly, the secondary host.

When an infected sand fly (*Phlebotomus* sp) bites a man, leptomonad forms (infective stage) enter the blood. Leptomonad forms enter into the cells of WBC, reticuloendothelial (R.E) cells of liver, spleen and bone marrow and transform into leishmanial forms. Thus, the parasite becomes intracellular. Gradually the number is increased by longitudinal binary fissions. It causes the disease kala-azar in man. When a sand fly sucks the blood of infected man, leishmanial forms enter the gut. There, they change into leptomonad forms and increase their number. Now they are ready for further transmission. The cycle is repeated in this way. (Fig. 4.4).

5. *Plasmodium*

Phylum	:	*Protozoa*
Class	:	*Sporozoa*
Order	:	*Haemosporidia*
Genus	:	*Plasmodium.*

It is the malarial parasite of man. There are 4 species that cause malaria to man. They are *Plasmodium vivax, P. malariae, P. falciparum* and *P. ovale. Plasmodium vivax* is studied as an example. The life cycle is completed in two hosts and hence it is known as digenetic parasite. Man is the primary host and the female Anopheles is the secondary host.

Life cycle in man (Primary host) (Asexual cycle)

P. vivax completes its life cycle in man in 3 stages.

1. Pre-erythrocytic cycle
2. Erythrocytic cycle
3. Exoerythrocytic cycle.

1. *Inoculation :* When an infected female Anopheles mosquito bites a man to suck his blood, then along with its saliva, it injects the Sporozoite stage of plasmodium into the human blood. The Sporozoite, the infective stage measures 11-12μ is length and 0.5-1μ in width. It is Sickle shaped with an oval nucleus. Within 30 minutes, they reach the liver cells where they multiply.

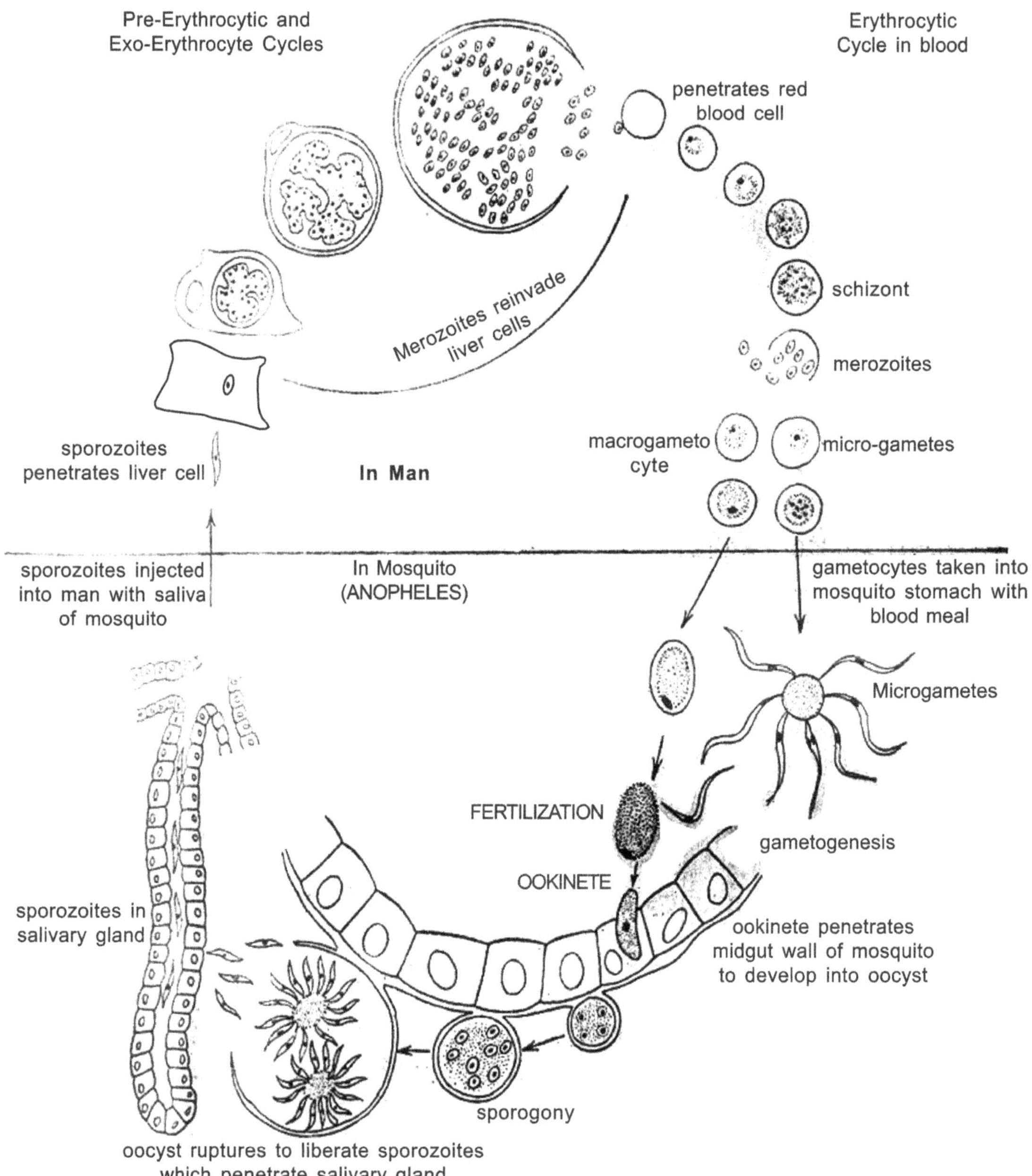

Fig. 4.5 Life cycle of *Plasmodium vivax* in man and mosquito

Pre-erythrocytic Schizogony in liver cells

In the liver cells the sporozoite grows to form a large round *schizont*. The schizont divides by multiple fission to form thousands of small spindle shaped cells called *Merozoites*. This multiple fission is called *Schizogony*. This phase of asexual multiplication is *pre-erythrocytic* Schizogony and the merozoites produced by it are called cryptozoites or cryptomerozoites.

Exoerythrocytic cycle

Cryptomerozoites attack fresh liver cells and start *exoerythrocytic cycle*. Cryptomerozoite attacks a liver cell and forms metacryptozoite or phanerozoite. It undergoes schizogony (multiple fission) to form nearly thousand metacryptomerozoites. The metacryptomerozoites are of 2 types. Smaller ones are micrometacryptozoites, which are many in number. They attack RBC. Bigger ones are called macrometacryptozoites. They are few in number and attack fresh liver cells. They survive in the liver for several years. Thus, liver works as a *storage organ of plasmodium in man*.

Erythrocytic cycle

The micrometacryptozoites, attack RBC and become Trophozoites. The trophozoite is feeding and growing stage. Inside the parasite, a non-contractile vacuole develops. Nucleus is pushed to one side. It is called signet ring stage. Then, the vacuole disappears and the parasite becomes irregular in shape. It is called *amoeboid stage*. It feeds on haemoglobin and digests globin part of the haemoglobin. Haematin part becomes haemozoin or melanin. The parasite becomes a Schizont. RBC shows Schuffner's dots at this stage on the surface. Schizont undergoes erythrocytic Schizogony (multiple fission) or merogony. Nearly 6-24 merozoites are formed. The merozoites are arranged like the petals of a rose. Hence it is called *Rosette stage*. Then RBC breaks, merozoites are liberated into the blood. Along with merozoites, haemozoin granules are also liberated. *Haemozoin granules cause malaria fever to man*. The liberated merozoites attack fresh blood cells and repeat erythrocytic cycle.

The duration between entry of sporozoites of plasmodium into man and the first appearance of malaria fever to man is called incubation period. It is about 14 days in *P. vivax*.

Some merozoites attack RBC and transform into sexual forms called gamonts or gametocytes. They take 96 hours to grow fully. They reach superficial blood vessels. The gametocytes are of 2 types. Smaller microgametocytes with bigger nucleus and larger macrogametocytes with smaller nucleus. They do not develop further in the human body because of high temperature. For further development they need the secondary host, the female *Anopheles* mosquito.

Life cycle in mosquito (secondary host) (sexual cycle)

When a female mosquito bites and sucks the blood of infected person, the gametocytes enter into the stomach where the RBC are dissolved and the gemetocytes are set free.

Gametogony

The microgametocytes after release in the stomach of mosquito, undergo the process of exflagellation. The nucleus of the microgametocytes divides into 6-8 haploid daughter nuclei. These nuclei migrate towards the periphery of microgametocyte. The cytoplasm pushes out forming a long flagellum like structure having one daughter nucleus in each. Thus, 6 - 8 flagellum like male gametes or microgametes are formed. Soon, these gametes separate and move actively in the stomach of mosquito. The *macrogametocyte* undergoes maturation process, thereby two polar bodies are pushed out and a female gamete or *macrogamete* is formed. The macrogamete is non motile and develops *cytoplasmic* or *receptive cone.*

Fertilization

The microgametes reach the macrogamete and only one will fertilize it and forms zygote. The fusion of male and female gametes is called *Syngamy*. As the gamates are of different sizes (i.e, anisogametes), the fusion is called *Anisogamy.*

The zygote remains rounded and motionless for 24 hours but soon elongates to form worm like and motile. Now the zygote is called *ookinete* or *vermicule.* The ookinete moves and bores through the wall of the stomach of mosquito and comes to lie beneath the outer epithelial layer. Then, it becomes spherical and secretes a thin elastic membranous cyst around it. Then it is called *Oocyst.*

Sporogony

The nucleus of oocyst first divides by meiosis and then by mitosis several times and its cytoplasm develops vacuoles forming faintly outlined cells called *Sporoblasts*. Then the cytoplasm forms slender spindle shaped haploid cells known as sporozoites. Each oocyst may have upto ten thousand sporozoites and groups of sporozoites get arranged around the vacuoles. This type of multiplication is called *Sporogony*. The oocyst bursts and the sporozoites are liberated into the haemolymph of the mosquito, from where they reach its salivary glands and enter the duct of hypopharynx. The sporozoites will infect a human host when the mosquito bites and the life cycle is repeated again. (Fig. 4.5).

6. *Balantidium coli*

Phylum : *Protozoa*
Class : *Ciliophora*
Order : *Heterotrichida*
Genus : *Balantidium.*
Species : *coli*

This parasite causes dysentery in man called ciliate dysentery or balantidiasis. It is the largest protozoan parasite inhabiting the large intestine of man. It is also found in pigs and monkeys. The pig is the common reservoir of infection.

The parasite exists in two stages :

1. Trophozoite stage, and 2. Encysted stage.

1. *Trophozoite :* It is oval in shape and measures 60-70μ in length and 40-50μ in width. The body is covered with a delicate pellicle showing longitudinal striations. The cilia are short and delicate and are of uniform length but those that line the mouth or cytostome appear to be longer and are called *adoral cilia*. Underneath the pellicle, a thin layer of ectoplasm and an inner granular endoplasm are seen. A large kidney shaped macronucleus and a small round micronucleus are seen in the cytoplasm. Two contractile vacuoles and many food vacuoles are also seen in the cytoplasm. At the anterior end is situated a groove known as peristome leading to a mouth (cytostome) terminating in a short funnel shaped gullet or cytopharynx extending upto one third of the body.

   At the posterior end is a permanent anus called the *cytopyge.*

2. *Cyst :* It is slightly smaller than the trophic form, measuring 50-60μ in diameter. The cytoplasm is granular and contains a macronucleus, a micronucleus and a refractile body. The contractile vacuoles may remain active for some time during encystment. The cyst is surrounded by a thick transparent double layered wall.

Life Cycle

It passes its life cycle in two stages but in one host only. The pig is the natural host and the man is the rare and incidental host. The cyst is the infective stage of the parasite (Fig. 4.6).

The entry of the parasite is through ingestion, via mouth.

Transmission occurs from pig to pig or pig to man, man to man or man to pig.

The cyst when ingested liberates trophozoites in the large intestine and from each cyst, a single individual is formed. The trophozoite may remain in the lumen or enter into the submucous coat of the large intestine where it grows and multiplies by *binary transverse fission*. With each division,

(a) (b)

20

10

0μ

Morphology of Balantidium coli.
(a), trophozoite stage; (b), cystic stage

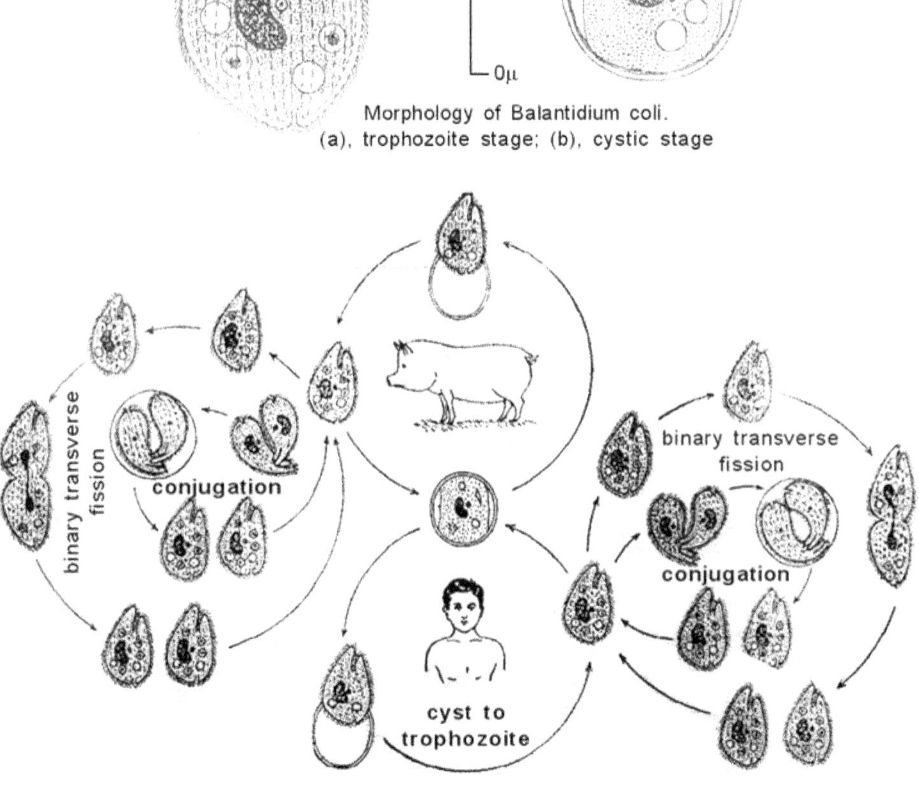

Fig. 4.6 Life cycle of *Balantidium coli* in pig and man

two daughter ciliates are produced and by successive divisions, large numbers of trophozoites are formed. First the micronucleus divides into 2, then the macronucleus and finally the body splits into two by a transverse partition. After a period of growth and multiplication, the trophozoite undergoes encystment particularly when it finds conditions unsuitable for vegetative existence.

Conjugation

During this process, the two trophozoites are enclosed in a cyst. An exchange of nuclear materials occurs, after which the individuals separate.

Because of the Balantidium infection, ulceration of colon occurs. Also, the passage of blood and mucus in the stool are noticed.

4.2 Biology of Helminth Parasites

The following is the account of the biology of some important helminth parasites like *Schistosoma haematobium, Fasciola hepatica, Paragonimus westermani, Diphyllobothrium latum, Taenia solium, Echinococcus granulosus, Hymenolepis nana, Trichinella spiralis, Ancylostoma duodenale, Wuchereria bancrofti, Ascaris lumbricoides, Dracunculus medinensis and Enterobius vermicularis.*

1. *Schistosoma haematobium*

Phylum : Platyhelminthes
Class : Trematoda
Order : Digenia
Genus : Schistosoma
Species : haematobium

It is a dioecious digenetic trematode being parasitic in the hapatic portal or pelvic veins of man. The common name of the parasite is “The Vesicular blood fluke”. The adult worms live in copula, in the pelvic venous plexus, vesical, prostatic and uterine plexuses of veins.

In schistosomes, the males and females are separate, but these two are found in pairs. The male is broad and flat having a gynacophoric canal formed by the folding of the ventral body wall. The long, slender female is lodged in the gynacophoric canal. Both the worms have oral sucker and ventral sucker or acetabulum.

Life Cycle

Schistosoma haematobium passes its life cycle in 2 hosts. Man is the definitive host and fresh water snails such as *Bulinus, Physopsis* or *Planorbis* being the intermediate hosts (Fig. 4.7).

The female held in the gynacophoric canal of the male deposits the eggs longitudinally, one at a time in the smallest venules one behind the other egg. In this way, the venules are filled with a number of eggs. The eggs are held in position by the spines and they slowly pass through the vessels and the mucosa of the urinary bladder and escape through the urine to the exterior. They enter water and the eggs hatch into free swimming ciliated larvae called *miracidia*. The miracidium searches for the intermediate host and on entering its proper larval host, penetrates into the soft tissues of the snail and ultimately makes its way into the liver. Here it loses its cilia and other organs

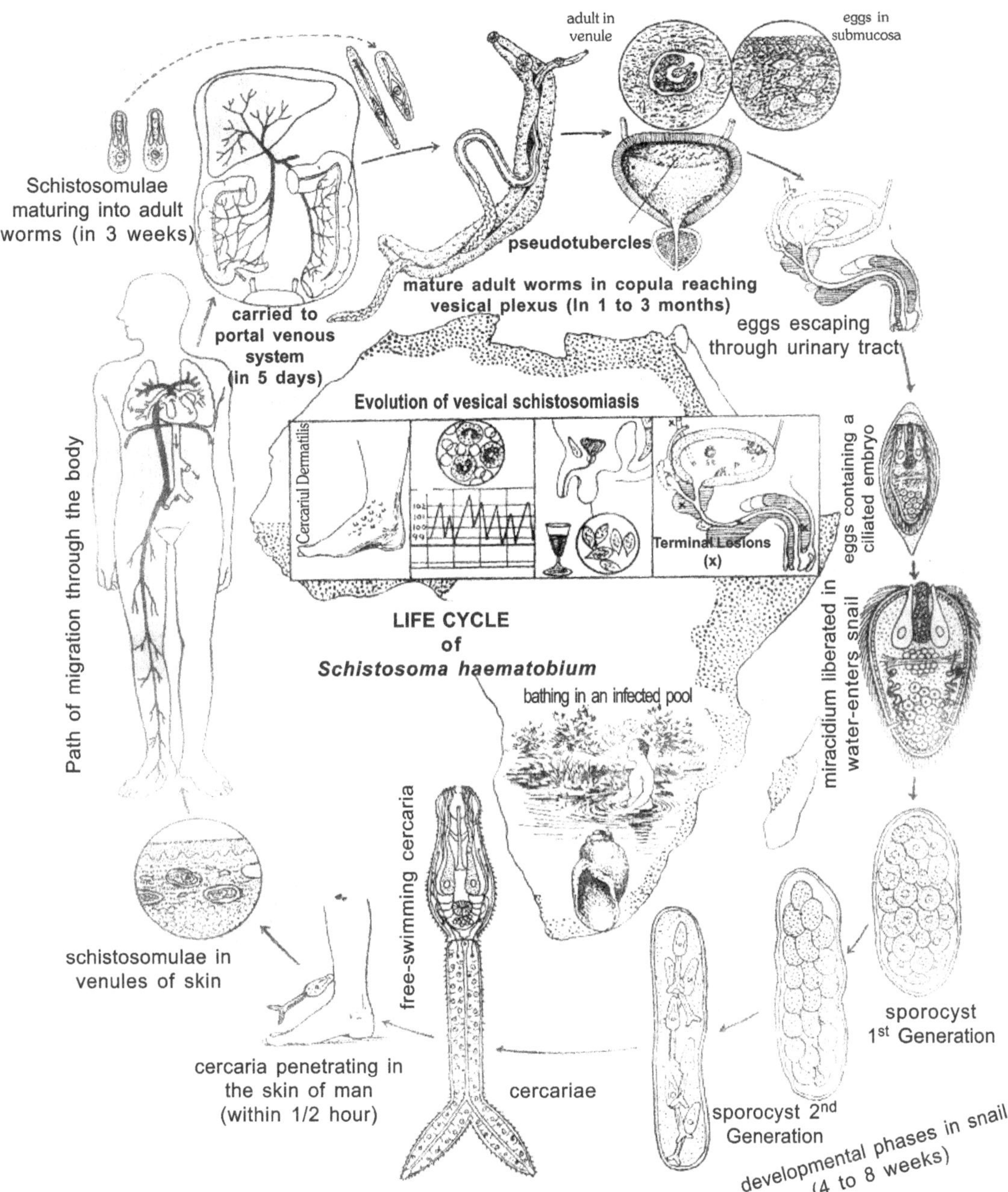

Fig. 4.7 *Schistosoma haematobium*

and is transformed into a tubular *Sporocyst* in the course of 4 - 8 weeks. It multiplies and forms second generation of sporocysts. Several weeks after the infection, when no further multiplication occurs, the sporocysts give rise to the *fork tailed cercariae*. These are the infective stage to the man. The cercariae break off from the sporocyst and escape from the snail into the water.

Infection results when human beings bathe or wash in this water. The cercariae penetrate the skin directly. On entry, the Cercariae cast off their tails and gain access to a peripheral venule. These forms called *Schistosomulae* are carried through the right heart into the pulmonary capillaries. From lungs they are carried through left heart into systemic circulation. The majority are pushed to the mesenteric artery, then through the capillary bed in the intestine to the portal circulation. There they become adults. After becoming sexually differentiated, they move out of the liver, migrate into the inferior mesenteric vein, rectal plexus, pelvic plexus and eventually enter the vesical plexus of veins. It takes about 1-3 months for the worms to reach to these plexuses from the initial infection. As the worms are sexually mature, they copulate (the females are enclosed in the males) and the fertilized females lay eggs which are ultimately released with the urine into the water. The cycle is thus repeated. (Fig. 4.7).

2. *Fasciola hepatica*

Phylum : *Platyhelminthes*

Class : *Trematoda*

Order : *Digenia*

Genus : *Fasciola*

Species : *hepatica*

It is commonly called the *sheep liver fluke* or *common liver fluke*. It is a parasite of herbivorous animals like sheep, goats and cattle living in the liver. It is occasionally found in man. These are the definitive hosts. Snails of the genus *Limnaea* are intermediate hosts. Liver flukes cause the disease liver rot in their definitive hosts.

The adult worm is a large leaf shaped fluke measuring 3 cm in length and 1.5 cm in width. It is brown to pale grey in colour and is provided with 2 suckers, the oral sucker and the ventral sucker. The oral sucker is situated at the anterior end forming a conical projection. The ventral sucker or acetabulum is situated midventrally about 3 - 4 mm behind the oral sucker. There are two permanent apertures on the body in addition to oral aperture or mouth. They are the gonopore and the excretory pore. The gonopore or genital aperture is situated midventrally in front of the ventral sucker. The excretory pore is situated at the posterior end of the body.

Life Cycle

Fasciola hepatia passes its life cycle in two different hosts. The definitive hosts are sheep, goat, cattle or man and the intermediate host is *Limnaea spp*. The eggs are passed out in the faeces of the definitive host and mature in water. A ciliated miracidium develops inside the egg in 2 - 3 weeks time. On escaping from the egg, the miracidium finds its way to its suitable intermediate host. Inside the lymph spaces of the molluscan host, the miracidium passes through the stages of sporocyst, two generations of rediae and finally to the stage of cercariae. The whole cycle takes a period of 30-60 days. The mature cercariae escape from the snail into the water, lose their tail, and encyst and are then called as metacercariae. They are lodged in the leaf blades of grass. These are swallowed along with the grass by herbivorous animals and occasionally by man. On entering the digestive tract, metacercariae lose their cyst in the duodenum, migrate through the intestinal wall into the peritoneal cavity, penetrate the capsule of liver, traverse its parenchyma and ultimately

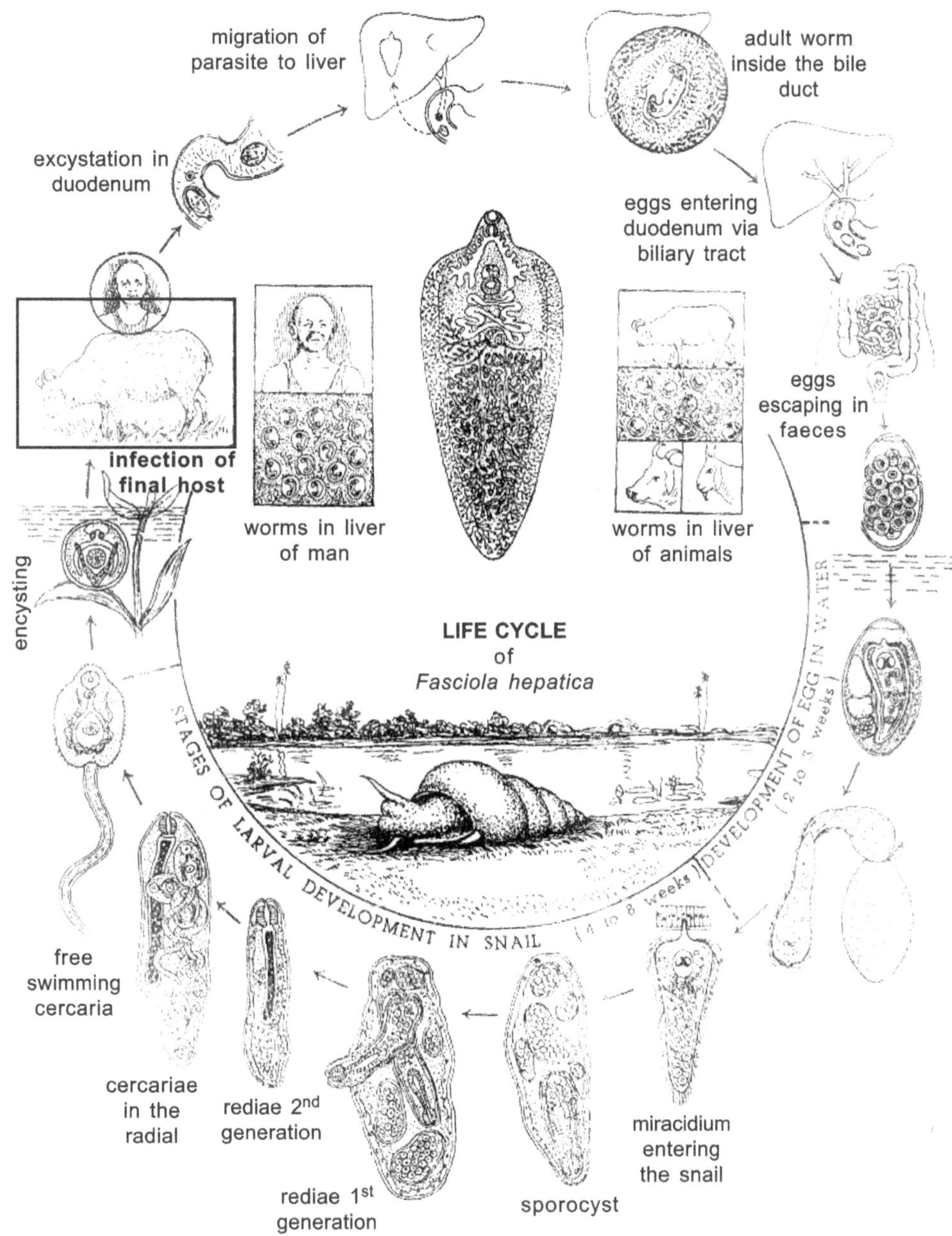

Fig. 4.8 *Fasciola hepatica*

settle in the biliary passages and grow to sexual maturity. It takes about a month to migrate from initial infection. Though the adults are hermaphrodites, cross fertilization is of common occurrence. Hence before fertilization, copulation occurs. When cross fertilization does not occur, self fertilization takes place and eggs are liberated in the faeces through the bile in about 3 - 4 months after infection. The cycle is thus repeated. (Fig. 4.8)

3. *Paragonimus westermani*

Phylum	:	*Platyhelminthes*
Class	:	*Trematoda*
Order	:	*Digenia*
Genus	:	*Paragonimus*
Species	:	*westermani*

The common name is Oriental lung fluke or the lung distome. The adult worms live in the respiratory tract or lungs of man and the infection with *P. westermani* is known as *Paragonimiasis*.

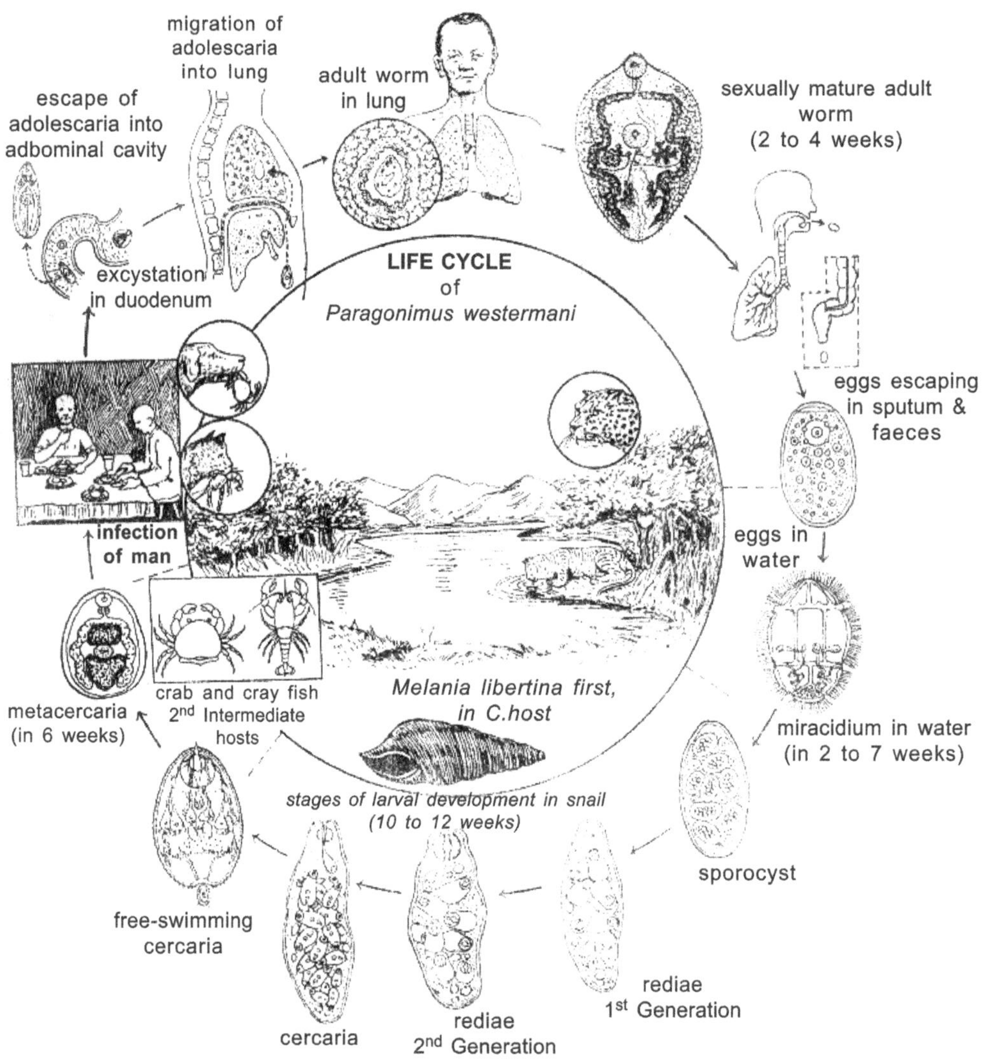

Fig. 4.9 *Paragonimus westermani*

Morphology

The adult worm is thick, fleshy and egg (oval) shaped. Its anterior end is slightly broader than the posterior end. It measures 8-12mm in length and 4-6mm in width. The ventral sucker is situated near the middle of the body. The life span of the adult worm is about 6-7 years.

Life cycle

P. westermani passes its life cycle in 3 hosts. One definitive host and two intermediate hosts. The definitive host is man or domestic animals. The first intermediate host is the fresh water snail, *Melania* and second intermediate host is a fresh water cray fish or a crab (Fig. 4.9).

The adult worms live in the respiratory tract of definitive hosts. The eggs generally escape through sputum and some are eliminated in the faeces. In water, the ciliated miracidium develops inside the egg in 2-7 weeks time and is liberated into water. It swims about in search of its snail host, *Melania* spp. Inside the soft tissue of the snail the miracidium casts off its cilia and passes through the stage of sporocyst and two generations of *rediae* and then finally transformed into cercariae. The whole process takes about 10-12 weeks. The mature *cercariae* escape from the snail into water and enter into its second intermediate host, a fresh water crab or a cray fish. Inside the crustacean host they become encysted in the viscera, muscles and gills.

When the raw flesh of the infected crab or cray fish is eaten by man or other host, the cyst wall is dissolved by the gastric juice and the worms are released in the duodenum. From there, they reach small intestine, penetrate its wall and enter the abdominal cavity. Later, they migrate upwards, piercing through thc diaphragm and the 2 layers of pleura, to gain entrance into the lungs where they finally settle and grow to sexual maturity. The adult worms copulate and the female releases the eggs which are discharged into a bronchiole and are coughed out with the sputum. The cycle is thus repeated.

4. *Diphyllobothrium latum*

Phylum	:	*Platyhelminthes*
Class	:	*Cestoda*
Order	:	*Pseudophyllidea*
Genus	:	*Diphyllobothrium*
Species	:	*latum*

It is commonly known as the fish tapeworm or the broad tapeworm. The infection of *D. latum* to man in known as Diphyllobothriasis. The symptoms are gastro-intestinal disturbances and anaemia. The adult worm lives in the small intestine (ileum) of man and also in dog, cat, fox and other fish eating mammals.

Morphology

The adult worm is yellowish grey in colour with dark central markings. It measures 3-10 metres in length. The individual worm may live for a period of 5-13 years.

The body of the worm is divided into the head or scolex, neck and proglottides or segments. The *head* or *scolex* is elongated and spoon shaped and measures 2-3mm in length and 1mm in width. It bears two slit like grooves or bothria on the dorsal and ventral surfaces respectively. There are no *rostellum* and hooklets.

The neck is thin and unsegmented and is much longer than the head.

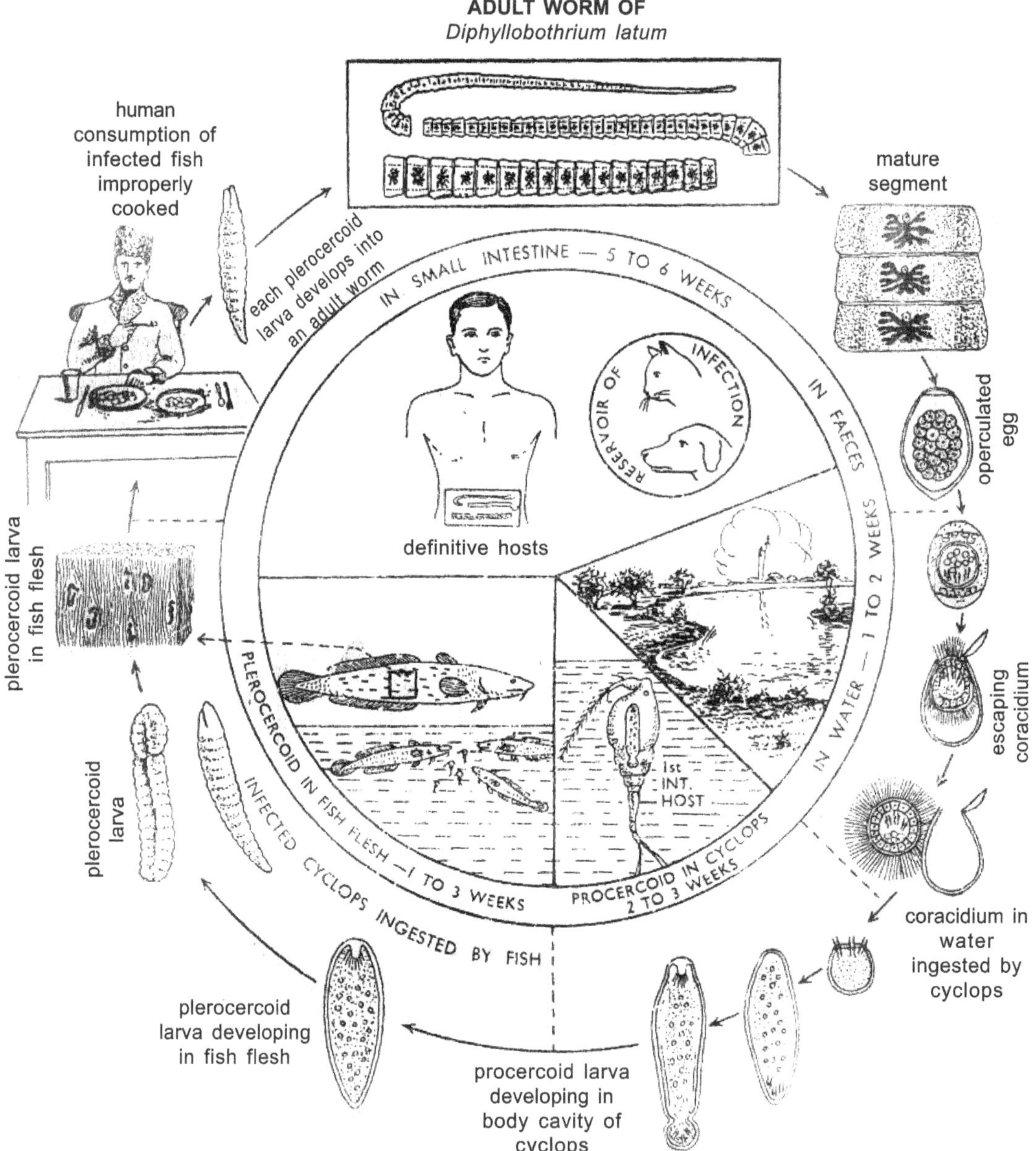

Fig. 4.10 Life cycle of *Diphyllobothrium latum*

The proglottides or segments are 3000-4000 in number. The segments are broader than length. For example, mature segment measures 2-4mm in length and 10-20 mm in width and is practically filled with male and female reproductive organs. Eggs are discharged through uterine pore. The posterior dried up segments break off from the body and are expelled through faeces.

Life Cycle

The worm passes its life cycle in one definitive host and two intermediate hosts. The definitive hosts being the man, dog and cat. The adult worms are found in the small intestine. The first intermediate host is a fresh water crustacean, a cyclops or a diaptomus and the second intermediate host is a fresh water fish, i.e., pike, trout, salmon, perch or other fish. (Fig. 4.10)

The eggs are liberated through the faeces of definitive hosts in water. A spherical ciliated embryo containing three pairs of hooklets, called Coracidium, after an incubation period of 1-2 weeks is released into the water and is then ingested by Cyclops.

Inside the intestine of cyclops, the coracidium loses its cilia and penetrates through the intestinal wall and settles in the body cavity. It is transformed into Procercoid larva with an elongated body. The cyclops containing the procercoid larvae is in turn fed by the second intermediate host, a fresh water fish.

In the intestine of fish the procercoid larva after freeing itself, passes through the gut wall and settles in the liver, muscles or fat tissue in the mesentery. In 1-3 weeks, the procercoid larva changes to a Plerocercoid larva or Sparganum. The body of the larva is white, somewhat flattened and marked by irregular unsegmented wrinkles. The plerocercoid larva is infective to man. It is not destroyed by insufficient cooking. When man consumes such fish, he is infected. Inside the intestine of man, the plerocercoid larva develops into an adult worm and after having attained sexual maturity in about 5-6 weeks, starts discharging eggs which are passed along with the faeces. The cycle is thus repeated.

5. *Taenia solium*

Phylum : *Platyhelminthes*
Class : *Cestoda*
Order : *Cyclophyllidea*
Genus : *Taenia*
Species : *solium*

It is commonly known as the *Pork Tapeworm* or *the armed tapeworm of man*. The adult worm lives in the small intestine of man. Because of *Taenia* infection, abdominal discomfort, chronic indigestion, anaemia and intestinal disorders such as diarrhoea alternating with constipation etc. may be noticed.

Morphology

The adult worm measures about 2-3 meters in length. The body is divided into the head or scolex, neck and the proglottides or segments.

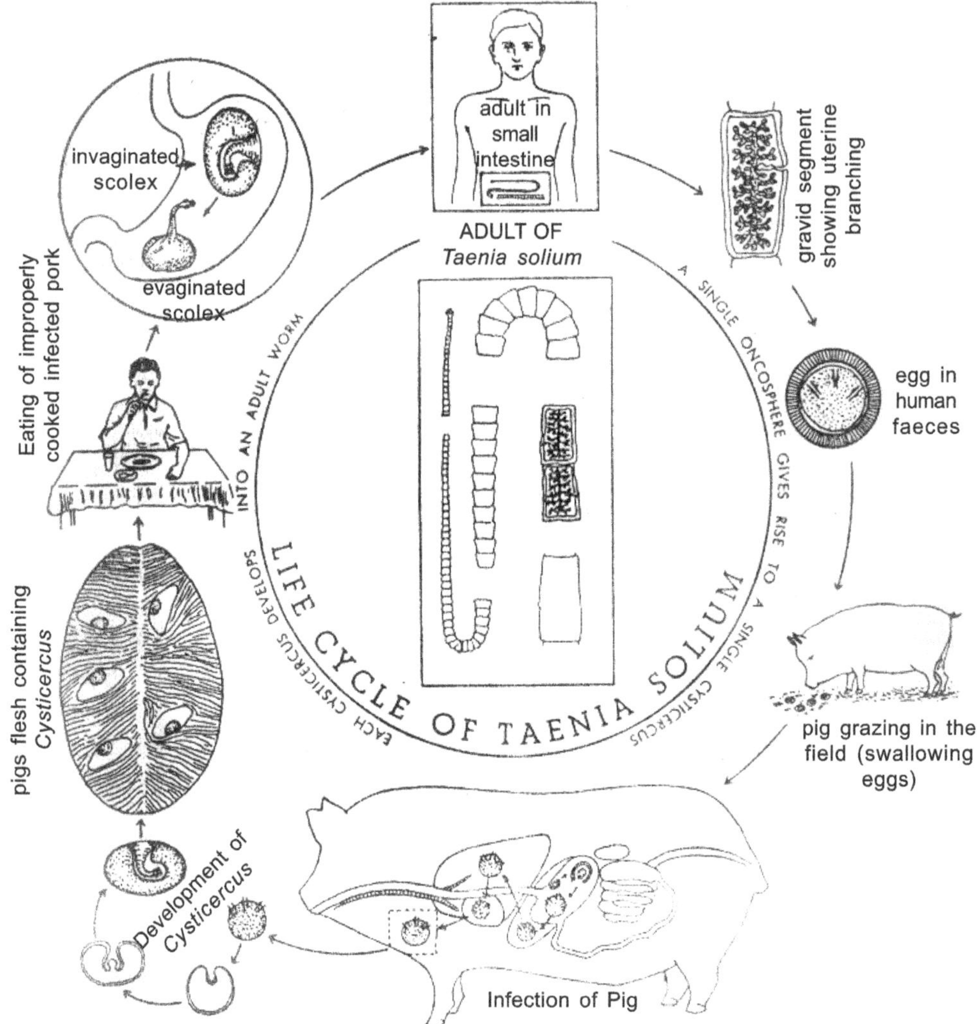

Fig. 4.11 *Taenia solium*

The head or scolex is about the size of a pinhead measuring about 1mm in diameter and has 4 circular suckers. The head is provided with a rostellum armed with a double row of alternating large and small hooklets.

The neck is short and measures 5-10mm in length.

The proglottides or segments are about 800-900 in number. A mature proglottid measures about 12mm in width and 6mm in length. The gravid segments are expelled in chains of 5-6 at a time. The worm has a life span of about 25 years.

Life cycle

The life cycle is passed in 2 hosts. The definitive host is the man and the intermediate host is the pig (Fig. 4.11).

The adult worm lives in the small intestine of man. The eggs or gravid segments are passed out with the faeces on the ground. The pigs eat the faeces along with the eggs. On reaching the alimentary canal, Onchospheres are liberated from the eggs. The Onchosphere is a 6 hooked (hexacanth) embryo (larva). These penetrate the gut wall with the aid of their hooks and gain entrance into the portal vessels and from there into the liver, right side of the heart, lungs, the left side of the heart and the systemic circulation. The Onchospheres, then reach the muscular tissues where they ultimately settle down and undergo further development. The Onchospheres lose their hooks and each forms an oval vesicle gradually increasing in size, containing at its bottom and the larva is known as Cysticercus. It takes about 60-70 days for the oncospheres to metamorphose into the cysticercus stage. Human beings are infected through the eating of undercooked pork containing cysticerci. Inside the alimentary canal of man, the scolex, on coming in contact with the bile, is everted and anchors to the gut wall by means of its suckers and develops into an adult worm by gradual strobilisation. The worm grows to sexual maturity in 2-3 months and starts producing eggs. The eggs are passed out along with the faeces, thereby repeating the cycle.

6. *Echinococcus granulosus*

Phylum : *Platyhelminthes*

Class : *Cestoda*

Order : *Cyclophyllidea*

Genus : *Echinococcus*

Species : *granulosus*

It is commonly known as the dog tapeworm or the Hydated worm. Man harbours the larval form and the adult worm is found in the small intestine of dog and other canine animals. The larval form in man causes Hydatid disease.

Morphology

The adult worm is a small tapeworm measuring 3-6mm in length. The body is divided into a head or scolex, neck and 3-4 segmented strobila. Of these segments, the first is immature, the second one is mature and the rest (3rd or 3rd and 4th) is gravid. The terminal segment is the biggest of all, measuring 2-3mm in length and 0.6mm in width. The head bears 4 suckers and protrusible rostellum with two circular rows of hooks. The neck is short and thick.

The egg is ovoid in shape and measures 32-36μ in length and 25-32μ in width. The hexacanth larva is with three pairs of hooks. The egg is infective to man, cattle, sheep and other herbivorous animals.

Life Cycle

The worm passes its life cycle in two hosts. The definitive host is dog, and also wolf, fox and jackal. The intermediate hosts are sheep, pig, cattle, horse, goat and man (Fig. 4.12).

The eggs are discharged with the faeces of the definitive host (i.e, dog). These are infected to the intermediate host i.e, man due to intimate handling of infected dogs.

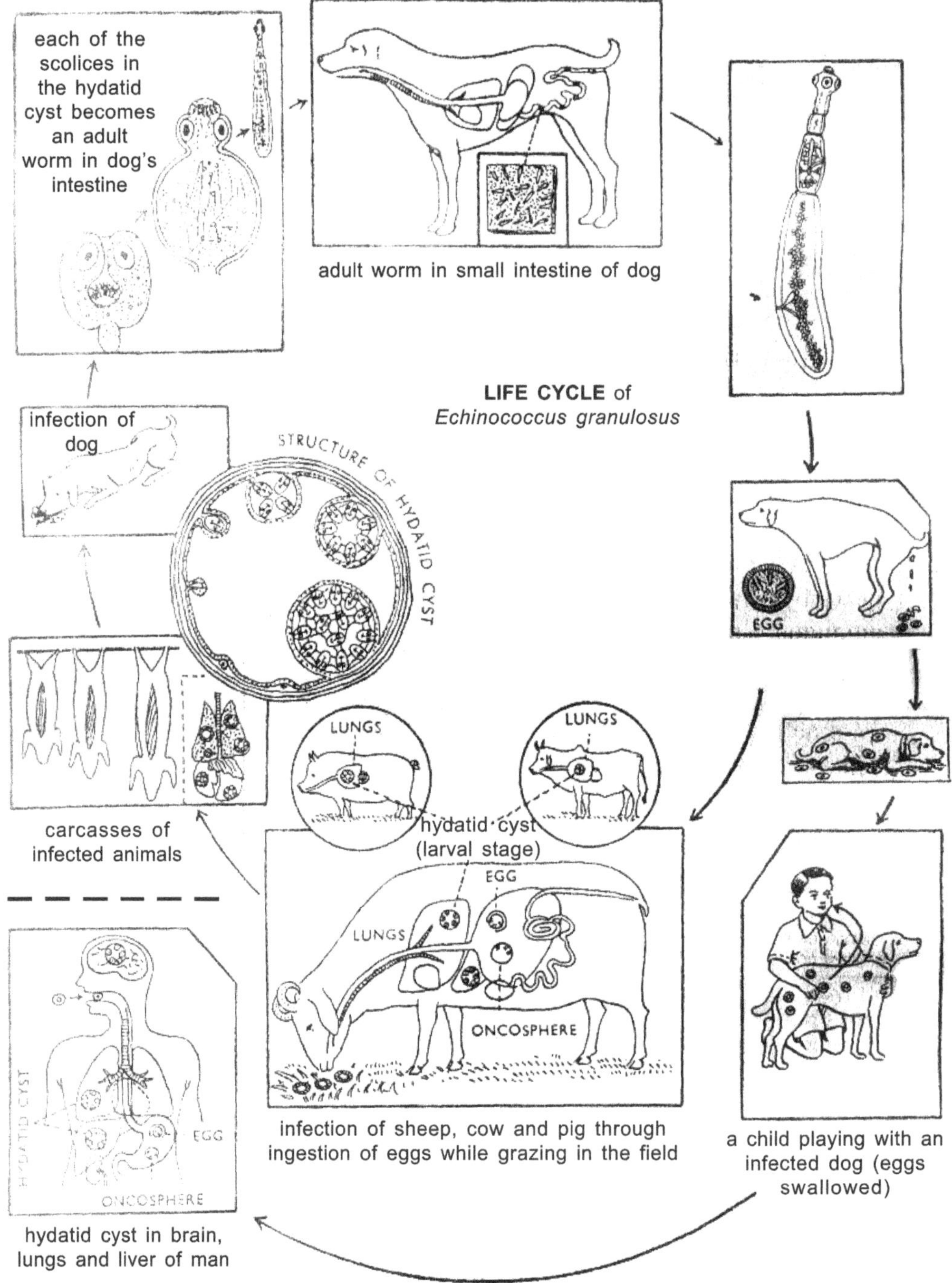

Fig. 4.12 *Echinococcus granulosus*

In the duodenum, the hexacanth embryos are hatched out. About 8 hours after ingestion, the embryos bore their way through the intestinal wall and enter the portal vein and thereby to the liver. The liver acts as the first filter. Some of the embryos may pass through the hepatic capillaries and enter the pulmonary circulation and enter the lung. Lungs act as the second filter. A few of the embryos may enter the general blood stream and are lodged in various organs. Thus the hexacanth embryos may be seen in the liver, lung or other organ. Wherever the embryo settles, it forms a hydatid cyst, the young larva being transformed into a hollow bladder. From the inner side of the cyst, brood capsules with a number of scolices are developed. A fully developed scolex is an end product and its presence inside the hydatid cyst is a sign of a complete biological development.

As the dogs have no access to the hydatid cyst developed in the viscera of man, the life cycle of the parasite comes to a dead end. The natural cycle is maintained by dog and sheep. The hydatid cysts from the viscera of sheep when ingested by the dog are capable of growing into adult worms in about 6-7 weeks in the intestine. Thus the cycle is repeated. The life span of the adult is about 5 months but the larval form lives for many years.

7. *Hymenolepis nana*

Phylum : *Platyhelminthes*

Class : *Cestoda*

Order : *Cyclophyllidea*

Genus : *Hymenolepis*

Species : *nana*

It is commonly called as the Dwarf Tape worm. The adult worm lives in the small intestine of man. It is also found in rodents i.e, mice and rats. In heavy infection, it causes abdominal pain and diarrhoea in man.

Morphology

The adult worm is small, thread like, measuring 1-4cm in length. The adult worm is short lived i.e, about 2 weeks. The body is divided into the head or scolex, neck and the segments.

The head is globular, has 4 suckers and provided with a short retractile rostellum armed with a single row of hooklets. The rostellum remains invaginated into the organ. The neck is long. The number of segments is about 200. A mature segment measures 0.3mm in length and 0.9mm in width. The eggs are liberated in the faeces by gradual distintegration of the gravid terminal segments.

Life cycle

No intermediate host is required and the entire development from the larval to the adult stage takes place in one host (Fig. 4.13).

With the ingestion of fully embryonated egg, a hexacanth embryo is liberated. It burrows into the villi of the anterior part of the small intestine and develops in about 4 days into a typical larval stage called *Cysticercoid*. After reaching maturity, the villus ruptures and the larva (Scolex) reenters the lumen of the small intestine. Later, it attaches to another villus further down and in the course of a fortnight develops into an adult worm. Strobilisation is rapid in about 30 days after the infection, and the eggs are observed in the faeces. In this way the cycle is repeated.

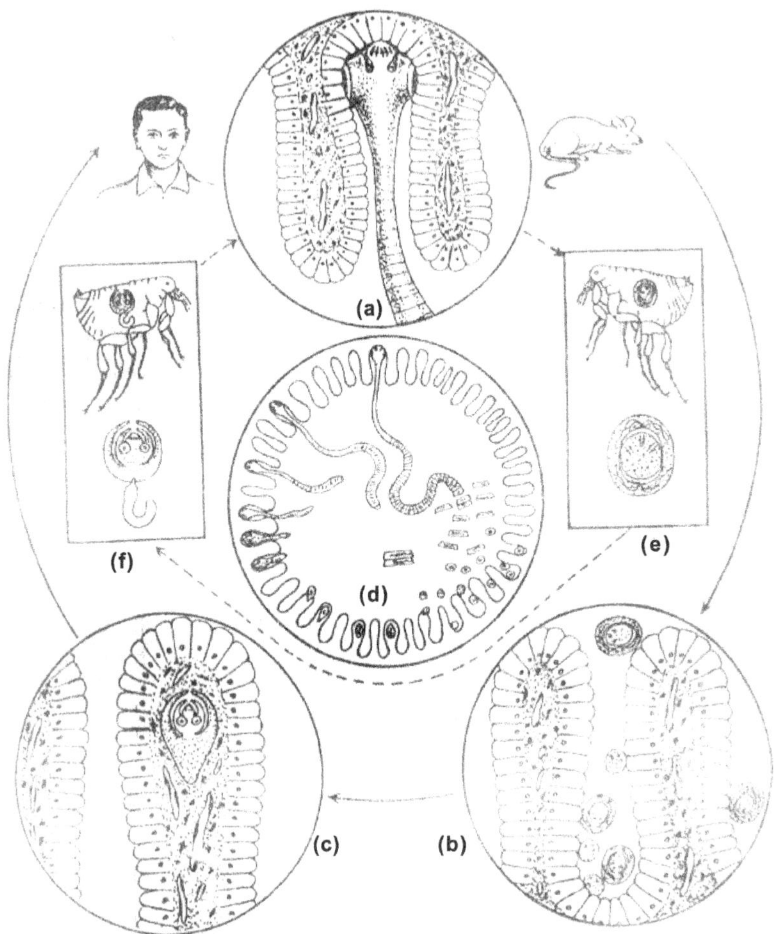

Fig. 4.13 Life cycle of *Hymenolepis nana.*

(a), (b), (c), (d) direct cycle; (a), (e), (f) indirect cycle. (a) head of the adult worm attached to the mucosa of the small intestine; (b) oncosphere penetrating intestinal villi; (c) cysticercoid in an intestinal villus; (d) stages of the evolution in the same host. (e) eggs in human faeces ingested by the rat flea; (f) cysticercoid in the haemal cavity of the rat flea (ingestion of these fleas causes infection).

8. *Trichinella spiralis*

Phylum : *Nemathelminthes*

Class : *Nematoda*

Order : *Enoplida*

Genus : *Trichinella*

Species : *spiralis*

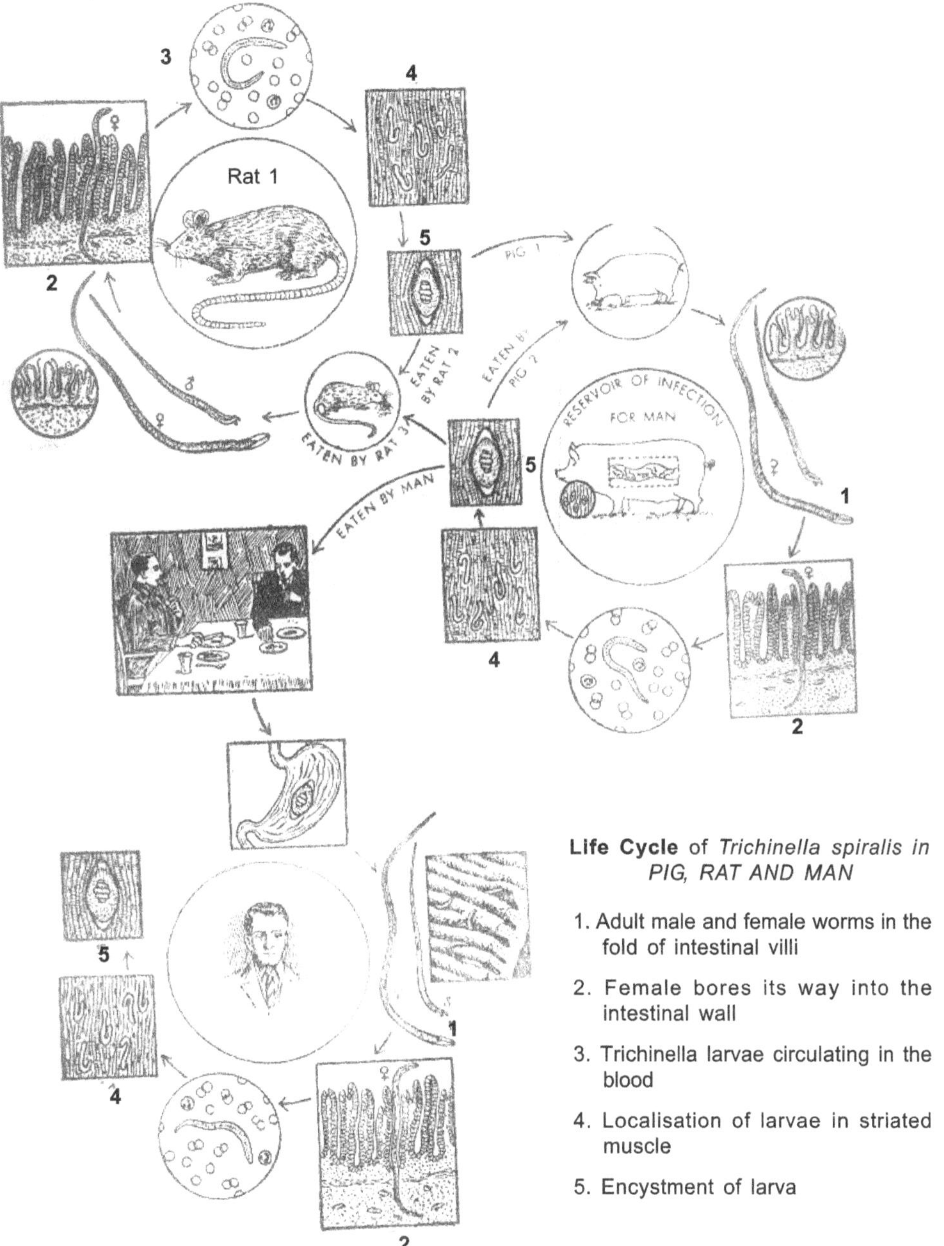

Fig. 4.14 *Trichinella spiralis*

It is commonly called as the trichina worm. The infection in man occurs through the consumption of infected pig's flesh, raw or insufficiently cooked, containing the viable larvae. In man it remains buried in the duodenal or jejunal mucosa and also in the striated muscles. Apart from man, pig and rat also are the hosts.

Morphology

The adult form is one of the smallest nematodes infecting man. The females are larger than males. The male worm lacks a spicule and the copulatory sheath but at the tail end there are two conspicuous papillae on either side. The females are viviparous and discharge embryos instead of eggs. The larvae are 100μ in length and 6μ in width.

Life Cycle

The whole life cycle is passed in one animal (pig, rat or man). Although one individual animal serves both as definitive and intermediate host, 2 hosts are required to complete the life cycle. The parasite entered into man is unable to complete its life cycle. The continuance of the species is however maintained by other animals. The primary host of *T. spiralis* is the pig, which serves as the reservoir host of man. Infection normally passes from pig to pig and from rat to rat. Infection of a new host is always brought about by ingestion of raw flesh of the trichonosed animal containing the viable encysted larvae (Fig. 4.14).

Larval *Trichinella* remaining encysted in hosts (pig or rat) muscle is eaten by a healthy animal (pig or rat or man). After ingestion, larvae develop into adults in the intestine in a very short time of about 48 hours. Male and female worms become sexually mature. Fertilization occurs and in another 24 hours, larvae are liberated. The larvae gain entrance into the blood stream and encystment occurs when they are taken up by another healthy animal such as rat or pig or man. When man eats the infected pig muscle, the larvae are set free in the duodenum and jejunum and grow into adult worms and within 5-7 days become sexually mature. The male after fertilizing the female dies. The fertilized female then burrows deeply into the mucosa of the intestine and discharges a large number of larvae into the lymphatics or blood vessels. This process continues for a period of 5-7 weeks. The female dies after discharging her progeny. The larvae reach the striated muscles where they undergo encystment. In man, the cycle ends there. The infection of *Trichinella* in man is known as trichinelliasis or trichiniasis or trichinosis. Because of this, fever, urticarial rash, leucocytosis with marked increase of eosinophils and myositis are noticed.

9. *Ancylostoma duodenale*

Phylum : *Nemathelminthes*

Class : *Nematoda*

Order : *Rhabditida*

Genus : *Ancylostoma*

Species : *duodenale.*

It is commonly known as the old world hook worm. The adult worm lives in the small intestine of man. The worm causes hook worm disease or ancylostomiasis in man characterized chiefly by anaemia.

Morphology

It is a small greyish white and cylindrical worm. The anterior end of the worm is slightly bent dorsally, hence the name hook worm. The oral aperture is not terminal but directed towards the

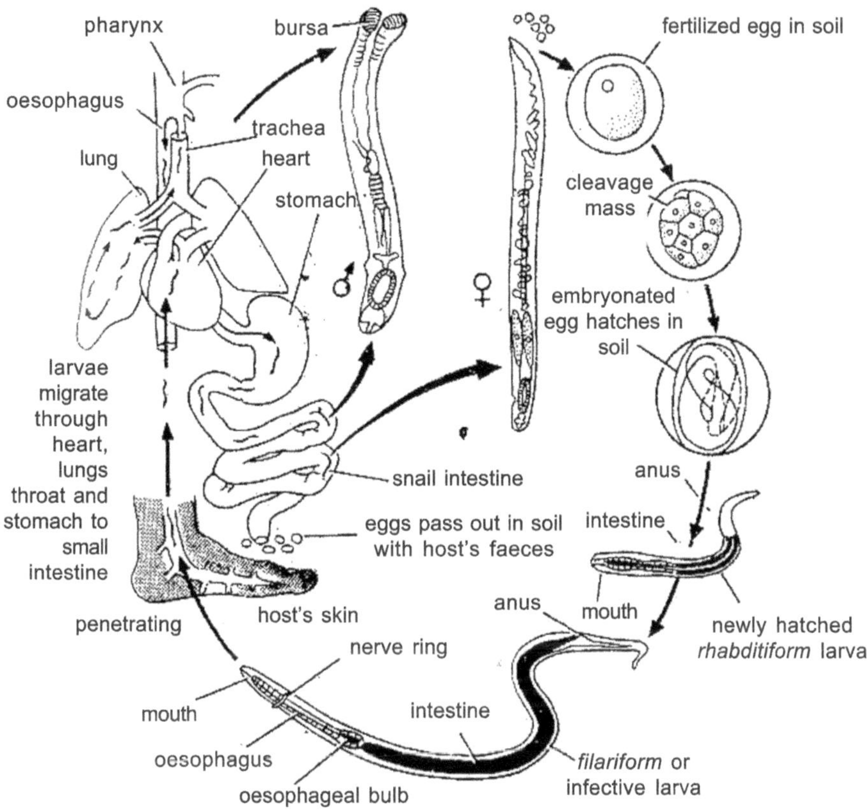

Fig. 4.15 Life cycle *of Ancylostoma duodenale*

dorsal surface. The females are larger than males. The posterior end is expanded in an umbrella like fashion in male while it is tapering in female. The life span of the adult worm in the human intestine is about 3-4 years.

Life Cycle

No intermediate host is required and the life cycle is passed completely in man. The eggs are passed out along with the faeces of the man. From each egg a rhabditiform larva (250μ in length) hatches out in soil in about 48 hours. The larva moults twice and then develops into a filariform larva (500-600μ in length). This is the infective stage of the parasite. The time taken for development from eggs to filariform larvae is on an average 8-10 days. The filariform larvae cast off their sheaths and enter into the host's body by penetrating the skin. On reaching the subcutaneous tissues, the larvae enter into the lymphatics or small venules. Through venous circulation they reach right heart, lungs, bronchi, trachea and larynx and then reach pharynx. On entering the cesophagus a third moulting takes place. The growing larvae settle down in the small intestine, undergo a fourth moulting and develop into adult worms. In 3-4 weeks time they are sexually mature and the fertilized females begin to lay eggs in the faeces. The cycle is thus repeated. (Fig. 4.15)

10. *Wuchereria bancrofti*

Phylum : *Nemathelminthes*
Class : *Nematoda*
Order : *Spirurida*
Genun : *Wuchereria*
Species : *bancrofti.*

It is commonly called as Bancroft's filaria. The adult worms are parasites of the lymphatic vessels and lymph nodes of man only. The infection to man is called filariasis. The transmitting agents are female mosquitoes such as *Culex fatigans* or *C. pipiens* or *C. guinque fasciatus.*

Morphology

The adult worms are long hair like transparent worms. They are filiform in shape and both ends are tapering. The male measures 2.5 to 4.0cm in length and 0.1mm in width. Its tail end is curved ventrally and contains 2 spicules of unequal length. The female measures 8-10cm in length and 0.2-0.3mm in width and its posterior end is pointed. The life span of adult worms is about 4-5 years.

Life Cycle

W. bancrofti passes its life cycle in two hosts i.e., the man and the mosquito (Fig. 4.16).

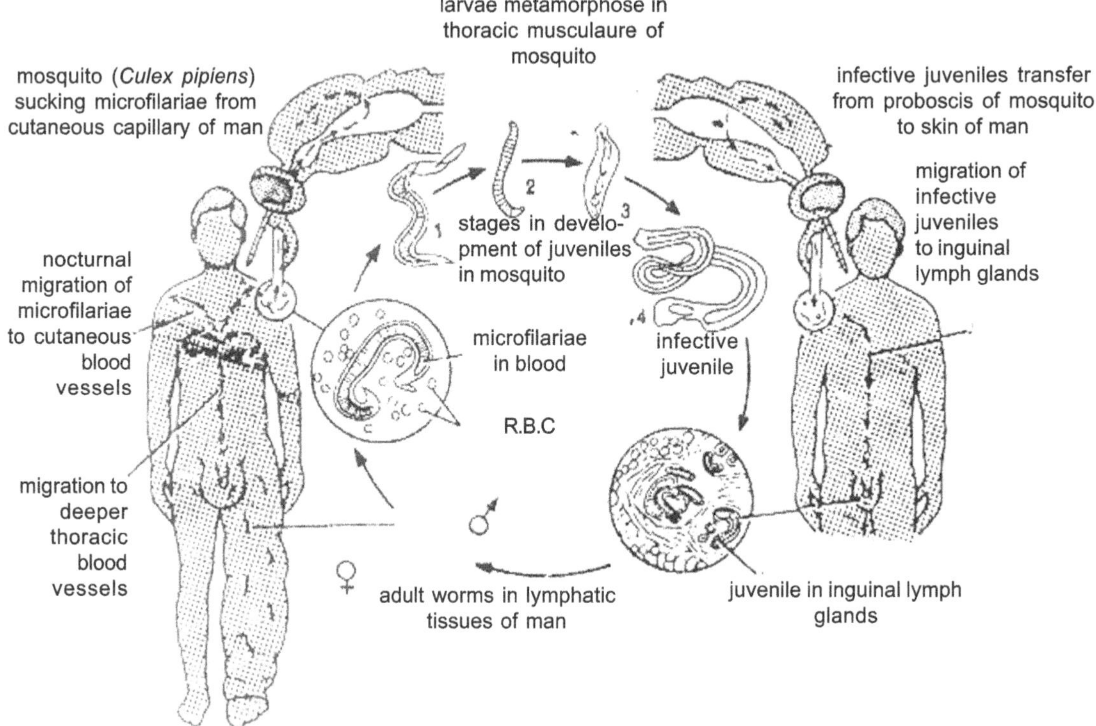

Fig. 4.16 Life cycle of *Wuchereria bancrofti*

The definitive host is man and the intermediate host is female mosquito.

The adult worms are harboured in the lymphatic system of the man. The embryos or microfilariae are discharged by the female. They undergo further development only when they are sucked along with blood by the intermediate host, *Culex fatigans* or *C. pipiens*. Hence they reach peripheral blood and remain there for a considerable time coinciding with the feeding habits of the mosquito. When the female mosquito sucks blood from an infected person, the parasites are ingested into the stomach of mosquito. The microfilariae come out of their envelopes, become active and migrate into the intestine. They bore the intestinal wall and enter the thoracic muscles of the mosquito. Then, they transform into short thick sausage shaped larvae (Ist stage larvae). Within 3-4 days, these larvae moult into second stage larvae and within 8-10 days moult into third stage larvae. These measure 1500-2000μ in length and 18-23μ in width. At this stage it is infective to man and enters the proboscis sheath of the mosquito by 14th day. When the infected mosquito bites a man, the parasites reach the site of the bite and make their way into the lymphatic tissue (such as inguinal, scrotal or abdominal lymphatics) of man, settle down and begin to grow into adult worms. Within 5-18 months, they become sexually mature. A male fertilizes the female and the gravid females give birth to the larvae. A new generation of microfilariae thus released reach the venous circulation, heart, lungs, systemic circulation and ultimately to the peripheral circulation and wait for the mosquito to receive them along with blood. In this way the parasite completes its life cycle.

11. *Enterobius vermicularis*

Phylum : *Nemathelminthes*
Class : *Nematoda*
Order : *Rhabditida*
Genus : *Enterobius*
Species : *vermicularis*.

It is commonly called as the thread worm, pin worm or seat worm. The adult worms live in the caecum and vermiform appendix of man. The infection of *E. vermicularis* in man is known as *enterobiasis*.

Morphology

The adult worm is small and white in colour. It is more or less spindle shaped and resembles a short piece of thread. A pair of wing like expansions called cervical alae are present at the anterior end. The male measures 2-4mm in length and the posterior one-third of the body is curved. The female measures 8-12mm in length and the posterior part is straight (Fig. 4.17).

Life Cycle

The life cycle is completed in one host only i.e, man. Infection is due to the ingestion of the eggs of the parasite from one person to another or via contaminated food and drink with eggs. The eggs newly laid on the perianal skin containing a tadpole like larva completes its development in 24 to 36 hours of time. Infection occurs by ingestion of these eggs. The egg shells are dissolved by digestive juices and the larvae escape to small intestine where they develop into adult worms. After the worms become sexually mature, the male fertilizes the female and then dies. The gravid female, then migrates from small intestine down to the caecum and colon (also the vermiform

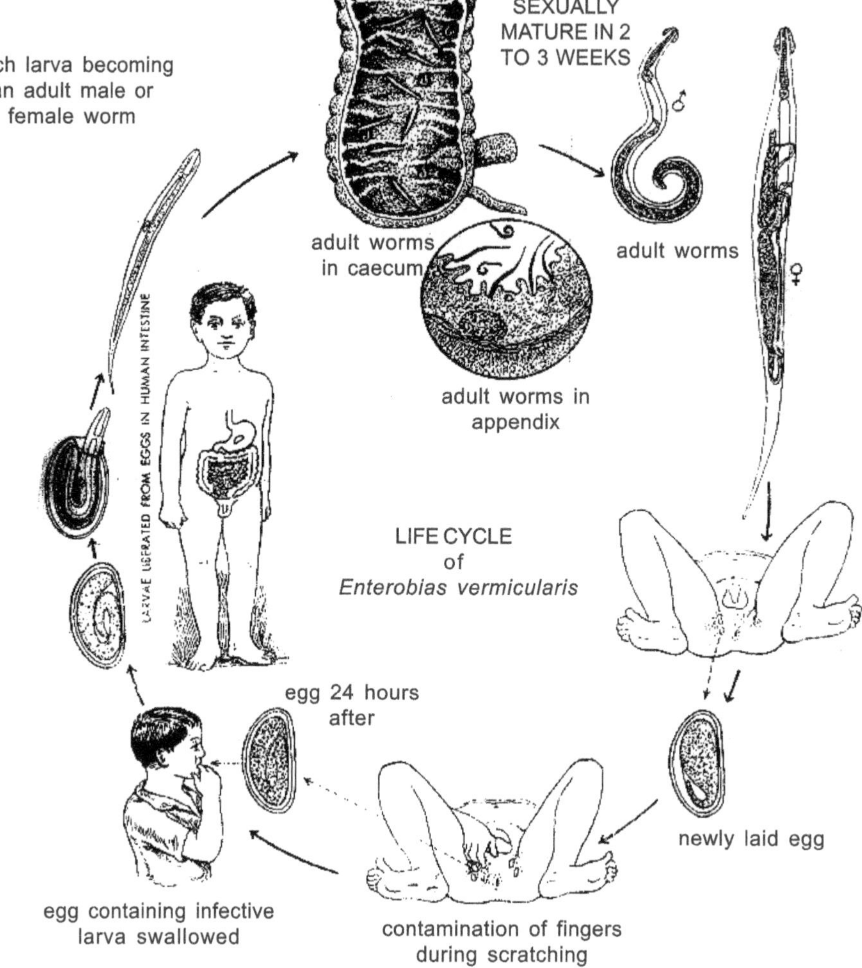

Fig. 4.17 *Enterobius vermicularis*

appendix) and remains there until the eggs develop. The fertilized female then passes down the rectum and works its way out of the anus during the night to deposit the eggs on the perianal skin. The movement of the worms at the time of egg laying causes intense itching, inducing the patient to scratch the affected part and thereby carrying the eggs on their fingers. These eggs are subsequently transferred to food and swallowed by the patient himself or the infection may occur direct from anus to mouth, a very common habit with children. The cycle is thus repeated.

12 *Ascaris lumbricoides*

Phylum	:	*Nemathelminthes*
Class	:	*Nematoda*
Order	:	*Ascaroidea*
Genus	:	*Ascaris*
Species	:	*lumbricoides.*

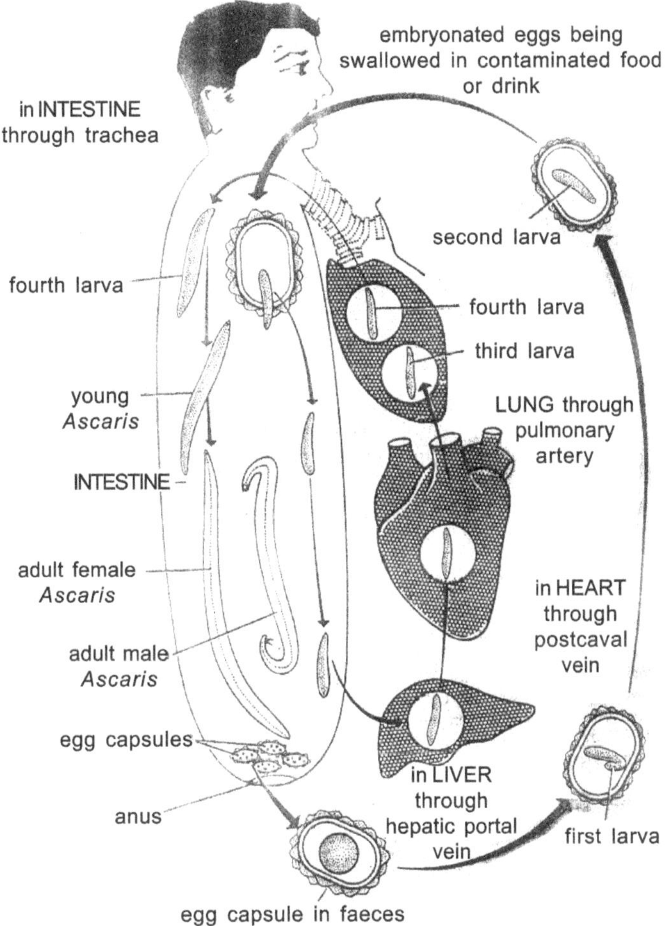

Fig. 4.18 *Ascaris lumbricoides* Life cycle

Ascaris is the most common roundworm. The adult worm lives in the small intestine of man. The infection of this parasite to man is known as Ascariasis.

Morphology

The adult worm is the largest intestinal nematode parasitising man. It is worm like, tapering at both ends. The posterior end of the male is curved ventrally in the form of a hook while that of the female is straight and pointed. The male worm measures about 15-25cm in length and with a maximum diameter of about 3-4mm while the female worm is longer and stouter than the male and measures 25 to 40cm in length with a maximum diameter of 5mm.

The life span of the adult worm is about 9-12 months.

Life Cycle

The worm passes its life cycle in one host only i.e, man. Fertilized eggs are released along with the faeces. A rhabditiform larva is developed within the egg shell in 10-40 days time depending upon the atmospheric temperature and humidity. This takes place in the soil. The ripe egg containing the larva is infective to man (Fig. 4.18).

When ingested with food, drink or raw vegetables, the eggs pass down the duodenum where the digestive juices weaken the egg shells and stimulate the larvae into activity. The rhabditiform larvae are liberated splitting the egg shell in the upper part of the small intestine. The newly hatched larvae burrow their way through the mucous membrane of the small intestine and are carried by the portal circulation to the liver. They live for about 4 days in liver. From there they enter the pulmonary circulation via heart. They grow in their size in lungs and moult twice on fifth and tenth day respectively. From the lungs, the larvae crawl up the bronchi and trachea and reach the larynx and pharynx. From here they pass down the oesophagus to the stomach and settle in the upper part of the small intestine. One more moulting occurs between 25th-29th day. The worms grow into adult worms and become sexually mature in about 6 to 10 weeks time. The gravid females begin to discharge eggs in the faeces within 60 days of infection. The cycle is thus repeated.

13. *Dracunculus medinensis*

Phylum	:	*Nemathelminthes*
Class	:	*Nematoda*
Order	:	*Spirurida*
Genus	:	*Dracunculus*
Species	:	*medinensis*.

It is commonly known as the *Guinea worm*, *serpent worm, dragon worm* or *Medina worm*. The adult females are usually found in the subcutaneous tissues especially in the legs, arms and back. The infection with this parasite in man is known as *guinea worm disease* or *dracunculosis*.

Morphology

The adult is a slender, long worm measuring 60cm to 1 metre in length with a diameter of 1.5-1.7mm resembling a piece of long twine thread. The body is cylindrical, smooth and milky white in colour. The males are smaller than females. The posterior extremity of the female is tapering and is bent to form a hook. The female is viviparous. The body fluid of the worm is toxic and causes a blister if it escapes into the tissues. The life span of the female is about one year and that of the male is less than 6 months (Fig. 4.19).

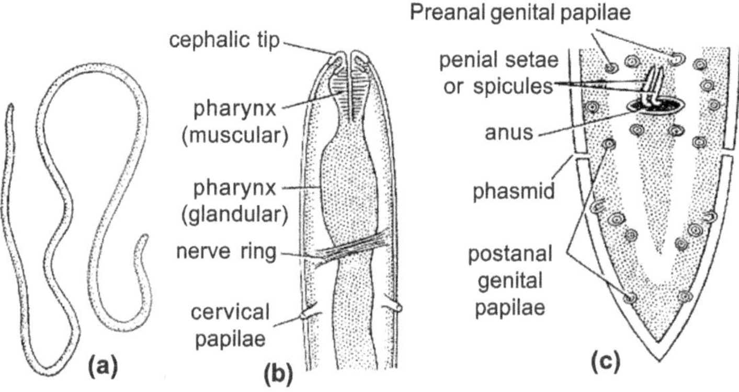

Fig. 4.19 *Dracunculus medinensis.* (a) Entire; (b) Anterior end of male; (c) Posterior end of male

Life Cycle

The worm passes its life cycle in two hosts ie., man and cyclops. The man is the definitive host and cyclops is the intermediate host.

The embryos are set free into the water when the affected part is submerged in water. Cyclops ingests these embryos from the water. The embryos penetrate the gut wall within 1-6 hours after ingestion and enter the body cavity of the cyclops where they undergo metamorphosis and increase in size to about 1mm. The cyclops containing the infective larvae are swallowed by man with raw drinking water. On reaching the stomach, the cyclops are digested by the gastric juice and the larvae are liberated. They penetrate through the gut wall and enter the retroperitoneal connective tissue where they grow to sexual maturity.

The males die after fertilizing the females. The mating of the adults takes place in the deeper connective tissues and not in the intestine. It takes about another 6 months for the gravid female to select its site for discharging the embryos in water. The gravid female migrates and selects only those parts of the skin liable to come in contact with water, such as arms, legs, back etc.

On reaching the skin surface, it secretes a toxin producing a blister which later ruptures and forms an ulcer. Contact with water stimulates the worm to protrude its head through the centre of the ulcer and causes a reflex discharge of a milky fluid containing a large number of coiled embryos ejected from the prolapsed uterus. Thus the embryos are back again in water and the cycle is repeated.

4.3 Biology of Arthropod Parasites

The following is the account of the biology of some important Arthropod parasites such as, Head louse, bed bug, mosquito, house fly, human flea, ticks and mites.

1. Head louse, *Pediculus humanus capitis*

Phylum	:	*Arthropoda*
Class	:	*Insecta*
Order	:	*Anoplura*
Genus	:	*Pediculus*
Species	:	*humanus*
Subspecies	:	*capitis*

Pediculus humanus capitis is commonly called as the head louse. It is an ectoparasite inhabiting the hair of the head and is a blood sucking pest. The infection with *Pediculus* to the man is called pediculosis.

Morphology

The head louse is wingless, with elongated dorso-ventrally flattened body and measuring 2.5 to 4mm in length. The insect has 3 pairs of legs of about equal length. The mouth parts are adapted for piercing the flesh and sucking the blood. The head louse undergoes incomplete metamorphosis i.e,. its life cycle consists of an egg, nymph and adult stages and both the nymph and adult stages resemble each other and have same habit and habitat (Fig. 4.20).

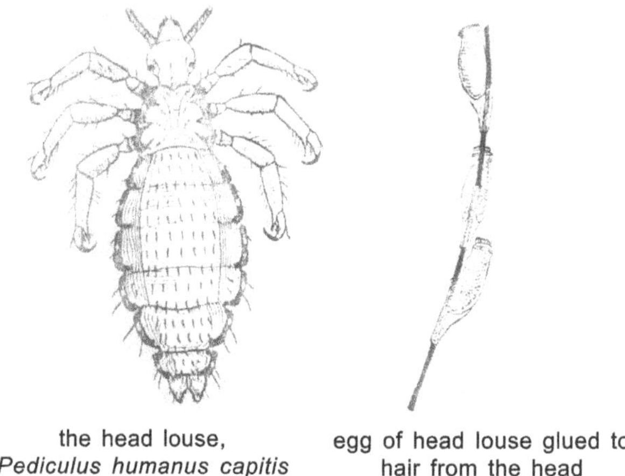

Fig. 4.20 Head louse, Pediculus

Life Cycle

The adult female lays upto 100 eggs known as nits in her life time. The female lays approximately 3 eggs per day and cements them on the hair of the head, sometimes on other hairy parts of the body also. The eggs hatch within about 7 days. There are 3 nymphal instars in the larval stage, which is completed within 10 - 14 days. The adult inhabits the hair of the head particularly behind the ears and around the occiput. Heavy infestations may force head lice to establish themselves on other hairy parts of the body. They are relatively less active feeding for hours at a time while clutching firmly to the hair shaft and they seldom leave the hairy regions voluntarily.

They are disseminated by physical contact or by sharing hats, combs or brushes.

Although the head lice have been shown to be capable of transmitting rickettsiae and spirochetes in the laboratory, they are usually not involved in the transmission of these organisms in natural conditions.

2. Bedbug, *Cimex hemipterus*

Phylum	:	*Arthropoda*
Class	:	*Insecta*
Order	:	*Heteroptera*
Genus	:	*Cimex*
Species	:	*hemipterus*

Cimex hemipterus is commonly known as the bedbug. Both the nymphs and adults suck blood during the night. The bites produce intensely itching and inflammatory reactions on the skin.

Morphology

The adult bug is a small, bristly, pear shaped, wingless and dark brown insect measuring about 10 -13 mm. They inhabit in a variety of human habitations such as houses, hotels, theatres, dormitories, barracks etc. They remain hidden in crevices in walls, floors, beds, seats etc (Fig. 4.21).

The insect has piercing and sucking type of mouth parts. The insect is active and flat before a blood meal but after it, it becomes oval and sluggish. The insect emits a typical smell which emanates from the stink glands.

Life Cycle

The adult female lays elongate white eggs singly at the rate of 3 to 4 per day in the cracks of the furniture, walls etc. The bedbug frequently returns to the same place to oviposit. About 500 eggs are laid by a female in her life time of about one year. The eggs hatch in about 6-7 days and the nymphs start feeding on blood. There are 5 nymphal stages and they resemble each other quite closely except in increasing size and deepening colour. The wing pads appear only after the last moult. Atleast one blood meal is required for each stage to moult into the next one.

The nymphal stage is completed in about 42-56 days.

Bed bug do not cling to the skin while sucking the blood, but prefer to remain on the clothing or bed linen. Feeding lasts for about 15 minutes. Although they may defaecate after the feeding, the deposits do not contaminate the bite wound. They can withstand starvation upto 6 months.

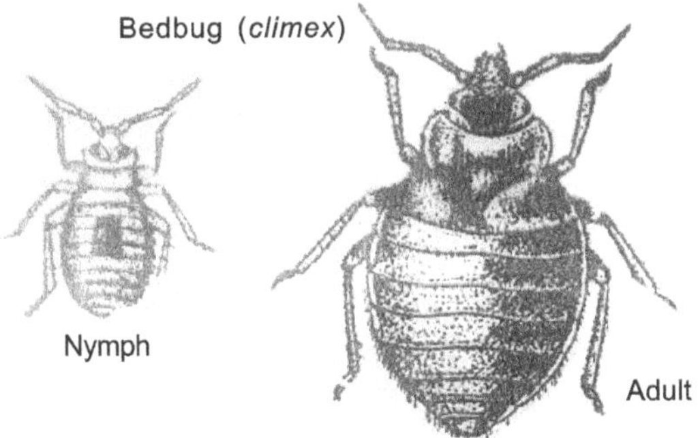

Fig. 4.21 The life cycle of Bedbug

Some persons especially children suffer severely from the bites of bed bug. During the biting, the saliva is injected which contains besides toxins an anticoagulin and an anaesthetic factor. The bites produce intense itching and inflammation. It is suspected but not yet proved of transmitting diseases.

3. *Mosquitoes*

Phylum : *Arthropoda*
Class : *Insecta*
Order : *Diptera*

There are mainly 3 genera of mosquitoes i.e., *Anopheles, Culex* and *Aedes* with several species under each genus. Female mosquitoes bite warm blooded animals including man and suck blood. The bite is painful, causes itching sensation and for sensitive people allergitic. Along with bites the mosquitoes also transmit pathogens responsible for the diseases like malaria, yellow fever, dengue fever, filariasis, encephalitis etc.

Anopheles, Culex and *Aedes* mosquitoes differ from each other by morphological features in the egg, larva, pupa and adult stages (Fig. 4.22). The details are given in Table 4.6.

Table 4.6

	Anopheles	*Culex*	*Aedes*
Eggs	1. Boat shaped, laid singly air float present	1. Cigar shaped, laid in rafts, no air float	1. Elliptical, laid singly, no air float
	2. Laid in flowing fresh water	2. Laid in stagnant, dirty and permanent bodies of water	2. Laid on moist ground likely to be flooded, in domestic containers, treeholes etc.
Larva	1. Lies parallel to water surface	1. Hangs at an angle from water surface	1. Hangs at an angle.
	2. Respiratory siphon absent	2. Respiratory siphon long and conical	2. Respiratory siphon long and conical
	3. Surface feeder	3. Bottom feeder	3. Bottom feeder.
Pupa	1. Green with short respiratory trumpet	1. Colourless with longer respiratory trumpet	1. Colourless with longer respiratory trumpet.
Adult	1. Resting posture inclined head down at 45° angle	1. Resting posture horizontal to ground	1. Resting posture horizontal to ground.
	2. Body grey coloured	2. Body grey coloured	2. Blackish brown with silvery stripes on thorax and abdomen.
	3. Maxillary palps equal to proboscis in both sexes but terminally clubbed in male and pointed in female.	3. Maxillary palps longer than proboscis in male and shorter in female.	3. Maxillary palps equal to proboscis in male, shorter in female.

Morphology

The adult mosquitoes are dipterous insects i.e., have one pair of wings only. The hind wings are modified into a pair of balancing organs known as halteres. They have piercing and sucking type of mouth parts. The females suck blood while males the plant juices (sap). The antennae are plumose in males (bushy) while they are pilose (thin-brush like) in females. They have 3 pairs of long legs. The egg, larval and pupal stages are aquatic while the adult stage is arboreal. The adults may have diurnal, crepuscular or nocturnal or both diurnal and nocturnal activities.

Life Cycle

The following is the account of *Anopheles* mosquito. The female *Anopheles* lays 40 to 100 eggs at a time in natural water such as ponds, swamps, rice fields, grassy ditches, flowing waters, wells etc. The eggs hatch in 24 to 48 hours into a larva. There are four larval stages in mosquito. The larval life lasts from 7 to 10 days in tropical climate while from 2 to 4 weeks or more in temperate climates. The larvae are called wrigglers because of their wriggling movement. They are voracious feeders of bacteria, yeasts, protozoa and other microorganisms.

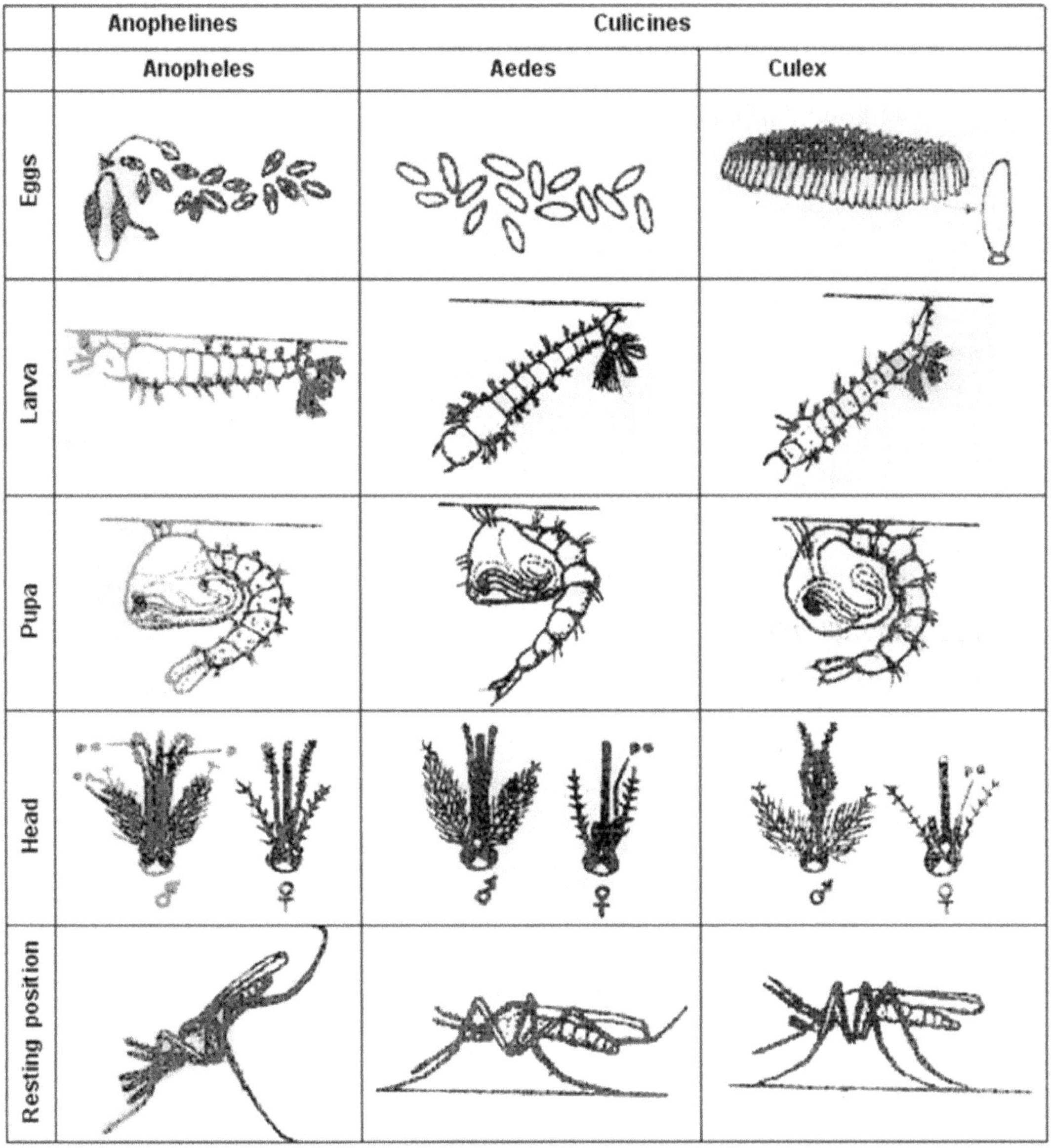

	Anophelines	Culicines	
	Anopheles	Aedes	Culex
Eggs			
Larva			
Pupa			
Head	p, pa ♂ ♀	pa ♂ ♀	pa ♂ ♀
Resting position			

Fig. 4.22 Distinguishing features between the developmental stages and the adult heads and resting postures of the anopheline and culicine mosquitoes (p = proboscis; pa = palp)

The pupa is comma shaped. It normally remains floating at the water surface but when disturbed, swims vigorously down to the bottom with characteristic jerky movements. The pupal period lasts for 2-3 days.

The adult mosquitoes are crepuscular or nocturnal in their activities. The blood feeding and oviposition normally occur in the evenings, at night or in the early morning around sunrise. Mosquitoes (male and female) copulate while flying. The longevity of adult males is about a week while the females live for about a month or more.

4. Housefly, *Musca domestica*

Phylum	:	*Arthropoda*
Class	:	*Insecta*
Order	:	*Diptera*
Genus	:	*Musca*
Species	:	*domestica*

Houseflies are cosmopolitan in distribution. The flies are a source of nuisance and great annoyance. Their bites are painful. The housefly transmits many diseases like typhoid, dysentery, cholera, yaws, T. B, leprosy, trachoma, anthrax, gonorrhoea and many helminthic diseases. It is a mechanical carrier of the pathogens of these diseases.

Morphology

The adult fly is about 6-7 mm long, grey in colour with four whitish lines on the thorax and yellowish tinge on the ventral surface. The body is divided into the head, thorax and abdomen. The head consists of a pair of compound eyes, aristate antennae and sponging type of mouth parts. The thorax consists of a pair of wings and a pair of balancers known as halteres and 3 pairs of legs. The abdomen consists of 10 segments but only 4 are conspicuous while the first two are atrophied and the last four are telescoped into the preceding segment.

The eggs are cigar shaped. The larvae known as *maggots* are apodous. The anterior end is narrow and pointed while the posterior end is broad (Fig. 4.23).

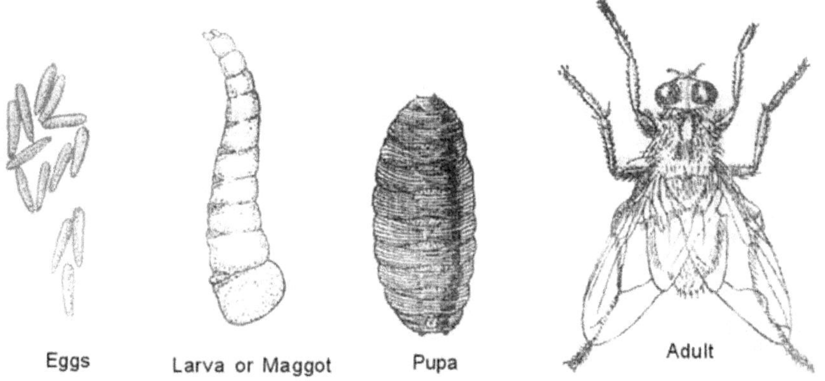

Fig. 4.23 Life history of *Musca domestica*

The pupa lies in a puparium which is amber coloured. The sexes can be differentiated in the adult stage by the following characters. In the male the compound eyes are very large and almost join obliterating the frons (the sclerite separating them) while in female frons is well developed and the compound eyes lie wide apart. The abdomen in male is cylindrical and thin while in the female it is broad. The females are larger in size than males.

Life Cycle

Copulation occurs on the ground 24 hours after adult emergence. Four days after mating, the female lays about 15-150 eggs in batches on any decaying matter such as cow dung, horse manure, garbage or decomposing animal or vegetative matter. During the breeding season the female lays in 4-6 batches a total of about 500 eggs.

The eggs are whitish and cigar shaped. They hatch in about 8-24 hours into tiny maggots (the larvae). There are 3 larval instars. The duration of these instars are 2-3 days, 1-2 days and 3-5 days respectively. The last larval instar when full grown is about 12mm long. It has 12 segments and there is an insignificant head.

When full grown, the larva seeks a drier place in the dung or garbage and pupates into a barrel shaped pupa. The larva changes into a pupa without moulting. The last larval skin hardens to form an outer covering called puparium.

The pupa is absolutely immobile and the pupal stage lasts for 4-5 days. The adult emerges out breaking the puparium with the help of a peg like ptilinum (a blood filled bag). The adults after about 10-12 hours take to flight. The adult longevity is about 20-30 days.

5. Human flea, *Pulex irritans*

Phylum : *Arthropoda*
Class : *Insecta*
Order : *Siphonaptera*
Genus : *Pulex*
Species : *irritans*

The fleas are ectoparasites of man sucking blood usually on ankles and legs, but in the nights a sleeping person is bitten on other parts of the body also. The irritation due to flea bite is felt after some time. They are also known to transmit the pathogen of plague from man to man or rat to rat or rat to man. Usually rat fleas *Xenopsylla cheopis* are well known to transmit plague pathogen.

Morphology

The adult fleas are relatively small measuring about 1-4 mm long and more or less oval insects, compressed laterally and are light to dark brown. Wings are absent but there are 3 pairs of well developed and powerful legs of which the hind pair is specialized for jumping. The head is small and triangular with a pair of small compound eyes, 3-segmented club shaped antennae and piercing and sucking type of mouth parts. Male fleas are identified by the upturned appearance of the abdomen. In the female, the tip of the abdomen is more rounded than in the male.

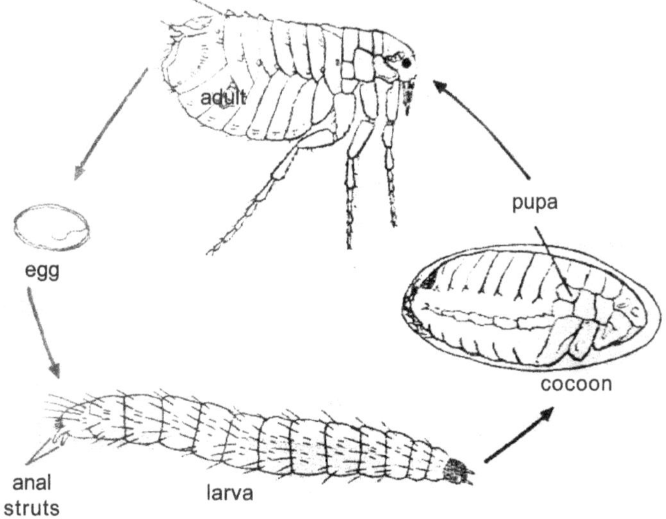

Fig. 4.24 The life cycle of *human flea*

Life Cycle

Both sexes of fleas take blood meals and are therefore equally important as vectors of disease. A female flea which is ready to oviposit may leave the host to deposit her eggs in debris which accumulates in host's dwellings. The females often lay their eggs in or near cracks and crevices on the floor or amongst dust, dirt and debris. The eggs are very small, oval, white or yellowish and are thinly coated with a sticky substance which helps the eggs to be covered with dust and concealed. Adult flea may live for upto 6-12 months or sometimes upto 2 years during which time it may lay about 300-1000 eggs in batches of 3-18 per day. The eggs hatch within about 2-14 days depending upon climatic conditions. A minute apodous larva hatches out from the egg. The larvae are characterized by setae on the body and a pair of finger like processes known as *anal struts*. The larvae are very active. They feed on almost any organic debris including hosts faeces. There are three larval instars. The larval period may be completed within 10-21 days. However in unfavourable adverse conditions it may be prolonged to upto 200 days. Mature full grown larvae are 4-10mm long (Fig. 4.24).

At the end of the larval period, the larva spins a whitish cocoon from silk produced by its larval salivary glands. As the cocoon is sticky, it is covered by dust and is difficult to be observed. After 2-3 days after spinning the cocoon, the larva pupates within the cocoon.

Adult emerges from the pupa in about 7-14 days. The freshly moulted adult needs a stimulus to come out from the cocoon. Adults may remain alive in their cocoons for about one year. The life cycle from egg to adult emergence may be as short as 2-3 weeks or sometimes as long as 20 months or more.

Adult fleas avoid light and are usually found sheltering among the hairs or feathers of animals or on man under his clothing or in his bed. Fleas feed several times during the day or night on their hosts. While feeding, fleas eject faeces composed of semidigested blood of the previous meal and then excess blood taken in during the feeding. Thus blood marks are found on clothing or bed linen of people heavily infected with fleas.

Although most species of fleas have one or more favourite hosts, they are not entirely host specific. For example rat, cat and dog fleas, *Xenopsylla, Ctenocephalides* sp readily feed on man in the absence of the hosts. Human flea feeds on pigs, rats etc.

6. *Ticks*

Phylum	:	*Arthropoda*
Class	:	*Arachnida*
Order	:	*Acarina*
Type	:	*Ticks*

Ticks are of two types the soft ticks (Argasidae) and hard ticks (Ixodidae).

The medically important soft ticks belong to the genus *Ornithodoros*. Among hard ticks, the important genera are *Ixodes, Dermacentor, Amblyomma, Hyalomma* etc. The ticks are ectoparasites feeding on blood of the host apart from transmitting tick borne relapsing fever, or different types of typhuses, encephalitis and tick paralysis.

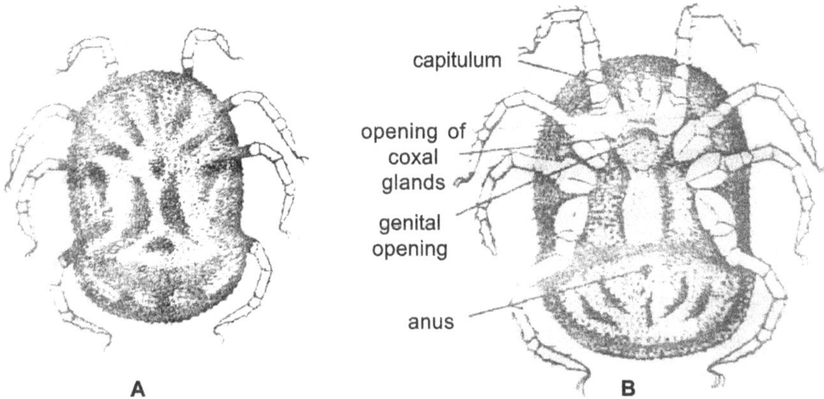

Adults of *Ornithodoros moubata*, (a) dorsal view; (b) ventral view.

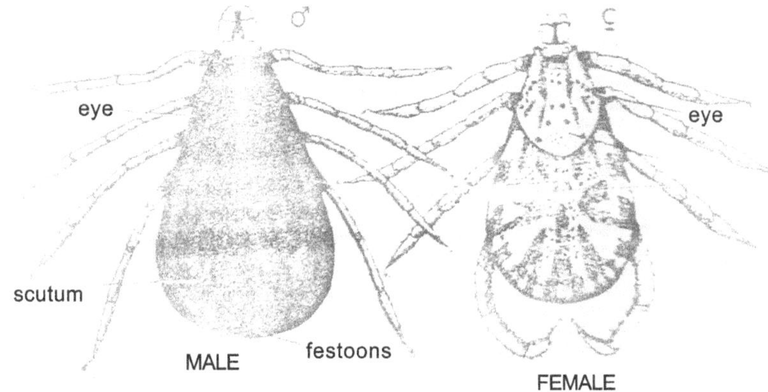

Fig. 4.25 Adults of male and female *Dermacentor species*; showing sexual differences in size of scutum. Festoons are present.

Morphology

In general, adult ticks are dorso-ventrally flattened and oval in shape. The integument is tough and leathery, wrinkled and there is no scutum in soft ticks but a scutum or dorsal shield is present in hard ticks. The mouth parts termed capitulum are situated ventrally in soft ticks and are projected forwards and visible in hard ticks. Four pairs of well developed legs terminate in a pair of claws. Males and females are very similar in external appearance and feed on blood and can consequently be disease vectors (Fig. 4.25).

Life Cycle

Both the soft ticks and hard ticks are hemimetabolous i.e., there is incomplete metamorphosis, involving a larval and nymphal stage.

Adult females of soft ticks take blood meal from their host and drop to the ground while hard ticks remain attached to their hosts for long periods as one blood feeding takes about 1-4 weeks. When feeding is completed the soft ticks lay their eggs in batches of 15-100 spherical eggs in or near the resting places of the adult ticks. As the adult tick lives for many years, thousands of eggs may be laid during her life time. Eggs hatch usually in 1-4 weeks. Hard ticks lay 100-8000 small spherical eggs in a gelatinous mass which is formed in front and on top of the scutum. The eggs are coated with a protective waxy secretion from gene's organ (a peculiar structure located in front of the midgut secreting gelatinous or waxy secretion and expels it through a small opening above the capitulum or mouth parts during oviposition) which protects the eggs for several months in adverse conditions. Hard tick female after laying one batch of eggs dies.

The eggs hatch to produce 6- legged larvae which superficially resemble the adults. They moult to produce 8- legged nymphs which resemble even more closely the adults.

The six- legged larva of soft ticks is very active and searches for a host for a blood meal and after a blood meal drops to the ground and after a few days, moults into eight legged nymph. There are 4-5 nymphal instars. Each nymphal stage requires a blood meal before it moults to the next stage. The duration of the life cycle from egg to adult ranges from 6-12 months. Adult soft ticks live for several years sometimes upto 15 years.

The six legged larvae of the hard ticks remain inactive for some days after hatching. Then they climb up the vegetation and cluster at the tips of the grasses and leaves. When suitable hosts pass, the larvae climb on to their hosts and crawl to their favoured site of attachment, commonly in the ears or on the eyelids where they start feeding on the blood. The larvae remain attached to the host for 3-7 days and then drop to the ground and take shelter in vegetation. After complete digestion of the blood meal, which takes about a week, they moult into eight legged nymphs.

The eight legged nymphs crawl and climb up the vegetation and behave similarly like the larvae. On getting the suitable host they take their blood meal for 5-10 days and then drop to the ground. There is only *one nymphal* instar in *hard ticks*. Then they moult to give rise to the male and female adult ticks.

In this way the life cycle is completed in both the soft ticks and hard ticks.

7. ***Mites***

Phylum : *Arthropoda*
Class : *Arachnida*
Order : *Acarina*
Type : *Mites*

Table 4.7 *There are several genera of mites the following table shows the details of the mites*

Common name	Scientific name	Host
1. Itch mite	*Sarcoptis scabiei*	man, domestic animals.
2. Hair follicle mite	*Demodex folliculorum*	man, domestic animals.
3. Chigger mites or trombiculid mites or red bugs	*Eutrombicula alfreddugesi*	man, other vertebrates, arthropods
4. Harvest mite	*Neotrombicula autumnalis*	man, vertebrates, arthropods
5. Rat mite	*Ornithonyssus bacoti*	rats
6. Mouse mite	*Liponyssoides (Allodermanyssus) sanguineus.*	mice
7. Chicken mite	*Dermanyssus gallinae*	poultry
8. Straw itch mite	*Byenotes ventricosus*	domestic animals.

The following is the account of 2 types of mites i.e, scabies or itch mites and chigger mites.

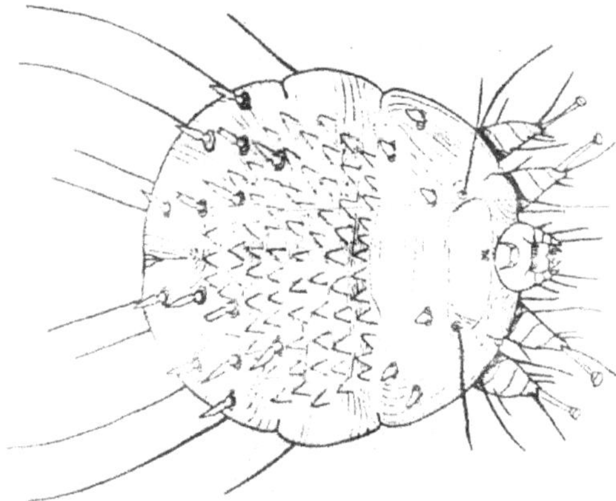

Fig. 4.26 Adult female *scabies mite, Sarcoptis scabiei*

Scabies Mite or Itch Mite

It causes scabies or itch in man and mange in animals. They are not vectors of any disease.

Morphology

The female mite is larger than the male and measures about 0.30-0.45mm. It is whitish and disc shaped. Dorsally the body consists of numerous small peg like protuberances and a few bristles. There are striations on dorsal and ventral sides of the body. The adults have four pairs of short and cylindrical legs. The first two pairs of legs end in short stalks called pedicels which terminate in suckers. In the female, the posterior two pairs of legs do not have suckers but end in long and conspicuous bristles. There is no real distinct head and the mouth parts protrude anteriorly from the body (Fig. 4.26).

Life Cycle

The female scabies mite selects places on the body where the skin is thin and wrinkled such as in between the fingers, wrists, elbows, feet, penis, scrotum, buttocks and axillae. The majority of mites are found on the hands. The mite digs and eats her way into the surface layers of the skin. Then they excavate winding tunnels @ about 1-5 mm per day and lay 1-3 eggs per day in the tunnels. The eggs hatch within 3-5 days and small six legged larvae emerge which look like adults. These larvae crawl out of the tunnels onto the surface of the skin and burrow into the stratum corneum or a hair follicle and produce small pockets called moulting pockets.

After 2-3 days, the larvae moult to produce 8- legged nymphs. A nymph destined to become a female mite moults to produce an adult female. It remains in the moulting pocket until she is fertilized by a male. Then she enlarges in size to become an egg laying (ovigerous) female. Mating may be accomplished by the male either burrowing through the surface of the skin into the moulting pocket containing the female, or on the surface of the skin. After 3-5 days after copulation, female starts laying eggs in the tunnel. The life cycle from the egg to adult takes about 11-20 days. Female scabies mite may live about 1-2 months on man.

Chigger Mites or Trombiculid Mites or Red Mites

The larvae of Trombiculid mites are called red bugs or chiggers. The important species are *Leptotrombidium* sp. Which cause Scrub typhus. Other trombiculid mites are *Neotrombicula autumnalis*, *Eutrombicula alfreddugesi* etc. These mites cause dermatitis in man known as scrub-itch or trombidiosis.

Morphology

The adults are small mites measuring 1-2 mm, reddish in colour and clothed with bushy hairs giving them velvety appearance. There are four pairs of legs ending in paired claws. The body is constricted between the third and fourth pairs of legs giving it the shape of '8'. The mouth parts project anteriorly and are clearly seen. The nymph resembles the adult but is smaller in size measuring 0.5 to 1 mm in length. The body is less densely covered with feathered hairs.

The adults and the nymphs are non parasitic and feed upon the eggs of insects and other arthropods. Only the 6- legged larvae are parasitic to man and hence responsible for the spread of diseases.

The larvae are very small 0.15 to 0.3 mm in length, usually reddish or orange and some may be pale yellow or straw coloured. There are 3 pairs of legs terminating in relatively large paired claws. The mouth parts are well developed and projected anteriorly. There is a rectangular or pentagonal scutum on which 3-6 setae are present.

Life Cycle

A female mite lays 1-5 spherical eggs each day on the damp soil or under leaves. After 5-7 days, the eggs hatch and the 6 legged larva crawls out. It becomes very active and climbs up grasses or other vegetation. The larvae attach themselves to birds or mammals including man. After reaching and selecting the suitable place on the body where the skin is soft and moist such as ears, genitalia and around the anus, they pierce the host's skin with their powerful mouth parts and inject saliva into the wound which causes disintegration of the cells. Lymph and other fluids and semidigested materials are fed by the larvae. They feed on the body fluids for about 2-10 days and after the feeding, they drop to the ground and bury themselves just below the surface of the soil. This stage is known as protonymph. After 7-10 days, it moults into an 8- legged reddish nymph. The nymphs feed on soil inhabiting arthropods. After about 2 weeks the nymph ceases feeding and becomes inactive. This stage is known as preadult. The preadult moults into an adult in about 14 days. Thus the life cycle takes about 2-3 months.

4.4 Parasitic Diseases of Fish and Shell Fish their Control

Water is the chief medium for fish culture. So the farmer should maintain proper hygienic conditions around the industry and he should take care about the quality of water. He should test water frequently for its purity and quality because most of the diseases in fishes are spread through water only. Polluted water contains harmful pathogenic organisms like bacteria, virus, fungi, protozoans, helminths, annelids and crustaceans etc. They cause heavy economic loss to the fish industry. If the disease outbreak is not identified and proper preventive measures are not adopted, mass mortality of fish takes place and this leads to heavy economic loss to farmer.

Reasons for parasitic infections

1. Environmental stress
2. Culture practices in unhygienic conditions
3. Wounds formed during transport to distant places
4. High stocking density

Types of diseases

Diseases which affect fish are parasitic and non-parasitic. The former includes those caused by bacteria, fungi, protozoans, worms, leeches and copepods and the latter comprise disorders associated with nutritional deficiency and sudden changes in abiotic and biotic factors. The common symptoms by which the affected fish are recognised are (a) changes from normal behaviour, (b) signs of reduced vitality, (c) lack of appetite and failure to feed, and (d) presence of lesions or sores.

Parasitic diseases

Organisms such as bacteria, fungi, protozoans, trematodes, cestodes, nematodes, acanthocephalans, hirudineans and copepods are mainly responsible for causing various parasitic diseases in fish.

Fungal Infections

Certain fungi infect the skin, gills, fins, lips and eyes of fish (Fig. 4.27). The formation of blisters, damage of tissues of the fins, decay of eyes, enlarged intestine and liver, blood capillaries damage, red spots due to clotting of blood are some of the diagnostic features of fungal infections.

Eg : Eclia, Branchiomysis demigrans, Ichthyophonus haferi etc.

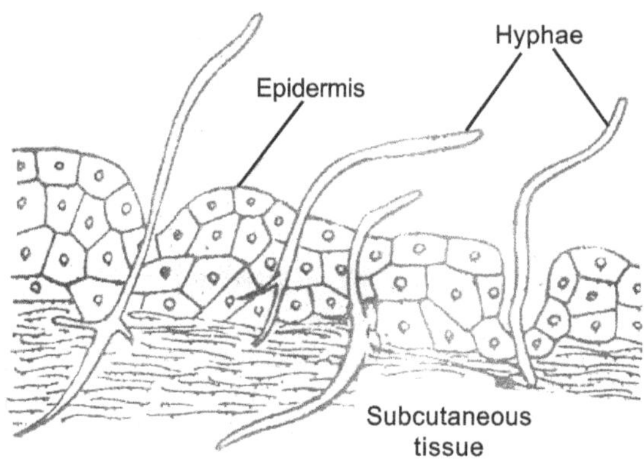

Fig. 4.27 Skin showing fungus hyphae

Bacterial Infections

Certain Bacteria infect skin, gills and intestine of fishes. The formation of blisters with water or pus or blood, black spots, bulging of belly, bulging of eye balls, decolouration of the body, redding of anal region and circular swimming movements on the surface of water are some of the diagnostic features of bacterial infections.

Eg : Vibrio cholera, Vibrio fisheri, Mycobacterium, Pseudomonas punctata. etc.

Protozoan Infections

Certain protozoa infect skin, scales, fins and gills of fishes. White spots on the skin and fins, light blue to red spots on the skin, scales, fins and gills, spiralling of the caudal end and deformity of the oral region are the principal diagnostic features of protozoan infections (Fig. 4.28).

Eg : Myxobolus, Ichthyophthirius, Costia, Trichodina, Myscosoma Cerebralis.

Helminth parasites

Certain helminth parasites infect skin, scales, gills, intestine, gall bladder etc. Decolouration of the body, mucous secretions near tail fin and gill region, withering of the skin, respiratory failure, black spots, deterioration of gall bladder are some of the symptoms of helminth infections (Fig. 4.28).

Eg : Diplostomulum, Dactylogyrus, Diplostomum, Gyrodactylus, Proteptus, Ligula etc.

Annelid parasites :

Two species of leeches like Hemiclepsis lead ectoparasitic sanguivorous life on the fishes. The wounds formed by the leeches are prone for secondary infections by bacteria, virus and fungi. Such infected fishes should be separated and treated for the removal of leeches (Fig. 4.28).

Crustacean Infections

Certain crustaceans like Copepods attack the eyes, skin, tissues, gills with the help of hooks and suckers. Fish lice attach near the eyes and suck the blood. Some parasites penetrate deep into the body and decay the tissue and gills (Fig. 4.28).

Eg : Fish, louse (Argulus), Anchor worm, (Lernaea), Ergasilus, Caligus, Pseudocycnus, Lernaenicus, Clavellisa etc.

Control Measures

The common sites of infection are skin, gills, eyes and fins. Parasites can be removed by :

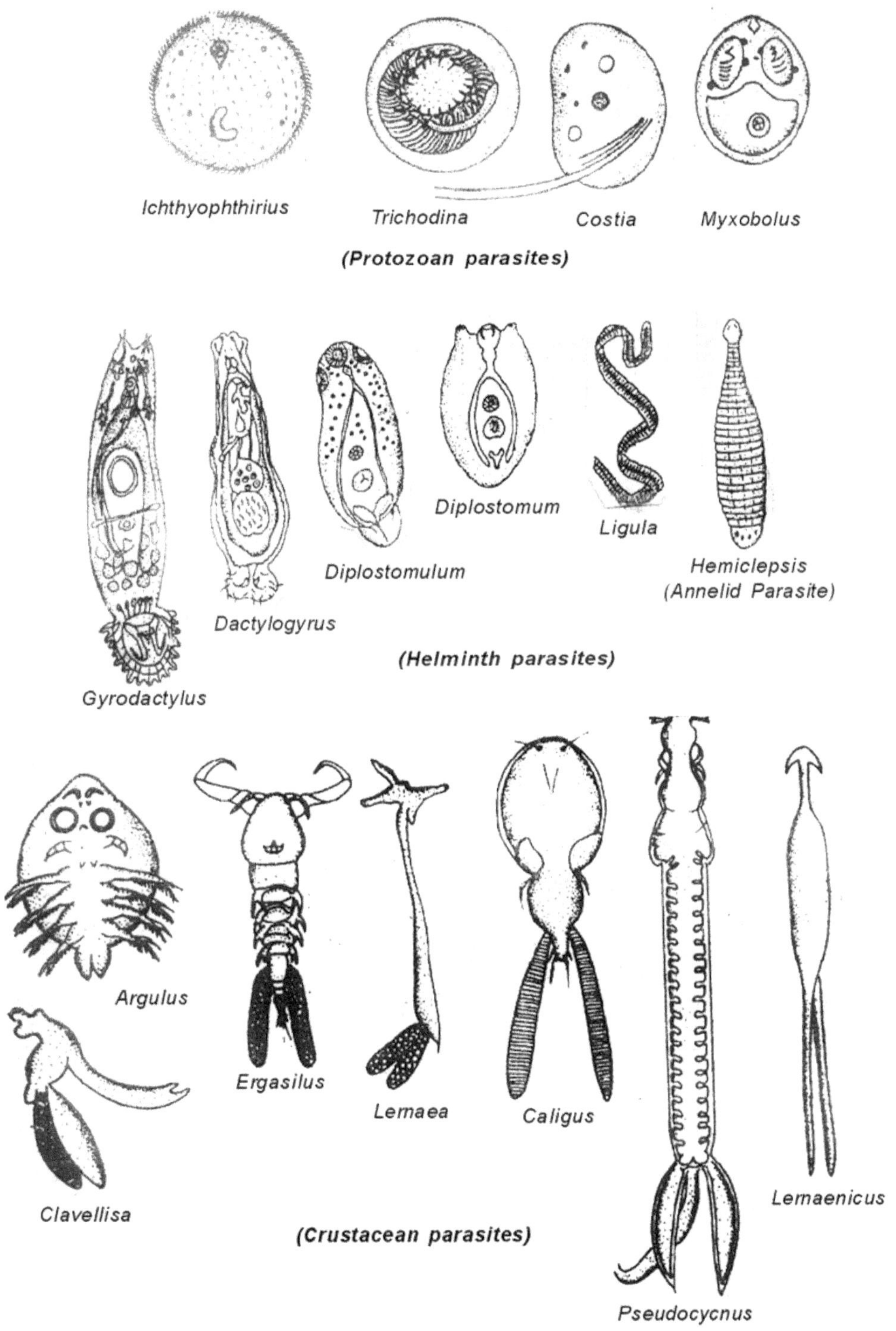

Fig. 4.28 Parasites of Fish and Shell fish

1. Dipping fishes in 3% NaCl solution
2. Keeping the fish for specific time in the dilute solutions of formalin, copper sulphate and potassium permanganate.
3. Pathogenic bacteria can be killed by keeping the fish for one hour in 1:50,000 Alkyl dimethyl benzyl ammonium chloride solution.
4. Chlorination of pond at regular intervals is necessary.
5. To destroy the eggs and young stages of leeches, ponds must be dried and sprayed with lime.
6. Lice can be easily taken out with the help of forceps.

4.5 Parasites of Domestic Livestock - Pig & Sheep

The man has domesticated the animals over the centuries to meet his requirements of food, clothing, protection or defense, companionship, transport etc. They include cattle, buffaloes, sheep, goats, pigs, horses, cats, dogs and poultry etc. They produce milk, meat, wool, skin and eggs for utilization by man. Domestication of animals began during the hunting and gathering phase of human civilization. As man realised the importance of domesticating animals for use as beasts of burden and as sources of milk, meat, fur, and leather etc, methods of improvement through selective breeding were used.

Livestock

The cattle (cows and buffaloes) constitute the most important species of livestock in India. They serve as the primary source of milk. Also, they play a significant role in agriculture in ploughing, harrowing, threshing, harvesting and transporting the produce to the market.

The cattle are used to draw water from wells. Their dung is used as fuel and as compost manure. Leather is used in the manufacture of footwear, bags, belts, coats etc. Horns, hooves and bones are used to prepare cattle feed and fertilizers. Compared to cows, buffaloes generally yield more milk of high fat content.

Cattle breeding

In the villages a few good pedigree bulls are selected and permitted to graze with the cows to bring about random breeding. About 10-60% cows are artificially inseminated by semen collected from high quality bulls with superior traits. Artificial insemination ensures good quality progeny and is also economical as semen from a single bull can inseminate several thousand cows or buffaloes.

An average cow or buffalo produces 8-10 calves during her lifetime.

To increase the milk yield, Indian cows are cross-breed with European breeds like Holstein, Brown Swiss, Jersey etc. The breeds developed through such cross breeding at the National dairy Research Institute, Karnal and in Kerala are Karanswiss and Sunandini.

Now a days, advanced technique is used for the production of super milch cows. The technique is called super ovulation and Embryo transplantation. In this technique, a pedigree bull and a high production cow are chosen to produce super milch cows. Super ovulation is induced by hormone injection. After artificial insemination 4-10 embryos are collected at a time. Each embryo is then transplanted into a carrier cow (surrogate mother).

By deep-freezing at -196 °C it is possible to preserve seven day old foetuses for several years for further use. An embryo can be cut into two to obtain homozygotic twins. The most beneficial outcome of embryo transplantation is the selection of high quality bulls for genetic upgrading.

Sheep

In our country sheep are reared for wool, skin, milk and meat, while droppings are used as a good manure. Sheep (*Ovis sp*) are classified according to the texture and quality of wool obtained from them. Sheep graze on the grass of pastures and hillsides.

The sheep rearers feed these animals on mineral mixture with oil cakes, pulses, sesame (til), corn and jowar. Sheep raising is an important occupation in the temperate Himalayan region (Kashmir, Himachal Pradesh, hill places of Uttar Pradesh, Sikkim and Arunachal Pradesh), Gujarat, Maharashtra and A.P. From April to October, shepherds migrate along with their sheep to pastures in high mountains. In winter, they are kept in houses and fed on foliage, dry fodder and grain.

Sheep that yield good quality wool and mutton are selected for breeding. Cross breeding with exotic breeds like Dorset, Horn and Merino (Australian) is being done to improve the local breeds. The wool obtained from the sheep of Kashmir is superior to that of other places. The Magra and Chokla are the best breeds from Bikaner and Kutchi and Joria are famous for the superior type of carpet wools.

Removal of wool from sheep at regular intervals is essential to promote the health of the sheep as it affords protection from ticks and mites which bite them for blood and transmits different disease causing pathogens. The recommended periods for shearing (removal) of wool from the sheep are winter (February-March) and rainy (August-September) seasons when rich grazing grounds are available.The manufacture of wool from sheep is a complicated process consisting of cleaning, drying, bleaching, dyeing, spinning and twisting (twining).

Pigs

The rearing of pigs is called piggery. The pig is used as a diet. i.e. it's flesh known as pork is taken by many people. The hairs of pig known as bristles are generally used in making the brushes for painting. Sausages are prepared by fresh minced pork. The fat of pig, squeezed from the body tissue, termed as lard, is used as a fine cooking medium and is also used in the manufacture or soaps, lubricants, greases, candles and waterproof materials. Pepsin, Thyroxin, pitutrin, insulin, liver extract, testosterone etc are derived from thyroid, pituitary, pancreas etc. of the pig which are used as pharmaceuticals. The skin of the pig is used as leather. The pig droppings are a good source of nitrogen, phosphorus and potassium for agricultural fields.

The important Indian breeds of pigs are Desi and Ghori and exotic pigs are Berkshire, large white Yorkshire, and Landrace etc.

Pigs feed on garbage, kitchen waste, vegetables and even human excreta. The domesticated indigenous pigs are fed on grass, straw and grains. They are also fed on roots and tubers.

In India, the pork industry is managed by traditional pig keepers and only recently commercial pig breeding farms and pork processing factories have been established at different places like Delhi, Aligarh, Kolkata, Mumbai, etc.

The flesh obtained from different parts of the body have been given different names i.e. Bacon (flesh obtained from the back and sides) and Ham (flesh obtained from the back of the thigh).

In India, the production of pork is 5% of the total production of meat.

The pork has to be cooked well to avoid possible tapeworm infections.

The details of the helminthes and arthropod parasites infecting man, cattle, pig, sheep and other domestic animals is already described earlier in the biology of helminth and arthropod parasites.

4.6 Parasites of Crops, Pathological Effects of Parasites - Disease Control

The crops are damaged by various organisms, i.e., the pests and diseases. The bacteria, viruses and fungi cause diseases in crops and some times cause huge yield and financial losses. Insect pests play a major role in crop losses being destructive at various stages of the crop starting from germination stage till harvesting and storage stages. Plant nematodes affect the root system and affect the growth of the plant and the yield. Likewise, rats and birds attack the crop in the final stages and cause considerable damage. Cattle, pigs, wild animals etc also sometimes attack the crops and inflict losses to the farmers. Of all the above, the insect pests occupy the first place as they cause huge crop lossess and affect the national economy significantly.

There are different types of insect pests damaging the crops depending upon their habits and habitats. Based on the mouth parts, the insect pests are differentiated into borers, sap sucking insects, nibblers, gall forming insects, leaf eating ones etc. In the same way, depending upon their habitat i.e., soil inhabiting ones damage the root system by nibbling, boring, cutting etc. Those insects living above the ground level feed upon the leaves, flowers, fruits, bore into the fruits, stems and feed on the internal tissues etc.

By observing the symptoms of the damage on the plant, the type of the insect is recognized.

For example, if a borer attacks the stem, the upper portions of the plant dry up and die, because of lack of nutrient supply from the root system and the lower portion of the site of attack will be normal. Such a condition i.e., upper dried condition of the plant is known as the 'dead heart'. In the same way when the root system is attacked by the pests, the growth of the plant is stunted and ultimately the plant dies. The leaf eating pests eat away or damage the leaves affecting photosynthesis and ultimately growth of the plant and the yields are affected. The flower or pod or head or fruit attacking insects directly feed upon the reproductive parts of the plant and affect the yield. The gall forming insects affect or block the transport of nutrients to other parts of the plants or affect photosynthesis and hinder the growth of the plant or kill it gradually.

Keeping all these points in view, the pest management practices have to be taken up for the plant protection.

There are different kinds of pest management practices in vogue. They are 1. Physical methods by exposing the pests to high or low temperatures, radiation, electricity, altering the other climatic conditions will reduce pest populations. 2. employing mechanical methods such as hand picking, using hand nets or bag nets, shaking or jarring, sieving, beating the pests with brushes, screening or meshing the doors and windows, using traps etc to trap or collect the insects or ward off the insects thus protecting our crops or houses or our ownselves from pest attacks. 3. Cultural methods such as tillage operations, weeding, crop rotation, trap crops, using resistant varieties from pests and diseases, inter cropping or mixed cropping, irrigation, manuring, using healthy seeds or cuttings or varieties etc. will minimise the pest attack or afford tolerance to the pest attack. 4. Legal methods or quarantine methods by which preventing the spread of pests from infested areas to uninfested areas i.e., from one place to the other or one country to the other etc is prevented. For this purpose, different laws have been enforced 5. Biological control in which the pests are controlled by using their natural enemies like parasites, predators, bacteria, viruses, fungi etc. 6. Chemical control in which different types of chemicals are used like inorganic pesticides (Sulphur, Arsenic, mercury, Fluorine, Chlorine, Zinc, Cyanide etc.), organic pesticides, like organo chlorinated compounds (DDT, BHC, Lindane, etc.) Organophosphorus compounds (Malathion, Parathion, Monocrotophos, demeton, Acephate, Carbamates (carbaryl, phorate) synthetic pyrethroids (Allethrin, permethrin, Decamethrin, fenvalerate), botanicals such as Nicotine, pyrethrum, Rotenone etc, antifeedants like Azadirachtin, repellents like Allethrin used in mosquito mats, bagon spray etc to repel mosquitoes. The pesticides are applied in different formulations like dusts, sprays, aerosols, fumigants etc to get optimum control and 7. Integrated pest management is the integration of all the control measures for pest management. Among these, the use of antifeedants, repellents, insect growth regulators, chitin synthesis inhibitors, pheromone traps etc is also practiced to give optmum plant protection from pest attacks.

The biology of protozoan, helminth and arthropod parasites has been already described.

A majority of protozoans are not pathogenic to man but a few can cause very serious diseases. *Entamoeba histolytica* causes amoebic dysentery, *Trichomonas vaginalis* infection severely damages vaginal membranes. It is spread through sexual intercourse. *Plasmodium* causes malaria. *Trypanosoma gambinse* causes sleeping sickness. Majority of protozoan parasites are intestinal or colic and cause dysentery or anaemia.

The helminths such as trematodes, cestodes and nematodes infect man at different parts of the body and cause different diseases. Most of the helminths harbour intestines and the rest attack various other parts like liver, brain, eyes, lungs, blood, lymph vessels and so on. Some of them are very serious parasites and kill their hosts in severe infections.

Similarly arthropod parasites attack man and cause severe pain by their bites, apart from transmitting pathogens such as protozoans, helminths, bacteria, viruses etc and pose serious problems, sometimes epidemics to human populations.

Control Measures

The diseases can be prevented by taking several precautionary measures such as clean surroundings, taking pure or uncontaminated water or sterilized water, wearing clean clothes, using curtains or meshes to the doors and windows and mosquito nets to beds, eating fresh and completely cooked food, keeping the cattle sheds, piggeries, poultry etc away from houses so that will take care of most of the diseases. When the man is affected by any of the parasites mentioned above he has to report to the health centres and get the disease diagnosed properly and the relevant medical attention must be taken till the disease is completely cured.

Table 4.8

	Pathogen	Treatment/drugs
		Protozoan Parasites
1.	*Entamoeba histolytica*	Iodoquinol, Metronidazol, Emetine, Dehydroemetine.
2.	*Trypanosoma*	Uramin, Pentamidine isothionate, melarsoprol, Difluoromethylornithine (DFMO), Nifurtimox, Allopurinol, Benznidazole.
3.	*Leishmania*	Antimony drugs. Stibogluconate etc.
4.	*Giardia*	Quinacrine, Dihydrochloride, Metronidazole and Furazolidine.
5.	*Trichomonas*	Metronidazole and Diodoquine .
6.	*Plasmodium*	Chloroquine, Quinine sulphate, Tetracycline or Sulfadiazine.
		Helminth Parasites
1.	*Schistosoma*	Praziquantel, Dxamniquine, Sodium Antimony Tartrate, Foudin, Anthiomaline, Miracil.
2.	*Fasciola hepatica*	Emetine
3.	*Paragonimus*	Emetine, chloroquine, Bithiono.
4.	*Diphyllobothrium*	Filix mas, mepacrine (atebrin), dichlorophen.
5.	*Taenia solium*	Filix mas, mepacrine (atebrin), dichlorophen, yomesan, hetrazan.
6.	*Echinococcus*	Filix mas, mepacrine (atebrin), dichlorophen, yomesan, hetrazan.
7.	*Hymenolepis*	Dithiazanine
8.	*Trichinella*	Corticosteroids
9.	*Ancylostoma*	Tetrachlorethylene, Bephenium hydroxynaphthoate, hexylbresorcinol.
10.	*Wuchereria*	Mel. w. (arsenical drug), Hetrazan (Diethylcarbamazine) para-melaminyl phenyl stibonate.
11.	*Enterobius*	Piperazine salts, pyrvinium pamoate, Dithiazanine iodide.
12.	*Ascaris*	Piperazine salts, Bephenium hydroxynaphthoate, hexylresorcinol, oil of chenopodium, hetrazan.
13.	*Dracunculus*	Hetrazan.
		Arthropod parasites
1.	*Pediculus*	Benzene hexa chloride (BHC), lindane, prioderm lotion.
2.	*Bed bug*	DDT, BHC, Kerosene, camphor, menthol or phenol lotions.
3.	*Mosquitoes*	Release of larvivorous fish, mosquito repellents (allethrin). Fumigants, etc.
4.	*Housefly*	Baygon baits, synthetic pyrethroids etc.
5.	*Fleas*	Malathion, DDT, BHC, Lindane, Chlorpyriphos etc.
6.	*Ticks and Mites*	r-BHC, insect repellents, acaricides.

Table 4.8 gives a list of drugs for the control of infections and diseases caused by protozoans and helminthes and pesticides for the control of arthropod (insects, ticks and mites) parasites. However, the diseases must be diagnosed and the treatments must be done under an expert physician supervision only.

4.7 Nematode Parasites of Crop Plants

The nematodes are world wide and are found in salt water, freshwater and in the soil. Besides, some are parasites of man and animals. Most of the species found in the soil are either saprophytic or predatory, whereas some of them are parasites of plants. The plant parasitic nematodes can be classified into three broad categories according to their feeding habits.

1. *Ectoparasitic nematodes :* These are free living forms that move about freely near or on the surface of the roots. Eg : *Trichodorus, Paratylenchus* etc.
2. *Endoparasitic nematodes :* These nematodes completely enter the root and lose physical contact with the soil or a large part of their body is lodged in the plant tissue. The normal habitat of these nematodes is inside the roots. Eg : *Meloidogyne, Heterodera, Pratylenchus* etc.
3. *Semiendoparasitic nematodes :* These nematodes penetrate partially in the root. The head and neck become permanently fixed in the cortex tissue and the remaining body portion remains outside in the soil. Once these nematodes penetrate the root, they become immobile. *Eg : Tylenchulus*, *Rotylenchulus*, etc.

The mouth parts of these nematodes are armed with a hollow protrusible stylet or spear which is used for puncturing the plant cells and extracting the sap. A spear is an important characteristic which is used for the identification of various groups of nematodes. For eg : in the order Tylenchida, it is known as stomatostyle and it develops in the pharynx. The following are some of the important plant nematodes presented in the Table 4.9.

Table 4.9 *The plant nematodes belong to the Phylum Nemathelminthes, Class, Nematoda and order Tylenchida*

S.No	Nematode	Scientific Name	Family
1.	Citrus nematode	*Tylenchulus semipenetrans*	Tylenchulidae
2.	Root knot nematode	*Meloidogyne incognita*	Heteroderidae
3.	The stem and bulb nematode	*Ditylenchus angustus*	Tylenchidae
4.	Cyst nematode	*Heterodera avenae*	Heteroderidae
5.	Burrowing nematode	*Radopholus similis*	Pratylenchidae
6.	Leaf and bud nematode	*Aphelenchoides spp.*	Aphelenchidae
7.	Root lesion nematode	*Pratylenchus spp.*	Pratylenchidae
8.	Lance nematode	*Hoplolaimus spp.*	Hoplolaimidae
9.	Seed gall nematode	*Anguina spp.*	Tylenchidae
10.	Golden nematode of potato	*Globodera rostochiensis*	Heteroderidae
11.	Spiral nematode	*Helicotylenchus spp.*	Hoplolaimidae
12.	Reniform nematode	*Rotylenchulus reniformis*	Nacobbidae
13.	Stunt or stylet nematode	*Tylenchorhynchus spp.*	Tylenchorhynchidae
14.	Rice root nematode	*Hirschmanniella oryzae*	Pratylenchidae
15.	Dagger nematode	*Xiphinema spp.*	Longidoridae
16.	Stubbyroot nematode	*Trichodorus spp.*	Trichodoridae
17.	Sting nematode	*Belonolaimus spp.*	Belonolaimidae
18.	Red ring nematode	*Rhadinaphelenchus cocophilus*	Aphelenchoididae

The importance of a nematode as a plant parasite depends largely on whether or not the population limit exceeds the level at which economic damage occurs to a crop plant. This concept of population threshold at which yield loss begins is often used in connection with crop pests to determine the tolerance level of a host crop. For any set of environmental conditions, each host crop has its own tolerance level for a nematode species. Thus, a nematode causes economic damage only if its population density exceeds the tolerance level of the plant grown in the field. The Table 4.10 shows the information on threshold densities of certain nematodes causing damage on specific crops.

Table 4.10 *Approximate threshold densities of certain nematodes causing damage on specific crops*

Nematode	Crop	Threshold density
Belonolaimus sp.	Cotton, Groundnut, Soyabean	25/500 cm^3 soil
Ditylenchus sp.	Onion	1-5/500 g soil
Globodera sp.	Potato	31 cysts/100 g soil
Hemicycliophora sp.	Vegetables	50-1300/100 ml soil
Heterodera sp.	Carrot	< 10 cysts/100 ml soil > 500 larvae/100 ml soil
Heterodera sp.	Peas	11-30 cysts/100 ml soil 700-2000 larvae/100 ml soil
Heterodera sp.	Sugar beet	4 -10 eggs/g of soil
Meloidogyne sp.	Tomato, cucumber	100-200/100 ml soil
Meloidogyne sp.	Strawberry	25/500 cm^3 soil
Meloidogyne sp.	Ground nut	100/500 cm^3 soil
Meloidogyne sp.	Tobacco, Tomato	25-50/500 cm^3 soil
Meloidogyne sp.	Cucumber	100-200/100 ml soil
Meloidogyne sp.	Tomato, Cucumber	100-200/100 ml soil
Pratylenchus sp.	Groundnut	25/500 cm^3 soil
Pratylenchus sp.	Tobacco	> 200/500 cm^3 soil
Pratylenchus sp.	Daffodil	1-5/500 g soil
Pratylenchus sp.	Potato	> 500/500 g soil
Pratylenchus sp.	Onion	> 5/500 g soil
Pratylenchus sp.	Roses	> 190/100 ml soil
Rotylenchus sp.	Beet, carrot, vegetables	100-1000/100 ml soil
Trichodorus sp.	Beet	20-240/100 ml soil
Tylenchorhynchus sp.	Grasses, grain crops, clover and peas	500-4500/100 ml soil
Tylenchorhynchus sp.	Azalea	> 100/500 cm^3 soil

(*Source : P. Parvatha Reddy 1987*)

The following is the account of some important plant nematodes.

1. The citrus nematode, *Tylenchulus semipenetrans* : This nematode causes slow decline disease of citrus. The symptoms include a general reduction of tree growth and vigour, yellowing and shedding of leaves and undersized fruits. The decline symptoms are more pronounced in the uppermost portion of the trees. The tree performance is lowered when the nematode population reaches about 4000 larvae and males per 10 grams of roots. In heavily infested roots, the cortex separates readily from the vascular tissue.

Life Cycle **:** The life cycle is completed is 6 to 8 weeks on *citrus aurantium*. The female nematode is saccate with a well developed stylet having basal knobs and with the excretory pore located just anterior to the vulva. It is a partially exposed sedentary endo-parasite. The young larva feeds on the outer cortical layer of the young roots. As the larva destined to be a female develops through successive moults, its head penetrates deep into the rootlet, while the body outside the root enlarges and becomes saccate. At maturity, the female deposits eggs in a jelly like mass. The eggs hatched in 12 to 14 days and sex differentiation could be made at second stage larvae. The second stage male larvae develop to maturity within 7 days without feeding while the female larvae required 14 days to locate the root, feed on the epidermal cells and moult (Fig. 4.29).

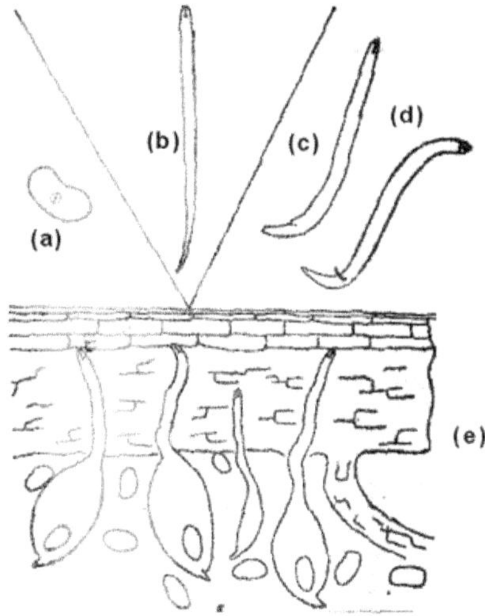

Fig. 4.29 Life cycle of *Tylenchulus semipenetrans.* (a) egg, (b) second stage larva, (c) young female, (d) male (e) adult females attached to root

2. Root lesion nematodes, *Pratylenchus spp. (Fam : Pratylenchidae) :* These nematodes are small, vermiform and have an annulated cuticle and a sclerotized lip region. Their stylets are well developed and bear basal knobs and the median oesophageal bulb is ovate to spheroid. These are the most important nematodes associated with the destruction of plant roots. They are endoparasites of fibrous roots on which they cause lesions, browning or necrosis. They move from root to root, feeding and laying eggs. The first moult takes place within the egg and three moults occur outside. The eggs hatch in 15-17 days, the larval life is 15-16 days and it takes 15 days from maturity to egg production (Fig. 4.30).

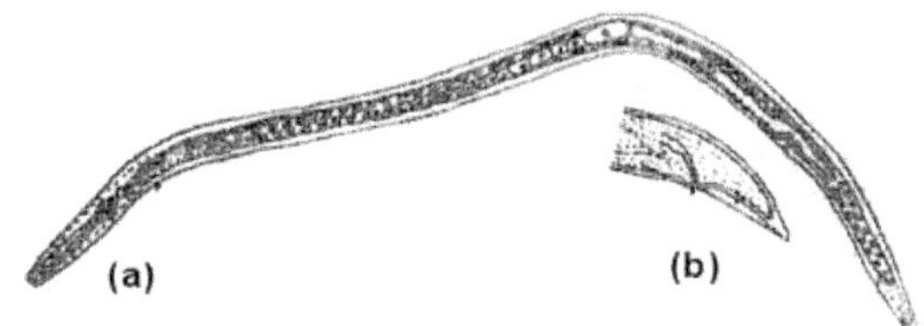

Fig. 4.30 *Pratylenchus coffeae,* (a) female (b) male tail

3. Root knot nematode, *Meloidogyne incognita (Fam : Heteroderidae) :* These nematodes show marked sexual dimorphism. Vermiform males have a well developed oesophagus and no bursa whereas the females are swollen and flask shaped, with two ovaries and an anus. The nematodes produce galls on roots which may range from a string of beads on smaller laterals to a heavily knotted root mass. With the formation of galls, the flow of water and food through the roots is blocked and the plants show symptoms of starvation. Several species of these nematodes are known to attack over 1700 species of plants and are most damaging in warm climates (Fig. 4.31).

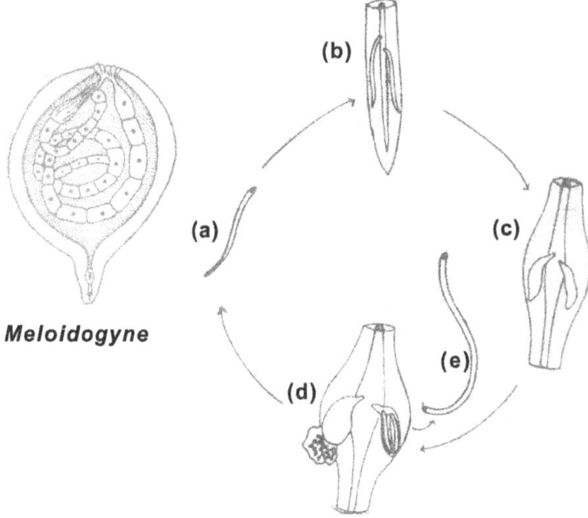

Fig. 4.31 *Meloidogyne spp*, (a) second stage larva (b) larvae inside the root, (c) development of larvae, (d) gall with mature female and egg mass, (e) male free in soil

The eggs are deposited in a mass within the root galls. The first stage larva which is found inside the egg shell and the second stage larva which is free are the infective stages. On entering a young root, the larva establishes itself and begins to feed and grow, hardly showing any movement, except that of the head while feeding. During its development, the larva swells. It becomes flask shaped and after several moults, develops into an adult.

4. Cyst nematode, *Heterodera sp. (Family : Heteroderidae) :* These nematodes resemble *Meloidogyne* in morphology. In addition, cuticle of the female *Heterodera* forms a tough, resistant, brown sphere or a pear shaped cyst containing embryonated eggs. These cysts may remain dormant in the soil for several years. The second stage larva constitutes the motile infective stage and the subsequent development takes place within the roots of the host plant (Fig. 4.32).

   Because of their attack, the roots proliferate into a shallow bushy system. The infested fields frequently have circular or oval patches of plants exhibiting poor growth. These patches increase year after year, if cropping continues.

   The fertilized female lays eggs inside the cysts. The hatching of the first instar larvae takes place within the eggs. The second stage larvae emerge from the eggs, move freely in the soil, attach with the host roots and penetrate the root mostly just behind the root tip.

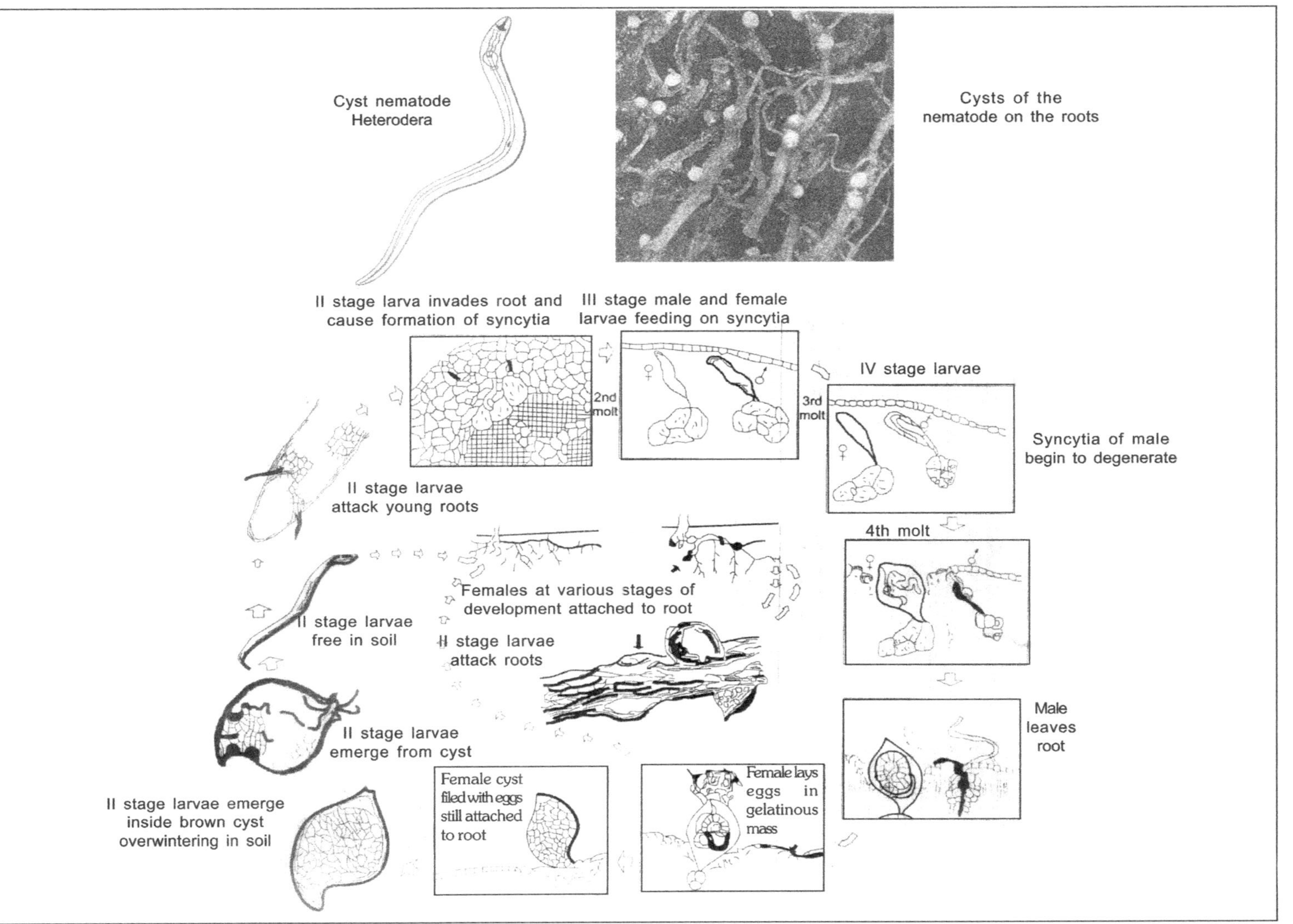

Fig. 4.32 Disease cycle of the cyst nematode *Heterodera glycines*

The second stage larvae undergo second, third and fourth moults and leave the host in the form of adult male. The female larvae are sedentary and attain adult stage after fourth moult. The adult female has its head lodged inside the root and the body projecting outside it. The complete life cycle takes one to two months.

5. Seed gall nematode, *Anguina spp. (Fam : Tylenchidae) :* These nematodes are obligate plant parasites giving rise to galls in leaves, stem and flowers of a number of cereals. *A. tritici*, which causes ear cockles in wheat, is the best known species in this group. The hard round galls containing the nematode larvae replace the seed in the ears. The nematode is characterized by a narrow, flattish head with irregularly shaped oesophageal glands. The ovary consists of cells arranged about a rachis and ending in a cap cell. The adults are slightly swollen and ventrally curved (Fig. 4.33).

   The second stage larvae liberated from the galls in moist soil infest wheat seedlings at an early stage. For some time, they live ectoparasitically between the young leaves and as the flower primordia develop, the nematodes enter the tissue. They moult successively to reach the adult stage. After mating, the females lay hundreds of eggs in the galls. The eggs develop into the second stage larvae and in this state, the nematode has been observed to resist desiccation for as long as 28 years.

6. Lance nematode, *Hoplolaimus spp. (Fam : Hoplolaimidae) :* These nematodes are known to attack more than 30 host plants. This nematode is widely distributed in different parts of India. Due to the nematode infection plant growth is retarded. The greatest nematode populations are found in the roots and soil of trees of 10 to 20 years old and smallest populations are associated with trees one to 5 years old (Fig. 4.34).

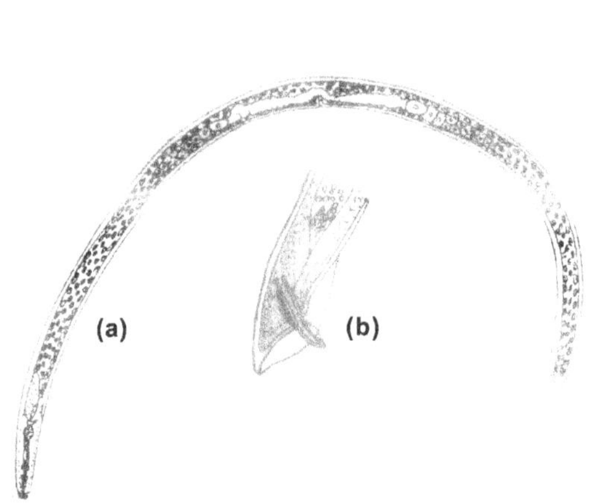

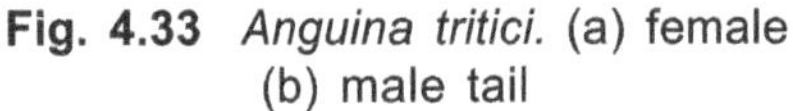
Fig. 4.33 *Anguina tritici.* (a) female (b) male tail

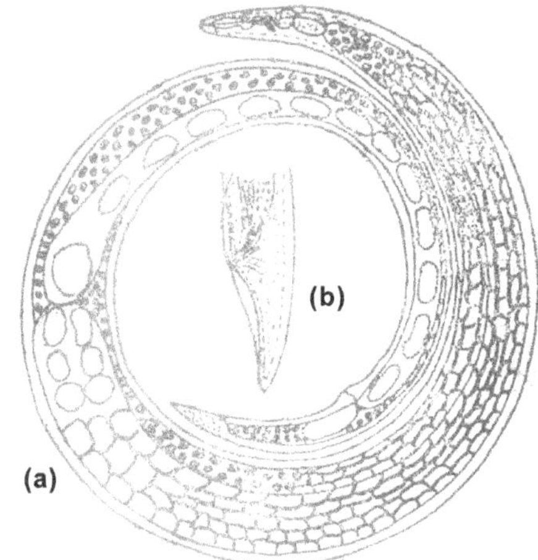

Fig. 4.34 *Hoplolaimus indicus.* (a) female (b) male tail

Life Cycle : The first moult occurs within the egg and the development outside the egg consists of 3 larval stages and the adult. Sex differentiation is indicated early in the second moult by the presence, in the female only, of 4 specialized ventral chord nuclei coinciding the position with that of the vagina.

7. Burrowing nematode, *Radopholus similis (Fam : Pratylenchidae) :* These nematodes are known to attack banana, sweet potato, pepper, arecanut, coconut, citrus, sugarcane, coffee, maize, avocado, tea, vegetables, ornamental trees, grasses, weeds, etc. About 244 host plants are attacked by this nematode.

   With the increase in nematode population, the roots are damaged completely, resulting in the debility and smaller fruits, finally the plants or trees 'fall' under wind pressure.

   Life Cycle : The burrowing nematode has a migratory endoparasitic habit and although the stages remain vermiform throughout, sexual dimorphism is apparent with adult males being somewhat degenerate and probably non-parasitic. Eggs are normally laid in infested tissue over 7-8 days at the rate of about 4 eggs per day. The life cycle from egg to egg extends over 20-25 days with eggs taking 8-10 days to hatch and the larvae 10-13 days to complete their development and mature. (Fig. 4.35)

8. The spiral nematode, *Helicotylenchus spp. (Fam : Hoplolaimidae) :* This nematode is known to infect banana, cocoa, sweet potato, citrus, sugarcane, cassava, coffee, maize, mango, oil palm, rice, rubber, tea, yam, avocado, grapevine etc.

   It causes root necrosis, die back and disfunction leading eventually to debility of the entire plant.

   Life Cycle : Groups of 8-26 eggs are laid in the cortical tissues of the roots. The eggs hatch in 48-51 hours at 30 ^{0}C in tap water and the first instar moults into second instar soon after hatching. The second stage larvae are recognized by their ventral digitate tail process. The female gonad primordium is 2 and 6 celled in the early third stage larvae and at the third moult respectively. During the fourth moult, the male and female gonads complete their development and the vulva and vagina are seen in females still within the roots and the entire life cycle is completed there within the root itself. (Fig. 4.36)

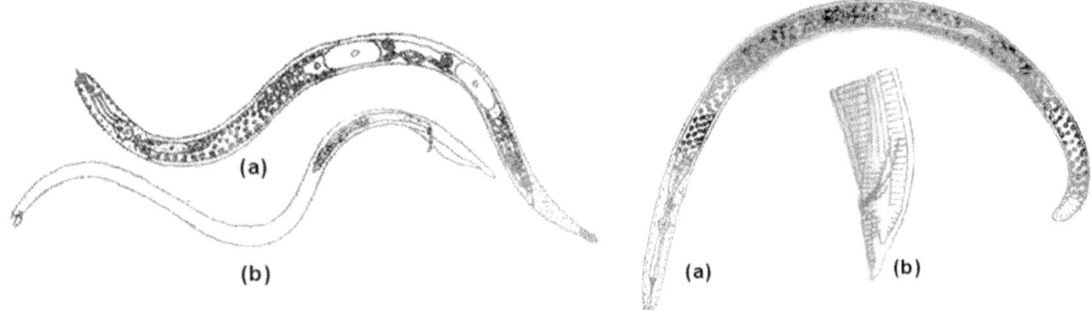

Fig. 4.35 *Radopholus similis.* (a) female (b) male

Fig. 4.36 *Helicotylenchus multicinctus.* (a) female (b) male tail

9. Golden nematode of potato, *Globodera spp. (Fam : Heteroderidae) :* It is known to attack, Potato, tomato and brinjal severely. Also a number of other solanaceous plants are recorded as hosts.

   Due to the nematode infection, plant growth is stunted and wilting occurs during hot parts of the day.

Globodera spp. is amphimictic, with sessile, globose females and motile, vermiform males. The second stage larva is the resistant, dormant and infective stage remaining inside the egg within the cyst formed by tanning of the female's cuticle. The cysts remain in soil after the death of the host, each containing about 500 eggs. After hatching, the second stage larva enters the root of a host plant, just behind the root tip and moves through the root to feed on pericycle, cortex or endodermis cells which are modified to form a large syncytial transfer cells on which the nematode feeds throughout the rest of the development upto the adult stage.

The adult female swells greatly, rupturing the root cortex and protrudes from the root, exposing the vulval region while the head and neck remain embedded in the root. As they develop, males become coiled within the earlier larval cuticles and on maturity they rupture them, leave the root and are attracted to females by a secretion. Both sexes mate several times. Males do not feed but remain active for upto 10 days, females normally mate soon after rupturing the root cortex and within 50 days of larval invasion (Fig. 4.37).

Fig. 4.37 Potato roots infected with *Globodera rostochiensis*

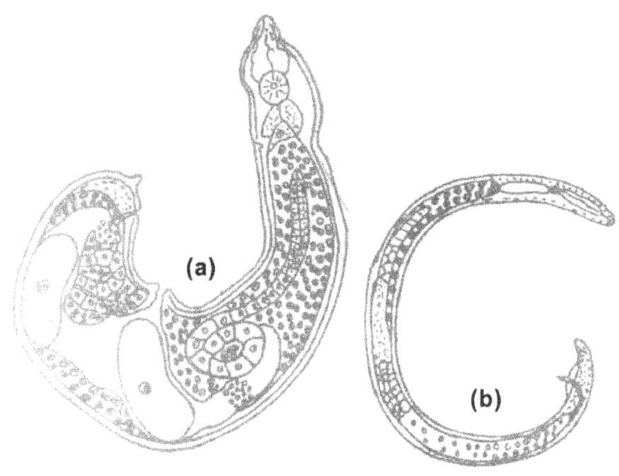

Fig. 4.38 *Rotylenchulus reniformis*, (a) female (b) male

10. The stunt nematode, *Tylenchorhynchus spp. (Fam : Tylenchorhynchidae) :* This nematode is known to attack a wide range of host plants like tobacco, maize, cereals, potato, sweet potato, legumes, cotton, tomato etc. The plants infected by these nematodes have stunted growth and have a much retarded root system.

    Life Cycle : Males appear necessary for reproduction. The life cycle is completed in about a month. Nematode populations increase rapidly between 30-34 ^{0}C.

11. The Reniform nematode, *Rotylenchulus spp. (Fam : Nacobbidae) :* This nematode is reported to attack 68 host plants belonging to 30 families. The important ones being cotton, banana, citrus, papaya, pineapple, mango, jack fruit, coffee, tea, tobacco, potato, tomato, vegetables, groundnut, sugarcane, rice, maize, etc.

    The nematode causes severe pruning of seedling roots and consequent stunted growth.

Life cycle : The adult female is an obligate, sedentary, semi endoparasite of roots while the male is nonparasitic. The species is hermaphrodite and reproduces by amphimixis.

The first moult occurs within the egg. The remarkable feature is that the eggs are able to hatch in water and the larvae develop to the preadult stage without feeding and growth, through a series of atleast three super imposed moults. The complete life cycle takes about 17-23 days. After infecting the roots, the young female feeds, grows, becomes obese and starts laying eggs in 7-9 days : An average of 66 eggs per egg mass is recorded on cotton (Fig. 4.38).

12. The White tip nematode, *Aphelenchoides spp. (Fam : Aphelenchidae) :* This nematode is a serious pest of rice. Besides rice, it also attacks strawberry, tuberose, chrysanthemum and others.

Infected seeds emerge late in seed beds and produce small seedlings. General stunting of the plant accompanies leaf injury. Panicles are shorter and spikelets reduced in number.

This nematode is carried beneath the hull of the rice kernel in a quiscent immature, usually pre-adult stage. In this stage the nematodes may lie dormant as long as 23 months. When the infected seed is planted and sprouts, the nematodes revive and move to the growing points of the leaves and stems, where they feed ectoparsitically. Apparently most eggs are deposited in leaf axils or in panicles, and several generations occur each season. Many nematodes find their way into the panicles and some eventually become located inside the hulls. On an average 5-6 nematodes are present in each seed. The life cycle takes 8 days at 23^0 C and 10 days at 21^0 C.

13. The rice stem nematode, *Ditylenchus. spp. (Fam : Tylenchidae) :* This is a serious parasite of rice. The symptoms of the attack are chlorosis of the leaves, branching of the stem in the infected portions, the panicle remaining enclosed in the leaf sheath or produces grains near the tip of the panicle while the rest remain unfertilized and do not give grains.

Cottony masses of preadult nematodes (4th stage larvae) are formed as the plants mature and dry. During humid periods, they become active, climb up on the stems and invade the growing point. They feed ectoparasitically on the epidermal cells of the young seed head, the peduncle, the part of the stems just above the upper nodes and the young leaves rolled about the bud above the growing point. Developing heads are found to contain enormous numbers of nematodes in all stages from eggs to adults. At no time they are found living endoparasitically in plant tissues (Fig. 4.39).

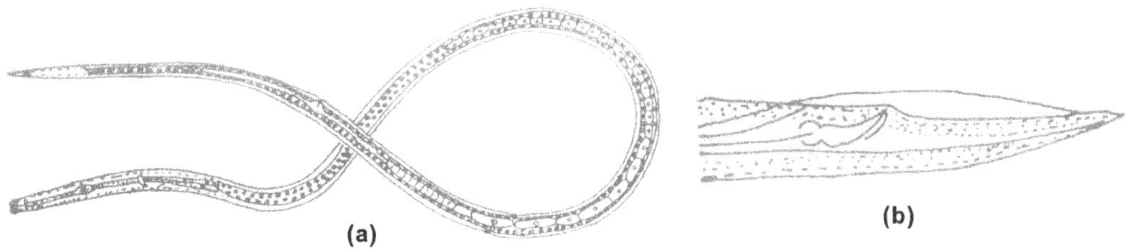

Fig. 4.39 *Ditylenchus angustus*, (a) female (b) male tail

General Methods of Control of Nematode Diseases

The plant parasitic nematodes are not as host specific as zooparasites, so a single parasitic nematode has a wide host range and each parasitic nematode may infect many host species. The pattern of life history is less variable in comparison to zooparasites. The adults are either parasitic or free living but in either case, the infective stages are found living freely in the soil. The main idea of plant parasitic control is to stop the life cycle whichever is possible in the particular species to protect the host species. The following control measures have been in practice for the plant parasitic nematode management.

1. *Chemical control :* Chemicals with nematicidal properties are known as nematicides. Many such chemicals are recommended for nematode control. There are two types of nematicides. Fumigants and nonfumigants. Chemically they can be categorized into halogenated aliphatic hydrocarbons, methyl isocyanate liberators, organophosphates and organocarbamates. The common pesticides of nematodes are given in Table 4.11.

Table 4.11 *Common nematicides*

Types of nematicides	Common name
1. Fumigants	
A. Halogenated aliphatic hydrocarbons	
Bromomethane	Methyl bromide
Trichloronitromethane	Chloropicrin
1, 3-Dichloropene	1, 3-D
1, 2-Dibromoethane	EDB
1, 2-Dibromo-3 Chloropropane	DBCP (Nemagon)
B. Methyl isothiocyanate liberators	
Methyl isothiocyanate	MIT
2. Non-fumigants	
A. Organophosphates	
O Ethyl S, S-Dipropylphosphorodithioate	Ethoprop
O, O-Diethyl-O- [(methylsulfinyl) phenyl] - Phosphorothioate	Fensulfothion
Ethyl 4-(methylthio)-m-tolyl-Isopropylphosphoramidate	Phenamiphos
B. Organocarbamates	
2, 3-Dihydro-2.2-dimethyl-7-benzofuranyl methylcarbamate	Carbofuran
2-Methyl-2 (methylthio) propionaldehyde O-(methylcarbamoyl) oxime	Aldicarb
Methyl N, N-(dimethylcarbamoyl) - N - (methylcarbamoyl) Oxy) - 1-thioformimidate	Oxamyl

Use of nematicides is the most important and reliable means for management of root knot nematodes. Nematicides are principally applied to control nematode population in the soil before the crop is grown. As the number of nematicides available for safe control of nematode diseases is declining drastically, the other methods of control are becoming increasingly important. Among the fumigants Carbon disulphides, Chloropicrin, DD, DBCP (nemagon), Vapam, methyl bromide, and others have been banned now due to their polluting effect. Now more stress is on systemic, non-fuming nematicides such as Carbofuran, Aldicarb, Oxamyl, Phenamiphos, and Phorate etc.

Chemical control of nematodes has not made much headway in India mainly due to the prohibitive cost of the nematicides, difficulty of application over large areas and the lack of awareness of the nematode problem. In spite of high initial costs of nematicides, there is increasing evidence that a judicious use of these chemicals would be economical on selected crops, such as cotton, sugarcane, tobacco, citrus, grape vine, potato, vegetables, and even wheat and barley in heavily infested fields.

Cultural Practices

They result in partial or complete control of nematodes and include :

Crop Rotation of Non-Host Plants

Several nematode species can infect only a few crops. Since all plant pathogenic nematodes are obligate, the absence of susceptible hosts from the soil for two to three years will result in the elimination of nematodes from that area through starvation and inability to reproduce. The value of crop rotation has long been recognized in the control of *Heterodera spp*. Thus continuous cropping with potatoes increases the cyst population of *H. rostochiensis* more than does three-year rotations with corn, green beans and red clover. Brown (1961) gave some idea of the effect of crop rotation on the yields of potatoes growing in soil infected with *H. rostochiensis*. Potato yield after three years of continuous cropping was 1.631 kg/hectare, and with a non-host in the second year, it was 4.515 kg/hectare.

In recent years a new term "cropping system" which is defined as sequence of growing crop may be spatial or temporal. Arrangement of different crops such as mixed cropping, row intercropping and strip cropping is known as spatial sequence. Distribution of crop over a period of time such as crop rotation relay cropping, discontinuous cropping and fallowing is called a temporal sequence. Both spatial and temporal sequencing have been found effective in controlling root knot nematodes on vegetable crops (Khan, 1997). A number of non-host or poor host of root knot nematodes like wheat, rice, onion and maize are known to increase yields of vegetable crops in the appropriate temporal sequence. Economic consideration is however a serious demerit of crop rotation. It can be successfully used when all the grown crops in the sequence are of economic value. Rotation of crops or cultivars with different genes for resistance against the species or races of root-knot nematodes has been suggested.

Some plants appear to be antagonistic to nematodes because they reduce populations to a greater extent than do non-hosts. *Crotolaria* is a good example of this. It has been suggested that the *Crotolaria* root system has a toxic effect on nematodes, and planted in advance of a crop, it may reduce subsequent damage.

There is evidence that *Tagetes* spp. may reduce the populations of some plant parasitic nematodes. Oostenbrink and his colleagues (1957) tested 16 varieties of *Tagetes patula* and *Tagetes erecta* placed between the rows or around other plants. They found that all varieties suppressed the populations of *Pratylenchus* in the roots of other plants as well as in the soil, resulting in a better growth of perennials in the second year of the succeeding crop. The *Tagetes* had to be grown for three to four months and it was suggested that the roots had a nematicidal action. In spite of many criticisms, the use of cover, trap or enemy crops provides a potentially useful control measure in crop rotations. Such an opinion is supported by the evidence that *Asparagus* exudes a nematicidal chemical from its roots.

In addition some grasses, varieties of mustard, Margosa (*Azadirachta indica*) root exudates, and *Asparagus* that have been tested as enemy plants to nematode population. In some vegetable plots and in greenhouse grown crops mulching of soil with straw, weeds, grasses, old newspaper sheets, dry pine needles, tree bark etc. have been found effective in reducing nematode attack.

Sanitary Practices

They include the cleaning of all machinery thoroughly before moving into an uncontaminated area, taking care not to bring nematodes into a field by means of contaminated nursery stock, seed, containers, and so on, keeping soil fallow free from host plants which deprive nematodes roots on which they feed. Thus regulatory methods (quarantine) can be the least expensive methods to control nematode diseases.

Flooding

Some species of nematodes have become adapted to living in the normal soil with ordinary moisture content and depend on a certain amount of aeration present in it. Flooding the land for a period of several months results in the death of the nematodes and thus frees the land of these pathogens. The applicability of this method is quite limited.

Fallowing

The fields or plants are deeply ploughed to keep them free from vegetation and exposed to dry summer heat of the sun. Starvation and desiccation kills the nematodes.

Predators

Although simple observation have shown that predaceous species of nematodes occur in the soil and can be utilized to eat away parasitic forms. It has been found that many nematophagous species of fungi are effective means of destroying the nematodes and these can be encouraged in the soil through organic soil amendments, such as the addition of decomposable vegetable matter.

Physical Methods

Plant parasitic nematodes can be easily killed in the laboratory by the application of heat irradiation, osmotic pressure etc. It is difficult, to employ such methods on large quantities of soil especially if they are under cultivation.

Irradiation

There is strong evidence that nematodes are killed or seriously affected by irradiation. Fassuliotis and Sparrow (1955) demonstrated that irradiation of potato tubers with X-rays inhibited sprouting and *Heterodera rostochiensis*. Cysts of this developing nematode exposed to 20,000 roentgens contained only brown and dead eggs; at 40,000 r. the eggs lost their contents completely. It is, however, difficult to irradiate nematodes. In fact it is impracticable because of the time needed to irradiate even small areas and because damage to plant roots occurs at levels of irradiation below those required to disrupt the nematode's life cycle.

Heat

Heat treatment is the most successful physical control measure developed so far. It is widely used for the killing of nematodes within plant tissues before planting and has proved useful on nematode-infested bulbs, corms and tubers, and roots of plants such as chrysanthemums, strawberries, bananas, and citrus. This process involves determining the correct time and temperature for killing the nematode but not the host. This entails immersing the nematodes (contained in as small a volume of water as possible) into thermostatically controlled water baths at different temperatures for different lengths of time. After such treatment the nematodes are examined for activity. Blake (1961) suggested that banana sets less than 13 cm in diameter should be immersed in water at 55^0 C for 20 minutes in order to kill *Radopholus similis*.

A new approach to heat treatment is the application of dry heat (Thomason et al., 1960). Sweet potato seed roots were treated at 45^0 C for 30 hours and these completely controlled root knot nematodes, with 60% of the roots surviving. The white tip nematode of rice *Aphelenchoides besseyi* is of widespread occurrence in India in most rice-growing areas. Hot water treatment of the seed at 52 - 55^0 C for ten minutes is the best way to ensure the complete destruction of the nematodes.

More than 100 fungus species have been designated as nematophagous, fungi that "feed" on nematodes either as predators or parasites. These nematohagous fungi, members of either the phycomycetes or the class Moniliales of the Fungi Imperfecti, are abundant throughout the world and are frequently encountered in soil samples.

Soil Solarization

Various methods of soil solarization have been found quite efficacious in controlling soil borne nematodes. For example *Hirschmanniella mucronata* population is reduced by black polythene mulching. Application is reduced by yard manure or de-oiled *Pongamia glabra* reduced *M.incognita* and *R.reniformis*.

Biological Control

According to Garrett (1965), "Biological control in any condition under which, or practice whereby, survival or activity of pathogen is reduced through the agency of any other living organism (except man himself), with the result that there is a reduction in incidence of the disease caused by the pathogen". Trap crops, resistant varieties and antagonistic plants that release nematocidal root exudates may be considered useful or potential biological controls for nematodes.

Biological control aims at increasing the parasites and predators of nematodes in the soil, to increase the mortality of plant nematodes. This can be done by changing the environment, adding organic amendments or introducing other organisms directly.

To increase the concentration of nematode-destroying organisms in the soil it is necessary to alter the environment so that some of the forces restraining the multiplication of the organisms are weakened. The only serious attempt to biologically control plant nematodes has been concerned with the use of nematode-trapping fungi.

Oil cakes of margosa, castor and peanut at 0.2% (W/W) three weeks before planting significantly reduce infestation of *M.javanica* on okra and tomato. In almost all these experiments a soil amendment was used, thereby altering the environment in favour of nematode-trapping fungi.

In the late 1930's Linford and his group successfully reduced the incidence of root-knot of pineapple by incorporating green plant material into the soil. It is generally assumed that the addition of this material stimulates the population of organisms antagonistic to nematodes, which in turn, reduces the parasitic nematode population. There are several nematode trapping fungi as *Arthabotrys robusta. A. oligospora*, and *A. superba*.

Duddington and Duthoit (1960) obtained statistically significant reductions in the mean number of larvae of *Heterodera avenae* on 60 cm square plots to which 675 g of chopped cabbage had been added as compared with untreated plots. Duddington, Everad and Duthoit (1961), in later experiments obtained further evidence that *Dactylaria thaumasia* reduces the infection of oat seedlings by *Heterodera avenae*. There is of course some doubt as to whether or not the fungi are entirely responsible for such reductions.

Resistance to nematodes may be due to the production of toxic root exudates as found in asparagus, lack of an attractant or a hatching factor in the exudates, barrier to penetration or failure of nematodes to develop within plant tissues.

Paecilomyces lilacinus has been found quite effective in controlling root-knot nematodes under field conditions. *Pasteuria penetrans*, a bacterial biocontrol agent for the control of *Meloidogyne* spp. has also shown promising results.

Breeding for Resistance

This is the best method of controlling any disease. However, breeding for resistance against soil nematodes is a difficult task. This is further made difficult by the presence of large host range of the pathogen and the occurrence of several races or biotypes of the pathogens. Several varieties of crop resistant to nematodes have been developed in our country. Some important ones are : Karnataka Hybrid and VNF-8 in tomato against root-knot nematode, Raj Kiran and RD-2052 in the barley crop against cereal cyst nematode, Anbeern-56 in cotton, Evergreen variety in onion and V-161 in cowpea against reniform nematode, TVAV 6464 in rice against rice lesion nematode, Troyer and Carrizo in citrus against citrus nematode, Gross Michel in banana against burrowing nematode, Branco Muniero in garlic against stem and bulb nematode, and SSM 591 in rice against *Aphelenchoides basseyi*.

Genetic Engineering in Nematodes

One of the major breakthroughs in the field of nematology is the identification of a dominant *Meloidogyne incognita* gene imparting resistance to the root-knot nematode and its linkage to an acid phosphatase gene (Rick and Fobes, 1974). The close association between an important phenotypic character and a relatively easily identifiable electrophoretic marker could be exploited for quick and more reliable methods for nematode resistance (Das gupta, 1998).

Biocontrol Strategies

Opportunities in recombinant DNA technology may extend to produce more effective biological control agents for plant parasitic nematodes. The toxin gene from the insect pathogen *Bacillus thuringiensis* has been transferred to the root colonizing bacterium *Pseudomonas fluorescens*. Similar techniques can be used to produce fungal strains with increased virulence, nematoxic or nematostatic metabolic production, lesser effects on non-target species.

The sophisticated sensory organs in nematodes provide a strong credence for the presence of well developed communication system in these organisms. The intervention in the communication system of nematodes offers several possibilities of biocontrol. Laboratory studies have revealed that when lectin binds to these moieties, host recognition phenomenon is disturbed in nematodes reproducing amphimictically. Intervention in the pheromone communication system offers possibilities of some control strategies. Saturation of worm's environment with a synthetic pheromone emanating from various sources such as slow release capsules, feed additives or soil amendments would disrupt reproduction continually so that gradients useful to the pest are eliminated.

There are certain bacteria that attract second stage *Meloidogyne incognita* while others repel them. Rhizosphere bacteria are positioned in the ideal location to intercept root feeding nematode. Seeds could be coated with an inoculum of strain that could colonize the root and produce nematode repellents. Bacteria that produce a nematode attractant, can be formulated into a nematicide pellet increasing its effectiveness.

This technique involves the production of monoclonal antibodies. The selected antigen is injected into a mouse. The antibody-synthesizing cells from the spleen are then mixed with fast-growing mouse cancer cells, called myclomas and a chemical "glue" PEG. The result is a hybridoma a fusion of an antibody making cell and a cancer cell. The hybridomas are separated, cloned, and tested to select those that make the desired monoclonal antibodies. The chosen hybridomas are injected into the pestonial cavity of mice, where they induce the accumulation of 2 to 10 ml of ascitic fluid containing large amounts of antibody. Or the hybridoma cells may be cultured in the laboratory. (They may be stored frozen until they are needed to generate additional supplies of monoclonal antibody). The final step is to separate the pure monoclonal antibodies from the hybridomas by differential centrifugation.

Antibodies specific for secretory granules could be used as immunocy-to-chemical probe to identify the gland products secreted through the nematode stylet and potentially important in pathogenesis. Antibodies specific for stylet secretions could then be used as probes to identify the subcellular sites in giant cells where secretions accumulate thereby providing some insight into the nature of the molecular and cellular mechanism by which secretions affect cell differentiation and growth.

Conclusions

During the past two decades, considerable progress has been made on plant parasitic nematodes, the damage they cause and their management. However, there is much more that needs to be done if we are to improve our agricultural production efficiency by effectively managing plant parasitic nematodes. By seeking help from technological advances in chromatography, high performance liquid chromatography, mass spectrophotometry, NMR spectrophotometry, electrophoresis, isoelectric focussing etc., the tools of biotechnology enable us to think in terms of single nematode the above opportunities to advance out discipline further so as to make this great revolution in applied biology and agricultural sciences a success.

4.8 Immunology

Immunology (Lt. immunitas : freedom from) is a branch of biology that deals with the immunity i.e. the inborn or acquired resistance of living organisms to infection of micro-organisms. It is concerned with the study of processes by which the body defends itself from the invasion and attack of foreign organisms. There are many micro-organisms like the bacteria, fungi, viruses and parasites that readily invade the body and cause disease(s). This requires 'external vigilance' on the part of the body defences to protect us against these invaders. The scientific basis of our defense mechanisms is the subject of immunology.

For decades immunology, the study of immune system was a back water in medicine. In reality we did not have the instruments to explore the 'battle fields' within us. Over the last 40 years, immunology has developed rapidly, thanks to the advent of molecular techniques. By 1980, it had become clear that immunology held a great promise for treating diseases as diverse as cancer and arthritis.

History

History of immunology shows its origin even about 2500 years back. During 15th century, dangerous and undesirable methods were practised in India and China. This method known as '*variolation*' (variola = small-pox), later came to be known as *innoculation*. In this method, small-pox lesions were scraped from the persons suffering from the disease and the material obtained was indroduced into a healthy individual with the help of needles.

The concept of immunity was first indroduced by Dr. Edward Jenner (1798) who first developed a scientific and safer method to small pox immunisation. But it was, Louis Pasteur who coined the term '*vaccine*' in honour of Edward Jenner who prepared the first vaccine from cow-pox (vacca=cow). A century later, Louis Pasteur with attenuated organism developed vaccines against anthrax, chicken pox, cholera and rabies. This method is now applied to protect against poliomyelitis, yellow fever, measles, small-pox and many other viral diseases.

In 1983, Buchner identified a serum factor called '*complement*' which was bactericidal. In 1908, *Emil Metchinikoff* shared the Noble prize with *Ehrlich* for their discovery that the phagocytic activity of certain cells play a role in immunity. This forms the basis of cellular immunity. *Wright* and *Douglas* (1903) demonstrated that the immune serum contained an active factor called 'opsonin', and its importance in phagocytosis. Its activity is called '*opsonization*'. *Portier and Richet* (1902) introduced the concept of hyper-sensitivity (Anaphylaxis), a phenomenon of increased immunological reactivity. *Zinsser* (1925) explained the distinction between humoral and cellular hyper-sensitivity reactions.

Definition of Immunity

Immunity is concerned with resistance to infection i.e., elimination of microbes or pathogens by antibodies. It is simply a process of recognition and disposal of non-self or foreign substances that enter the body. The non-self substances may be infectious, micro organisms, or tissue grafts. It also includes defence mechanisms against cancer, allergies and auto-immune diseases.

For example, man as a model-the first contact with antigen say small-pox (non-self) develops memory with the immune system, provides the ability to recognize the same antigen during subsequent contact (*recognition*) but not other antigens (*specificity*).

Thus, '*recognition*', '*memory*' and '*specificity*' constitute the three fundamental features of immunity or immune response.

Therefore, immunity may be defined as "all those physiological mechanisms that endow the animal with the capacity to recoginize materials as foreign to itself and to neutralize or eliminate or metabolise them with or without injury to its own tissues".

Classification of Immunity

Immunity is broadly, classified into two types '*Innate immunity*' and '*Aquired immunity*'. (Table 4.12)

Table 4.12

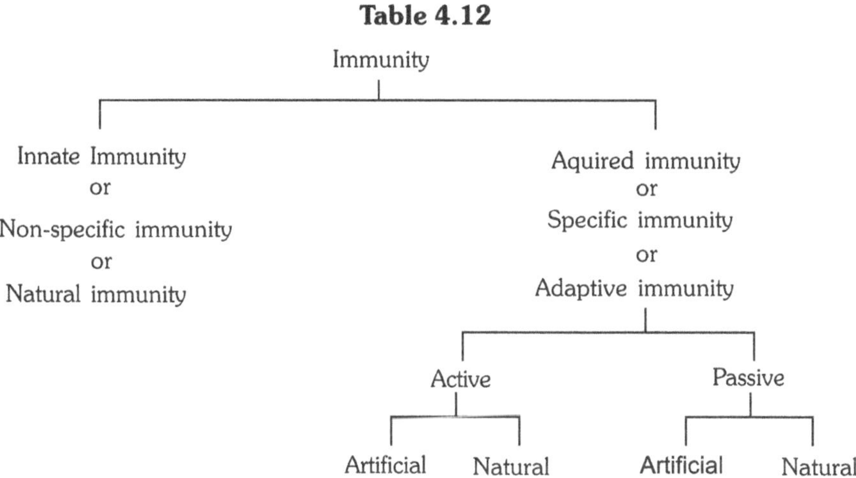

Innate Immunity

All living organisms possess natural resistance to infections from birth and this native or natural defense mechanism is known as "*innate immunity*" or "*natural immunity*". Innate immunity provides the general protective reactions of organisms against any invasion but not against any specific micro-organism. Hence, it is also known as *non-specific immunity*. Non-specific immunity operates through many factors and is effective against a wide range of infectious organisms.

Aquired Immunity

The resistance that an individual acquires during one's life is known as 'acquired immunity or adaptive immunity'. In contrast to innate immunity, adaptive immunity takes longer time to develop and is highly specific for antigens including those associated with microbes and shows memory. Hence, it is also known as '*specific immunity*'. This aquired specific immunity may be 'active' and 'passive' types. Both active and passive immunity again may be '*natural* or *artificial*'.

Though, innate and adaptive immunity are considered separately for convenience and understanding, it is important to know that both the systems work together through direct cell contact, through interactions involving chemical mediators, cytokines and chemokines. More over many of the cells of innate immune system are the same cells used by aquired immune system also. Let us consider briefly the components and operation of both 1. innate 2. aquired immunity.

1. *Innate Immunity :* Operates through several factors such as physical, mechanical, biochemical, cellular, genetic and other factors. The innate immune system is considered as the *first line of defense* against micro-organisms i.e the physical and mechanical barriers together with host secretions constitutes the body's first line of defense against micro organisms. Physical and mechanical barriers include the *unbroken skin* and *mucous membranes* which are generally impervious to infectious agents. It is possible for micro-organisms to enter through hair follicles, openings of sweat glands or abrasions but mostly are inhibited by lactic acid and fatty acids in sweat and sebaceous glands and low pH. *Mucous secretions* of the respiratory tract, the digestive tract, the urinogenital tract and other tissues form a protective covering over mucous membranes and collect and hold many micro-organisms and make them ineffective. *Peristalsis* of intestine, cilia lining body cavities sweep bacteria from susceptible surfaces. *Coughing, sneezing, shedding tears, perspiring, salivating* provide mechanical flushing or clearing of micro-organisms.

Some of the secretions of the mechanical barriers contain chemical substances that inhibit or destroy the micro-organisms (Fig. 4.40). These *chemical factors* are *secretions of the skin* i.e., sweat glands and sebaceous glands, produce sweat and sebum which are bactericidal, *HCl secreted by digestive tract* is microbicidal, *human milk* (lacto ferritin), *nasal secretions*, *saliva* and lysozyme (enzyme- anti-microbial action) certain hormones, *interferons* act as antiviral agents, *complements* help in destruction of microbial organisms by phagocytosis, semen contains spermine and a bactericidal chemical and are polyamines.

Cellular factors include *cells involved in phagocytosis* (micro and macro phages), *natural killer cells* (NK cells).

Genetic factors include immunity at the level of *species, races* and *individuals.*

Other factors include *temperature, inflammation, fever* etc.,

The second line of defense consists of body's internal mechanisms which can be mobilized against invading micro organisms. These mechanisms can be non-specific in their action eg. phagocytosis or specifically directed against pathogens eg. antibodies and sensitized cells. These two types are called *non-specific defense mechanisms* and *specific aquired immunity* respectively (Table 4.13).

The body can mobilize many factors which act non-specifically to the invasions of foreign organisms. These factors include *complement, phagocytic cells, cytotoxic lymphocytes*, and *interferons* among others. These shall be discussed in detail later, whereas *the specific aquired immunity* is a result of an infection or artificial immunization and is directed to the specific causative organisms.

During infection both the mechanisms act together to rid the body of the invading micro-organisms. This interrelationship and the interrelationships between the defense mechanisms discussed here, are illustrated in Table 4.13.

Aquired or Specific Immunity

Unlike innate responses, acquired immune responses require a lag period during which the immune system gears up for an attack against a foreign agent. Unlike innate responses, acquired responses are highly specific, that is, they are directed only against specific molecules present in the foreign material. For example, the blood of a person who has just recovered from measles, but not with a related virus, such as that which causes mumps.

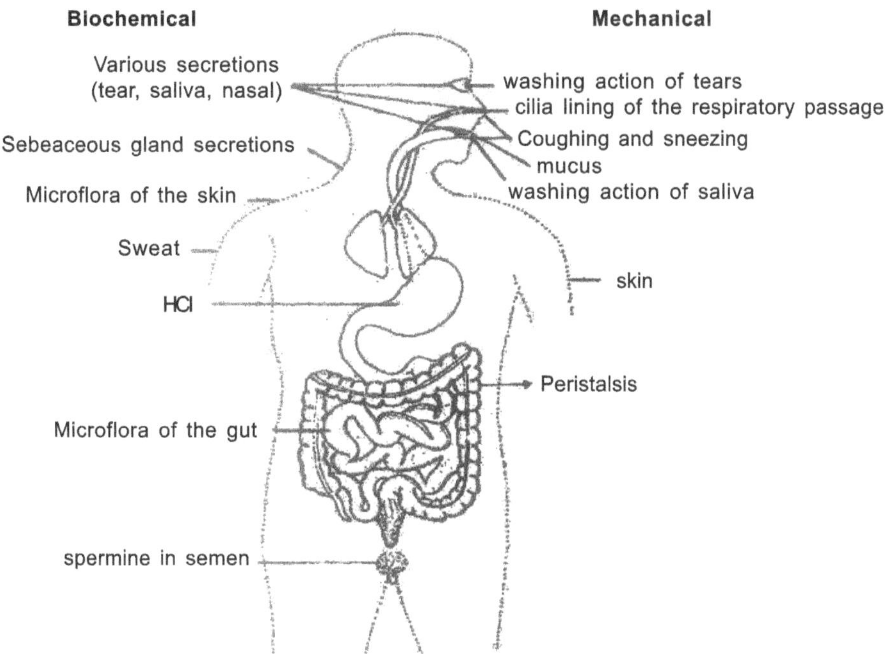

Fig. 4.40 Different biochemical and mechanical barriers in humans.

Table 4.13

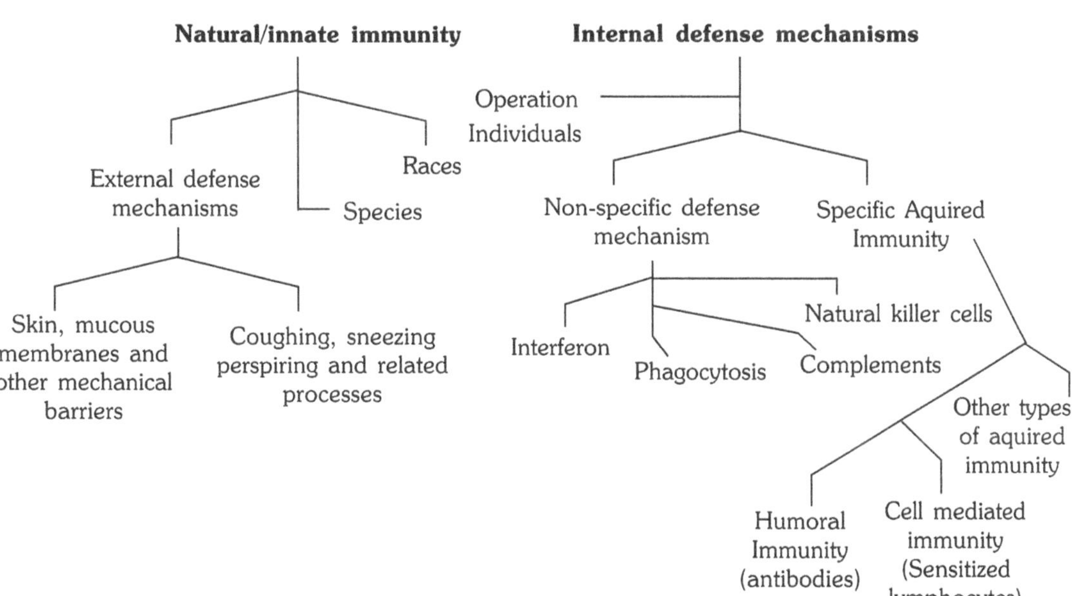

The acquired or specific immunity is of two types namely **active** and **passive**. Both active and passive immunity may again be natural or artificial.

Active Immunity

Active immunity is the resistance developed by an individual in response to antigenic stimulus. Active immunity involves the active functioning of a personal immune system, resulting in the synthesis of specific antibodies (humoral immunity) or production of immunologically active cells (cell-mediated immunity).

Active immunity may be *natural* or *artificial* based on the source of antigen.

(a) *Natural Active Immunity :* Here immunity is developed by the host in response to the antigen that enters by natural infections. *Eg :* measles, small pox, influenza, common cold, etc.

(b) *Artificial Active Immunity :* Artificial active immunity is attained by the host in response to the antigens obtained by vaccination. *Eg :* Anthrax vaccine, BCG, measle vaccine etc.

Passive Immunity

The immunity that a non-immune individual acquires by receiving antibodies or white blood cells from another immune individual is known as *passive immunity*. It differs from active immunity in the following :

* There is no production of antibodies and so there is no participation of the recipient's immune system.
* The protection got is temporary, usually lasting for a few days or weeks, till the passively transmitted antibodies are metabolized and eliminated.

Thus, the immunity caused by passive immunization is less effective and inferior than that caused by active immunization. The main advantage of passive immunization is that it is immediate in its action of producing immunity and so this method would be used when immediate immunity is needed.

Passive immunity is also of two types namely *natural* and *artificial.*

(a) *Natural Passive Immunity :* The immunity transferred from the mother to the child passively is known as natural passive immunity. Eg : colostrum (milk produced by mother after the birth)

(b) *Artificial Passive Immunity :* Transfer of immunity from an immunized donor to a non immune recipient by transferring antibodies or immunized lymphocytes is known as artificial passive immunity. Artificial passive immunity is therapeutically used in the treatment of *tetanus, diptheria, snake bite* and *gas gangrenes.*

Organs of Immune System

The immune system is a complex network of organs, cells and secretions of these cells. The organs of immune system are located throughout the body. The cells that are responsible for the immune responses in the body are made or differentiated in certain organs. These organs are *lymphoid organs* and contain *lymphoid cells*, together form *lymphoid system*. Lymphoid organs of immune system include blood (or reticulo-endothelial system) and *lymphatic vessels* that carry leucocytes to and fro from structures like bone marrow, thymus, lymph nodes, the spleen, tonsils, adenoids, appendix,

Peyer's patches located in the small intestine. The lymphoid organs are grouped into 1. *Primary Lymphoid organs* and 2. *Secondary lymphoid organs*. In adults, the primary lymphoid organs are the *thymus* and bone marrow (in birds it is Bursa of Fabricius) and the secondary lymphoid organs are the *spleen*, the *lymph nodes and tonsils*.

The mature, immuno competent and non-functional lymphocytes of primary lymphoid organs migrate through blood or lymph and accumulate in peripheral and secondary lymphoid organs. In these organs, the cells are exposed to antigen-stimulus, become functional by producing specific antibodies or sensitized cells.

Primary Lymphoid Organs

1. *Bone - marrow :* It is the primary source of pleuripotent stem cells that give rise to all haemopoietic cells including lymphocytes. Bone marrow is the major organ for B-cell maturation and gives rise to precursor cells of thymic lymphocytes.
2. *Thymus :* Thymus is the major organ for T-cell maturation and development. Immature T-cell precursors migrate from the bone marrow to the thymus (to become thymocytes) where they are processed into mature T-cells which develop cell-mediated immunity. The term *T-cell means thymus derived cell* and should be used only to describe mature T-cells. These T-cells are non-functional within the thymus but migrate to sites through blood and lymph and become functional on coming in contact with antigens. (Fig. 4.41)

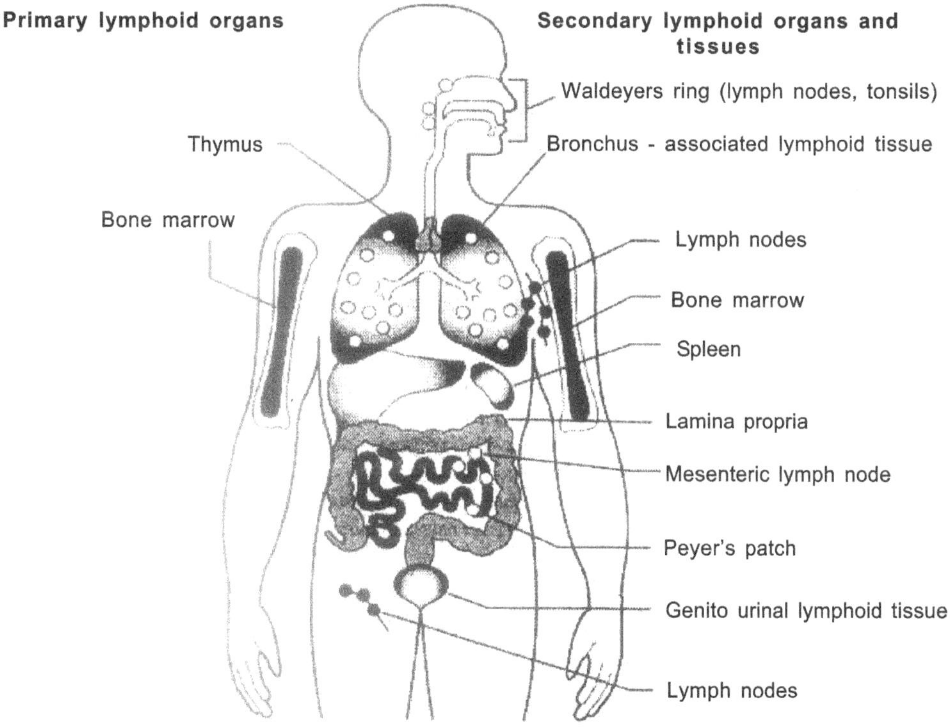

Fig. 4.41 Lymphoid organs

Secondary Lymphoid Organs

1. *Spleen :* It is a solid, large encapsulated bean shaped organ with a spongy interior, situated on the leftside of the body below the diaphragm. It is divided into an outer red pulp and inner white pulp. The red pulp contains large amount of blood containing *phagocytes* and *plasma cells* whereas the white pulp is made up of lymphoid tissue containing T and B-lymphocytes. The main immunological function of spleen is to filter the blood by trapping blood borne micro-organisms and producing an immune response i.e. humoral and cellular immunity. It also removes damaged blood cells and immune complexes.
2. *Lymph nodes :* Lymph nodes are small, solid structures found at varying points along the lymphatic system i.e. groin, arm pit and mesentery. It is spherical in shape and encapsulated one. The interior is divided into three zones : *cortex*, a middle *para cortex* and inner *medulla*. The cortex contains *B-cells*, the para cortex contains *T-cells* and medulla contains *T-and B-cells*. These help in phagocytosis and responsible for cell-mediated and humoral immunity (Fig. 4.41).

Cells of the Immune System

The majority of the cell types involved in the immune system are produced from a common haemopoietic stem cell (HSC). HSC's are found in the foetal liver, foeal spleen and neonate and adult bone marrow. They differentiate into functionally mature cells of all blood lineages. The organs and properties of the cells (and molecules) of the immune system (both innate and adaptive), inter actions and modulations of cellular components as well as the mechanism by which effective immune responses develop i.e. the way in which the immune system distinguishes from self from non-self in order to mount an immune response are described below.

The haemopoietic stem cell (HSC) develops through the process of differentiation into functionally mature blood cells of different lineages. The lineage of cells different from the HSC is determined by the micro environment of the HSC and requires contact with stromal cells (including epithelial cells, fibroblasts and macrophages) and also cytokines (small signal molecules between cells induce growth, differentiation, cytotoxicity and regulation of immunity). Cytokines if produced by leucocytes are *interleukins*, *monokines* if produced by myeloid cells and *lymphokines* if produced by lymphocytes. (Table 4.14) (Fig. 4.42 & Fig. 4.43) (Table 4.15 & 4.16).

The stem cells proliferate and differentiate into Mast cells, Antigen presenting cells, Mega karyocytes and platelets under the influence of colony stimulating cell factors (CSF).

Antigen (Ag), Antibody (Ab), Antigen-Antibody (Ag-Ab) Complex

Antigen (Abbreviated as 'Ag')

The first stage of removing an invading organism is to recognize it as being foreign i.e. non-self. The immune system recognizes the invader as having antigens. An antigen may be defined as "any substance which induces an immune response in the form of proliferation of lymphocytes and production of antibodies, specific for the antigen introduced". These foreign substance are described as *antigens* and are normally proteins, carbohydrates, lipids and nucleic acids. The ultimate aim of immune response is to inactivate, destroy and eliminate the antigen from the body. Immune response occurs through a complex series of interactions that stimulate the production of antibodies. This shows two properties of an antigen i.e. *specificity* and *immunogenicity*. The *specificity* is the ability

Table 4.14 *Haemopoietic Stem Cell (HSC)*

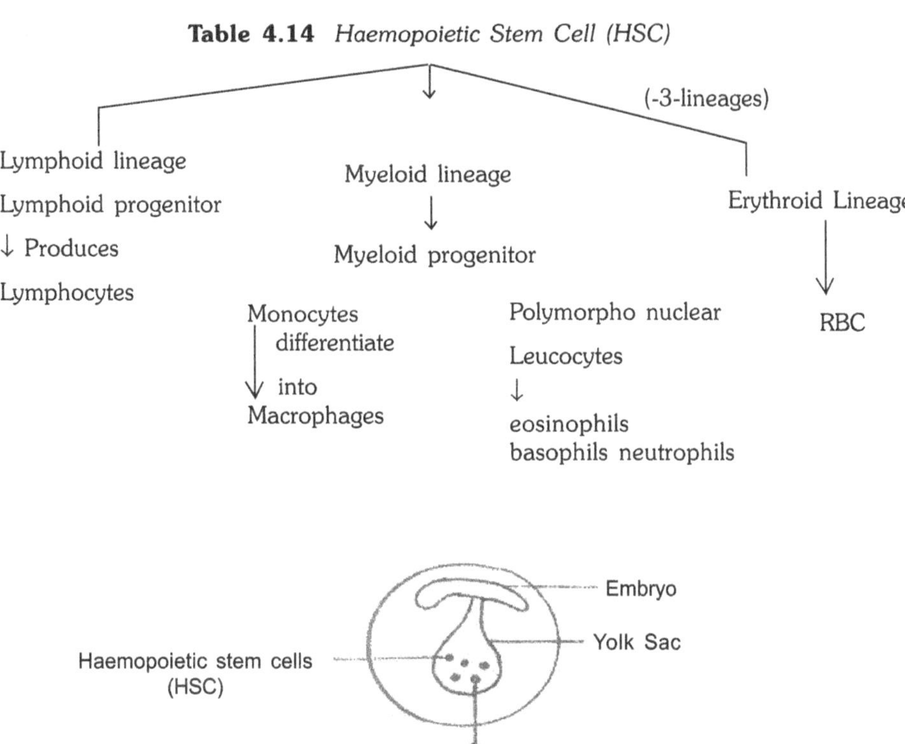

Fig. 4.42 Origin of stem cells and their differentiation.

to react with specific immune antibodies or cells whereas *immunogenicity* is the capacity to stimulate an immune response. The term *antigenicity* and *immunogenicity* differ in that the former exhibits "*foreignness*" of a substance while the latter refers to the ability of a substance to stimulate an immune response. The specific properties of an antigen are due to certain parts of the antigen, separable into low molecular weight *antigenic determinants* or *haptens* or *epitopes* normally linked to the larger supporting portion of the molecule called *carrier*. The hapten is the specificity determing portion or single antigenic determinant, incapable of eliciting antibody response and hence, is also known as *incomplete antigen*. High molecular weight antigens possess both antigenic and immunogenic properties and are called *complete antigens* or *immunogens*. Antibodies present in albumins. Some substances may act as haptens in some animals while the same substance acts as a *complete antigen*.

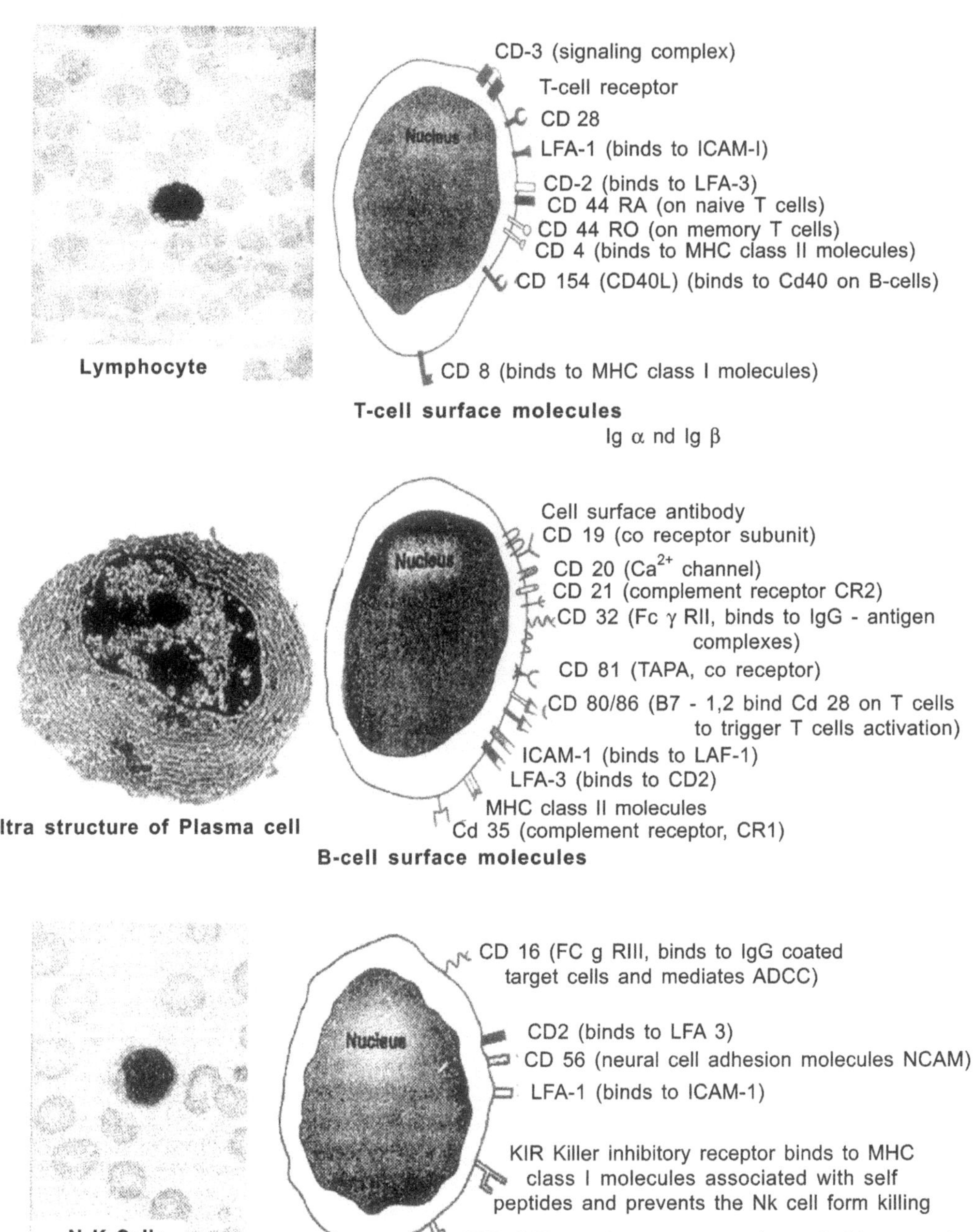

Fig. 4.43 Cells differentiated from lymphoid lineage.

Table 4.15 *Lymphoid Lineage*

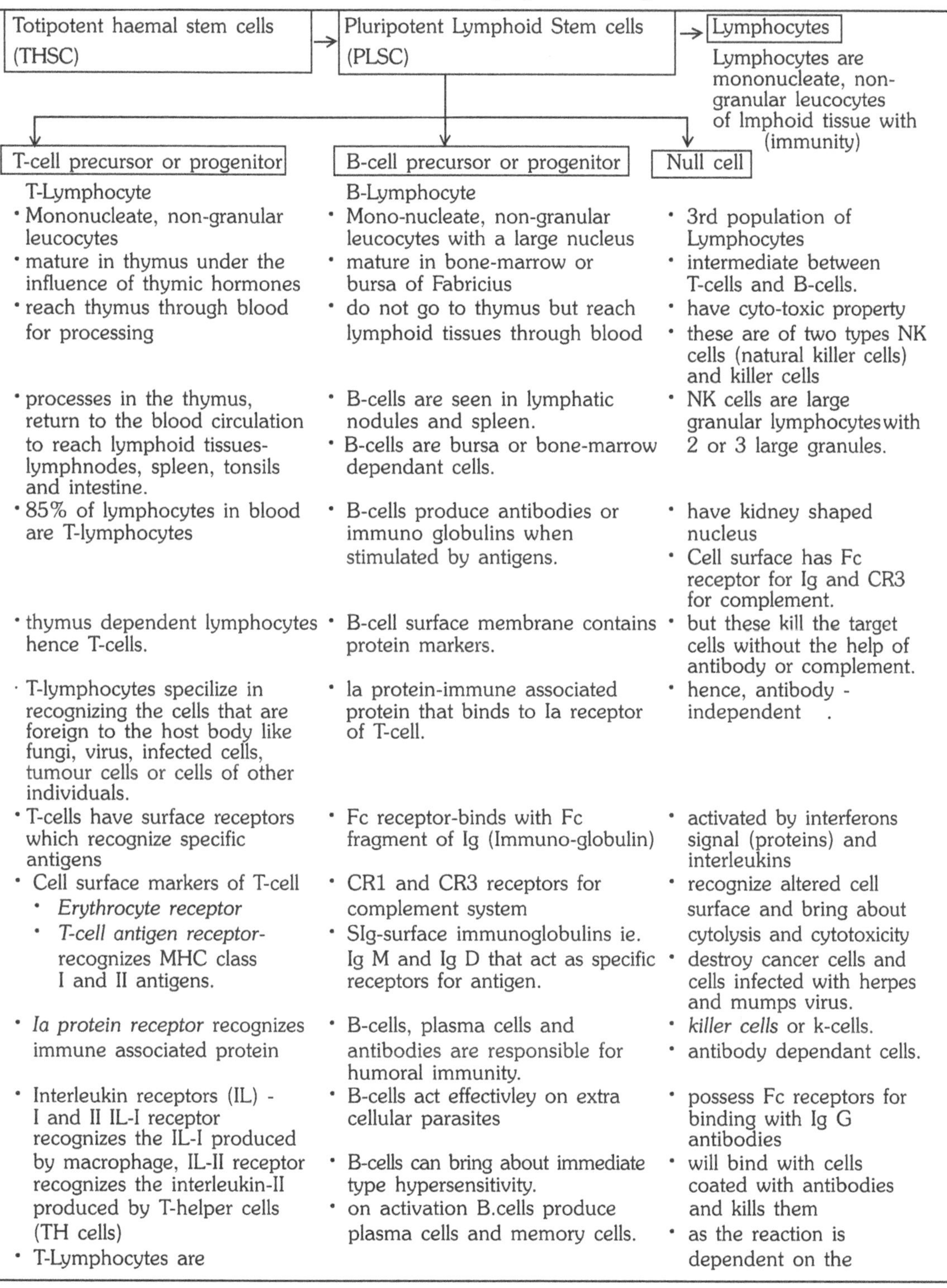

Totipotent haemal stem cells (THSC) → Pluripotent Lymphoid Stem cells (PLSC) → Lymphocytes

Lymphocytes are mononucleate, non-granular leucocytes of lmphoid tissue with (immunity)

T-cell precursor or progenitor	B-cell precursor or progenitor	Null cell
T-Lymphocyte	B-Lymphocyte	
• Mononucleate, non-granular leucocytes • mature in thymus under the influence of thymic hormones • reach thymus through blood for processing	• Mono-nucleate, non-granular leucocytes with a large nucleus • mature in bone-marrow or bursa of Fabricius • do not go to thymus but reach lymphoid tissues through blood	• 3rd population of Lymphocytes • intermediate between T-cells and B-cells. • have cyto-toxic property • these are of two types NK cells (natural killer cells) and killer cells
• processes in the thymus, return to the blood circulation to reach lymphoid tissues-lymphnodes, spleen, tonsils and intestine.	• B-cells are seen in lymphatic nodules and spleen. • B-cells are bursa or bone-marrow dependant cells.	• NK cells are large granular lymphocyteswith 2 or 3 large granules.
• 85% of lymphocytes in blood are T-lymphocytes	• B-cells produce antibodies or immuno globulins when stimulated by antigens.	• have kidney shaped nucleus • Cell surface has Fc receptor for Ig and CR3 for complement.
• thymus dependent lymphocytes hence T-cells.	• B-cell surface membrane contains protein markers.	• but these kill the target cells without the help of antibody or complement.
· T-lymphocytes specilize in recognizing the cells that are foreign to the host body like fungi, virus, infected cells, tumour cells or cells of other individuals.	• la protein-immune associated protein that binds to Ia receptor of T-cell.	• hence, antibody - independent .
• T-cells have surface receptors which recognize specific antigens	• Fc receptor-binds with Fc fragment of Ig (Immuno-globulin)	• activated by interferons signal (proteins) and interleukins
• Cell surface markers of T-cell • *Erythrocyte receptor* • *T-cell antigen receptor*-recognizes MHC class I and II antigens.	• CR1 and CR3 receptors for complement system • SIg-surface immunoglobulins ie. Ig M and Ig D that act as specific receptors for antigen.	• recognize altered cell surface and bring about cytolysis and cytotoxicity • destroy cancer cells and cells infected with herpes and mumps virus.
• *Ia protein receptor* recognizes immune associated protein	• B-cells, plasma cells and antibodies are responsible for humoral immunity.	• *killer cells* or k-cells. • antibody dependant cells.
• Interleukin receptors (IL) - I and II IL-I receptor recognizes the IL-I produced by macrophage, IL-II receptor recognizes the interleukin-II produced by T-helper cells (TH cells) • T-Lymphocytes are	• B-cells act effectivley on extra cellular parasites • B-cells can bring about immediate type hypersensitivity. • on activation B.cells produce plasma cells and memory cells.	• possess Fc receptors for binding with Ig G antibodies • will bind with cells coated with antibodies and kills them • as the reaction is dependent on the

Table 4.15 *Contd....*

responsible for cell-mediated immune response. • also influences the immune responses of B-cells, other T-cells and non-lymphocytic cells.	–	presence of bound anti body, it is called antibody dependant cell-mediated cytotoxic reaction (ADCC).
• act on intracellular parasites hidden in cells.	–	• If the surface of the target cell has specific antigen-antibody complex lymphocytes become activated and kill or destroy the target cell.
• destruction of foreign cells by T-cells is responsible for the rejection of tissues or organs grafted from one person to another	–	• K-cells kill a variety cells such as tumour cells, bacteria, virus, fungi and parasites.
• Such rejection is one of the major problems in organ transplantation. Sophisticated techniques established several catergories of T-cells or sub-populations of T-cells.	–	
They are functionally classified into 2 types.	–	• Antigenic stimulation (viruses, bacterial infection, and toxins) of B-cells makes it to divide into two types of daughter cells namely 1. Plasma cells and 2. B-memory cells (secondary B-cells)
• regulator cells 1. T- helper cells (T_H cells) 2. T-suppressor cells (T_S cells) • effector cells 3. T-cytotoxic cells (T_C cells) or (T_K cells) 4. T-delayed type hyper-sensitivity cells (T_D cells)	–	
• T_H cells help both B-cells and T-cells in immune responses	–	• Plasma cells are the end cells of B.lymphocytes
• T helper cells can be activated even by small quantity of antigen • activated T_H cells secrete lymphokines.	–	• secrete antibodies or immunoglobulins in response to specific antigen.
• also secrete a lymphokine called macrophage migration inhibition factor which causes the accumulation of macrophages around the antigen.	–	• Plasma cells are very rarely seen in plasma of blood but are found in the secondary lymphoid organs like lymph nodes and spleen.
• activates macrophages to perform phogocytosis 2. T. supressor cells (T_S cells) • suppress the activity of B-cells other T-cells	–	• The nucleus gives a *certaied* appearences with well formed protein synthesis machinery i.e., it has rough ER with ribosomes.
• inhibit antibody production by B-cells • suppress the function of T. helper cells.	–	• Rough ER is the site for protein synthesis namely synthesis of immunoglobulins.
• responsible for immune tolerance by limiting the ability of the immune system to attack a person's own body tissues.	–	• two types of polysomes are found-one for synthesis of light chains and the other for synthesis of heavy

Table 4.15 *Contd....*

3. T.cytotoxic cells (T_C) or		chains.
T.killer cells (T_K)	–	• no surface receptor but have Ig in cytoplasm
• kills microorganisms or body's own cells	–	• do not divide with a half life of 2 or 4 days.
• can directly attack a cell and capable of killing microorganisms	–	• Though plasma cells are major antibody producing cells, lymphocytes, and transitional cells also may synthesize antibodies.
• receptor protein on the surface of T_C cells help them in binding to the micro-organisms	–	
• binding results in swelling of the T_C cells and release cytotoxic substances directly into the attacked cell. Cytotoxic substances may be lysozymal enzymes prepared by T-cells.	–	
• capable of attacking many different organisms one after the other.	–	–
• T_C cells are lethal to tissue cells that have been involved by viruses.	–	–
• virus particles entrapped in the membrane of T_C cells, attract the T_C cells by antiviral antigenicity ultimately damaging the virus infected tissue cells.	–	–
• destroy cancer cells, heart transplant cells or other type of cells, foreign to persons own body.	–	–
4. T. delayed type hypersensitivity cells (T_D)		
• bring macrophages to the sites where delayed hypersensitivity reactions occur.	–	–
• T_D cells are similar to T_H cells	–	–
• secrete macrophage haemotoxin and macrophage migration inhibition factor.	–	–
• by secreting lymphokines are directly responsible for delayed hypersensitivity reactions.	–	–
• macrophage chemotoxin attracts monocytes and macrophages whereas macrophage migration inhibition factor inhibits migration of macrophages and monocytes, thereby infiltration of these cells occur into tissues where a delayed type hypersensitivity reaction is occurring.	– –	– –

Table 4.16 *Totipotent hemel stem cell → Pleuripotent stem cell → Myeloid progenitor*

Microphages (Polymorphonu- clear Leucocytes)	**Monocytes Macrophages (Mononuclear phagocytic system)**	**Mast cells**	**Antigen presenting cells**	**Megakaryotic cells Platelets**
1. Neutrophils 40-75% of total leucocytes multilobed nucleus with granular cytoplasm play a role in allergic reactions and in inflammaton processes. 2. Eosinophils Acidophilic leucocytes, granular cytoplasm, granules are rich in hydrolytic enzymes, nucleus elliptical or bilobed about 2.5% of leucocytes, phagocytic, detoxify and ingest antigen-antibody complexes, play a role in allergic reactions. 3. Basophils constitute 0.4% of total leucocytes, nucleus bilobed, granular cytoplasm granules contain, platelet activating factor, play a role in systemic allergic reactions and anaphylaxis.	large mono nuclear phagocytes. Also known as reticuloendothelial cells. Derived from bone-marrow promonocytes differentiate into monocytes. Monocytes from blood migrate into tissues to become macroph-ages. These are long-living and usually found engulfing bacteria, viruses and protozo-ans by a process of phagocytosis. Found in lymph nodes, spleen, bone-marrow and liver. Primary function is phagocytosis. Help in processing of antigen before they are presented to T or B-cells. Present antigen to T-cells and is very specific. Macrophages are secretory cells and secrete components of complement system i.e., hydrolytic enz-ymes toxic forms of O_2 and monokines.	Mast cells are found in connective tissues through the body, close to blood vessels and mucosal surfaces. Produced in the bone-marrow and have similar morphology and functions of basophils. Large cells with baso-philic granular cytop-lasm with polymorph-ous nucleus. The granules of the cells secrete substa-nces like seratonin, haparin, histamines, leucotrienes which bring about the inflammatory and allergic reactions.	These cells process and transfer antigen to lymphocytes. Three groups of these cells are present in the immune system. They are macrophages, dendritic cells and Langer, hans cells. Macrophages are phygocy-tic and contain class I&II MHC molecules and rece-ptors of F_C and complemet; present antigens to T or B-cells. Dendritic cells are present in spleen and lymph nodes. Contain F_C receptor and class II MHC molecules. Present antigens to T_H cells and B-cells. Langerhans cells are present in the skin, migrate to spleen and lymph nodes. Contain Class I & II in HC molecules, F_C and C_3.	Non-nucleated spherical large cells derived from megakaryocytes of bone marrow. Contain I MHC molecules and receptors of I_gG&I_gE. Besides blood clotting, also help in immune response, inflammation and hyper-sensitivity reactions.

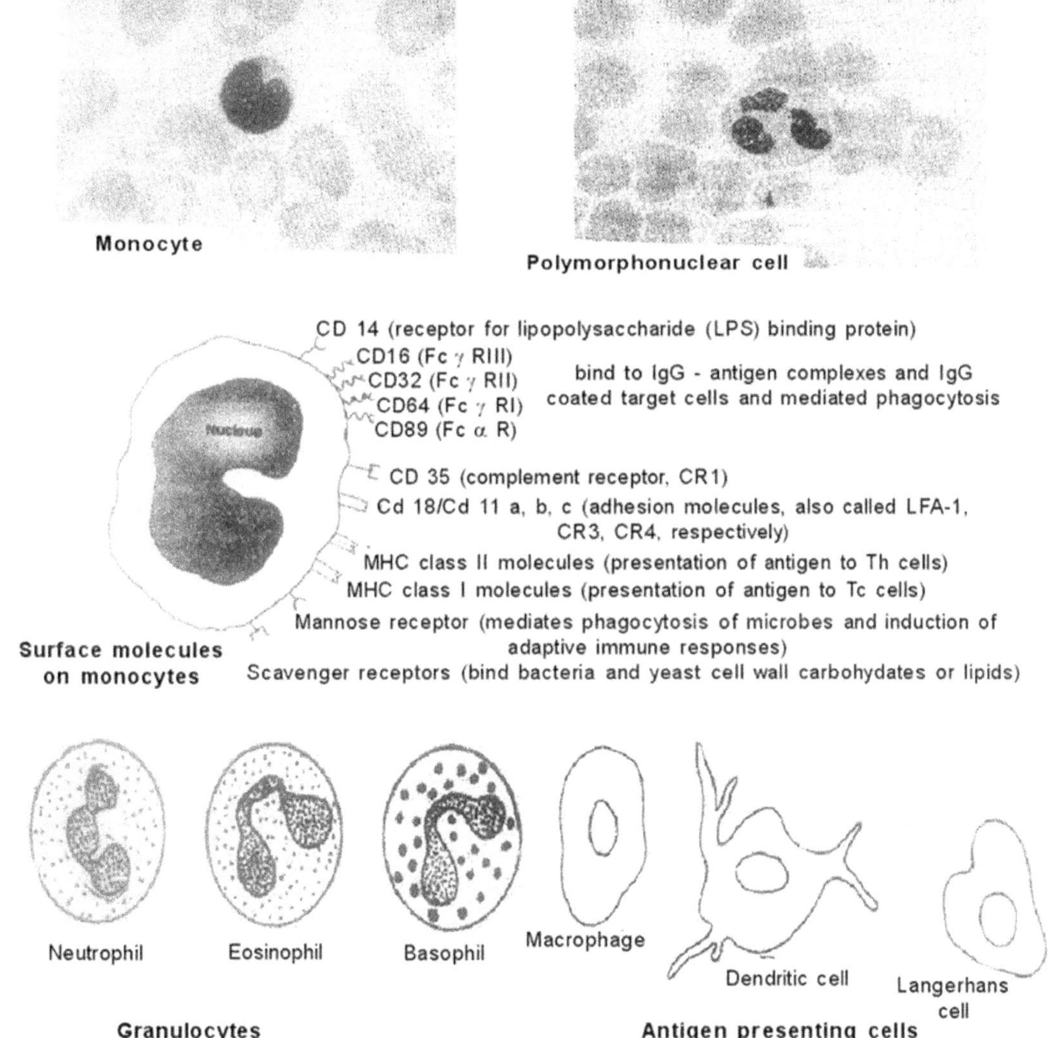

Fig. 4.44 Cells differentiated from Myeloid Lineage

Chemically, antigens are proteins or carbohydrates with molecular weight greater than 10,000 daltons and may be nucleo-proteins, glycoproteins, lipoproteins or synthetic polypeptides or polysaccharides. All these function as complete antigens whereas heavy metals like nickel, small organic compounds function as *haptens*. In these macromolecules, the antigenic determinant or epitope is only a small area with only 6-10 aminoacid residues.

When a potential antigenic molecule is encountered early in the development of a T cell or B-cell, the cells die. No immune response occurs during the second encounter of the same molecule. Then the molecule or antigen is said to be *tolerogen* and the host is said to be *tolerant* to it. The immune system considers this *tolerogen as "self"*.

Immunogenicity of an antigen depends on many factors :

1. Foreignness
2. Chemical complexity
3. Molecular size
4. Antigen dose
5. Route of immunization
6. Adjuvants.

1. *Foreignness :* In general, the immune response is greater when the antigen is more different from 'self'. A person will not normally respond to his or her own (same species) antigens or to those of an identical twin but will react to proteins from (unrelated species) different persons or any foreign source.
2. *Chemical complexity :* A chemically complex molecule is more immunogenic than to a simple one. Good immunogens are proteins (> 20 amino acids), lipo-proteins (endotoxins), glycoproteins (blood groups), nucleic acids in combination with small proteins. Small polypeptides *eg :* insulin can be weak immunogens. Pure lipids or pure polysaccharides are also not good immunogens. Some polymeric antigens can trigger an IgM - B cells response without T-cell help. These are called T-independent antigens.
3. *Molecular size :* The molecular size of an antigen has a direct relation to antigenicity. The size is usually correlated to the complexity of the molecule and the number of antigenic determinants. Very large molecules such as haemocyanin (Mol. wt 6,00,000), tetanus toxin (Mol.wt. 55,000) are excellent antigens whereas molecules of low molecular size are poor antigens *eg.* insulin (5,700 mol.wt) and histones (6,000 Mol. wt)
4. *Antigen dose :* In general, the higher the dose of antigen the greater is the immune response. Repeated exposure to an antigen leads to activation of large numbers of immuno-competent cells. This activation allows a faster response each time the antigen is encountered. This is called anamestic or secondary immune response.
5. *Route of Immunization :* The immune response will be different depending on the route of immunization. Antigens are more effective via subcutaneous or intra muscular and intraveinous routes.
6. *Adjuvants :* (Lt. adjuvare-to help). These are the substances which enhance the immune response to an antigen. Usually the adjuvants are administered along with antigens to produce antibodies vigourously. *eg.* silica particles, beryllium sulphate, mineral oils, endotoxins etc.

Antibody (-ies) (Ab)

Antibodies (abbreviated as Ab) are immunoglobulin proteins formed in response to antigen (Immunogen) and specifically react with the immunogen. Immunoglobulins are immunologically active serum proteins. These are represented as 'Ig'. These are synthesized by B-lymphocytes and plasma cells and are also called as antibodies (Ab). Immunoglobulins or antibodies are "immunological missiles" produced by the body. These are produced by vertebrates only and are found in the serum, body fluids and tissues. All antibodies have two basic functions. 1. antigen binding and 2. participation in effect or functions depending upon the physical properties of the antibody.

Antibodies have binding sites to interact specifically with antigen molecules, having the same structure as the antigen which induce the Ab. It is this heterogenicity that gives the host a wide range of Ab-mediated mechanisms that destroy the potential pathogens.

Structure of Antibody or Immunoglobulin

Gerald Edelman and Rodney Porter (1962) have proposed the basic structure of an immunoglobulin (Fig. 4.45). It is a glycoprotein, 'Y' or 'T' shaped symmetric dimer molecule. Antibodies have a basic unit of four polypeptide chains two identical pairs of short light chains and longer heavy chains. Each light chain is made up of 214 amino acids whereas each heavy chain is made up of 400 to 700 amino acids. Based on the structure and amino acid sequences, the light chains are of two types namely *kappa* (k) and *lambda* (λ) chains. The immunoglobulins with kappa chains are called k-type and with lambda chains are called L-type (60% k-type and 40% L-type Igl). The heavy chains are of five types : 1. *Gama chain* (γ) in IgG 2. *Alpha chain* (α) in IgA 3. *Mu chain* (μ) in IgM 4. *Delta* (δ) *chain* in IgD and 5. *Epsilon* (ε) *chain* in IgE.

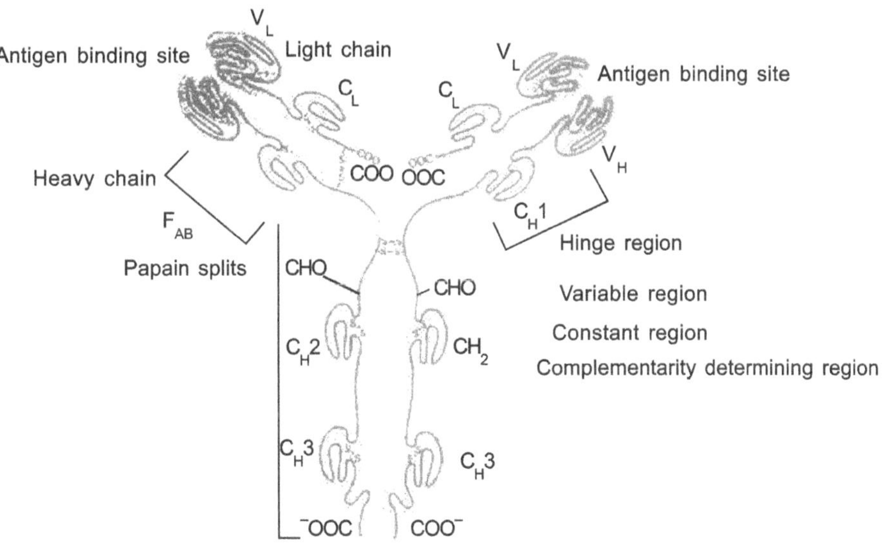

Fig. 4.45 Structure of Human IgG molecule

One end of each chain is called *aminoterminal end or N-terminal end* and the other end is called *carboxyl terminal end or C-terminal end.* All four chains are bound together by interchain disulfide bonds. In this basic unit, one L-chain is bound to one H-chain by single interchain disulfide bridge and non-covalent interactions. Similarly, the two H-chains are bound together by 1 to 13 interchain covalent disulphide bridges as well as by non-covalent hydrophilic and hydrophobic interactions. The heavy chain (H-chain)has central flexible region called *hinge.* Both H and L chains are divided into two regions namely a *variable region (V-region)* and a *constant region (C-region).* In the constant region, amino acid sequences remain constant whereas in variable region, the amino acid sequence shows variability, hence the name. The variable region is located at the extremity of the arms of Y in the N-terminal end of the immunoglobulins and the constant region is located at the base of the stalk of Y in the C-terminal end.

Each polypeptide chain consists of an aminoterminal portion called variable (V) region and carboxyterminal portion called constant (C) region.

The polypeptide chain do not exist 3-dimensionally as linear sequences of amino acids but are folded disulfide bonds into globular regions called *domains*. The domains of H-chain are called V_H (variable) and C_H1, C_H2, C_H3, C_H4 (constant) and those L-chains are designated V_L and C_L.

Porter (1959), showed that immunoglobulin (IgG) is cleaved through limited proteolysis with papain, into three fragments : two identical F_{ab} *fragments* and *one* F_C *fragment*. F_{ab} is antigen binding (ab) fragment and F_C is a crystallisable fragment. F_{ab} forms the arms of Y-shaped Ig molecule whereas F_C is derived from stem of the Y. F_{ab} reacts with the antigenic determinants and its N-terminal binds with the antigen. The F_C fragment is involved in complement fixation, attachement to phagocytic cells, degranulation of mast cells etc.

The variable regions contain certain highly variable zones called *hyper variable regions (HV) or hot spots*. Three such zones exist in each V_L domain and four in each V_H domain. These zones are involved in the antigen binding site and are called *idiotypic regions*.

Classes of immunoglobulins

There are five classes of immunoglobulins occurring in human serum and are named based on the nature of heavy chains and are as follows (Table 4.18).

Name of IG	Nature of heavy chain
1. IgG - Immunoglobulin	G - Gamma (γ) H-chain
2. IgA - Immunoglobulin	A - Alpha (α) H-chain
3. IgM - Immunoglobulin	M - Mu (μ) H-chain
4. IgD - Immunoglobulin	D - Delta (δ) H-chain
5. IgE - Immunoglobulin	E - Epsilon (ε) H-chain

Biological Activities of Differnt Classes of Immunoglobulins (Fig. 4.46)

IgG

In normal human adults, IgG constitutes approximately 75% of the total serum immunoglobulins. There are 4 subclasses of IgG i.e., IgG1, IgG2, IgG3 and IgG4. IgG is the only immunoglobulin that can cross the placenta. IgG is also capable of fixing serum complement.

IgA

IgA is the predominant immunoglobulin in body secretions comprising 15% of the total immunoglobulins. Secretory IgA provides the primary defence mechanism against some local infections owing to its abundance in saliva, tears, bronchial secretions and mucous secretions of small intestine.

IgM

It is prominent in early immune responses to most antigens and predominates in certain antibody responses like natural blood group antibodies. IgM is the most efficient complement fixing immunoglobulin.

IgD

The main function of IgD has not been determined, though there are isolated cases reported of IgD activity against certain antigens like insulin, penicillin, milk proteins, diphtheria toxoid and thyroid antigens.

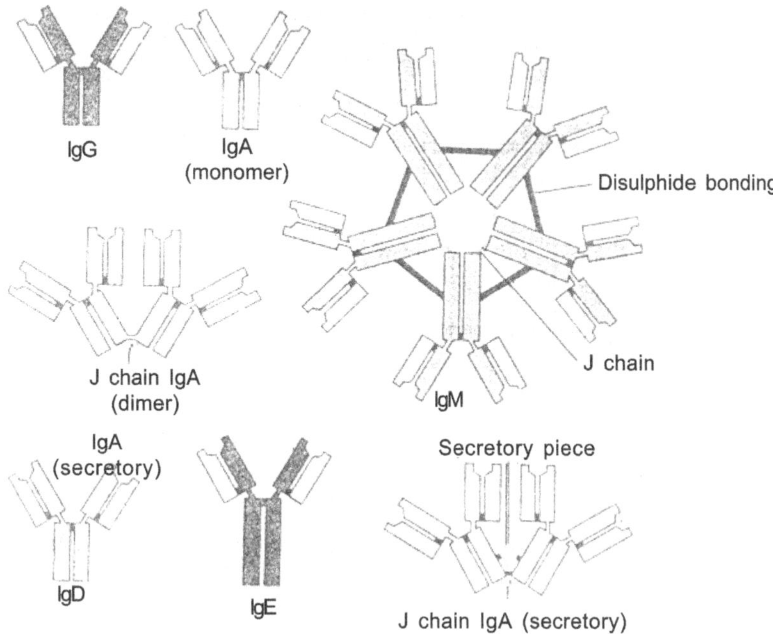

Fig. 4.46 Different types of antibodies (Igs)

IgE

Its main function lies in allergic reactions.

Functions of Immunoglobulins

The main function of the antibodies or Immunoglobulins is to protect the body against the invading organisms. This function is carried out in different ways, such as

1. Agglutination of antigens
2. Precipitation of antigens
3. Neutralization of antigens
4. Lysis of antigen
5. Opsonization
6. Tissue fixation
7. Phagocytosis
8. Chaemotaxis
9. Activation of mast cells and basophils.

Antigen (Ag) - Antibody (Ab) complex (Ag - Ab complex)

The interaction between antigen and antibody is called antigen-antibody reaction. It is abbreviated as Ag-Ab reaction. Antigen-antibody reaction is the basis for humoral or antibody mediated immunity response.

$$\text{Ag} + \text{Ab} \rightarrow \text{Ag - Ab complex}$$

The reaction between Ag and Ab is highly specific i.e. it is the discriminating ability of the antibody to combine with only one type of antigen. For example, the antibody produced against lens antigen will react with only lens antigen which is the cause for its production.

In antigen-antibody reaction, antigen combines with the antibody through *antigenic determinants or epitopes* and the antibody combines with antigen through *antigen-binding site or paratope.*

When antigen and antibody are closely fit, the strength of the binding is great. The binding is due to closeness between Ag and Ab, intermolecular forces and antibody affinity. Between Ag and Ab, inter molecular non-covalent forces operate like electrostatic forces, hydrophobic bonding, hydrogen bonding and Vanderwaals bonding.

The strength of binding between a particular antibody and a single antigenic determinant is called *affinity* and if the binding is strong it is called *high afftinity antibodies*. Avidity refers to the capacity of an antiserum containing various antibodies to combine with the whole antigen that stimulated the production of antibodies. Avidity is the capacity of antibodies to combine with multivalent antigen.

The antigen-antibody reaction leads to visible manifestations like precipitation, flocculation, opsonization, complement fixation, cytolysis, agglutination and immuno florescence. Here, as an example of Ag-Ab reaction detection, only opsonization is discussed.

4.9 Opsonization

Opsonization is a process of making microbe easier to phagocytose (Fig. 4.47). This process is facilitated by molecules called *opsonins* (Gr : to make more tasty) which do this by coating the microbes. They help in attachment of the microbe to the phagocyte and also trigger activation of phagocytosis i.e. opsonin is an antibody when combined with a particulate antigen (insoluble) increases the susceptibility of the antigen to phagocytosis. This antigen is named as opsonizing antibody.

Opsonins include the complement component C3b and antibody itself, and the latter acts as a bridge between the innate and adaptive immune systems.

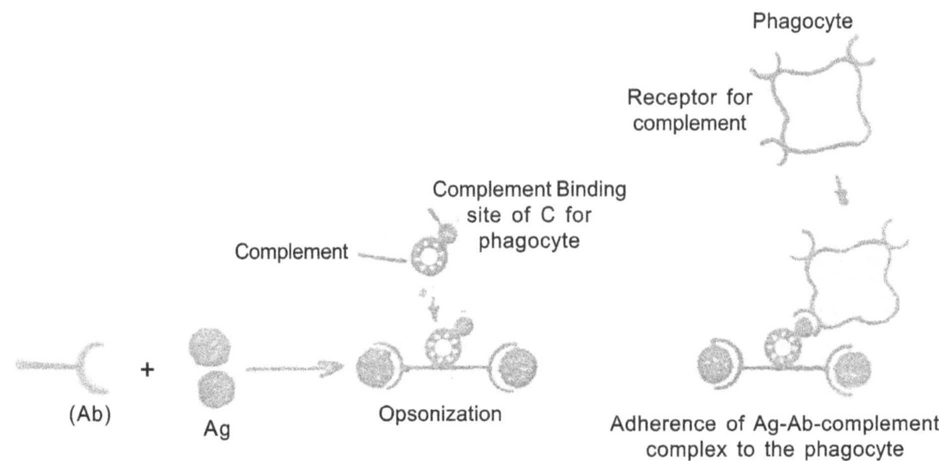

Fig. 4.47 Opsonization

In opsonization, the antibody combines with the surface antigen of bacteria, activating the complement system. The activated complement C3b is attached to Ag - Ab complex to form Ag - Ab complement complex. Phagocytes use their surface receptors for C3b and Fc fragment of the antibody. This results in coating or adherence of Ag-Ab complement to phagocytic cells.

Thus, the microbe or antigen fixed on the phagocytic cell is killed by phagocytosis or lysis.

Immune Response

Immunity in vertebrates is conferred by certain types of white blood cells collectively known as lymphocytes. They arise like other blood cells from common precursor cells i.e stem cells in the bone marrow. Lymphocytes, however, leave the blood vessels and patrol the inter cellular spaces for foreign intruders. They eventually return to the blood via the lymphatic vessels but not before interacting with specilized lymphoid tissues such as the thymus, the lymph nodes and the spleen, the sites of immune response (Fig. 4.48).

The defender white blood cells fall into three groups : the *phagocytes* or *cell-eaters* of which the stalwart is *macrophage* and two kinds of lymphocytes, called *T* and *B cells*. All share one common objective to identify and destroy all substances, living and inert, that are not part of the human body, that are '*not self*'. These include cancer cells, which have turned from '*self*' to '*non-self*'.

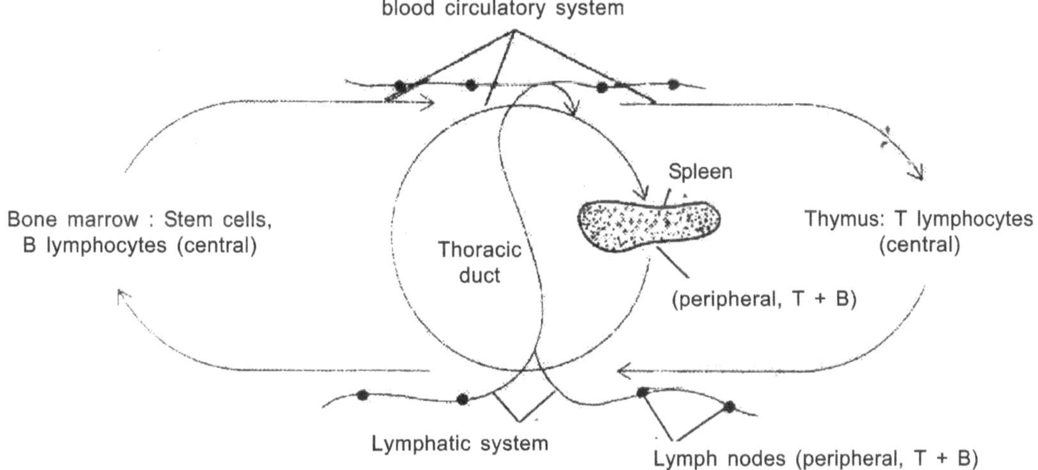

Fig. 4.48 Basic Anatomy of Immune System

Any foreign macromolecule or non-self enters the body of an organism, the organism reacts or responds to it and triggers a response called *immune-response*. These foreign molecules or antigens may be proteins, carbohydrates or nucleic acids.

There are four critical phases to each immune response : recognition of the enemy, amplifications of defenses, attack and slowdown. Each immune response is a unique local sequence of events, shaped by the nature of the enemies.

The immune responses may be primary and secondary immune responses.

The rates of appearance of antibodies in the blood serum following primary (1°) immunization on day 0 with antigen A and secondary (2°) immunization on day 28 with antigens A and B. Antigen B is included in the secondary immunization to demonstrate the specificity of immunological memory for antigen A. Note that the secondary response to antigen A is both faster and greater than the primary response. T-cell mediated responses exhibit similar immunological memory.

Primary and Secondary Immune Responses

The immune response produced by the entry of antigen for the first time is called *primary immune response*. The first injection of the antigen producing primary immune response is called *priming dose*. Primary response is slow and has a long latent period. Antibody titre is minimal in primary response.

If the same animal is exposed to the same antigen for the second time, it elicits an immune response called *secondary* immune response. The injection of the antigen for the second time is called *secondary dose* or *booster dose*. The secondary response is very rapid, more intense and long lived with short or negligible latent period. Antibody titre is more, about 10 fold than in the primary response.

Immune responses can be represented by a graph by plotting antibody titre against time, that gives a sigmoid curve called *immune response curve* (Fig. 4.49).

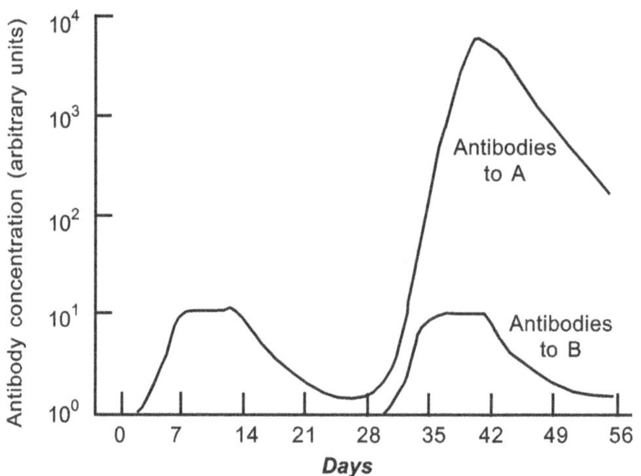

Fig. 4.49 Primary and secondary immune responses.

Two types of Immunity (Fig. 4.50 & 4.51) have been distinguished :

1. Cellular or Cell-Mediated Immunity (abbreviated as CMI)
2. Humoral or Antibody-Mediated Immunity (abbreviated as HI)

Cell Mediated Immunity (CMI)

In this type, the body produces large number of activated lymphocytes that are specially designed to destroy the foreign organisms hence called cellular or cell-mediated immunity. It is mediated by T-lymphocytes or T-cells and are so called because of their development and processing occurs in the Thymus. 'T' in T-cells stands for thymus. B-lymphocytes or B-cells and antibodies have

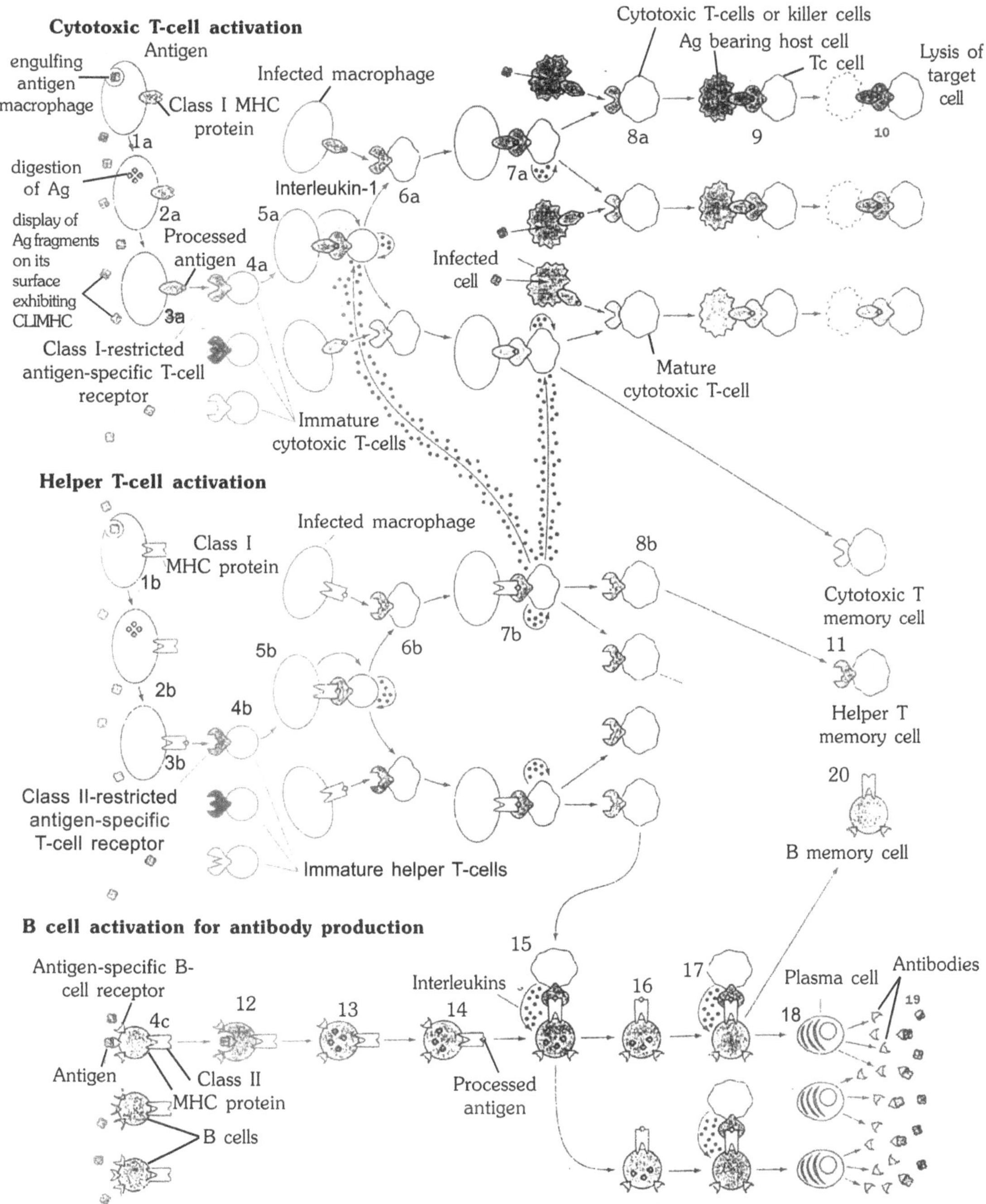

Fig. 4.50 Generations of cell mediated and humoral immune responses
(Cytotoxic T-cell activation and Helper T-cell activation are cell mediated immune responses and B-cell activation for antibody production is humoral immune response)

no role to play in cell mediated immunity. Here, cellular antigens are handled, cells and tissues are damaged, in contrast to the antibody which can not handle intracellular antigens.

CMI is the function of two distinct T-cell sub-populations. In particular, T helper cells (T_H) are required for activation and differentiation of certain other cells (eg., B-cells) of the immune system while *cytotoxic T-cells* (T_C) directly kill the cells infected with microbes. Both T_C (also known as *cytotoxic (CD8) T-lymphocytes, (TL'S)* and T_H cells (helper (*CD*) *4 cells*) need to interact directly with the cells, they are going to kill/help, through specific recognition mechanisms.

The cellular immune response leads to the destruction of offending cells. This process occurs through a complex series of interactions among various types of T-cells that specifically bind a particular antigen.

It all begins when a *macrophage* (white cell) engulfs (1a,1b) and partially digests a foreign antigen (2a, 2b) and displays the resulting antigenic fragments (3a, 3b) on its surface. There, these fragments bind to one of the two types of cell surface proteins known as Major Histo Compatibility Complex (MHC) proteins (*histo* refers to tissues). *MHC proteins* are the antigen presenting markers through which the immune system distinguishes body cells from invading antigens (class I MHC proteins) and immune system cells from other cells (class II MHC proteins). Class I MHC proteins are displayed by all the nucleated vertebrate cells on their cell-surface. Millions of helper T-cells circulate in the blood stream and a select few are programmed to 'read' that antigen.

Macrophages displaying class II MHC proteins are recognized by proteins known as *T-cell receptors* present on the surfaces of immature cytotoxic T-cells (T_C).

In order to bind to the antigen-displaying macrophage, these receptors must specifically complex the antigen together with the class II MHC protein.

In the same way, macrophages displaying antigenic fragments complexed to class II MHC proteins are bound by immature helper T-cells (T_H) bearing the cognate receptor. Binding to the macrophage, the T-cells become activated. Once activated, helper T-cells begin to multiply. These then stimulate the multiplication of few killer T-cells and also B-cells that are sensitive to the invading antigen (virus).

T-cells that bind to a macrophage-displayed antigen-MHC protein complex are induced to proliferate a process known as *clonal selection*. Only those T-cells that can specifically recognize the intruding are produced in quantity. Clonal selection occurs because a macrophage bound to a T-cell releases protein growth factors named lymphokines by which immune cells communicate with each other. It is Interleukin 1 (IL-1) (5a, 5b) which activates the helper T-cells (T_H) and also stimulates the brain to raise the body's temperature causing fever which enhances the activity of immune cells. Activated helper T-cells produce autostimulatory secretion, the interleukin-II (IL-II) which stimulates the other helper T-cells and killer T-cells to grow and divide (7a, 7b).

Helper T-cells also produce a lymphokine called *gamma interferon* (IF) which has multiple effects. Like IL-2, it helps activated killer T-cells or cytotoxic T-cells (T_C) (8a) to attack invading organisms. The cytotoxic T-cells (T_Cs) bind to antigen-bearing host cells (9) and at the point of contact release *70 KD protein, perforin* that lyses or punctures the membrane of target cells (10), letting the contents to spill out, thus disrupting the viral replication cycle. Thus, with each exchange of signals between macrophages and T-cells, a *lymphokine cascade* amplifies the immune response until the enemy is over powered.

As the infection is contained, *Suppressor T-cells (T_S cells)* half the entire range of immune responses preventing them from spiraling out of control.

Memory T and B cells remain in the blood stream and lymphatic system ready to move rapidly, should the same virus once again invades the body.

Thus, the cell mediated immune system functions mainly to prevent the spread of a viral infection by killing virus infected host cells. It is also effective against viral infections, parasites, certain types of cancers and rejection of tissue and organ grafts. Another vital function of cell-mediated immune system in recent years is spread of *aquired immune deficiency syndrome (AIDS)* whose causative agent, *human immuno deficiency virus (HIV)*, acts by specifically attacking helper T-cells (T_H cells).

Humoral Immunity : abbreviated as HI or Antibody Mediated Immunity

(humor is an archaic term for fluid)

This is most effective against bacterial infections and extra-cellular phases of viral infections. This immunity is mediated by an enormously diverse collection of proteins known as *antibodies (Ab)* or *immunoglobulins (Ig)*. Antibodies are produced by B-lymphocytes or B-cells which in mammals mature in bone marrow.

The destruction of antigens by producing antibodies is called *antibody mediated immunity* and as the antibodies are present in *body fluids (humors)*, it is called *humoral immunity*.

B-cells cannot bring about humoral immune response but needs the co-operation of other cells such as *macrophages* (f), *T. helper cells* (T_H) and *dendritic cells*-possess high levels of surface MHC class II molecules that process and present antigens to T-cells.

Humoral Immune System (Fig. 4.51)

- B-cells (B-lymphocytes) display both immunoglobulin and class-II MHC proteins on their surfaces. (4c)
- If a B-cell encounters an antigen that binds to its particular immunoglobulin, it engulfs the antigen (12), partially digests the antigen (13) and displays the antigen fragments on its surface in complex, with the class II MHC protein.
- Mature helper T-cells T_H (8b) bearing receptors specific for this complex bind to the B-cell (15) and releases interleukins (IL) that stimulate the B-cell to proliferate and differentiate (16).
- As the number of B-cells increases, helper T-cells signal them (17) to start producing antibodies. (16-8b)
- Most of the B-cells progeny are plasma cells (18) that are specialized to secrete large amounts of *antigen-specific antibody*.
- Then these antibodies bind to available antigen (19), neutralize, thereby marking it for destruction either by *phagocytosis* or by activating the complement system. It also prevents them from attacking other cells.
- Most T-cells and B-cells live only for a few days unless stimulated by their respective antigen. B-cell proliferation is limited by its interaction with suppressor T-cells (T_S cells) which have the opposite function of helper T-cells.

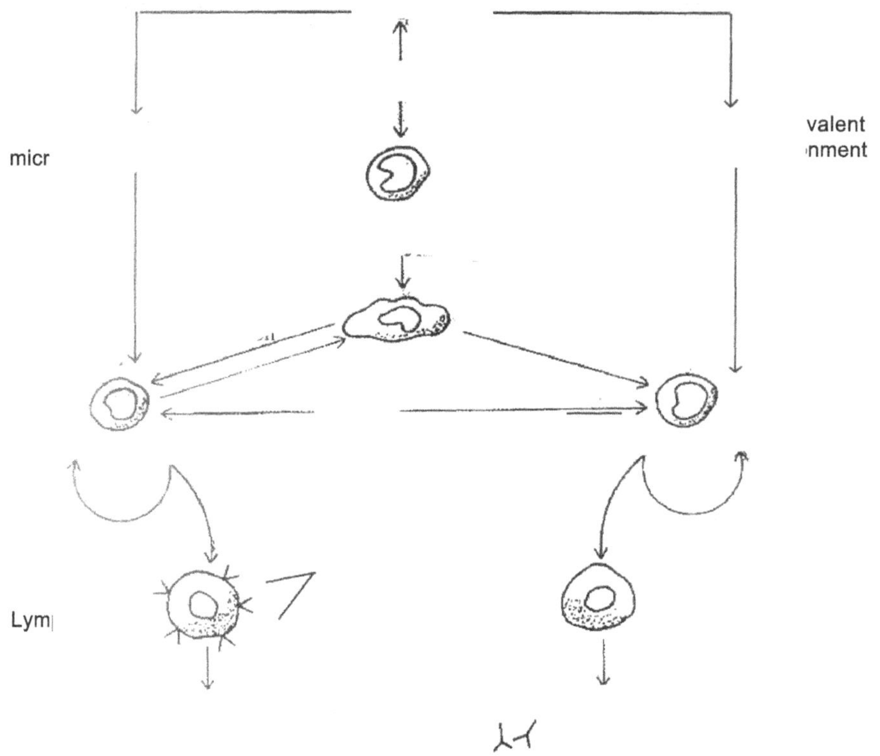

Fig. [illegible] Generations of cell mediated and humoral immune responses

- One of the hall-marks of the immune system is that an animal is rarely infected twice by exactly the same type of pathogen i.e., recovery from an infection by a pathogen renders an animal immune from that pathogen.

This, so called *secondary immune response* is mediated by long lived memory T-cells and memory B-cells.

Mediators of Immune System

These include the *complement system, lymphokines* etc.

The complement system is the primary humoral mediator of antigen-antibody reactions. It consists of atleast 20 chemically and *immunologically distinct thermolabile plasma proteins* capable of interacting with each other abbreviated as C and numbered C1, C2, C3, C4, C5 etc. The various functions they carry out are increased vascular permeability, chemotaxis, opsonization prior to phagocytosis and lysis of target organisms.

Activation of complement system occurs via two pathways :

(i) *Classic pathway :* Initiated by antigen-antibody complexes.

(ii) *Alternate pathway :* Initiated by a variety of largely non-immunological stimuli.

The Lymphokines

When antigen is injected intradermally into an appropriate sensitized host, a delayed skin reaction may develop over the course of 24-28 hours, characterized by mononuclear cell infiltrate. The initial reaction of antigen with a few specifically sensitized lymphocyte results in the production of soluble mediators called *lymphokines*.

The functions of lymphokines are :

1. Recruit host inflammatory cells
2. Activate the inflammatory cells
3. Keep them at the site of inflammation
4. Amplify the cell mediated hypersensitivity reaction.

While these substances exert marked biological effects on the cells, they are produced in only minute quantities by activated lymphocytes.

A variety of lymphokines, detected usually by diverse functional assays has been described. It is not yet clear, how many chemically distinct molecules are involved, nor how many of the lymphokine activities detected in vitro are relevant in vivo.

Some of the important lymphokines are :

1. *Macrophage migration Inhibition factor :* This substance inhibits the normal active migration of macrophages.

   It is of low molecular weight and lacks immunological specificity. It is a glycoprotein which can be generated in vitro by interaction of immune T-cells with antigen.

2. *Interferons :* They represent a family of molecules that can also exert immuno modulatory effects in addition to its antiviral activity.

   It is of three types α, β and γ.

   The interferon produced by sensitized T cells is called *gamma* (γ) *interferon* which is chemically distinct from alpha (α) and beta (β) interferons induced by viruses.

   Gamma interferon can activate macrophages, levels of metabolic activity accompanied by greater ability to kill tumour cells and ingest microbes.

3. *Chemotactic factors :* There are chemotactic factors for neutrophils, eosinophils, basophils, monocytes and other lymphocytes.

4. *Interleukin-2 :* This substance also called T cells growth factor is produced by antigen of mitogen activated T cells. It, in turn, causes proliferation of the T cells involved in mediating cellular immunity.

5. *Transfer factor :* This is an ill defined substance present only in humans, which has the ability to transfer delayed hypersensitivity and enables to prepare the non sensitized lymphocytes to respond to specific antigen without preparing the host to make antibody against the same antigen. The exact mechanism is not known, but it is postulated to be a single-stranded polynucleotide which either is informational or may provide some part of a receptor for antigen.

Major Histo-Compatibility Complex Proteins (MHC)

The membrane bound proteins encoded by the major histo compatibility complex or MHC (histos refers to tissues) are the antigen presenting markers through which the immune system distinguishes body cells from invading antigens (class I MHC proteins) and immune system cells from other cells (class II MHC proteins).

Generally, MHC refers to a cluster of genes responsible for immune response, transplantation antigens and proteins of the complement system. MHC produces a set of proteins called histo compatible molecules. MHC molecules are located on the cell surfaces of all nucleated cells of the body as well as in the blood serum.

These molecules are responsible for allograft rejection, immune recognition, complement levels etc.

The gene segment is named as *major histocompatibility complex* because of its original discovery was based on transplantation experiments. Transplantation is nothing but implantation of a tissue from one individual to another. The tissue implant is called *graft*. The person who donates the tissue is called the *donor* and the person who receives is called *recipient*. If both the donor and recipient are identical in their genotype and phenotypes (antigen pattern) then the graft remains alive and healthy. Now the graft is said to be *histo-compatible*, and if the graft is non-compatible, it dies and is *rejected* (graft). The rejection of the graft is due to an immunological reaction.

The graft tissue releases antigen into the recipient. The immune system of the recipient recognizes the antigen as 'non-self' and elicits the immune response which rejects the graft. The antigen responsible for graft rejection is called *transplantation antigens* or *histocompatibility antigens*. It is now known that the genes are responsible for production of antigens, associated regulation of immuno response, components of complements and certain proteins.

Histo compatible molecules are produced by MHC genes and are classified into four types :

1. Class I MHC molecules (Lymphocytes, platelets and in man, human leucocyte antigens (HLA)
2. Class II MHC molecules (B-cells, macrophages, monocytes, antigen presenting cells and activated T cells)
3. Class III MHC molecules (complements like C2, C4 and factor B)
4. Class IV MHC molecules (T-cells leukemia (T*la*))

MHC functions are 1. production of HLA (human leucocyte Antigen) 2. Production of immune associated antigens (Ia) 3. controls the level of components of complement i.e C4, C2 of classical pathway and factor B alternate pathway 4. helps in T-cell recognition (Cytotoxic T cells require Cl I MHC whereas helper and suppressor T cells require Cl II MHC).

Hyper Sensitivity

May be defined as "the violant reaction of the immune system to specific antigen or pathogen leading to severe symptoms and even death in sensitised animal when it is reexposed to the same antigen for the second time".

Hyper sensitivity in clinical terms is called *allergy* and the factors causing hyper sensitivity are called allergens.

Traditional classification of hypersensitivity is based on the time required for a sensitised host to develop clinical reactions upon reexposure to the antigen. These are '*immediate*' and '*delayed*' (hypersensitivity) types. Both these are further sub divided into several clinical types.

Coombs and Gell (1963) classified hyper sensitivity reactions into four types based on the different mechanisms of pathogenesis or mechanisms of immunologically mediated disorders.

Type I, *Type* II, *Type* III *and Type* IV *Hypersensitivities*

These occurs as a Systemic reaction or a Local reaction.

Systemic Reaction

It usually follows an intravenous injection of an antigen to which host has already become sensitized. This often leads to a state of shock, which is sometimes fatal,

Local Reaction

It depends on the portal of entry of the allergen and takes the form of localized cutaneous reaction, conjunctival discharge, nasal discharge or bronchial asthma or food allergy.

Type I *Anaphylaxis : (Anaphylactic hypersensitivity)*

Mediators of Type I *Reaction*

Type I reactions are mediated by IgE antibodies produced by lymphocytes and plasma cells in response to an allergen. Once released, IgE antibodies attach to mast cells and basophils through cell surface receptors. Once attached, these cells release a variety of powerful vasoactive mediators after a series of reactions which are shown as follows :

Primary mediators released	Secondary mediators
1. Histamine	1. Arachidonic acid metabolites
2. Eosinophil chemotactic factor of anaphylaxis (ECF-A)	2. Leukotrienes
3. Neutrophilic chemotatic factor	3. Platelet activating factors (PAF)
4. Neutral Proteases	

Reactions of Systemic Anaphylaxis

After administration of heterologous proteins e.g. antisera, penicillin, enzymes etc., within minutes there is itching, hives, skin erythema followed shortly by contraction of respiratory bronchioles leading to respiratory distress. There is vomiting, diarrhoea, laryngeal obstruction leading to shock and death.

Reactions to Local Anaphylaxis

These are exemplified by atopic reaction which is defined as a genetically controlled predisposition to the production of specific IgE antibodies upon inhalation or ingestion of minute amounts of antigen like pollen, dust, animal dander etc. Specific diseases like urticaria, allergic rhinitis, angio-oedema and bronchial asthma are also caused.

Type II *Hypersensitivity (Antibody dependent cytotoxicity)*

This type is mediated by antibodies directed toward antigens present on the surface of cells or other tissue components. The antigenic determinants may be intrinsic to the cell membrane or extrinsic absorbed on the cell surface. In either case the reaction takes place by the binding of anitbody to normal or altered cell surface antigens.

There are two antibody dependent mechanisms :

Complement mediated cytotoxicity

In this reaction, antibody (IgM or IgG) reacts with an antigen present on the surface of the cell, causing activation of complement system and resulting in direct membrane damage and lysis. At the same time, the antibody coated cells become susceptible to phagocytosis. Example of such reaction is seen in Good-Pasteur's disease, Autoimmune haemolytic anaemia and erythroblastosis foetalis.

Antibody dependent cell mediated cytotoxicity (ADCC)

In this, the target cells coated with IgG antibody can be killed by a variety of non sensitized cells that have Fc receptors. This is carried out by K cells, eosinophils, neutrophils and monocytes which bind to the target cell by the receptor for Fc fragment of IgG and carry out lysis without phagocytosis.

This mechanism has been studied in vitro and whether it plays a role in vivo is not known.

Type III *Hypersensitivity (Immune complex mediated)*

This type of hypersensitivity reaction is induced by antigen-antibody complexes that produce tissue damage as a result of their capacity to activate a variety of serum mediators, principally complement system and biologically active fragments e.g. kinins. The toxic reaction is initiated when antigen combines with antibody, either within the circulation (circulating immune complexes) or at extra vascular sites (in situ immune complexes).

There are two types of antigens that cause such reaction.

Exogenous antigen : eg. foreign protein, bacteria, viruses.

Endogenous antigen : In which the individual can produce antibody against its own body complement.

Type IV *Hypersensitivity (Cell mediated delayed)*

This type of hypersensitivity is initiated by specifically sensitized T-lymphocytes. It induces the classic delayed hypersensitivity reactions and cell mediated cytotoxic reactions.

1. *Delayed Hyper-sensitivity :* This type of reaction is initiated by specifically sensitized T-lymphocytes generated during the initial contact with antigen, some of the memory cells remain in the circulation for a long time. When the individual is re-exposed to the specific antigen, e.g. tuberculin, T-lymophocytes are stimulated to divide and release a variety of biologically active molecules K/a lymphokines. The lymphokines amplify the inflammatory response by recruiting inflammatory cells, activating them and keeping them at site. The most important result is stimulation and activation of macrophages within the lesion which further cause tissue damage and inflammatory reactions. Example of this reaction is Tuberculin Reaction.

2. *T-cell mediated cytotoxicity :* The process of CTLs is complex and not properly understood. However, once sensitized, they kill the antigen bearing target cells. This group of T cells are characterized by the presence of OK T5/T8 antigens on their surface. The CTL after causing lysis of the target cell, survives the encounter and is then recycled to kill other larger cells.

   This CTL mediated lysis is highly specific and the cells are not affected.

It is now clear, that this specificity is due to the presence of cell surface HLA antigens by which CTLs recognize their target cells. It is suggested that the infection of the cell with the antigen e.g. virus, changes the cell surface HLA antigen, which is different from the normal self HLA antigen so as to evoke the CTLs, capable of recognizing altered self and thereby kill the virus infected target cell. e.g. Transplant Rejection Reaction.

Immune Disorders Classified by Source of Antigen

Exogenous

Atopic disease (e.g. poison ivy contact dermatitis; reactions to plant pollen, sera and drugs).

Homologous

Reactions to isoantigens (e.g. transfusion reaction, erythroblastosis fetalis, transplant rejection)

Autologous

Autoimmune diseases (e.g. systemic lupus erythematosus, erytheumatoid arthritis, Siogren's syndrome).

4.10 Production and Screening of Monoclonal Antibodies against Parasites

Antibody produced by a single clone of cells is a "*monoclonal antibody*". All the antibody molecules are identical and bind to the same antigenic site with identical binding affinities.

Monoclonal antibodies can be generated in large quantities by creating a cell fusion called *hybridoma* between an antibody producing cell and a myeloma cell (cancer).

Substances, foreign to the body such as disease causing bacteria/viruses/infectious agents known as antigens are recognized by the body's immune system as invaders. Our natural defenders against these agents are antibodies (proteins) that destroy them. There are cells in the body, the immune lymphocytes, which can produce millions of different antibodies. But a single cell can only produce antibodies with a certain predetermined specificity. Hence, for the formation of a multitude of antibodies, occurrence of excess of lymphocytes is a prerequisite. If the body is exposed to a certain foreign antigen, there may be stimulation of a lymphocyte that can identify the particular antigen. This lymphocyte then starts to divide and form a clone of monospecific cells which produce identical 'mono-clonal antibodies' (mAb). Thus, a single antibody forming (lymphocyte) cell or clone, produces antibodies specifically directed against a single antigen or antigenic determinants.

Earlier it was thought that homogeneous antibodies or immunoglobulins could be obtained in quantity by simply cloning a single lymphocyte and harvesting the antibodies, the clone produced. Unfortunately lymphocytes donot grow continuously in culture. In 1970, Cesar Milstein and George Kohler developed a technique for immortalizing such clones, named *hybridoma technology*. Hybridomas are somatic cell hybrids produced by fusing antibody forming spleen cells with myeloma (is a cancer of the antibody forming lymphocytes/cells) cells. The resultant '*hybrid*' cells express the properties of both parental cell lines i.e. the myeloma provides properties of unrestricted proliferation and production of vast quantities of monoclonal antibody whereas the *immune lymphocyte* provides information for the specificity of the antibody (i.e. an antibody with known antigen binding properties). Such hybridoma-derived monoclonal antibodies are becoming increasingly important in diagnostic and therapeutic medicine.

Production and Screening of Monoclonal Antibodies (mAbS) (Fig. 4.52)

Steps involved in the production of monoclonal antibodies using hybridoma technology are :

1. A rabbit or mouse is immunized through repeated injections of specific antigen for the production of specific antibody and for proliferation of desired B-cells.
2. Tumors (or myelomas) are produced in a mouse or rabbit (myeloma is a cancer of the antibody forming B-lymphocytes).
3. Spleen cells that produce specific antibodies and myeloma cells that produce tumours are cultured separately from the above two sets of mice.
4. The myeloma cell line used stops synthesizing antibodies as it is a mutant called HGPRT that can not synthesize enzyme hypoxanthine-guanine phsophoribosyl transferase or HGPRT in HAT medium (Hypoxanthine Aminopterin Thymidine).
5. The antibody synthesising cells from the spleen (specific) are then fused with fast growing mouse cancer cells called myelomas (immortal) and a chemical 'glue', polyethylene glycol.
6. The result is a 'hybridomas', a fusion of antibody making cell and a cancer cell. The hybrid cells are grown in a selective HAT medium.
7. HAT medium contains a drug aminopterin one pathway for nucleotide synthesis, making the cells dependent on another pathway that needs HGPRT enzyme absent in myeloma cells. Therefore the myeloma cells that do not fuse with B-cells will die since they are HGPRT. B-cells that do not fuse will also die as they lack tumorigenic properties of immortal growth.
8. HAT medium, therefore allows selection of hybridoma cells which inherit HGPRT gene from B-cells and tumorigenic property from myeloma cells.
9. The hybridomas are separated, cloned and tested to select those that make the desired monoclonal antibodies.
10. The chosen hybridomas are injected into the peritoneal cavity of the mouse where they induce the accumulation of 2 to 10 ml of ascitic fluid containing large amounts of antibody or these hybridoma cells may be cultured in the laboratory and may be stored frozen to generate additional supplies of mAbs.
11. The final step is to separate the pure monoclonal antibodies from the hybridomas by differential centrifugation.

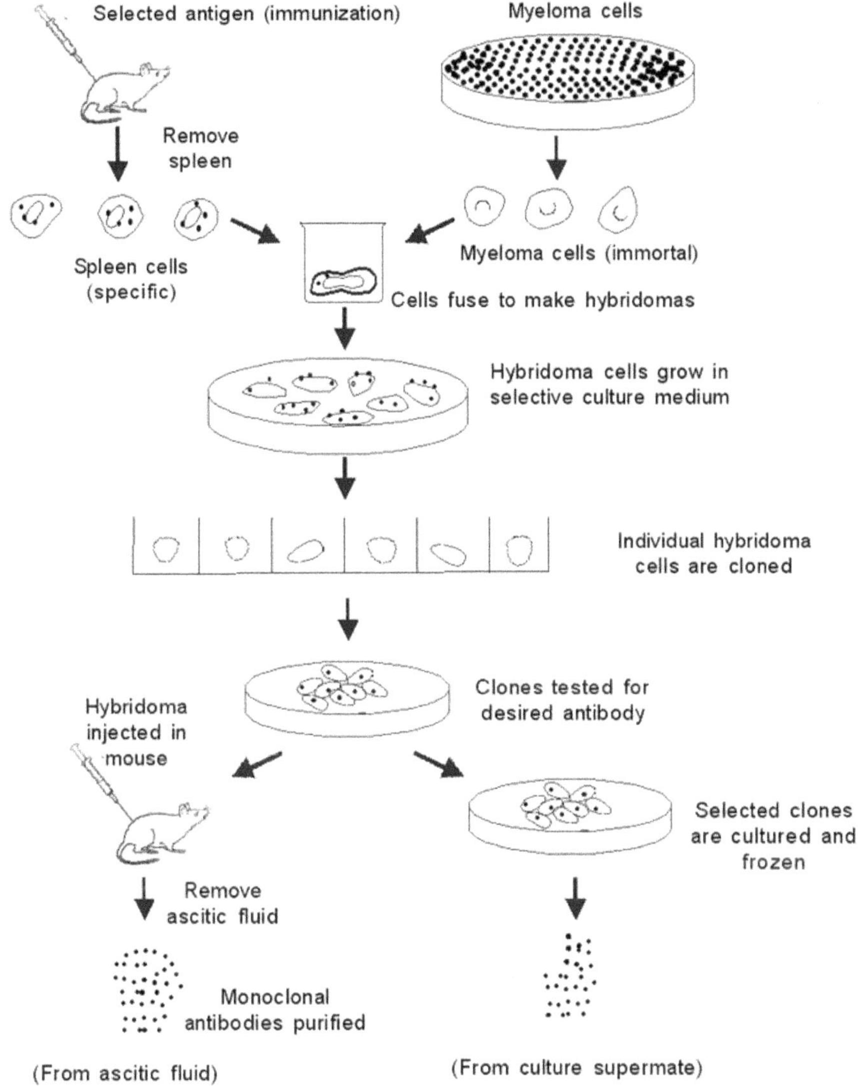

Fig. 4.52 Production of Monoclonal antibodies by Hybridoma technology

The discovery of hybridoma technology for production of monoclonal antibodies has revolutionised in immunology by opening up new possibilities for basic as well as applied biomedical research. For example, they have made possible the routine testing of blood for the presence of HIV (AIDS virus), thereby protecting the public blood supply.

Monoclonal Antibodies Against Parasites

More than 1000 million people living in the tropical areas face a threat of painful and devastating parasite diseases which kills millions of people each year. Even more distressing is the fact that in many countries these diseases are becoming more and more common.

Biotechnology today is undergoing a resurgence in a wide range of applications and the tremendous increase allover the world of all the new technologies, those that utilize monoclonal antibodies (mAb) are likely to have the most rapid and wide spread impact on medicine. There is a great possibility that genetic engineering will produce new and effective vaccines for many parasitic diseases but in the near future monoclonal antibodies are likely to be of much greater importance.

4.11 Enzyme Linked Immuno sorbent Assay (ELISA)

ELISA (enzyme linked immuno sorbent assay) was initially developed in 1971 independently by two groups : 1. Engvall and Perlmanna and 2. Weeman and Shuurs.

ELISA is a simple and versatile technique, widely employed variants of enzyme immuno assay. Enzyme immuno assays have emerged as quantitative techniques for detection of antigens, haptens and antibodies by coupling enzymes to them. ELISA has found application for the detection of a variety of antibodies and antigens such as hormones, toxins and viruses.

ELISA is a very sensitive (i.e Ag and Ab are detectable at levels of about 10^{-10} gm/ml or 1 part in 10 billion), less expensive, safer, reliable and accurate as radio immuno assays.

ELISA is so named because it employs various enzymes linked to either antigen or antibody as a label which can easily be detected by measurement of enzyme activity. So it is the enzyme activity which is the measure of the quantity of antigen (Ag) or antibody (Ab) present in the test sample.

Enzymes used in ELISA include β-galactosidase, glucose-oxidase, peroxidase and alkaline phosphatase.

At least three different ELISA techniques are available that have significant clinical value :

1. Double antibody sandwich-ELISA abbreviated as DAS-ELISA for the detection and measurement of antigen.
2. Direct antigen coating-ELISA (DAC-ELISA) for the detection of enzyme-Ig conjugates.
3. Indirect microplate ELISA for detection and measurement of antibody.

The test may be done in polystyrene tubes or polyvinyl micro litre plates. If it is done in tubes, it is called *macro-ELISA* or if in plates, it is called *micro-ELISA*.

In double antibody sandwich DAS-ELISA technique (Fig. 4.53)

- The wells or depressions in polystyrene plate receive antiserum, the antibodies in the antiserum adhere to the surface of each well.
- The test antigen is added and if the antigen is homologous it attaches to the antibody immobilized on the well surface.
- Enzyme-labeled specific antibody is then added, it will bind to the antigen already fixed by the first antibdoy.
- This results in "*antibody* (with enzyme) antigen *antibody sandwich*".
- Finally enzyme substrate is introduced for reaction with the enzyme.
- The rate of enzyme action is *directly proportional to the enzyme-labeled antigen-antibody* and this in turn is *proportional to the amount of test antigen.*

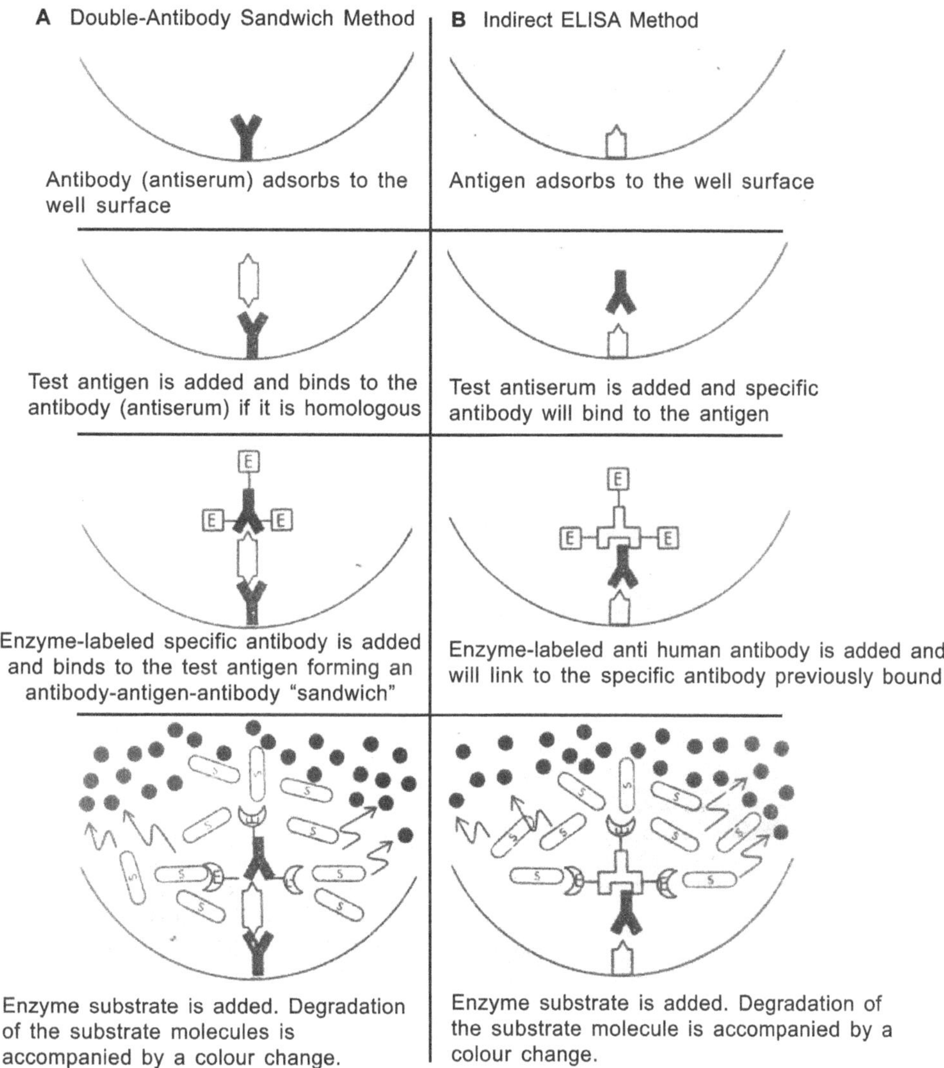

Fig. 4.53 Enzyme linked immuno Sorbent Assay (ELISA) techniques (A) double-antibody sandwich DAS.ELISA and B. Indirect microplate ELISA (Pelezaretal).

- Enzyme activity may be observed visually by colour change (yellow colour) brought about by hydrolysis of the substrate or it can also be measured by using a spectro photometer.
- This method is highly successful in the assay of B-Hepatitis.

Indirect - Microplate - ELISA

- Employed for detection and measurement of antibody.
- The first step in this technique is coating of polystyrene wells with antigen by passive adsorption.
- Test anti serum is added and allowed to incubate

- The antibodies (homologous) in the antiserum will bind to the immobilized antigen.
- Enzyme labeled antihuman antibodies (antiglobulins or antibodies) are added to the system that will fix to the antibody-antigen complexes formed in the previous step.
- Finally, the enzyme substrate is added and the rate of its hydrolysis/degradation is associated with a colour (yellow) change.
- The coloure change is proportional to the concentration of antibody present in the test sample.
- The colour is read in a spectrophotometer or a micro ELISA reader.
- This method has been used for the immuno diagnosis of many infectious pathogens such as viruses, parasites and fungi.
- For example, one antiserum sample from a pregnant woman can be screened simultaneously for infections (rubella virus and type II herpes virus) and other infections that can affect the foetus.

For the convinience of the readers, a procedure for the ELISA is given below.

ELISA Test (Enzyme Linked Immuno sorbent Assay)

Principle

The ELISA assay can be used to measure either antigen or antibody. To measure antigen, the antibody is bound to solid phase, incubated with test serum and then a second enzyme labelled antibody is added, substrate is then added and enzyme activity is related to antigen concentration.

Clinical Applications

In detection of hepatitis B surface antigen (HBsAg), human chorionic gonadotropin (HCG) levels in urine or blood, CEA (Carcino Embryonic Antigen) levels, Steroid hormones, immunoglobulins, antibodies to bacteria, viruses, DNA and allergens.

Procedure

Material and Methods

1. *Test sample :* Serum sample stored at -75^0 C till the time test is performed.
2. *Enzyme conjugate :* Antibody to the specific antigen is conjugated to enzyme. The commonly used enzyme being horse-radish peroxidase. Others are alkaline phosphatase, Lysozyme and glucose-6 phosphate dehydrogenase. The enzyme is coupled to antigen or antibody by various cross-linking agents e.g. glutaraldehyde and dimaleimide.
3. *Substrate :* Ortho phynylene diamine. (OPD) or 5-aminosalicylic acid. Or PNPP (p-nitrophenyl phosphate disodium salt tablets).
4. *Antibody Microtest Plate :* Precoating of antibody carried out in a flat bottomed, 96 well microtest plates with specific dilution.
5. *Diluent :* Phosphate buffer saline (PBS) pH 7.2, 0.1 M and Tween (0.5 ml Tween-20 to 1 litre of PBS) or carbonate buffer pH 9.5 or normal saline with 0.1% sodium azide.

Method

1. The antibody is diluted to 1:500 in PBS at pH 7.2 and 75µl is added to the bottom of each well.
2. Incubate the plates at 4^0 C in a moist chamber overnight for 18-20 hrs.
3. Wash the plate three times with PBS supplemented with 0.05% Tween 20 (BS-T) for 5 minutes each.
4. Add 200µl of 1% bovine serum albumin (blocking solution) in each well and incubate overnight at 4^0 C in a moist chamber.
5. Add 25µl of the test samples (primary antibody) positive and negative control in to the test plates coated with antibody.
6. Incubate the plate for 2 hours at room temperature.
7. Wash three times with PBS-T and add 50µl of peroxidase conjugated antibody (secondary antibody) to each well.
8. Incubate the plates at room temperature for 2 hours and wash five times with PBS-T.
9. Add 100 µl of freshly prepared substrate i.e. *orthophenyl diamine* in citrate buffer pH 5.0 and hydrogen peroxide.
10. After 30 minutes incubation at room temperature in a dark box, add 75µl of 2 M sulphuric acid or MNaOH to stop the reaction.
11. Add 75µl of PBS to make final volume of reaction mixture as 250µl.
12. A positive test is indicated by the production of yellow colour as a result of the product of enzyme reaction. This is read by the naked eye against a white background or by measuring the optical density at 492 nm in a spectro photometer using a microcuvette or a micro ELISA reader with automated photometer preferably within half an hour to one hour of the addition of stopping solution.

Calculations

$$= \frac{\text{O.D.Test} - \text{O.D.BlankII}}{\text{O.D.Blank} - \text{O.D.BlankII}} = \frac{\text{O.D. of sample}}{> \text{O.D. mean}}$$

Interpretation of Results

A sample is labelled positive if the yellow colour is stronger than that of the negative control and if the the optical density, measured as the ratio of test to negative control is more than 2:1 (O.D. of blank I is the average optical density of 5-7 negative controls and O.D. of blank II is the optical density of 100µl substrate, 75µl of 2M sulphuric acid and 75µl of PBS).

A test sample with a ratio between 1 : 5 and 2 : 1 is repeated and if the ratio is still less than 2 : 1 the test should be considered negative.

Advantage of ELISA Technique

1. Simple and easy to perform
2. Degree of sensitivity at low cost
3. Stability of reagents
4. Lack of radiation hazards
5. Potential for automation
6. Relatively inexpensive equipment.

ELIS Assays Currently Available

1. Hepatitis	:	HBs Ag, anti-HBS Ag, Anti – HBc Ag, IgM anti – HBc Ag HBe/anti HBe, anti – HA V anti-Delta
2. Cancer	:	CEA, AFP, Beta HCG, Estrogen Receptor, TdT, PAP
3. STD	:	N. Gonorrhoeae Chlamydia Trachomatics
4. Infections	:	Rubella, Rotavirus cytomegalovirus, IgE, Toxo – G
5. Metabolic	:	Ferritin, Prealbumin
6. Thyroid	:	TSH
7. AIDS	:	HTLV – III

Essay Questions

1. Describe in detail the parasitic adaptations.
2. Enumerate different Helminths.
3. Arthropod parasites pathogenic to man.
4. Describe the structure and biology of *Entamoeba.*
5. Describe the structure and biology of *Trypanosoma.*
6. Describe the structure and biology of *Plasmodium.*
7. Describe the structure and biology of *Balantidium.*
8. Describe the morphology and life cycle of *Schistosoma.*
9. Describe the morphology and life cycle of *Fasciola.*
10. Describe the morphology and life cycle of *Diphyllobothrium.*
11. Describe the morphology and life cycle of *Echinococcus.*
12. Describe the morphology and life cycle of *Dracunculus*.
13. Describe the structure, nature of damage and life cycle of head louse/bed bug/mosquito/housefly/ticks/mites/fleas.
14. Enumerate the differences between *Anopheles*, *Culex* and *Aedes* mosquitoes.
15. Write in detail the parasitic diseases of fishes and their control.
16. Give a detailed account of the livestock breeding (i.e., the breeding of cattle, sheep & pigs).
17. Describe in detail the different control measures of the crop pests.
18. Enumerate different plant parasitic nematodes.
19. Write in detail the structures nature of damage and life cycle of *Meloidogyne/Heterodera.*
20. Give a detailed account of the methods of control of plant parasitic nematodes.
21. Give the definition of Immunity and types of Immunity.
22. Give an account of organs of Immune system.
23. Describe in the factors responsible for immunogenicity of an antigen.
24. Write about different Immunoglobulins and their biological activities.
25. Describe the Antibody-Antigen reactions.
26. What is Opsonization ? How does it help in Phagocytosis.
27. Describe on the Cell-mediated immune responses.
28. Write the humoral immunity responses.
29. Write an essay on Hyper sensitivity and its types.
30. Write about Hybridoma Technology.
31. Write about ELISA and its applications.

Short Questions

Explain the Following :

1. Effects of Parasitism.
2. Life cycle of *Entamoeba.*
3. Life cycle of *Giardia.*
4. Different forms of *Trypanosoma.*
5. *Leishmania* morphology.
6. Preerythrocytic/Exocrythrocytic/Erythrocytic cycle of plasmodium.
7. Gametogony/Sporogony of plasmodium in mosquito.
8. Structure of *Schistosoma* or *Fasciola* or *Paragonimus* or *Diphylobothrium* or *Taenia solium* or *Echinococcus* or *Hymenolepsis* or *Trichinella* or *Ancylostoma* or *Ascaris* or *Dracunculus.*
9. Structure of Head louse.
10. Structure of Bed bug.
11. Sex differences in Adult mosquitoes..
12. Structure of house fly.
13. Structure of flea.
14. Itch mite.
15. Chigger mite
16. Soft tick.
17. Hend tick morphology.
18. Control measures of fish parasitic diseases.
19. Cattle breeding.
20. Sheep breeding.
21. Piggery.
22. Pathogenicity of *Tylenchulus semopenetrans.*
23. Root knot nematodes.
24. Seed gall nematodes.
25. Lance nematodes.
26. Radopholus.

27. Globodera.
28. Aphelenchoides.
29. Ditylenchus.
30. Chemical practices for the control of plant nematodes.
31. Biological practices for the control of plant nematodes.
32. Cultural practices for the control of plant nematodes.
33. Physical practices for the control of plant nematodes.
34. Genetic Engineering of Nematodes.
35. Brief History of Immunology.
36. Definition of Immunity.
37. Innate and Aquired Immunity.
38. Interferons.
39. Phagocytosis.
40. Active Immunity.
41. Antibody/Antigen.
42. Antibody-Antigen complex.
43. IgG / IgA / IgM / IgD / IgE.
44. Functions of Immunoglobulins.
45. Opsonization.
46. Primary Immune response.
47. Secondary Immune response.
48. Humoral Immune system.
49. Complement system/Lymphokines.
50. Major histocompatibility complex (MHC) proteins.
51. Type I/Type II/Type III/Type IV hyper sensitivity.
52. Monoclonal Antibodies.
53. Production of Monoclonal antibodies.
54. DAS - ELISA Technique.
55. Indirect Microplate - ELISA Technique.
56. ELISA applications.

5 Animal Biotechnology

5.0 Transgenic Animals

Animals are a major source of food, milk, wool, leather and many other byproducts. Obviously, there has been growing interest in improving the efficiency of livestock production. Until recently, selective breeding was the only way to enhance the genetic quality of desired animals. However, selective breeding method is time consuming and can introduce limited genetic variability. Although progress in various gene transfer techniques like transduction, transfection and introduction of cloned genes into fertilized eggs, implantation of such genetically modified eggs into receptive female and obtaining progeny which carry cloned gene has resulted in the rapid development of a new era of animal biotechnology.

It is now well known that eukaryotic genes can be isolated, modified, amplified and can be expressed. Thus, it is also possible to introduce new foreign or synthetic genes into eukaryotic cells. One of the widely used strategies to accomplish this goal is to incorporate the DNA into the genome of a non replicating virus and allow the virus to infect. To date, the application of genetic engineering technologies to animals has focussed on three major goals : namely,

1. The introduction or enhancement of desirable trait in domestic animals.
2. The production of valuable products by transgenic animals and
3. To investigate basic biological processes such as gene expression.

1. The mouse has been the choice organism for developing gene transfer methodologies and transgenic mice are now routinely produced and studied in laboratories throughout the world. Over 60 different genes have already been transferred into recipient mouse eggs or embryos and stably maintained in adult mice and their progeny.

2. The foreign introduced genes are referred to as *transgenes* and the animals that contain the foreign or introduced genes are called *transgenic animals*.
3. Transgenic animals are useful both to study the expression and control of foreign eukaryotic genes and also as recipients of successful therapy and manipulation.
4. One of the most encouraging results of these studies on transgenic mice is that the transgenes have usually exhibited their normal patterns of expression (tissue-specific or developmental stage specific expression) and regulation (induced or constitutive).
5. For example a mutant mouse lacking the gonadotropin-releasing hormone, and thus sterile, was restored to normalcy by being transfected with an intact gonadotropin-releasing hormone gene that was incorporated into its genome and its germ line.
6. Transformation of the germ line in mammals can be carried out in several ways and the following methods are routinely used to produce transgenic animals.
7. (a) Microinjection of DNA into the pronuclei of fetilized eggs and (b) Infection of preinplanted embryos with retroviral vectors and (c) Stem cells.

A complete discussion of the various methods is beyond the scope of this book but a few commonly used methods and some applications of recombinant DNA technologies in animal breeding have been outlined below.

Methods for Production of Transgenic Animals

A number of standard procedures are now available to introduce naked DNA into mammalian culture cell system through a process called *transfection*. Most often the cells are suspended in the presence of calcium precipitate of the DNA.

It is estimated that only about one in 10 to 10^{16} cells takes up the DNA and incorporates it stably into the recipient chromosomes. It is not fully known why this small fraction of cells in the population get transfected, but those that are transfected pick up several DNA fragments. One way to select for those cells that have taken up the foreign DNA is to induce a gene that allows the transfected cells to grow in a particular medium in which nontransfected cells cannot survive. Because transfected cells typically take up more than one fragement, the gene used for selection need not be located on the same DNA fragment as the gene whose role is being investigated called *the transgene*.

Electroporation Method

In this method the animal cells are incubated with DNA in special vials that contain electrodes that deliver a brief electrical shock. This causes the plasma membrane to become transiently permeable to DNA molecules some of which find their way in to the nucleus and become integrated into chromosome thus bringing about transfection.

Microinjection of DNA into fertilized egg

This is the most commonly used method to introduce foreign genes into an animal cell. The nuclei of oocytes and eggs are ideally suited for this approach. In this technique (modifications of nuclear transplant techniques developed in the 1950s, the eggs are surgically removed from the female parent and are fertilized *in vitro*. The DNA is then microinjected into the male pronucleus (the haploid nucleus of sperm) prior to nuclear fusion of the fertilized egg through a fine tipped glass capillary tube as shown in Fig. 5.1. The infected DNA can be linear or circular, but the former integrates more efficiently and several copies (from few hundreds to few thousands) are injected into each egg. The integration of the injected DNA molecules occurs at random sites in the genome. Anti sense RNA gene strategy is the right technology with many applications.

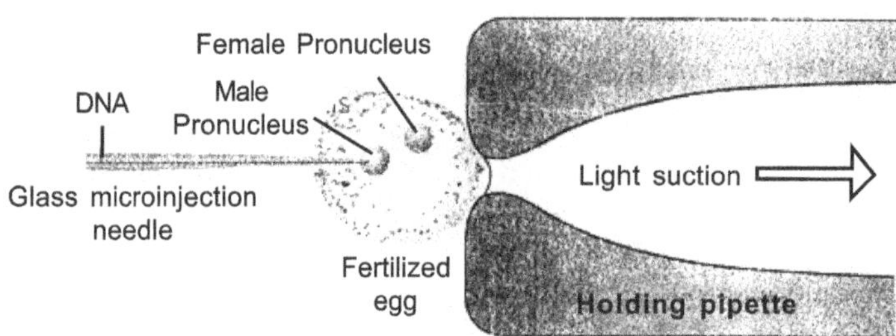

Fig. 5.1 The production of transgenic animals by microinjection of DNA into fertilized eggs.

Injection of DNA into the nucleus (germinal vesicle) of a mouse oocyte. The oocyte is held by a pipette.

Because the DNA is inserted early in development even before the haploid nuclei to form the diploid nucleus of the zygote, integration of the injected DNA molecules occurs early during embryonic development. As a result some germ line cells usually carry the transgene. When the zygote is two celled it is implanted in a pseudo pregnant foster mother for further development. The animals that develop from the injected egg in foster mother are called the G_0 generation and are almost mosaics with some somatic cells carrying the transgene and others not carrying it. The initial G_0 transgenic animals must be mated and the G_1 progeny produced to obtain animals in which all cells carry the transgene. In most of the cases where progeny studies have been performed, the transgenes were transmitted to progeny. Thus, *transgenic animal* is obtained. The success with this method is 15-20%.

Embryonic Stem Cells

Another important, effective and recent method for transforming mammals involves the use of *embryonic stem cells* present in the blastocyst that gives rise to embryo proper. These are derived from embryo, can be cultured and transfected with a vector containing the desirable gene. Mutations in them can be selected or introduced by the use of recombinant DNA vectors. These stem cells can then be injected into the cavity of another blastocyst where they become incorporated into the developing embryo and often participate in forming the germ line. In this way, any mutations introduced into the embryonic stem cells while they are in culture may become incorporated into the germ line of living animals. The method was first used to create mice with a deficiency of the enzyme HPRT used in purine biosynthesis in order to have an animal model for the human disease. Surprisingly, the HPRT-deficient mice are virtually normal, whereas the enzyme deficiency in human beings results in a serious disease.

Genetic Engineering with Animal Viruses

The most common method is direct injection of vector DNA into the nucleus of a fertilized egg, which is then transferred to the uterus of a foster mother for development. The vector is usually a modified retrovirus (SV40). This animal virus is icosahedral in shape with 5224 bp in its RNA, first isolated from monkey and named Siamian Vacuolating virus. This virus replicate in the host without making active virus particles and make circular plasmids in host cell or the viral genome gets integrated into the host genome. In view of this, the virus is ideal for mammalian genome studies and r-DNA studies. Transfection can also be mediated by retroviruses (RNA viruses containing the gene for reverse transcriptase i.e., Retroviruses have RNA as their genetic material and code for a reverse transcriptase that converts the retrovirus genome into double-stranded DNA that becomes inserted into the genome in infected cells. Genetically engineered retrovirus containing inserted genes undergo the same process. These features of retroviruses make them convenient vectors for the genetic manipulation of animal cells, including those of birds, rodents, monkeys, and human beings (Fig. 5.2).

Genetic engineering with retroviruses allows the possibility of altering the genotype of animal cells. Because a wide variety of retroviruses known, including several that infect human cells, genetic defects may be corrected by these procedures in the future. However, many retroviruses contain a gene that results in uncontrolled growth of the infected cell, thereby causing a tumor. When retrovirus vectors are used for genetic engineering, the tumor-causing gene is first deleted. The deletion also provides the space needed for incorporation of foreign DNA sequences. The recombinant DNA procedure employed with retroviruses consists of synthesis in the laboratory of double stranded DNA from the viral RNA, through use of reverse transcriptase. The DNA is then cleaved with a restriction enzyme and by the technique already described foreign DNA is inserted. Transformation yields cells with the recombinant retroviral DNA permanently inserted into the genome. In this way, the genotype of the cells can be altered (Fig. 5.2).

Transgenic Animals - the success story

The first transgenic animal was produced in USA by Ralph Brinster in 1981 by introducing a gene for rat growth hormone (GH) into the fertilised eggs. In 1988, a transgenic mouse that is prone to cancer was the first genetically engineered animal to be patented. This mouse provides an excellent model for studying cancer. Since then many transgenic animals such as cattle, sheep, goats, pigs, rabbits, poultry, fish and silk worm have been produced. Following is an account of few selected transgenic animals. Such animals are useful both to study the expression and control of foreign eukaryotic genes and also in their own right as recipients of successful therapy and manipulation.

Transgenic mice are produced routinely in laboratories all over the world as they provide valuable tools for the study of gene expression in mammals and an excellent model system to test various gene-transfer vectors and methodologies for possible use in humans. Hundreds of different transgenes have been studied in mice and in most cases, the transgenes show normal patterns of inheritance, indicating that they have been integrated into the host genome.

One of the early significant effects of a transgene was the increased growth rate that occurred when rat, bovine, or human growth hormone genes were expressed in mice (Fig. 5.3). This prompted animal breeders to ask whether the introduction of either 1. Extra copies of the homologous (same-species) growth hormone gene or 2. Copies of heterologous growth hormone genes from related species might result in domestic animals with enhanced growth rates. *Transgenic pigs* were produced

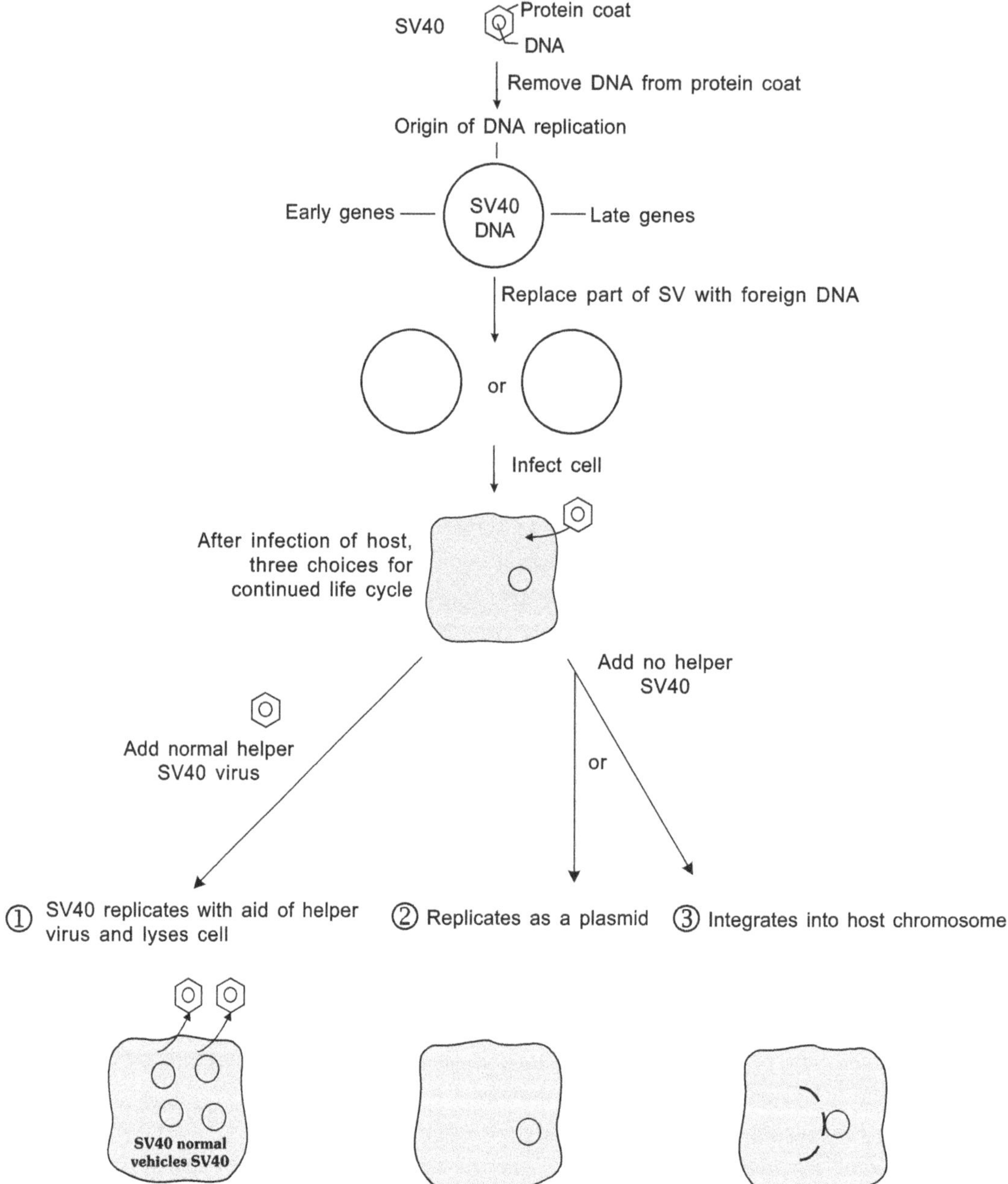

Fig. 5.2 SV40 virus can be used as a gene-cloning vehicle. Although part of the virus is replaced by inserted DNA during cloning, it can still replicate with the aid of normal helper viruses (non recombinant SV40). Without the aid of helper viruses, it can either replicate as a plasmid or integrate into the host chromosome.

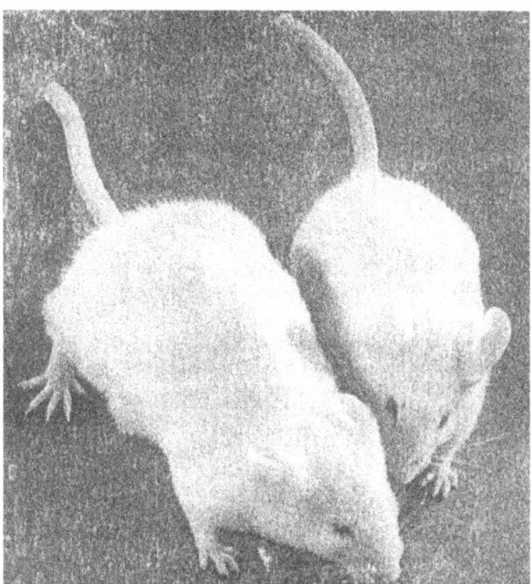

Fig. 5.3 Transgenic mouse on the left, carries a chimeric human growth hormone gene, is double the size of the control mouse (on the right).

with the hope that enhanced growth hormone levels might result in leaner pigs with improved meat quality and with faster growth. Other researchers introduced growth hormone transgenes into fish and chickens with similar objectives. The experiments with transgenic pigs indicated that growth rates were not increased on standard diets, but were increased on high-protein diets. However, the transgenic pigs were found to be leaner than the controls, because growth hormone favors the synthesis of proteins instead of fats.

Unfortunately, the transgenic pigs also exhibited several undesirable side effects of the higher growth hormone levels. Notably, the female transgenic pigs were sterile. And both sexes were lethargic with weak muscles and were highly susceptible to arthritis and ulcers.

In 1988, a transgenic mouse that is prone to cancer was first genetically engineered animal to be patented. This mouse provides an excellent model for studying cancer.

Transgenic Sheep

In 1997, Ian Wilmut at the Roslin Institute near Edinburgh and his colleagues created clones of sheep using somatice cells. As you may have read in papers they named the first *cloned lamb, Dolly*.

Fig. 5.4 illustrates how Dolly was created. The researchers removed mammary cells from an adult female sheep and grew them in the laboratory. The researchers then extracted the nucleus from a sheep oocyte and fused a diploid mammary cell with the enucleated oocyte cell. Fusion was promoted by electrical pulses as mentioned earlier. After fusion, the zygote was implanted into the uterus of an adult sheep. One hundred and forty-eight days later, Dolly was born.

Thus, Dolly is a genetically identical copy of the sheep that donated a mammary cell to create Dolly. Dolly and the donor sheep are genetically identical in the same way that identical twins are since they carry the same set of genes and look remarkably similar.

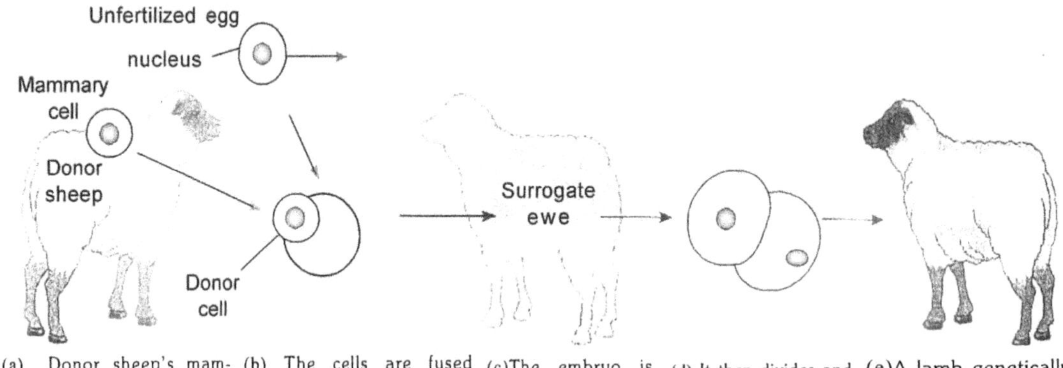

(a) Donor sheep's mammary cell is extracted and grown in a tissue culture flask. Another sheep's unfertilized egg is extracted and the nucleus is removed.

(b) The cells are fused together. The donor nucleus and the maternal proteins initiate development of the egg into an embryo.

(c)The embryo is transferred into a surrogate ewe.

(d) It then divides and grows like a normal embryo.

(e)A lamb genetically identical to the donor sheep is then born.

Fig. 5.4 Protocol for the successful cloning of sheep

The rate of transgenics in sheep is very low. The first transgenic sheep carried about 10 copies of human anti-hemophilic factor IX gene fused with BLG gene (b - lactoglobulin) as it was necessary for specific expression of gene in mammary glands. The sheep consequently secreted human factor IX into their milk.

Genetic engineering has also been profitably done to increase wool production in sheep. This was done by producing sheep that were transgenic for two bacterial enzymes converting Serine to cysteine. Growth hormone (GH) genes have also been introduced and can be used to promote body weight.

In another work, transgenic sheep which gave a_1-antitrypsin upto 35 gms/litre of milk were produced.

Human tissue plasminogen activator and urokinase have also been produced in transgenic sheep.

Mammalian cloning is still at an early stage of development. Nevertheless, the breakthrough of creating Dolly has shown that it is possible. This provides the potential for many practical applications. With regard to *livestock*, cloning would enable farmers to use the somatic cells from their best individuals to create genetically homogeneous herds. This could be advantageous with regard to agricultural yield, although such a genetically homogeneous herd may be more susceptible to rare diseases.

Apart from practical uses, however, people have become concerned with the possibility of human cloning. This prospect has raised serious ethical questions. In the future, society will have to make decisions regarding the application of cloning to humans.

Transgenic Chicken

Other transgenic animals are being tested for resistance to viral infections. Avian leukosis virus (ALV) is a major viral pathogen of chickens, causing losses to the poultry industry of $50 million to $100 million per year. Obviously, the availability of an ALV-resistant strain of chickens would be of major commercial value. Therefore, reserchers have produced transgenic chickens that carry a defective ALV genome. These chickens produce viral RNA and the viral envelope protein, but no progeny viruses. Most importantly, they are resistant to infection by ALV. The synthesis of large amounts of the retroviral envelope protein somehow blocks the reproductive cycle of intact, pathogenic ALV viruses. That ALV resistance has been transmitted to several generations of progeny indicates that the trait is stable. These encouraging results suggest that the introduction of defective viral genomes into domestic animals may be useful tool in the production of virus-resistant genotypes.

Major albumen proteins (60 g/egg) are as follows :

(a) Ovalbumin (54%)

(b) Ovotransferrin (12%)

(c) Ovomucoid (12%)

(d) Lysozyme (3.4%)

From the commercial point of view, the production of large quantity of foreign proteins especially *Immunoglobulins* for clinical application, prophylactic agents and in diagnostic kits which are used as structure.

Transgenic chick with anti-HIV protein have been developed by Roslin Institute in December 2000.

Transgenic Cows

In the production of transgenic cows, oocytes were collected from slaughter house killed cows, followed by desired *invitro* manipulation of oocytes, and the oocytes were fertilized in vitro. Then, their pronuclei were microinjected with an engineered segment of DNA encoding the antibacterial human iron binding protein; lactoferrin. The embryos were cultured to morula, blastula stage and were non surgically implanted into recipient females. Since cows produce approximately 10,000 litres of milk in a year and if 35g of protein is produced per litre, then cows are best suited as bioreacters.

Transgenic Cow Clone for Mastitis Disease Resistance

The disease Mastitis costs U.S. dairy farmers about $1.7 billion annually, including lost milk revenues. In 30% cases it is caused by *Staphylococcus aureus* and destroys milk-secreting cells in udder. Antibiotics are only effective in about 15% of cows infected with *S. aureus*. *Annie* will resist such cellular attacks by secreting an added protein. *Lysostaphin* will offer alternative defense bioengineered right into the animal's cells.

Transgenic Fish

The genes that have been introduced by microinjection in fish include, human or rat gene for growth hormone, chicken gene for della-crystalline protein, E.coli gene for b-galactosidase; E.coli gene for neomycin resistance, winter flounder gene for antifreeze protein; rainbow trout genes for growth hormone.

Transgenic Mosquito

Scott and his colleagues, California USA. have recently, unraveled the genetic code of malarial parasite (*Plasmodium*) and its host the *Anopheles* mosquito. Each year about 2.7 million people die of malaria. New genes could be introduced into laboratory mosquitoes that would block transmission of deadly *Plasmodium* parasites.

Molecular Pharming

A novel avenue of research involves the production of medically important proteins in the mammary glands of livestock. This approach is called *molecular pharming*. As shown in (Table 5.1) several human proteins have been successfully produced in the milk of domestic livestock. Compared with the production of proteins in bacteria, one advantage is that certain proteins are more likely to function properly when expressed in mammals. This may be due to post-translational modifications (E.g., attachment of carbohydrate groups) that occurs only in eukaryotes. Further, certain proteins may be degraded rapidly when expressed in bacteria. Therefore, as an ideal alternative to recombinant bacteria, the expression of human proteins in the milk of transgenic livestock is an area of great promising interest.

Table 5.1 *Proteins that can be produced in the milk of domestic animals*

S.No	Protein	Host	Use
1.	Lactoferrin	Cattle	Used as an iron supplement in infant formula
2.	Tissue plasminogen	Goat	Dissolves blood clots
3.	a-1- antitrypsin	Sheep	Treatment of emphysema
4.	Factor IX	Sheep	Treatment of certain inherited forms of hemophilia
5.	Insulin-like growth factor	Cattle	Treatment of diabetes

To introduce a human gene into an animal so that the encoded protein will be secreted into its milk, the strategy is to clone the gene next to a *milk-specific promoter* since eukaryotic genes are expressed in a tissue-specific fashion. In mammals, certain genes are expressed specifically within the mammary gland so that their protein product will be secreted into the milk. Examples of milk-specific genes include genes that encode the milk proteins such as lactoalbumin, β-lactoglobulin, casein, and whey acidic protein. In order to express a human gene in domestic animal's milk, the promoter and regulatory sequences for these milk-specific genes are linked to the coding sequences for the human gene (Fig. 5.5). In this way, the protein encoded by the human gene will be expressed within the mammary gland and secreted into the milk. The milk can then be obtained from the animal, and the human protein isolated in higher quantity and purified. The mammary glands, thus, serve as bioreactors and the production of proteins is much cheaper than other methods and the proteins are of clinical importance.

Thus transgenic animal species can provide important information regarding the functional roles of genes. In addition, transgenic animals as model organisms are becoming invaluable tools for investigating the mechanisms and treatment of human diseases. Trangenic animals serve as models for the study of Neurodegenerative disorders such as Alzheimer's disease, Parkinson's disease, Hutington's disease. Muscular dystrophy, arthritis/rheumatitis, cataract (Tau protein phosphorylation) and also can be used for drug testing; either classical drugs, macromolecular drugs such as vaccines, antibodies, growth factors, hormones, cytokines etc.

1. Using recombinant DNA technology, clone a human hormone gene next to a sheep β-lactoglobulin promoter. This promoter is functional only in mammary cells so that the protein product is secreted into the milk.
2. Inject this DNA into a sheep oocyte. The plasmid DNA will integrate into the chromosomal DNA, resulting in the addition of the hormone gene into the sheep's genome
3. Implant the oocyte into a female sheep, which then gives birth to a transgenic sheep offspring.
4. Obtain milk from female transgenic sheep. The milk contains a human hormone
5. Purify the hormone from the milk and use as medicated milk

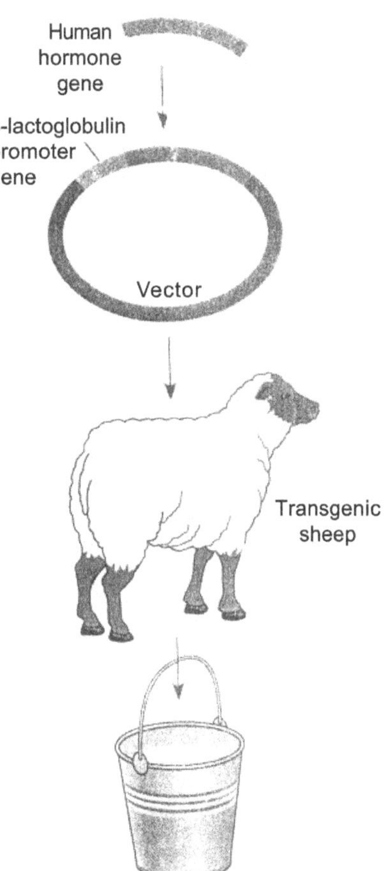

Fig. 5.5 Strategy for expressing human genes in a domestic animal's milk

5.1 Recombinant DNA Method in Vaccine Development

Vaccination is the simptest and the most effective means of controlling outbreaks of diseases in both animals and humans. Still, about five million infants die each year from diseases that can be prevented by existing vaccines. Morever, at present no effective vaccine is available against diseases like influenza, malaria etc. Almost 200,000 persons are infected each year with hepatitis B virus. On the other hand, malaria which is one of the greatest health problems of the world, afflicts an estimated 300 million persons is tropical areas annually killing 1-2 million of them. The acquired immunodeficiency syndrome (AIDS) is also rapidly increasing for want of a vaccine for its control. The first major vaccine was developed against small pox by active immunization with the attenuated agent of the disease, namely the living cowpox virus. All of the successful vaccines developed since then for the prohylaxisis of contagious diseases, which require making the vaccine with causative agent itself such as poliomyelitis, typhoid etc have always had defects in that they caused adverse reactions in some patients because of the extraneous materials they contained.

An Effective Vaccine

An effective vaccine must stimulate the immune system to make antibodies that can attack and neutralize the infective agent. Vaccines which are now available mostly use live attenuated organisms, killed inactived organisms and toxoids (detoxified toxins). Vaccination using whole organisms (live, attenuated or killed) includes risk of unanticipated active replication of the pathogen in the vaccine. Live poliovirus, vaccines, for example, led to infection and inadvertent spread of the virus in non-vaccinated individuals.

Vaccination induces about twice as many cases of encephalitis as does the natural disease (5.0 cases/million of postvaccinal encephalitis as compared with 2.5 cases/million postinfection encephalitis). In case of vaccination against flavivirus such as Dengue, there is a chance of induction of immune responses that facilitate infection of target cells which bear surface receptors for antibodies or antibody receptors.

A major problem faced by doctors, especially those working in ill-equipped hospitals and clinics, is to make accurate diagnoses of diseases. An array of individual monoclonal antibodies which can be used to identify specific parasitic antigens would be of great help. The production of such specific antibodies against parasites is now beginning and their use in simple tests should become possible in few years.

The United Nations Development Programme, the World Bank and the World Health Organization, in 1975 set up a special programme for Research and Training in Tropical Diseases. Six diseases malaria, leprosy, filariasis, leishmaniasis, schistosomiasis and tryponosomias which pose threat human life have been identified and research in biotechnology is showing signs of success in the treatment of some of these diseases.

Malaria

Until 1970's it seemed that malaria would be conquered if not eradicated completely. But, it did not happen and malaria fought back against all that science and medicine can throw against it. It is now estimated that there are on an astonishing 300 million cases of malaria and in Africa alone there be as many as one million deaths a year.

People were optimistic about treatment of malaria initially as the malarial parasite was being killed by the drugs and simultaneously by DDT and insecticides were controlling the mosquitoes which spread the diseases. However, both the parasite and mosquitoes have now become resistant to chemicals which previously killed them. A approach which can overcome this situation is urgently required, which can be possible with the aid of biotechnology.

One of the reasons for the malarial parasites 'success' is that while inside flow blood cells it is hidden from patients immune system. A good vaccine which guides the body how to make antibodies against malarial parasites would make all the difference.

Recently Ruth Nussenzcoeig and her colleagues at New York University have made monoclonal antibodies which recognize a protein on the surface of *Plasmodium*, Mice and monkeys injected with these monoclonal antibodies resisted malarial infections. Such immunisation is called passive because the protection depends upon foreign antibodies injected into the body, rather than on antibodies made by the individuals own immune system, as in the case of active immunization.

Trypanosomes

The term trypanosomiasis includes several tropical diseases, each caused by a different species of the group of protozoans called *Trypanosomes*. Two most common forms of these diseases are Chagas disease and African sleeping sickness disease which claims around ten thousand fresh victims each year. *Trypanosomea cruzi*, the parasite responsible is often transmitted by blood transfusion. *T. Cruzi* infections are wide spread but are often undetected because they are usually chronic and involve relatively low number of parasites. These parasites attack the nervous system, and unless treated, the disease may be invariably fatal.

Recently an anti-cancer drug, daunorubian was found to be active against trypanosomes if the drug is linked to albumin or ferritin, two proteins found in human blood. The supply of these proteins is limited at present. However required amounts can be produced, probably by inserting the appropriate genes into bacteria and purifying the proteins. The proteins thus manufactured caused adverse reactions in some patients because of the extraneous materials they contained.

5.2 Preparation and Functions of Viral Vaccines

The new technologies for preparing vaccines employ recombinant DNA technology, peptide synthesis, internal image (idiotype) antibodies, as well as the use of inactivated intact virus, or their subunits replicated in continuously propagated cell limes.

Subunit vaccines are produced by recombinant DNA technology which helps to express the heterologous genes in various prokaryotic and eukaryotic systems. Therefore, large quantities of purified immunogenic substances can now be obtained at low cost for the use as a subunit vaccine.

Recombinant DNA vaccine was first reported for the virus of foot and mouth disease. In these live recombinant viral vaccines, the virus can express antigens in the same way as they are expressed during a natural infection. It can also replicate in the host, thus amplifying the antigen load and activating both the humoral and cell mediated responses.

If such responses prevail, protection is not achieved and in addition, disease is aggravated upon exposure to the vaccine. Finally, certain pathogens have evolved even more complicated mechanisms to induce inappropriate immune response. These mishaps and concerns emphasize the need to develop better defined and safer vaccines. The recent developments in immunology, molecular biology and biotechnology have provided modern tools for the development of such new vaccines.

The immunogenic B-subunit of the cholera toxin, glycoprotein D of herpes simplex virus (HSV) type 1, the circumsporozoite (CS) protein and other antigens from the human malarial parasite *Plasmodium falciparum* and the immunogenic protein antigen of *Mycobacterium leprae* etc. are among the several microbial protein genes which were cloned and expressed in *E. coli*. The surface antigen of human hepatitis B virus (HBS Ag) has been cloned and expressed in the yeast, *Saccharomyces cerevisiae*. Laboratory animals immunized with viral recombinant DNA vaccines are protected against infections by producing neutralizing antibodies. However, the ability of recombinant vaccines to induce cell mediated immunity has not been fully known as yet.

Immunization with a genetically engineered *vaccinia virus* (VSV) the anti small pox agent, which has been successfully used for a long time as an immunizing agent against small pox is an additional means for making vaccines. Using vaccina virus would not have to be refrigerated and thereby would be highly cost effective as well as applicable for the developing countries. The amount of exogenous DNA that can be inserted into the vaccina virus genome is unknown. An addition of 25 - kilobase pairs of foreign DNA has no significant effects on virus growth, hence multivalent vaccines can be produced by using a single vaccina virus. In this method a single vaccina virus recombinant has been generated which expresses several foreign genes, namely the influenza virus, hemagglutinin, the hepatitis B surface antigen and the herpes simplese virus type 1 glycoprotein D. Vaccination with appropriate recombinant vaccina virus protects at least partially the following species.

(a) Hamsters against infection with influenza virus
(b) Mice against lethal intraperitoneal inoculation with herpes simplex viruses and against the development of a latent trigeminal HSV-1 infection, or against lethal intravenous infection with VSV.
(c) Cattle against lingual infection with VSV
(d) Chimpanzees against hepatitis B virus.
(e) Sheep against intradermal infection with rift valley fever virus.
(f) Primates and guinea pigs against lassa virus.

Recombinant vaccina virus can stimulate a specific cytolytic T. lymphocyte (CTL) response, but such a response depends on secondary immunization of T cells in vitro with the native virus.

However, vaccination with live vaccina virus have several drawbacks. Sometimes it causes complications even death. The other major problem is that, an anaphylactic response due to existing immunity to vaccina is likely to inhibit the replication of live recombinant vaccina virus in the host.

Molecular genetics offers enormous possibilities both for elucidating microbial pathogenesis and for developing new vaccines. Because BCG is currently the most widely used vaccine in the world and a superb adjunant, Bloom and his co-workers undertook to develop BCG into a multivaccine vehicle capable of immunizing simultaneously against antigens of multiple pathogens. Since BCG grows so slowly a strategy to introduce and express foreign genes using shuttle vectors was developed that could have general application. Shuttle plasmids or plasmid vectors can be readily genetically manipulated in *E. coli* and then shuttled to *Mycobacteria*. Foreign antigens have now been successfully expressed in BCG vaccina strains.

Considerable work have also been carried out regarding the construction of recombinant avirulent *Salmonella* for oral immunization to induce mucosal immunity to bacterial pathogens. A group of microorganisms of often collectively called *Streptococcus* mutants constitutes the principle etiologic agents of dental caries. Avirulent derivatives of *S. tymphimurium* harbororing plasmids expressing *S. mutans* colonization protein antigens are constructed with the help of gal E mutations along with other mutations. The gal E mutations originally employed by Germanier and Furer for live oral typhoid vaccine blocks synthesis of a normal LPS core and O-antigen side chain in *S. typhimurium*. In the absence of galactose strains become rough and are now susceptible to non-specific host defences. Based on studies in humans, rabbits and monkeys, the titers of secretory IgA achieved in saliva of mice immunized with these bivalent vaccines should be protective and prevent colonization of teeth by *S. mutans*. A similar immunization stratergy should be effective in contending with bacterial, viral and mycotic pathogens that colonize on or invade through mucosal surface. Lot of studies are also being carried out using adenovirus as expression vectors and recombinant vaccines an approach that would have additional benefit of producing novel veterinary vaccines.

5.3 Stem Cell Biology

Dr. James A. Thomson, University of Wisconsin, Madison USA, (November 1998) first isolated cells from the inner mass of several days old human embryo. This marked the begining of stem cell Biology as these cells theoretically differentiate into virtually any type of human cells from blood to skin and used for repair of damaged tissue. In January 1999, U.S. Department of Health and Human Service approved public funding for research on embryonic stem cells followed by approval by President Bill Clinton in September 1999. On the basis of recommendation by National Bioethics Advisory Commission on August 9, 2001, President G.W.Bush approved Federal funding of *Embryonic stem cell research.* This led to hectic research the world over and Human Embryonic stem cell Registry followed in April 27, 2001, the National Institute of Health reported 64 cell lines world wide; interestingly 19 cell lines are from USA, 24 cell lines are from Sweden and 6 cell lines from Australia. The National Centre for Biological Sciences, Bangalore and Reliance Life Science group, *Mumbai*; *has* several cell lines where active research is being carried out.

In a parallel discovery and at the same time John Gearhart (1998) from John Hopkins, USA reported first derivation of human embryonic germ cells from a isolated population of cells in foetus gonadial tissue known as *Primordial germ cells* which later becomes the eggs and sperms. Cell lines were developed from these cells also.

Stem cells have received much attention from the scientists, world over and stem cell biology is in the forefront of science and public policy. It's fascinating and equally promising science as this will lead the scientists to develop cell based therapies to treat disease and to develop tissues and organs from such cells.

Definition

Stem cells are "undifferentiated embryonic cells (ES)". ES cells are derived from a group of cells called *the inner cell mass* which is part of the 4-5 day embryo called the blastocyst. Once removed from the blastocyst, the cells of the inner cell mass can be cultured into embryonic stem cells. Embryonic stem cells and embryonic germ cells are pluripotent and have self-renewal property.

Stem cells are one of the most fascinating and promising areas of modern biology today and leading biologists are investigating the possibility of cell-based therapies to treat disease and how an organism develops from a single cell and how healthy stems cells replace damaged cells in adult organisms.

The history of research on stem cells began about 40 years ago but however, in reality stem cell research gained importance only in the past five years. However, in the 1960's researchers discovered that the bone marrow contained at least two kinds of *stem cells or precursor cells*, one called *hematopoietic stem cells* from which all types of blood cells in the body develop and secondly bone marrow stromal cells, that generate bone, cartilage, fat and fibrous connective tissue.

Stem cells play a crucial role in the normal growth and development of animals and humans including the maintenance of tissues and organs in the body. However, certain specialised cells such as blood and muscles cannot divide and produce new cells. Such type of cells are regenerated from stem cells, which have the unique ability to divide to produce both copies of themselves and other cell types. Stem cells, thus play an important role in supporting tissues such as blood, skin, and gut that undergo continuous cell replacement, and muscle that can be built up according to the body's need.

Properties of Stem Cells

Some of the important properties of stem cells which make them so important are :

1. *Stem cells are unspecialized :* One of the fundamental properties of a stem cell is that it does not have any tissue-specific structures that allow it to perform specialized function. Unlike a red blood cell on a nerve cell, the stem cell does not have any physiological property. However, unspecialised stem cells can give rise to specialised cells like heart muscle cells, blood cells and nerve cells.
2. *Stem cells are capable of replication and produce copies of themselves :* Stem cells are totipotent and produce identical copies of themselves, again and again. This process called self-renewal continues throughout their life. In contrast heart muscle cell, and blood cells do not replicate, when damaged or injured they can not be replaced.
3. *Stem cells give rise to specialised cells :* A unspecialised stem cell on differentiation gives rise to a specialised stem cell such as blood or muscle cell.
4. Stem cells have been grouped according to how many different types of cells they have the potential.

   A fertilized egg is said to be *totipotent* meaning that its potential is total; it gives rise to all the different types of cells in the body (fully functional organism). *Pluripotent* stem cells can give rise to any type of cell in the body and has the potential to develop into more than 200 cell types except those needed to develop a foetus (not fully functional organism). Stem cells that give rise to a small number of different cell types are generally called *multipotent* i.e. less plastic, hence, limited cell types are derived.
5. Stem cells can be isolated and maintained in a unspecialised state for many generations and on differentiation give specialised cells and tissues for the lifetime of the organism.
6. Stem cells can be grown in culture and maintained in a defined medium for several years and can be manipulated in culture and later injected into normal embryo to produce a chimera.
7. Stem cells have normal chromosome number i.e. $2n = 46$.
8. Stem cells generate three cell types with primary germ layers the meso, endo and ectoderm layers.
9. Cells lack immuno response but this requires further study
10. Stem cells do not age
11. Stem cells do not mutate, their cell lines are therefore stable
12. Stem cells do not show malignancy
13. Stem cells are ideal target cells in view of its replicative potential and long in vivo half life. Introduction of the gene into stem cells can result in the expression of transferred gene in a large population of daughter cells. At present, bone marrow is the only tissue for which stem cells could be used as recipients of transferred genes.

Classification of Stem Cells

Stem cells are classified based on their properties :

1. Embryonic stem cells
2. Adult stem cells
3. Haemopoietic stem cells

1. *Embryonic stem cells :* Embryonic stem cells, as their name suggests, are derived from embryos. These are isolated from embryos that are 5-6 days old. At this early stage the embryo is a hollow ball of cells the size of pinhead, called a blastocyst.

   The Embryonic cells are derived from a small group of pluripotent cells within the blastocyst, called the inner cell mass, which gives rise to all the highly specialized cells needed to produce an adult. Technology for producing animals from undifferentiated embryonic cells grown in culture is an outergrowth of studies carried out on mouse and human teratocarcinomas. These unusual tumors were originally recognized because they consist of both undifferentiated cells and multiple types of differentiated cells. Undifferentiated cells from terato carcinomas can be grown in culture and are pluripotent because multiple tissues reappear after the cultured teratocarcinoma cells are injected back into embryos of adult animals (Fig. 5.6)

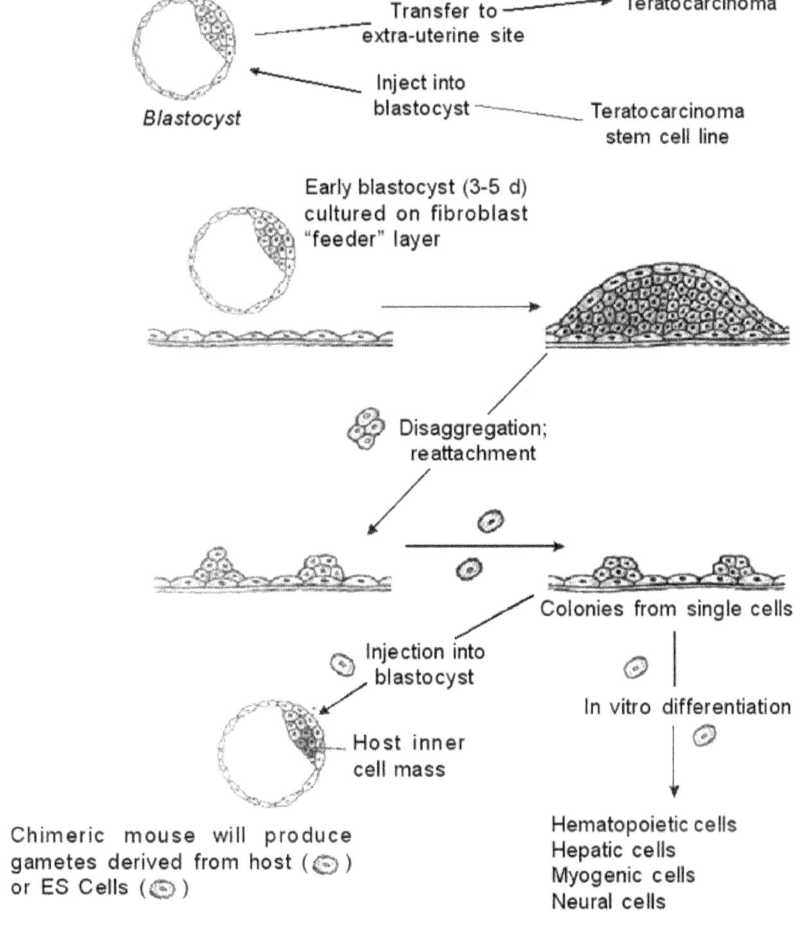

Fig. 5.6 Teratocarcinomas and embryonic stem cells. Cells from teratocarcinomas can be grown in culture and injected into a blastocyst and will give rise to many different tissue types. Embryonic stem cells share many features with teratocarcinoma cells but are derived directly from the inner cell mass or primitive ectoderm and, in contrast to teratocarcinoma cells, are often euploid and can therefore contribute to the germline. Cultured embryonic stem cells differentiated in vitro can give rise to a variety of different cell types.

2. *Adult stem cells :* An adult stem cell is an undifferentiated cell found among differentiated cells in a tissue or organ, can renew itself and can differentiate to yield the major specialized cell types of the tissue or organ. The primary routes of adult stem cells in a living organism are to maintain and repair the tissue in which they are found unlike embryonic stem cells, which are defined by their origin (the inner cell mass of the blastocyst), the origin of adult stem cells in mature tissues is unknown to date, adult stem cells have been found in bone marrow, blood, skin, muscle, liver, brain, the corneo and retina of the eye, the lining of the gastrointestinal tract, and pancreas.

3. *Haemopoietic cells :* Embryonic Stem Cells Vs Adult or Tissue Stem Cells. Human embryonic and adult stem cells have their advantages and disadvantages regarding their potential use of cell-based regenerative therapies.

   Embryonic stem cells can become all cell types of the body because they are pluripotent and hence have a wide applications and tremendous therapeutic potential. Adult stem cells are generally limited to differentiating into different cell types of their tissue of origin. However, recent studies suggests that adult stem cell may exhibit plasticity, a phenomenon where an adult stem cell from one tissue may under the right conditions give rise to cell types of another tissue.

For some cell types, acquisition of a normal differentiated phenotype is associated with a loss of proliferative ability. Neurons, erythrocytes, and platelets, for example, are highly differentiated cells that carry out essential functions but do not dividie. Replacement of erythrocytes and platelets occurs on a regular basis, however, and depends on stem cells. Undifferentiated cells that are capable of extensive proliferation, stem cells may give rise to additional stem cells (self-renewal) or to progenitor cells comitted to forming differtiated derivatives. (Fig. 5.7). Most stem cells are said to be pluripotent, meaning that they give rise to multiple different types of cells after subsequent rounds of division. The entire program of blood cell development is based on a lineage that makes use of stem cells and pluripotent precursors. Besides their relevance to the pathogenesis of genetic disease, the programs that create and maintain hematopoietic stem cells have provided the basis for using bone marrow transplantation to treat diseases such as leukemia and lymphoma.

Applications

Biomedical research on stem cells is still in its infancy, but advancing very rapidly. Embryonic stem cells as research tools have opened many doors for biomedical research.

The ability of the stem cells to replace damaged cells in the body that would otherwise not be replenished has led to investigate the possible use of stem cells in regenerative medicine. Currently, organs and tissues donated are used to replace a damaged/destroyed tissue but the need for transplantable tissues and organs far out weights the available supply. Stem cells directed to differentiate into specific cell types of offer unlimited possibility of a renewable source of replacement cells and tissues to treat Parkinson's and Alzheimer's diseases, spinal cord injury, strokes, burns, diabetes, arthritis etc.

Scientists in academic and industrial research laboratories are vigorously pursuing all possible avenues of research including ways to direct the specialisation of adult and embryonic cell to become pancreatic islet like cells that produce insulin and can be used to control blood glucose level.

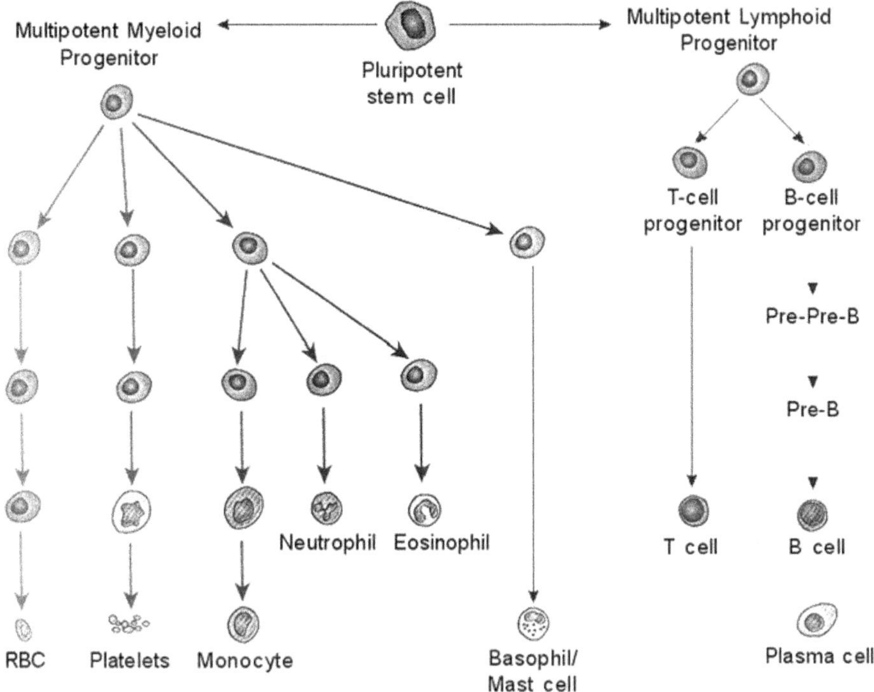

Fig. 5.7 Development of hematopoietic cells

There are many ways in which human stem cells can be used in basic medical research and in clinical research. However, there are many technical hurdles between the promise of stem cells and the realization of these applications which will only be overcome by continued intensive stem cell research. Stem cells can be used to treat many disorders to repair a damaged heart, to study congenital birth defects and placental abnormalities, to explore the effects of chromosomal abnormalities in early development, to test subjects therapeutic drugs in preclinical tests to screen potential toxins, for devising better methods for gene therapy and genetic engineering, in cancer chemo therapy, graft verses tumour disease treatment of cancer, disorders of nervous system and in the treatment of auto immune disease like diabetes.

Stem cell therapies, like other tissue transplants, face the problem of being rejected by the patients immune system. The immune system protects the body against disease by recognising foreign substances and destroying them. This has resulted in the failure of many organ transplants, and is a major hurdle that must be overcome if stem cell therapies are to succeed for routine used.

In people who suffer from type I diabetes, the cells of pancreas that normally produce insulin are destroyed by the patients own immune system. Recent studies suggest the probability of obtaining insulin producing cells from embryonic stem cell culture that could eventually be used in transplantation therapy for diabetics. (Fig. 5.8)

Studies of human embryonic stem cells may yield information about the complex events that occur during human development.

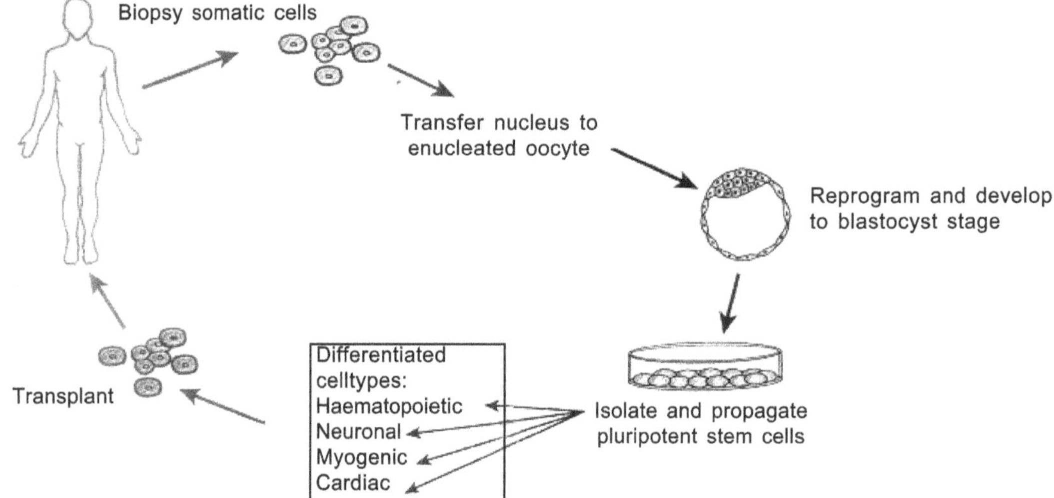

Fig. 5.8 Potential application of cell cloning and embryonic stem cell technology to organ and tissue transplantation.

Human stem cells could also be used to test new drugs for example, new medications could be tested for safety on differentiated cells generated from human pluripotent cell lines.

Assessing Human Stem Cell Safety

The isolation of human stem cells offers the promise of a remarkable assay of novel therapeutics. Biologic therapies derived from such cells (through tissue regeneration and repair as well as through the targeted delivery of genetic material) are expected to be effective in the treatment of wide range of medical conditions. Efforts to analyze and assess the safety of using human stem cells in the clinical setting are vitally important to this endeavour.

Donor Screening

Whether human stem cells are of embryonic, fetal or adult origin, donor sources must be carefully screened. Routine testing should be done to guard against the inadvertent transmission of infectious diseases. In addition, pedigree assessment and molecular genetic testing should be done.

The purpose of pedigree evaluation and for genetic testing is to establish whether the human stem cells in question are suitable for use in the context of a particular clinical situation. For example, genetic analysis could detect a mutation in the gene for alpha synutein. This gene is known to be responsible for the rare occurrence of early onset of Parkinson's disease. Detecting such a genetic abnormality in neuronal progenitor cells derived from embryonic stem cell line could block the use of those cells as a treatment for a number of neurodegenerative conditions, including Parkinson's disease.

Stem cells are being used to develop specialised liver cells to evaluate drug detoxifying capabilities and represents a new type of early warning system to prevent reactions in patients.

Research on stem cell biology and developmental biology are closely related and therefore it is necessary to understand the genes and molecules, such as growth factors and nutrients that function during development of the embryo so that they can be used to grow stem cells in the laboratory and direct their development with specialised cell type. ES cells are key tools for understanding fundamental events in embryonic development that may one day help explain birth defects and approaches to correct them.

Human stem cells must be transplanted into animal models of human disease. Transplantation of neural stem cells should demonstrate measurable evidence of efficacy in models of neurodegenerative disease, such as Parkinson's disease, Huntington's disease and Amyotrophic Lateral Sclerosis (ALS) and this area offers tremendous potential for treating a variety of human diseases, such as Parkinson disease, type 1 diabetes mellitus, and cirrhosis. At present, the only stem cell of clinical relevance is the hematopoietic stem cell, which can reconstitute the blood system after bone marrow transplantation. Although bone marrow has long been the major source of transplantable hematopoietic stem cells and progenitor cells, the discovery that placental blood (which is readily available) represents a rich source of hematopoietic stem cell is begining to make a substantial impact on the treatment of malignancy and genetic disease.

The use of placental blood has several advantages over bone marrow as a source of hematopoietic stem cells.

1. recipients are more tolerant of histoincompatible placental blood than of other donor cells. Thus, engraftment occurs even if as many as three HLA antigens are mismatched between the donor and the recipient.
2. the easy availability of placental blood, together with the increased tolerance of histoincompatible donor cells, greatly expands the number of potential donors for any recipient.
3. The risk of graft-versus-host disease is substantially reduced using placental blood cells as the donor source. It is likely that the isolation and characterization of stem cells from other tissues, including the nervous system, will eventually make cell replacement feasible in a wide variety of malignancies and genetic diseases.

In addition to efficacy, evidence for anatomic and functional integration of transplanted human stem cells should be assessed.

From the perspective of toxicology, the proliferative potential of undifferentiated human embryonic and embryonic germ cells evoke the greatest level of concern. A characteristic of human embryonic stem cells is their capacity to generate teratomas when transplanted into immunologically incompetent strains of mice. It is essential that careful toxicology studies are performed that are of the appropriate duration and that involve transplantation in immunocompromised animals of undifferentiated or partially differentiated embryonic stem cells, as well as adult stem cells.

Since ES do not show chromosomal abnormality and malignancy, studies on chromosome abnormality in early development might include the ability to monitor the development of early childhood tumors, many of which are embryonic in origin.

Embryonic stem cell technology provides a mechanism for creating animal models of important human genetic diseases and, in many cases, has provided unexpected insight into the role such genes play in normal development. For example, the breast cancer susceptibility gene BRACI is required both for early embryonic development and for proper branching and architecture of developing mammary ducts. Creating animal models for diseases like hemophilia or cystic fibrosis is a prerequisite for developing and testing gene testing therapy; in addition, embryonic stem cell technology provides a testing ground for efforts to correct genetic defects in utero.

The world's first embryonic stem cells bank was developed at Kings College, London and the Centre for Life in New Castle, England on 19th May, 2004. The bank aims to store and supply stem cell lines for research and possible treatment of dreaded diseases.

5.4 Gene Therapy

Deliberate alteration of human genome for alleviation of disease is called *Gene Therapy* or *Gene therapy* is correction of an inborn error of metabolism by the insertion into the afflicted person of a normal gene. Genetic disorders may however, be amenable to treatment, either symptomatic or potentially curative. Treatment may range from conventional drug or dietary management and surgery to the future possibility of *gene therapy/molecular/genomic medicine*!

The speed with which the ideas of gene therapy have caught the medical scene and achieved such a high level of scientific and medical acceptane is remarkable. Even more is the impact that new genetic thinking will have on the effectiveness of therapy for many kinds of human disease. Medicine for the first time in its history, can now imagine genetic manipulations of human patients that are truly curative.

The prospect of curing genetic disorder with gene therapy is being investigated and debated. Several experimental strategies have shown that gene transfer is possible.

Guidelines

Human gene therapy as per guidelines of NIH (USA) are as follows :

1. The gene must be cloned and well characterised i.e. must be available in pure form.
2. An effective method must be available for delivering the gene into derived tissue (s) or cells.
3. The risk of gene therapy to the patient must have been carefully evaluated and shown to be minimal.
4. The disease must not be treatable by any other strategies.
5. Data must be available from preliminary experiments with animal models or human cells and must indicate that the proposed gene therapy should be effective.

Minimal Requirements for Gene Therapy for a Genetic Disorder

1. Identification of the affected locus or at least of the biochemical basis of the disorder.
2. A complementary DNA (cDNA) clone of the gene, the gene itself (particularly if it is not large), or a functional version of the gene from which noncritical components have been removed to reduce its size.
3. A substantial disease burden and a favorable risk-benefit ratio in comparison with alternative therapy.
4. Sufficient knowledge of the molecular basis of the disease to be confident that the gene transfer is likely to ameliorate or correct the biochemical pathology and to prevent or reverse critical phenotypic abnormalities. Whereas loss of function mutations will require replacement with a functional gene, some dominant alleles, such as dominant negative alleles, will require inactivation of the mutant gene or its products.

5. Appropriate regulatory components for the transferred gene. Tight regulation of the level of gene expression is relatively unimportant in some diseases and critical in others. In thalassmia, for example, overexpression of the transferred gene would cause a new imbalance of globin chains, whereas low levels of expressions would be ineffective. In contrast in some enzymopathies, abnormally high levels of expression may have no adverse effect.
6. An appropriate target cell with, ideally, a long half-life or good replicative potential in vivo.
7. Adequate data from cultured cell and animal studies to indicate that the vector, gene construct, and target cell are both efficacious and safe.
8. Protocol review and approval by an Institutional Review Board and, in most countries oversight governmental agency.

Models for Gene Therapy

In most cases, '*gene therapy*' means the introduction of foreign genes into defective target cells and the subsequent introduction of the genetically modified cells to complement the genetic defect and correct a disease phenotype. There are two ways in which the gene transfer might theoretically be carried out (Fig. 5.9). The *first*, well developed method involves identifying suitable target cells in an affected patient, removing them, carrying out the gene transfer *in vitro*, and returning the genetically and phenotypically corrected cells to the patient. A *second* approach involves the introduction of a vector directly into a patient that is targeted to the appropriate tissue or organ in the patient's body. While little work has been carried out so far on the second technique, it would seem in the long run to be the simpler and more easily delivered of the two potential approaches of gene therapy.

To introduce new genes efficiently and functionally into mammalian cells without damaging or killing the host cell, the various agents include several classes of tumour viruses, those with DNA genomes (papovaviruses such as SV40, polyoma, and papilloma viruses) and those with RNA genomes (retroviruses such as the murine and avian leukaemia and sarcoma viruses).

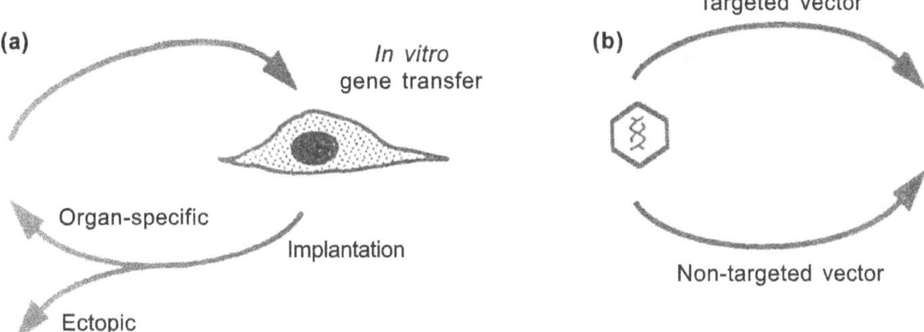

Fig. 5.9 Two theoretical routes for the introduction of foreign genes into a human patient for the purpose of gene therapy. (a) target defective cells are removed from the patient, genetically modified in vitro and returned to the patient into a site that allows the newly expressed gene product to correct the disease phenotype. (b) is the direct introduction of the gene transfer vector into the patient. The vector might be targeted specifically to the tissue expressing the abnormal phenotype or delivered more indiscriminately to the entire body. In the latter case, therapy might be achieved by body-wide expression of the foreign gene or by regulation of its expression in selected target through the inclusion in the vector of tissue-specific expression signals.

To be useful for clinically applied gene transfer procedures, viral vectors must be capable of transferring foreign genes efficiently and of stable expression of the newly introduced gene in target cells without bringing about any metabolic or genetic damage in those cells. Because retrovirus vectors satisfy the above criteria, they have been particularly useful in the early stages of the development of many gene therapy models. The mechanisms by which they infect cells, integrate into the genome of host cell, and express foreign genes in a stable manner ('*transgenes*') refer to any DNA construct that contains the coding sequences of the gene in question, under the control of appropriate regulatory elements. Very often, a transgene gene consists of a cDNA under the control of a promoter that may not necessarily be the natural promoter of the gene, are now well understood. They are capable of infecting many kinds of target cells and they do little or no damage to infected cells.

Somatic Cell / Non Heritable Gene Therapy

This will treat the disease symptoms of the patient but will not cure the disease i.e. defective gene will be present in the germ line and may be transmitted through the germ line. Somatic cell gene therapy is generally viewed as being acceptable as this is seen as similar to existing treatment such as organ transplantation since this involves removal of patients own cells, repaired and reimplanted in the patient. *Somatic cell Gene Therapy* is less extreme manipulation of an individual than an organ transplant.

Gene therapy may be undertkan to replace or inactivate a dominant mutant gene whose abnormal product causes the (generally dominant) disease. Gene replacement of all or part of a mutant gene at the normal locus is much more difficult. This strategy would be required, for example, to replace a Huntington disease gene containing an expanded CAG repeat or at least most of the CAG expansion itself. Alternatively, gene "surgery" of this type could also be achieved by modifying or degrading the mutant RNA rather than the gene that encodes it. For example, selective degradation of a mutant mRNA encoding a dominant negative pro α_1(I) collagen that causes osteogenesis imperfecta should, in principle, prevent the bony abnormalities of the disoder. Therapeutic genes that encode RNA enzymes, or ribozymes, designed to degrade only the mRNA from the mutant allele, have shown much promise in laboratory studies.

Currently, somatic cell gene therapy protocol are essentially gene addition procedure, i.e. functional copies of the gene defective in patient to the genome of recipient cells. The introduced gene are added at random sites in the chromosome of the host cells. The ideal gene therapy protocol would replace the defective gene with a functional gene. Gene therapy may be widely used to achieve a pharmacological effect to counteract the effects of a mutant cellular gene or genes or to counteract disease pathogenesis in some other way. Patients with acquired diseases, including cancer, are likely to benefit from this strategy.

In the future, somatic cell gene therapy will probably involve gene replacement rather than gene additions, which would be mediated by homologous recombination and would place the introduced gene at its normal location in the host genome. Human gene replacements are usually referred to as targeted gene transfers.

Germ line/heritable Gene Theory is universally viewed as ethically unacceptable as it affects gonads. This amounts to tinkering with nature or '*Playing God*'. This theory has been tested on animals but not on human.

Ethical Issues

1. Somatic gene therapy risks are not different from the risks associated with vaccines or antineoplastic agents. Somatic gene therapies must be subjected to critical assessment of safety and efficacy as any other therapies.
2. Somatic gene therapy are designed to avoid germ line transfer/inadvertent germ line. Therapy may still be possible. Risk of genetic damage transmitted to offsprings is not unique to gene therapy. This concern is not different from cancer/radio and chemotherapy.
3. Patients with somatic gene therapy still carry genes in their gonads, therefore this is called *Euphenics* (and not Eugenics or Egenics).
4. Somatic gene therapy has potential risk of
   (a) Induction of neoplasia within target tissue by actions of vector/inactivation of protooncogenes.
   (b) Damage/inactivation of other genes in target cells.
   (c) Altertion of germ line should not occur because therapeutic intervention is limited to somatic cells only.

As many as 4000 inherited human disorders are now known but only few are presently curable. In July 1990, two committees of National Institute of Health, USA, approved gene therapy treatments on people.

The first gene therapy in human was performed in September 1990 on a 4 year old girl suffering from ADA deficiency. She was treated with transgene-adenosine deaminase-deficient severe combined immunodeficiency disease (ADASCID). This is a rare autosomal disease of the immune system. Several patients have since been treated in USA and are school going and receive ADA treatment on regular basis. The somatic cell gene therapy holds great promise with the identification and cloning of several genes that cause serious human diseases, gene therapy is on the horizon for sufferers of the diseases like. PKU, thalassemia, Duchnene muscular dystrophy, sickle cell anemia, (Lesch-Nyhan syndrome) (defect in protein metabolism). Hurler syndrome (defect of polysaccharide metabolism). Current gene therapy involves gene additions i.e. simply adding functional copies of the gene that is defective in the patient to the genome of recipient cells. The inserted gene get randomly on the chromosome of the host cell. This ideal protocol for gene therapy is to replace the defective gene with a functional one i.e. targeted gene replacement.

Use of Stem Cells in some Cell Based Gene Therapies

Several of the early gene therapy studies using stem cells were carried out for therapeutic purposes but to track the cells fate after they were infused back into the patient. The studies aimed to determine where the stem cells ended up and whether they are producing the desired gene product, and if so, in what quantities and for what length of time.

The major reason for using stem cells in cell based gene therapies is that they are a self renewing population of cells and this may reduce or eliminate the need for repeated administration of the gene therapy.

Human gene therapy is still at an early stage of development. Relatively few patients have been treated with gene therapy and initial results are promising. Table 5.2 describes several types of diseases that are being investigated as potential targets for gene therapy.

Table 5.2 *Future Prospects in Gene Therapy*

Type of disease	Treatment of
Blood :	Sickle cell disease, hemophilia, severe combined immuno-deficiency syndrome (SCID)
Cancer :	Brain tumors, breast cancer, ovarian cancer, colorectal cancer, malignant melanoma.
Cardiovascular :	Atherosclerosis, essential hypertension.
Infections :	AIDS, latent viral fevers
Lung :	Cystic fibrosis
Muscular :	Duchnene muscular dyotrophy, myotonic muscular dyotrophy
Metabolic :	Glyocogen storage disease, lysosomal storage disease black urine disease.
Neurological :	Alzheimer's disease, Parkinson's disease.

Recently, *recombinant DNA technology* is being used to combat human diseases via gene therapy. In this method, a cloned gene is introduced into the somatic cells as a way to compensate for a defective gene. This technique may become an important way to alleviate many human disorders.

5.5 Genetic Counselling

Genetic counselling is defined as "*a process in which patients or their relatives at the risk of a genetic disorder are made aware of the consequences of the disorder, its transmission and the ways by which this can be prevented*". Genetic counselling is a communicative rather than a directive process and is non judgemental.

Genetic counselling includes the following aspects :

1. An accurate diagnosis of the disorder is essential for genetic counselling
2. Advising the patient/the family members with survey of relatives with similar complaints for disease confirmation or in otherwise normal relatives for carrier detection.
3. Management of the disorder either curative (if possible) or supportive.

To enable accurate diagnosis the following three procedures should be followed :

1. *History :* A proper record of the history of the patient is necessary.
   (a) This includes both present and relevant past history.
   (b) Family histroy includes sibs and other relatives also. If there is any other person in the family with a similar problem also should be noted.
   (c) Obstetric history of the mother includes exposure to teratogens (drugs X-rays) in pregnancy. History of abortion or still-birth if any, should also be recorded.
   (d) Enquiry should be made about consanguinity (inbreeding), as it increases the risk especially in autosomal recessive disorders.

2. *Pedigree charting :* This offers in a concise manner the state of disorder in a family. Constructing a pedigree with proper interrogation though time-consuming, is ultimately rewarding. It forms an indispensable step towards counselling. Standard symbols are used in pedigree charting in human beings (Fig. 5.10).
3. *Estimation of genetic risk :* This depends on the pattern of inheritance of a disorder.

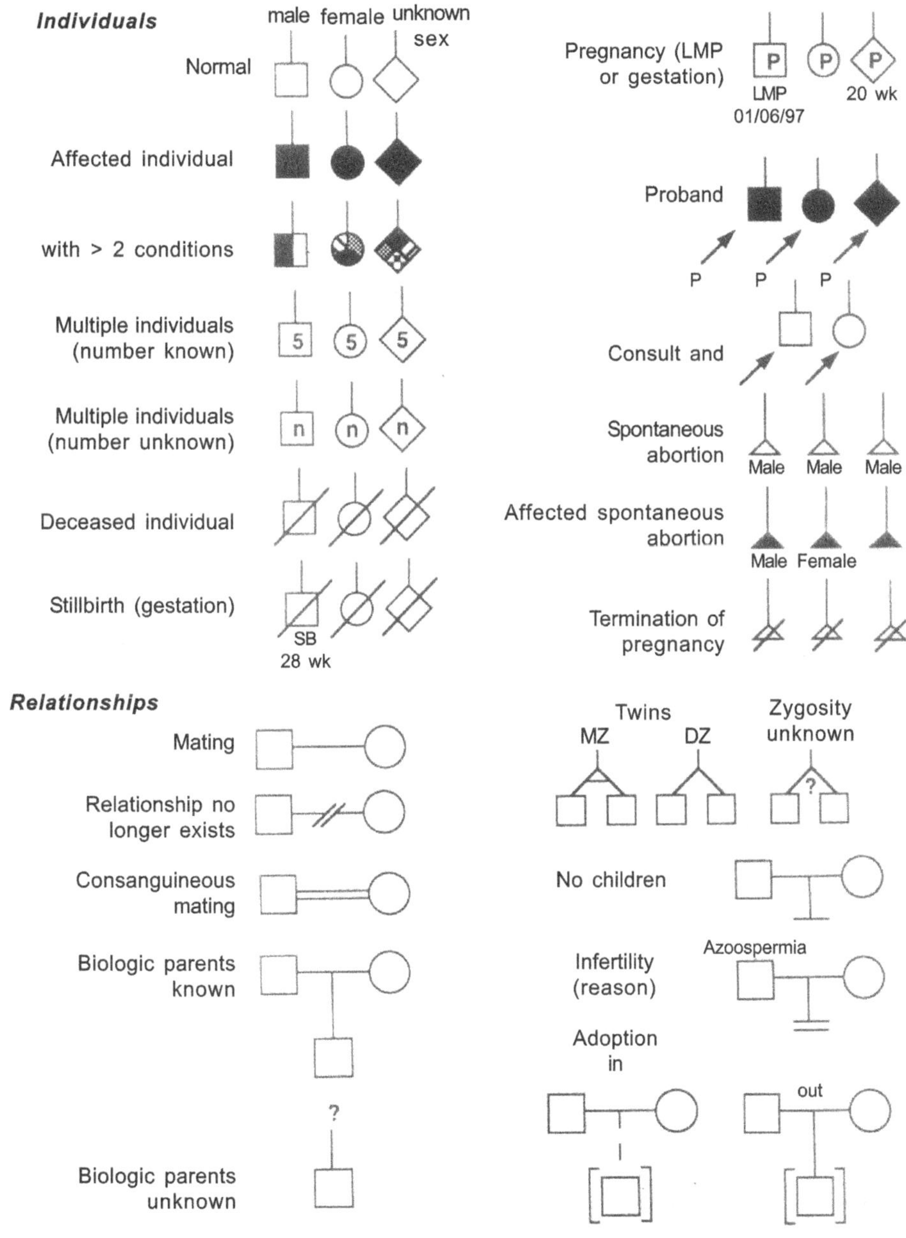

Fig. 5.10 Standard Pedigree Charting Symbols

This forms one of the most important aspects of genetic counselling called recurrence risk. To estimate it, one requires to take into account the following :

1. Mode of inheritance (Mendelian, chromosomal disorder etc)
2. Analysis of pedigree/family tree
3. Results of various tests such as linkage studies.

In order to arrive at a risk one has to work out the probability. The probability of an outcome is defined as the number or more precisely the proportion of times it occurs in a large series of events. Routinely, the probability is indicated as a proportion/fraction of one. Probability 0.25 or 1/4 indicates that, on average the event will be observed on 1 in 4 or 25% of occasions.

According to Andrew Poole (1975) genetic counselling is virtually nonexistent in dental practice and this may be because of lack of training or awareness. However, many parents would appreciate counselling, especially for dental defects such as *dentinogenesis* imperfect or cleft palate lips which are not ordinarily dealt with in a general counselling device service.

Although most of the bleeding disorders are inherited as per known patterns, genetic counselling probably receives the least attention by those who are involved in treating these diseases. Because of this, families do not have a clear understanding of risks to further offspring, parents carry an unnecessary burden of guilt, marital problems develop because of blame being assigned to a spouse, and the fear of conception disrupts normal husband-wife relationships.

In any genetic disease situation it must be stressed that :

1. Genetic counselling is an integral part of patient management.
2. It is the responsibility of the physician to see that genetic counselling is provided by all appropriate source.
3. Genetic counselling is based on an accurate diagnosis.
4. To help determine the patterns of inheritance family history in pedigree form should be taken.
5. Other health workers (nurses, social workers, hygienists) should know about the disease and follow through with valid information as they assist the family.
6. Genetic counselling is not decision making for families, but the supplying of current, correct information on which parents can make decisions which they will not regret.
7. The management of emotional and psychological problems is just as important as the physical and medical problems.
8. Genetic counselling should be initiated early and reinforced several times so the family can be supported and feel as secure as possible about their knowledge of the disease. Appropriate referrals for financial, emotional, sexual and family-planning problems should be made.

From the foregoing it is apparent that is pre-requisite of good family counselling is a thorough family history. This should include not only the genealogy of the particular defect but also a record of all pregnancies regardless of outcome, stillbirths, and all other defects. In addition, a good prenatal history is essential and every effort should be made to follow up on drugs, illnesses, and other problems which the mother experienced during pregnancy. This aspect of the counselling may help to eliminate those clefts which have a "pure" genetic etiology or have a pronounced environmental or teratologic involvement and which will therefore not follow the counselling based on multifactorial inheritance. The following facts should be remembered with respect to cleft lip and cleft palate.

1. An affected female has a greater chance of having an affected offspring than an affected male, although both have 40 times greater risk than the population risk, i.e., 0.001 for Caucasians.
2. The more severe the defect in the parent the greater the risk for an affected offspring. For example, a parent with bilateral cleft of the lip and palate is more likely to have an affected offspring than a parent with a unilateral cleft of the lip. The greatest risk in this sense is of the severely affected female parent.
3. A first degree relative (son, daughter, brother, sister) has the highest risk (40 times the population incidence), second degree (aunt, uncle, niece, nephew) an intermediate risk (7 times the population incidence), and the third degree the least risk (3 times the population incidence). The risk thus decreases rapidly with decreased degree of relationship as distinct from the single gene inheritance which merely halves. Inbreeding in USA is forbidden by law.
4. The risk for a second child to be affected increases rapidly if one child is already affected. This rises to 4 percent for one affected child (unaffected parents) to 9 percent for two affected children. For an affected parent with one affected child the risk is 17 percent.

It should be also emphasized that recurrence risk is significantly lower than it would be if the condition was single gene inherited. In addition, the risk does not change from child to child in single gene inheritance. This again indicates the need for a good family history to eliminate a single gene inheriance pattern.

Genetic Counselling in Mitochondrial diseases inheritance (from Mothers side).

- Some disorders of mitochondrial function are due to nuclear gene mutations.
- Many disorders caused by mitochondrial mutations are sporadic.
- It is not known whether the degree of heteroplasmy in the mother determines the risk to offspring.
- Severity is very variable and difficult to predict.
- It is difficult to advise asymptomatic relatives who carry the mitochondrial mutaiton.

Clinical Genetic Services

Genetic disorders place considerable genetic load on population and hence associated economic burdens not only to the affected people but their families and the community as a whole. Many diseases in adult life have a genetic predisposition like diabetes, cleft-lip/palate, familial cancer and coronary heart diseases. Genetic diseases though rare are largely incurable and often severe. With advanced medical technology, there is growing awareness and concern by the people and the doctors of the role of genetics in clinical practice. Genetic counselling covers a wide spectrum of risk, diagnosis and possible preventive cures. The role of clinical genetic service is to advice the patient/family as to how the disease can be prevented and its implication in the family/society. This service cover several aspects and diagnosis is the first pre-requisite for genetic counselling without a defined diagnosis, appropriate genetic advice may be given to the needy in terms of the mode of inheritance etc. Drawing a pedigree is the best way to record genetic information as per the international norms. Once this is done, estimation of gender risk depends on the pattern of inheritance of the disorder, with emphasis to identify the carrier. Mendelian disorders due to mutant genes carry high risk of recurrence where as chromosome disorders are at low risk. Interpretation of risk varies depending on severity of the disorder and this aspect need to be discussed with the doctor vis-a-vis the family.

The risk of curing a disorder, the severity and nature of disorder, pre-natal diagnosis, its implication on the couple, religious considerations/taboos etc have to be considered.

The psychological basis of diagnosis involving emotional stress affecting the counselling though supplementary, is very vital in decision making and requires sensitive counselling over a period of time. Department of clinical genetics in teaching hospitals are rich in academic and technically trained staff with access to genetic services for patients/families. Diagnostic and genetic counselling provided by medical genetic counsellor who have nursing or other paramedical background, associated labour services is integral in providing clinical genetic service. These provide biochemical, psychological, cytological, cytogenetical, pathological, dermatoglyphic and even molecular genetic services. Finally clinical use of registers is aimed at ascertaining is completely as possible all people at risk of developing/ transmitting a particular disorders. A register permits a long term follow up of family history. Genetic registers are best preserved in computers and can be accessed for incidence and natural course of diseases, effect of counselling and preventive measures etc.

Reasons for Referral to a Genetics Clinic

- Genetic disease diagnosed, counselling requested.
- Testing carrier state of family members for Mendelian disorders.
- Investigation and diagnosis of possible genetic disease.
- Diagnosis of mentally handicap or physical abnormality
- Diagnosis of malformation in neonates or stillbirth including amniocentesis
- Genetic investigation of recurrent pregnancy loss.
- Genetic management of high risk pregnancies
- Interpretation of abnormal pre-natal tests.
- Inbreeding etc.

Brain Users of Clinical Genetic Services

- Paediatricians
- Obstetricians
- Other hospital specialists, pyschiatry, dentistry, neurology
- General practitioners
- Community child health services
- Others (self referrals, adoption services)

Organisation of Clinical Services

Medical staff : Consultant clinical geneticists

- Specialist registrars
- Clinical assistants or clinical medical officers

Field workers

Genetic associates (scientific officers or counsellors)

- Specialist health visitors
- Social workers

Clerical supporting staff

- Computer personel
- Photographer
- Medical record assistant

Laboratory services

- Biochemical testing service
- Cytogenetic testing service
- Molecular genetic testing service
- Pathology testing service
- Dermatoglyphic service etc.

There are a number of counselling clinics, in the advanced countries but for too few in India to take care of the needs for this service and to have maximum eugenic value. To be effective such clinics must involve the cooperation of medical profession as well as geneticists. There must be both a correct diagnosis of the trait being studied and a knowledge of the complicated reactions of genes may be involved. Since many of the tests for heterozygous carriers of harmful recessive genes are difficult to perform and interpret, they must be done by experts. Also, it seems that such clinics must be supported by funds from the government or foundations, since people are generally unwilling to pay the heavy cost involved for such counselling although they can pay for the services of a physician diagnosing and treating a disease.

5.6 Pre-Natal Diagnosis

A doctor who cares for a child with a birth defect or genetic disorder may be in the best position to alert families to the possibility of a recurrence of a genetic disorder in future offspring. The family may wish to know and may benefit from methods that convert probability statements about recurrence risks into more precise knowledge.

The information gained from prenatal diagnosis may be helpful in the management of pregnancy, labour and delivery and in some instances may improve the outcome of pregnancy. The availability of prenatal diagnosis gives couples options they might not otherwise have, including preparation for the birth of a child with an abnormality, termination of an affected pregnancy, or use of fetal treatment, which though now experimental, may be an option in the future.

The various techniques currently used in investigation of prenatal diagnosis include the following.

1. *Ultrasonography*
2. *Amniocentesis*
3. *Chorionic villus sampling*
4. *Fetal blood cells in maternal blood*
5. *Maternal serum alpha-fetoprotein*
6. *Maternal serum beta-HCG*
7. *Maternal serum estriol.*

An account of the various pre-natal diagnosis procedures is given as under.

Ultrasonography

Ultrasound has become the primary method for imaging foetal anatomy. This is a non-invasive procedure that is harmless to both the foetus and the mother. High frequency sound waves are utilized to produce visible images from the pattern of echos made by different tissues and organs, including the foetus in the amniotic charity. The developing embryo can first be visualized at about 6 week gestation.

During the early second trimester, ultrasound is used to date the pregnancy, identify twins, diagnose some foetal structural anomalies, and examine the placenta and the amount of amniotic fluid. Many foetal organ systems and anatomic lesions can be visualized by ultrasound, including some genitourinary, gastrointestinal, skeletal and central nervous system abnormalities. When a structural abnormality is found in the foetus by ultrasound, further diagnostic tests may be indicated. Ultrasound is also used to guide invasive sampling such as amniocentesis, CVS, and various biopsies.

Although an ultrasound examination can be quite useful, there are certain limitations to this procedure. Subtle abnormalities may not be detected until later in pregnancy, or may not be detected at all. A good example of this is Down syndrome (trisomy for chromosome 21) where the morphologic abnormalities are often not marked, but only subtle.

Amniocentesis

This is an invasive procedure in which a needle is passed through the mother's lower abdomen into the amniotic cavity inside the uterus.

Amniocentesis is best performed at 14 to 16 week gestation (menstrual age). However, an ultrasound examination always proceeds amniocentesis in order to determine gestational age, the position of the foetus and placenta, and determine if enough amniotic fluid is present.

Within the amniotic fluid are foetal cells (mostly derived from foetal skin) which can be grown in culture for chromosome analysis, biochemical analysis, and molecular biologic analysis. (Fig. 5.11)

In the third trimester of pregnancy the amniotic fluid can be analyzed for foetal lung maturity. This is important when the foetus is below 35 to 36 weeks gestation, because the lungs may not be mature enough to sustain life. This is because the lungs are not producing enough surfactant. After birth, the infant may develop respiratory distress syndrome from hyaline membrane disease. The amniotic fluid can be analyzed by fluorescence polarization (f poll), for becithin : sphinogomyelin (LS) ratio, and/or for phospatidyl glycerol (PG).

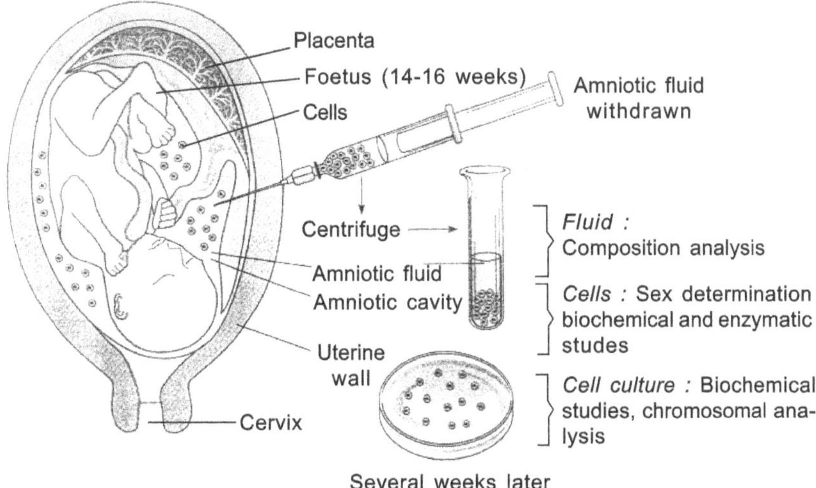

Fig. 5.11 The techniques of amniocentesis. The position of the fetus is first determined by ultrasound, and then a needle is inserted through the abdominal and uterine wall to recover fluid and fetal cells for cytogenetic and or biochemical analysis

Amniotic fluid is bloody in about 21 of amniocentesis usually, the blood does not affect amniotic cell growth and is maternal in origin, however, if the blood is foetal, it may falsely elevate amniotic fluid. Dark red or brown fluid indicates previous intra-amniotic bleeding and is associated with adverse pregnancy outcomes.

Amniocentesis rarely results in significant morbidity (e.g. symptomatic amnionitis). The probable risk of foetal loss from amniocentesis is about 0.5%. Needle injuries to the foetus are rare.

Foetal sex can be determined before birth by constructing the karyotype from cultured foetal cells. When the mother has been identified as a carrier of an X-linked mutation and a specific test for the foetus is unavailable, foetal sex may be determined with the option to abort males, although half of them will be normal. DNA analysis can identify the affected male for many X- linked disorders.

Chorionic Villus Sampling

Chorionic villus sampling (CVS) is used for pre-natal diagnosis in the first trimester. In this procedure, a catheter is passed via the vagina through the cervix and into the uterus to the developing placenta under ultrasound guidance, and chorionic villi aspirated and cultured short-term and long-term in the laboratory. Karyotypes are obtained from both cultures, although based on the long-term specimen are more accurate. Most, of the enzymes in cultured amniotic fluid cells can be measured in extracts of chorionic villi or cultured villus cells, and DNA can be extracted for molecular genetic studies. However, CVS cannot be used for tests that require amniotic fluid (eg, amniotic fluid-foetoprotein levels for neural tube defect screening). (Fig. 5.12).

The primary advantage of CVS is that results are available much earlier in pregnancy. Thus, if results are abnormal, simpler methods can be used to terminate the pregnancy, or the foetus can be treated prenatally (*eg.* dexamethasone to prevent virilixation in a female foetus with 21-hydroxylase deficiency). Normal results decrease prenatal anxiety earlier.

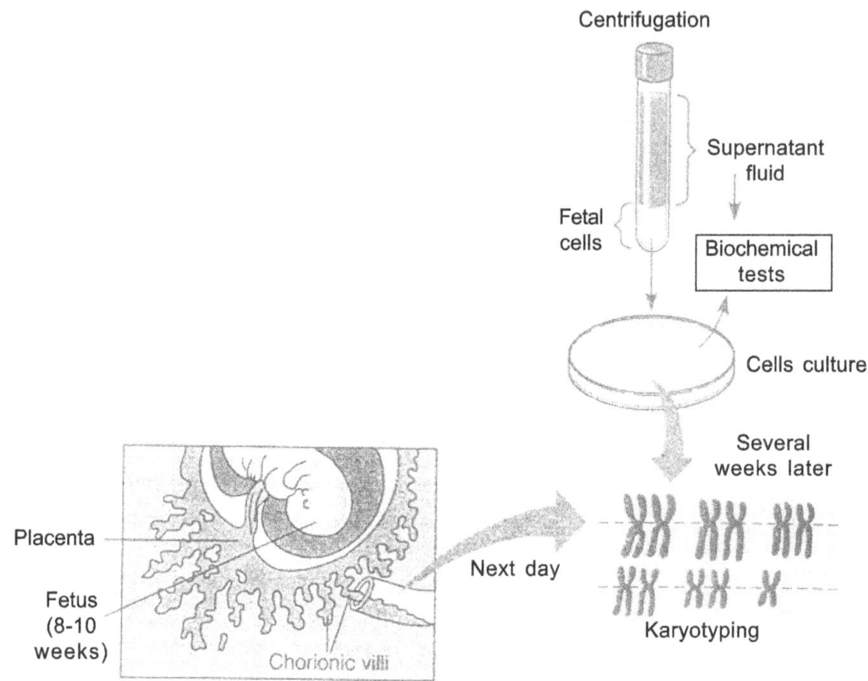

Fig. 5.12 Chorionic Villus Sampling

There are two CVS approaches. Transcervical CVS is performed at 10 to 11 weeks of gestation. Transabdominal CVS may be performed from the 10th week of gestation to term. The total foetal loss rate after first trimester CVS (including spontaneous and induced abortions and losses after 20 weeks) is not statistically different from that after amniocentesis.

Placental location determines which approach is used Low-lying or posterior placentas are usually more easily sampled transcervically. The transabdominal approach is best for fundal placentas or those located anteriorly.

After CVS, foetal heart rate is verified by ultra sonography Rh_0 (D) immune globulin 300mg is given to unsensitized mothers with Rh-negative blood. A minimum of 5mg of villi is usually necessary for analysis; 10 to 25mg is optimal. After incubation overnight, cytotrophoblast cells are harvested directly for cytogenic analysis.

An error in diagnosis because of maternal cell contamination is rare when laboratory techniques are good. When the diagnosis is uncertain, amniocentesis may be necessary to make a definitive diagnosis.

CVS has the disadvantage of being an invasive procedure, and it has a small but significant rate of morbidity for the foetus; this loss rate is about 0.5% to 1% higher than for women undergoing amniocentesis. Rarely, CVS can be associated with the limb defects in foetus. The possibility of Rh sensitization is present. There is also a possibility that maternal blood cells in the developing placenta will be sampled instead of foetal cells and compound chromosome analysis.

Maternal Blood Sampling for Foetal Blood Cells

This is a new technique that makes use of the phenomenon of foetal blood cells gaining access to maternal circulation through the placental villi. Ordinarily only a very small number of foetal cells enter the maternal circulation in this fashion (not enough to produce a positive test for fetal maternal hemorrhage). The foetal cells can be sorted out and analyzed by a variety of techniques to look for particular DNA sequences, but without the risks that these latter two invasive procedures inherently have. Flourescenece in-situ hybridization (FISH) is the best technique that can be applied to identify particular chromosomes of the foetal cells recovered from maternal blood and diagnose aneuploid conditions such as the trisomies (2n + 1) and monosomy.

This technique has limitations since it is difficult to get many foetal blood cells.

Maternal Serum Alpha-Fetoprotein (MSAFP)

The developing foetus has two major blood proteins-albumin and alpha-fetoprotein (AFP). Since adults typically have only albumin in their blood, the MSAFP test can be utilized to determine the levels of AFP from the foetus. Ordinarily, only a small amount of AFP gains access to the amniotic fluid and crosses the placenta to mother's blood. However when there is a neural tube defect in the foetus, from failure of part of the embryologic neural tube to close, then there is a means for escape of more AFP into the amniotic fluid. Neural tube defects include anencephaly (failure to closure at the cranial end of the neural tube) and spina bifida (failure to close the caudal end of the neural tube. Also, if there is an omphalocche or gastroschisis (both are defects in the fetal abdominal wall), the AFP from the foctus will end up in maternal blood in higher amounts

In order for the MSAFP test to have the greatest utility, the gestational age must be known with certainty. This is because the amount of MSAFP increases with gestational age (as the foetus and the amount of APP produced, increase in size). Also, the race of the mother and presence of gestational diabetes are important to know, because the MSAFP can be affected by these factors. The MSAFP is typically reported as multiples of the mean (MOM). The greater the MOM, the more likely a defect is present. The MSAFP has the greatest sensitivity between 16 to 18 weeks gestation, but can still be useful between 15 and 22 weeks gestation.

However, the MSAFP can be elevated by a variety of reasons which are not related to foetal several tube or abdominal wall defects, so this defect is not 100% specific. The most common cause for an elevated MSAFP is a wrong estimation of the gestational age of the foetus.

Using a combination of MSAFP screening and ultrasonography, almost all cases of anencephaly and most cases of spina bifida can be detected.

Neural tube defects can be distinguished from other foetal defects (such as abdominal wall defects) by use of acetylcholinesterase test performed on amniotic fluid obtained by amniocentesis. If the acetylcholinesterase is elevated along with MSAFP then a neural tube defect is likely. If the acetylcholinesterase is not detectable, then some of the the fetal defect is suggested.

The MSAFP can also be useful in screening for Down syndrome and other trisomies. The MSAFP tends to be lower when Down syndrome or other chromosomal abnormalities are present.

Maternal Serum Beta - HCG

This test is commonly used as a test for pregnancy. Beginning at about a week following conception and implantation of the developing embryo into the uterus, the trophoblast will produce enough detectable beta-HCG (the beta subunit of human chorionic gonadotrophin) to diagnose pregnancy. Thus, by the time the first menstural period is missed, the beta-HCG will virtually always be elevated enough to provide a positive pregnancy test. The beta-HCG can also be quantified in serum from maternal blood, and this can be useful early in pregnancy when threatened abortion or ectopic pregnancy is suspected; because the amount of beta-HCG will be lower than expected.

Later in pregnancy, in the middle to late second trimester, the beta-HCG can be used in conjugation with the MSAFP to screen for chromosomal abnormalities, and Down syndrome in particular. An elevated beta-HCG coupled with a decreased MSAFP suggests Down syndrome.

Very high levels of HCG suggest trophoblastic disease (molar pregnancy). The absence of a foetus on ultrasonography along with an elevated HCG suggests a hydatioform mole. The HCG level can be used to follow up treatment for molar pregnancy to make sure that no trophoblastic disease, such as a choriocarcinoma persists.

Maternal Serum Estriol

The amount of estriol in maternal serum is dependent upon a viable foetus, a properly functioning placenta, and maternal well-being. The substrate for estriol begins as dehydroepiandrosterone (DHEA) made by the foetal adrenal glands. This is further metabolized in the placenta to estriol. The estriol crosses to the maternal circulation and is excreted by the maternal kidney in urine or by the maternal liver in the bile. The measurement of serial estriol levels in the third trimester will give an indication of general well-being of the foetus. If the estriol level drops, then the foetus is threatened and delivery may be necessary urgently. Estriol tends to be lower when Down syndrome is present associated with adrenal hypoplasia and anencephaly.

5.7 Animal Tissue Culture

The artificial rearing of animals cells in a chemically defined medium under controlled conditions i.e. in vitro culture of cells and tissues is called animal cell culture. It is also otherwise known as in vitro culture of animal cells. Tissue culture is a method of research and involves removing from animals or plants small fragments of tissues called explants and placing them in an environment in which they can survive outside the parent organism. Thus, a tissue in culture may consist of population of cells or parts of organs in which different cell types retain natural relationships to each other (organ culture). The cells in culture may either, increase in size and number or they may differentiate from simple cells and grow into cells with specialized functions or they may even perform specialized functions.

In 1839, Theodur Schwann,a German Zoologist suggested that individual cells could probably be grown outside an organism. However, such demonstration of growth outside an organism was dependant on the development of suitable techniques. In 1887, it was observed that the living white blood cells or leucocytes collected from a frog undergoing mitosis outside an organism. These cells were placed in glass paradises for microscopic study. In 1902, first attempt was made to grow cells from higher plants in nutrient media. But it was Dr. Ross G. Harrison; an American Zoologist in 1907 who was the first to devise a cell culture technique and successfully demonstrated the growth of processes from frog nerve cells in medium of clotted lymph. This a technique known as hanging drop method; made it clear that the tissue culture was not only feasible but it could be useful in biological research. A student of Dr. Harrison developed a cell culture technique and applied it to cells of birds and mammals using a medium of clotted blood plasma. He later, collaborated with Alexis Carrel, a Nobel prize winning environmentalist and developed many refinements in cell culture methodology particularly in tissue of embryonic extracts combined with blood plasma.

Cell culture methods were quite improved after the discovery of antibiotics during 1940's. The significance of animal cell culture was increased when viruses were used to produce vaccines using animal cell culture in late 1940's. Similarly during 1950's polio vaccine production on cultured cell enhanced the significance of animal cell culture technology. In recent times, animal cell culture has become an important component of many scientific disciplines. Advance in medical and veterinary science would have not been possible, without the availability of cultured animal cells. Animal cell and tissue culture is not considered as a science, rather a method of approaching problems in experimental biology and medicine. Animal cell cultures have been used extensively in the diagnosis and characterization of viral agents, the production of vaccines and immunotherauptic agents studying the structure and function of various tissues and organ systems, in the analysis of harmful and beneficial effects of natural and artificial compounds, in bioremediation and monitoring of toxic waste sites. Animal cell culture also plays an important role in molecular biology and biotechnology. The significant biotechnology applications of animal cell culture is the production of monoclonal antibodies and large scale production of biochemicals, reproductive technologies and the production of transgenic animals.

Requirements for Animal Cell and Tissue Culture

The desirable requirements for animal cell and tissue culture are various types of culture vessels (Fig. 5.13 & Fig. 5.14) and the following facilities :

1. Air-conditioned room.
2. Hot room with temperature recorder.
3. Microscope room with different types of sophisticated research microscopes with camera devices.
4. Sterilization room for sterilization (autoclave) of glassware and culture media.
5. An osmometer to test the osmoliability of cells during tissue culture.
6. Dark room for photography.
7. Service room.
8. A preparation room for media.
9. Glassware shelving.
10. Plastic shelving.
11. Specialised equipment.

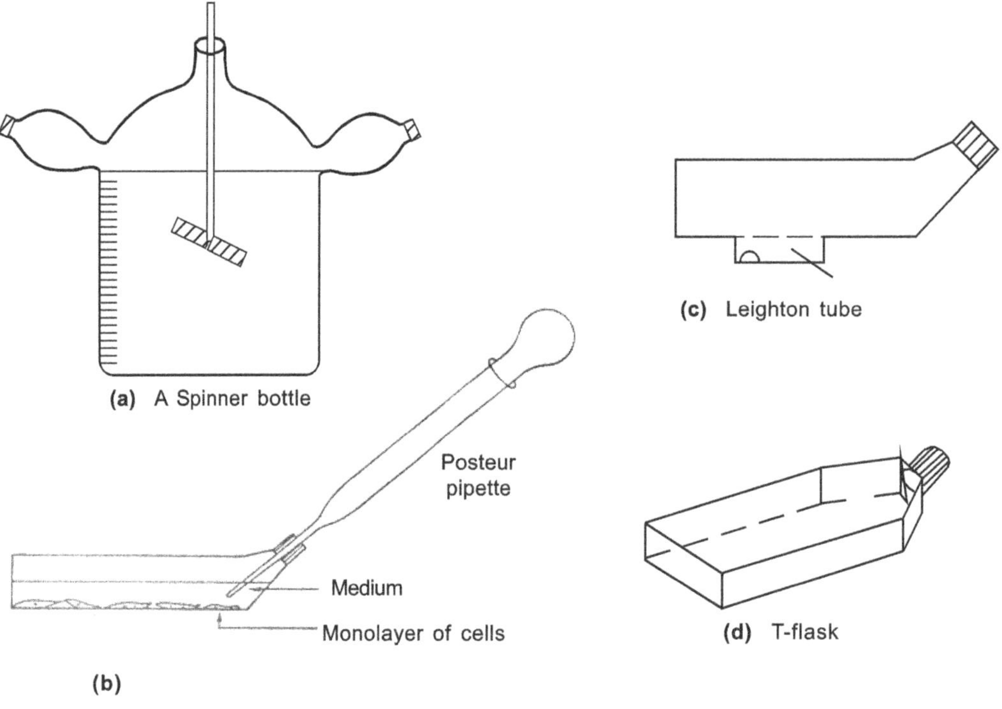

Fig. 5.13 (a) Spinner bottle (b) Pasteur pipette, and (c) Leighton tubes used for animal cell culture. Petri dishes may also be used (d) T-flask

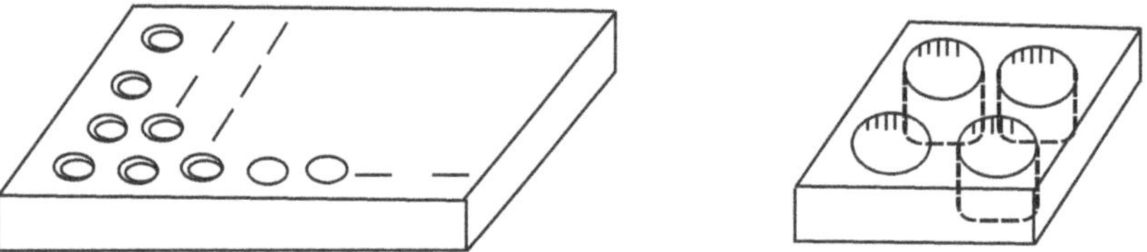

Fig. 5.14 Multiwell plastic culture plates in different sizes and with different numbers of wells per plate

Basic Requirements in Cell-culture

(a) *Culture medium :* The food material on which the organisms are grown in the laboratory is known as *culture medium* and the growth itself is called *culture* or in other words, the *nutrient preparation* on or in which a culture (a population of organisms) is grown is called a culture medium. The basic requirement of any culture medium is that it should serve as a source of energy and should contain all essential organic and inorganic ingredients required for the growth of animal cells. The pH of the culture medium should be well suffered for maintaining a pH between 7 and 7.3 i.e. slightly alkaline, a requirement for most of the animal cell cultures. The pressure exerted by media used for cultures of vertebrate animal cells should not cause/the cells to shrink or swell. It should be sterile and isotonic to cell cytoplasm. Animal culture media are broadly divided into two types, namely

1. Natural media
2. Synthetic media

Natural media for tissue cultures are composed of natural biological fluids and are of three types such as

1. Coagulants or plasma dots
2. Biological fluids and
3. Tissue extract of animal cells.

Synthetic media or artificial media consisting of salts, carbohydrates, vitamins, proteins and nutrients were made and in 1911, first attempts were made to culture animal cells in this media. Basically synthetic media are of two types: (i) *Serum-containing media* ie. media containing serum and (ii) *Serum-free media* (media devoid of serum)

This was followed by studies on the effects of variations of certain components of the medium on the cells. Later it was found that the study of cell behavior could be best carried out in a controlled environment i.e. completely defined medium. These initial studies resulted in the development and use of numerous chemically defined nutrient media containing salts, amino acids, vitamins and other components necessary for the growth of specific cells eg. Hormones,fatty acids and trace metals. The technique of growing cells in a blood plasma-embryonic tissue extract medium made possible the successful culture of the organs both whole embryonic organs such as bone and also fragments of larger organs such as pituitary or mammary glands.There are many types of vertebrate cells that require support for their growth in vitro otherwise, they will not grow properly. Such cells are called an charge-dependent cells. Therefore a large number of substratcs are used which may be adhesive such as glass.

(b) *Substrate used in Tissue Culture :* Glass, plastic, treated plastic surfaces, micro carriers or non-adhesive substrates like liquid gel interface (agar, agarose) hollow fibers. (Fig. 5.13)

(c) *Optimal Temperature in Tissue Culture :* The optimal temperature for tissues in culture depends in body temperature of the animal from which cells are obtained (the optimum temperature) eg. for humans the optimum temperature is 37^0 C, avian cells need higher temperature, insect cells need lower temperatures. Generally, cultured cells can tolerate considerable drop in temperature but not more than 2^0 C above optimal temperature of growth.

(d) *Gas Phase in Tissue Culture :* In artificial preparation of synthetic media in addition to organic and inorganic nutrients O_2 and CO_2 gas phases are to be added. About 90 to 95% oxygen and 5 to 10% of CO_2 is required. Dissolved CO_2 play a role in maintaining pH and bicarbonate concentration. At low cell density, CO_2 needed is low while in fully-grown cultures, cells produce high acid, hence is need to allow the CO_2 to escape.

(e) *Media sufficient in Tissue Culture :* 1. The balanced salt solution media is a mixture of inorganic and organic salts at pH 7.2 to 7.4. 2. Serum is supplemented at 5 to 50% concentration with balanced solution. 3. Serum free media with (a) balanced self solution (b) Known serum concentration like albumin. (c) Transferine and growth factor

(f) *pH and cell maintenance :* The pressure exerted by media used for culture of vertebrate animal cells should not cause cells to shrink or swell. So for the maintenance of cells in culture, the media should have appropriate alkaline or acid conditions. Most cells grow well at pH 7.4. A drop in pH greater than 0.4 unit per day will require media change every 24 to 48 hours.

For routine maintenance of cells in tissue culture, initiated cultures and non-proliferating cells require periodic media change. The replacement of media is indicated by drop in pH, cell concentration, cell type and cell morphology i.e. granularity around the nucleus, cytoplasmic vaccuolation, rounding of cell and cell detachment.

(g) *Cell feeding :* For a slow growing culture add half medium and leave half spent medium. Cell lines that do not grow to confluency within 10 days need to be trypsinized and repeated. Cell cultures need to be sub cultured at least once in a week.

(h) *Sub-culturing of Cell Lines :* This is necessary when the available substrate is occupied by the cells or available nutrient is used up by the cells. Sub-culturing is of two types 1.Dilution of suspension culture to fresh medium and 2. Dissociation of cells in the monolayer followed by dilution in the fresh medium.

(i) *Monitoring growth of cells in Tissue Culture :* The cell count is determined *by try pan blue dye exclusion method* or by *determination of total protein* content by dye method (*SRB ASSAY*) or *by metabolic labelling of cells* and determination of viable cells by conversion of soluble dye to insoluble crystals (MTT assay). Live (viable) cells actively pump out the dye by efflux mechanism and dead (non viable) cells do not. White transparent cells are viable cells and blue cells are dead cells.

Culture of Animal Tissues

Cultures are usually grown either as single layers of cells (monolayers) on a solid surface such as glass or plastic or a suspension in a fluid medium. Cellulose sponges have been used to grow cells in order to maintain normal three-dimensional characteristics of tissue growth; similarly a foam prepared from a plasma protein, the fibrin responsible for normal blood clotting is also used for their purpose. Monolayer cultures may also be grown on semi-solid supporting materials i.e. gels, blood clots or surfaces coated with the protein collagen. Cells in suspension culture are kept from setting to the bottom either by stirring or shaking or rotating the cells in their fluid medium. As kind of media that support cell growth are also good nutrients for other contaminant organisms, the glassware, the equipment and culture media are to be sterilized by heat, steam or Millipore filter paper (Autoclaving).

The culture of *human fore skin tissue* is an ideal example for in vitro culture of animal tissues. Human fore skin tissue is collected from male infants, and preserved in balanced salt solutions some times the fore skin is sterilized directly to establish cell or tissue cultures.

1. *Sterilization :* The fore skin tissue is first kept dipped in 70% isopropanol for 15-30 seconds. Then it is placed in Dulbecco's phosphate buffer saline (DPBS) for a few seconds. It is once again transferred to DPBS solution in a petri dish and swirled to remove the blood and debris. Finally it is washed in DPBS solution and used for tissue culture.

   For long time preservation, the sterilized fore skin tissue is maintained in a flask containing Eagle's medium with penicillin, steptomycitrin and amprotericin and the flask is stored in a refrigerator at 4 to 5^0 C.

2. *Culture of explants :* The sterilized fore skin tissue is cut into small pieces fragments with sterile surgical apparatus. Each of these pieces fragments is an *explant*. The explants which are bigger than 2 to3 mm or smaller than 1 mm size are not suitable for tissue culture.

   For culturing explants, two methods were suggested (i) sandwich method and (ii) etched 'X' method.

   (i) *Sandwich method :* With the help of a fine tipped forceps, an explant is aseptically transferred into a sterile petridish and a cover slip of 22 × 22 mm size is placed over the explant. Sufficient volume of medium is transferred into petridish and is incubated at 37^0 C until the cell growth appears (Avian cells require higher temperatures where as insect cells require low temperatures)

   (ii) *Eteched 'X' method :* 'X' shaped etches are made in the sterile petridish with the help of a sterile scalpel. Then, an explant is placed on each 'X'-mark and is gently pressed into it. The raised lines of etches prevent the floating of explant while adding the nutrient medium to the petri dish. It is left undisturbed for 30 to 40 mts to keep the explants compacty with the etch mark. Then a suitable nutrient medium is poured into the petri dish and is incubated at 37^0 C for a few days. The explants usually produce viable young daughter cells within 24 - 48 hrs. Some tissues take 7 days to produce daughter cells. Once a daughter cell is produced, it migrates away from the explant and grows freely in the medium. This daughter cell undergoes repeated mitotic divisions to form a single layer of cells (monolayers).

Culture of Isolated Cells

A culture containing isolated individual cells is called isolated cell culture. Cell culture is an inevitable tool in modern biology. The *in vitro* culture of animal cells involves the following steps.

1. *Isolation of animal material (tissues) :* The dissociation of cells from an explant tissue from a target organ is done by two methods i.e. (a) Mechanical method and (b) Biochemical method

   (a) In mechanical method or physical disaggregation method, small pieces of explants tissue is forced through a silk cloth are forced through syringe and needle or repeated pipetting. Although the physical method is quick and cheap yet it damages many live cells. In another method of physical disaggregation, the explants are placed in sterile petri dish containing buffered medium with balanced salt solution these cells are now alternately passed through sieves of decreasing pore size (50 mm and 20 mm mesh). The process is repeated to get more disaggregation of cells. The debris remaining on sieves are discarded and medium containing cells is collected. The cells are counted using a haemocytometer. The viable disassociated cells are now termed as 'primary cells'.

   (b) In biochemical disaggregation or enzymatic disaggregation method undamaged viable cells in large numbers are obtained by using enzymes for dislodging the cells of tissues. In embryonic tissue, a high number of undifferentiated cells with least extra cellular matrix is found. Therefore, disaggegation of embryonic tissue readily occurs then in the tissues of adult or newborn. In fragile tissues, such as tumours the chances of cell death and cell recovery are more than normal tissues. There are two important enzymes used in tissue disaggregation ie. collaginase and trypism.

Collaginase is used for disaggregation of embryonic or normal as well as malignant tissues. The intra cellular matrix contains collagen and thus collaginase disaggregates the normal and malignant tissues. The epithelial cells can be damaged by it but the fibrous tissues remain unaffected. The crude collaginase also contains non-specific proteases. The trypsin is also used for disaggregation of tissue into cells and the process is called trypsinization. The enzyme trypsin is commonly used for embryonic tissue but many kinds of cells can tolerate it and different types of tissues are significantly affected. In enzymatic disaggregation methods, the explant tissue is cut into pieces and transferred to a trypsiniation flask containing nutrient media. Then the medium is treated with a enzyme solution containing collaginase, trypsin or proteases. The flask is continuously agitated by keeping it on a magnetic stirbar and the cells get disassociated from the explant and go into suspension. This is called trypsinization and the cell suspension is filtered through a fine nylon cloth to separate out the cell clusters or clumps. The filterate is centrifuged to discard the excess liquid. The concentrated cell suspension is used for cell cultures. The cells are plated on medium and incubated for 48 - 72 hrs for cell growth.

2. *Primary cell culture :* A *primary culture* is the culture of cells taken directly from a living animal. It is also known as *primary cell line.*

A small volume of cell suspension is aseptically transferred to a flat-bottomed culture flask containing nutrient medium and the culture flask incubated at 37^0 C for few days. During incubation, the cells attach to the bottom of the flask and undergo mitotic divisions. This results in the formation of a continuous layer of cells at the bottom of the flask. This layer is of one cell thickness and is called monolayer. Secondary cultures are established from the *monolayer.*

The viability and rate of growth of cells differ from cell to cell. For e.g. blood cells die and disappear within two or three days, whereas neurons and muscle cells can live for thirty days. Fibroblasts also grow faster and survive well. The shape and size of cells vary greatly depending upon the composition of cell culture medium. The primary culture has its own advantages and disadvantages.

The main advantages are :

1. No need to establish stock cultures.
2. A high number of cells can be produced within a short time for physiological studies.
3. Primary cultures can be grown in relatively simple media.
4. Primary cultures are more sensitive to viral infection than established cultures and thus are suitable for production of viral vaccines and interferons.

The main disadvantages are :

1. Primary cultures may have contaminants (latent viruses)
2. Not suitable for long term studies.
3. Primary cultures have mixed populations of cells, experimental findings may often lead to confusion for different cell types.

3. *Secondary cultures :* Sub cultures of primary culture leads to secondary cultures or sub-cultures. The monolayer at the bottom of the flask is made to detach from it by adding trypsin/ EDTA containing balanced salt solutions without calcium and magnesium (to loosen them). EDTA chelates calcium and magnesium, where as trypsin digests and clears the adherence proteins. It is then cut into small fragments and are transferred to a flask containing nutrient medium and is incubated at 37^0 C for few days. As a result monolayers are formed and are called sub-cultures. The scare sub-cultured several times on fresh medium and propagated accordingly until most of the sub-cultures are identical in their morphology and growth rate. These cell cultures are called secondary culture. Generally, animal cells can divide only for a few times before aging and death. For e.g. Human cells divide 50 to 100 times before ageing. Therefore, the secondary cultures cannot be maintained indefinitely.

Establishment of Cell Cultures

There are many types of animal cells that can be grown *in vitro* such as tumor cells, pigmented melanoma cells, neuroblastoma cells, steroid producing adrenal cells, growth hormone prolactin secreting cells, teratoma cells capable of differentiating in artificial conditions, pigmented or cartilage cells. Depending upon the purpose, a continuous cell line can be established from cultured tissues. Sometimes, a few cells in secondary cultures get a different morphology, higher growth rate and indefinite life. Such cells are transferred to a fresh medium and cultured. These proliferate and grow as a definite cell lines. This type of culture is called established cell line or continuous cell lines. These cell-lines possess an unusual chromosome number. These may produce tumours if transplated in the body of related animals. A cell culture can be designated as established cell line, until it is sub-cultured at least 70 times at regular interval of 3 days.

The classic and international examples of cell lines are :

1. Hela cell line (established from sub-cultures of epithetical like cells of cervical carcinoma of a negro woman, Henrietta Lach 1952 and abbreviated as Hela Cells).
2. WISH (raised from epithetical cells of huamn amnion).
3. HEP-2 (raised from sub-cultures of epithetical cells of carcinoma of human larynx).
4. 3T3 AND L (raised from fibroplast sub cultures of connective tissue of mouse.
5. CHO (raised from sub-cultures of epithetical cells of every Chinese hamster.
6. WI-38, raised from normal embryonic lung tissue of a woman (1961).
7. MRC-5 raised from normal embryonic lung tissue, (1966) etc.

Mass cultivation of cells or cultivation of animal cells enmasse in bioreactor systems

The cells are cultured in two different phases as :

1. Freely suspended cells in liquid medium, and
2. Immobilized cells on the solid phase or embedded in suitable materials.

The choice of cultivation technique depends mainly on whether the cells are capable of growth in an adherent or anchorage independant condition. In suspension culture, the cells grow freely dispersed in a liquid medium which is free of any other solids sedimentation of cells is prevented by medium agitation. In contrast, adherent cells are attached to and spread on a solid surface and thus cultivated as an immobilized culture. Cells normally grown in free suspension may be induced to attach and spread on a surface after a suitable pre-treatment of the surface with attachment factors such as fibroneetin or vitroneetin.

The most suitable bioreactor used for suspension culture is compact-loop bioreactor which relies on the integrated monitoring and control of physical, chemical, biological and biochemical parameters. (Bioreactor is a large vessel in which microbes or plant cells or animal cells are cultured under aerobic conditions to produce a novel products).

Online monitoring of chemical and biological parameters in suspension culture is carried out by online flow injection analysis (FIA) by using gas chromatography (GC), high performance liquid chromatography (HPLC) etc.

In immobilization, the adherent cells are allowed to get attached on solid support and they grow as immobilized cell culture. In order to increase surface area for mass cell cultivation, different materials such as glass spheres for cell attachment are packed in different types of bioreactors. A bioreactor operated with support material packed with in it, through which medium is perfused is called a packed bed reactor and is most suitable for immobilized cells. The properties of micro carrier beads used for cell immobilization are that they develop charged surfaces, either positive or negative charges for e.g. DEAE Sephadex (dietrylaminoethyl dextran) developed positive charges, polysystem negative charges, gelatin beads slightly positive or negative charges. The beads require surface coating for adhesion of sufficient number of cells, as energy is required for cell adhesion. In addition to this glass and ceramic materials are also used as good carriers as they contain high surface energy.

Organ culture

Organ culture refers to the rearing of an organ or parts of the organ in vitro from an appropriate tissue. The whole organ cannot be cultured in artificial medium, as it is made up of different types of cells. The parts of the organ with identical cells can be grown by tissue culture and care should be taken that the tissue should not be damaged while handling the organ.

Therefore, organ culture technique demands more tactful manipulation than tissue culture. The culture media on which organ is cultured are the same as described for cell and tissue culture. However, it is more easy to culture embryonic organs than the adult animals. Methods of culturing embryonic organ and adult organs differ. Besides, culture of whole or part of animal organ is difficult because these require high amount of O_2 (about 95%). Special serum-free media (eg.t8) and special apparatus (Towell's type II culture chamber) are used for adult culture. In addition, the embryonic organs can be cultured by applying any of the following three methods.

Organ culture on plasma clots

A plasma clot is prepared by mixing five drops of embryo extract with 15 drops of plasma in a watch glass placed on a cotton wool pad. The cotton wool pad is put in a petridisho. Time to time cotton is moistened so that excess evaporation does not occur. Thereafter, a small piece of organ tissue is placed on the top of plasma clot present in the watch glass. In the modified technique the organ tissue is placed into raft of lens paper or rayon. The raft makes easy to transfer the tissue, excess fluid can also be removed.

Organ culture on Agar

Solidified culture medium with agar is also used for organ culture. The nutrient agar media may or may not contain serum. When agar is used in medium, no extra mechanical support is required. Agar does not allow to liquefy the support. The tumours thus obtained from adults fail to survive on agar media, whreas embryonic organs grow well media consist of ingredients: agar (1% in basal salt solution), chick embryo extracts and horse serum in the ratio of 7:3:3.

Organ culture in liquid media

The liquid media consists of nutritional ingredients except agar. When liquid media are used for organ culture, generally perforated metal gauge or cellulose acetate or a raft of lens paper is used. This possibly provides support.

The best example for organ culture is artificial skin. To get an artificial skin, a small fragment of skin is excised from a person and cultured on a suitable mesh of fibres in a flask. The artificial skin is used as skin transplants for severe burns. In bone marrow transplantation, the bone marrow tissue is grown in vitro to get a large amount of tissue. It is used to treat cytokine deficiency and for genetic engineering.

Whole embryo culture

During 1950s, Spratt studied how metabolic inhibitors affect the development of embryo in vitro. Old embryo (40h) were studied up to another 24 - 48 hrs in vitro until died. For embryo culture a suitable medium prepared is poured into watch glasses which are then placed on moist absorbent cotton wool pad in petri dishes. For the culture of chick embryo, eggs are incubated at 38^0 C for 40 - 42 hrs so that a dozen of embryos could be produced. The egg shell sterilized with 70 percent ethanol is broken into pieces and transferred into 50 ml of BSS. The vitelline membrane is removed. The embryo is observed by using a microscope so that the developmental stage of blastoderm could be found out. The blastoderm is placed on the medium in watch glass placed on sterile adsorbent cotton wool pad in petri dishes. Excess of BSS is removed from the medium and embryo culture of chick is incubated at 37.5^0 C for further development.

Animal cell cultures-storage

The storage of cultures in liquid nitrogen (–196 ^{O}C) will ensure zero metabolic activity and an almost indefinite period of storage. This long term cold storage is called cryopreservation. It is very useful for laboratories working with cells whose life span is relatively short. Cryopreservation involves the following steps :

1. The cell culture is treated with trypsin to dissociate cells from the monolayers.
2. The cell suspension is centrifuged to remove the excess medium.
3. Cryoprotective agents like diethyl sulfoxide (DMSO) is added to the concentrated cell suspension to decrease cryodestruction.
4. The culture flask is capped with metal foil or thick plastic and is kept in a large container containing liquid nitrogen which keeps the temperature at -196^0 C.
5. The container is first kept at 5^0 C (for one hour), at 20^0 C for the next one hour and then at -80^0 C for the next 24 hrs. This treatment of tissue at low temperatures makes

the cells to ooze out all intra cellular freezable water. Later, the temperature is raised to 27^0 C to melt extra cellular ice crystals on the surface of the cells. This temperature treatment is called *thawing*.

6. Then the container is placed in -196^0 C freezer to preserve the culture for indefinite time.
7. This followed by washing and reculturing washing is done to remove the toxic cryoprotectants. Washing follows the following procedure i.e. dilution (nutrient medium), resuspension, centrifugation and removal of cells. The final viable cell suspension is used as starter culture for reculturing.

Advantages of cell culture

- Easy propagation of cells.
- Homogeneity in cell populations.
- Short-term cultures possible.
- Cells can be frozen for future use.
- Long term culture of cells is possible.
- Establishes cell lines with specific gene(s).
- Cloning of cells with specific phenotype.
- Expression of recombinant proteins by cell culture.

Applications of animal tissue culture

Animal tissue culture is an important tool in modern biology and biotechnology. From cultured animal cells, several valuable products such as human monoclonal antibodies and biochemicals can be produced on a large scale. More interestingly, the genetically engineered cells have revolutionized the cell culture industry. Several specific promoters of human origin are utilized for high expression of foreign genes. For large scale production of certain biochemicals, genetically engineered baculovirus infected animal cells are also in use in a bioreactor. Master cell banks (MCB's) of cell culture system are established to meet the demand of commercial production of products. Thus large scale cultures are the source of all valuable products (Table 5.3).

Some of the important products which are produced from animal cell cultures are :

(i) enzymes (asperagenase,collagenase,urokinase,pepsin,hyaluronidase,rennin,trypsin,tyrosin hydroxylase), (ii) hormones (leutinizing hormone, follicle stimulating hormone, chorionic hormone and erythropoietin), (iii) vaccines (foot and mouth disease vaccine, vaccines for influenza, measles and mumps, rubella and rabies), (iv) monoclonal antibodies, v.interferons, etc. Tolbert et al. (1982) succeded in producing large quantities of human interleukin-2 or T-cell growth factor by culturing a permanent lymphoblastoid T-cell line in a large batch suspension culture in a bioreactor.

Production of Vaccines

Viruses can grow and multiply only in the living cells. Therefore, animal cell cultures are used to propagate viruses to make vaccines. For example, poliomylitis virus is grown in mammalian cell cultures to make polio vaccine.

Table 5.3 *Some products of medical use derived from animal cell cultures.*

Products	Application
Erythropoietins	
Erythropoietin-α	Anaemia resulting from cancer and chemotheraoy
Erythropoietin-β	Anaemia secondary to kidney disease
Human growth hormones	
hGH	Human growth deficiency in children renal cell carcinoma
Somatotropin	Chronic renal insufficiency. Turner's syndrome
Monoclonal antibodies (therapeutic)	
Anti-lipopolysaccharide	Treatment of sepsis
Murine anti-idiotype/human	
B-cell lymphoma	B-cell lymphoma
Monoclonal antibodies (diagnostics)	
Anti-fibrin 99	Blood clot
Tcm-Fab(breast)	Blood cancer
Pr-356 CYT-356-IN-111	Prostate adinocarcinoma
Plasminogen activator	
Urokinase type plasminogen activator	Acute myocardial infarction, acute
Tissue type plasminogen activator	stroke, pulmonary embolism,
Recombinant plasminogen activator	deep vein thrombosis
Vaccines	
HIV vaccines (gp120)	AIDS prophylaxis and treatment
	Malaria prophylaxis
	Poliomyelitis prophylaxis

The virus is allowed to infect animal cell cultures. The virus multiplies and produces numerous virus particles in the cells. The mature virus particles come out of the animal cells by breaking their cell membranes.

The mature culture is treated with a hypertonic solution to lyse the cells. The lysate is then centrifuged to remove the cell debris. The supernatant is filtered through a high quality filter to get virus particles for vaccination.

Somatic cell fusion

The method of fusing two different cells types is called somatic cell fusion or somatic hybridization. The fusion product is called somatic hybrid. Polyethylene glycol(PEG), electric field, Japanese hemagglutination virus(JHV) and Sendivirus(SV) are used to induce the fusion between somatic cells. A typical example for somatic hybrid is hybridoma. It is the product of fusion of lymphocyte and myeloma cells. It shares the antibody producing property of the lymphocyte and the fast growth of the myeloma cell. Hybridoma clones produce identical antibodies with a single type of epitope called monoclonal antibodies. They have wide applications in the field of medicine, pure and applied research as well as industries.

Genetic Engineering

In genetic engineering, recombinant viruses are constructed and allowed to infect eggs or zygotes growing in a nutrient medium. After reaching certain globular stage, the embryo is implanted into the uterus of a female. The female gives birth to transgenic individuals. The transgenic animal has the insert gene in all its cells. Therefore, cell culture is believed to be an inevitable step in animal genetic engineering.

Sperm Bank

Sperm bank is a collection of semen from superior farm animals, fishes, oysters or even man; stored in liquid nitrogen. It supplies semen timely when and where it is in need.

Semen is collected from farm animals by using artificial vagina (AV), electro-ejaculation or massage method and diluted properly by adding a proper dilutent. The diluted semen is transferred to tubes and the tubes are capped tightly with a metal foil. The closed tubes are then kept in a can fitted in a large container having liquid nitrogen. The can be preserved at -196^0 C for indefinite period and can be used even after several years. For this, the semen tube is kept at 37^0 C for a few hours. Then the semen is introduced into the uterus of a female by mechanical means without casual mating. This method of insemination is called artificial insemination(AI). The artificial insemination extends the service of superior sires in breeding. And it decreases the semen wastage through mounting.

In vitro fertilization

The process of fertilizing an egg with a sperm in a nutrient medium in the culture plate is known as *in vitro* fertilization (IVF). This method is practiced in farm animals and also in women having damaged fallopian tubes. It overcomes the difficulties in fertilization in such individuals.

Unfertilized eggs are collected from the female and placed in the nutrient solution in culture plate. Semen is collected from a male and added in the plate to fertilize the eggs. The culture plate is then incubated at 37^0 C for a few days till the fertilize eggs attain 8 to 16 cells stage. The globular embryo is then implanted into the womb of the female. After the suitable period of conception the female produces a young one as ususal and normal manner.

Cattle breeding

In vitro fertilization is used in :

1. Cattlle Breeding.
2. Breeding rare and endangered species using another species of surrogate mother.
3. Developing human test tube babies.

Raymond's Embryo Research Centre (Madhya Pradesh) has bred high yielding calves through inferior quality cows. Holstein-Freeian variety cows produce a large quantity of milk but they suffer with frequent abortion. Malvi breed, a native variety of Madhya Pradesh, is poor in milk yield but resistant to abortion. Shyam Zawar and co-workers had taken the egg from Holstein-Freeian variety cows and fertilized in-vitro with semen collected from the bull. Having attained 8-16 celled embryo, it was implanted in the womb of the Malvi breed. The local breed produced Holstein-Freeian variety calves. This method is known as embryo transfer technology.

This technique is also used in horse breeding programme.

Breeding rare species

IVF technology is used to save rare species of animals from extinction. Many wild animals have been bred by this technique. In a London zoo, wild Indian bison and guar have been bred by using Freeian cow as a surrogate mother. Horse-mares were made to give birth to Britain zebra.

Test tube babies

Babies raised from in vitro fertilized eggs are called test tube babies. In this technique, the woman desirous of having a test tube baby is first treated with gonadotrophin hormone, after menstrual cycle. This stimulates the development of several follicles.

A small slit is made just below her naval and an instrument called laparoscope is inserted into the slit. The laparoscope has a light source, a lens system and a micropipette. The surface of the ovary is viewed through the lense system for searching mature follicles and the mature follicles are taken out through the micropipette. The eggs (follicles) are inoculated into a nutrient medium, containing bovine serum albumin and pyruvate. Semen is collected from the woman's husband or any other male donor and added into the medium. The culture plate is incubated at 36^0 C for a few days to notice the fertilized eggs.

At the 8 - 16 cell stage, an active embryo is implanted into the uterus of the woman using a micropipette. In 38 weeks she will produce a baby.

This method helps immensely the infertile woman with badly damaged fallobian tubes. (Fig. 5.15)

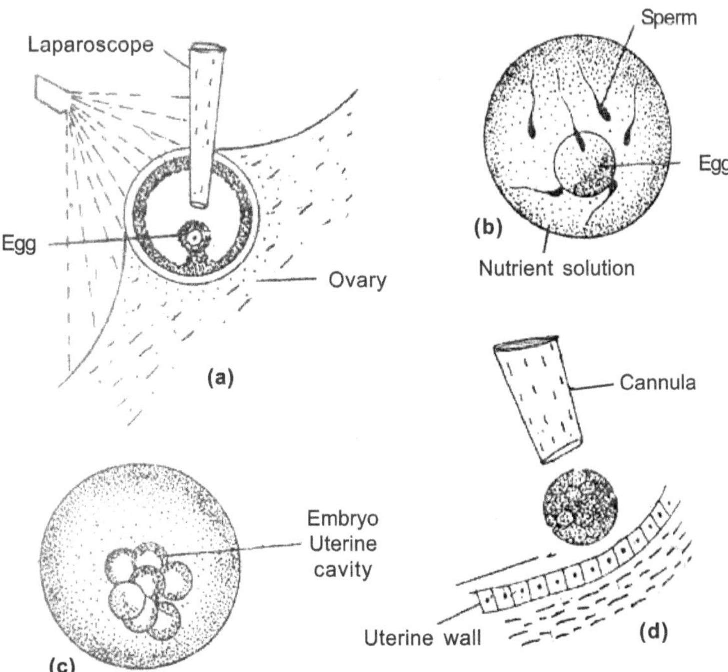

Fig. 5.15 Protocol for *in vitro* production of Test tube baby through IVF.

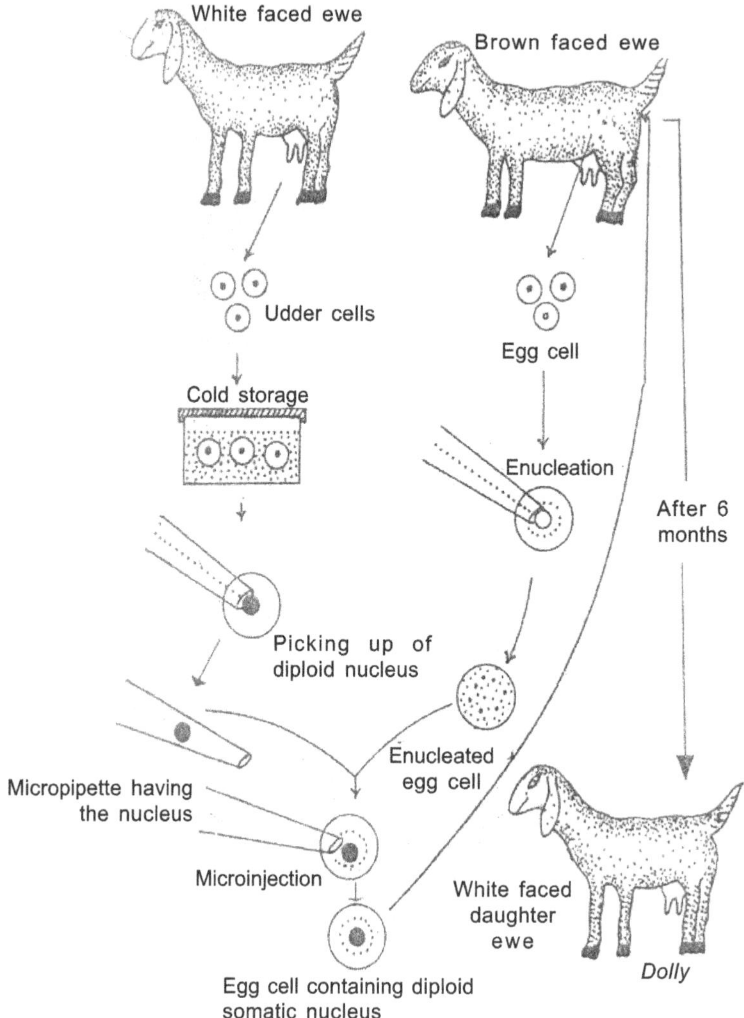

Fig. 5.16 Animal cloning technique to make xerox copies

Xerox Animals

The offspring which are the exact copies of their parents, are called xerox animals or copy animal or clones. The method of making xerox animals is referred to as animal cloning. For the first time xerox frogs were made by Robert Bricks in 1952. Later it has been practiced in fishes, poultry, rats, pigs, goats and cows.

Ian Wilmut and his co-workers at Roslin institute in Scotland for the first time succeeded in cloning a mammal in 1997, they created a cloning goat named Dolly. Dolly is the xerox copy of her mother.

Some mature somatic cells were taken from the udder of a donor ewe having whitish face. They were placed in a nutrient solution in a petri dish and incubated at cold temperature (less than 5^0 C).

Then an ewe having brownish faced was treated with gonadotrophin injection to stimulate the development of eggs. Mature eggs were taken from the ewe using a laparoscope and placed in a nutrient solution. From the mature egg cell, the haploid nucleus was sucked out using a micropipette under a microscope. Later, a diploid nucleus was taken from the somatic cell and introduced into the enucleated egg cell. The egg cell was then allowed to grow in the medium till it reached 8 - 16 cells stage. That cell mass was implanted into the uterus of the brown faced ewe.

At the end of the sixth month the ewe delivered a white faced female goat (Dolly). Dolly is the exact copy of the white faced adult ewe (Fig. 5.16).

This method can also be adopted for human being and is called human cloning. However, it has not been practiced so far because of some social and ethical issues involved in it.

Embryo splitting

The embryo of monkey at 8-celled stage is split into two partially separated units using a nylon thread. This embryo when implanted into uterus of a female monkey produces identical twins. This technique is known as embryo splitting.

5.8 Monoclonal Antibodies (MCA)

A single type of antibodies having the same antigenic determinant produced by a single hybridoma clone,is called monoclonal antibody. The hybridoma is made by fusing a lymphocyte (B cell) with a myeloma cell. Presence of a single antigenic determinant is the useful feature of the monoclonal antibodies. MCA's bind with only one type of epitope on the antigens.

Applications of monoclonal antibodies

Monoclonal antibodies are very pure and specific to antigenic substances. Hence they have been put into many practical uses in research, clinical diagnosis of diseases and in the treatment of the disease. Their overall applications are discussed briefly here under.

1. *Monoclonal antibodies as research tools :* Monoclonal antibodies have been used as research tools in the fields of medicine, immunology,biochemistry, genetic engineering, carcinobiology etc. Some of their research applications are given below :
   (a) In genetic engineering,monoclonal antibody is used to screen recombinants.
   (b) In immunological studies, MCA's are used to identify various cell types involving in immune responses and to detect interactions among them. For this purpose MCA's are raised against the cell surface antigens and used as markers to identify the cells.
   (c) The rejection of transplanted organs, especially kidney transplants, is associated with OKT-3 antigen present on the surface of T-cells. Monoclonal antibody against OKT-3 antigen is made and injected into the patient to suppress the transplant rejection. The MCA named orthoclone OKT-3 is mainly used for this purpose.
   (d) Monoclonal antibodies are used to determine the structure of cell membranes.
   (e) They are employed in serological classification of closely related bacterial, viruses and protozoans.

(f) They are used in radioimmuno assay, ELISA and immunoflourescence assay in research to identify and detect some target products.

(g) MCA's are used to detect the exact position of enzymes in the cells by using immunocytochemical methods.

(h) MCA's are being used to detect and classify various enzymes.

(i) Monoclonal antibodies are used in affinity columns to pick up proteins from the solution. They help to isolate and purify proteins and enzymes from solutions.

2 *Clinical diagnosis :* Since 1979, monoclonal antibodies have been put into the clinical diagnosis of certain severe diseases. MCA's are used as diagnostic kits for the following diseases :

(i) Childhood diarrohea

(ii) Malaria

(iii) Cancers of gastro-intestinal tract, pancreas, breast, uterus, kidney, stomach, larynx, liver, bone, and connective tissue.

(iv) Leukaemias

(v) Veneral diseases caused by *Nisseria gonohoreae* and chlamydias

(vi) Infections by herpes virus 1 and 2.

(vii) Pregancy can be detected in woman as early as one or two weeks after conception using MCA's.

3 *Treatment of several diseases :* The cancers of lungs, breast, pancreas, etc are cured by using monoclonal antibodies. Marx in 1982, in the US National Cancer Institute, made monoclonal antibodies against some surface antigens of cancer cells. That institute has cured several cancer cases by injecting the MCAs.

Simpson and Cioli (1982) have prepared MCAs against the surface antigens of some parasites. These MCAs are used to control the spread of *Troponema pallidom*, *Mycobacterium leprii*, *Haemophilus influenzae*, etc.

Hazards of MCA's

Most of the mouse myeloma cells contain certain genes related to retroviruses. If monoclonal antibodies were prepared from such meyloma cells, the MCA's may have the dangerous viruses. The hazards of the viruses should be eliminated before using the MCA's. In order to prevent the viral hazards, the regulatory authorities approve MCA's for medicinal use only after confirming that they are virus free.

5.9 Drugs and Chemicals from Animals

The Greek physician Galen (about 130-200 AD) mentioned opium, hyocyamus, sqill and viper toxin among other drugs in his apothecary shop. Holland (1994) in his paper mysticism to science, the use of medicines from animals etc.

Recently Lev and Peri (2001) in their paper gave Historical evidence of the beginning of modern medicine in the holy land (Jerusalem) and provided the list of medicinal preparations in various drugs found in stores and dispensary (1857). The huge archives of London Society in England lists 268 substances, 147 of which are of plant origin and 117 chemical and mineral and 4 of animal origin.

Modern medicines from animals and animal products and their success in treatment of various diseases is the result of extensive collaborative research by scientists and pharmaceutical industries. For the proper development of pharmaceuticals of plant/animal origin pharmaceutical industries involve number of specialists; pharmacologists, pathologist; physiologists, toxicologists, endocrinologists, neurologists, veterinarians, medical advisors who work in collaboration with organic and physical chemists before a drug is finally approved and marketed.

Before the World War I, the drugs were mainly prepared from synthetic chemicals but as the supply of drugs was cut off; the use of domestic animals for pharmaceutical industry was started since then thanks to new researches and developments of industrial technologies the use of animal medicine has an important place in modern medicine.

The drugs prepared from animals and clinical products showed good response in the treatment of several diseases without affecting the longivity of life. New drugs are being derived from the cardiac glycosider/digitoxins protein hormones like insulin and many others. The animals commonly used for experiments in pharmaceutical industry include cats, guinea pigs, dogs, mice, albino rats, monkeys, Chinese hamsters etc.)

The pharmaceutical industry is taking the full advantage of modern trends of development and research in the fields of biological sciences, chemistry and engineering and by applying these trends in formulating new drugs and making them available to doctors and patients. Although number of animal drugs are used in various systems of medicines in India, the following is the account of famous animal drugs.

1. *Musk : Kasturi : Moschus moschiferous*

Musk, an aromatic secretion from the male Himalayan musk deer is used to prepare medicines to treat asthma, nervesedative in epilepsy, hysteria, convulsions and other disorders. It is obtained from the round glands situated beneath the abdomen of the deer and fetch Rs. 3,00,000 per kilogram in the international market. Musk is an inspissated and dried secretion (testicular extract) from the preputial follicles of the male deer. Fresh musk is milky with strong diffusible smell.

It is almost impossible to extract musk without killing the animal. In view of its occurrence in Himalayas, the animal is included in schedule I of the Wildlife Protection Act (WPA) which is a list of endangered species. Musk deer farms and their breeding is done in Uttaranchal Pradesh and Shimla in Himachal Pradesh.

In Ayurveda and Siddha musk is used as aphrodissiac, antispasmodic, vomiting, cold, cough etc.

In Unani hot or dry musk is used as a cardiotonic tonic to brain, increases vitality, for curving nervous diseases, paralysis, apoplexy etc.

The musk substitutes at present known as trinitro-meta-tertiarybutyl-tolune and the corresponding compounds obtained from the homologues of toluene and dinitro derivatives of the ketones which are formed by the interaction of acyl chlorides on derivatives of toluene, of these, Trinitrobutyltoluol (C_6 H NO_3 CH_3 C_4 H_7) is considered to be the best. Its odour is very similar to the natural musk and is sold in perfumery under the name of artificial musk.

2. *Cockroach glands :* A preparation from cockroach glands in homeopathy is said to be very effective in the treatment of chronic cough and asthma.
3. The sheep brain is used in the preparation of anti-rabies vaccines.
4. Hirudin : *Hirudo medicinalis* (Leech)

   Hirudin is the secretion of the salivary glands of leech which prevents coagulation of blood during its feeding. The anti coagulant properties of hirudin can be made use of by employing it an anticoagulant in surgical operations. It is also recommended for prevention of phlebitis and post operative pulmonary inflammations. Employing leeches is a natural and cheap therapy for arthritis. According to a Russian team of doctors (Dr. I. G. Salikhov and colleagues from Kazan State Medical University Kazan, Russia) they were able to successfully cure patients both rheumatoid and osteoarthritis.
5. *Bats :* There is a popular belief that bats have medicinal value and cure diseases especially respiratory illnesses.
6. *Bee venom :* This is secreted by poison glands of sting of bees. Clinically it is used in the treatment of rheumatism, kerato conjunctivitis, iritis and in the control of blood pressure.
7. Honey is used as a general tonic and is the base for most Unani and Ayurvedic formulations. It is also used as a demulcent and laxative.
8. *Shark, cod liver oil and halibut liver oils :* The oils are rich source of vitamins A and D are used in the manufacture of number of medicines while the whale liver oil, which is rich in vitamin 'A' is routinely recommended for proper maintenance of eyes and eye diseases.

   Vitamin A and D are both extracted by passing steam into tanks containing the livers (shark & cod) suspended in water, until the temperature reaches 70 - 80^0 C ($150^0 - 176^0$ F). The tissues disintegrate, and the oils float to the surface and are skimmed or centrifuged off, purified and marketed.
9. Mulberry silkworm pod. This is used as styptic, tonic and astringent. It helps in controlling profuse menstruation, leucorrhoea and chronic diarrhoea.
10. *Heparin :* Heparin is a natural coagulant and is extracted from the liver of higher animals and is frequently used against coronary thrombosis.
11. *Snake Venoms :* Black cobra venom is used in the treatment of leucoderma botulism. Pain injury to nerves, rheumatism and sciatica.

    Indian cobras (*Naja tripudens, N. bungarus*). Cobra venom (cobroxin) is used as hepatic stimulant in small doses, it is used in neuritis, neuroheprosy, spinal tuberculosis, migrine, intracetable pain due to cancer, arthritis epilepsy and haemorrhage of retina and uterus. The medicinal property of venom is due to non toxic protein component of venom which destroys certain types of tumor cells. Cobra cardiotoxic is recommended in bronchial asthma and pain injury to nerves, rheumatism and sciatica. Black cobra venom is used in the treatment of leucoderma botulism.

Indian Vipers (Daboia russelli and Viper aspis)

Viper venom can be used as hemostatic factor because its coagulating quality. It is used in haemophilia and haemorrhage. A non tonic protein 'stypren' is isolated from viper venom and is recommended for anti-coagulant therapy.

1. Snake venom is made up of a complex mixture of enzymes, proteins, cellular debris, fats and certain salts such as calcium, ammonium and magnesium chloride and phosphate. It also contains toxic substances like haemotoxin and neurotoxin accordingly there are two types are known, neurotoxin which attack the central nervous system and haemotoxin which targets the circulatory system. Snakes with neuro toxic venom include cobras, mambas, sea snake, kraits and coral snakes. Snakes with haemotoxic venom include rattle snakes copperheads and cottonmouths. Some snakes contain venom that contains combinations of both neurotoxins and haemotoxins. Snake venome is used to cure heart attacks and strokes.
2. The British Heart Foundation has expert team of doctors at Oxford, Birmingham and Liverpool Universities with experience in collection, storage and analysis of snake venoms from around the world.
3. Snake venom is used for the preparation of 'antivenin' and is being used in the number of diseases. In Ayurveda different kinds of pills containing different proportions of snake venom are used in the collapse stage of fever, cholera and many complaints from time immemorial.
4. Manospecific anti venoms are found to be more effective and are less likely to cause side effects than the bivalent and polyvalent antivenoms.
5. A number of *enzymes* have been isolated from cobra venom which help in neuro toxin paralysis, respiratory movements, and act mostly as *accelerators*. Many other enzymes isolated are hyaluronidase, ribonulease, lecithinase, phospho-diesterases etc., for research purposes.

   Antisnake venoms are non being routinely manufactured by several pharmaceutical companies, notably Haffkine Institute, Serum Institute Bhart's serum.

Frogs

Phyllobates bicolor and *Dendrobates sp.* (Fam : Dendrobatidae) are poisonous dart frogs of central and South America, secrete a powerful toxin from glands on their back which is potent pain killer. Drs. Daley & Meyers working with *Epibedobates tricolor* (Ecuqdorian frog) produces epibatidine, a toxin 200 times more effective in suppressing pain, promotes alertness and without side effects unlike opium. Chemists of Abbott laborataries have synthesized this and several novel molecules resembling epibatidine. Clinically it is ABT-594 that depresses pain and is non addictive.

Animal Origin Drugs - The Modern era

Thyroxine is normally produced by the thyroid gland. If insufficient amounts are produced (hypothyroidism) patients become mentally slow, lethargic, and cold. L - Thyroxine, obtained from pork or beef thyroid glands or by synthesis.

Chlorpropamide and insulin are both drugs for diabetes, a condition characterized by a deficiency or reduced effectiveness of insulin in the body. Insulin, obtained from the pancreas of cows and pigs or by a novel biotechnological process involving recombinant DNA, is an agonist and supplements body insulin. It is a large molecule consisting of a sequence of amino acid joined by bridges between chains. Insulin must be injected but chlorpropamide can be taken orally. It is structurally similar to the sulfonamide drugs but has no antibacterial action. Instead, it stimulates the pancreas to produce more insulin.

Insulin

Other drugs of animal origin primarily include glandular extracts containing hormones, such as thyroid and parathyroid, and purified hormones, such as insulin, obtained from the pancreas of a pig, sheep, or ox. For insulin production, the pancreas must either be absolutely fresh or frozen immediately after removal from the animal. Otherwise, enzymes will cause decomposition of the insulin. The glands are minced and immediately stirred with cold acidified alcohol, which extracts the insulin and prevents its destruction by the enzyme trypsin. The gland residue is separated by centrifugation and further extracted with more acidified alcohol. The final alcoholic extract is passed through filter presses to remove the remaining fine particles of pancreas tissue. The extract is now made sufficiently basic to cause precipitation of the inert proteins and allow their removal. The extract is removed quickly, concentrated, and cooled, fat separates and is filtered off. The crude insulin is then precipitated by adding common salt and is purified by repeated crystallization. For medical practice, insulin is almost always made up in the form of a purified solution for injection.

Sea creatures have long been screened for medically useful chemicals. U.S. researchers reported in *Journal of the American Chemical Society* that the microbes that live on these animals might be a more fruitful place to look for new drugs. Dr. Haiyin He of the pharmaceutical company Wyeth-Ayerst in Pearl River, New York, and colleagues have found a bacteria that produces an anti-cancer agent on the body of the Fijian sea squirt *Polysyncraton lithostrotum*.

The researchers were looking for the source of the chemical namenamicin, thought to come from the sea squirt. Namenamicin damages DNA, so can be used against cancer cells that multiply rapidly. This chemical has shown some promise in treating tumours.

Many living organisms, especially sea creatures, make a wide range of poisons, and these are constantly being screened for possible use as anti-cancer agents. Sea squirts – a group of marine invertebrates produce a range of nasty chemicals, some of which have pharmaceutical potential. Namenamicin resembles the calicheamicins, chemicals made by actinomycetes bacteria that are also being studied as anti-cancer drugs. So the group suspected that namenamicin might not be made by the sea squirts at all, but by bacteria living in or on their bodies.

When the researchers extracted actinomycetes from sea squirts, and grew them in fermentation vats, they discovered a previously unidentified strain that generates a substance highly damaging to tumour cells. They name the bacteria *Micromonospora lomaivitiensis* and its two related active agents lomaiviticin A and B. These compounds also act as antibiotics, and may be produced by the actinomycetes cells to ward off other bacteria. *Lomaiviticin* kills *Staphylococcus aureus*, one of the virulent, antibiotic resistant bacteria that causes infection and subsequent death.

The researchers failed to identify the source of namenamicin, they stumbled across another class of potentially powerful drugs. Their discovery suggests that researchers might profitably search for new pharmaceuticals, not in squishy sea creatures, but in the bacteria that live on them.

Animals have served humans in their search for drugs in a number of ways. They have not only yielded to drug testing and biological assay procedures but also have provided drugs from their own tissues or through their biological processes. Hormonal substances such as thyroid-extract insulin and the pituitary hormone obtained from the endocrine glands of cattle, sheep and swine, are life saving drugs employed daily as replacement therapy in human body. The urine of pregnant woman is a rich source of estrogens. The use of animals in the production of various biological products including serums, anti-toxins and vaccines, has been of life-saving significance ever since the pioneering work of Dr. Edward Jenner on the small pox vaccine in England 1796. Today poliomyelitis vaccine is prepared in cultures of rhesus monkey tissues; the mumps and influenza vaccines in fluids of chick embryos, the rubella vaccine in duck embryos and the small pox vaccine from the skin of bovine calves.

Immunoglobulins are preparations containing antibodies against infections microorganisms and are usually prepared from human plasma or serum. Normal immunoglobulins are prepared from material from blood donors and contain several antibodies against infections diseases prevalent in the general population. Specific immunoglobulins contain minimum specified levels of one antibody.

Tetanun immunoglobulin (TIG), Rabies immunoglobulin, Hepatitis B immunoglobulin, human anti-D immunoglobulin are available for the treatment of various medical emergencies. Gammaglobulin is routinely used as a intravenous preparation for replacement therapy for patients with congenital agammaglobulinemia, idiopathic thrombocytopenic purpurea and kawasaki syndrome etc.

In future vaccines for diseases like AIDS and cancer may be developed through utilization of cell and tissue cultures technologies.

Role of Biotechnology in Animal Medicines

During past 20 years biotechnology has revolutionised the era of biological research particularly in the field of medicine, agriculture, pollution control, food processing, biochemistry and genetically engineered micro-organisms. Medicines like hormones and various types of vaccines (mentioned elsewhere) production from genetically designed micro-organism has begun. About 25% of total drugs produced in the world are derived with the help of biotechnological methods.

Genetic engineering of the '*double helix*', the spiral DNA chain of life has made possible to produce more abundant, much purer antibiotics, vaccines and biological products to combat human diseases. There are two basic technologies in the genetic field of drug development. They are recombinant DNA (r-DNA) and monoclonal antibody (MOAB) production. Both techniques has the ability to manipulate and produce proteins which are an infinite source of drugs. These techniques can influence the cells in their ability to produce proteins. The more fundamental of the two techniques is the recombinant DNA. It can potentially produce any protein. It is possible to transplant genetic material from higher species such as man into lower species like bacterium and vice versa. Drugs such as human insulin, 'Humulin', human growth hormone hepatitis B vaccine, epoietinalpha and interferon interleukins, enkephalins, endorphins are being produced in this manner.

While r DNA technique involves manipulation of proteins within the cells of lower animals, the monoclonal antibody production is conducted entirely within the cells of higher animals including man (the patient). The technique exploits the ability of the cells to produce a described antibody and stimulates pure antibody production. These antibodies then have the capacity to combat the specific target. Monoclonal antibodies have enormous potential to change the face of medicine and pharmacy in the next decade and applications for their are already in progress.

In medicine, monoclonal antibodies are being used to localise malignant cancer cells and infuture MOABS may be used to combat diseases like juvenile onset of diabetes. Other uses of MOAB include diagnosis of ABO blood grasp, cancer, pregnancy, allergies, viral diseases and certain hormones, organ translantation etc.

Enzymes are of great significance in the treatment of a number of human diseases. The enzymes can not be produced in a laboratory but enzymes of medicinal importance have been extracted from human urine (Urokinase), organs, blood or from micro-organisms. These enzymes can be produced biotechnologically and can be replaced for deficients enzymes in human body. A protein tissue, "*plasminogen activator*" *(t - PA)* has been produced biotechnologically, which is found in minute quantities in human blood system. This enzyme dissolves blood clots in the coronary artery and restores the normal circulation of the blood in the heart of human-beings. *Superoxide dismatase* (SOD) is another biotechnologically produced enzymes which prevents tissue perfusion damage.

Several drugs are now routinely produced using recombinant DNA, which has increased the purity and decreased the cost of production. Cortisone, interferon and cyclosporin are all produced with the help of r-DNA technology.

ESSAY QUESTIONS

1. Write an account of Transgenic animals and their applications.
2. What are transgenic animals ? Describe the various methods employed in producing transgenic animals. What are their uses.
3. Describe the essential procedures involved in animal tissue culture.
4. Differentiate between cell, tissue and organ culture, add a note on their application.
5. What is molecular farming ? Describe some useful products so obtained by transgenic animals.
6. What the various techniques used for prenatal diagnosis.
7. Differentiate between amniocentesis and chroionic villus sampling.
8. What is gene therapy ? What are its prospects and limitations ?
9. What is gene therapy ? Differentiate between the somatic and germ line therapies.
10. What are the conditions and aims of gene therapies.
11. What is genetic counselling ? How it is useful in disease diagnosis and carrier detection.
12. Write an account of stem cell biology and its applications.
13. Write an essay on drugs and chemicals from animals.
14. Describe r-DNA methods in vaccine development.
15. What are vaccines ? What is the role of r-DNA technology in vaccine production ?
16. Describe the retro virus mediated gene transfer in animals.
17. What are the criteria for pedigree charting

SHORT QUESTIONS

Explain the Following :

1. Transgene
2. Transgenesis
3. Transfection.
4. Microinjection.
5. Test Tube Baby.
6. Amninocentesis.
7. Pedigree charting.
8. Gene therapy
9. Eugenics.
10. Pluripotent stem cells.
11. Animal cell culture.
12. Stem cells.
13. Animal toxins.
14. Cell lines and their applications.
15. Animal viruses.
16. Ultrasonography.
17. Somatic cell therapy
18. Germ lime cell therapy
19. Xerox copies of animals
20. Retrovirus
21. SVS virus
22. Karyotype
23. Snake vemoms
24. Insulin
25. Musk
26. Sperm bank
27. Stem cell bank

6 Microbial Biotechnology

6.0 Introduction

Microbiology is a branch of biology that deals with the study of microorganisms. Microorganisms are microscopic in size and most of them are unicellular and include bacteria, fungi, algae, protozoa, viruses and is concerned with their form, structure, reproduction, physiology, metabolism, classification, culture, economic importance and pathogenicity etc.

Microorganisms are essential to life. Among their many roles, they are necessary for geochemical cycling and soil fertility. They are used to produce food as well as pharmaceutical and industrial compounds. On the negative side, they are the cause of many diseases of plants, animals and human and are responsible for the spoilage of food etc (Table 6.1 & 6.2). Microbes are also extensively used in research laboratories all over the world to investigate cellular and allied processes.

Microorganisms include a great diversity of living forms. Algae, fungi and bacteria were once placed in the plant kingdom, while animal like organisms, protozoans and some bacteria were kept in the animal kingdom. Viruses, are extremely minute "living things" that did not fit in animal or plant kingdom, because they are neither cells nor organisms. Haeckel (1866) proposed that all these microscopic unicellular forms into a third kingdom '*Protista*'. A more recent and comprehensive system of classification the five kingdom systems was proposed by Whittaker (1969), based on three principal modes of nutrition : Photosynthesis, absorption and ingestion. Those are *Plantae* (multicellular green plants and higher algae), *Animalia* (multicellular animals), *fungi* (yeasts, molds and multinucleate higher fungi), *Monera* (bacteria and cyanobacteria) and *Protista* (micro algae and protozoa).

Microorganisms were the first live organisms on Earth. The chemical evolution led to primitive forms of prokaryotic life that originated on the planet Earth around 3.5 billion years ago. The recent studies on microbial evolution and diversification revealed three lines of approach : 1. *Archaea* (Prokaryotes with cell walls that lack murien, having either bonds in their membrane phospolipids - methanogens, extreme halophiles, thermoacidophiles which diverged from other bacteria at a very early stage of evolution); 2. *Bacteria* (members of a group of diverse and ubiquitous prokaryotic, single celled organisms, organisms with Prokaryotic (pro = before, karyote = nucleus) single celled organisms, organisms with Prokaryotic cells i.e. cells lacking nucleus but nucleoid; 3. *Eucarya* (cellular organisms having a membrane bound nucleaus within which the genome of the cell is stored as chromosomes composed of DNA; include algae, protozoa, plants and animals. All these biological systems have the following characteristics :

(a) the ability to reproduce
(b) the ability to ingest or assimilate food substances and metabolise them for energy and growth.
(c) the ability to excrete waste products.
(d) the ability to react the changes in their environment.
(e) the susceptibility to undergo mutation.

Table 6.1 *Important Landmarks in the History of Microbiology*

Antony Van Leeuwenhoek (1632-1723)	First to observe, record and report microorganisms.
Edward Jenner (1749-1823)	Vaccination for small pox using cowpox vaccine.
Justus Van Liebig (1803-1873)	Phytochemical theory of fermentation
Jocab Henle (1809-1885)	Established principles of germ theory of diseases.
Louis Pasteur (1822-1895)	Germ theory of fermentation and diseases. Immunisation techniques.
Florence Nightingale (1820-1910) infection	She organised hospitals to minimises cross
Joseph Lister (1827-1912)	Isolated bacteria in pure culture
John Tyndall (1820-1893)	Developed fractional sterilization to kill spores (Tyndallization)
Fanny Hesse (1850-1934)	Aagar as solidfying material for microbial media used
Robert Koch (1843-1910)	Pure culture technique Koch's postulates and discovered a causative agent of anthrax and tuberculosis.
Paul Ehrilich (1854-1915)	Modern concept of chemotherapy
Hans Christian Gram (1853-1933)	Developed gram stain for differential bacterial staining.
August Von Wassermann (1866-1925)	Introduced complement fixation test for syphilis
Marhinus Willem Beijerinck (1851-1931)	Utilised principle of enrichment culture confirmed first finding of viruses.
Frederick W. Twort (1877-1950) and *Felix M. D. Merelle* (1873-1949)	Independently discovered bacteriophages and viruses that destroy bacteria.
Rous (1966)	Viruses and Cancer
Delbruck, Khorana & Nirenberg (1968)	Genetic code
Milstein, Kohler and Jerne (1984)	Immunology
Bishop and Varmus (1989)	Genetic basis of Cancer
Mullis (1993)	Polymerase chain reaction
Doherty and Zinkernagol (1996)	Recognition of virus infected cells by immune defenses
Prusiner (1997)	Prions

The fundamental microbiological work was contributed by Louis Pasteur (1822 - 1895). Pasteur took up the problem of souring of wine and beer and found that it is due to the growth of undesirable organisms. Pasteur further showed that wine did not get spoiled if it were held for a few minutes at 50 - 60^0 C i.e., Pasteurisation. Roberk Koch (1843-1910) established that germs were the causes and not the end products of diseases. Koch's postulates provided the method to identify the real causative agents of a disease. Koch's postulates are :

1. A specific organism should be found constantly in association with a disease.
2. The organism should be isolated and grown in a pure culture in the laboratory.
3. The pure culture when inoculated in to a healthy suceptible animal it should produce symptoms of the same disease.
4. From the inoculated animal it should be possible to recover the organism with the same characteristics as the original isolate.

Microbiology has emerged as one of the most important branch of the life sciences. It made significant contribution to basic biological sciences as well as in several applied areas, particularly medicine.

Some important applied areas of microbiology are listed as under :

1. *Medical Microbiology* : Causative agents of diseases, diagnostic procedures, preventive measures.
2. *Food Microbiology* : Food preservation and preperation of food borne diseases and their prevention
3. *Industrial Microbiology* : Production of medicinal products such as antibiotics andvaccines, fermentcd beverages, industrial chemicals, production of proteins and harmones by GEMs (Genetically Engineered Microorganisms).
4. *Geochemical Microbiology* : Coal, mineral and gas formation; prospecting for deposits of coal, oil and gas; recovery of minerals from low grade ores.
5. *Environmental Microbiology* : Microbial distribution in air, water and soil microbial role in recycling of biological elements-water purification microbiological examination-biological degradation of wastes - contamination of air dissemination of diseases.
6. *Agricultural Microbiology* : Rhizosphere–Effects as plant growth. Root nodulation. Decomposition of organic and inorganic substances. Root pathogenic substances. Root pathogenic microbes, plant diseases.
7. *Microbial Biotechnology* : Microbial role in synthesis of important chemicals, acetone and acetic acid. Genetic engineering techniques for cloning of pharmaceutically important polypeptides in to microbes, which may be then produced on a large scale millions of copies of the same single cell in shorter duration at low cost.
8. *Dairy Microbiology* : Milk as bacteriological medium. Milk proteins, carbohydrates (lactose), fats, minerals, and vitamins relation to enzymatic degradation by microbes preservation of milk free from pathogenic microbes preperation of curds, yogurt, suncreem, butter and cheese by microorganisms.

Table 6.2 *Important Microbial diseases their causative agents and discoverers*

Year	Disease /infection	Causative microscope	Discoverer
1876	Anthrax	*Bacillus anthracis*	Koch
1879	Gonorrhea	*Neisseria gonorrhoeae*	Neisser
1880	Typhoid fever	*Salmonella typhi*	Eberth
1880	Malaria	*Plasmodium sp.*	Laveran
1881	Wound infections	*Staphylococcus aureus*	Ogostan
1882	Tuberculosis	*Mycobacterium tuberculosis*	Koch
1882	Glanders	*Pseudononas mallei*	Loeffler and Schytz
1883	Cholera	*Vibrio cholerae*	Koch
1883	Diphtheria	*Corynebacterium diphtheriae*	Klebs and Leffler
1885	Tetanus	*Clostridium tetani*	Nicolaier
1886	Bacterial Pneumoniae	*Streptococcus pneumoniae*	Fraenkel
1887	Meningitis	*Neisseria meningitidis*	Weichselbaum
1887	Malta fever	*Brucells Sp.*	Bruce
1888	Equine strangles	*Streptococcus Sp.*	Schutz
1892	Gas gangrene	*Clostridium perfringens*	Welch and Nuttall
1894	Plague	*Yersinia pestis*	Kiotasato and Yersin
1895	Fowl typhoid	*Salmonella gallinarum*	Moore
1896	Botulism	*Clostridium botulinum*	Van Ermengem
1898	Dysentery	*Shigella dysenteriae*	Shiga
1905	Syphilis	*Treponema pallidum*	Schaudin and Hoffman
1906	Wooping cough	*Bordetella pertussis*	Bordet and Gengou
1909	Rock mountain spotted fever	*Rickettesia Rricketsii*	*Rrickets.*

6.1 Bacteria

Bacteria are the prokaryotic microorganisms and are far distinct from Eukaryotic organism structurally and functionally. The concept of a bacterium as a prokaryotic organism was established in microbiology in 1962 by Stamir van Niel. Whittakar (1969) proposed five kingdom systems of microorganisms. Recently, the evolutionary relationships of bacteria have been defined on the basis of ribosomal RNA sequences. In a modern sense, bacteria, cyanobacteria, actinomycetes are distributed in the domain of bacteria e.g, methanogenic bacteria and other extremophiles. Although the bacterial structures are quite simple, but their classification is very difficult. Currently, the bacterial nomenclature is regulated by a committee on systematic nomenclature. The bacterial taxonomy has a rapid and radical change due to involvement of modern molecular biological techniques. The use of computers in bacteriology in general and taxonomy in particular gave an added advantage to restore the impact of the techniques of numerical taxonomy. "Bergey's (957) Manual of Systematic Bacteriology" 7th edition in four volumes is internationally recognised book containing names and description of bacterial species.

Characteristics

The size, shape of and arrangement of bacterial cells vary with species to which they belong. Bacteria are about 0.1 to 60 × 6 nm in size. It has the rigid cell wall that determine the shape of bacterial cell. The bacterial cells are spherical, elongated helical rods, pear shaped, lobed sphere, and also rods with squared ends. Arrangement of cells are more complex in cocci than bacilli. It also depends on adherence of cells together following the cell division Fig. 6.1.

eg : Monococcus (single); diplococcus (pairs); streptococcus (chains); tetracocci (4 celled) sarcinae (8 celled), staphylococcus (bunches).

Similarly *Bacillus* also show the cell arrangements of mono, diplo, strepto bacillus, coccobacillus, vibrioid (curved) spiral (helical) forms. The cyanobacterial cells are comparativety larger than the bacterial cells.

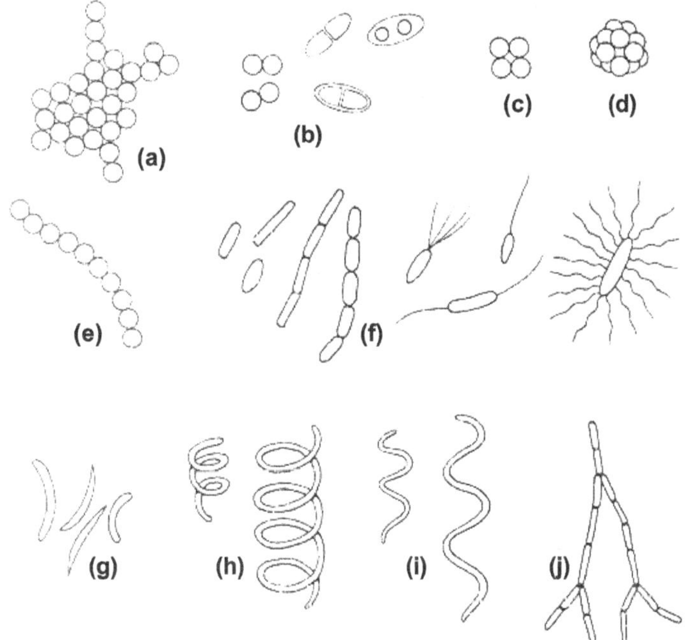

Fig. 6.1 The shapes of bacteria
(a) Various Staphylococci (b) Diplococci (some with capsules) (c) Tetrads (d) Sarcinae (e) Streptococci (f) Bacilli, some with flagellae (g) Vibrios (h) Spirillae (i) Spirochaetes (j) Actinomycetes

Cell Structure

Microscopic observation reveal that the structural components outside and inside the cell wall differ considerably from species to species hence characteristic (Fig. 6.2).

Capsule

Some of bacterial cells are surrounded by Extra Cellular Polymeric (ECP) substances which are called capsule. Capsule forms an envelop around the cell wall and can be observed under a light microscope after special staining technique.

The capsule is gelatinous polymer wall made up of polysaccharides, protein etc. Capsule in certain species is specific and can be used for immunological differentiation of related species. It is sticky in nature and secreted from the inner side of cell wall, which gets firmly attached to the surface of the cell wall. The marine bacteria form trichome, and are enclosed with gelatin matrix.

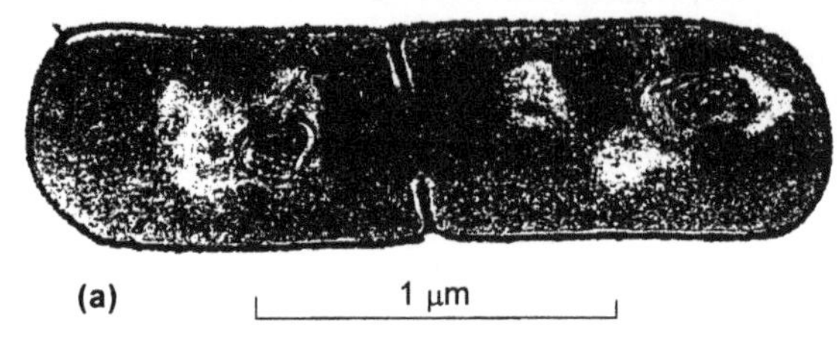

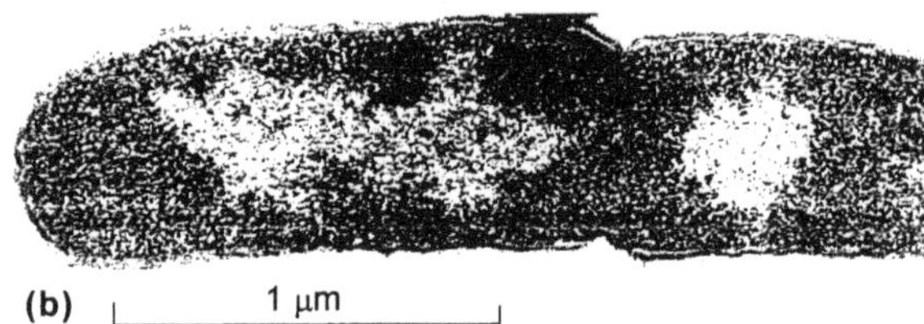

Fig. 6.2 Electron micrographs showing sections of (a) *Bacillus subtilis* and (b) *Escherichia coli* during transverse fission. *B. subtilis* is a gram-positive bacterium possessing a thick electron dense cell wall. Fission is by the ingrowth of the cell wall. *E. coli*, a gram-negative bacterium, has a thinner and more flexible cell wall, permitting fission to occur by constriction. In both species these are relatively electron transparent regions occupied by nuclear material. In addition, mesosomes are clearly visible in the *B. subtilis* cells.

Functions of Capsule

1. Prevent the attachment of bacteriophages (viruses eating bacteria).
2. Protect from environmental barriers.
3. Inhibit the engulfment by White Blood Corpuscles (WBC).
4. Capsule protects the cell from dessication, maintains the viscosity and inhibit the movement of nutrients from cell.

Flagella

It is a hair like, helical and surface appendage. It is of 20 - 30 nm in diameter and 15nm long. It provides the motility to the bacterium.The arrangement of flagella may be

monotrichous, ; diplotrichous ; peritrichous ; lophotrichous ;

amphitrichous .

A flagellum consists of 3 basic parts (a) basal body, (b) hook and (c) filament. Ultra structure of flagellar basal bodies and hooks in Gram-negative and Gram-positive bacteria are shown in (Fig. 6.3).

There are 3 types of movement in bacteria.

(a) Flagellar movements,

(b) Spirochaetal movement,

(c) Gliding movement,

Pili

It is a hair like appendage found on Gram negative bacteria. (Fig. 6.4)

They are distributed on the surface of the cell. Pili measure 0.2 - 20 nm long with 250A^0 diameter. Based on their functions, pili are of two types :

I - pili → for attachment

F - pili → for formation of conjugation tube and transfer of DNA from donor to recipient cell during conjugation.

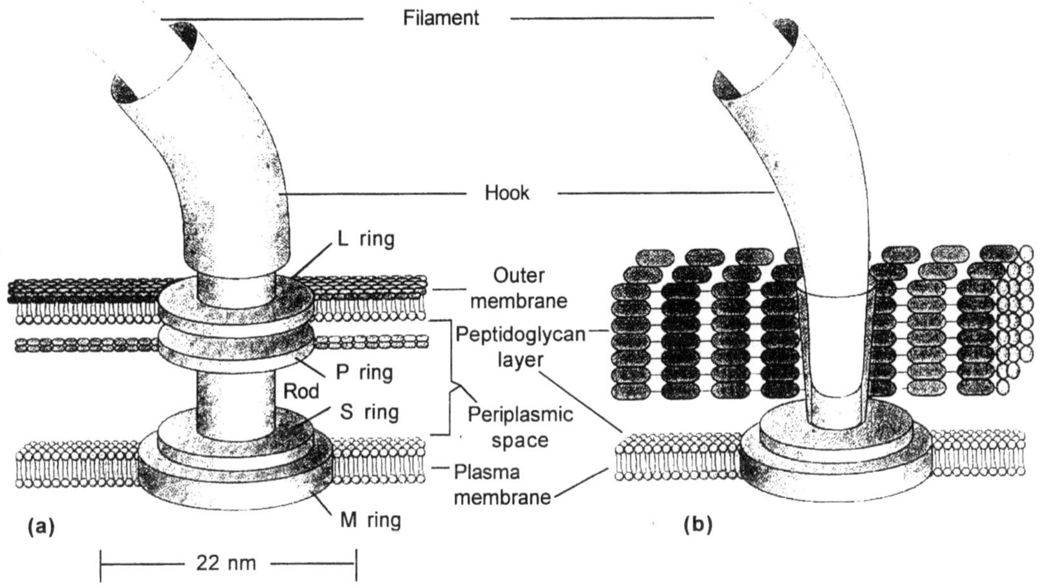

Fig. 6.3 The Ultrastructure of Bacterial flagellae. Flagellar basal bodies and hooks in (a) Gram-negative and (b) Gram-positive bacteria.

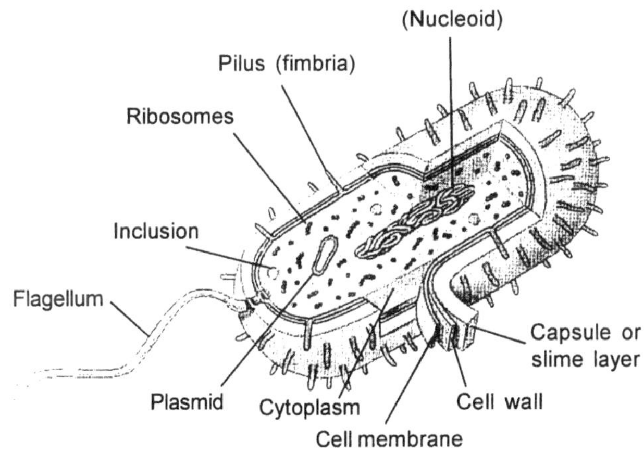

Fig. 6.4 Gram-negative Bacterial Cell with Pilus (fimbria)

Cell Wall

Cell wall is semi rigid with complex structure beneath the capsule.

The cell wall protects the cell from environmental barriors, like osmatic pressure, temperature, dessication, etc. It provides attachments to the flagella. It rescues the cell from antibodies and harmful chemicals. Cell wall of Gram negative bacteria is comparativety thinner than Gram positive bacteria (Table 6.3).

Table 6.3 *Differences between cellwalls of Gram positive and Gram negative bacteria*

Character	Gram positive	Gram negative
Gram staining	Retain crystal violet	Pass crystal violet
Peptidoglycan	Several layer thick	Thin
Lipid and lipo-proteins	Low	High
Lipopolysaccharides	Absent	High
Teichoic acid	Present	Absent
Periplasmic space	Absent	Present
Susceptibility to penicillin and sulfonamide	High TN	Low
Nature of Toxin Production	Exotoxin	Endotoxin

Chemical Nature

The cell wall of bacteria is made up of network or peptidoglycan except in halobacterium (marine bacterium).

Peptidoglycan determines the shape of a cell. Peptidoglycan consists of alternately arranged sub-units of NAM (N-acetyl muramic acid) and NAG (N-acetyl glucosamine) (Fig. 6.5). The Gram negative bacterial cell wall also contain peptidoglycan with low concentration 5-10% and remain with lipopolysaccharides and lipoproteins. Lipoproteins functions for development of diffusion channels. Lipid A is toxin, when present in blood is responsible to induce fever and shocks. Polysaccharide is composed of 3 components. O-antigens region, inner core and outer core regions and these chains have antigenic property and therefore distinguished serologically. The outer membrane consists of two molecules of glucose sandwitching a molecule of galactose. The structural variation in gram-positive and gram-negative cell walls is shown in Fig. 6.6.

Function

Peptidoglycon provides the structural integrity to the cell.

Cell envelop acts as a barrier and matrix portion as receptor sites (Fig. 6.7). O-antigen determines the specificity of Gram negative bacteria. There are few groups of microorganisms that have no cell wall. eg. Myloplasma sps.

Plasma Membrane

Also called cell membrane, is a structure internal to the cell wall, and no cell could live without a plasma membrane.

It is situated beneath the cell wall and is about 7.5 nm thick. It consists of proteins (20-70%), lipids (20-80%), polysaccharides (1.51%) water 20 - 21%. The plasma membrane consists of bilayer of phospholipid in which globular proteins are embeded.

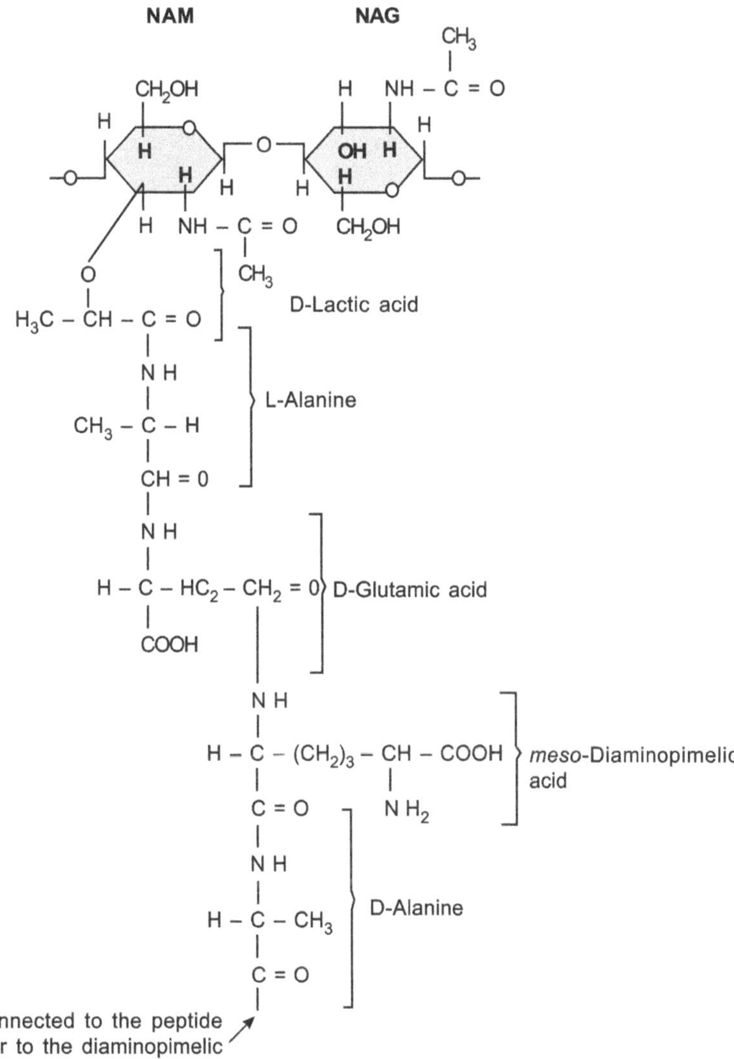

Fig. 6.5 Peptidoglycan subunit composition. The Peptidoglycan subunit of *Escherichia coli*, most other gram-negative bacteria, and many gram-positive bacteria. NAG is N-acetylglucosamine. NAM is N-acetylmuramic acid (NAG with lactic acid attached by an ether linkage). The tetrapeptide side chain is composed of alternating D- and L-amino acids since meso-diaminopimelic acid is connected through its L-carbon.

Function

Transportation of organic and inorganic nutrients is the chief function.

It consists of enzymes for biosynthetic pathways. It provides permeability barrier. The inner membrane invaginates to form mesosomes.

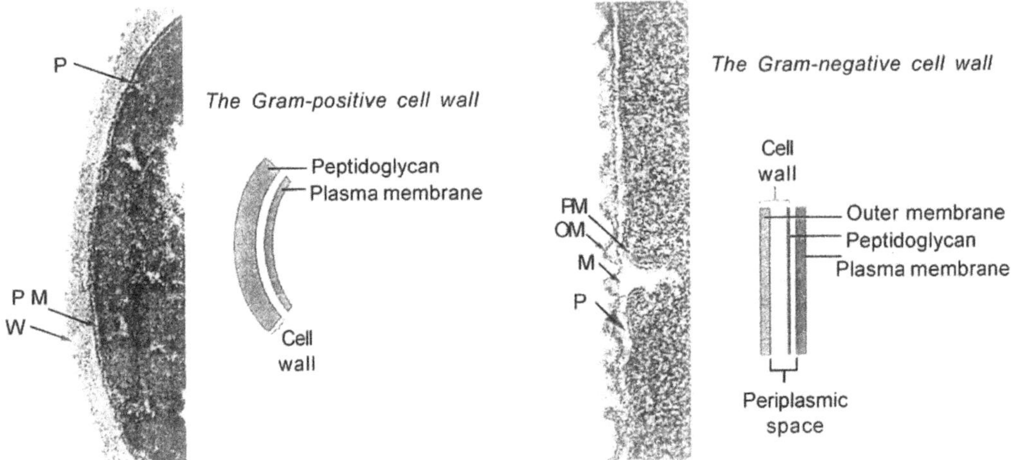

Fig. 6.6 Gram-positive and Gram-negative cell wall. The gram-positive envelope is from *Bacillus licheniformis* (left), and the gram-negative micrograph is of *Aquaspirillum serpens* (right). M; peptidoglycan or murein layer; OM, outer membrane; PM, plasma membrane; P, periplasmic space; W, gram-positive peptidoglycan wall.

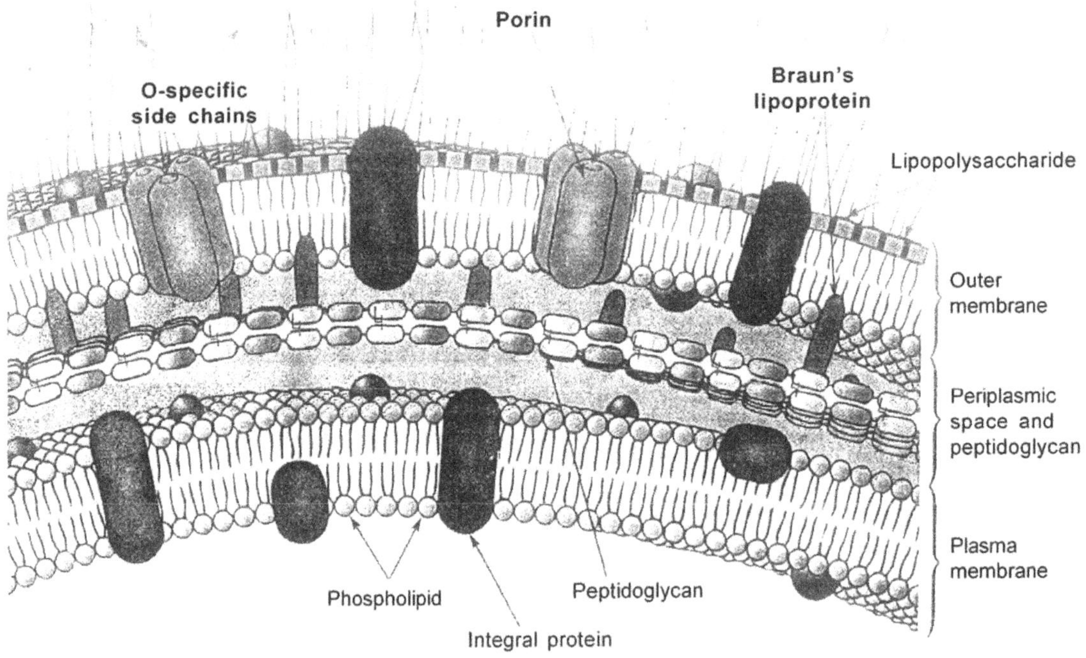

Fig. 6.7 The Gram-Negative Envelope

Mesosomes

These are the invaginated structures formed by localized infoldings of the plasma membrance. It comprises of vesicles, tubules, lamellar whorls. Generally mesosomes are found in association with nuclear area near the site of cell membrane. Mesosomes take part in respiration but they are not analogous to mitochondria. Mesosomes play a role in reproduction during fission. It also separates the bacterial DNA into each daughter DNA.

Cytoplasm

The cytoplasm consists of water, proteins, carbohydrates, lipids, inorganic salts and ions.

Ribosomes

Bacterial cells contain ribosomes of 70S, and acts as site of protein synthesis. Cytoplasm of prokaryote contain about 10,000 ribosomes which is upto 30% of dry weight, 70S ribosome have the subunits of 30S (smaller) and 50S (larger) later they get divided into RNA and protein.

Nucleoids

Bacterial chromosome is called nucleoid and lack mitotic apparatus. DNA is the genetic material not enclosed by any covering. Some bacterial cells consists of extra chromosomal DNA called P-DNA. There are R plasmid, Col plasmid, S plasimids etc.

Cytoplasmic inclusions

Prokaryotic cells contain several reserve deposits called inclusions : These are volutin granules, Polysaccharide granules, Lipid inclusions, Polyhedral bodies, Sulphur granules, Gas vacuoles and chromotophore vesicles.

Spores and Cysts

Certain bacteria produce metabolically dormant structures known as cysts. *eg : Endospores.*

Nutrition, growth and metabolism of Bacteria

The major biological function of the microbial cells is growth. Growth is the orderly increase in mass and number of all components of the cell. All living organisms require nutrients for their growth as well as for cellular maintenance and multiplication. The nutrients provide the energy required by the microbial cell to synthesize new cellular components and reproduce. Bacterial cell consists of protein, chromosomes, enzyme membranes, cell wall and flagella. These components are made up of large (macro) molecules. The main function of the bacterial metabolism is the synthesis of macromolecules.They are derived from their respective subunits. Proteins from amino acids ; nucleic acids from nucleotides; polysaccharides from simple sugars and lipids from glycerol, alcohols and fatty acids. Bacteria differ greatly in their needs of nutrients, basically, they obtain their nutrients from the environment in which they live. Bacteria are called osmotrophs because they obtain their food in solutions that diffuse through their plasma membranes. They cannot however, transport macromolecules such as proteins or polysaccharides through their membranes. These molecules first digested outside the cell by exoenzyme that are excreted by the cell to the internal environment.

For, *eg* : enzyme protease will degrade protein into amino acids which are then absorbed. The absorbed nutrients are metabolised to produce energy rich compounds. eg : Adenosine-tri-phosphate (ATP) is chemically converted to meet the specific nutritional requirements of the bacterium. The construction of the cellulose components requires the ATP as the source of the energy.

Nutritional Requirements of Bacteria

Nutritional requirements of bacteria are often satisfied in different ways. Some bacteria are capable of synthesizing their food from simple elements within their environment. Most of the pathogens do not able to do this. They therefore depend on a free supply of ready made or organic compounds for their growth and metabolic activity.

The essential nutritional requirements of bacteria are :

Water

All metabolic functions of a bacterial cell are performed in aqueous environments. Bacterial cultivation and growth requires water.

Energy Source

Bacterial cellular functions are synthesis of new components, concentration of nutrients in the cell.

Cell mobility require energy. Some of them obtain from sunlight, oxidation of chemical compounds are referred phototrophs and chemotrophs respectively. The phototrophic bacteria are divided into cyanobacteria and photosynthetic bacteria ; the former are able to produce O_2 as a by product of photosynthesis. Chemotrophic bacteria get their energy from the oxidation of organic and inorganic compounds, such as glucose. These oxidations may be by fermentation or respiration.

Carbon requirement

Carbon (C) atoms are essential in the nutritional requirements of bacteria since they take part in cellular metabolism. They participate in the biodegrading (catabolic) activities which result in the production of ATP and the biosynthetic (anabolic) activities that result in the synthesis of cellular components. Some bacteria derive their requirements of carbon from carbon dioxide of the atmosphere. These organisms are called autotrophs. They are mainly free living bacteria. Those bacteria that depend on freely supplied organic molecules for their source of carbon are called heterotrophs. These are parasitic and evolved to adapt to the environment and most of the human pathogenic bacteria belong to this group.

Nitrogen requirements

Nitrogen is necessary for the synthesis of essential amino sugars, nucleotide and vitamins. Microorganisms are able to utilize the various nitrogenous compounds which are present in the structural components of cell wall, to satisfy their needs.

Growth

Bacterial growth is obtained usually by binary fission.

Based on the nutritional status, the bacteria will grow on the surface of culture medium by utilizing the ingredients and preferable sources.

Growth of a bacterium is studied under the different phases.

(a) lag phase - initial phase
(b) log phase - exponential phase
(c) stationary phase - static phase
(d) decline phase - death phase

Growth may also depends on several environmental conditions like pH, temperature, osmosis, diffusion etc. The metabolic activities of bacteria are signified with membrane. Several enzymes and proteins are present in it. Respiratory chain consists of several metabolic enzymes, co-enzymes that perform the respiration, fermentation and some times nitrogen fixation.

Important Applications of Bacteria

1. Bacteria are widely employed in fermentation technology.
2. Biodegradations, Bacteria are widely used in various industrial processes.
3. Dairy industry depends on various bacteria for milk and by-products.
4. Nitrogen fixation is carried out by number of soil bacteria.
5. Biofertilizers; several algal and fungal species are used as biofertilizers.
6. Genetic Engineering is accompanished by several bacterial species.

6.2 Viruses

"An infectious agent having a simple acellular organisation with a protein coat and a single type of nucleic acid, lacking independent metabolism, and reproducing only within living host cells virus"

A complete virus particle or virion consists of one or more molecules of DNA or RNA enclosed in a coat of proteins, and some times also in other layers. Viruses differ from living cells in three ways, (i) their simple cellular organisation, (ii) the absence of both DNA and RNA in the same virion, (iii) their inability to reproduce independent of cells and carry out cell division as prokaryotes and eukaryotes.

Characteristics of Viruses

1. Virions range in size from about 10-300 or 2100 nm in diameter (Fig. 6.8).
2. Virions are constructed around a nucleocapsid core. The nucleocapsid is composed of a nucleic acid, either DNA or RNA, held within a protein coat called the capsid, which protects viral genetic material and aids in its transfer between host cells.
3. Some capsids are icosohedral or helical in shape. Many viruses have an envelope, an outer membranous layer, other structures (eg. many bacteriophages) or have complex multilayered walls surrounding the nucleic acid (small pox virus).

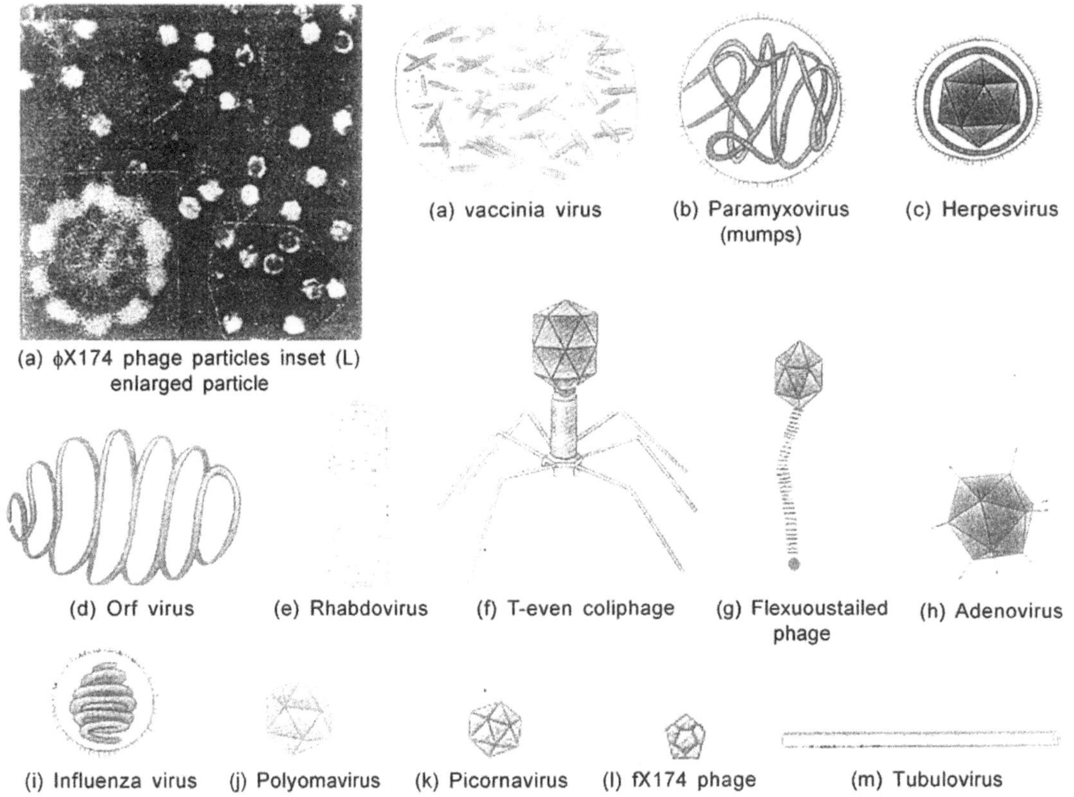

(a) ϕX174 phage particles inset (L) enlarged particle
(a) vaccinia virus
(b) Paramyxovirus (mumps)
(c) Herpesvirus
(d) Orf virus
(e) Rhabdovirus
(f) T-even coliphage
(g) Flexuoustailed phage
(h) Adenovirus
(i) Influenza virus
(j) Polyomavirus
(k) Picornavirus
(l) fX174 phage
(m) Tubulovirus

Fig. 6.8 The size and morphology of selected viruses.

In Fig. 6.9, Tobacco Mosaic Virus (TMV) first isolated by Dr. W. M. Stanley in 1953. It is pure RNA and forms crystal. Electron microscopic studies have shown that around RNA coil proteins of units are arrange as knobs called capsids (Fig. 6.9).

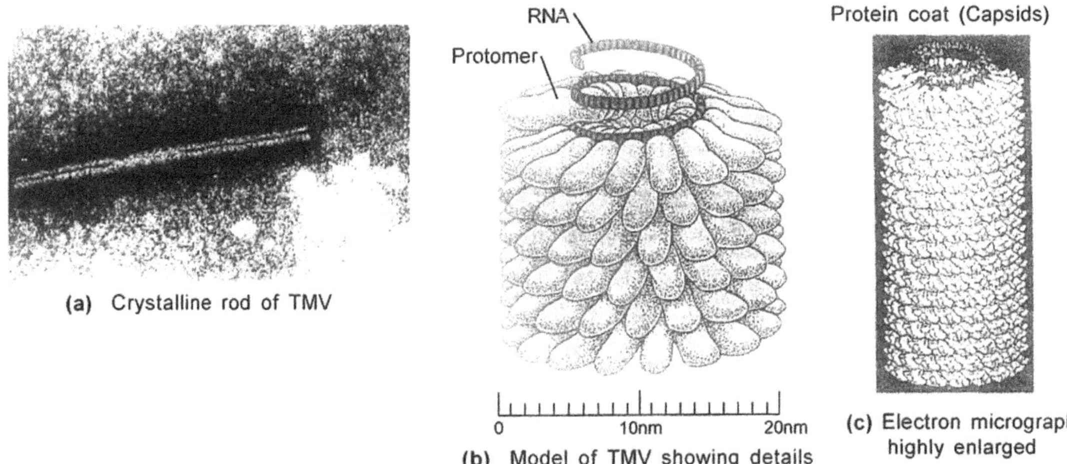

RNA
Protomer
Protein coat (Capsids)
(a) Crystalline rod of TMV
0
10nm
20nm
(b) Model of TMV showing details
(c) Electron micrograph highly enlarged

Fig. 6.9 Tobacco mosaic virus (TMV) structure

4. *Bacteriophage is a virus whose host is a bacterial cell. Phages are of several kinds.*
   (a) T-phages, numbered T_1 to T_7 are *E.coli* phages are tad pole like with a hexagonal head and a tail. The T phages may be even (T_2, T_4, T_6) or uneven (T_1, T_3, T_5, T_7). (Fig. 6.10).
   (b) P_{22} is a strain of phage that infects *Salmonella*.
   (c) Rhizobiophages are associated with root nodule bacteria, *Rhizobium*.
   Phages may be virulent (intemperate) causing death of the phage releasing 100-1000 phages. This is a lysogenic cycle. Phages may be temperate, with a prophage and are lysogenic, lamda is a temperate phage.
   (d) Phages $\phi \times 174$ also have been studied, has a single stranded DNA and its protein coat isicosahedral (with 20 plane surfaces) and is devoid of tail.

The *virulent bacteriophage* : The bacteriophage particle is more complicated than the poliovirus. It has a "*head*" and a "*tail*". The head is a hexagonal prism, composed of a ball of DNA surrounded by a protein coat. The tail is a long cylindrical structure, which is composed of (a) a contractile outer sheath, (b) an internal tube, and (c) a tip.

The bacteriophage is not motile. But when the tip of its tail comes into contact with the bacterial wall, the bacteriophage becomes attached to the bacterium (Fig. 6.8). An enzyme present at the extremity of the phage tip attacks the bacterial wall. The external sheath of the tail contracts. The internal tube penetrates through the membrane inside the bacterium. The genetic material is injected through the tube into the bacterial cytoplasm. All this happens within a minute. Then the vegetative phase takes place. The synthesis of phage is initiated and 5 minutes after the infection the genetic material of the bacteriophage starts multiplying. After 12 to 20 minutes, the phage DNA is condensed into a ball, and the subunits of the head are assembled around the ball of nucleic acid. Finally, by a process that is not fully understood, the tail appears where it should appear. The infectious particle is formed.

In the meantime, a peculiar phage protein has been produced, the endolysine, which in the particle will be localized at the tip of the tail. This endolysine depolymerizes the mucopolysaccharides of the bacterial wall. The bacterium is lysed and liberates some hundred bacteriophage particles. This completes the "*lytic cycle*" of the bacteriophage. Some bacteriophages, called *virulent* phages, are multiplied only by the lytic cycle, others, called *temperate* phages, may evolve in two directions.

The temperate bacteriophage; Lysogeny. After a bacterium has been infected with a temperate phage, the vegetative phase may be started, which culminates in the production of infectious particles and bacterial death. Sometimes, however, the vegetative phase does not started, and the bacterium survives. The survivors multiply normally (Fig. 6.11). But they are now endowed with a remarkable property. Each surviving bacterium possesses and perpetuates the potentiality of producing bacteriophage particles in the absence of infection. When this potentiality is not expressed, the bacterium grows and divides. When it is expressed, it dies. The bacteria that perpetuate the possibility of producing bacteriophage in the absence of infection are called *lysogenic* (Fig. 6.13).

The Prophage. A given bacterial species can be infected by a number of different strains of temperate bacteriophages, A or B, which might differ by a number of properties such as host range,

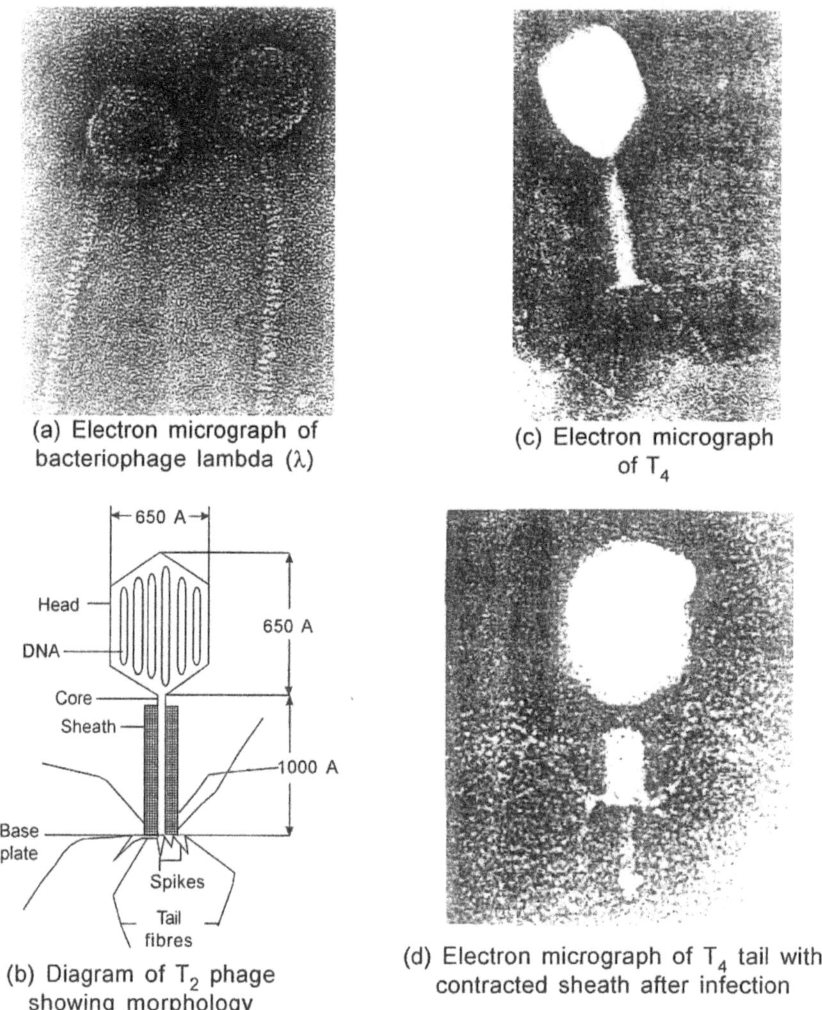

(a) Electron micrograph of bacteriophage lambda (λ)

(c) Electron micrograph of T_4

(b) Diagram of T_2 phage showing morphology

(d) Electron micrograph of T_4 tail with contracted sheath after infection

Fig. 6.10 Electron micrograph of bacteriophage lambda (λ), T_2 and T_4

immunological specificity, size, shape, etc. A given bacterium infected and lysogenized by a temperate bacteriophage A will perpetuate the potentiality to produce bacteriophage A. When infected with B, it will perpetuate the potentiality to produce B. The lysogenic bacterium multiplies into a specific structure: the prophage (etymologically, "*before the phage*").

A number of experimental data have led to three essential conclusions.

1. The prophage is the genetic material of the bacteriophage. It carries the information for the production of bacteriophage particles.
2. There is only one prophage per bacterial chromosome.
3. A given prophage is always attached to the same site, or locus, of the bacterial chromosome.

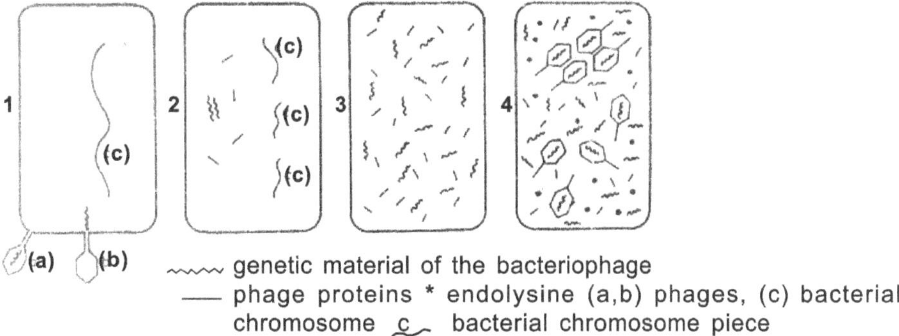

Fig. 6.11 The Life Cycle of a Bacteriophage

Table 6.4 *Types of viral nucleic acids*

Nucleic Acid Type	Nucleic Acid Structure	Virus Examples
DNA		
Single - Stranded	Linear single strand	Parvoviruses
	Circular single strand	ϕ X 174, M13, fd phages
Double - Standard	Linear double strand	Herpesviruses (herpes simplex viruses, cytomegalovirs, Epstein - Barr virus), adenoviruses, T coliphages, lambda phage, and other bacteriophages
	Linear double strand with single chain breaks	T5 coliphage
	Double strand with cross - linked ends	Vaccinia, smallpox
	Closed circular double strand	Papovaviruses (polyomavirus, human papilloma viruses, SV40), PM2 phage, cauliflower mosaic
RNA		
Single - Stranded	Linear, single stranded, positive strand	Picornaviruses (polio, rhinoviruses) togaviruses, RNA bacteriophages, TMV, and most plant viruses
	Linear, single stranded, negative strand	Rhabdoviruses (rabies), paramyxoviruses (mumps, measles)
	Linear, single stranded, segmented, positive strand	Brome mosaic virus (individual segments in separate virions)
	Linear, single stranded, segmented, diploid (two identical single strands), positive strand	Retroviruses (Rous sarcoma virus human immunodeficiency virus)
	Linear, single stranded, segmented, negative strand	Paramyxoviruses, orthomyxoviruses (influenza)
Double - Stranded	Linear, double stranded, segmented	Reoviruses, wound-tumor virus of plants, cytoplasmic polyhedrosis virus of insects, phage j6, many mycovirusess

The host range, immunologic relationships, phage morphology and nucleic acid properties used in classifying phages.

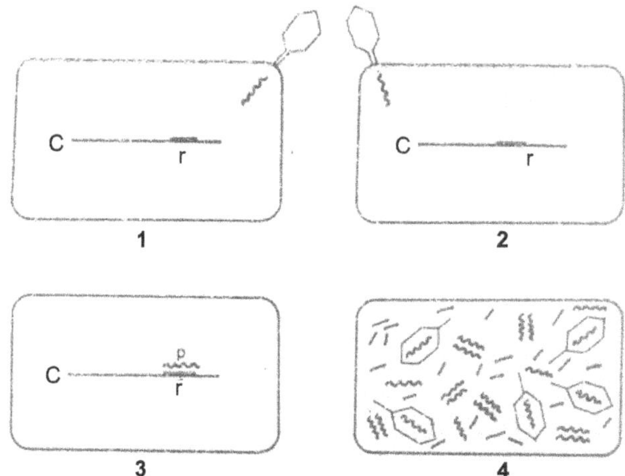

Fig. 6.12 The fate of an Infected Nonlysogenic Bacterium

Types and groups of virus : - (Table 6.4 & 6.5)

1. Viruses are flexible with respect to the nature of their genetic material. They employ four possible nucleic acid types : ss DNA, ds DNA, ss RNA and ds RNA.
2. After DNA bacteriophages have reproduced within the host cell many of them are released when the cell is destroyed by lysis. A phage life cycle that cumulates with the host cell bursting and releasing virion is called a lytic cycle (Fig. 6.12).
3. Lysogeny is reversible, and prophage can be induced to become active again and lyse the host.
4. Although most temperate phages exists as integrated prophages the phage P_1 can lysogenise *E.coli* even though its DNA remains separate from the bacterial chromosome.
5. The important step in animal virus life cycle is adsorption of the virus to a target cell receptor site, often special capsid structures are involved in this process.
6. Virus penetration of the host cell plasma membrane may be accomplished by capsid removal from the nucleic acid or uncoating.
7. Almost all plant viruses are RNA viruses, either single stranded or double stranded. Caulimo viruses and gemini viruses with their DNA genomes are exception to this rule.
8. The uses of phages in molecular cloning are : Cloning vehicles, enzymes, DNA sequencing, gene expression and intrabacterial manipulations.
9. Viroids are a unique class of acellular self replicating entities that differ viruses in not processing a protein coat. They consist of only naked RNA and therefore lack the dormant phase of characteristic of viruses.
10. Prions represent the first exception of the rule that all self-replicating entities contain nucleic acid as the genetic material, as these contain only proteins.
11. Prions may specify their own amino acid sequence by reverse translation to yield a nucleic acid that is normally translated by the cellular system.

Table 6.5 *Some common virus groups and their characteristics*

Nucleic Acid	Strandedness	Capsid symmetry[a]	Presence of Envelope	Size of Capsid (nm)[b]	Number of Capsomers	Virus Group	Host Range[c]
RNA	Single	I	–	22-30	32	*Picornaviridae*	A
		I	+	40-75(e)	32	*Togaviridae*	A
		I	+	100(e)		*Retroviridae*	A
		H	+	9(h) 80-120(e)		*Orthomyxoviride*	A
		H	+	18(h), 125-250(e)		*Paramyxoviridae*	A
		H	+	14-16(h), 80-160(e)		*Coronaviridae*	A
		H	+	18(h), 70-80× 130-240 (bullet shaped)		*Rhabdoviridae*	A
		I	–	25		*Bromovirus*	P
		I	–	30		*Cucumovirus*	P
		H	–	18 × 300		*Tobamovirus*	P
		I	–	26-27	32	*Leviviridae [QB]*	B
RNA	Double	I	–	70-80	92	*Reoviridae*	A.P
		I	+	100(e)		*Cystoviridae*	B
DNA	Single	I	–	20-25	12	*Parvoviridae*	A
		I	–	18 × 30 (paried particles)		*Geminivirus*	P
		I	–	25-35		*Microviridae*	B
		H	–	6 × 900 1,900	–	*Inoviridae*	B
DNA	Double	I	–	45-55	72	*Papovaviridae*	A
		I	–	60-90	252	*Adenoviridae*	A
		I	+	130-180		*Iridoviridae*	A
		I	+	100, 180-200(e)	162	*Herpesviridae*	A
		C	+	200-260× 250-290(e)		*Poxviridae*	A
		H	+	40×300(e)		*Baculoviridae*	A
		C	+	28 (core), 42(e)	42	*Hepadnaviridae*	A
		I	–	50		*Caulimovirus*	P
		I	–	60		*Corticoviridae*	B
		Bi	–	80×110, 110[4]		*Myoviridae*	B

(a) Types of symmetry : I = Icosahedral; H = Helical; C = complex; Bi = Binal

(b) Diameter of helical capsid (h); diameter of enveloped virion

(c) Host range : A = animal; P = plant; B = bacterium.

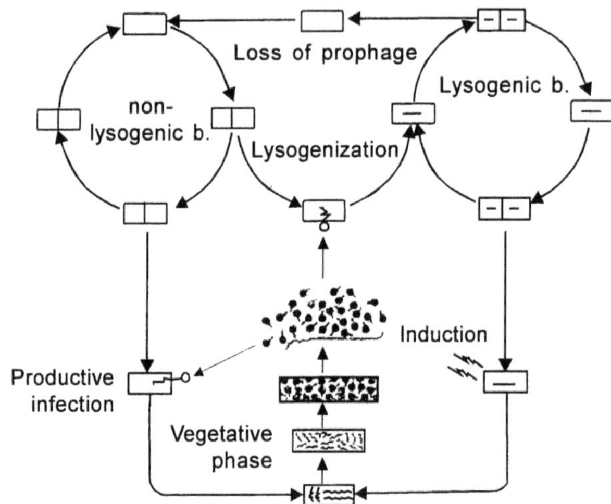

Fig. 6.13 Digrammatic Representation of Lysogeny

Prions

Prions are proteinaceous infectious agents. Prions cause several types of neuro degenerative diseases affecting humans and livestock.

According to Stanley Prusiner (1982), University of California, USA, prions can act as infectious agents composed entirely of proteins. In 1987 he won Nobel Prize for this discovery.

The idea ofprion-based disease is still somewhat controversial, though gaining much experimental support. According to the prion hypothesis, such diseases arise from the ability of the prion protein to exist in two conformational states : a normal form PrP^{C}, which does not cause disease, and an abnormal form PrP^{SC}, which does. The normal prion protein is encoded by normal individuals genomes, and the protein is expressed at low levels. The abnormal conformation of the prion protein can come from two sources. An individual can be infected with the abnormal protein by direct contact with an affected individual or by eating raw meat from an organism with the disease. Alternatively, some people carry alleys of PrP gene that cause the normal prion protein to convert spontaneously to the abnormal conformation at a very low rate. These individuals have an inherited predisposition to developing a prion-related disease. An example of an inherited prion disease is fatal familial insomnia.

Neurodegenerative diseases caused by prions are as follow :

1. *Infectious diseases :*
   - (a) *Kuru :* Human disease once common in New Guinea leads to loss of co-ordination dementia. Infection spreads by cannabolism that ended in 1958.
   - (b) *Scrapie :* Disease of pigs and sheep, characterised by intense itching, the animals scrape against trees, finally leading to neuro degeneration and death.
   - (c) *Mad cow disease :* Changes in posture and temperment, loss of co-ordination and neuro degeneration.

From the molecular mechanism by which prions cause disease, the prion protein is predicted to exist in two conformations, PrP^{SC} and PrP^{C}.

2.*Human inherited diseases :*

(a) Creutz feldt Jacob disease leads to loss of co-ordination and dementia (loss of memory)

(b) Gerstmann - Straussler Seheinker disease leads to loss of co-ordination and dementia.

(c) Familial fatal insomnia begins with sleeping and autonomic nervous system disturbance followed by insomnia and dementia.

Plasmids : Plasmids are extrachromosomal, autonomous, self replicating genetic particles and are double stranded when in host. These were first discovered in Japan in 1950 when bacteria simultaneously acquired resistance to several antibiotics. R 222 was the first plasmid studied. There are various types of plasmids; 1. R plasmid confers resistance to antibiotics called as RTF, 2. Col plasmid confers resistance to toxins called colicin and kappa 3. F plasmid (episome) carries fertility factor called F. All plasmids are circular DNA molecules occurring in microorganisms measuring 1kb (=1000 base pairs) to 200 kb in size. They carry an operation for transfer and insertion sequences for attachment to cell chromosome and each other and are highly mobile segments of genetic material. The first plasmid used in genetic engineering (cloning) was *E.coli* pSC101 which contains single Eco R1 cleavage site with tetracycline resistance gene. The plasmid pBR 322 is widely used cloning vector with ampicillin and tetracycline gene and numbers of restriction enzyme cleavage sites. Derivatives of this called second and third generation vectors are generally used in cloning experiments.

Cosmids : Charon phages, plasmids are hybrids between plasmids and a lamda chromosome. These vectors or chimeric plasmids contain Cos sites at each end (Cohesive sites). The Cos has 12 base complementary single stranded termini on the mature lambda (λ) chromosome. The Cos site is recognised (λ) DNA packaging to produce complimentary cohesive ends of mature lamda chromosome. The advantages of cosmids are : their ability to replicate autonomouly in *E. coli* cells and in vitro packaging capacity of the lambda chromosome. These are useful for cloning large segments of foreign DNA (2.5 kb to 50 kb).

(Details of recombination in microorganism and related aspects are not covered for want of space).

6.3 Actinomycetes

The Actinomycetes are a large group of filamentous bacteria, usually Gram positive, that forms branching filaments. As a result of successful growth and branching, a ramifying network of filaments is formed called mycelium. Although it is of bacterial dimensions, the mycelium is in some way analogous to the mycelium formed by the filamentous fungi. Most actinomycetes form spores; The manner of spore formation varies and is used in separating sub groups (Fig. 6.14).

The DNA base compositions of most members of the actinomycetes fall within the range of 63–78 % G +C. Organisms at the upper end of this range have the highest GC percentage of any bacteria known. Phylogenetically, the filamentous actinomycetes form a coherent group; thus the mycelial spore forming habit is of both phylogenetic as well as taxonomic importance.

Some bacteria in this group such as of *Nocardia sps*. exhibit only a limited degree of growth. *Actinoplanes* and *Streptomyces sps*. exhibit filamentous growth so extensive as to form hyphae.

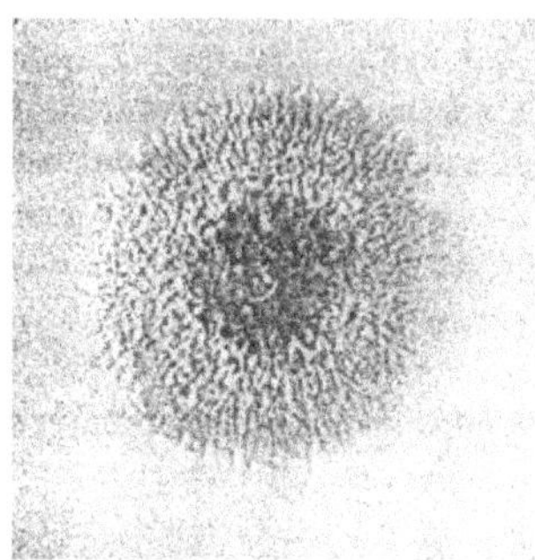

Fig. 6.14 *Actinomycetes*, Characteristic growth of Streptomyces griseus on an agar plate

Growth of actinomycetes is very different than the process of cell enlargement and binary fission that occurs in unicellular bacteria.

Reproduction occurs by several mechanisms of fragmentation of hyphae to form spherical cells in the non spore - forming genera such as *Nocardia*. The genomes of actinomycetes may be circular or linear in diverse species or may bear large up to the genome of *E. coli* (Fig. 6.15).

Physiologically produce various enzymes and thousands of metabolic products synthesize and excrete. Many of these products are antibiotics.

Actinomycetes are major antibiotic producers in the pharmaceutical industry.

Some actinomycetes are pathogens and some cause allergic reactions. Spores cause farmers lung infection, Pneumonitis and hypersensitivity.

Actinomycetes sometimes cause biodeterioration of materials and are often responsible for spoilage of hay, straw, cereal grains, seeds, wood, paper, wool, hydrocarbons, rubber, and plastics.

In nature, biodegradation by actinomycetes plays an extremely useful role in waste removal and is an integral part of the recycling of materials in nature.

Many actinomycetes live in aerobic soils, where they biodegrade organic substrates.

Number of actinomycetes in the soils often exceed 1 million per gram. Soil actinomycetes decompose lignocellulose from plant residues.

They are prolific producers of extracellular enzymes and degrade the complex material. Spores of actinomycetes are especially resistant to dessiccation. Actinomycetes have great biotechnological potential for the production of pharmaceuticals and converting waste material into useful chemicals. The most beneficial activities of actinomycetes related to agriculture and forestry include their activity as biological control agents in the regulation of fungal disease and the capacity of *Frankia* sps carry out symbiotic nitrogen fixation. The cell wall chemistry is especially important feature and used to differentiate actinomycetes. The presence or absence of mycolic acid, whether the peptidoglycan contain L - diaminopimelic acid or meso-diaminopimelic acid and the presence of specific amino acids or carbohydrate residue in the cell wall are used to define the different types.

Phenotypic characteristics based on variation of pure culture. Growth obtained on specialized medium. The mycelium penetrates into the agar.

In many species the colony becomes fluffy due to aerial mycelial growth. Sometimes mycelia are covered by sheath.

Colonies are grey to white often turn to range of colored spores. There is also a characteristic "*earthy*" aroma associated with most cultures of actinomycetes. It involves intentionally inhaling bacteria that may be pathogenic or cause allergic reactions.

At least seven separate taxonomic groups of actinomycetes can be recognised, based on phenotypic characteristics.

eg : Nocardioform : They are characterized by their formation of hyphae that fragment into shorter elements, cell wall composition, and whole cell sugar patterns. Many undergo morphological changes.

eg : Maduromycetes : It represent a heterogenous assemblage of genera. The cell wall contains meso-diaminopimelic acid.

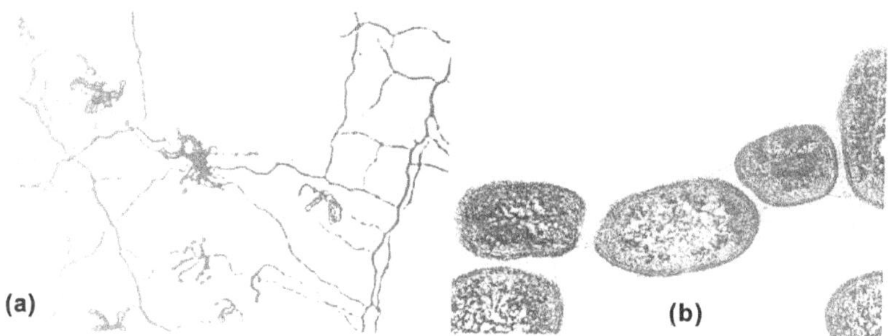

Fig. 6.15 *Nocardioform Actinomycetes*, The hyphae of nocardio form actinomycetes fragment to form arthrospores. (1,250×) (a) micrograph of aerial hyphae and chains of conidia of *Nocardia asteroides*. (b) Development of arthrospores still connected to the aerial hyphae in *Nocardiodes albus*. (50,000×)

6.4 Protozoa

The subkingdom protozoa which was originally included in the kingdom of protista includes unicellular, non-photosynthetic eukaryotic microorganisms that are considered single-celled animals (Gr. protos primitive; zoon - animal). There are more than 65,000 known species of protozoa including several fossil forms. (Fig. 6.16 to 6.24).

Many of the protozoans are parasitic causing serious diseases of humans such as amoebiasis, giardiasis, trichomoniasis and malaria. The free living (non-parasitic) protozoans play an important role in maintaining the ecological balance in nature as many of them feed on fungi and bacteria.

Protozoa consists of a large group of organisms that exhibit a wide diversity of size and morphology. These are unicellular colonial organisms with enkaryotic like cellular structure. The size and shape of protozoa show considerable variation. Leishmania donovani 1-4 mm in length, *Amoeba proteus* 600 mm; ciliates - 200 mm - 2 cm; Formaniferida - 15 cm.

Most protozoa have an optimum temperature between 16-25^0 C; the maximum is between 36-40^0 C. Protozoa is divided in to free living forms and those living on or in other organisms (symbiotic). Some of the symbiotic ones are parasitic and cause various diseases.

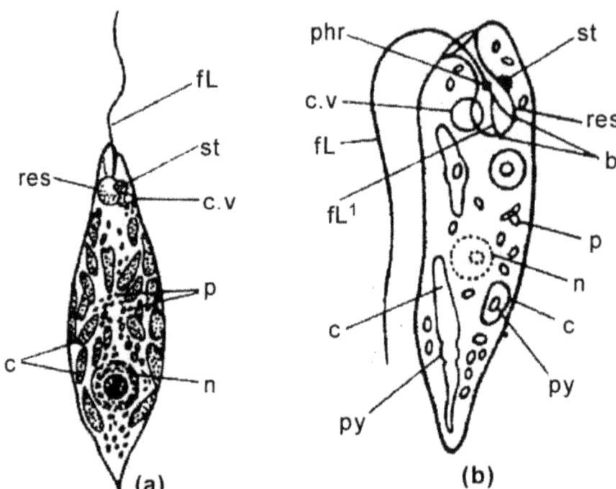

Fig. 6.16 (a) *Euglena viridis*; (b) *Euglena gracilis* (diagrammatic). b, basal body; c, chloroplast; *c.v*, contractile vacuole; *fl*, flagellum; *fl*, short accessory flagellum; n, nucleus; p, paramylum; phr, photoreceptor; py, pyrenoid; res, reservoir; st, stigma.

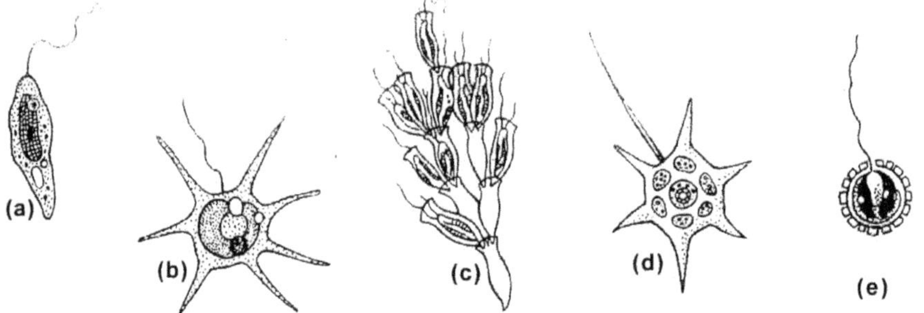

Fig. 6.17 Chrysomonads. (a) *Chrysamoeba*; (b) *Chrysamoeba*, amoeboid phase; (c) *Dinobrvon*; (d) *Distephanus*, a silicoflagellate; (e) *Pontosphaera*, a coccolithophorid. (After various authors).

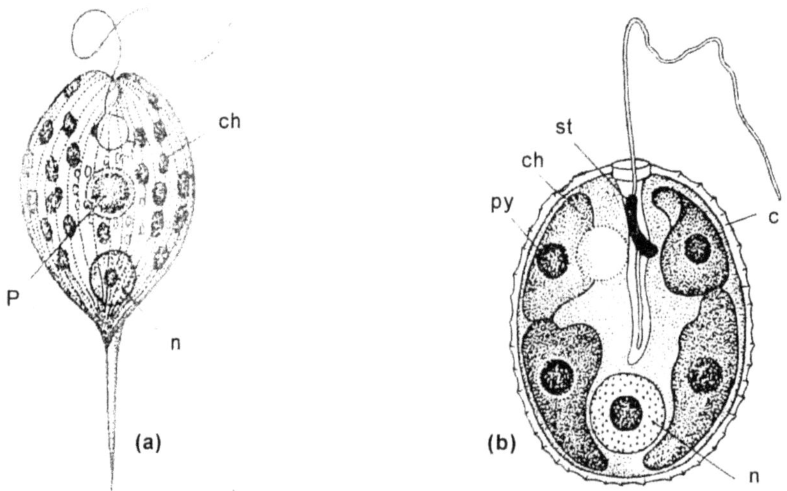

Fig. 6.18 Euglenoids. A, Phacus; B, Trachelomonas. ch, chloroplast; I, lorica; n, nucleus; p, paramylum; py, pyrenoid; st, stigma.

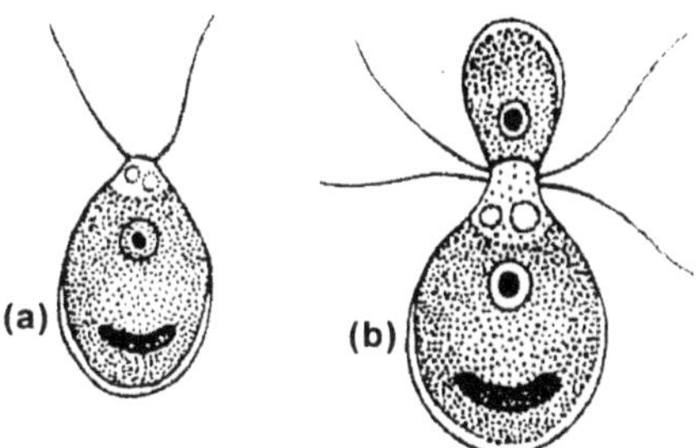

Fig. 6.19 Phytomonads. A, *Chlamydomona*; B, fusion of unequal gametic individuals in *Chlamydomona*.

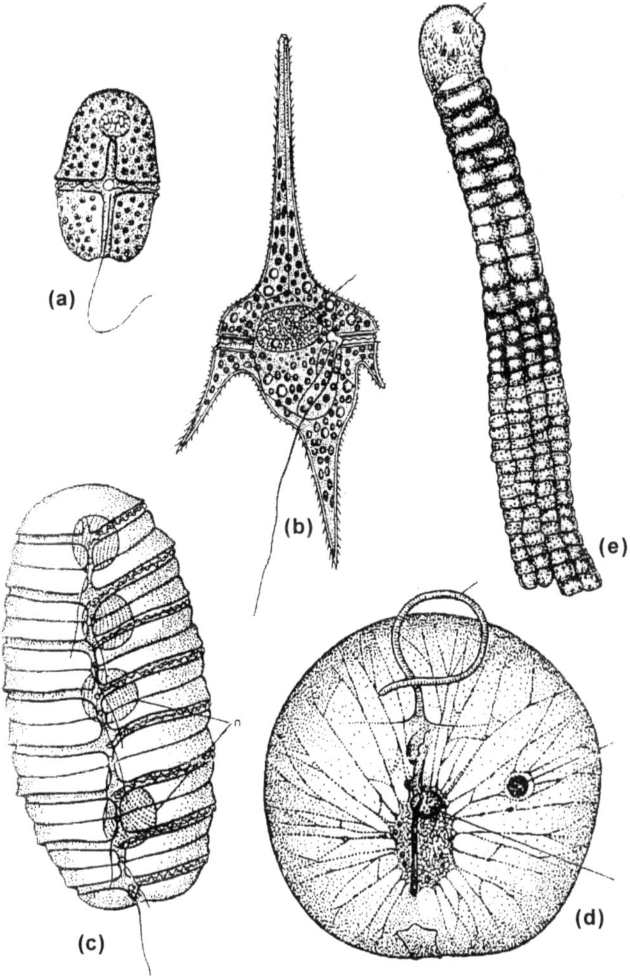

Fig. 6.20 Dinoflagellates. (a) *Gymnodinium*; (b) *Ceratium*; (c) *Polykrikos*; (d) *Noctiluca*; (e) *Haplozoon* (parasitic in many marine polychaetes; the unflagellated colony products many small swarmers).

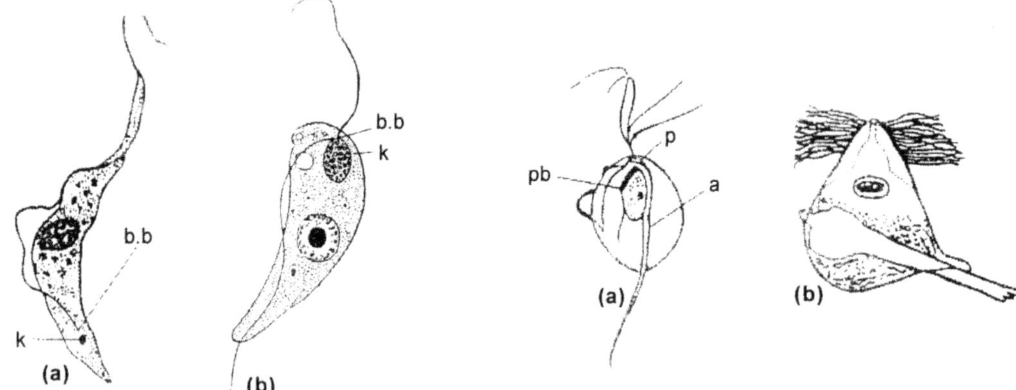

Fig. 6.21 Simpler monadarians. (a) Trypanosoma rhodesiense, one of the causative agents of African sleeping sickness; (b) Bodo, a common flagellate found in stagnant water. b.b, basal body; k, kinetoplast.

Fig. 6.22 More complex monadarians. (a) Trichomonas vaginalis, from the human vagina; (b) Trichonymha, ingesting a large particle of wood. a, axostyle; p, pelta; pb, parabasal body.

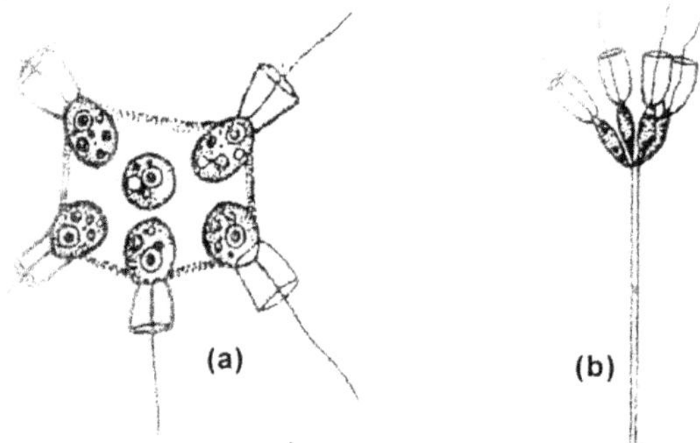

Fig. 6.23 Choanoflagellates. (a) Proterospongia; (b) Codosiga. (From Kudo, op. cit.)

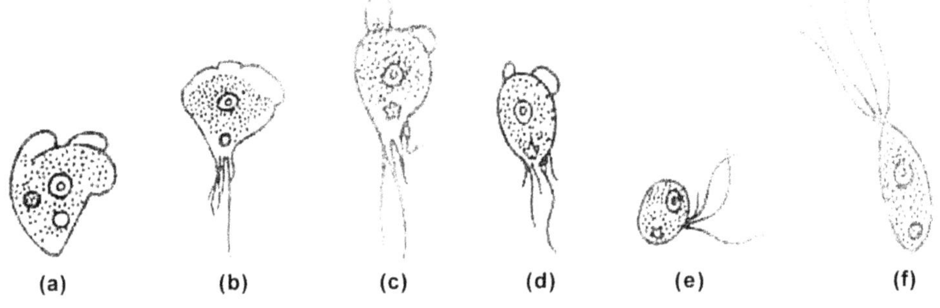

Fig. 6.24 *Naegleria*, (a-c), amoeboid phase (b and c are characterised by filiform posterior extensions); (d) formation of flagella among the filiform extensions; (e) migration of the flagella, anteriorly; (f) typical flagellated phase.

Structure

Plasma membrane forms the outer boundary of most protozoans. The DNA in protozoa is arranged in chromosomes, contained with in nuclei. The flagellum is a filamentous and moves in undulating manner and propels the cell. Cilia are smaller than flagella (5 - 20mm) in length and propel the cell by beating. Contractile vacuoles are found in free living protozoa, particularly in fresh water species and functions to regulate osmotic pressure.

Protozoa like other eukaryotic cells, respond to environmental stimuli such as chemicals (chemotaxis), gravity (geotaxis) and light (phototaxis). These sensory responses are mostly linked to motility. Ameboid moment results from cytoplasmic streaming within the cell and formation of finger like extentions (pseudopodia).

This type of movement requires that the cell is in contact with a solid surface.

Feeding

Protozoa have a wide range of feeding mechanisms. They live in organic rich environments and assimilate dissolved nutrients by absorption through the plasma membrane. Protozoa that possess photosynthetic capacity utilise light for the fixation of CO_2 hence phototrophs.

Respiration

Aerobic protozoa obtain oxygen by diffusion. The utilization of oxygen by respiration drives oxygen in to the cell. In anerobic environments protozoans exhibit a fermentative metabolism.

Osmoregulation

Osmoregulation is the vital function as they lack a rigid outer wall. Most fresh water protozoa are hyperosmotic and hence accumulate water by osmosis. Most marine organisms lack a contractile vacuole as the cells will be hypoosmotic.

Metabolism

Protozoa inhabit a wide range of habitats and can utilise a large number of substrates. Most aerobic protozoa possess a typical glycolytic pathway. Some anaerobic organisms (eg. *Entamoeba sp.*) utilise inorganic pyrophosphate instead of ATP for some of the reactions. Free living protozoa can utilise inorganic as well as organic sources of nitrogen when available.

Reproduction

Protozoa multiply by asexual reproduction and some are able to carryout both asexual and sexual process. The simplest method of binary fission is seen in the amoebas, forming two equal daughter cells. In multiple fission, a single mother (parental) cell divides to form many daughter (filial) cells (eg. *Plasmodium*). Budding is also found in protozoa, is a variation of unequal cell division. Mother cell remains sessile and releases one or more swarming daughter cells.

Sexual fusion of two gametes (syngamy) occurs in various groups of protozoa. Conjugation is a sexual process found exclusively in the ciliates. After exchange of nuclei, the conjugants seperate and each of them give rise to its progeny by fission or budding.

Classification : Details of classification of protozoa are summarised in Table 6.6.

Table 6.6 Classification of Protozoa

	Taxonomic group	**Characteristics**
Phylum I	*Sarcomastigophora*	*Single type of nucelus syngamy*
Sub phylum	*Mastigophora*	One or more flagella typically with chromatophores
Class	*Phytomastigophorea*	Plant like flagellates typically with chromatophores
Class	*Zoomastigophorea*	Animal like flagellates
Sub phylum	*Opalinata*	Binary fission occurs between rows of flagella which cover the entire body surface.
Sub phylum	*Sarcodina*	Pseudopodia - Binary fission
Sub phylum	*Rhizopoda*	Locomotion by pseudopodia includes the naked (*Amoeba*) and testate amoebae (*Arcella*).
Super class	*Actinopoda*	Spherical - planktonic.
Phylum II	*Labyrinthomorpha*	Spindle shaped or spherical vegetative cells. Capable of producing a network of mucus tracks.
Phylum III	*Apicomplexa*	Sporozoa (spore forming stage in their life history)
Class	*Sporozoea*	Reproduction both sexual and asexual - oocysts contain infective sporozoites.
Class	*Gregarinia*	Trophozoites large and extra cellular
Sub class	*Coccidia*	Trophozoites small and typically intracellular
Sub class	*Piroplasmia*	Parasites of vertebrate blood cells - small, pyriform, round rod shaped or amoeboid.
Phylum IV	*Microspora*	Intracellular parasites of invertebrates particularly arthropods.
Phylum V	*Acetospora*	Spores with one or more sporoplasms - parasitic.
Phylum VI	*Myxozoa*	Spores of multicellular origin with one or more polar capsules.
Phylum VII	*Ciliophora*	Organisms possess cilia atleast in one stage of life cycle
Class	*Kinetofragminophorea*	Oral infraciliature cytosome often apical or mid ventral on surface of body.
Class	*Olifohymenoporea*	Oral apparatus generally well defined - oral ciliature clearly distinct from somatic ciliature.
Class	*Polymenophorea*	Well developed adoral zone of membranelles - paroral ciliature.

Protozoa with beneficial and detrimental effects

Autotrophic endosymbionts share some of their photosynthetic products with the host. Ectosymbionts provide useful products for the higher animals. Protozoa that cause detrimental effects are listed in Table 6.7.

Table 6.7 *Protozoa with detrimental effects (human diseases)*

	Pathogen	Disease
1.	*Entamoeba histolytica*	diarrheal diseases
2.	*Sarcocystis sp.*	
3.	*Plasmodium*	malaria
4.	*Trypanosoma brucei*	African trypanosomiasis
5.	*T. cruzi*	Americean trypanosomiasis
6.	*Leishmania sp.*	Leishmaniasis (Kala Azar)
7.	*Trichomonas vaginalis*	Sexually transmitted systemic infection
8.	*Toxoplasma gondii*	during pregnency
9.	*Giardia intestinalis*	Giardiasis (intestinal diseases).

6.5 Microorganisms In Waste Water Treatment

The waste water consists of the following components :

1. Domestic water-borne waste including human wastes, wash waters and everything that goes down from home drains to sewage system.
2. Industrial water-borne wastes such as acids, oils, grease and organic animal and vegetable wastes from factories, and
3. Ground and surface water that enter the sewerage system carrying agricultural wastes and live stock wastes.

The types of microorganisms and their quality will vary depending on the native of the waste water. Raw sewage contain millions of bacteria per millilitre. These bacteria include coliforms, faecal streptococci, anaerobic spore forming bacilli, proteus group etc.

The major steps involved in the waste water treatment are :

1. Primary treatment to remove coarse and settleable solids
2. Secondary treatment (biological) to absorb and oxidise organic constituents of waste water thus reducing BOD (Biochemical Oxygen Demand).
3. Advanced treatment to remove additional unwanted chemicals.
4. Final treatment to disinfect and dispose of liquid effluent, and
5. Solid processing to stabilise solids removed from wastewater and disposal of solids.

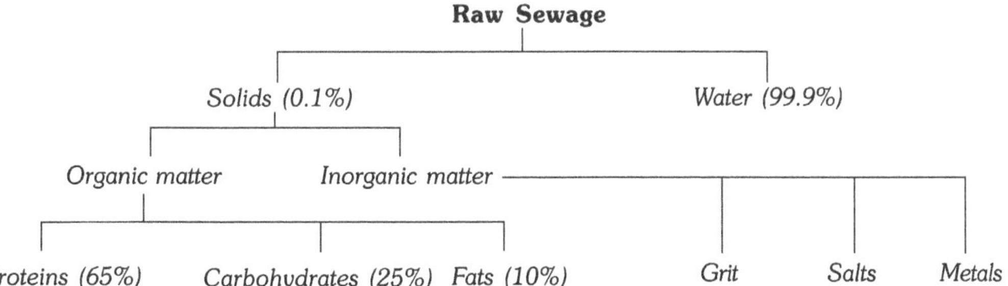

Typical composition of domestic water

Fatty Acids	:	23.8 %
Carbohydrates	:	17.8 %
Proteins	:	10.0 %
Fatty acid esters	:	9.1 %
Soluble acids	:	6.8 %
Anionic surface active agents	:	4.5 %
Amino acids	:	1.6 %
Creatinine	:	1.1 %
Amino sugars	:	1.6 %
Amides	:	0.5 %

Ken P. Flint (1986)

Lipids	-	33 %
Protein	-	25 %
Ash	-	20 %
Cellulose	-	8 %
Starch	-	8 %
Lignin	-	6 %

Sanitary Sewage

Spent water supply of the community.

Domestic Sewage

Wastewater from kitchen, bathrooms, lavatories, laboratories, etc.

Aim of Sewage Treatment

(i) disposal safely without causing pollution, and

(ii) to conserve natural resources

Aim is achieved by three methods

(i) The reduction in BOD

(ii) The reduction in the numbers of pathogenic microorganisms

(iii) The reduction in the levels of inorganic nutrients (phosphates, nitrates, sulphates).

Sewage treatment consists of

1. Separation of solid and liquid phases
2. Treatment in aerobic and anaerobic conditions.

Aerobic treatment gives

- Activated sludge, and
- Active film

Anaerobic treatment gives

- Septic sludge and
- Digested sludge.

Aerobic and Anaerobic Treatment Consists of

1. Adsorption of pollutants on activated sludge biological film and septic sludge (10 - 15 min.)
2. Mineralisation of pollutant with microorganisms (long time)

These two depends on

1. Conditions of the plant operation
2. Temperature
3. Mixing fresh material with mineralising agents.
4. Oxygen supply

1. *Biological Filters* (Fig. 6.25)

Microbiological Process : During commissioning period, a biological film is formed on the lumps of the packing, the main agent (microbial population) which oxidises organic substances.

Biocenosis of the biological film : Biological film consists of algae (blue, green, and diatoms), protozoa, insect larvae, beetles, worms and fungi. In the upper layer of film algal members are dominant and in lower layers all other members exists. The approximate population of each individual is as follows :

Bacterial (α, β, meso saprobes) innumerable; Green algae - 226 $\times$ 10^9 cells/cu.m; Blue green - 143 $\times$ 10^9 cells/cu.m; Diatoms - 24 $\times$ 10^9 cells/cu.m; Protozoa - 66 $\times$ 10^9 cells/cu.m; Worms - 181 $\times$ 10^9 cells/cu.m; Insects - 9 $\times$ 10^9 cells/cu.m.

The Bacteria mineralise organic material, Protozoa feed on bacteria (infusaria and rhizopoda) Algae produce O_2 and phytoncides (substances detrimental to microbes), Worms dig passage between particles to loose the biological film to ensure O_2 to it, and feed on organic substances to digest and destroy some stable compounds chitin and cellulose.

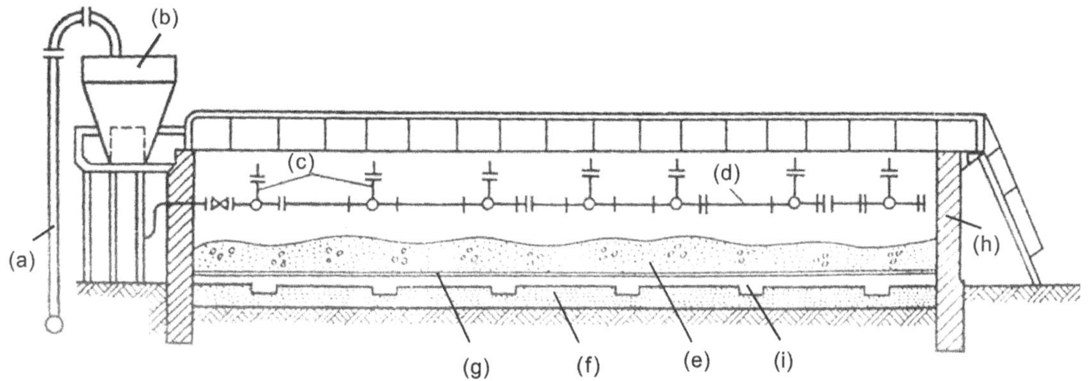

Fig. 6.25 A Biological filter
(a) inlet pipe; (b) batcher; (c) sprinklers; (d) distribution network; (e) filter packing (f) solid bottom (g) perforated bottom (drains) (h) walls (i) troughs in solid bottom

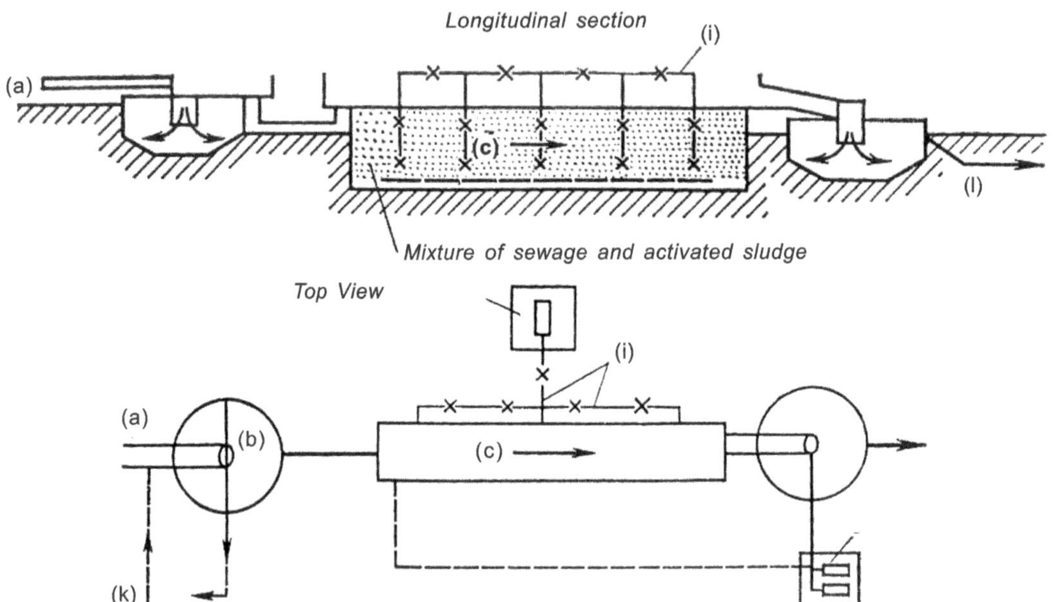

Fig. 6.26 Structure of Aeration tank for complete purification of sewage : (a) inlet pipe; (b) primary settling tank; (c) aeration tank; (d) secondary settling tank; (e) activated sludge pumps (f) air blowers (g) return activated sludge (h) excess activated sludge (i) air ducts; (j) fresh sludge (k) excess activated slude in compacting section; (l) treated sewage

2. *Aeration Tanks* (Fig. 6.26)

Microbiological Process : Activated sludge is formed by adsorption of colloids, suspended particles contained in sewage, multiplication of microorganisms on the substrate.

The activated sludge comprises the whole complex of microbes contained in sewage including the pathogenic flora of man. The microbes are absorbed on activated sludge and depending on their physiological properties, become adapted to the surrounding medium to enter the complex of mineralising agents. Part of flora which cannot adapt to new environmental conditions is killed and the mineralising microbes feed upon it.

Sludge contain

More of bacteria (1×10^{12}) i.e., nitrifying bacteria, unicellular bacteria, filamentous bacteria, yeasts, moulds, protozoa, sarcoids, flagellates, rotifers etc. Activated sludge has tremendous adsorbing power due to developed surface area, and the adsorption rate decreases with time.

Main mineralisers are Bacteria

Sarcoids - feed on sludge convert complicated substances to simpler ones which can be consumed by other microbes.

Infusaria are regulators of bacterial growth.

Protozoa promote accumulation of nitrogen in surrounding medium.

Rotifers are indicators of work of sewage treatment plant.

Following changes occur in the liquid treated in aeration tank :

1. The concentration of pollutants decreases due to the dilution with the liquid carrying activated sludge.
2. Pollutants are adsorbed on activated sludge during the first 15 - 30 mts (I oxidation state)
3. The amount of organic material dissolved in water and adsorbed on activated sludge gradually decreases (II oxidation state).
4. The nitrogen of ammonium salts and nitrites gradually diminish due to their oxidation to nitrates (III oxidation state) - nitrification.

The quality of activated sludge is characterised by :

V/g (m1/g)

V = volume of sludge in ml.

g = dry weight of sludge in grams.

Sludge index

60 ml/g - well precipitating sludge

80 - 90 ml/g - loose sludge

300 ml/g - inadequate work to treatment plant.

- Minute organisms forms dense, quickly precipitating activated sludge.
- Long filaments and branched microbes form loose difficultly precipitating sludge (bulking).

Chemical composition of activated sludge : (in %)

Ash	-	15.58	organic nitrogen	-	3.61
Silica	-	3.95	crude fat	-	2.54
Ironoxide	-	1.58	Fatty acids	-	0.39

$P_2 O_5$ - 2.14

Crude sewage contain : large number of non - pathogenic faecal bacteria, eg :

Escherichia coli
Proteus sp
Serratia sp.
Bifido bacteria sp.

Smaller number of pathogenic bacteria, eg :

Clostridium perfringens
Streptococcus faecalis
Salmonella sp.

Gram negative bacteria of sludge :

Pseudomonas
Flavobacterium
Nocardia
Achromobacter
Alkaligenus
Mycobacterium

Bulking in which the sludge is very difficult to settle due to the presence of filamentous organisms. Like Sphaerotilus natans, Beggiatoa, Leptothrix, Leucothrix and Thiothrix

Chaemolithotrophs :

Nitrosomonas sp. and *Nitrobacter* sp.

3. *Biological Ponds*

Used for final purification of the biologically treated sewage in tropical and subtropical conditions. The basic reactions, design construction, operation, maintenance, disposal, recycle and reclamation are important in the management of biological ponds. (Fig. 6.27)

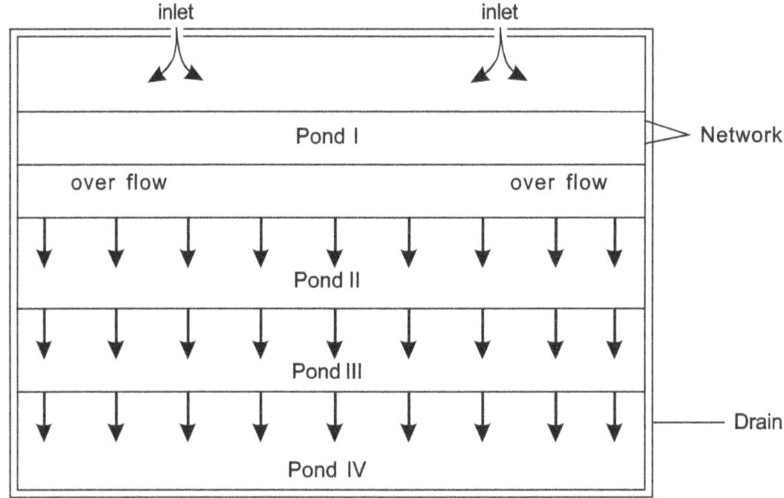

Fig. 6.27 A Biological Pond

1. ***Primary Treatment***
   - Removal of floatable and settleable materials
   - Removal of grit, sludge and greese etc.
   - Enhancement of sedimentation } by algal treatment
   - Heavy metal precipitation } by algal treatment
   - Greese destruction } by algal treatment
2. ***Secondary Treatment***
   - Removal of the fraction of the dissolved organic matter
   - Disinfection of residual water with Cl_2 and O_3 by chemical coagulation of suspended solids.
   - Growth of an enriched microbial culture.
3. ***Tertiary Treatment***
   - Removal of plant nutrients (P, NH_3, NO_3)

4. Quaternary Treatment

- Removal of any hazardous substances or agents such as virus, resistant bacteria refractory substances (which are not oxidised) toxic substances.

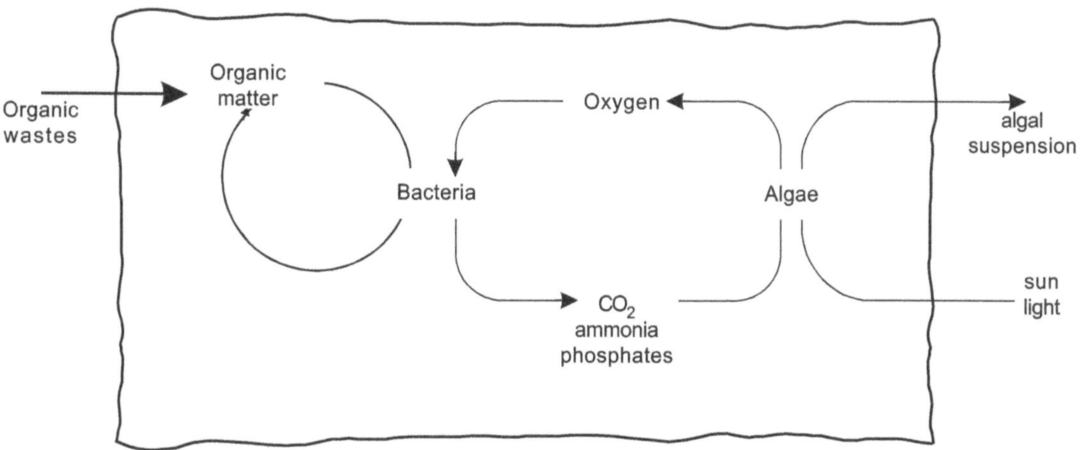

Fig. 6.28 Cycle of controlled Photosynthesis in organic wastes

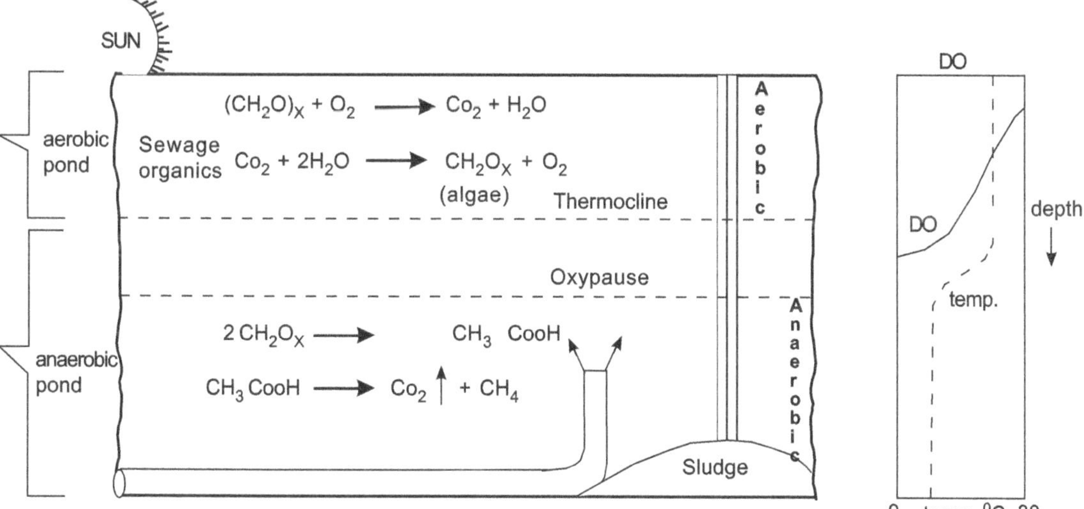

Fig. 6.29 Schematic cross section of a stabilization pond

Based on Microbiological activity ponds are classified to

1. Oxidation pond
2. Acid fermentation pond
3. Methane fermentation pond
4. Algae pond

Based on discharge

Terminal pond (no discharge by evaporation and percolation)

Intermittent pond (discharge only in wet season)

Overflowing pond (discharge at all times)

Percolation pond (evaporation and percolation exceeds in flow).

Anaerobic Treatment

Process Microbiology

Consortia of microbes (mostly bacteria) ⇒ Transforms the high molecular weight organic compounds to methane.

Organic matter $\rightarrow CH_4 + CO_2 + H_2 + NH_3 + H_2S$

Bacteroides, Bifidobacterium, Clostridium, Lactobacillus, Streptococcus, involved in hydrolysis and fermentation of organic compounds.

There are four categories of bacteria that are involved in the transformation of complex materials in to simple molecules. They operate synergistically.

Group I : Hydrolytic bacteria (Breakdown Proteins, cellulose, lignin, lipids).

Group II : Fermentative acidogenic (acid - forming) bacteria (converts sugars, AA, fatty acids to organic acid, alcohols, ketones, acetate, CO_2 and H_2).

Group-III : Acetogenic Bacteria (Accetate and H_2 producing)

Syntrobacter wolinii

Syntrophomonas wolfei : Converts fatty acids and alcohols to acetate, H_2, CO_2 (these are used by methanogens)

$$CH_3.\ CH_2\ OH + CO_2 \rightarrow \underset{\text{acetic acid}}{CH_3\ COO\ H} + 2H_2O$$

$$\underset{\text{Propionic acid}}{CH_3.\ CH_2\ COOH} + 2H_2O \rightarrow CH_3\ COOH + Co_2 + 3H_2$$

$$\underset{\text{Butyric acid}}{CH_3.\ CH_3\ CH_2\ COOH} + 2H_2 \rightarrow 2CH_3\ COOH + 2\ H_2O$$

Symbiotic relation between acetogenic and methanogenic bacteria.

Group-IV : Methanogens :

Anaerobic digestion of organic matter in the environment release 500 - 800 million tonnes of methane per year.

This represents 0.5 % of the organic matter from photosynthesis.

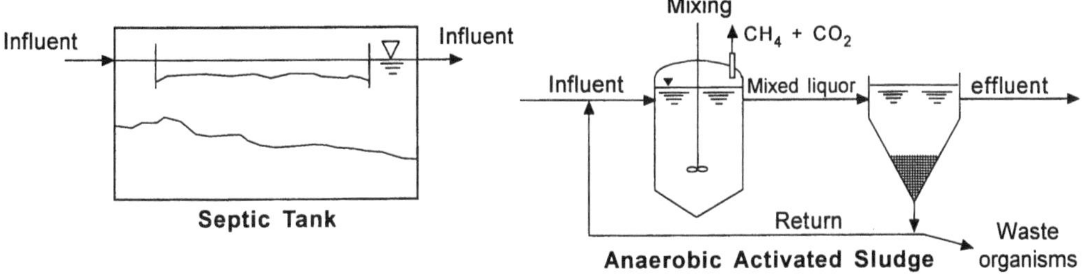

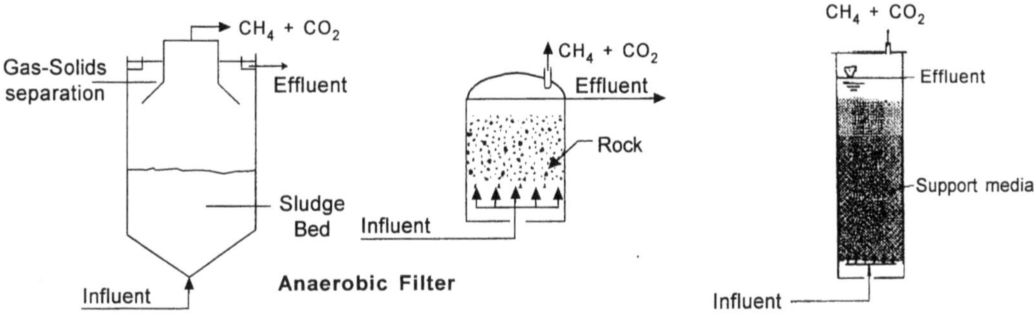

Fig. 6.30 Scheme of microbiological processes used for anaerobic treatment of domestic sewage

Methanogens are of two categories :

1. *Hydrogenotrophic methanogens :* (Hydrogen utilizing chaemolithotrophs)

$$CO_2 + 4H_2 \rightarrow CH_4 + 2H_2O$$

2. *Acetotrophic methanogens :* (Acetate splitting bacteria) (Acetoclastic)

$$CH_3\,COOH \rightarrow CH_4 + CO_2$$

Methanosarcina mazei

Methanothrix sp.

Methanobacterium formicicum

M. bryanti

M. thermoautotrophucum

M. ruminantium

Methanococcus voltae

Methanogenium cariaci

M. marisnigri

Factors controlling anaerobic digestion :

- Temperature
- Retention time
- pH
- Chemical composition of waste water
- Competition with other bacteria
- Toxicants

Anaerobic digestion of waste water and sludge :

Anaerobic digestion consist of series of microbiological processes that convert organic compounds to methane. It has long been used for the stabilisation of wastewater sludges.

Advantages

1. It uses readily available CO_2 as an electron aceptor. It requires no oxygen.
2. It produces lower amount of sludge (3 - 20 times less than aerobic process). 50% organic carbon is converted to biomass under aerobic process, in anaerobic process only 5 % is converted in to biomass.
3. Produce useful gas methane. This gas contains about 90 % of the energy and has a calorific value of 9000 Kcal.
4. Energy requirement for wastewater treatment is reduced.
5. Useful for high strength industrial wastes
6. Can biodegrade xerobiotic compounds.

Disadvantages

1. It is a slower process.
2. It is more sensitive and upset by toxicants,
3. Start up of the process requires long periods,
4. Requires relatively high concentration of primary-substrates.

Process Description

Anaerobic digesters are large fermentation tanks provided with :

- Mechanical mixing
- Heating
- Gas collection
- Sludge addition
- Withdrawl ports
- Supernatant outlets

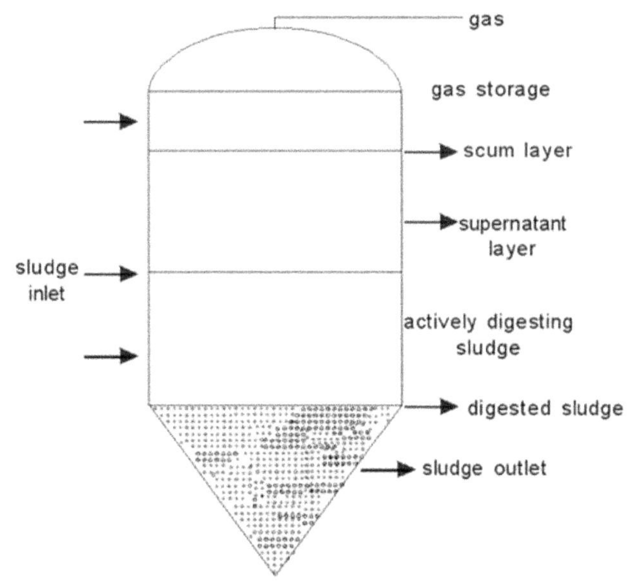

conventional single stage anaerobic digester

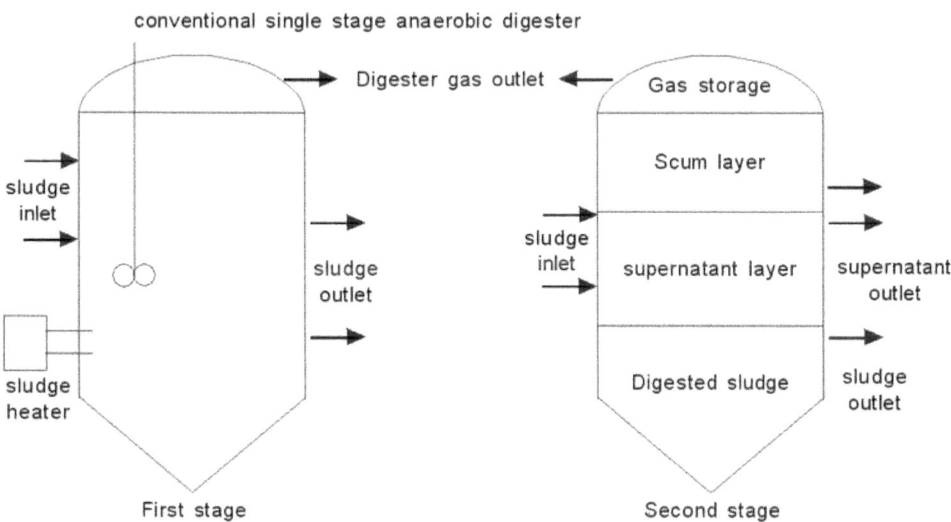

Fig. 6.31 A Two-Stage anaerobic digester

First stage : Continuously mixed and heated for sludge stabilization.

Second stage : Thickening and storage prior to withdrawl and ultimate disposal.

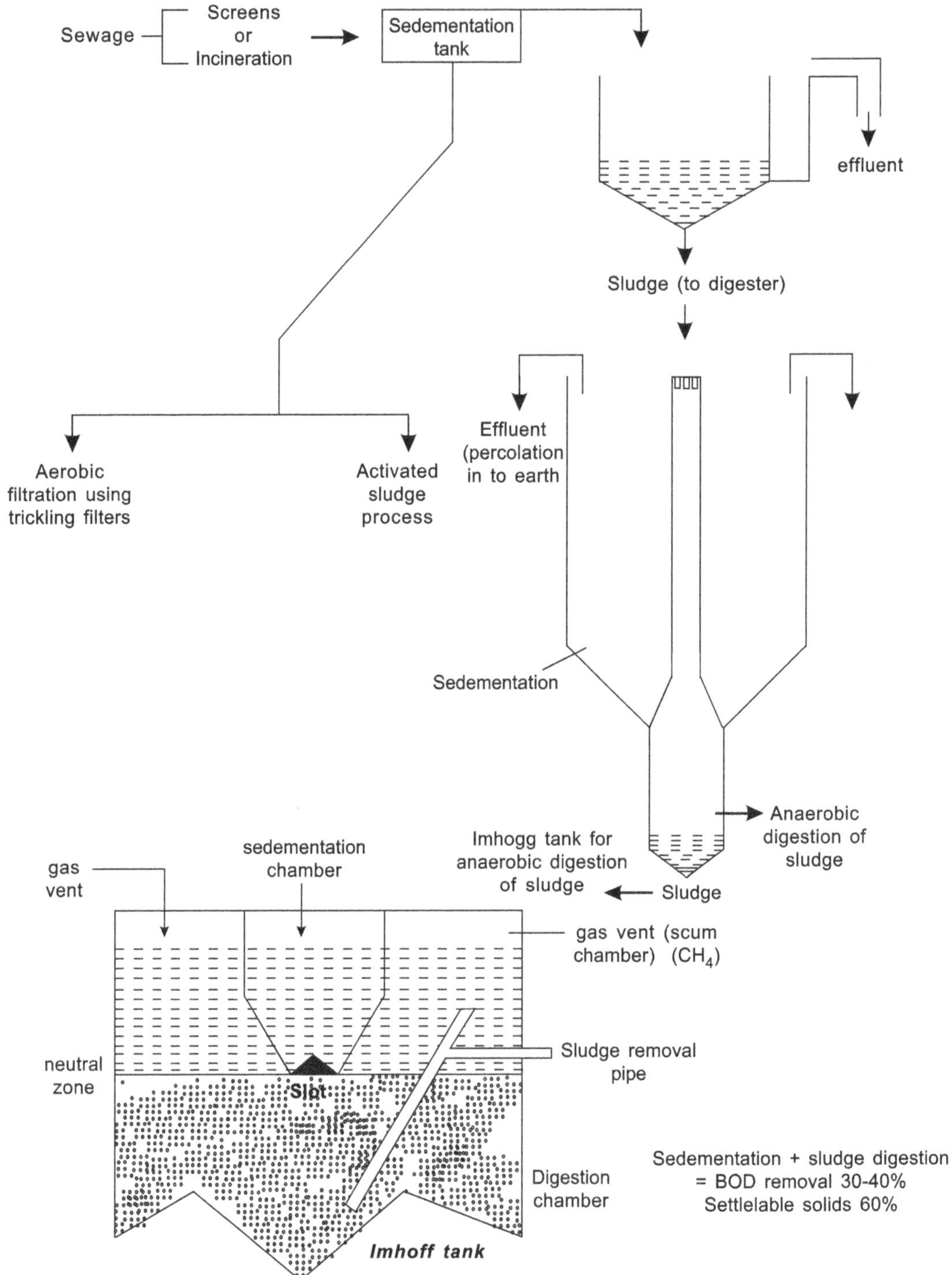

Fig. 6.32 A Primary sewage treatment plant

Biological Treatment : Technological Innovations

1. *Anaerobic Wastewater Treatment :* The direct treatment of high strength waste waters by methenogenic consortia occurs in a range of biofilm and sludge - blanket reactors.
2. *Biological drinking water treatment :* Removal of low concentrations of biodegradable material by a range of biofilm processes reduces disinfectant demand and problems of microbially produced tastes, odour, and corrosion.
3. *High - rate biofilm processes :* Excellent cell retention is combined with a specific surface area to make volumetric loadings very high.
4. *Biofilm activated sludge systems :* Adding biofilm attachment media to activated sludge increases biomass accumulation and process stability.
5. *Membrane bioreactors :* Microorganisms are held inside the membrane, while substrate are fed with the wastewater and across the membrane.
6. *On-line control with expert systems :* Automation of on-line control with expert systems increases reliability when influent loadings vary greatly and when processes have short liquid detention times.
7. *In situ bioreclamation of contaminated aquifers :* Biologically active zones in which an increased mass of microorganisms biodegrade the contaminants in place are stimulated by injecting required nutrients and substrate into the aquifer.

Reuse of Waste Water

Excreted load.

- latency
- multiplication
- persistence
- treatment survival

Infective dose applied to land / water

- persistence
- intermediate host
- type of use practice
- type of human exposure

Infective dose reaches human host

- human behaviour
- pattern of human immunity

Risks of infection and disease

alternate routes of transmission

public health importance of excreta and wastewater use.

Factors influencing the sequence of events between the presence of a pathogen in excreta and the occurance of measurable disease attributable to wastewater reuse.

Options for health protection (with regard to wastewater irrigation)

1. Waste water treatment,
2. Restriction of the crops grown,
3. Choice of irrigation methods, and
4. Control of human exposure to infection.

6.6 Bioindicators

The microorganisms which exist in any environment are direct indicators of the conditions of that area. The types of micro organisms are all a direct result of the nutrient concentration and environmental conditions. The functions of microbes in the environment, nutrient status and direct microscopic examination is used as a bioindication of the system. Bioindicators are considered to be more reliable for the assessement of the status of system over some chemical tests.

In the event of pollution, the physical and chemical structure of the system undergoes a change resulting in change in biota. Some species are sensitive to the changed situations while some are tolerant. The sensitive organisms, therefore, disappear and their place is taken by the tolerant species. This mode of identification of organisms indicates the degree of pollution.

The basic interest of pollution is due to its effects on living organisms and hence the assessment of pollution is a biological problem (Sastry, 1988). The physical and chemical data gives quantitative information while biological data only qualitative in nature, hence in the establishment of pollution status the physico-chemical was considered as ideal base. Further, the biological analysis has been tedious and expensive. The biological data was never investigated in their simplified forms. So, the bioindication has been omitted in major situations. But there are a number of advantages with biological examinations :

1. The organisms can integrate the situation with their presence, number and behaviour, which chemicals may not do.
2. Many chemical pollutants produce adverse biological reactions.
3. Living organisms can preserve any ecological perturbation.

The degree of ecosystem imbalance is measured only by biological information. Biological indices provide the information based on the following criteria :

1. Species composition
2. Diversity of the species
3. Distribution pattern
4. The presence or absence of indictor organism.

Kolukurtz and Marsson (1908) introduced the concept of indicator organisms for the assessment of water pollution. In the course of self-purification of water body shows distinct zones of decreased pollution or improved water quality. Biologically each zone offers optimal conditions for certain species or communities or organisms known as the ‘indicator organisms’ or indicator community. The saprobic zones are classified as (a) Polysaprobic (gross pollution) (b) Alpha-mesosaprobic (heavy pollution) (c) Beta - meso saprobic (moderate pollution) and (d) Oligo saprobic less pollution (clean water).

Liebmann (1962) developed four quality levels of water and showed in maps of rivers and lakes with colour indication of blue, green, yellow and red with increased levels of pollution and eutrophication. Palmer (1969) made the attempt to identify and prepare a list of genera and species of algae tolerant to organic pollution (Table 6.8). He prepared a list of 60 genera and 80 species tolerant to organic pollution.

Table 6.8 *Pollution index of various algal genera*

Genera	Pollution index	Genera	Pollution index
Anacystis	01	*Micractinum*	01
Ankistrodesmus	02	*Navicula*	03
Chlamydomonas	04	*Nitzschia*	03
Chlorella	03	*Oscillatoria*	04
Closterium	01	*Pandorina*	01
Cyclotella	01	*Phacus*	02
Euglena	05	*Phormidium*	01
Gomphonema	01	*Scenedesmus*	04
Lepocinelis	01	*Stigeoclonium*	02
Melosira	01	*Synedra*	02

If an algal sample has the genera *Chlorella, Oscillatoria, Stigeoclonium, Synedra, Nitzschia, Chlamydomonas* and *Navicula*, the score according to above list is 3 + 5 + 2 + 2 + 3 + 4 + 3 = 22. A score of 20 or more for a sample is an indication of organic pollution. This confirms the high organic pollution of this water sample. A score of 15-19 is taken as probable evidence of high organic pollution. Lower figures indicate that the organic pollution is not high.

The sequention comparison index, proposed by Cairns *et al* (1968) used by professional biologists for quick surveys and by the non-biologists for getting an idea of the biological effects of pollution.

Saprobic index (Pantle and Buck, 1955)

The frequency of the indicator organisms is used as a criterion as also its saprobic valency for numerical expression. The frequency ratings (h) are : Random occurrence - h - 1; Frequent occurrence - h - 2; Massive development - h - 3;

The saprobic zones (S) are expressed as :

1. Oligosaprobic - S - 1
2. β - Mesosaprobic - S - 2
3. α - mesoyraprobic - S - 3
4. Polysaprobic - S - 4

The product of frequency (h) and saprobic zones (S) expresses the saprobic value.

The saprobic index for all indicator species in a sample is obtained from the formula :

$$S = i = \frac{\sum_{i}^{n} = 1(S_1 h_i)}{\sum_{i}^{n} = 1(h_1)}$$

Where i is the index for different species. Saprobic index can be calculated only when the number of indicator species exceeds nine.

Biotic Index (Beck, 1955)

Biotic indices are developed on the basis of tolerance of the organisms to various zones of pollution.

$$B_1 = 2 \text{ (n class I)} + \text{(n class II)}$$

Where, n represents the number of organisms in each class. In class I, the organisms are little tolerant to organic pollution. Class II organisms are tolerant to organic pollution but not to anaerobic conditions. The index value range between 0 - 10.

Diatoms taken from surface waters of different degree of pollution are found to differ in quality and population diversity. These findings can be represented graphically by plotting abundance and diversity.

Gabriel (1946) used an index (I) based on the number of decomposers, B (bacteria), producers (P) algae etc., and consumers, C (ciliates etc.)

$$I = \frac{2P}{B+C}$$

In heavily polluted waters, the decomposers prevail, with increasing self-purification, the number of heterotrophic consumers increases and finally autotrophic producers flourish.

Biological index of pollution developed by Morasawa, is estimated with the following formula:

$$BIP = \frac{A}{A+B} \times 100$$

Where,

A is the number of chlorophyll bearing organisms,

B is the number of non-chlorophyll bearing organisms.

This concept of bioindication has originated from the recognition of the fact that each species in the animal and plant community has specific and optimum ecological requirements for its best survival, propagation and growth.

6.7 Biosensors

Biosensors are compact, low-cost disposable devices which circumvent the need for frequent recalibration procedures. The biosensor technology enables rapid, sensitive measurement to be made using a simple procedure, with a device that is relatively easy to operate.

Definition : Compact analytical device incorporating a biological or biologically derived sensing element either integrated within or intimately associated with a physico-chemical transducer. The sensing element incorporated may be an enzyme, antibody, nucleic acid, microorganism or cell.

The basic aim of a biosensor is to produce either discrete or continuous digital electronic signals which are proportional to a single analyte or related group of analytes.

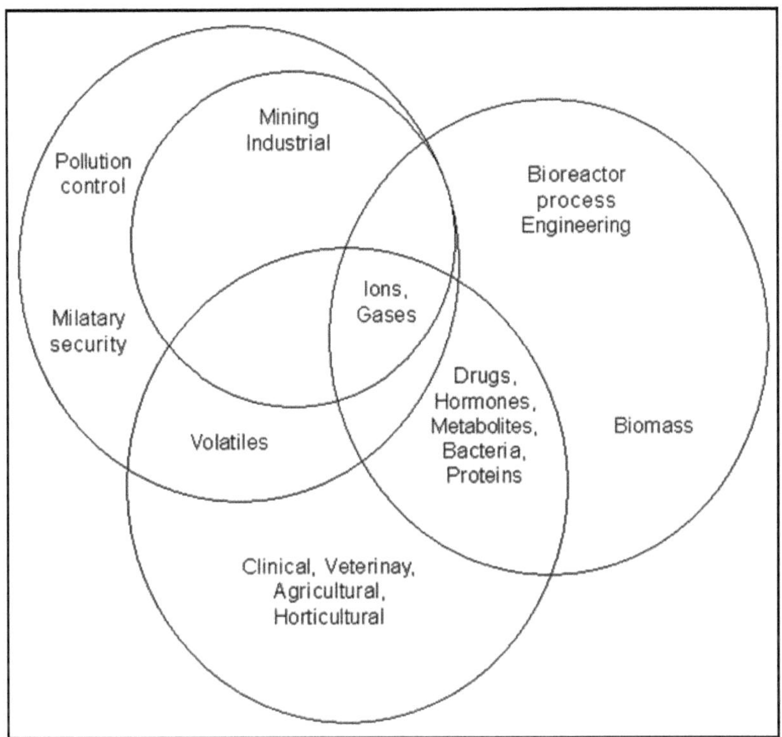

Fig. 6.33 Some potential analysis and their applications

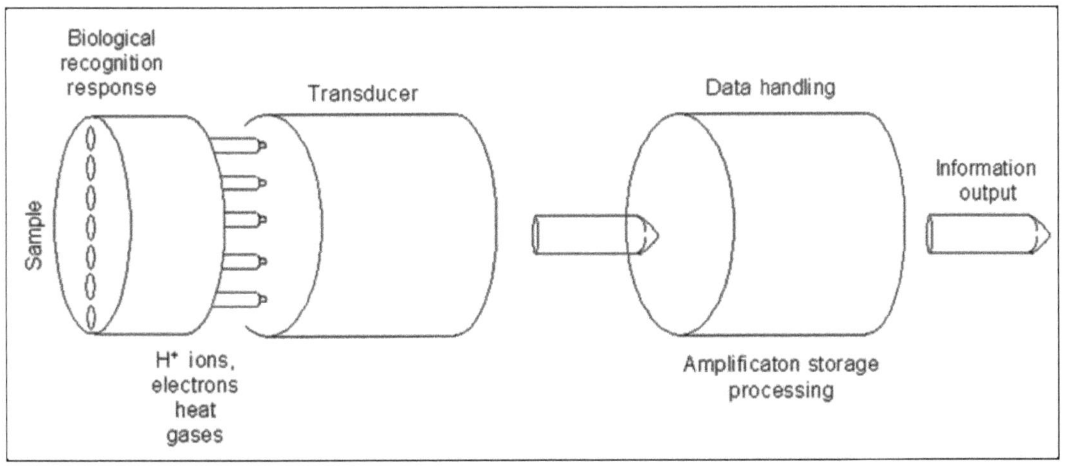

Fig. 6.34 Plan of a generalised biosensor

In the mechanism of process, the biosensor upon making contact with an appropriate sample converts the presence of the desired analyte in to an electrical signal. Then the concentration of the sample is interpreted as the electrical signal via, a suitable combination of a biological recognition system and an electrochemical tranducer. Usually the enzymes, organelle, microbial, plant and animal cell, tissue, slice, section, or immune binding or receptor protein acts as biological sensing systems in the biosensor. (Fig. 6.34).

The organic polymers are the most important parts of the biosensor mainly used in the preparation of enzyme membranes and redox membranes. Polymers enhance the stability, biocompatibility, and selectivity and are used to support for enzymes, antibodies, microorganisms, organelles and dyes as components for the construction of membranes in the transducer.

Boretos and Eden (1984) identified certain important polymeric materials. The following are the polymers and their functions are mentioned in the parenthesis :

Polyacetylene (Redox polymer); Polyacrylamide (Polymer gel); Polyaniline (Redox polymer); Polyethylene (membrane); Poly ethyleneimine (membrane); Poly ethylene phthalate (enzyme membrane); Polymethacrylate (enzyme membrane); Poly oxyethylene ferrocenes (mediator); Poly propylene (membrane); Polypyrrole (Redox polymer); Poly styrene (optical fibre); Poly thiophene (Redox polymer); Poly Urethane (enzyme membrane); Poly vinyl alcohol (polymer matrix); Poly vinyl butyral (polymer gel); Poly vinyl chloride (polymer matrix); Poly vinyl pyrrolidone (enzyme membranes)

Methods commonly used for the immobilisation of enzymes on these polymeric materials are :

1. Adsorption
2. Cross linked adsorption
3. Covalent fixation
4. Cross linking
5. Entrapment
6. Encapsulation.

Advantages of Polymers in Biosensors

The use of polymers improves the properties and facilitates the construction of biosensors. This is interms of their sensitivity, stability, selectivity and measuring time. Food technologists and environmentalists have an interest in increased sensitivity and limits of detection for a range of analytes.

Many strategies have employed to restrain or modify the structure of biological receptors to enhance their *longevity*. Improvement in the *selectivity* of biosensors is sought at two levels (i) the interface between the transducer and the biological receptor be made more exclusive then reducing interference (ii) new receptors can be developed with improved or new affinities. The use of PVC membranes in urease sensors decreases the measuring time and enlarge the linear measuring area.

Transducers and their Applications

The immobilizations of the bio active materials with a transducer makes it possible to convert their bio chemical signals in to measurable electrical signal. (Table 6.9).

Table 6.9

Transducer		Applications
Ion selective electrodes	..	Ion in biological media, enzyme electrodes
Thermistors	..	Enzyme, organelle, or cell sensors for substrates, gases, antibiotics, pollutants.
Fibre optic and opto electronic devices	..	pH, enzyme substrates, immunological systems,
Piezoelectric crystals	..	Volatile gases

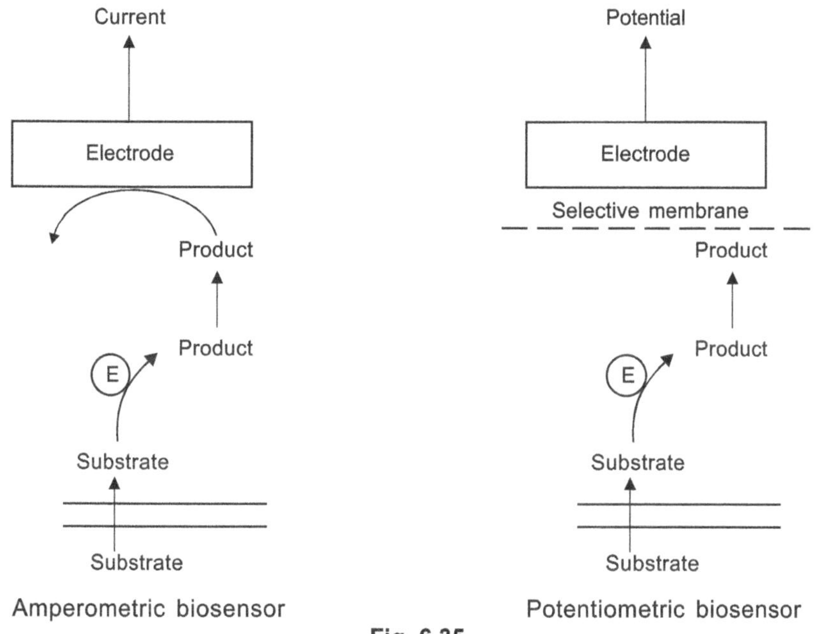

Fig. 6.35

Classification of Biosensors

The classification of biosensors is based on the type of detector. The majority of the biosensors have at this time a detector based on an electrode.

There are two principal types of electrodes employed in biosensors, namely the amperometric and the potentiometric electrode (Fig. 6.35).

In the amperometric electrode a potential is imposed between the sensitive part of the electrode and a reference electrode. At this imposed potential, a molecule intervening in the enzymatic reaction (substrate or product) can transfer an electron with the electrode surface to be oxidised or reduced. The current through the electrode reflects the concentration of this molecule, but because this molecule is consumed at the electrode surface, the current is also function to the mass-transfer to the surface.

In the potentiometric electrode, no current passes through the electrode and the signal is then the potential obtained between the sensitive part of the electrode covered with a selective membrane and a reference electrode. No molecule is consumed at the electrode surface to the signal is independent of the mass transfer to the active surface.

The optical electrode (optrode), a new technology of fibre optics, an enzymatic catalysis by the change of the spectrophotometric characteristics of the solution. It is used to measure the absorbance of light of a set wavelength in a mixture of substrate, the absorbance being proportional to concentration of the product.

The prospective applications optical biosensors include; clinical and diagnostic medicine; monitoring of chemical, food and drug processes and fermentation; monitoring of effluents, and pollutants and monitoring chemical and biological warfare.

Immunosensors

Biosensors find useful applications in immunology. A combination of immobilized antibody (or antigen) with an electrode device yields an immunosensor. The immunoglobulins (antibodies) are produced in the animal body in reaction to the presence of foreign substances (antigens). It is possible to produce a batch of antibody (monoclonal antibody) which reacts specifically to one individual antigen. If the substance to be measured is itself antigenic, then specific antibody against it can be produced and can be incorporated in to the biosensor system.

Biosensor for Environmental Control

BOD Sensor : The Biochemical Oxygen Demend (BOD) is one of the most widely used and important test in the measurement of organic pollution. The conventional BOD test requires a five day incubation period and values of the test results depend on the skill of an operator. Therefore, rapid and reproducible methods are desirable for the BOD test. Yeasts, *Trichosporon cutaneum* which was utilised for waste water treatment was employed for the sensor (Hikuma et al. 1979). The microorganisms immobilised on a porous acetyl cellulose membrane were sandwiched between an oxygen permeable Teflon membrane and a porous membrane. Then the membrane was directly fixed on the surface of the platinum cathode of an oxygen probe. A continuous flow system using a new microbial sensor was developed for automatic estimation of 5 - day BOD.

Ammonia Sensor

In ammonia biosensor a nitrifying bacterium *Nitrosomonas* sp. utilises ammonia as a sole source of energy and oxygen is consumed by the respiration.

$$2\ NH_3 + 3O_2 \xrightarrow{\textit{Nitrosomonas}\ \text{sp.}} 2\ HNO_2 + 2\ H_2O$$

Here, ammonia is determined by microbial sensor using immobilised nitrifying bacteria such as *Nitrosomonas* sp. and an oxygen probe.

Nitrite Sensor

Nitrobacter sp. utilise nitrite as the sole source of energy and oxygen is consumed by the respiration.

$$2\ NO_2^- + O_2 \xrightarrow{\textit{Nitrobacter}\ \text{sp.}} 2\ NO_3^-$$

Oxygen uptake by the bacteria can be directly determined by the oxygen electrode attached to the immobilised bacteria. Therefore, NO_2 generated in the buffer (pH 2.0) can be determined by the microbial sensor using immobilised *Nitrobacter* sp. and an oxygen electrode.

Sulphite Ion Sensor

The determination of sulphite and sulphur dioxide in wastewaters and atmosphere is important in environmental analysis. In this process sulphite is modified to sulphate by enzymatic oxidation. An amperometric organelle sensor for the determination of sulphite, composed of immobilised microsome particle, a gas permeable Teflon membrane and an oxygen electrode is used.

Advantages of Biosensor

1. Very low concentrations of biological substances (even about10^{-15} m) may be measured.
2. Wide range of sensitive and selective analyses
3. Rapid response and low cost
4. Continuous real time assay
5. Disposable, reagentless nature and operation in opaque, turbid solutions
6. Computer compatibility

Current and Future Challenges of Biosensors

The key issues in the biosensor field that are currently being addressed by biotechnology include :

1. Novel techniques for the realistic mass production of biosensors
2. Molecular engineering of the biosensor interface and biorecognition components for optical performance
3. Direct transduction of biorecognition events
4. Micro fabricated biosensors, microysystems incorporating biosensors and array based biosensors.

6.9 Microbial Enzymes

Enzymes are defined as protein catalysts that have great specification for the reaction catalysed and the molecules acted on (Table 6.10). A catalyst is a substance that increases the rate of chemical reaction without being permanently altered itself. Thus enzymes speed up cellular reactions. The reacting molecules are called substrates, and the substances formed are the products. Many enzymes are pure proteins, and consist of a protein, the apoenzyme, and also a non-protein component, a cofactor, required for catalytic activity. The complete enzyme consisting of the apoenzyme and its cofactor is called the holoenzyme. If the cofactor is firmly attached to the apoenzyme, it is a prosthetic group

Table 6.10

Type of Enzyme		Reaction Catalysed by Enzyme
Oxidoreductase	:	Oxidation - reduction reactions
Transferase	:	Reactions involving the transfer of groups between molecules
Hydrolase	:	Hydrolysis of molecules
Lyase	:	Removal of groups to form double bonds or addition of groups to form double bonds
Isomerase	:	Reactions involving isomerisations
Ligasae	:	Joining the two molecules using ATP energy.

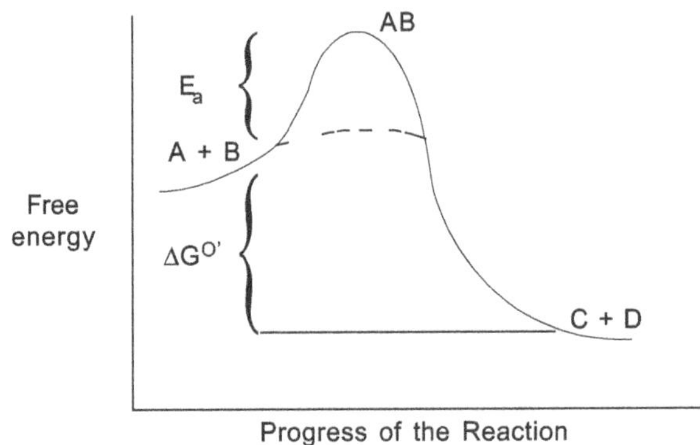

Fig. 6.36 Enzymes lower the energy activation.

In this chemical reaction A and B are converted to C and D. The transitional state complex is represented by $AB^{\pm}$, and the activation energy required to reach it, by Ea. The line represents the course of the reaction in the presence of an enzyme. The activation energy is much lower in the enzyme catalysed reaction. The difference in free energy level between reactants and products is $\Delta G^{0'}$ (Fig. 6.36).

This Michaelis-Menten Kinetics explains the dependance of enzyme activity upon substrate concentration (Fig. 6.37). The maximum velocity (Vmax) is the rate of product formation when the enzyme is saturated with substrate and making products as fast as possible. The Michaelis constant (Km) is the substrate concentration required for the enzyme to operate at half of its maximam velocity. It relates the reaction velocity (V) to the substrate concentration (S).

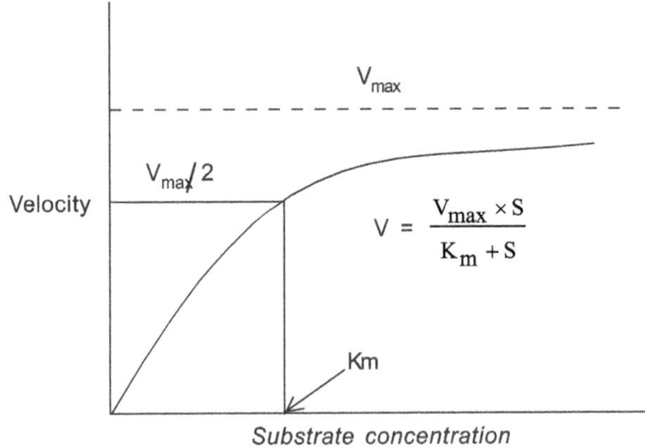

Fig. 6.37 Michaelis - Menten kinetics

Applications of Enzyme Technology

Major targets of modern enzyme technology continue to be preservation of foods and food components (eg. vitamins), more efficient use of raw materials and improvement of food quality such as texture and taste (Table 6.11). Enzymes are also used in the food industry for process optimization to reduce process cost. Other current objectives include the utilization of new raw materials for feeding humans and animals.

Biotechnological methods have also replaced some of the traditional chemical processes. Enzymatic methods, which include the most recent procedures with immobilized enzymes and immobilised microorganisms constitute one aspect of modern biotechnology. Biotechnology encomposses such as amino acids and antibiotics as well as processing waste water.

New Biotechnological Processes Include :

1. The use of enzymes for synthesis of amino acids, peptides and antibiotics.
2. The use of enzymes for recycling food wastes and in waste water treatment.
3. The use of enzymes in non - aqueous media for the production of chiral compounds and the synthesis of special polymers.
4. The production of sweetner, such as aspartame, using combined microbiological and enzyme methods.
5. The production of cyclodextrins from starch with specific enzymes.
6. The use of celluloses and lipases as active components of detergents.
7. The production of rare and expensive mammalian enzymes by microorganisms.
8. The large scale production of lignolytic enzymes in biobleaching and biopulping of wood based industries.
9. The production of tailored enzymes to serve as specific, process - adapted catalysts.
10. Large scale production of enzymes by gene technology is to be expected in future years.

Table 6.11 *Survey of Technical Enzyme Applications*

Process or Product	Enzyme Preperation
Alcohol	α - anylases, β - amylases, β - glucanases, protease, cellulase.
Amino acids	Amino acid acylase, proteases with peptidases.
Animal feed	Protease, amylase, hemicellulase
Beer	α - amylases, amyloglucosidases, proteases
Berries	α - amylases, β - amylases, fungal proteases, pentosanases.
Cheese sipening	Lipases, peptidases
Chicken breakdown	Chitinase
Citrus juices	Pectinases
Citrus slicing	Cellulases
Confections	Invertase
Detergents	Proteases, amylases
Egg Products	Glucose oxidase, catalase
Flovours	Petinases, lipases, cellulases
Flax Processing	Pectinases
Fruit, vegetable pulps	Macerases
Lactose	Yeast lactase
Leather manufacture	Proteases, Lipses
Lipid hydrolysis	Lipases
Meat tenderizing	Proteases, especially papain.

Table 6.11 Contd...

Process or Product	Enzyme Preperation
Milk	Cold sterilization : H_2O_2 / catalase
Oils	Lipases
Phytic acid	Phytases
Rubber	Proteases
Silage	Grasses - Cellulases, b - glucanases, lignases.
Tea	Tannases
Waste water	Prosease, cellulase. Complex enzyme preperations.
Wine making	Maceration - Pectinases - proteases, β. gluconases
Wood based - Rayon, Paper manufacturing	Lignolytic (lignin peroxidase, laccase), cellulases.
Yeast hydrolysis	β. gluconases.

Economic considerations for the use of Techinical Enzymes

The economic benefit of using technical enzyme preparations lies in lowered process costs, often increasing product quality, reducing environmental impact and making use of renewable resources in technology and food possible (Table 6.12).

Table 6.12 List of Some Industrial Microbial Enzymes

Name	Type of Enzyme	Origin
50 : 50	Rennet	Animal
Agidex	Amyloglucosidase	*Aspergillus niger*
Alfamalt VT	Fungal α - amylase	*Aspergillus oryzae*
Alfamalt PRO	Bacterial Protease	*Bacillus subtilis*
Alkalase	Bacterial Protease	*Bacillus lichemiformis*
Ambazyme	Amyloglucosidase	*Aspergillus niger*
Amylase P	Fungal α - amylase	*Aspergillus oryzae*
Amyliq	Bacterial α - amylase	*Bacillus subtilis*
Anthocyanase YAZ	Anthocyanase	*Aspergillus niger*
Arazyme	Protease	*Microbes*
Aquazyme	Bacterial α - amylase	*Bacillus sp.*
Asgo 1500	Glucose oxidage	*Aspergillus niger*
Avizyme	Cellulase	*Trichoderma - reesei*
BAN	Bacterial α - amylase	*Bacillus subtilis*
Beta - Glucanase	β - Glucanase	*Arthobacter sp.*
Beta - Glucanase	β - Glucanase	*Rhizopus solani*
Bergazyme H	Cellulose, hemicellulase	*Trichoderma reesei*
Biobake	Fungal α - amylase	*Aspergillus oryzae*
Biocellulase A	Cellulase	*Aspergillus niger*
Biocellulase T	β - Glucanase	*Trichoderma reesei*
Biodiastase	Fungal α - amylase	*Aspergillus oryzae*
Bioinvert	Invertase	*Saccharomyces sp.*

Table 6.12 Contd...

Name	Type of Enzyme	Origin
Biokleistase	Bacterial α - amylase	*Bacillus subtilis*
Biopectinase	Pectinase	*Aspergillus niger*
Bioprase	Bacterial protease	*Bacillus subtilis*
Biotamylase	Bacterial α - amylase	*Bacillus subtilis*
Biotex	Bacterial α - amylase	*Bacillus subtilis*
Gabo	Rennet	Animal
Canlpha	Bacterial α - amylase	*Bacillus amyloliquefaciens*
Capalase	Lipaesterase	Goat
Capalase KL	Lipaesterase	Goat, Lamb
Catalase LA 3	Catalase	*Aspergillus niger*
Catalase U 5L	Catalase	*Micrococcus sp.* *Lysodeikticus sp.*
Cellulast	Cellulase	*Trichoderma reesei*
Cellulase Onozuka	β - Glucanase, cellulase	*Trichoderma viride*
Cellulase CP	Cellulase	*Penicillium funiculosum*
Cellulosin AP	Cellulase	*Aspergillus niger*
Cellulosin HC	Hemicellulase	*Aspergillus niger*
Celluzyme	Cellulase	*Humicola insolens*
Cereflo	β - Glucanase	*Bacillus subtilis*
Ceremix	β - Glucanase	*Bacillus subtilis*
Clampzyme	Cellulase	Fungal species
Collupulin	Papin	Papaya
Coralase L 10	Papin	Papaya
Corolase N	Bacterial proteinase	*Bacillus subtilis*
Corolase PK	Pancreatic protease	Animal
Corolase PN	Fungal alkal. protease	*Aspergillus sojae*
Cyclodextrinase	Glucanotransferase	*Bacillus sp.*
Degomma	Bacterial amylase and protease	*Bacillus subtilis*
Denapsin	Fungal acid protease	*Aspergillus sp.*
Deoxin	Glucose - oxidase	*Penicillium amagasakiense*
Dex - Lo	Bacterial α - amylase	*Bacillus subtilis*
Dextranase 50 L	Dextranase	*Penicillium lilacinum*
Dextrozyme	Glucoamylase, pullulanase	*Bacillus acidopullulyticus*

Table 6.12 Contd...

Name	Type of Enzyme	Origin
Diase	Amyloglucosidase	*Aspergillus niger*
Emporase	Rennet	*Mucor pusillus*
Esperase	Bacterial alkal protease	*Bacillus subtilis*
Extractase	Pectinase	*Aspergillus niger*
Fermcolase	Catalase	*Aspergillus niger*
Filtrase B	β - Glucanase	*Bacillus subtilis*
Filtrase NL	β - Glucanase	*Aspergillus* sp.
Finizym	β - Glucanase	*Aspergillus niger*
Flavourzyme	Protease, peptidase	*Aspergillus oryzae*
Fromase	Rennet	*Mucor miehei*
Funcellase	β - 1.3 Glucanase	*Trichoderma viride*
Glucanex	β - Glucanase	*Aspergillus niger*
Glucoseoxidase L	Glucose - Oxidase	*Aspergillus niger*
Glucoseoxidase P	Glucose - Oxidase	*Aspergillus niger*
Glucox	Glucose - Oxidase	*Aspergillus niger*
Glucozyme	Amyloglucosidase	*Rhizopus* sp.
Glucozyme NL 3	Amyloglucosidase	*Rhizopus* sp.
Glutase	Amyloglucosidase	*Rhizopus* sp.
Godo - AGI	Glucoseisomerase	*Streptomyces griseofuscus*
Godo - BSP	Bacterial neutral protease	*Bacillus subtilis*
Godo - MSD	Dextranase	*Aspergillus niger*
Godo - YNL	Lactase	*Kluyveromyces lactis*
Grindamyl A	Fungal β - amylase	*Aspergillus niger*
Grindamyl S	Xylanase, hemicellulase	*Aspergillus niger*
Hannilase	Rennet	Microbiell
Hazyme	Fungal β - amylase	*Aspergillus oryzae*
Hitempase	Bacterial β - amylase	*Bacillus* sp.
Hydrolact	Lactase	*Khuyveromyces lactis*
Inulinase MJ	Inulinase	*Aspergillus* sp.
Kitalase	β - 1,3 Glucanase	*Arthrobacter* sp.
Kleistase	Bacterial α - amylase	*Bacillus amyloliquefaciens*
Klerzyme	Pectinase	*Aspergillus niger*
Lactase Y - AO	β - Galactosidase	*Aspergillus niger*
Lactase Y - L	β - Galactosidase	*Kluyveromyces lacts*
Lecitase	Phospholipase A 2	Hog pancrease
Lipase AP	Lipase	*Aspergillus* sp.
Lipase CE	Lipase	*Humicola lanuginosa*
Lipase MAP	Lipase	*Mucor* sp.
Lipase N	Lipase	*Rhizopus* sp.

Table 6.12 Contd...

Name	Type of Enzyme	Origin
Lipase P	Lipase	*Pseudomonas* sp.
Lipase Seiken	Lipase	*Rhizopus* sp.
Lipolact K1	Lipaesterase	*Aspergillus niger*
Lipolase	Lipase	Microbial
Lipozyme L	Fungal lipase	*Mucor miehei*
Lysozyme	N - Acetylmuramoyl-hydrolase	Hen's egg
Macerozyme 2S	Pectinase	*Rhizopus* sp.
Maltamyl	Fungal α - amylase	*Aspergillus niger*
Maturex	Diaacetyl reductase	*Bacillus* sp.
Maxamyl	Bacterial α - amylase	*Bacillus licheniformis*
Maxatase	Bacterial alkal. protease	*Bacillus licheniformis*
Maxinvert	β - fructofuranosidase	*Saccharomyces lactis*
Maxiren	Rennet 100 %	*Kluyveromyces lactis*
GB Megafresh	Bacterial α - amylase	*Bacillus subtilis*
Meicellulase	Cellulase	*Trichoderma reesei*
MKC - Fungal Amylase	Fungal α - amylase	*Aspergillus oryzae*
Molsine	Mix of hydrolases	*Aspergillus* sp.
Multifresh	Fungal α - amylase	*Aspergillus*. sp
Mycolase	Fungal α - amylase	*Aspergillus oryzae*
Neutrase	Bacterial neutral protease	*Bacillus amyloliquefaciens*
Newlase	Fungal acid protease	*Rhizopus* sp.
Novamyl	Maltogenic amylase	*Bacillus* sp.
Novoren	Rennet	*Mucor miehei*
Novozyme 188	Cellobiase	*Aspergillus niger*
Novozyme 230	Inulinase	*Aspergillus* sp.
Novozyme 234	Cell-wall-lysing hydrolase	*Trichoderma harzianum*
Optamyl - L	Bacterial α - amylase	*Bacillus licheniformis*
Optidex - L	Amyloglucosidase	*Aspergillus niger*
Optimase	Bacterial alkal.protease	*Bacillus licheniformis*
Optimash	Bacterial α - amylase	*Bacillus subtilis*
Optisp rit	Amyloglucosidase	*Aspergillus niger*
Optisweet S	Glucoseisomerase	*Streptomyces olivaceus*
Optisweet Immob	Glucoseisomerase. immobilized	*Streptomyces rubiginosus*
Optitherm	Bacterial α - amylase	*Bacillus licheni formis*
Orientase 90	Bacterial neutral protease	*Bacillus subtilis*
Oropan	Proteases	Pancreatic, fungal
Ovazyme	Glucoseoxidase, catalase	*Aspergillus niger*

Table 6.12 Contd...

Name	Type of Enzyme	Origin
Palatase	Lipaesterase	*Aspergillus niger*
Pancellase	Cellulase	*Aspergillus niger*
Pandicase XP	Fungal neutral protease	*Aspergillus oryzae*
Panprosin	Fungal acid protease	*Aspergillus niger*
Panzym	Pectinase	*Aspergillus niger*
Pectinex	Pectinase	*Aspergillus niger*
Pectinol	Pectinase	*Aspergillus niger*
Pectolase	Pectinase	*Aspergillus niger*
Pentopan	Pectinase	*Aspergillus niger*
Phytase	Fungal phytase	*Aspergillus niger* *Rhizophus sp.* *Trichoderma sp.*
Piccantase	Lipaesterase	Microbial
Piexazyme LA 1	β - Galactosidase, immobilized	*Aspergillus niger*
Proctase	Fungal acid protease	*Aspergillus* sp.
Prolysin	Bacterial neutral protease	*Bacillus subtilis*
Promen	Bacterial neutral protease	*Bacillus subtilis*
Pronase	Protease	*Streptomyces griseus*
Proteinase 18	Bacterial neutral protease	*Bacillus subtilis*
Prozyme	Fungal protease	*Aspergillus* sp.
Promozyme	Pullulanase	*Bacillus acidopullulyticus*
Pulluzyme	Pullulanase	*Klebsiella aerogenes*
Rapidase AG	Amyloglucosidase	*Aspergillus niger*
Rapidase C	Pectinase	*Aspergillus niger*
Rapidase CPE	Penctinesterase	*Aspergillus niger*
Rennilase	Rennet	*Mucor* sp.
Rhozyme P 11	Fungal protease	*Aspergillus oryzae*
Rhozyme P 53	Bacterial protease	*Bacillus subtilis*
Rhozyme P 64	Bacterial alkal. protease	*Bacillus licheniformis*
Rohalase A	Bacterial α - amylase	*Bacillus amyloliquiefaciens*
Rohalase M	Fungal α - amylase	*Aspergillus oryzae*
Rohamalt MG	β - Glucanase	*Trichoderma reasei*
Rohamalt P	Bacterial protease	*Bacillus subtilis*
Rohament CA	Hemicellulase	*Aspergillus niger*
Rohament PC	Macerase	*Aspergillus niger*
Rohament P	Pectinglycosidase	*Aspergillus* sp.
Rohapect D, MA, MB, C	Pectinases	*Aspergillus niger*
Samptase	Fungal acid protease	*Aspergillus niger*

Table 6.12 Contd...

Name	**Type of Enzyme**	**Origin**
Sanactase	Fungal α - amylase	*Aspergillus* sp.
Sanzyme	Fungal α - amylase	*Aspergillus oryzae*
Sanzyme SS	Fungal acid protease	*Aspergillus* sp.
Savinase	Bacterial alkal. protease	*Bacillus* sp.
Sclase	Pectinase	*Aspergillus usamii*
Speedase	Bacterial α - amylase	*Bacillus subtilis*
Spezyme	Glucoseisomerase	*Streptomyces* sp.
Spezyme GI	Glucoseisomerase	*Streptomyces rubiginosus*
Starzyme AG	Amyloglucosidase	*Aspergillus niger*
Sternzyme H	Hemicellulase	*Trichoderma reesei*
Sterzyme B 2000	Fungal protease	*Aspergillus niger*
Sternzyme BE	Protease, hemicellulase	*Bacillus*. sp., *Aspregillus* sp.
Sternzyme BK	Bacterial protease	*Bacillus subtilis*
Sumizyme AC	Cellulase	*Aspergillus niger*
Sumizyme AP	Pectinase	*Aspergillus niger*
Sumizyme C	Cellulase, hemicellulase	*Trichoderma reesei*
Sumizyme FP	Fungal acid protease	*Aspergillus oryzae*
Sumizyme MP	Fungal alkal. protease	*Aspergillus melleus*
Sumizyme MC	Macerase	*Rhizopus oryzae*
Sweetzyme	Glucoseisomerase, immobilized	*Bacillus coagulans*
Taka - Sweet	Glucoseisomerase, immobilized	*Bacillus olvaceus*
Taka - Therm	Bacterial α - amylase	*Bacillus licheniformis*
Termamyl	Bacterial α - amylase	*Bacillus licheniformis*
Thermoase	Protease	*Bacillus thermoproteolyticus*
Thermolysin	Protease	*Bacillus thermoproteolyticus*
Tunicase	Cellwall lytic enzyme	*Arthrobacter* sp.
Ultrazyme	Pectinase	*Aspergillus niger*
Unicase	Amyloglucosidase	*Rhizopus* sp.
Veron A, AV, AC	Fungal α - amylase	*Aspergillus oryzae*
Veron F 25	Bacterial α - amylase	*Bacillus* sp.
Veron HE	Hemicellulase	*Trichoderma reesei*
Veron P	Bacterial protease	*Bacillus subtilis*
Veron SX	Xylanase	Microbial
Veron GX	Xylanase	*Aspergillus niger*
Xylanase 250	Xylanase	*Trichoderma viride*
YL 5	Cellwall lytic enzyme	Microbial

Source - (Helmut Uhlig, 1998)

In Industrial Enzymes and their applications. John Wiley and sons. Inc.

The increasing yields and improving raw-materials utilization, lowering costs, altering technical properties, and improving preservation, flavour, and cleansing action are the prime requisites in the economic considerations.

6.10 Biosorption

Biosorption processes using live and dead organisms to remove toxic heavy metals or recovery of rare (costly) metals from effluents of various industries is an interesting and useful application of biotechnology. It is proved to be more efficient than physical and chemical methods in many respects. Thus, copper, zinc, iron, mercury, lead, nickel, silver, gold, uranium, thorium can be removed or recovered from effluents by biosorption.

Definition

'*Biosorption*' is defined as the removal of metals from solution as a result of physical and chemical actions of living or dead biomass, excreted and derived components and products. Among all living systems, microorganisms gained significance due to easy availability in nature and artificial cultivation. Biosorption is also considered as a property of certain types of inactive, dead microbial biomass to concentrate heavy metals from even very dilute aqueous solution.

The removal of heavy metals is by the following process :

1. *Extra cellular precipitation and complexation :* Many extracellular fungal products can complex and precipitate heavy metals. Acid can be an efficient metal ion chelator and oxalic acid is a effective insoluble metal oxalaters.
2. *Metal binding to cell wall :* It includes the phenomenon, such as ion exchange, adsorption, complexation, precipitation and crystallisation. A variety of potential sites may be involved in metal sequestration include carboxyl, amine, hydroxyl, phosphate and sulphydryl groups.
3. *Transport of toxic metals :* The plasma and vacuolar membranes are the main transport membranes of fungi. Metabolism - dependant intracellular uptake or transport of metal ions in to cell can be inhibited by low temperature, absence of energy source, metabolic inhibitors etc.
4. *Intra cellular fate of toxic metals :* Metallo thioneins are small, cystine rich polypeptides that can bind essential metals like Cu and Zn and non - essential metals like Cd.
5. *Metal Y - glutamyl peptides :* Another metal binding molecules are short, cysteine containing Y-glutamyl peptides, encompassed by the trivial name phytochelatinins.

   Schizosaccromyces is known to synthesise at least seven different homologons phytochelatins in response to metal exposure.
6. *Vacuolar compartmentation :* The fungal vacuole is a highly important organelle with functions that include, macro molecular degradation, storage of metabolites, and cystolic ion and pH homeostasis. The electro chemical proton gradient generated across the vacuolar membrane energises transport of monovalent and divalent cations in to vacuole as well as other substances including basic amino acids.

7. *Metal transformation :* Microorganisms can effect chemical transformations of metal by oxidation, reduction, methylation and dealkylation. Some enzymatic transformations may be involved in survival, since certain transformed metals are less toxic and more volatile than the original.

Biosorption – Non living biomass : Advantages

1. Toxicity effects on the absorption are less.
2. No need to supplement with nutrients.
3. Process is not physiologically controlled.
4. Toxicity limitation or thermal inactivation not concerned to immobilization.
5. Biomass acts as ion exchanger, so rapid and efficient in metal uptake.
6. Metals are desorbed and recovered easily.

Biosorption – Living Biomass : Advantages

1. This system using living biomass is self replinishing due to growth.
2. The metal is deposited in an altered chemical state and less sensitive to spontaneous desorption.
3. Metabolic activity is the only economic way of degradation of organo-metallic compounds.
4. It is a microbial property of removal of metal toxicity rather than a product under exploitation.
5. Two or more organisms can be employed in a synergistic way.

Biosorption is very cost effective system when compared to any other prevailing treatment methods like electrodialysis, ion exchange, distillation, reverse osmosis, reduction, precipitation, freezing, liquid - liquid extraction etc.,

6.11 Bacterial Leaching and Biomining

It is relatively recently, the bacterial activity on weathering and leaching of mineral ores was established. The relations between biotechnology and metallurgy has created a new branch of applied science, biohydrometallurgy. This area of research, using bacteria for extraction of minerals is popularly known as bacterial leaching and biomining.

In 1947, the bacterium, *Thiobacillus ferrooxidans* was used to leach the copper ores in USA. These acidophilic autotrophic bacteria *T. ferrooxidans* and *T. thiooxidans* followed the two biochemical reactions for the extraction of copper and uranium : (i) Redox processes at the mineral surfaces and (ii) Biosorption of metals on bacterial cell surfaces. The important activity of the bacteria for its survival and for leaching of minerals is the oxidation of sulphur that binds copper, zinc, lead and uranium. Mining operations using bacteria are becoming popular, as the conventional methods are being costly, involves lot of energy and releases toxic gases. The bacteria like, *Leptospirillum, ferooxidans*, *Sulfolobus* sp and *Thiobacillus acidophiles* were isolated from biobleaching sites and their role in biomining was understood.

Leaching reactions from direct bacterial attack are :

$4\ Fe\ S_2$ (Pyrite) + $15\ O_2 + 2\ H_2O \rightarrow 2\ Fe_2\ (SO_4)_3 + 2\ H_2\ SO_4$

$4\ Cu\ Fe\ S_2$ (Chalcopyrite) + $17\ O_2 + 2\ H_2\ SO_4 \rightarrow 4\ Cu\ So_4 + 2Fe_2\ (So_4) + 2\ H_2O$

$2\ Fe\ As\ S$ (arsenopyrite) + $7\ O_2 + 2\ H_2O \rightarrow 2\ Fe\ \ Aso_4) + 2\ H_2\ So_4$

CuS (covellite) + $2O_2 \rightarrow Cu\ So_4$

The alternate, indirect bacterial leaching :

$2\ Fe\ S_2 + 2\ Fe_2\ (So_4)_3 \rightarrow 6\ Fe\ So_4 + 4S$

$Cus + Fe\ (So_4)_3 \rightarrow 6\ Cu\ So_4 + 2\ Fe\ So_4 + S$

$Uo_2 + Fe\ (So_4)_3 \rightarrow Uo_2\ So_4 + 2\ Fe\ So_4$

$Uo_3 + H_2\ So_4 \rightarrow Uo_2\ So_4 + H_2O$

The bacterial leaching can be followed by three methods : Vat leaching (in aerated tanks), heap method (natural heap formation) and *in situ* method (without removing the ore from underground i.e. by applying bacterial leaching solution).

In the bioleaching process of Uranium, the microbes do not interact with uranium but act on iron oxidant. Pyrite converts in to ferric sulphate and sulphuric acid by *Thiobacillus ferooxidans*. The dissolved uranium is extracted from the leached solution with organic solvents as tributyryl phosphate and the uranium is then precipitated.

Process of Biomining of Copper

Preparation of piles of ore and ground permeability

⇓

The pile is sprayed with leaching solution that contains Fe 3+ ions, sulphate ions and *T. ferooxidans* (The sulphate ion in the leaching solution form $H_2\ SO_4$ giving suitable acidic solution for growth of bacteria) ⇓

Copper released due to catalytic chemical reaction is drained in the form of solution.

⇓

Copper is then removed from the solution with the help of another solvent

⇓

The remaining leaching solution flows in to an open pond, where *T. ferrooxidans* catalyses a reaction that oxidises Fe 2+ in to Fe 3+ ions and sulphide into sulphate ions - now the leaching solution is recharged.

⇓

The recharged leaching solution is pumped back to the top of the pile for the cycle to begin again.

⇓

Then the copper is extracted from the solution by '*bioabsorption filters*' made up of algae.

As these biological methods, though cheap and ecofriendly, delays cash flow due to slower production and therefore hindered its adoption. Hence, genetically engineered bacteria are constructed with modern recombinant –DNA technology to enhance the speed of reaction. The function of some chromosomal genes has been identified in *T. ferooxidans* by cloning, sequencing and complementation of the corresponding gene delections in *E. coli*. A small broad host range (BHR) plasmid has been discovered in *T. ferooxidans*. A natural mutant, *Alcaligens eutrophus* var. *metallo-tolerans* was isolated from soils containing high levels of heavy metals.

Members of genera *Botrytis, Aspergillus, Cephalosporium, Spicaria, Fusarium, Hormodendrum, Mucor, Penicillium, Trichoderma* have been observed to dissolve the ores and their use in biomining is now established.

6.12 Bioremediation

'*Bioremediation*' encompasses biological methods for clean-up of contaminated soil and ground water. It involves establishing the conditions in contaminated environment, so that appropriate microorganism flourish and carry out the metabolic activities to detoxify the contaminants.

Bioremediation can be defined as stimulation of microorganisms to rapidly degrade hazardous organic contaminants to environmentally safe levels in soils, sub surface materials, water, sludges, residues etc.

Biodegradation of target pollutants is achieved with the following environmental conditions :

1. Bio availability
2. Physical and chemical factors (pH, oxygen, moisture, redox potential, electron acceptors)
3. Concentration of pollutant
4. Recalcitrance
5. Xenobiotic nature of pollutant.

Bioremediation is becoming a potential treatment technology as it is less costly and toxic chemicals are removed from the environment.

There are some constraints in the bioremediation process, they are :

1. The process of bioremediation is relatively slow,
2. Heavy metals are not removed,
3. It does not remove all quantities of contaminants, and
4. Comprehensive pollutant audits are required.

The natural bioremediation related to an increase in the contaminant - degrading microorganisms by adding nutrients or removal of toxins or elimination of predators. It also enhanced by induction of enzymes responsible for the degradation of contaminants. Here natural microorganisms have

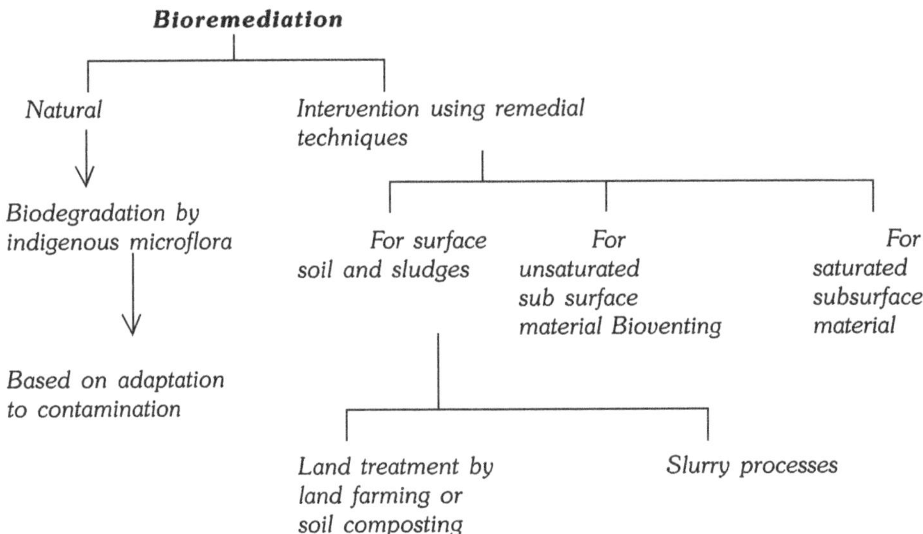

not evolved genetic competence to utilise synthetic compounds. In this regard, the application of genetic selection and gene manipulation techniques, can make the process of capacity development fast.

In the bioremediation of oil polluted soils, microbes are used in a transportable bioreactor which help to degrade the oil components within weeks. The final products are reusable soil, CO_2 and H_2O. The soil can be reused after cleaning in horticulture or landscape culture.

The *Pseudomonas acidovorans, P. fluorescence, P. mendocina, P. aeruginosa* are proved to be efficient in degrading 2, 4, 6 trinitrotoluene (TNT) by cometabolism. The two intermediate compounds generated are 4 - amino. 2, 6 dinitrotoluene and 2 - amino 4, 6 dinitrotoluene. This biological degradation of TNT may provide safe and cost-effective approach to bioremediation of contaminated soil. Two important things are needed is bioremediation of TNT.

1. Foreign organisms must survive in new environment
2. Inoculum must remain in contact with TNT.

   Several criteria must be met before bioremediation can be effective :

   (i) Microbes that can degrade must be present at the site or brought to the site, and

   (ii) Microbe must attack the toxicant and convert them in to harmless products.
3. Environmental conditions (temperature, pH, nutrients, depth, diffusion, spatial variability) must permit microbial growth and activity. Where poor conditions exist, the environment can be modified to support or accelerate microbial growth by the addition of an electron acceptor (O_2) and essential nutrients such as N_2, P, and trace elements.
4. If the contaminant is present in the subsurface, that sub surface must be permeable enough to allow perfusion with a solution of oxygen and nutrients.

6.13 Bio Polyesters

Poly (β-Hydroxyalkanoates)

Structures : Microorganisms are popularly known to produce intracellular energy and carbon storage products, called as poly (β - hydroxy butyrate) (PHB). This polymer is fixed conveniently in the group of poly (β - hydroxyalkanoates) (PHA). These polymers are formed as intracellular inclusions under stressed growth conditions. The presence of excess carbon and energy sources; limiting nutrient or growth factors are the unbalanced growth conditions. These results in reduction equivalents, which originate from metabolic oxidation processes, are stored in water insoluble, chemically and osmotically inert form of the following structure :

$$-\left[O-\overset{\overset{\large R}{|}}{C}H-CH_2-\overset{\overset{\large O}{\|}}{C}\right]-m$$

R = n - alkyl pendant group of variable chain length

HB, β - hydroxy butyrate, where R = methyl

HV, β - hydroxy valerate, where R = ethyl

HC, β - hydroxy caproate, where R = n - propyl

HH, β - hydroxy haptanoate, where R = n - butyl

HO, β - hydroxy octanoate, where R = n - pentyl

IIN, β - hydroxy nonanoate, where R = n - hexyl

HD, β - hydroxy decanoate, where R = n - heptyl

HUD, β - hydroxy undecanoate, where R = n - octyl

HDD, β - hydroxy dodecanoate, where R = n - nonyl.

Distribution

The different groups of eubacteria and archaebacteria, and cyanobacteria are capable of accumulation of PHA in the form of storage products. Among them *Alcaligens eutrophus* is known to contain high amount (96 %) of PHA with fructose as PHA producing substrate. The *Rhodobacter*, phototrophic bacteria is also efficient in production of PHA with acetate as its substrate and reached around 80% (Table 6.13).

Table 6.13

S.No	Group	Genus	PHA Content (% dry weight)	PHA Producing substrate
1.	*Phototrophic Bacteria*	*Chloroflexus*	<1	Yeast extract, glycylgycine
		Ectothiorhodospira	ND	NS
		Rhodobacter	80	Acetate
		Rhodospirillum	47	Acetate

Table 6.13 Contd...

S.No	Group	Genus	PHA Content (% dry weight)	PHA Producing substrate
2.	*Gliding bacteria*	*Beggiatoa*	57	Acetate
3.	*Sheathed bacteria*	Leptothrix *Sphaerotilus*	67 45	Pyruvte Glucose / Peptone
4.	*Budding and Appendaged bacteria*	*Caulobacter* *Hypomicrobium*	36 ND	Glucose / Glutamate Methanol, glucose
5.	*Spiral or curved bacteria*	*Aquaspirillum* *Azospirillum* *Spirillum*	ND 75 40	Malate Malate Lactate
6.	*Gram negative, aerobic rods and cocci.*	*Azotobacter* *Derxia* *Methylobacterium* *Methylosinus* *Pseudomonas* *Rhizobium* *Zoogloea*	73 26 47 25 67 57 ND	Glucose Glucose Methanol Methane Methanol Mannitol Yeast extract, Casaminoacids
7.	*Gram negative facultative anaerobic rods.*	*Chromobacterium*	37	Glucose / Peptone
8.	*Gram negative anaerobic bacteria*	*Syntrophomonas*	30	Crotonate
9.	*Gram negative cocci and coccal bacilli.*	*Acinetobacter* *Moraxella*	< 1 20	Glucose NS
10.	*Archelebacteria Gram positive cocci*	*Halobacterium* *Micrococcus*	38 28	Glucose Peptone / Tryptone
11.	*Endospore forming rods and cocci*	*Clostridium* *Bacillus*	13 25	Tryptone/Peptone/Glucose Glucose
12.	*Actinomycetes*	*Mycoplana* *Nocardia* *Streptomyces*	ND 14 4	Methanol Butane Glucose
13.	Cyanobacteria	*Aphanothece* *Chlorogloea* *Microcoleus* *Spirulina*	< 1 10 < 1 6	NS Acetate, Co_2 NS Co_2

ND - *Not Determined* **NS - *Substrate not Specified***

The Physiological and ecological conditions generally controls the quantitative formation of PHA. Low concentration or total absence of a variety of different nutrients can induce or stimulate the formation of PHA. Usually ammonia, carbon, iron, magnesium, manganese, oxygen, phosphate, potassium and sulphate are the preferred nutrients to induce the PHA formation. The intracellular presence of PHA seems to play a significant role in the survival of the microorganisms under other conditions of environmental stress such as when subjected to osmotic pressure, or desiccation. In general, the cells containing this polymer has a higher survival rate than PHA-free cells.

The physical and mechanical properties of these copolymers such as stiffness, brittleness, melting point, grass transition temperature, or resistance to organic solvents can change considerably as a function of the monomer composition.

Analytical Methods

Earlier, the PHA quantities were determined by the digestion of cells containing granules with hypochlorite and then subsequently measuring the turbidity caused by the remaining polymer granules. PHA amounts were also determined by gravimetric methods after extraction of the polymer with a solvent such as chloroform or propylene carbonate and subsequent determination of the polymer in either.

Another method of the quantification of PHA was by μv-spectrophotometric technique after the acidic conversion of PHB to crotonic acid which was measured at 235 nm. Recently, the gas chromatography, high pressure liquid chromatography (HPLC) and nuclear magnetic resonance spectroscopy (NMR) are used to quantify PHA.

PHA as Biodegradable Plastics

PHA production has gained considerable industrial interest and environmental importance to evaluate it as polyesters for the use in either biodegradable and biocompatible plastics for a wide range of applications.

Medical Applications

- Surgical pins, sutures, staples, and swabs
- Wound dressing
- Blood vessel replacements
- Bone replacements and plates
- Stimulation of bone growth by piezoelectric properties.
- Biodegradable carrier for long term dosage of drugs and medicines.

Industrial Applications

- Biodegradable carrier for long term dosage of herbicides, fungicides, insecticides or fertilizers.
- Packaging containers, bottles, wrappings, bags and films.
- Disposable items such as diapers or feminine hygiene products.

For these medical and industrial applications the PHA are commercially available under the trade name of '*Biopol*'. Biochemically, there are two different ways of achieving polymer formations in microbes :

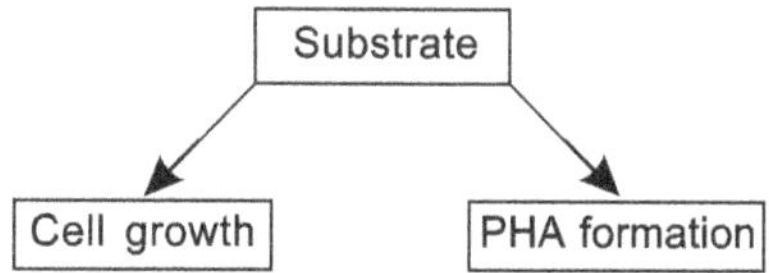

1. Substrate is utilised simultaneously for both cell growth and PHA formation.

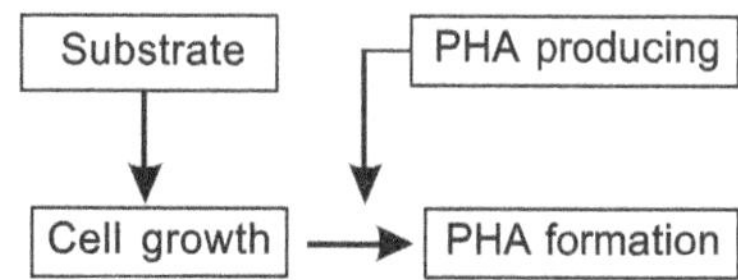

2. A polymer producing substrate is added to the culture after the growth

Biodegradation of PHA

A series of parameters can influence the rate of biodegradations and the life span of a plastic material in nature, which includes :

(i) The type of environment,

(ii) The presence of a microbial population and its microbial activity,

(iii) The availability of water,

(iv) The temperature,

(v) The section thickness of the plastic material,

(vi) The surface texture and porosity of plastic material, and

(vii) The presence of second components in the plastic such as fillers or colouring agents.

Two mechanisms are responsible for the degradation of PHA. Under sterile or aseptic conditions PHA is degraded by a hydrolytic mechanisms, especially at high pH. This type of degradation is important for medical applications. In natural environments, the polymer is degraded enzymatically by the action of depolymerase or esterase. PHA's with longer chain length pendant groups undergo slower biodegradation rates in environments such as sewage sludge or sediments for the following reasons :

(a) Low molecular weight carbon compounds are mineralised more rapidly, where as more complex structures have much lower rates of mineralisation.

(b) Long chain repeating units increase the hydrophobicity of the polymer and it generally inhibits or prevents microbial growth on the polymer surface.

6.14 Bio Fuels

'*Bio fuel*' is the general term for sustainable fuels because the source is harvestable. Solid biofuels, such as firewood are abundant. Liquid biofuels are convenient and safer. Therefore, biofuels, which are non - conventional, renewable are of high priority for a fast developing country like India. Tropical countries like India are previlaged with a high rate of biomass production, stores up enormous quantities of convertable solar energy. Agricultural and forest wastes abounding in villages, and municipal solid waste heaping in urban centres can be used to yield high quantities of eco-friendly biofuels.

In practice, we acquire two-fifths of the energy from burning of oil and the balance from other sources, such as coal, gas and nuclear energy. Indeed, almost 90 percent of our energy comes from fossil fuels which will one day run out. What will we do then ? What are the alternative fuels to which we can turn ? The convenient answer comes from biofuels. Ethanol, methanol, esters and hydrogen are the principal renewable bio fuel contenders.

Among many important biofuels used for energy generation purposes in various parts of the world; in view of the importance the two biofuels, Ethanol and Hydrogen are described as under.

Ethanol

From early days ethanol has been produced from sugar by fermentation. Brazil, Zimbabwe and the U.S.A. produce this as a fuel. Ethanol does not release as much energy as petrol. But in terms of volume, the fuel almost match, ethanol releasing 23,000 kilojoules per litre as against 25,000 for petrol. By 1990 Brazilians were growing 25 million tones of sugarcane a year on four million hectares (8 % of their cultivated land) and turning three quarters of it is to 12 billion litres of ethanol, and another million which burn a mixture of 20 percent ethanol and 80 percent petrol by volume.

In terms of cost, ethanol can no longer compete with petrol. This is despite the Brazilian government buying the electricity generated by burning sugarcane residues, farmers increasing sugarcane yields and manufacturers boosting alcohol production by 15 percent. Ethanol is still used because, as a so called oxygenated fuel, it produces less pollution. Ethanol enabled Brazil to be one of the first countries to phase out leaded petrol, and to eliminate photochemical smog caused by unburnt hydrocarbons. The U.S. meanwhile, produces more than 3.5 billion litres of ethanol a year by fermenting the starch of surplus corn and grain. The plan is to double this by the year 2004, and much of it will end up in reformulated blends of gasoline.

Steps involved in the manufacture of ethanol :

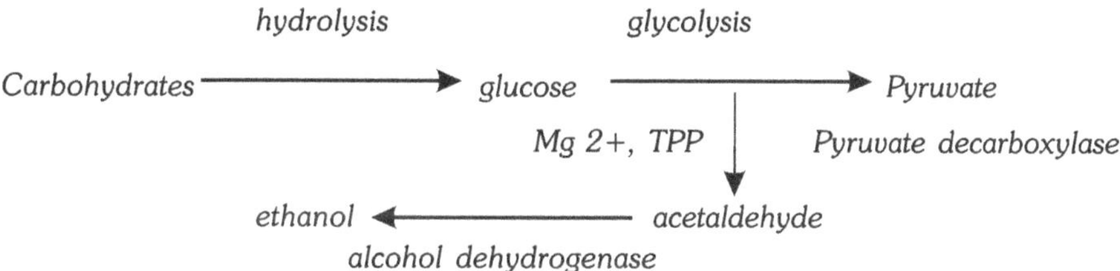

The three important substrates for the production of ethanol are :

(i) Roots, tubers, or grains that are rich in starch

(ii) Molasses or juice derived from sugarcane, palm and sugar beet, and

(iii) Wood or wood products from processed wood. In India, ethanol is produced mainly from fermentation of molasses. The flow scheme is as follow :

Powdered seed meal in water (pH 6.0 – 6.5).

heat 60 – 80^0C cool

⇓ + small amount of α - amylase

Liquefied substrate (50 – 60^0C)

⇓ + glucoamylase and α - amylase

digested starch

⇓ + air + yeast culture

Fermentation of starch

⇓ (30 – 35^0C) + yeast cells

Fermented broth

Ethanol

Hydrogen

Besides ethanol and esters, there are other fuels produced from fossil fuels which could be made from sustainable resources. Among them very important and highly accepted biofuel is hydrogen. When hydrogen burns one gram yields more energy than the same amount of conventional fuels. In the future energy plans, the hydrogen economy is emerging on first priority, but the problem is its storage. A tonne of hydrogen occupies 11 million litres, which is about the size of a big house. More over, hydrogen is less dense than the natural gas, methane, and requires three times of the volume of gas to produce the same amount of heat.

Condensed hydrogen to a liquid at – 253^0 C and a tonne of the gas becomes a manageable 14,000 litres. This is how large quantities of hydrogen are transported and stored for powering rockets in the American space programmes the only vehicle which currently run on hydrogen.

Hydrogen is already manufactured on a large scale for the chemical industry, and piped over hundreds of kilometers around Europe and U.S. World production of hydrogen is around 350 billion cubic metres per year. There are two natural sources of hydrogen : water and hydrocarbons such as Methane.

Production of Hydrogen

(i) When an electric current passes through water it releases hydrogen. The electrolysis of steam inside porous electrodes of zirconium oxide produce hydrogen even more efficiently. Even so, the process would not make the hydrogen economy viable. It is cheaper to produce electricity than to make hydrogen. But the surplus electricity produced by nuclear power stations at night could provide a cheap way of electrolysing water to make hydrogen.

(ii) The second way of generating hydrogen is to blow steam through white-hot coal to produce a mixture of hydrogen and carbon monoxide gas, known as synthesis gas. Charcoal, a renewable fuel, could also be used.

(iii) Another way of making hydrogen is from alkanes such as methane and propane by treating them with steam at about 1000^0C. This gives a mixture that is 75 percent hydrogen by volume.

(iv) The recent *biotechnological method* of hydrogen production is through photosynthetic machinery by green plants. The chlorophyll helps in trapping solar energy in the form of ATP, which is utilised in photosynthesis. Here, the water is subjected to photolysis, which splits the water molecule in to oxygen, electrons and hydrogen (H^+) ions. These hydrogen ions can be converted in to hydrogen gas (H_2) and collected and used as fuel. The important enzymes for the conversion of H^+ to H_2 are hydrogenase and nitrogenase.

The algal members like, *Scenedesmus, Chlamydomonas, Dunaliella, Porphyridium, Oscillatoria,* and some Photosynthetic bacteria like, *Rhodospirillum, Chromatium, Clostridium, Thiocapsa* can be employed for hydrogen production. The *hydrogenase* produced by these organisms combine the H+ ions and electrons liberated during photolysis of water to form hydrogen. The hydrogen gas thus produced will bubble out of the solution and can be easily collected.

Nitrogenase is the chief hydrogen producing enzyme in cyanobacterial species. This enzyme reduces nitrogen to ammonia with associated release of hydrogen gas. Compared to hydrogenase, the production of hydrogen with nitrogenase is efficient, because the former is sensitive to oxygen, and in heterocyst of cyanobacteria the oxygen concentration is low or completely absent. Hydrogen production depends on light, which provides electron donors and also activates nitrogenase and ATP formation in heterocysts.

Modern trends in Hydrogen production using Blue - Green Algae (BGA)

Hydrogen production using cyanobacteria (BGA) is a clean technology. But more research is needed to identify the right type of blue - green algae or genetically engineer them for higher activity so that the efficiency of the hydrogen production system will be more economical.

Research on hydrogen production by photobiological methods was to illuminate chloroplasts in sun light and use the photosynthetically derived energy and reducing power to produce hydrogen in the presence of the enzyme hydrogenase. However, the system had a short life due to the instability of some of the chloroplast components when exposed to long periods of light and oxygen.

The components of chloroplasts were continuously synthesised in the presence of light worked or whole organisms (cells), which are much more stable when exposed to light. Cyanobacteria, especially the nitrogen fixing ones contain the enzymes nitrogenase and hydrogenase, which can generate hydrogen under particular conditions.

Theoretically this is an easy process of splitting of water in to hydrogen and oxygen; cyano -bacteria are easy to grow; do not require expensive chemicals; main substrates are water, CO_2 and air; however, in practice it is not economical to produce hydrogen.

Another problem during the process of splitting of water is the production of oxygen. This evolved oxygen inhibits enzymes that catalyse the hydrogen production. So the reactions should under go in microaerobic conditions (partial vacuum).

6.15 Biofertilizers

Nitrogen is a major element required by the plant for growth and development. The nitrogen is provided in the form of chemical fertilizer. Such chemical fertilizer pose health hazards and pollution problem in soil besides these are quite expensive. Therefore, biofertilizers are being recommended in place of chemical fertilizers

Definition

The term Biofertilizers or microbial inoculants can be defined as preparations containing live or latent cells of efficient strains of nitrogen fixing, phosphate solubilizing or cellulolytic microorganisms used for application to seed, soil or composting areas with the objective of increasing the numbers of such microorganisms and accelerate certain microbial processes to augument the extent of the availability of nutrients in a form which can be easily assimilated by plants.

Further, the microbial process is,

1. Quick
2. Consume less energy than industrial processes,

3. Diversified in to small units to meet the demands of the specific problems
4. Recycling of farm wastes (organic resources)
5. Work in fixing nitrogen, mobilizing other plant nutrients, and degrading lignocellulosic wastes.

Approaches to the problem

There are both novel and conventional approaches to the problem related to biofertilizers; they are:

1. To increase the area under cultivations of grain legumes by the introduction of legumes in inter, multiple and relay cropping.
2. To produce efficient strains of rhizobia for inoculating legumes and other non symbiotic nitrogen fixers for inoculating cereals.
3. To evolve a technology enabling the newly introduced nitrogen fixers to successfully compete with the strains of nitrogen fixing microorganisms already present in the soil.
4. To evolve nitrogen responsive varieties of plants.
5. To overcome the inhibition of fertilizer nitrogen on biological fixation of nitrogen.
6. To define agronomic practices heading better fixation and conservation of nitrogen on the farm.
7. To develop inexpensive denitrification inhibitors such as neem cake, and
8. To evolve simple practices to conserve water on the farm because optimum moisture is needed for successful nodulation and hence biological nitrogen fixation in legumes.

Potential Organisms for Biofertilizers

Many microorganisms, both symbiotic and non symbiotic are being identified and exploited as potential source of biofertilizers. Attempts are being made to large scale production and mass inoculation of these organisms to the seed or soil. Some of these are:

I. *Nitrogen Fixing Organisms*

A. *Symbolic Organisms*

(i) *Rhizobium*

(ii) *Azolla*

(iii) Associated organisms - *Azospirillum*

B. *Non symbiotic nitrogen fixers (Free - Living)*

I. *Heterotrophic*

(i) *Aerobic - Azotobacter, Azomonas, Beijerinckia, Derxia, Mycobacterium,*

(ii) *Anaerobic - Clostridium, Desulfovibrio, Chlorobium, Chromatium*

(iii) *Facultative anaerobic - Bacillus, Enterobacter, Escherichia, Klebsiella, Rhodopseudomonas, Rhodospirillum*

2. *Phototrophic*

(i) *Blue green algae : Anabaena, Anabaenopsis, Aulosira, Cylindrospermum, Nostoc, Calothrix, Scytonena, Tolypothrix, Fischerella, Mastigocladus, Stigonema, Westiella, Westiellopsis, Campylonem, Michrochate*

(ii) *Phototrophic bacteria : Rhodomicrobium, Rhodopseudomons, Rhodspirillum, Rhdocyclus, Thiospirillum, Thiocapsa, Chromatium, Ectothrioshodopsira,*

II. *Phosphate solubilizing or phosphate augumenting organisms*

(i) *Thiobacillus. Bacillus, Pseudomonas* sp.etc.

(ii) Mycorrhizae.

Rhizobium

Rhizobium forms symbiotic association in the roots of leguminous crops, leading to the formation of nodules, which help in improving the fertility of soil through fixation of nitrogen from the atmosphere. Several groups of Rhizobia have been recognised for inoculating legumes in India. They are, *R. leguminosarum, R. melilotii, R. trifoli, R. lupinli, R. japonicum* and *Rhizobium* sp.

The species of *Rhizobium* form nodules in the roots of leguminous plants and establish themselves inside the root tissue of the host plant. The bacteria invade the roots by forming are infection thread which enters through root hairs which curl and invaginate. The infected root cells undergo proliferation and cell enlargement leading to the formations of nodule (Fig. 6.38).

Bacteroids are the seats of nitrogen fixation and nitrogenase is the key enzyme to convert N_2 to NH_3. This nitrogen fixation is essentially an anaerobic process and nodule developed a mechanism to exclude oxygen from the bacteroid which is the site of fixation.

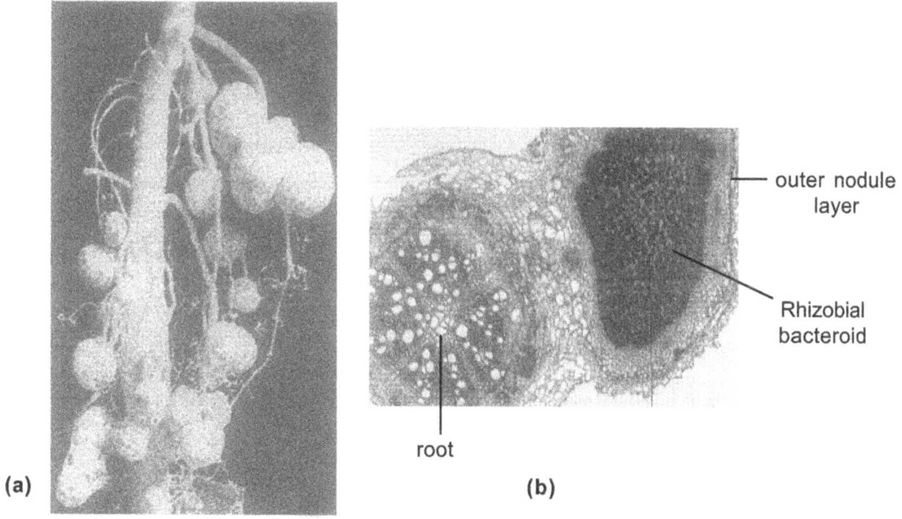

Fig. 6.38 (a) bunch of N2 fixing root nodules on a legume root and (b) T. S. of root and nodule

Cultivation and mass production of Rhizobium

Rhizobia are maintained on Yeast Extract Mannitol (YEM) agar medium. The selection of suitable strains of *Rhizobium* species is with a single strain or more than one strain for a particular cultivar or a group of cultivars one strain for a particular cultivar or a group of cultivars or crops. The selected strain is grown for 3-4 days on YEM agar slants. The culture is then transferred to a large flasks containing sterile solid or liquid medium for 4-9 days. By this time, a large quantity of a liquid broth is formulated in the fermentor, pH adjusted to 6.5 to 7.0 and sterilised. After cooling to 30^0 C, inoculum from the seed tank fermentor is transferred asceptically to the production fermentor at the rate of 1 percent by volume (Fig. 6.39).

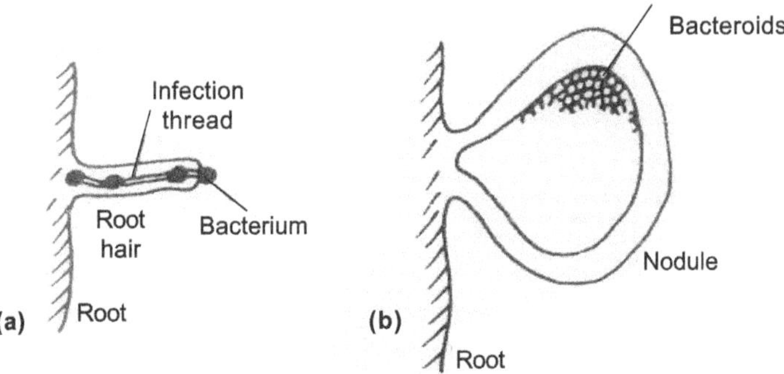

Fig. 6.39 Diagram showing of root nodule formation, (a) Infection by *Rhizobium* bacteria via the root hairs by the invagination of the cell wall to form an 'infection thread', (b) release of the bacteria from the infection thread leads to the formation of the nodule, packed with nitrogen-fixing bacteriods.

Procedure for inoculating seeds with Rhizobium

Sugar 50 g + 500 ml water
⇓
Heat the solution for 15 mts + gram aerobic 200 g.
⇓
Cool to room temperature
⇓
Add legume inoculant in peat
⇓ (150g)
Mix inoculum to form slurry
⇓
Add seeds to inoculum slurry
⇓
Mix the seeds with incolum slurry by hand
⇓
Dry the seeds in shade on paper.

Azolla

Azolla is a free floating aquatic fern used as an organic manure. It has an algal symbiont, *Anabaena azollae,* with in a cavity and fixes nitrogen. The common species of *Azolla* occuring in India are, *A caroliniana, A. nilotica, A.filiculodes, A.mexicana,A.microphyllg,A pinnata. Azolla* has become a common organic input in rice cultivation in Thailand, Indonesia, China and Philippines

Cultivation of Azolla

Azolla is seeded in flooded rice fields
⇓
After 5 –10 days the water is drained and *Azolla* is ploughed.
⇓
Before/after transplantation the *Azolla* plants are buried with hands in soil
⇓
The process is repeated as and when crop needs nitrogen fertilizer.

Azospirillum

Nitrogen fixation has been shown in two species, *Azospirillum brasilese* and *A. lipoferum* from rhizosphere soils of orchards and plantation crops based on their growth in nitrogen free medium. Nitrogen fixation was found in tomato plants as an associative effect of the combined influence of VAM fungi and free living nitrogen fixers.

Azospirillum is a gram negative and contains poly 'β' hydroxy butarate granules. For fixations of molecular nitrogen, it needs microaerophilic conditions. This organism is also known to produce growth substances such as IAA, kinetins and gibberellins. The response of sorghum and pearl millet to A. *brasilense* inoculant under several agroclimate conditions are successful.

Azotobacter

Bacteria belonging to the genera *Azotobacter* (family Azotobacteraceae) constitute the majority of free living nitrogen fixing bacteria. Many of species of *Azotobacter* are recognised, and among them most intensively investigated and commercially exploited species are:

A. chroococcum, A. agilis, A vinelandii, A. beijerincki, A.insigais, A. macrocytogenes, A.paspali.

Inoculation of soil or seed with *Azotobacter* is effective in increasing yield of crops. It is also synthesises various substances such as, vitamin - B, IAA, and gibberellins. Its application was proved in improving seed germination, plant growth and yield of crops of wheat, barely, maize and sugar cane.

Blue Green Algae

Biofertilisers belonging to blue green algae play a vital role in solving the problems of soil fertility, soil productivity and environmental quality. *Anabaena, Aulosira, Calothrix Scytonema,* and *Tolypothrix* are considered to be very important genera for the spontaneous fertility of the tropical rice fields. The application of blue green algae enhance the organic mater to the soil and secrete growth promoting substances like auxins, and vitamins. It also helps in mobilization of insoluble phosphates and improve the physical and chemical nature of the soil. This algal amendment to the soil also improved the water holding capacity, reduce soil compaction and narrow C:N ratio. In the process of soil conservation the role of blue green algal is well established (Fig. 6.40).

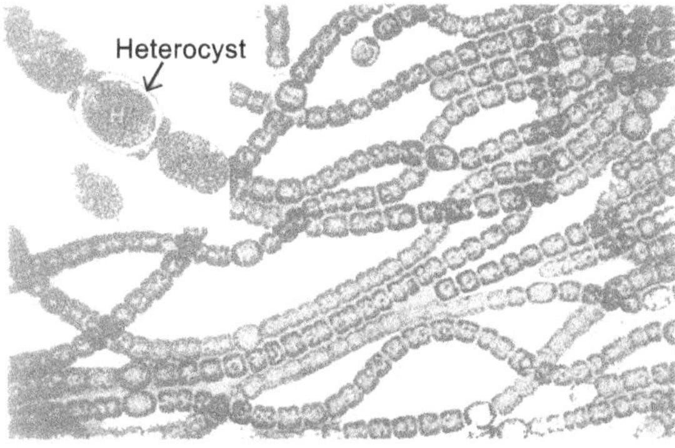

Fig. 6.40 Blue-green algal filaments with enlarged heterocysts amidst vegetative cells

Mass Production of Cyanobacterial biofertilizers

Preparation of cemented tank/ shallow trays/ polythene lined pits with width not more than 1.5m for proper handling of culture.

⇓

Transfer of 2-3 kg soils to 100g super phosphate

⇓

Water the pit to up to about 10 cm height, and add 2 ml insecticide to protect the culture from mosquitoes.

⇓

Sprinkle 100^0 g of starter culture on the surface of water.

⇓

Maintain the temperature 35.4 $\leftarrow$ 0^0 C for optimum growth, and water level always maintained about 10 cm during the growth period.

⇓

After drying, the algal mass (mat) is separated from the soil that forms flakes.

⇓

Further, it is collected, powdered, and packed in polythene bags for marketing.

⇓

The algal flakes can be used as starter inoculum again.

Photosynthetic bacteria

Photosynthetic bacteria synthesise carbohydrates by the chlorophyll in the presence of sunlight from CO_2 and reductants taken from air and oxygen.

$$2\,H_2A + CO_2 \xrightarrow[\text{bacterial chlorophyll}]{\text{light}} (CH_2O)\,x + 2A + H_2O$$

All the photosynthetic bacteria are divided in to 35 groups. The anoxygenic group (no evolution of oxygen) has purple and green bacteria, while oxygen evolving group has only cyanobacteria. The green bacteria utilise H_2S, H_2 or thiosuplhate as an electron donor and CO_2 as an carbon source. The purple bacteria contain two groups : The purple sulphur bacteria (Thiorhodaceae), which utilise H_2S as an electron donor, and the purple non sulphur bacteria (Athiorhodaceae) which depends upon organic compounds for their metabolism. All these groups contain major storage material and supplied to the plants as fertilising agents. *Rhodopseudomonas* are known to fix the atmospheric nitrogen in photo-autotrophic manner, and established themselves as effecient biofertilizers.

Phosphate solubilizing Organisms

Phosphate dissolving soils microorganisms play an important role in correcting phosphorus deficiency of crop plants. They release soluble inorganic phosphate into the soil through decomposition of phosphate rich organic components. Among microorganisms fungi, bacteria and mycorrhizae observed to be promising source of biofertilizers in phosphate supply. Many fungi *(Aspergillus, Penicillium)* and bacteria (*Bacillus megaterium, Pseudomonas striata*) are the potential solubilizers of bound phosphates. As the 50 per cent of the soil biomass is occupied by fungal mycelium, fungi is playing an essential role as phosphate solublizers. In the process of commercial preparation of these organisms as biofertilizers large flasks and rotary shakers are used in batch fermentation. A commercial preparation of phosphobacterin containing bacterial cells of *Bacillus megatherium* is widely used in USSR. Vegtable crops respond better than cereal crops to the application of phosphate dissolving organism.

Mycorrhizae

Bernhard Frank (1885) coined the term '*mycorrhizae*' means, root fungus. Physiologically mycorrhizae represents symbiosis and not parasitism. On the basis of infection anatomy two major groups of mycorrhizae have been recognised viz., '*Ectomycorrhizae*' and '*Endomycorrhizae*'. In ectomycorrhizae association the infection is intercellular while in endomycorrhizal association the infection is intracellular. The endomycorrhizae are of four major groups viz. Arbutoid mycorrhizae, Ericoid mycorrhizae, orchid mycorrhizae and Vesicular Arbuscular (VA) mycorrhiza. The development of VA mycorrhizae in roots can be divided in to four stages. (i) Spore germination and hyphal growth from infective propagules of VAM fungi, (ii) Growth of hyphae through soil to host roots, (iii) Penetration and successful initiation of infection in roots, and (iv) Spread of infection and development of internal hyphal system, arbuscules and vesicles. Ectomycorrhizae can be mass multiplied *in vitro* by using routine microbiological techniques. VAM fungi are obligate endosymbionts and scarcely grow in pure culture. Large scale innoculum production is hampered by this obvious limitation and presently root based bulk innocula is the only means of using VAM fungi in agriculture. In this methods of inoculum production there are every chance of contamination by unwanted airborne microflora. Also the chances of dissiminating root borne pathogenic infections can not be ruled out. To overcome these defects, root cultures under sterile conditions either as organ cultures or solution cultures have been worked out. Fungal cultures of VAM genera are, *Glomus sp*, *Gigosopora, Sclorocystis Aculospora, Scutellospora, Entrophospora* etc (Fig. 6.41).

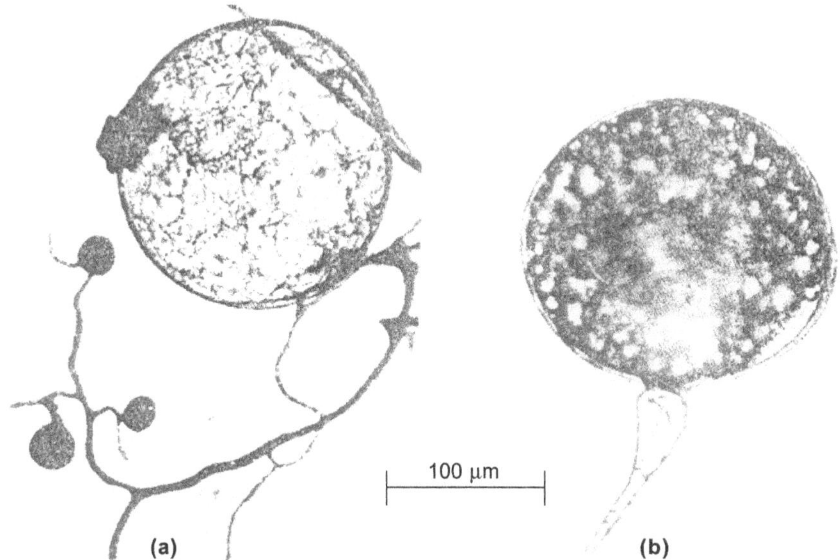

Fig. 6.41 Vesicular arbuscular mycorrhizae (a) a spore of the 'yellow vacuolae' Glomus mosseae, (b) a spore of the 'bulbous reticulate' type, *Gigaspora calospora.*

The beneficial effects of VAM on plant growth are numerous. Mycorrhizae improve the phosphatous nutrition of the host particularly in low fertility due to exploration of the soil by the external hyphae beyond the root hairs and phosphorous depletion zone. Mycorrhizal fungi also stimulate plant uptake of zinc, copper sulphur, potassium and calcium. They play an important role in water economy of plants and import the drought tolerance to the plants. VAM infection alleviates heavy metal toxicity and increase the tolerance of the crops to high acidity and temperature. Mycorrhizae also improve seedling survival and growth by enhancing uptake of nutrients and water by lengthening root life and by providing production against pathogens. Due to these beneficial actions, mycorrhizae are widely used as biofertilizers for improving the growth of agricultural and horticultural plants. They have been exploited to save the costly phosphate fertilizers. They are also proved to be successful organisms in wasteland reclamation and afforestation.

6.16 Biopesticides

Biological control of insect pests was in practice since last several decades even without the knowledge of active principles of biopesticides. Similarly farmers were not aware of the biological mechanisms behind those biocides. With rapid advancement of science and technology in developed countries, chemical pesticides were manufactured and frequently used to protect crop plants and other vegetation. Prompted by the consequences of environmental build up of toxic chemicals in the environment, environmental scientist and biologists started looking for a environmentally safer alternative to the chemical pesticides. This notion resulted in researches in the area of biopesticide development throughout the world.

Problems of pest control is changing constantly because of development of resistance against chemical pesticides in insect pests. Newer populations well equipped with resistance mechanisms are taking over replacing the older insect populations. This situation has resulted in continuous increase of chemical pesticides.

Biopesticides have emerged as potential means to control insect pests without adversely affecting the environment and ultimately the food chain. Biopesticides from bacteria, fungi and viruses have been developed and successfully used to combat insect pests involved in devastation of our agriculture. Various types of biopesticides, their production and their targets have been studied in order to understand the whole scenario of biopesticides development and their application in crop protection.

The two characteristics of chemical pesticides, i.e. their high efficiency and ease of use have eclipsed the modern development of biopesticides. But unforeseen long term consequences of introducing new chemicals to the environment prompted many researchers to find solutions in microorganisms in whose presence man has evolved. Therefore, studies aimed at replacing chemical pesticides with environmentally safer methods are currently being conducted at many research centres.

Characteristics of Biopesticides

The principle of biological control has been in practice in the control of various plant and animal pests resulting in a class of products that of by name biopesticides containing microorganisms antagonistic towards particular pest population. These biopesticides are.

(i) Potential means to control pests
(ii) Environmentally safer or non-hazardous
(iii) Easy to produce in large quantities
(iv) Specific in their insect control activity

Biopesticides utilises the pathogenic microbial strains isolasted from diseased insects during natural epidemics. These are also considered to be potentially complementary alternative to chemical pesticides possibly by prolonging their commercial life span by reducing the amount used. The active ingredient of microbial pesticides include, the spores or vegetative cells of pathogenic microorganisms.

Objective of microbial pest control

"*To bring forward the time when the insect pathogen controls the population, rather than to introduce a new material in to the pest management system*".

Over 400 fungal species are known to attack insects and mites. The first mass production attempts were made in 19th century. However, significant commercial development has occurred only in last several years(8–10 years). More than 90 bacterial pesticides have been described in 1971 by Falcon. This includes *Bacillus thuringiensis*, varieties of this bacterial pesticide are manufactured and sold throughout the world which are used

(i) to control caterpillar pests, and
(ii) to control mosquitos and black flies.

The first biopesticide product from *B.thuringiensis* was marketed in 1979. In the year 1983 western world has used worth 30 million dollor product from this particular bacterium. This represented 90–95% of the total market for biolgical control agents.

Microorganisms used to produce Pesticides

(i) *Bacteria as Pesticides :* The endospore forming bacteria belonging to genera *Bacillus* and *Clostridium* are very good candidates for pesticides. Among the non-spore formers, species of *Pseudomonas*, *Enterobacter, Proteus, Serratia*, and *Xenorhabdus* have potentials for field use.

Bacillus thuringiensis is being used in control of various insect pests of cotton, beans, cabbage, tobacco, tomato, oranges, grapes, shade trees and ornamentals. It has been considered as most commonly used and economically important strain successfully tested against 140 insect species. The important facts about this bacterium are :

(i) Crystalliferous bacterium
(ii) Sporulated cells possess crystals of high molecular weight (230 KD).
(iii) These crystals are taken up by the insects with their food and become insecticidal after digestion by proteinases in an alkaline environment.
(iv) Toxic subunits cause disintegrations of epithelial cells at the gut surface, resulting in paralysis and death.
(v) Depending on the dose caterpillars that have eaten *B. thuringiensis* crystals stop feeding with 10–15 minutes and die after 3 – 9 days (Fast,1981).
(vi) The beta exotoxin (water soluble and thermostable) known as 'thyringiensin" has broad host range for insects also effective against house flies.
(vii) 19 Serotypes and 320 isolates of B. *thuringiensis* have been identified and among them B. *thuringiensis israelensis* (serotype 14) has relatively specific activity against aquatic *Diptera* and now being produced commercially.

There are certain other *Bacillus* species used as pesticides. They are *Bacillus popillae* and *B. lentimorbus* pathogenic to Japanese Beetle, *B. moritae* pathogenic to fly pathogens and *B.sphaericus* promising pathogens to mosquitoes.

Fungi as Pesticides

Fungal insect parasites exist in all classes of Mycota. Some fungal genera that come under entomogenous fungi are, *Beauveria, Metarrhizium, Entomophthora* and *Coelomyces*. Deuteromycetes (Moniliales, Sphareopsidales) and Phycomycetes (Blastocladiales, Chytridiales, Entomophthorales) contain very important and useful species as biopesticides. The important Deuteromycetes member, *Verticillium lecanii* is pathogen of aphids and white flies and *Hirsutella thompsonii* is very useful pathogen to kill eriophyid mites. Fungi infects not only through mouth parts and gut like bacteria, but also through cuticle, and referred as contact insecticides. Recently, there is an addition of many more fungi as potential bioinsecticides, they are:

Normurea rileyii, Phytophthora palmivora, Colletotrichum gloeosporioides, Cercospora rodmanii, Trichoderma viride.

Viruses as Pesticides

Insect pathogenic viruses are common and are known to cause epidemics known as epizootics in insect populations. Nuclear polyhedrosis virus (NPV) and some granulosis viruses possess insecticidal properties and can successfully control Boll worms of cotton crops. '*Elcar*' is the popular product from Sandoz company with NPV as insecticide.

6.17 Single Cell Proteins

Due to population explosion, protein food stuffs becomes insufficient in quantity and high in price. Scientific and technological explosion development of modern agriculture is not considered to be effective to solve the protein problem in developing nations. Hence, the development of a creative sources of proteins such as SCP is badly needed.

Definition

The term single cell protein (SCP) was coined by group of scientists at Massachussetts Institute of Technology, USA in 1966. It refers to isolated cell protein and any microbial biomass from both uni and multicellular bacteria, yeasts, filamentous fungi or algae which can be used as food or feed additives.

SCP as source of food and feed

Compared to traditional methods of producing proteins for food or feed, large scale production of microbial biomass has the following advantages :

1. Conventional sources are not adequate to meet the requirement of growing population.
2. Microorganisms have a very fast growth rate. A protein doubling time as short as two hours has been recorded in *Neurospora crassa.*
3. They can be easily grown on a variety of substrate including wastes of agriculture, urban and industrial.
4. They can be easily genetically modified for growth on particular substrata and under particular cultural conditions.

5. Their protein content is quite high varying from 35 - 60 %.
6. They can be grown in slurry or solid state fermentation.
7. Their nutritional value are as good as those of other conventional food rich in proteins.
8. SCP production can be carried out in continuous culture and thus be independent of climatic changes.

Characteristics of SCP

1. SCP should have satisfactory nitrogen and protein contents, aminoacid profile, lipid contents, mineral and vitamins.
2. SCP should have satisfactory biological value.
3. SCP product must be free from toxic substances in the cells themselves or introduced in to the cells through residues in the substrate.
4. SCP should have a satisfactory flavour, aroma, colour and texture.
5. Methionine supplementation is required to obtain satisfactory feeding performance in non ruminants.
6. SCP should have high protein efficiency ratio (PER).
7. Toxic residues should not accumulate in the tissue of animals to be subsequently consumed by man.
8. SCP should not show adverse affects on reproductive performance in animals.
9. SCP should have acceptable physical characteristics such as formulation, blending and storage stability.

SCP Microorganisms

A large number of bacteria, actinomycetes, yeasts, filamentous fungi and algae have been studied as SCP sources. Among these major groups the following genera are most promising : (Table 6.14)

Table 6.14 Major source of single cell proteins

Major Group		Taxa
A. Bacteria		
(i) Phototrophs	:	*Rhodospirillum, Rhodopseudomonas*
(ii) Gram negative aerobes	:	*Alcaligenes, Pseudomonas*
(iii) Gram negative facultative anaerobes	:	*Aeromonas, Chromobacterium, Flavobacterium, Klebsiella, Hydrogenomonas, Methanomonas, Methylomonas.*
B. Actinomycetes	:	*Actinomyces, Mycobacterium, Nocardia, Streptomyces*
C. Yeasts	:	*Rhodotorula, Saccharomyces, Candida, Rhodosporidium Debaromyces, Endomyces, Saccharomycopsis, Torulopsis.*
D. Filamentous fungi	:	*Cunninghamiella, Morteriella, Mucor, Aspergillus, Aureobasidium, Botrytis*
E. Algae		
(i) Green algae	:	*Chlorella, Scenedesmus*
(ii) Blue green algae	:	*Spirulina*

Raw materials for SCP production

Production of SCP has acquired importance in several countries in recent years, since it is considered to be a delicacy for human food. These are cultured on solid substrates which range from straw to composted straw to pieces of wood and saw dust. The important and mostly used raw materials for SCP production are :

Agricultural wastes

Straw (rye, wheat, barley, rice, oat, grass) stover (sorghum, soybean, corn), sugarcane, bagasse, cotton gin, husk from grains, vegetable and fruit peelings.

Industrial wastes

Paper mill sludge, lignocellulose based industrial by products and wastes.

Municipal wastes

Municipal garbage, sludge.

Hydrocarbons

Methane, Ethane, Propane, Butane, Liquid hydrocarbons (gas oil, paraffins, petroleum, kerosene) Alcohols (methanol, ethanol).

Hydrocarbons Utilising Microorganisms

From early days hydrocarbons from various sources have been used for large scale production of SCP. Following are some of the examples of hydrocarbons as substrates that support the growth of microorganisms in the product on of SCP :

Table 6.15

Hydrocarbon		Microorganisms
Methane	:	*Methanomonas methanica* Methanomonas methanoxidans Methylococcus capsulatus Pseudomonas methanica *Graphium penicillioides.*
Propnae and Butane	:	*Arthrobacter simplex* *Nocardia paraffinica*
LPG gas	:	*Candida rigada*
Liquid hydrocarbons	:	*Achromobacter delvacata* Acinitobacter cerificans Mycobacterium phiei *Pseudomonas aeruginosa*
Alkanes	:	*Candida guillimondia* Candida intermedia Candida lypolitica *Candida tropica*
Methanol	:	*Hyphobacterium sp.* Methanomonas methanica Streptomyces sp. *Candida sp.*
Ethanol	:	*Candida ethanothermophilum*

Nutritional Value of SCP

SCP is mainly produced as an animal feed. There are problems with using SCP for direct human consumption because of :

(i) High content of nucleic acid (6-11%) leading to development of kidney stones and gout if consumed in high quantity,

(ii) Possibility of the presence of toxic secondary metabolites,

(iii) Poor digestibility and stimulations of gastrointestinal and skin reactions.

Yeast protein is the only SCP approved for human consumption but other food grade SCPs may soon be produced. When SCP is used as animal feed protein, the two most important factors are their nutritional value and safety (Table 6.16, 6.17, 6.18 and 6.19).

Table 6.16 *Nutritional Value of Food Protein (Dubey, 1993)*

Source	Analytical composition(%)		Essential amino acids (g/100g crude protein)		Biological coefficient		Digestibility of crude protein(%) (K.cal/kg)	Metabolisable energy for poultry
	Total	Crude protein	Lysine	Threonine	NPU	NPV		
Algae	8.0	45-71	5.7	5.2	-		82	-
Dried skimmed milk	5.7	35.90	8.0	-	87	31.2	-	2510
Soya bean meal	7.0	44.00	6.4	4.0	64	65.0	-	2240
Alakae yeast	11.2	7.00	7.4	4.2	91	96.0	92	2540
Bacteria from methanol	11.5	7.2	6.2	4.6	84	88.4	91	3468

NPU - Net protein utilization

NPV - Net protein value.

Table 6.17 *Comparison of Single Cell Proteins (%)*

Name of the Produce	Raw Protein	Fat	Nucleic acid	Mineral salts	Amino acids	Moisture
Milk powder	34	1.0	-	8.0	-	5.0
Soya meal	45	1.8	-	6.0	40	12.0
Isolated leaf protein	91	0.5	1.0	3.5	102	4.0
Yeast alkaline	60	9.0	5.0	6.0	54	4.5
Filamentous fungi	31.7	4.9	2.5	2.0	-	13.5
Algae	72.6	7.3	4.6	4.7	-	3.6

Table 6.18 *Nutritive Value of SCP - Mushrooms*

Nutrient	*Agaricus bisporus*	*Pleurotus ostereatus*	*Lentinus edodes*	*Volvariella volvacea*
Protein	3.9	2.8	2.8	5.0
Fat	0.2	0.6	0.4	0.7
Crude fibre	1.1	1.1	1.2	1.4
Vitamins				
Thiamine	1.1	4.8	7.8	1.2
Riboflavin	5.0	4.7	4.9	3.3
Niacin	44.7	108.7	54.9	91.9
Ascorbic acid	81.9	-	-	20.2
Minerals				
Calcium	23.0	98.0	33.0	71.0
Phosphorus	1429.0	476.0	1248.0	677.0
Sodium	-	61.0	837.0	374.0
Potassium	4762.0	3756.0	3793.0	3455.0
Iron	0.2	8.5	15.2	17.1

Table 6.19 *Comparison of Nutritive Value of SCP Mushroom with Various Foods (Li and Chang, 1982).*

Essential amino acid indexes			Nutritional indexes		
Pork, chicken, beef	-	100	Chicken	-	59
Milk	-	99	Beef	-	43
* Mushroom (high)	-	98	Pork	-	35
* *Volvariella volvacea*	-	96	* Mushroom (high)	-	28
Potatoes, kidney beans	-	91	* *Volvariella volvacea*	-	27
* *Pleurotus ostreatus*	-	90	Spinach	-	26
Corn	-	88	milk	-	25
* *Agaricus ostreatus*	-	87	* *Agaricus bisporus*	-	23
Cucumbers	-	86	Kidney Beans	-	21
Spinach, soyabeans	-	76	Peanuts	-	20
* Mushroom (low)	-	72	Cabbage	-	17
Turnips	-	69	* *Pleurotus ostreatus*	-	15
Carrots	-	53	Cucumbers	-	14
Tomatoes	-	44	Corn	-	17
* SCP			Turnips	-	10
72 - 98			Potatoes	-	09
			Cenots	-	06
			* Mushrooms (low)	-	05
			* SCP **5 - 28**		

Cultivation and Production Process of SCP

The nutrients, other than carbon sources, employed in this process are many of those commonly used for yeast cultivation. Any nitrogen source that can be assimilated may be employed (soyabean meal, yeast extract, urea etc.). Inoganic salts, eg. salts of phosphoric acid such as magnesium, potassium, or iron salts may be used. Cultivation is advantageously effected by aeration and agitation, and may be batchwise, semi-batchwise or continuous. Cultivation temperature around 40^0C are advantageous.

SCP Process Includes the Following Stages

1. Sterilisation - air and aqueous solution medium
2. Fermentation
3. Seperation and water washing of biomass
4. Drying and granulation.

(i) *Production Process of Yeast :*

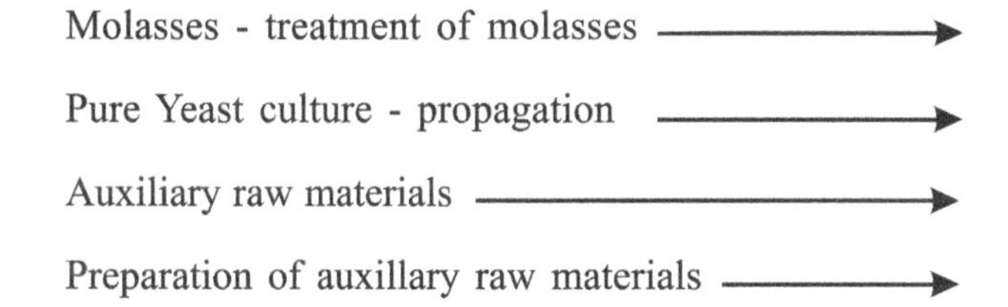

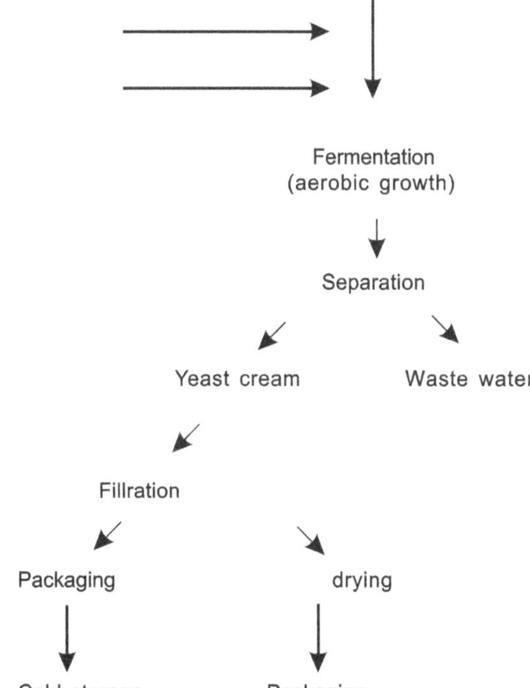

(ii) *Production Process of Spirulina :*

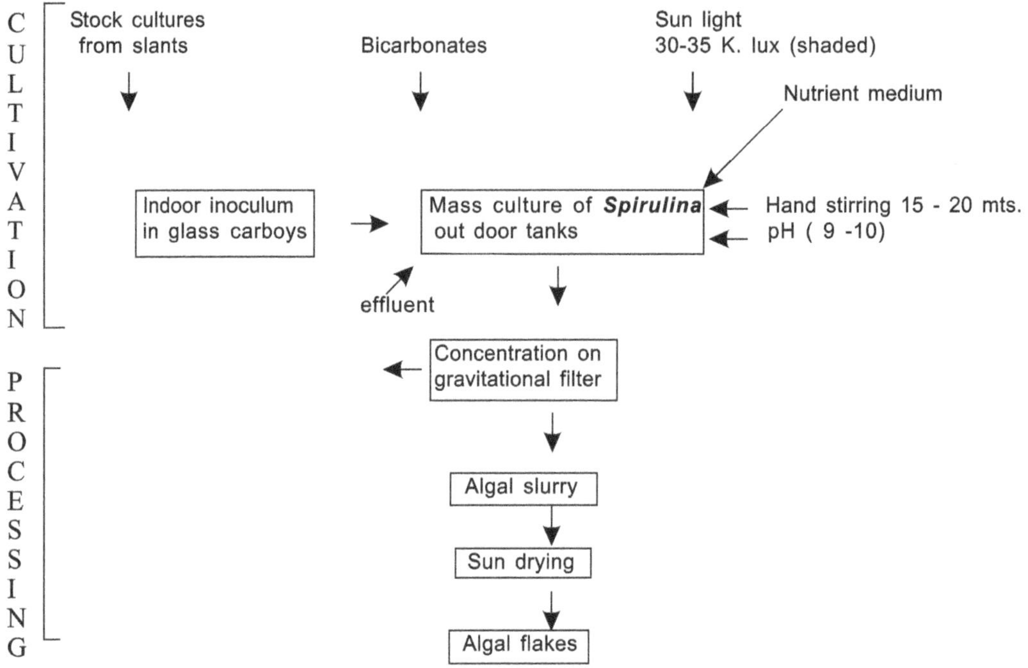

(iii) *Production Process of White Button Mushroom (Agaricus bisporus)*

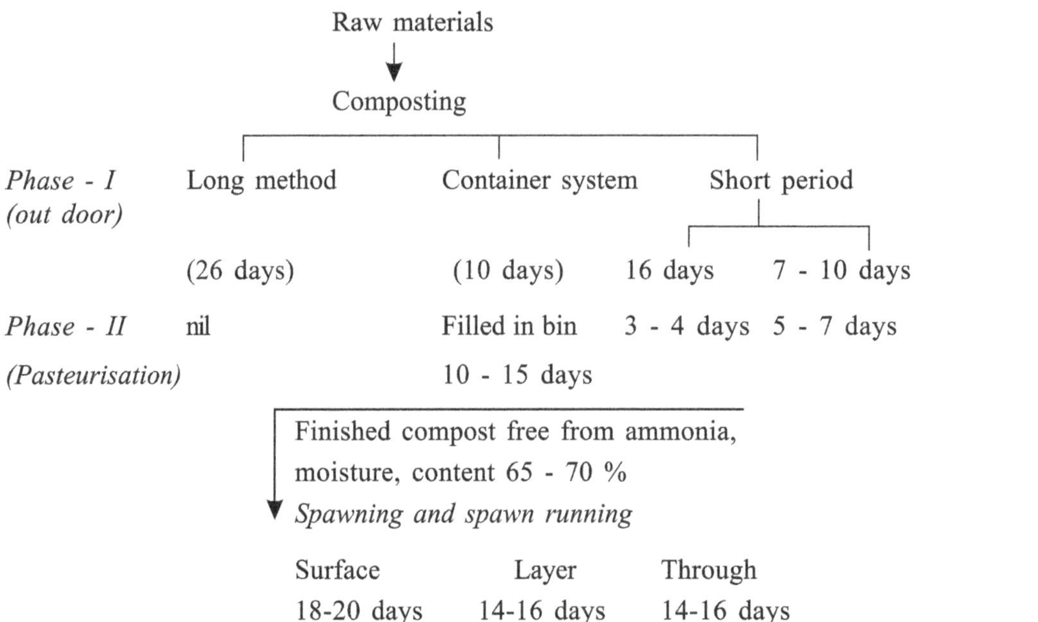

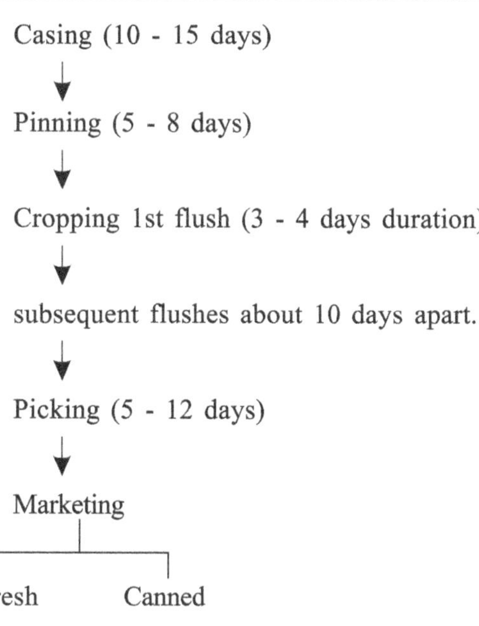

(iv) *Production Process of Oyster Mushroom (Pleurotus sofarcaju) :*

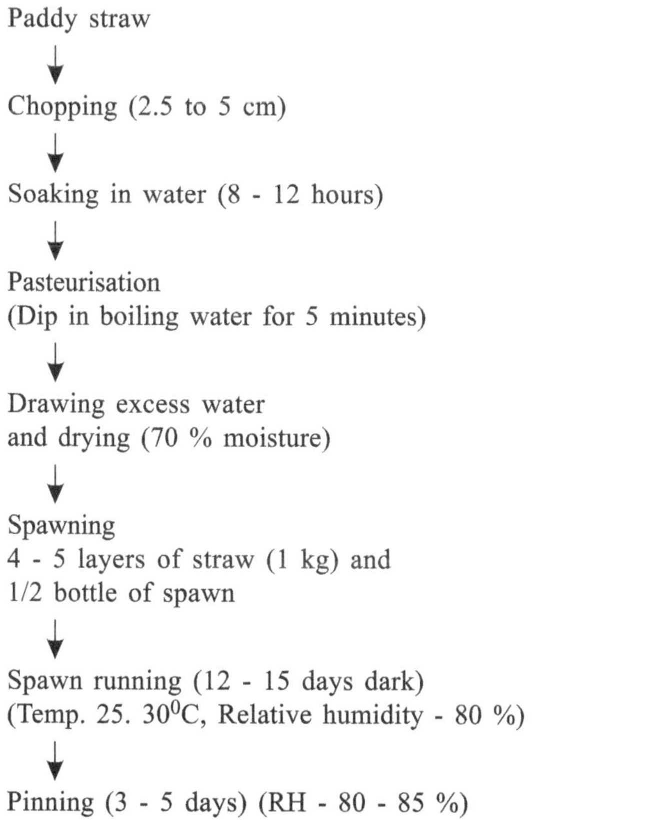

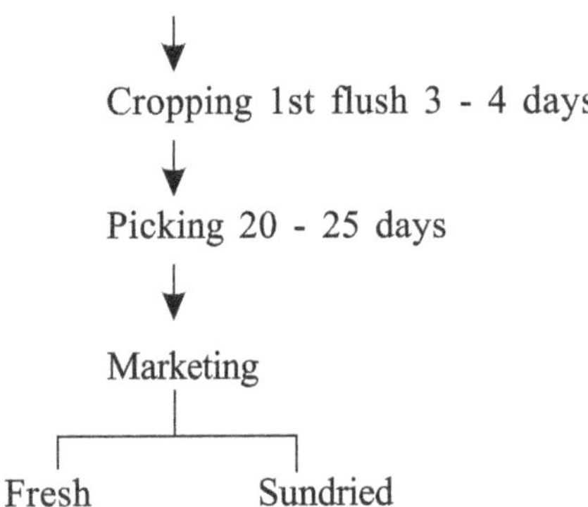

Factors Affecting Production of SCP

A. Yeast Biomass

1. Organic substrate and nitrogen ratio
2. pH of nutrient medium
3. Temperature (30 - 34^{0}C)
4. Oxygen
5. Maintenance of sterile conditions throughout the process
6. Suitable strain of yeast.

B. Algal Biomass

1. Illumination time
2. Light intensity
3. Supply of Co_2 concentrations
4. Nitrogen sources
5. Agitation of growing cells to maintain cell in suspension

C. Bacterial and Actinomycetous Biomass

1. Suitable strains of bacterial cultures
2. Genetic stability
3. Absence of bacteriophage
4. Suitable pH (5-7) of growth medium and temp. (15-35^{0}C).
5. Oxygen agitation to create aerobic conditions
6. Organic substrate and nitrogen concentrations (C : N ratio)
7. Maintenance of sterile conditions.

6.18 Mushroom Cultivation

Among the 80 crore undernourished people of the world 20.8 crore people alone line in India. Around 25 crore people live below the poverty line. To think of a hunger free India and translate this thinking into real action, the mushroom can play an important role to enrich human diet. As majority of the Indian population are vegetarians, mushrooms substitute the protein requirements. The mushroom cultivation is an ideal mode of recycling of agro-wastes which are generated by various agricultural practices. India has vast resources and potential for large scale production of mushrooms for almost daily consumption.

The edible mushrooms are of three important types (Table 2.20) :

1. Humus inhabiting (Humicolous)
2. Wood inhabiting (Lignicolous)
3. Dung inhabiting (Coprophilous)

Table 6.20 *Edible Mushrooms*

A. *Humicolous (humus inhabiting)*		
(i) Saprophytic	:	*Lepista nuda*, *Volvariella* sp., *Marasmium* sp., *Polyporus tuberastus*
(ii) Symbiotic	:	*Boletus* sp., *Lacterius* sp., *Tricholus* sp., *Tuber* sp., *Morchella* sp.,
B. *Lignicolous fungi (Wood inhabiting)*		
(i) Saprophytic	:	*Agrocybe* sp., *Pleurotus* sp., *Auricularia* sp., *Lentinus edodes*
(ii) Parasitic	:	*Armillaria mellea* *Cyttaria* sp.,
C. *Coprophilous (Dung inhabiting)*		
		Agaricus sp., *Coprinus* sp.,

=	0.5 kg meat 2½ eggs 1.5 kg. potato 1.0 kg. pulses 0.5 kg. soyabean

The mushroom cultivation was started as early as 17th century in France, Japan, China, etc. Mushroom cultivation began in India in 1943, with the cultivation of paddy straw mushroom (*Volvariella volvacea*). Subsequently, the cultivation of button mushroom (*Agaricus bisporus*) and oyster mushroom (*Pleurotus ostreatus*) was started in 1966 and now it is popular industry for small and big farmer all over the country.

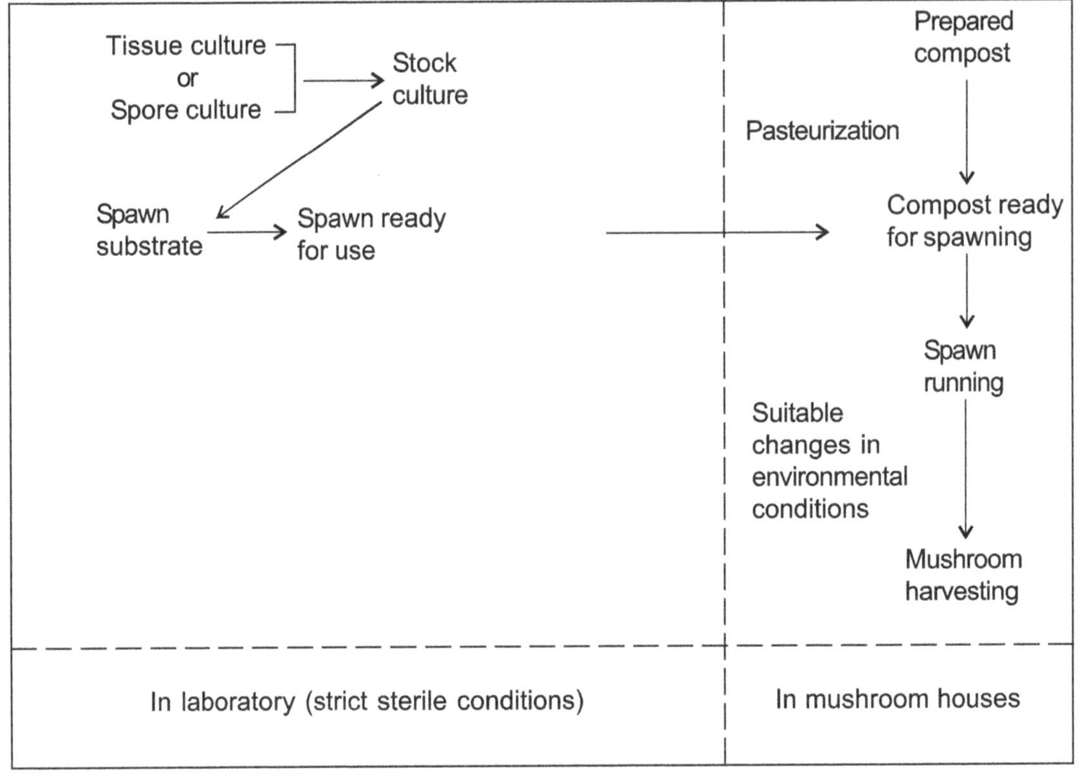

Fig. 6.42 Flow Chart of Mushroom production

The important steps involved in the mushroom cultivation, are shown in Fig. 6.42 *and are as follows :*

(i) Pure culture (obtained by sub-culture/tissue culture/spore culture)

(ii) Spawn production (seed material of mushroom fungi)

(iii) Preparation of beds (composting)

(iv) Spawning (planting spawn on to the bed materials)

(v) Cropping

Major Steps in Mushroom Cultivation

Cultivation of Paddy Straw Mushroom

This mushroom is originally grown on paddy straw and popular in tropical and sub-tropical countries. The important species are : *Volvariella volvacea* and *V. diplasia*. The important steps involved in the paddy straw mushroom cultivation are as follows :

Cultivation of Oyster Mushroom

Oyster mushroom (*Pleurotus ostreatus, P. florida, P. sajor-caju*) can be grown on various substrates viz. wheat/paddy straw, maize stalks, maize cobs, cotton waste, wooden logs, saw dust and vegetable plant residues. The various steps involved in the cultivation of oyster mushroom are as follow :

Cultivation of Paddy Straw Cultivation
(Flow Chart)

Paddy straw

↓

Bundles (1 kg)

↓

Soaking (12 hours)

↓

Draining excess water
(Pasteurization if needed)

↓

Layering and spawning
(20 kg straw 4 bottles spawn)

↓

Spawn running (10 days)

↓

Cropping (7-10 days)

↓

30 - 35^0 C. 80 - 90% RH

↓

Harvest

↓ ↓

use on the same day | Drying

Cultivation of Oyster Mushroom (*Pleurotus florida*)
(Flow Chart)

Paddy straw

↓

Chopping (2.5 to 5cm)

↓

Soaking (8-12 hours)

↓

Pasteurization

↓

(Dip in boiling water for 5 mts)

↓

Draining excess water
and drying (70% moisture)

↓

Spawning
4-5 layers of straw (1 kg) and ½ bottle of spawn

↓

Spawn running (12-15 days dark)
(Temp. 25 - 30^0 C; RH - 80%)

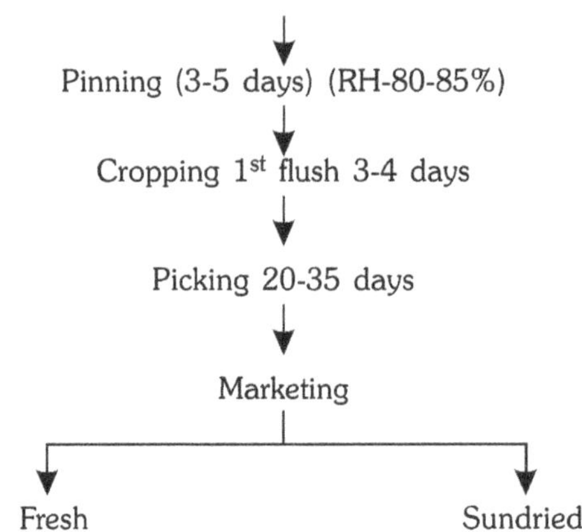

White Button Mushroom (*Agaricus bisporus*) cultivation (Flow chart)

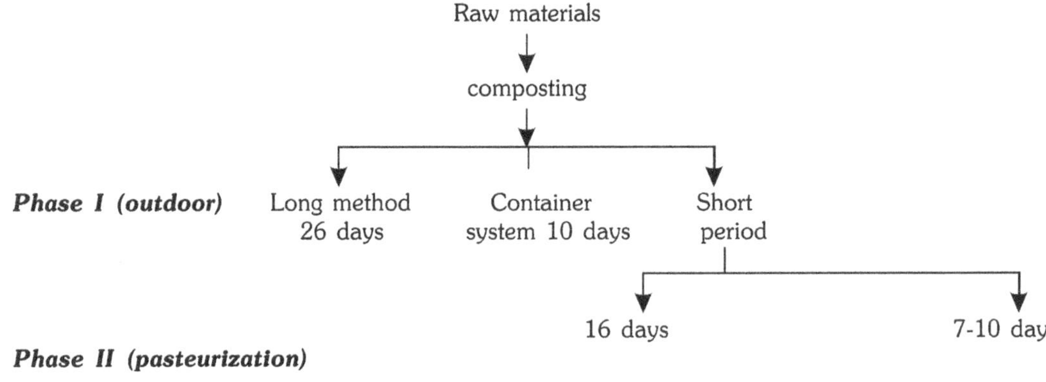

Finished compost free from ammonia, moisture content 65-70% spawning & spawn running

Surface	Layer	Through
18-20 days	14-16 days	14-16 days

Casing (10-15 days)

↓

Pinning (5-8 days)

↓

Cropping 1st flush (3-4 days duration)
Subsequent flushed about 10 days apart.

↓

Picking (5-12 weeks)

↓

Marketing

Fresh — Canned

White Button Mushroom Cultivation

White button mushroom (Agaricus bisporus) is popular in USA, France, China etc. The compost made out of straw of grain crops are widely used for the cultivation of this mushroom paddy straw supplemented with compost, cotton seed and molasses showed higher yield hence recommended.

The flow chart is shown Fig. 6.43.

The canning of mushrooms is also an important process when consumer is located in far off places. The important steps involved are as follow.

Pre-cleaning → washing → Blanching → Cooling → Grading → Slicing → Filling → Brining and exhausting → can seaming → Retorting → Labelling and Packing.

In Table 6.21 the comparative nutritive value of four common edible mushrooms is summarised.

Table 6.21 *Nutritive value of some common Mushrooms*

Nutrient	*Agaricus bisporus*	*Pleurotus osterreatns*	*Lentinus edodes*	*Volvariella volvacea*
Protein	3.9	2.8	2.8	5.0
Fat	0.2	0.6	0.4	0.7
Crude fibre	1.1	1.1	1.2	1.4
Vitamins				
Thiamine	1.1	4.8	7.8	1.2
Riboflavin	5.0	4.7	7.8	1.2
Niacin	44.7	108.7	54.9	91.9
Ascorbic acid	81.9	–	–	20.2
Minerals				
Calcium	23.0	98.0	33.0	71.0
Phosphorus	1429.0	476.0	1248.0	677.0
Sodium	–	61.0	837.0	374.0
Potassium	4762.0	3756.0	3793.0	3455.0
Iron	0.2	8.5	15.2	17.1

Nutritive value of some common Mushrooms

In Table 6.21 & 6.22 comparison of nutritive values of mushrooms with various vegetables and non vegetarian food shiff is given. The data clearly indicates that in terms of essential amino acids, amino acid score and nutritional indexes are very high, hence mushrooms are ideal food for good healthy life style. In Table 6.23 world production of various cultivated mushrooms representing various genera, their common market names other details are summarised.

Table 6.22 *Comparison of Nutritive value of mushroom with various foods. (Li and Chang, 1982).*

Essential Amino acid		Amino Acid Scores		Nutritional indexes	
100	Pork, chicken, beef	100	Pork	59	Chicken
99	Milk	98	Chicken, beef	43	Beef
98	Mushrooms (high)	91	Milk	35	Pork
89	Mushroom (high)	28	Mushroom (high)	96	*Volvariella volvacea*
71	*Volvariella volvacea*	27	*Volvariella volvacea*	91	Potatoes, kidney beans
63	Cabbage	26	Spinach	90	*Pleurotus ostreatus*
56	Potatoes	25	Milk	88	Corn
55	*Pleurotus ostreatus*	23	*Agaricus bisporus*	87	*Agaricus ostreatus*
53	Peanuts	21	Kidney beans	86	Cucumbers
50	Corn	20	Peanuts	87	Peanuts
46	Kidney beans	17	Cabbage	76	Spinach, Soyabeans
42	Cucumbers	15	*Pleurotus ostreatus*	72	Mushrooms (low)
40	*Lentinus edodes*	14	Cucumbers	69	Turnips
33	Turnips	17	Corn	53	Carrots
32	Mushrooms (low)	10	Turnips	44	Tomatoes
31	Carrots	9	Potatoes	28	Spinaches
6	Tomatoes	18	Tomatoes	5	Mushroom (low)

Table 6.23 *World Production of cultivated Edible Mushrooms*

SI.No.	Common Name	Botanical Name	Substrate	Production in $\times 10^3$ tonnes	Contri bution (%)	Purpose
1.	Button mushroom	*Agaricus bisporus* *A. bitorquis*	Straw	1424	37.8	Food
2.	Oyster mushroom	*Pleurotus sojar-caju*	Straw	909	24.2	Food
3.	Straw mushroom	*Volvariella volvacea*	Straw	207	5.5	Food
4.	Shiitake mushroom	*Lentinus edodes*	Shil and oak tree	393	10.4	Food & Medicine
5.	Nameka	*Pholiota nameka*	Squl dust	53	1.4	Food
6.	Inkcap	*Coprinus comatins*	Straw	–	–	Food
7.	Wood ear or Jews ear	*Auricularia polytricha* and other species	Wood	400	10.6	Food
8.	Winter mushroom	*Flammulina velutipes*	Wood	143	3.8	Food
9.	Silver ear	*Tremella fusiformis*	Wood	105	2.8	Medicine
10.	Bracket	*Ganoderma luciderma*	Wood	–	–	Medicine

Essay Questions

1. Describe the salient features of bacteria and their economic importance.
2. Give an account of the structure of bacteria their life history and importance in the economy of nature.
3. Write a general account on protozoa.
4. Define enzyme. Write the applications of microbial enzymes in industrial technology.
5. What is biofertilizer ? Explain the role of *Rhizobium* in agricultural biotechnology.
6. Describe the functions of phosphate solubilizing microorganisms as biofertilizers.
7. Enumerate the characteristics of biopesticides and explain the role of bacteria, fungi and viruses as biopesticides.
8. Write a detailed account of microbial formation of poly b - hydroxyalkanoates.
9. Define single cell protein. Explain their source, significance and characters.
10. Describe the nutritional value of single cell protein.
11. Write the cultivation and production process of some important SCP.
12. What are biofuels ? Explain the steps involved in the large scale production of Ethanol.
13. Write an account on the role of aerobic microorganisms in the wastewater treatment.
14. Explain anaerobic digesters in wastewater treatment.
15. Describe the role of bacteria in leaching and mining of copper.
16. Write the characteristics of biosorption and explain its role in bioremediation of heavy metals.
17. What are bioindicators ? Explain their role in biomonitoring and conservation.
18. Explain the plan and process of biosensor.
19. Write the advantages and current and future challenges of biosensors.
20. Write an account of edible mushrooms.
21. What are mushrooms ? Give an account of their culture and importance.
22. Compare the food value of mushrooms with other foods.

SHORT QUESTIONS

Explain the Following :

1. Chemotrophs
2. Organotrophs
3. Flagella
4. Endospores
5. Gram negative bacteria
6. Bacterial cell wall
7. Peptidoglycan
8. Condiospores
9. Yeast (*Saccharomyces*)
10. *Neurospora*
11. Antibiotics
12. Actinomycetes
13. Virus
14. Phages
15. Reproduction in phages
16. Life cycles of T_4 phages
17. Detrimental effects of Protozoa
18. Protease enzyme
19. Catalases and Peroxidases
20. *Azolla*
21. *Azospirillum*
22. Cyanobacteria
23. Leghaemoglobin
24. *Azotobacter*
25. Mycorrhizae
26. *Bacillus megaterium*
27. *Bacillus thuringiensis*
28. Nuclear polyhidrosis virus (NPV)
29. *Rhodospirillum*
30. *Thiobacillus*
31. *Spirulina*
32. Prions
33. *Rhizobium*
34. Symbiosis
35. Plasmids

36. Cosmids
37. Lytic cycle
38. Lysogenous cycle
39. Hydrocarbon utilising microbes
40. *Candida rigada*
41. Net protein utilization
42. Net protein value
43. Mushrooms
44. Essential Aminoacid index
45. Oyster Mushroom
46. Production of hydrogen
47. Nitrogenase and hydrosenage
48. BGA in hydrogen production
49. *Thiobacillus ferrooxidans*
50. Activated sludge
51. Biological ponds
52. Trickling filters
53. Biological films
54. AAFEB (Anaerobic attachment film expanded bed)
55. Septic tank
56. Sewage
57. Bulking
58. High rate oxidation ponds
59. Bacterial indicators
60. Biomonitoring
61. Transducer
62. Optrode
63. Immunosensors
64. BOD sensor
65. Sulphite ion sensor
66. *Nitrobacter* sp.
67. Biosensors
68. Biomagnification
69. Nif genes
70. Heterocyst
71. Single Cell Protein (SCP)
72. World production of edible mushrooms

References

Alkinson B. and Mavitane F. 1991, Biochemical Engineering and Biotechnology.

Anke T. 1997, Funal Biotechnology, Chapman & Hall, London

Arnold L. Demain, J.E. Davies, Atlas R.M. 1986, Industrial Microbiology & Biotechnology.

Bains W. 2004, Biotechnology from A to Z. 3rd Edition, Oxford University Press.

Baker HG. 1970, Plants and Civilisation, MacMillan, London.

Began M. Harper J.L, Towns end. Cr. 1986, Ecology, Blackwell, London.

Bell P R and Hemsley A R 2000, Green Plants : Their Origin and Diversity. Cambridge University Press.

Biotechnological Innovations in Animal Producting Biotechnology, 1992, Butterworth-Heineman.

Biotechnological Innovations in Crop Improvement Biotechnology, 1991, Butterworth-Heineman.

Blackwell WH. 1990, Poisonous and Medicinal Plants, Prentice Hall. N.Y.

Black JG. 2002, Microbiology, Principles and Explorations, John Wiley, N.Y.

Bold HC, Wynne M.J. 1985, Introduction to Algae, Prentice Hall, N.Y.

Bold HC. Alexopoulus CJ and Delevoryas T. 1980, Morphology of Plants and Fungi, Harper and Row, N.Y.

Brown. T.A. 1990, Gene Cloning : An Introduction, Champman and Hall, London.

Chari M and Reddy P.J. 2004, Principles of Environmental Studies, B.S. Publications, Hyderabad.

Chatterjee K.D. 1962, Parasitology, Personal Publication.

Chauhan AS. 2004, Environmental Studies, Jain Brothers, New Delhi.

Cheng T.C. 1986, General Parasitology, Academic Press Inc. London.

Chopra.R.N.,1956. Glossary of Indian Medical Plants, CSIR, New Delhi.

Dodds JH and Roberts L.W. 1995, Experiments in Plant Tissue Culture, Cambridge University Press.

Esan, K 1977, Anatomy of Seed Plants, J. Wiley, N.Y.

Evans. DE 2002, Plant Biology, Viva Books Pvt. Ltd., New Delhi.

G.S Omen & M. Alexander. 1990, Genetic Control of Environment

Gautam. P.L.,Rana.R, Srivastava .U., Ray Choludhuri. S.P. and Singh.B.R. 1998. Prospects of Medicinal Plants. ISPGR, New Delhi.

Gifford EM and Foster AS. 1989, Morphology and Evolution of Vascular Plants, WH. Freeman, USA.

Gilbert S. Omenn. 1987, Environmental Biotechnology.

Glazer, A.N & Nikaido 1995. Microbial Biotechnology.

Gray P (Edit) 1971, Encyclopedia of Biological Sciences, Von Nostrand, N.Y.

Grierson D. (ed.) 1991, Plant Genetic Engineering Blacikie, Chapman and Hall, London.

Grime J.P. Hodgson JG and Hunt RP. 1988, Comparative Plant Ecology, Unwin Hyman, London.

Gupta PK. 1994, Biotechnology, Rastogi Publications, Meerut.

Holland K. 1994, Medicines from Animals : from mysticism to science Pharm Fish, London, 24 : 9-12.

Hood. L.V. et al. 2002, Immunology 5th Edition, Benjamin, John Wiley, California.

Jadhav H and Bhosale V.M. 1995, Environmental Protection and Laws, Himalaya Publishing House. New Delhi.

Jhingran VG. 1994, Fish and Fishery of India, Hindustan Publishing Co., Delhi.

Jordon EL & Varma PD. 1995, Invertebrate Zoology, S. Chand & Co., Jalandhar.

Khan.I.,Khanum, 2001-2003, Role of Biotechnology in Medicinal and Aromatic Plants, vol 1 to 4. Ukaaz pulications, Hyderabad, AP.

Kochar SL. 1998, Economic Botany in the Tropics, McMillan India Ltd., New Delhi.

Kotpal R.L. 2002, Modern Textbook of Zoology - Invertebrates, Rastogi Publications, Meerut.

Kumar HD. 1977, Modern Concepts of Ecology, Vikas Publishing House Pvt. Ltd., New Delhi.

Kumar HD. 1999, Introduction to Phycology, Affiliated East West Press, New Delhi.

Lack, A. J. and DE. Evans. 2001, Plant Biotechnology-Instant Notes, Viva Books. New Delhi.

Lev. E and Peri Y. 2001, Historical Evidence of the begining of modern medicine in the holy land ! List of medicinal preparations found in preparations in stores and dispensary on July 21, 1857, Harefush 140 : 1109-1102.

Lindsey K. and MGK. Jones. 1989, Plant Biotechnology in Agriculture, Open University Press, London.

Mantell. SH, Mathews. JA and RA. Mckee. 1985, Principles of Biotechnology, Blackwell, Oxford.

Mauseth JD 1988, Plant Anatomy, Benjamin, California.

May E. & Ashlock PD. 1991, Principles of Systematic Zoology, Tata McGraw Hill Inc. New Delhi.

Meagher RB. 2002, Phytoremediation of Toxic Elememtial and Organic Pollutants, Curr Opin Plant Bio. 3 : 15-162.

Miller S.A & J.B. Harley. 2002, Zoology, Tata McGraw-Hill Publishing Company Ltd., New Delhi.

Nayar.M.P.,Sastry.A.R.K.,1987 Red Data Book of Indian plants. Botanical survey of India, Calcutta.

Rana. A.K. 2003, Indian Medical Aromatic Plants Special, Indian Foresters (1 & 2) Dehradun.

Rao, M.N and Datta A.K. 1987, Waste Water Treatment, Oxford & IBH, New Delhi.

Rao. P. S 1999 Cultivation of Medicinal and Aromatic Plants Research and Development Circle, A.P.Forest Dept, Hyderabad.

Ratledge C & Krestiansen B (ed) 2001, Basic Biotechnology, Cambridge Union Press.

Ravikumar,2000, 100 Red Listed Medicinal Plants FRLHT, Bangalore.

Roit, Brostoh J and Male D. 1998, Immunology, Times International Publishers, London.

Sambamurthy V.S.S. and V.S. Subramanyam, Economic Botany of Crop Plants, Asiatech Publications, New Delhi.

Santhanam R. Sukumaran N. & Natarajan P. 1987, A Manual of Fresh Water Aquaculture, Oxford & IBH Publishing Co. Pvt. Ltd. New Delhi.

Schery RW. 1972, Plants for Man, Prentice Hall, N.Y.

Service M.W. 1986, Lecture Notes on Medical Entomology, Blackwell Scientific Publications, London.

Sharma B.K 2001, Environmental Chemistry, Goel Publishing House, Meerut.

Sharma PD. 2000, Environmental Biology and Toxicology, Rastogi Publications, Meerut.

Shukla G.S & Updhyay V.B. 2003, Economic Zoology, Rastogi Publications, Meerut.

Sivarajan 1991, Introduction to the Principles of Plant Taxonomy, Oxford & IBH, New Delhi.

Sing V. & DK. Jain 1999, Taxonomy of Angiosperms, Rastogi Publications, Meerut.

Slater A. Scott. N and Fowler N. 2004, Plant Biotechnology, Cambridge University Press.

Smith I.D. 1994, Animal Parasitology, Cambridge University Press.

Sobti RC. 1996, Medical Zoology, Shoban Lal Nagin, Chand & Co., Jalandhar.

Srivastava C.B.L. 1999, A Textbook of Fishery Science and Indian Fisheries, Kitab Mahal, Allahabad.

Srivastava K.P. 1996. A Textbook of Applied Entomology, Kalyani Publishers, New Delhi.

Stern KR 2000, Introductory Plant Biology, McGraw Hill, Boston.

Stewart C.N. Jr. 2003, Transgenic Plant, Horizon Scientific Press.

Smith, G. 1988, Cryptogami Botany Vol. I & II, Tata McGraw Hill, Indian Reprint.

Steward FC. and AD Krikorian. 1971, Plants Chemicals and Growth, Academic Press, London.

Transgenic Plants as Edible Vaccines, Curr Top, Microbial, Immunology, 240.159-176. 1999.

Mosses V., 1994, and RE. Cape. 1994, Biotechnology, The Science & Business, Harwood, Academic Press, N.Y.

Vines AE. & N. Rees 1964, Plant and Animal Biology, Pitman, London.

Walden R. 1988, Genetic Transformation in Plants, Open University Press Milton Keynes, London.

Zaman V. & L. Keong. 1989, Handbook of Medical Parasitology, Keong Publishing Pvt. Ltd., Singapore.

Index

A

B

C

D

E

F

I

M

N

O

P

T

U

V

W

X

Y

Z

www.ingramcontent.com/pod-product-compliance
Ingram Content Group UK Ltd.
Pitfield, Milton Keynes, MK11 3LW, UK
UKHW061534190726
13845UKWH00005B/104